THE PANTANAL OF POCONÉ

MONOGRAPHIAE BIOLOGICAE

VOLUME 77

Series Editors

H.J. Dumont and M.J.A. Werger

The titles published in this series are listed at the end of this volume.

The Pantanal of Poconé

Biota and Ecology in the Northern Section of the World's Largest Pristine Wetland

by

CHARLES W. HECKMAN

KLUWER ACADEMIC PUBLISHERS
DORDRECHT / BOSTON / LONDON

A C.I.P. Catalogue record for this book is available from the Library of Congress.

ISBN 0-7923-4863-X

Published by Kluwer Academic Publishers,
P.O. Box 17, 3300 AA Dordrecht, The Netherlands.

Sold and distributed in the U.S.A. and Canada
by Kluwer Academic Publishers,
101 Philip Drive, Norwell, MA 02061, U.S.A.

In all other countries, sold and distributed
by Kluwer Academic Publishers,
P.O. Box 322, 3300 AH Dordrecht, The Netherlands.

Printed on acid-free paper

Printed in the Netherlands

Dedication

This book is dedicated to my wife, Wai-Yuen Sylvia Heckman, whose patience and understanding, in spite of very difficult living and working conditions, encouraged me to complete this book.

Cover: The inundated floodplain of the northern Pantanal, still largely in a pristine state, is typically covered by great expanses of herbal vegetation dominated by few plant species. A nearly monospecific stand of *Pontederia lanceolata* in the foreground is traversed by a stork jabiru, which has come to be a symbol of the Pantanal. The native herbal vegetation is interspersed by stands of woody plants on higher ground along the banks of the water courses and on hummocks as well as by temporary and permanent streams and rivers, various kinds of lakes, and a strongly modified vegetation on cattle ranches.

Contents

Color Plates

Color Plates

The production of the color plates was financed through the kind support of the Johanna and Fritz Buch Gedächtnisstiftung in Hamburg.

Plate 1a: View of the system of cliffs that divide the lowland plain in which the Pantanal is located from the *alto plano* to the north and east. Not far from these cliffs separating the low plain of the Pantanal from the vast plateau to the north and east is the continental divide separating the Paraná and Amazon Watersheds. The air is typically smoky during the dry season due to the many fires deliberately and illegally set to produce pasture land.

Plate 1b: Some of the extremely eroded, rocky outcroppings near Chapada dos Guimarães, evidence of ancient mountain building activity.

Plate 2a: A drying pool of water during the *seca* dominated by *Microcystis aeruginosa* and with a surface bloom of *Euglena sanguinea*, producing the strong color contrast between the red surface and green water below.

Plate 2b: The aquatic fern, *Azolla* cf. *mexicana*, grows among other water plants, such as *Salvinia auriculata* and *Hydrocleys* sp. as shown here.

Plate 3a: The thick masses of *Salvinia auriculata* rhizoids form the substate in which the interrhizon community develops.

Plate 3b: Because they bloom intensely during the dry season, the color of the *Tabebuia heptaphylla* flowers (a) make the abundance of the trees in the Pantanal evident (b).

Plate 4a: The deep red leaves of *Myriophyllum mattogrossense* beneath the extremely clear water that covers the floodplain during the *cheia*.

Plate 4b: *Ludwigia sedoides* is a typical species of the floodplain aggregation during the period of flooding, which sometimes becomes dominant in small areas. The flower in the center belongs to *Nymphoides grayana*, while the other two yellow flowers are those of *L. sedoides*. The lancet-shaped emergent leaves belong to *Pontederia lanceolata*.

Plate 5a: The flower of *Passiflora cincinnata*, one of several species of its genus that have apparently been introduced along the highways of the Pantanal by man.

Plate 5b: Typically, the ecotone between the trees on the hummocks and the semiaquatic plants at the high water level on the floodplain is dominated by *Bromelia balanse*, which is very abundant in the Pantanal of Poconé. Its hard, spined leaves form a barrier to large animals.

Plate 6a: A typical member of the interrhizon community, *Suphisellus grammicus* is invariably found concealed among the rhizoids of *Salvinia auriculata*.

Plate 6b: *Hyla raniceps* is one of the small predators in the interrhizon community. These nocturnal frogs often rest in the leaves of *Eichhornia azurea* during the day.

Plate 7a: Neotropic ormorants, *Phalacrocorax brasilianus,* fish for considerable periods of time, continually diving until they have caught enough to eat. They then roost on trees to dry their feathers.

Plate 7b: An anhinga or snake-neck bird, *Anhinga anhinga,* drying its feathers beside the Rio Pixaim.

Plate 8a: The capped heron, *Pilherodius pileatus*, in a bed of *Eichhornia azurea*. This species remains either solitary or together with its mate, each pair in its own small territory.

Plate 8b: The rufescent tiger heron, *Tigrisoma lineatum*, is a common bird in the Pantanal, where it usually remains solitary near the water's edge. The individual in the photo displays the feathers of a young adult. The older individuals lose the striated coloration and develop uniform reddish brown feathers on the head and upper body.

Plate 9a: The roseate spoonbill, *Ajaia ajaja*, shown together with a great egret, is fairly common in the Pantanal. During the study, it was more numerous in the lowland marshes near Porto Jofre, although it was occasionally seen near Poconé, as well.

Plate 9b: The bare-faced ibis, *Phimosus infuscatus*, is often seen in small flocks patrolling along the edges of the water bodies and on partially inundated areas of the floodplain throughout the year.

Plate 10a: The southern screamer, *Chauna torquata*, is a large bird generally observed in pairs throughout the Pantanal. They do not permit persons to approach them too closely, and before anyone can get within about 50 m of a pair, the birds fly to a low tree where they scream raucously until the person leaves.

Plate 10b: The crested caracara, *Polyborus plancus*, is by far the most photogenic bird in its family, posing along the Transpantaneira Highway and allowing vehicles to approach within a few meters. These birds flee only when persons leave the vehicles. Early in the morning, the caracaras patrol the highway to feed on the frogs, reptiles, and small mammals that were killed by trucks during the night.

Plate 11a: The sunbittern, *Eurypyga helias*, is so plain looking that it is often overlooked as it moves along the shores of the water bodies in the Pantanal (a). When it takes off, however, it reveals a brilliant sunburst (b). This "startle effect" is apparently effective against the wild felids that frequent the Pantanal.

Plate 11b: The ringed kingfisher, *Ceryle torquata*, generally remains on its personal lookout post on a section of the power lines, a bridge, or a fence overlooking the water, where it waits for fishes of a suitable size to come near enough to the surface to catch. Although these birds are often observed fairly near others of the same species, each one has a small territory which it defends against intrusion by other kingfishers.

Plate 12a: The yellow-billed cardinal, *Paroaria capitata*, is a species that seasonally forms very large flocks in the Pantanal. One large flock was encountered year after year near the Rio Pixaim.

Plate 12b: A very large group of long-nosed bats, *Rhynchonycteris naso*, lived under a bridge along the Transpantaneira Highway. On cool mornings, large groups of the more than 50 individuals from the colony would emerge to sun themselves on the sides of the bridge. After being taken in the hand early in the morning, these bats were reluctant to fly away when released until they had warmed themselves from body heat.

Plate 13a: The Pantanal deer, *Blastocerus dichotomus*, often enters the water to feed on *Eichhornia azurea* and other water plants.

Plate 13b: A crab-eating fox, *Dusicyon thous*, running along the edge of the ditches bordering the Transpantaneira Highway during the dry season.

Plate 14a: Giant otters, *Pteronura brasiliensis*, concealed in the shadows along the shore of the Rio Bento Gomes, are leaving the water to rest in the gallery forest.

Plate 14b: A ranch house on the site of a hummock surrounded by water during the *cheia*. The boat at the upper right is used to cross the vast flood plain during this season. During the *seca*, the floodplain becomes completely dry.

Plate 15a: The tailings of the Dois Paderes gold mine flow over a large area. They are contaminated by mercury.

Plate 15b: As large areas of the floodplain dry up during the *seca*, the fishes are forced into small, shallow pools of water in which they are nearly defenseless against flocks of water birds that prey upon them.

Plate 16a: Palm fruit is one of the favorite foods of the hyacinthine macaw, the largest species in the parrot family. It is presently threatened by excessive hunting to satisfy the house pet trade.

Plate 16b: During the dry season, the shores of the Rio Bento Gomes and other rivers in the Pantanal are typically lined with dozens of well-fed caimans, which gorge themselves on the trapped fishes early in the morning and then sun themselves side by side for the rest of the day.

Plate 1a

Plate 1b

Plate 2a

Plate 2b

Plate 3a

(a)

(b)

Plate 3b

Plate 4a

Plate 4b

Plate 5a

Plate 5b

Plate 8a

Plate 8b

Plate 9a

Plate 9b

Plate 10a

Plate 10b

(a)

(b)

Plate 11a

Plate 11b

Plate 12a

Plate 12b

Plate 13a

Plate 13b

Plate 14a

Plate 14b

Plate 15a

Plate 15b

Plate 16a

Plate 16b

1. Introduction

There are many famous wetlands in the world that have been deemed important for the homeostasis of the biosphere and accorded some form of protected status in view of the richness of their flora and fauna. The Pantanal, located almost directly in the center of the South American continent, is among the largest such wetlands in the world. It is, in fact, the largest wetland that is still in a nearly natural state and has not been extensively modified by man for agriculture. The extent of the region covered by flood waters at the end of each rainy season varies from year to year, sometimes considerably, so estimates of the exact area it occupies have varied enormously. The size of the Pantanal has been reported to be between 80,000 km^2 (Bonetto, 1975) and 250,000 km^2 (Tundisi and Matsumura-Tundisi, 1985). The most commonly accepted estimates at the present time are approximately 130,000 to 140,000 km^2, supported by estimates made from Nimbus-7 satellite observations (Hamilton *et al.*, 1996). However, Por (1995) suggested that the area extending into Bolivia and Paraguay has not been satisfactorily surveyed, leading him to accept the figure of 200,000 km^2 as the area of the entire Pantanal, including all of its extensions.

The main reason for the year to year variation in the extent of the flooding is the considerable difference in timing of the rainfall on the watershed. The total amount of rain varies less than does its temporal distribution among the months during which most of it falls on the drainage area of the upper Rio Paraguay. The irregular local rainfall throughout the Pantanal watershed results in a pattern of flooding that always leaves part of the floodplain dry. Thus, over a nine year period, a maximum of only 110,000 km^2 was found by Hamilton *et al.* (1996) to be under water at any one time, although 131,000 km^2 was subject to inundation at some time during their satellite observations. A dry season minimum of 11,000 km^2, reported by the same authors, is indicative of the extreme fluctuations in the extent of the area under water.

To make any conservation program effective, not just the wetland itself but also the watershed feeding it must be protected. Its area is somewhat easier to calculate because the kind of terrain, its elevation above sea level, and its annual flooding need not be taken into consideration. The wetland and its entire catchment area were estimated to occupy nearly a half a million square kilometers (Walter and Breckle, 1984). Por (1995) accepted a figure of nearly 490,000 km^2, of which 393,600 km^2 is located in Brazil.

The Pantanal may well become the last major wetland in the tropical wet-and-dry climatic zone that has been modified only slightly by man. Plans for the "economic development" of the wetlands in the upper Nile Valley may totally drain the largest remaining wetlands in the wet-and-dry zone of Africa. The largest periodically inundated region in the world is located in Southeast Asia in the valleys of the Mekong, Chao Phya, and Irrawaddy Rivers. This wetland, however, has been almost entirely modified for the culture of rice, and scarcely any pristine regions remain. In fact, agriculture is thought to have first been practiced in that region of the world (Solheim, 1971) with rice, a grain native to periodically flooded wetlands and the first crop of any kind cultivated on a large scale.

To bring the region of Southeast Asia under cultivation, nearly the entire area was divided into paddy fields with low dikes, and deeper ditches and canals were constructed to drain off excess water during the rainy season and maintain a water supply during the dry period. In these ditches, enormous natural fish production is possible (Heckman, 1974, 1979). To protect the fishes, the native fauna of crocodilians and water birds was largely eliminated, and the smaller seed-eating birds were also driven away to protect the rice. The native turtle species were used for food faster than they could reproduce, and they are rare today. The native mammals were also driven from the floodplains and survive today only in mountainous regions and near a few temples where they find protection.

Finding similarities between present-day Southeast Asia and the Pantanal is rather difficult, but on close examination of the flora and of the fish and invertebrate fauna, numerous similarities can be discovered (Heckman, 1997a). Furthermore, ancient ruins, such as those of the temple complex of Angkor Wat in Cambodia, are decorated by carvings in stone depicting scenes from centuries long past. They show fisherman on the lakes catching turtles and crocodilians in addition to fishes, while others can be found depicting large wading birds, recognizable as egret species. Similar wildlife scenes, although no longer a part of life in Southeast Asia, could well be encountered in the Pantanal today.

There are other wetland regions climatologically and ecologically similar to the Pantanal in northern Australia (Taylor and Tullock, 1985; Finlayson, 1993), but these are considerably smaller than those on the other continents. It is therefore safe to assume, as stated above, that the Pantanal is now the largest flooded tropical savanna in the world in a near natural state.

The human impact on the Pantanal has been greatly limited by various factors, including the inaccessibility of the region, the low human population density in Central Western Brazil and adjacent regions of Bolivia and Paraguay, and the lack of natural resources, except for gold.

Various native American tribes inhabited the region to the north of the Pantanal in prehistoric times, and many of them penetrated into the wetland itself (Bertelli, 1988). However, they apparently took little more from the land than they needed to survive. The earliest European visitors to the region of Cuiabá to the north of the Pantanal arrived only as late as 1713 looking for slaves. Several bitter armed conflicts largely eliminated many of the native settlements during the next few years, and when gold was discovered, attention turned to prospecting. In addition to the city of Cuiabá, the permanent settlement of Nostra Senhora de Livramento was founded in 1731 followed by Cáceres in 1778, and São Pedro del Rei, now known as Poconé, on January 21, 1781, by Captain General Luiz Albuquerque de Melo Pereira e Cáceres. The search for gold was the main occupation of the population of these towns during the 18th and early 19th centuries. To the south, Corumbá was founded in 1775 and became a bastion to defend the border against the neighboring Spanish colonies when European wars spread over into the New World from the Old. In 1831, the name São Pedro del Rei was changed to Vila da Nossa Senhora do Rosario de Poconé, and in 1863, the town became known as Comarca de Poconé (Costa e Silva & Ferreira, 1994).

As time went on, more stable forms of income were developed, the most important of which was ranching. Cattle descended from a race brought by the Spanish to Paraguay came to Mato Grosso, and these were captured and raised on the ranches. Later, other breeds of cattle were introduced, as well. In the Pantanal, a race of horse was developing that was particularly suited to wetlands. These horses have particularly strong fore-legs and can carry riders easily for long distances through rather deep water. They are also descended from a stock that had originally been brought to Paraguay by way of Argentina.

Paraguay had been developed agriculturally well over a century earlier than Mato Grosso by the Jesuits, who had founded settlements, called reductions, in which the native Guarani people had been encouraged to live in order to escape the slave traders who regularly raided the interior of South America, in spite of various Spanish and Portuguese laws to discourage the enslavement of the native people. These settlements prospered, and cattle and horses were plentiful in the regions to the south of the Pantanal by the early 18th century. In the second half of that century, when settlers from the eastern parts of Brazil were reaching Mato Grosso in increasing numbers, the Jesuit order was suppressed in Europe, and the Guarani towns were soon destroyed by joint efforts of Spanish and Portuguese troops. Apparently, cattle and horses were there in abundance for the taking.

Until recently, the human impact on the Pantanal has been limited mainly to the traditional activities of gold mining and cattle ranching. The region now supports a small human population, probably no greater than that of the native Americans who resided in the region during prehistoric times. Somewhat more numerous are the horses, cattle, and other domestic animals that the ranchers have introduced. A diversified and extensive form of agriculture has been practiced that had scarcely modified the land until fairly recently, when attempts to improve the pastures resulted in the introduction of plants from other continents that are not liked by the cattle and have become noxious weeds.

Recent developments are threatening to change the traditional situation in the Pantanal. These include a resurgence of gold mining employing large quantities of mercury to amalgamate the gold dust. This activity not only causes local heavy metal pollution, it also leaves deep excavations and large fields of mud to mar the landscape (Fig. 1.1). Another serious problem is the increased poaching directed at various species that are either marketable for their skins, such as the local caiman, or at smaller birds, mammals, and reptiles that bring high prices on the international pet market. Other, much more serious dangers are apparent in proposals for regional economic development, including the construction of an inland seaway to Cuiabá and Cáceres through the Paraná, Paraguay, and Cuiabá Rivers. A deepening and regulation of this river system would certainly reduce the area of the Pantanal and might even drain it completely for most of the year. While the economic feasibility of such projects is more than highly questionable, the danger of severe financial losses has never discouraged the "technological development" lobby in the past.

Figure 1.1: An open pit gold mine scars the landscape at the borders of the Pantanal. After it is abandoned, it will become a pond with mercury in its sediment.

At the beginning of the research on the Pantanal in 1991, it rapidly became apparent that physical, chemical, and biological information on this wetland did not exist. The publications on the region were few with little specific physical and chemical data and definitive descriptions of only a few groups of the flora and fauna. A few more extensive publications on the regional biota have recently begun to appear, such as works on the terrestrial plants (Pott and Pott, 1994) and the birds (Dubs, 1992). While a few of the most attractive plants and animals had already been amply illustrated on countless postcards, brochures, posters, and guide books to the region, the great majority of the species inhabiting the wetland had never been catalogued. Systematic collection of meteorological data had never been initiated anywhere in the wetland, and there are still few weather stations in the entire state of Mato Grosso. A small number of general or regional works on the Pantanal provide the names of the bird species and some of the mammals and reptiles, but information about their ecology is still sketchy.

In organizing the research work, it was quickly noted that the Pantanal is markedly heterogeneous in its structure, and general studies on the region must include information about a great variety of habitats (Fig. 1.2). Studies concentrated at any one location are likely to provide results similar to those obtained by the blind men who felt various parts of the elephant in the fable by Aesop. It would be wrong to present the features of biotic communities in a river, an oxbow, a lake, a permanent lowland marsh, or temporarily flooded areas as those of "the Pantanal." Similarly, it would be wrong to designate the characteristics of this wetland during any one season as "typical" for this region. Nearly all of the habitats are in a state of constant change that follows a complex seasonal cycle. Thus, the ecosystem must be investigated as a dynamic system in a constant state of change regulated by the seasonal climatic and hydrological cycles. The great differences in these cycles in the northern and southern parts of the wetland, particularly those related to the hydrological regime, produce totally different conditions that seem to have promoted the development of a richer flora and fauna in the South. This greater number of species is mainly encountered in the terrestrial and littoral habitats, as indicated by comparing lists of the terrestrial plant species in the northern and southern parts of the wetland (Pott *et al.*, 1989; Pott and Pott, 1994; Section 7) and range information on the birds (Pinto, 1938; Sick, 1988; Dubs, 1992) with data on the biota recorded during the field studies (Section 8).

Figure 1.2: Although a low-lying plain, the Pantanal is interspersed by a great variety of different aquatic and wetland habitats, which became distinct from one another mainly due to slight differences in terrain elevation.

The limnological study of the Pantanal was limited by the sponsoring agencies to a period of only about 44 months. During this period, much work was required to purchase equipment and set up laboratories where they had not existed before.

Considering that the main part of this vast wetland in Brazil extends from north to south about the same distance as Germany does, it was fully unrealistic for the planners to expect general information on the wetland as a whole to emerge from the study. The goal of the limnological research was therefore focused on providing a preliminary overview of the general limnological features of the northern part of the Pantanal, and particularly the conditions prevailing in the Municipal District of Poconé, located in the State of Mato Grosso. This book provides a summary of the data that have been collected to date on its physical and chemical features, its fauna and flora, and the interrelationships among the climatic, abiotic, and biotic characteristics of this wetland. Comparisons with other regions can then be made, and the peculiarities of the northern part of the wetland can be better understood.

Although the study was undertaken primarily to elucidate the limnology of the Pantanal, it is scarcely possible to ignore the terrestrial communities that are intricately interwoven with the wetland and aquatic ecosystem, both spatially and functionally, in such a vast, heterogeneous wetland. The Pantanal is situated in the middle of a great area of shrub savanna, generally referred to as *cerrado* sensu lato in Brazil (Rizzini and Heringer, 1962). Since this kind of ecosystem completely surrounds the Pantanal (Goodland and Ferri, 1979), it determines to a great extent which terrestrial species will be encountered in the drier parts of the wetland. Although the enormous variety of terrestrial species in the *cerrado*, including its extensions into the Pantanal, can only be touched upon in this work, a considerable amount of information could be obtained on some of its more common plants and animals, especially those that are best adapted to develop in the Pantanal in spite of the flooding. General phytogeographical reports on the relationship of *cerrado* vegetation to that in the wetlands were provided by Adámoli (1981, 1986a). In addition, general botanical works on the tracheophyte communities of Mato Grosso have been published periodically (Hoehne, 1923; Guarim Neto, 1992; Dubs, 1994), and floristic works on terrestrial vegetation of the Pantanal itself are also available (Nascimento and Cunha, 1989; Pott & Pott, 1994; Scheßl, 1997).

Also touched upon are the upland streams and water bodies which are fed by springs and flow into the Pantanal throughout the year. Although they occupy only a tiny fraction of the area encompassed by the vast wetland, these proved to be much more diverse in species than the Pantanal itself, apparently because the species present do not have to cope with the extreme seasonal changes in water quality and periodic desiccation. The Pantanal, in contrast, is noted for vast numbers of relatively few species, which have seasonal phases of active growth and dormancy.

A word must also be said about the state of systematics and taxonomy in the region of the Pantanal. Without doubt, many of the names used in this work will be revised in the years to come by taxonomists. Some of the species referred to will be spit into several different ones according to minor morphological differences, while others will be renamed. This is especially true for the tracheophytes, annelids, molluscs, arthropods, and vertebrates, which are represented overwhelmingly by species endemic to small regions within the Neotropics. The taxonomy of the South American species within these groups is still in its pioneer stages. Several of the species identified only to the level of genus are with certainty still undescribed. It is hoped that some of the serious errors and omissions of the past can be avoided in future work. During the preparation of this book, an enormous amount of taxonomic literature had to be reviewed, and much of it was chaotic, to say the least.

From the publications on many groups studied in the past, a general pattern of taxonomic progress can be seen. The flora and fauna of Mato Grosso, like that in much of the rest of South America, is still in the phase of being described, too often in short publications in which single species are treated with insufficient reference to how they differ from congeneric and confamilial forms and little or nothing about their behavior and ecology is reported. As this phase progresses, small differences between individuals are often used to justify the recognition of distinct species. In the parts of the Neotropical Region where more work has been done than in Mato Grosso, this has produced a large number of nominal taxa that are almost impossible to tell apart. Eventually, it is recognized that the differences between them fall within the natural variability of single species, and the lists of synonyms begins to grow. For example, in his summary of the leech fauna in South America, Ringuelet (1981) provided a long list of species names that he himself had given to specimens he had described as new leech species, all of which he later admitted should be regarded as junior synonyms of *Helobdella stagnalis*, a widespread species throughout both South America and Europe.

The taxonomy of the pilid snail fauna has been all but abandoned since the 1920s, when different taxonomists had selected different morphological features for distinguishing the species, and their keys proved to be incompatible. This occurred mainly because many of the morphological features of the soft parts, which the authors had used as a basis for their systematic work, are extremely variable. As a result, almost all of the systematic schemes proved faulty (Pain, 1972). What is presently accepted for want of anything better is a rather loosely applied nomenclature based on the general shape and coloration of the shell. In the long run, this taxonomy will probably prove closer to the phylogenetic realities than any of the highly specialized morphological or biochemical methods that have been attempted.

The taxonomy of many South American arthropod groups is in an extremely confused state. Unfortunately, in some cases, the more intensively certain taxonomists have worked on a group, the more difficult it has become to distinguish the taxa and to decide which of the forms described are actually valid species. There is often a great gap between the way in which species are supposed to be identified and the actual options open to the ecologist with specimens he is trying to classify. In theory, a specimen should first be tentatively identified using a key. However, with the exception of those prepared for a few taxa for which exemplary literature is available, published keys are locally limited to regions far from Mato Grosso, hopelessly out of date, or non-existent. To complete the work in the Pantanal, new keys had to be produced.

After tentative identification using a key, the next theoretical step is comparing the specimen with a complete description of the species it seems most likely to belong to. In some cases, this is possible, but in others, the morphological characterizations are so cursory that it is pointless to try to recognize the specimen from the description. A negative example is the genus *Gyretes* in the family Gyrinidae. This genus included 225 nominal Neotropical species when Spangler (1981) advised leaving identification to the "specialists." Unfortunately, there seem to be no specialists at the present time who are familiar with these species. Therefore, the literature had to be searched exhaustively to find the isolated descriptions and fragmentary keys developed mainly by Ochs (1964, 1965) for regions far from the Pantanal. The literature yielded little useful information, however. One of the two taxonomists responsible for most of the nomenclature was able to describe new species in as little a two printed lines. He continued his work through the 1920s before leaving the field to the second prolific author, who described dozens of new *Gyretes* species without ever even bothering to prepare a single illustration. Furthermore, the genitalia, which have come to be recognized as organs of primary importance in insect taxonomy, were not described at all until relatively late in the second author's career.

After about 40 years of describing *Gyretes* species with nothing more than a few lines of text, the second of these authors started to provide satisfactory illustrations, beginning about 1960. Thereafter, he produced a few works with minimum satisfactory descriptions, but he never reviewed many of his own previous publications or those of his predecessor in a way to permit other workers to distinguish the new species he was still regularly describing from those he had named in earlier years.

In theory, the remedy for poor published descriptions is a comparison of the specimen with the types maintained in museum collections. Even if scientists in South America had unlimited funds for flights to Europe and the United States, where many of the types have been deposited, they would still not come any farther with their work because many of the types have been lost or are severely damaged. In addition, just locating the museums where the types are most likely to be deposited is a difficult task. The first of the authors mentioned above with reference to *Gyretes* stated that his types were in his personal collection, which has had plenty of time since the 1920s to get lost. The pre-World War II specimens of the second author were apparently deposited in the

Senckenberg Museum in Frankfurt am Main, where they were destroyed by bombs during the war. In general, South American species have been described from few type specimens, and only a single type was designated for a great many. Therefore, there are few species represented in more than a single museum collection, and if that collection is not well maintained, the unique type may never be found again. Often, the single types that can be found lack essential parts or had been dissected or macerated, leaving little remaining for future examination. Thus, there are a great many species, and not just gyrinids, that cannot be distinguished with certainty from the literature and probably cannot be found in museum collections in a condition suitable to compare with fresh specimens. This leaves too much to guesswork and will continue to hamper ecological studies until taxonomy work is intensified to provide lectotypes of the local species to South American museum collections.

There are similar problems with the tracheophytes and even with the vertebrates. Hence, the names used in the text represent a state of the art taxonomy, which, however, must be accepted at the present time as a matter of necessity in lieu of more precisely defined species. Without basic taxonomic work, ecological studies are severely limited in scope and accuracy. It must be bluntly stated that the present policy of many government agencies and private organizations to withhold funds for fundamental biological research, especially systematics and taxonomy, has created insuperable barriers for the ecologists working in South American and other regions of the world in which the flora and fauna are still largely uncatalogued and inadequately described. The various manipulations using indices or counts of unknown organisms are little more than ecological pseudoscience when the basic knowledge of the species present is unavailable. Ecosystems, particularly those in the tropics, are far too complicated to be reduced to a single index number or placed in any meaningful numerical category.

After surveying the flora and fauna of the Pantanal, which for the above reasons required a great deal of time and effort, the functioning of the ecosystem was investigated. The interactions between species is regulated to a large extent by the seasons. Thus, analyses of community structure, food webs, survival strategies, and life cycles were undertaken during all seasons. Direct observation in the wetland yielded a great deal of factual information, which was enhanced by special studies undertaken by doctoral and master of science candidates. Some of these have still not been completed, and hopefully they will bring still more detailed information about specific processes in the Pantanal to light.

The overall purpose of the research project being carried out while this book was being prepared was to provide a coherent strategy for ensuring the conservation of this great wetland. The conclusions reached with this goal in view are presented in the final sections of the work. Presenting an approach to the environmental protection of a great wetland like the Pantanal is be no means sufficient. Past experience has repeatedly shown that gaining knowledge, no matter how well substantiated, is no guarantee that responsible leaders will reach reasonable decisions. There is still a long way to go before an effective plan for the conservation of this unique resource is implemented and the region achieves some protected status against selfish interests that threaten to grab quick profits and rob this and future generations of a great natural treasure.

2. Geography

In order to obtain enough data to reach and substantially support meaningful conclusions about the ecological conditions prevailing in the Pantanal during the short period allowed for the research project, the study had to be limited mainly to an area small and uniform enough to study intensively. In addition, however, adjacent regions had to be visited in order to determine how representative the findings are for the wetland as a whole. The sites selected for intensive sampling are located in the Municipal District of Poconé, near enough to the main roads to preclude excessive loss of time to reach them. For comparison, sites in the Municipal Districts of Cáceres and Barão de Melgaço were also visited during all of the seasons, and comparative data on the water chemistry, flora, and fauna were assembled. In addition, studies were conducted along streams in the adjacent upland regions of southern Mato Grosso, and observations were made of the terrestrial biotic communities, as well. This section is intended to provide basic geographical background information about the Pantanal and about the limited study area selected for intensive investigation.

2.1. Regional characterization

2.1.1. Location of the Pantanal

Most of the Pantanal is located in the Brazilian states of Mato Grosso and Mato Grosso do Sul. A small part of the region belongs to Bolivia, and an even smaller portion is located in Paraguay (Fig. 2.1). The region of the wetland itself lies roughly between 16° and 20°S and 50° and 58°W, but its borders are irregular, ill-defined, and variable from year to year. Furthermore, the river system to which it belongs tends to overflow onto extensive floodplains in other locations downstream from the main part of the Pantanal throughout Paraguay and into northern Argentina, and smaller but still rather extensive wetlands are encountered along the system as far south as the Rio de la Plata. As in the case of the Amazon System (Sioli, 1984), the part of the Paraná River System that lies in the tropical wet-and-dry climatic zone is responsible for considerable annual water level changes even in those parts of the river that lie far outside of the zone.

The borders of the Pantanal cannot be precisely shown on a map because, as already mentioned in the introduction, the area covered by water changes greatly during the course of a year. The terrain elevation of the region varies from about 75 m to about 200 m above mean sea level. A major part of the floodplain area is just a few meters above or a few meters below the average annual high water level, making a demarcation between the permanently dry and temporarily flooded areas is difficult. Furthermore, the time and amplitude of the rise in the water level also varies considerably from year to year. In any case, part of the wetland is always under water, including the courses of the main rivers feeding and draining the region throughout the entire year and the deeper-lying streams and small tributary rivers that always contain water of various depths, although that water may be lentic for part of the dry season and lotic for the rest of the year.

As Figure 2.1 shows, the Pantanal is located near the geographic center of South America, far from the sea. It is situated on a lowland plain that is bordered to the north and east by the uplands of the Central Brazilian Shield and to the west by foothills of the Andes. The rainfall over millennia has leached the soil and rocks in the drainage area of their soluble substances, and the water on the floodplains during the rainy season is similar in most physical and chemical properties to rainwater, as discussed in subsequent sections of this book. It is therefore an inland freshwater system without any influence from the sea, either in historic or recent prehistoric times. From the locations of the headwaters of the streams feeding the Pantanal (Fig. 2.2), it is also clear that the geologically younger Andes Mountains do not influence the water quality in any

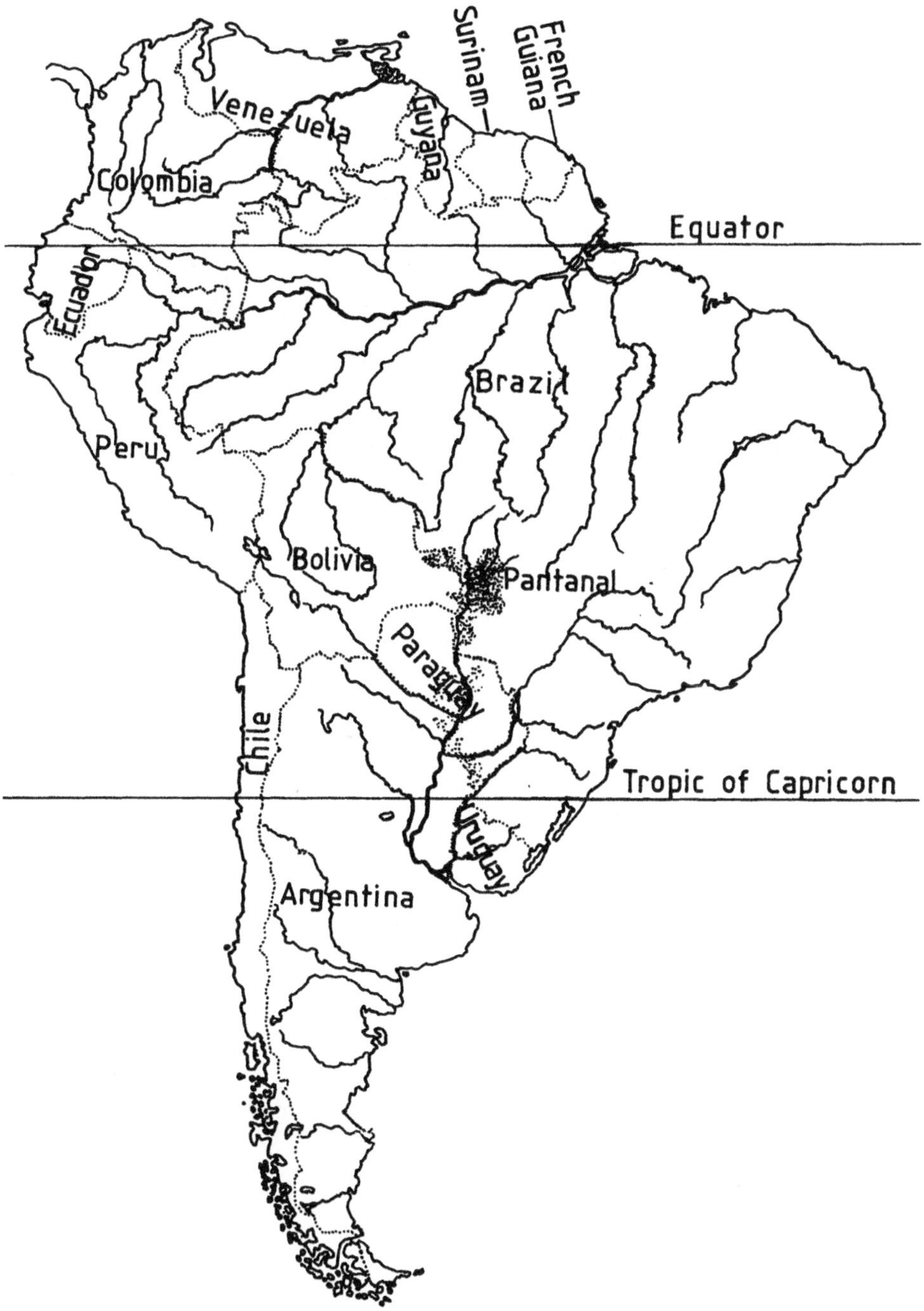

Figure 2.1: The location of the Pantanal, near the geographic center of South America, and its adjacent wetlands along the Rio Paraguay to the south (stippled areas).

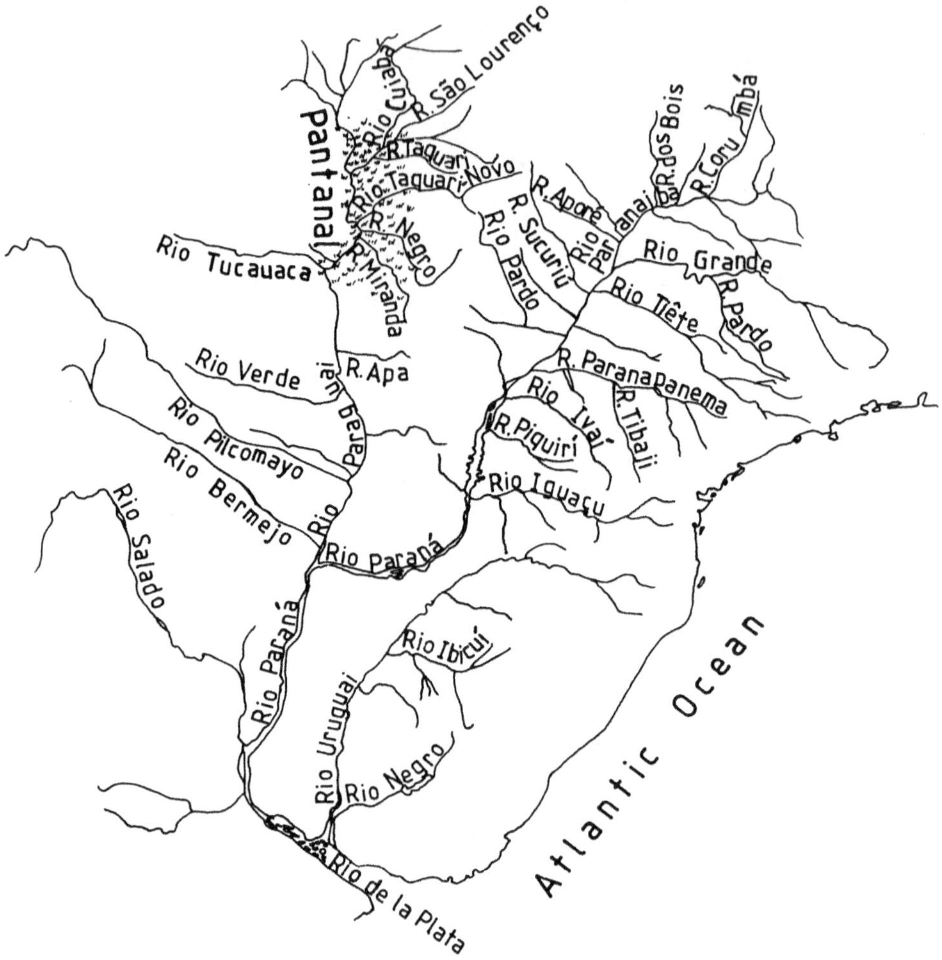

Fig. 2.2: A map of the Rio Paraná System, showing the location of the rivers feeding the Pantanal near the headwaters of the Rio Paraguay.

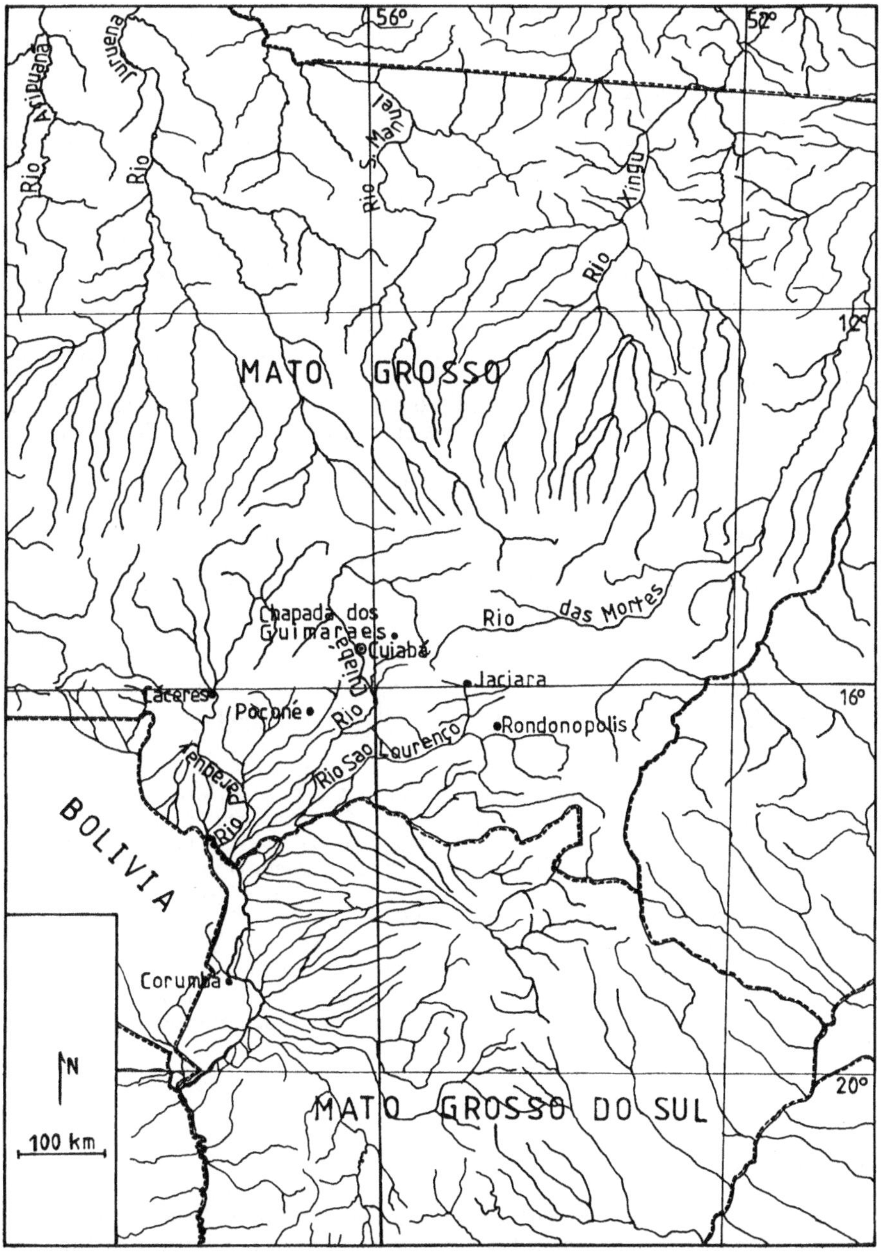

Figure 2.3: General map of the Pantanal Watershed in Brazil showing most of the states of Mato Grosso and Mato Grosso do Sul. The approximate location of the highlands dividing the watersheds of the Rio Paraná and the Amazon System is apparent to the north and east of Cuiabá, where the headwaters of the streams feeding both systems are located close to one another. The Rio das Mortes flows into the Rios Araguaia and Tocantins and empties into the ocean at the mouth of the Amazon. The Pantanal occupies most of the area between Poconé in the North and Corumbá in the South and stretches from just west of Rondonopolis to just west of the Bolivian border.

extensive part of the system, as they do in the Amazon Basin. The largest tributaries flowing into the Pantanal from the west join the Rio Paraguay south of the wetland.

The main river along which the Pantanal is situated is the Rio Paraguay, which is joined by the Rio Cuiabá from the east near the borders of Mato Grosso, Mato Grosso do Sul, and Bolivia. Most of the smaller rivers crossing the northern part of the wetland flow directly into the Rio Paraguay rather than joining the Rio Cuiabá before the confluence. To the south of the Rio Cuiabá, several other large rivers pass through the Pantanal to empty their water into the Rio Paraguay as it passes through the Brazilian state of Mato Grosso do Sul (Fig. 2.2). Between the Rio Cuiabá and Rio São Lourenço, a large river draining the area to the west of the wetland, a considerable number of anastomosing channels are formed near their confluences with the Rio Paraguay. Because the area through which they flow is flooded for much of the year, it is not easy to distinguish which is which on the maps. In fact, this is a common problem with the smaller rivers as well, which may not always be depicted at exactly the same locations on different maps. During the high water period, they are all confluent anyway, and during the dry seasons, it is often difficult to decide what should be considered a main channel and which of the low laying areas should be considered as oxbows or blind arms.

To the north of the Pantanal, the Rio Paraguay is joined from the west by the Rios Sepotuba and Cabeça, as well as a few smaller streams. These flow into the main river not far from their sources. As the river flows southward into the Pantanal, it tends to break up into smaller, parallel channels flowing past large islands and forming anastomosing branches among islands of various sizes, some of which completely or nearly disappear during the high water season. The entire region is located in the greater drainage area of the Rio Paraná fed through the Rio Paraguay and its own extensive system to the east. Downstream from the confluence of these two rivers, the Rio Paraná becomes a major water course, emptying into the broad Rio de la Plata, which separates Argentina from Uruguay. This is the second largest river system in South America.

The highlands located north of Cuiabá at a point just north of Chapada dos Guimarães and passing along the northeastern side of the São Vincente Mountains separate the Paraguay-Paraná and Amazon-Tocantins Watersheds. Thus, the drainage area upstream from the Pantanal in the state of Mato Grosso is considerably smaller than the area of that state drained through the rivers of the Amazon Basin (Fig. 2.3). The relatively short distance from the headwaters of these streams and rivers to the Pantanal accounts for the close relationship between rainfall and the water level in the Municipal District of Poconé (Section 5). Simply stated, the rainfall on the watershed does not have to flow very far to reach the northern borders of the Pantanal (Fig. 2.2). Of course, this is not the case in the southern part of the wetland. Thus, in the region of the Pantanal near the city of Corumbá, the high water level is usually recorded after the season of heavy rainfall has ended (Section 5).

2.1.2. Industry and agriculture

Another geographical feature of the Pantanal is its location far from the population centers of South America. Mato Grosso was one of the last regions in Brazil to be settled, and according to the estimate of the Instituto Brasileiro de Geografia e Estatística in Cuiabá for 1992, its present area of 901,420.7 km^2 was populated by only 2,107,698 people, most of them inhabiting the cities, the largest of which is the capital of Cuiabá. Because other regions are easier to exploit for agriculture, the Pantanal has remained relatively free from human activity, except that of the few cattle ranchers, fishermen, and gold miners, who generally concentrate around the edges of the great wetland or at least on relatively high ground. Although the population has been increasing rapidly in recent years through migration from more crowded regions in the East, the immigration has served mainly to increase the population of Cuiabá, the state capital, rather then the small towns or rural areas, except for those in which gold mines are located. Hence, the Pantanal is not an area currently subject to increased human activity, although there has been a slight increase in the number of tourists visiting the region and the numbers of fisherman traversing the Transpantaneira Highway to reach the Rio Cuiabá at Porto Jofre. This traffic has been encouraged by attention called to the natural wonders of this wetland by the Brazilian news and entertainment media.

Recent demographic growth in rural areas of Mato Grosso has resulted from a renewed interest in mining the gold dust, which has been deposited with alluvium in various regions near the Pantanal. As in the 18th century, gold has been the main factor motivating many settlers to come to Mato Grosso. Activities of this kind have recently been especially vigorous in the Municipal District of Poconé and near the town of Cangas. This represents a "false growth", in that no concurrent industries are developed, and as soon as the richest deposits have been exploited or the world market price of gold falls, the mining will be abandoned, and the many people working in the mines will leave the region. The gold from the mines leaves Brazil, legally or illegally, and no long-term economic gains are achieved from this activity.

Agriculture represents a more stable basis for the settlement of southern Mato Grosso. In the Pantanal, cattle ranching has long been the primary form of food production in the wetland. Along the rivers, fishing has provided a livelihood for a limited number of inhabitants. Fruits and vegetables are raised to a moderate extent, but the production has never become intensive because of the limited productivity of the land caused by the high temperatures and soil deficiencies as well as the limited market for the products, which cannot be shipped rapidly or economically enough to reach the markets in the East while still fresh. Recently, intensive sugar cane and soybean production has enjoyed a brief popularity in the upland regions of the watershed, but this form of agriculture has little prospect for long-term success for economic reasons, as discussed in Section 13.2.4. It promises, however, to cause severe damage to the soil and biota before it is abandoned.

The typical cattle ranches in the Pantanal are enormous in size, but only part of their vast areas are suitable for grazing. An intensification of the production would cause the already low market prices for beef to fall even more, and a considerable number of cattle in the Pantanal would die before they could be sent to market, especially during the dry season. The ranchers do not usually concern themselves much about mortalities among the cattle because they know that more beef is being produced than the present market requires. Beef is extremely cheap in the restaurants of Mato Grosso, but it is still not possible to market more than a small proportion of the total production. Vultures feeding on dead cattle and horses are therefore a familiar sight in the Pantanal. The frequent mortality of livestock reduces the overproduction but does not provide the ranchers with any more income for the beef produced on their land.

While there are many hungry people in the eastern part of Brazil, the economic conditions prevent them from earning enough money to afford beef. Thus, under the rules of the present world economic system, they are denied food that is in superabundance. Formerly, the beef from Mato Grosso, Argentina, and other parts of South America was exported in large amounts to Europe and North America. Now, however, an oversupply of high quality beef produced under agro-industrial conditions in North America and as a by-product of an intensive dairy industry in Europe has saturated those markets and induced various governments to make it more difficult to import beef. Thus, the production of soybeans for export in Mato Grosso is actually in direct competition with the traditional ranching since the soybeans are purchased as feed for the cattle raised in North America and Europe, and the meat produced there has driven the South American beef from the market. Taking only the energy costs into consideration, it would be cheaper to produce beef in South America and ship it to other regions of the world than to ship much greater tonnages of feed long distances to produce the cattle nearer to the local markets.

The largest area of the Pantanal is located in the State of Mato Grosso do Sul, which has an area of only 357,471.5 km^2. Since the area of the Pantanal in this state is roughly double that in Mato Grosso, the wetland covers a much greater percentage of the state:

roughly 25%. In contrast, only about 5% of Mato Grosso is accounted for by the Pantanal. Human influence on the wetland might be expected to be considerably greater in Mato Grosso do Sul, which had a population of 1,943,069 inhabitants in 1990 and a demographic density of 5.43 inhabitants per km^2. These figures may be misleading, however, because the population of Mato Grosso is overwhelmingly concentrated in the region of Cuiabá and other cities at the northern edge of the wetland. So although the demographic density is only 2.44 inhabitants per km^2, the human impact on the Pantanal in Mato Grosso may well be greater than that in Mato Grosso do Sul, where a considerable portion of the human population is concentrated to the south and east of the capital, Campo Grande, and hence at the opposite end of the state from the Pantanal.

2.1.3. Transportation

The surface transportation system in Mato Grosso has always been extremely limited. Until the present century, Cuiabá had most easily been reached through the river system. Overland travel from the states in eastern Brazil was arduous as no reliable system of roads existed. As a result, the port of Cuiabá was quite well developed. Since the completion of a paved highway, however, the river steamer traffic has been discontinued, and the port has lost its importance.

Today, in the entire region to the north of the Pantanal, there are still no railroads. In contrast, the capital of Mato Grosso do Sul, Campo Grande, is connected to the cities in the east by a railway, which continues westward into Bolivia. This system was part of an early attempt at constructing a transcontinental railroad in South America. Although it does not represent an important means of transportation at the present time, it could be improved whenever there is a demand. There are presently no definite plans to extend the railroad northward to Cuiabá, although this would be the most sensible and economical way to supplement the highway system.

The road traffic between Cuiabá and eastern Brazil is limited to a single main highway running southward toward São Paulo via Campo Grande with a branch about 100 km east of Cuiabá toward Goiania, Brasilia, and the state of Minas Gerais. In the opposite direction, this highway continues north and westward from Cuiabá toward Porto Velho in Rondônia. From a point along this road not far from Varzea Grande, the sister city of Cuiabá, the road to Poconé branches off and runs generally southward. Another paved road leads from Cuiabá to the town of Chapada dos Guimarães, a favorite weekend resort on the plateau overlooking the lowlands in which Cuiabá and the Pantanal are located. Cooler temperatures prevail there, permitting the Cuiabá residents to escape the heat of the city. There is also a paved road leading from Cuiabá to Santo Antônio de Leverger, a city farther downstream along the Rio Cuiabá (Fig. 2.4).

In addition to these paved highways, there are many unpaved roads leading to smaller cities in the region of Cuiabá. Some of these lead to Barão de Melgaço and smaller towns near the Rio Cuiabá, while others connect various towns in the upland regions with the state capital.

As the name implies, the Transpantaneira Highway was constructed to cross the Pantanal from Poconé in Mato Grosso to Corumbá in the state of Mato Grosso do Sul. This highway has remained unpaved and crosses about 130 wooden bridges along its 147 km to the hamlet of Porto Jofre on the Rio Cuiabá. From this point, construction was abandoned, and a bridge was never constructed to cross the main river into the neighboring state. The road now serves various ranches located along its route and a few tourist hostels. Traffic is normally light, but some of the vehicles that the wooden bridges have to support are remarkably heavy, carrying cattle from the ranches, tourists with much baggage to the ranches and various hostels, and fishermen with boats and supplies to the Rios Cuiabá and São Lourenço at Porto Jofre. The bridges are frequently in bad condition because of a shortage of funds for repair, and the unpaved surface of the road develops many deep holes during the rainy season.

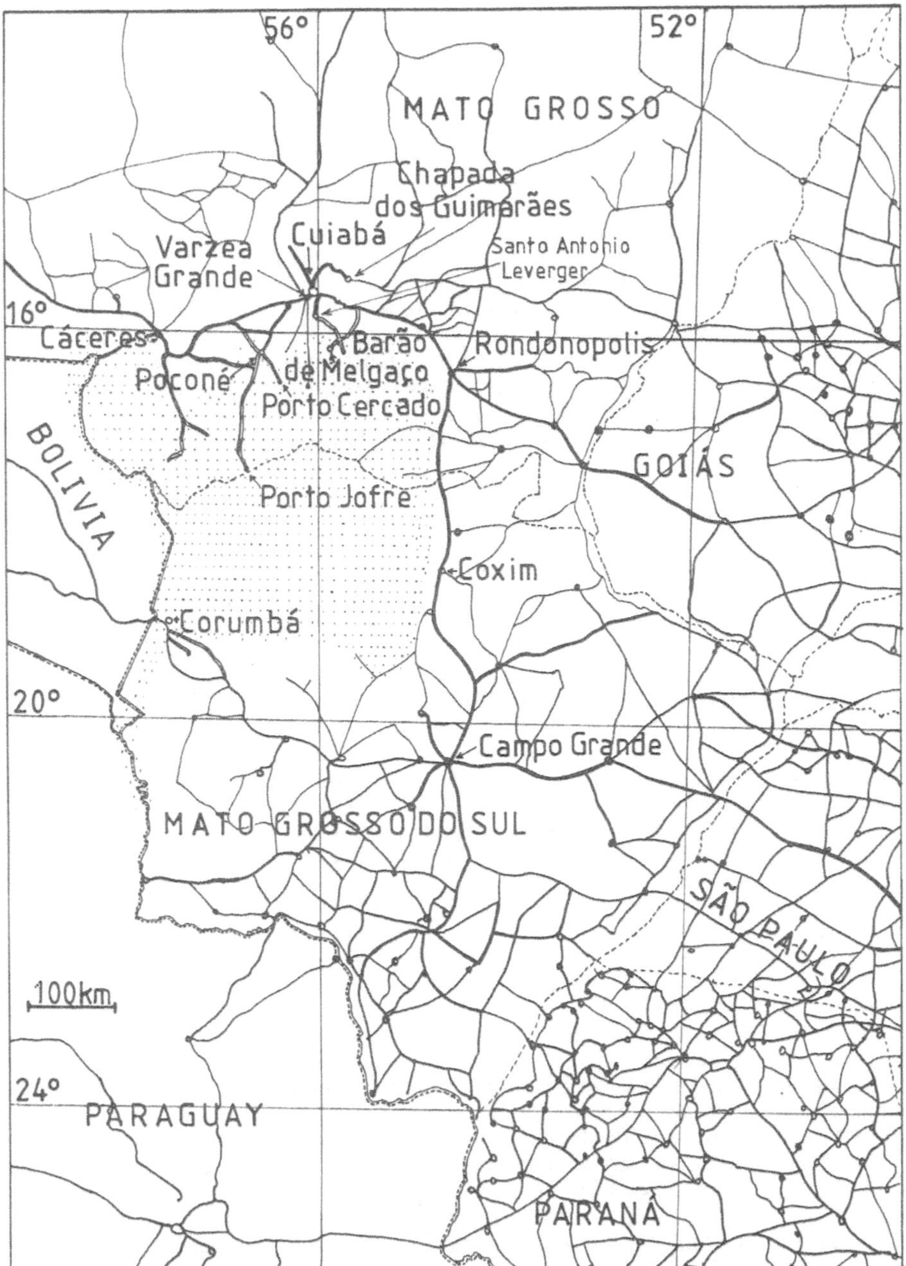

Figure 2.4: Map of the southern part of the Central Western Region of Brazil showing the main roads and a few of the unpaved roads used to reach sampling sites in the wetland. The general sparseness of the road network in the Pantanal and the region to the north is apparent. Thick solid lines represent well-traveled paved roads, while thinner solid lines are used for less traveled roads, which may not all be paved. Unpaved roads in and near the Pantanal are depicted as thin parallel lines.

There are a few short sections of the Transpantaneira leading northward from Corumbá, but construction in Mato Grosso do Sul was abandoned early. There are some trails that are passable quite far northward during the dry season (A. Pott, pers. comm.), but there are no means available to cross the Rio Cuiabá should anyone come that far.

Presently, a trip over paved highways from Cuiabá to Corumbá by way of Campo Grande covers roughly 1200 km. While the Transpantaneira would have shortened this distance considerably, the time saved driving over a road in the condition that the Transpantaneira is usually in would be negligible, and considerable damage would be sustained by the vehicles through wear and tear on the suspension and steering. While the cost of constructing this highway would have been great, the cost of maintaining it properly would have been astronomical, and the economic value of such a road would certainly not cover these costs, especially considering the extremely sparse human population in this part of Brazil.

Many ranchers oppose improvement of the Transpantaneira because a better road would bring more unwelcome visitors and change the infrastructure of the region. Some of the "pantaneiros" have lived in the Pantanal for many generations, and they have mixed feelings about the augmentation of the tourist industry.

2.1.4. Subdivisions of the Pantanal

The Pantanal is not a wetland that can be easily studied as a unit. It is rather a patchwork of many kinds of habitat that must all be understood in order to learn why the ecosystem functions the way it does. The habitats are discrete but constantly interact in ways that change with the season. The Pantanal can be broken down into many kinds of units, each of which deserves individual attention. Some of these are peculiar to the Pantanal, while other are continuations of systems that exist far from this wetland. Among examples of the habitats peculiar to the Pantanal and a few small wetlands in tropical South America are the vast temporarily flooded plains, which occupy more area than any other kind of habitat and undergo complete transitions from wet to dry and dry to wet every year; the permanently marshy lowlands, which are not notably widespread in the Pantanal and occur chiefly near the large rivers; the small rivers in the Pantanal, which flow rapidly during the late rainy and early dry seasons and become standing water bodies during the late dry season.

The geographic subdivisions of the Pantanal are depicted in Figure 2.5. Although these are mainly political divisions, they display considerable geographical differences because they are drained by different tributaries. The findings of this study apply specifically to the northern part of the Pantanal located in the Municipal District of Poconé and State of Mato Grosso. The approximately 2/3 of the Pantanal located in the State of Mato Grosso do Sul can be considered distinct, and the findings for Poconé cannot necessarily be carried over to the regions farther south. The reasons for the differences are caused to a large extent by the interaction of climatic and hydrological factors, which will be discussed in subsequent sections.

Only three subregions of the Pantanal and a small part of a fourth are located in the Brazilian state of Mato Grosso (Pott and Pott, 1994). These are the Districts of Poconé, bordered to the south and east by the Rio Cuiabá and to the west by the Rio Paraguay; Cáceres, located between the valley of the Rio Paraguay to the east and the Bolivian border to the west; Barão de Melgaço, forming a wedge between the Rio Cuiabá to the northwest and Rio Correntes to the south; and Paraguai, a narrow strip of land on both sides of the Rio Paraguay, the northernmost part of which is in Mato Grosso, and the rest, in Mato Grosso do Sul. Although the three main northern districts are located between the same lines of latitude, their prevailing physical, hydrological, and biotic conditions are distinct enough to give each a character of its own.

The subregions located in the state of Mato Grosso do Sul are even more distinctive, being located at latitudes progressively farther south and farther downstream along the river system. Other than the narrow Paraguai Subregion running north-south along the

Figure 2.5. The ten generally accepted subdivision of the Pantanal, according to Magalhães (1992). In Portuguese, these are sometimes referred to as *pantanais*, suggesting that each one is a separate and distinct wetland.

western edge of the Pantanal, the subregion of Mato Grosso do Sul located farthest to the north is Paragua's, a vast area between the Rio Correntes in the north and Rio Taquari to the south. Another vast area is occupied by the Nhecolândia Subregion to the south of the Rio Taquari and north of the Rio Negro. The southeastern corner of the Pantanal is occupied by the Aquidauana Subregion, bordered in the north by the Rio Negro and to the west by the Rio Aquidauana. The Abobral Subregion is a narrow strip of land to the south of the Rio Negro, west of the Rio Aquidauana, and east of the Rio Paraguay. To the south of this is the Miranda Subregion. The southwest corner of the

Pantanal is occupied by the Nabileque Subregion, with a narrow extension along the Paraguay border extending south of the 22° parallel. All of these regions share characteristics with the northern part of the wetland but are sufficiently distinct to demand separate ecological studies, some of which have already been undertaken by different groups of Brazilian scientists. Generally speaking, until the project to investigate the Pantanal of Poconé was undertaken, the southern part of the Pantanal was much better covered in the scientific literature than the northern part. Many authors still refer to the southern part of the wetland as the "Pantanal of Mato Grosso" because the region had been studied before Mato Grosso do Sul became an independent state of Brazil. Previously, both this state and Rondônia to the north were included in the vast state of Mato Grosso. Today, however, it would be less confusing to use the term Pantanal of Mato Grosso to refer to the northern part of the wetland and Pantanal of Mato Grosso do Sul to designate the southern part. Geographical as well as political and administrative differences would justify this. Sometimes, the subregions of the Pantanal are referred to as "*pantanais*" (Magalães, 1992), giving each the status of an independent "Pantanal" or wetland.

Turning again from the districts and returning to the differences among the discrete kinds of habitat encountered in the wetland, we should distinguish more precisely between those endemic to the Pantanal and those that are extensions of the even more extensive systems bordering them. The habitats that are continuations of those also commonly encountered outside the Pantanal include the large rivers with their fish fauna, encompassing species that can be found far upstream and downstream from the Pantanal as well as in the stretches of river traversing the wetland. The gallery forests along these rivers, where they have not been eliminated, are also the homes of certain species that can be found both within and outside the wetland, wherever such kinds of habitat exist. The lakes, called *baias*, which occur along the rivers in the Pantanal, mainly at the edge of the wetland areas, are also similar to such water bodies at somewhat higher elevations that do not have contact with the Pantanal floodplain. Their physical, chemical, and biotic features are quite different from those in the rivers and floodplain areas and are also generally distinct from one another. Finally, there are the species of the savanna woodland or *cerrado*, which penetrate into the Pantanal wherever the terrain elevation is sufficiently high to keep them from being drowned during the high water period.

2.1.5. Selection of Sampling Sites

In selecting the sites from which samples were taken, emphasis was placed on the temporarily flooded areas and small rivers draining them because these account for the greatest part of the surface area in the Pantanal. However, other kinds of habitat were also studied according to a regular but less frequent schedule in order to avoid a one-sided presentation of the limnology of the region. It must be emphasized that the Pantanal is a composite of all of the different kinds of habitat and their resident biotic communities, and many species move freely from one kind to another, interacting with the resident species in both. Figures 2.3 and 2.6 show the general locations of various kinds of discrete habitat types within the Pantanal of Mato Grosso and some of their biotic features.

The ecological research was conducted mainly along the Transpantaneira Highway, on ranches that can be reached from the Highway, and at various other sites in the Pantanal that can be reached by car or boat without undue hardship. The stations visited most frequently are shown in Figure 2.6.

The Rio Cuiabá and smaller rivers to the north and south flow westward generally parallel to each other. To obtain information about the main river system, samples were taken during all seasons from two sites along and near the Rio Paraguay. One of these sampling sites was located on the Fazenda Santo Antonio das Lendas along the Rio Paraguay between Cáceres and its confluence with the Rio Cuiabá, while the second was located in the city of Cáceres itself (Fig. 2.6).

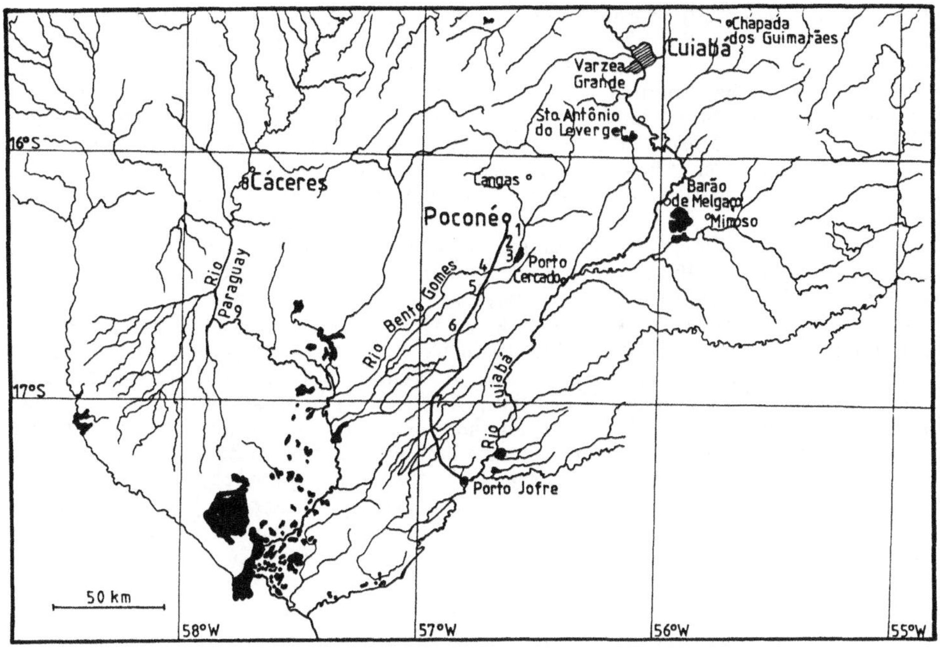

Figure 2.6: A map of the northern Pantanal showing the Transpantaneira Highway along which most of the sites systematically sampled during the study are located. It runs from Poconé to Porto Jofre. The dark areas are permanent lakes. 1 = exposition park of Poconé beside the Rio Bento Gomes; 2 = Birici, a slightly depressed area through which the water drains from the flood plain during the *vazante*; 3 = the west shore of the Baia Piuval, a large lake along the Rio Bento Gomes; 4 = Rio Bento Gomes where it passes beneath the Transpantaneira Highway; 5 = Rio Novo where it crosses the Transpantaneira; 6 = Rio Frio or Rio Claro where it crosses the Transpantaneira; 7 = Rio Pixaim and a shallow pond that are confluent during the high water period and separate during the dry season; 8 = Rio Paraguay at Cáceres; 9 = arm of the Rio Paraguay on the Fazenda Santo Antonio das Lendas.

For the routine, periodic sampling and analyses, sites were chosen at the Exposition Park of Poconé (Fig. 2.7), situated on the Rio Bento Gomes at the edge of the Pantanal; the Baia Piuval on the Fazenda Ypiranga (Fig. 2.8), a large permanent lake along the course of the Rio Bento Gomes at the point where it enters the main part of the Pantanal floodplain; a stream at a point locally called Birici on the Fazenda Ypiranga about 15 km south of Poconé (Fig. 2.9); the Rio Bento Gomes where it crosses the Transpantaneira Highway (Fig. 2.10); the point where the Rio Novo crosses the Transpantaneira Highway (Fig. 2.11); a site where the Rio Frio crosses the highway (Fig. 2.12); and the Rio Pixaim and an adjacent permanent pond near the highway that form a single water body during the high water period but are isolated from one another during the dry season (Fig. 2.13). The floodplain is almost entirely under water during the season of flooding and is contiguous with all streams and rivers crossing the wetland and with the southern part of the Baia Piuval. At that time, the rivers and streams contain water that flows rather rapidly, while the water on the floodplain is lentic. During the rainless seasons, the floodplain is parched, and the streams and small rivers contain standing water. Only the large rivers remain lotic throughout the year.

Samples for comparison were taken from other water bodies in different parts of the Pantanal. One site on the Rio Paraguay at an abandoned state research station in the city of Cáceres and another along the same river near its confluence with the Rios Cuiabá and São Lourenço on the Fazenda Santo Antonio das Lendas were visited once during each season for sampling and water analysis (Fig. 2.14). The Baias Acorizal, Recreio, and Porta de Fora in the Pantanal of Barão de Melgaço were also visited periodically for sampling. In addition, analyses were conducted at various times at sites along the Rio Cuiabá, including Porto Cercado to the east of Poconé and Porto Jofre at the terminus of the Transpantaneira Highway. When these sites were visited, samples from the streams and wetlands draining the Pantanal near these sites were also analyzed. Furthermore, large areas were subjected to regular observation, and the flora and fauna they supported were determined and catalogued throughout the entire annual cycle. These areas included a wide strip of land on both sides of the Transpantaneira Highway, a large area of the floodplain on the Fazenda Ypiranga, and the shores of the Rio Bento Gomes from the exposition park of Poconé to the dense thickets that separate the navigable section of that river from the Baia Piuval.

Figure 2.7: Aerial view of the exposition park of Poconé beside the Rio Bento Gomes during the season of flooding. The sampling site is located at the end of a road, mainly obscured by trees, which runs to the edge of the river at a point just to the right of the large tree left of center in the photograph.

Figure 2.8: Looking west across the Baia Piuval. Most of the samples taken from this lake were obtained at the boat dock shown near the center and from a boat in the area of the lake shown in the photo.

Figure 2.9: The shallow drainage channel called Birici on the Fazenda Ypiranga during the dry season, showing the aquatic plants as they are dying off. *Eichhornia azurea* forms a dense bed at the lowest-lying sites, where it dies off only during the driest years.

Figure 2.10: The bridge along the Transpantaneira Highway crossing over the Rio Bentc Gomes shown during the season when the flood waters are deepest.

Figure 2.11: The bridge along the Transpantaneira Highway crossing the Rio Novo during the *vazante*, the season of rapidly falling water levels.

Figure 2.12: The bridge along the Transpantaneira Highway crossing the Rio Frio, a branch of the Rio Claro, during the dry season.

Figure 2.13: Looking southward over the Rio Pixaim and the shallow pond along the Transpantaneira Highway during the season of maximum water depth, when these water bodies form one continuous flooded system. During the dry season, they remain separate. The sampling sites were located near the palm tree to the left of the bridge.

Figure 2.14: The sampling site beside a channel of the Rio Paraguay on the Fazenda Santo Antonio das Lendas north of its confluence with the Rios Cuiabá and São Lorenço.

Aerial surveys were undertaken during the rainy and dry seasons to photograph the region from the northern border to Porto Jofre, and a photographic record was made of the kinds of vegetation covering much of this area.

In addition to the program of sampling and analysis in the Pantanal, parallel studies of streams and rivers in the watershed upstream from the wetland were undertaken (Heckman, 1995). These include the Rios Cuiabá, Arica Mirim, and Coxipó and the Corrego Guanandi. In addition, a study was made of the headwaters of the Rio das Mortes, a river arising near the streams feeding the Pantanal but flowing northward toward the Tocantins Basin and reaching the Atlantic Ocean through the Rios Araguaia and Tocantins, over a distance totaling more than 2000 km (Fig. 2.3). In addition, a few data were obtained from upland streams to the east, including the springs at Aguas Quentes and the Corregos Filipa, Ajudicaba, Correia, and Formoso (Heckman, 1995).

To the south of the area of the watershed investigated, the Rios São Lourenço, Itiquira, and Taquari flow into the Pantanal, joining the Rio Cuiabá from the south upstream from its confluence with the Rio Paraguay. In the southern part of the Pantanal in the State of Mato Grosso do Sul, the Rio Paraguay is joined by the Rios Taquari Novo, Negro, and Miranda from the east, and farther south by the Rio Apa at the border between Brazil and Paraguay. All of the main rivers in the North and many in the South were visited, but examination and analysis of the water to the south of the Rio Cuiabá were limited to spot samples taken from the Rio Miranda, Rio Paraguay, and the wetlands north of Corumbá. Other information on these southern regions of the Pantanal was obtained from the literature for comparison, although the volume of data available is still not at all satisfactory, especially considering the vast area covered by the wetland.

2.1.6. The drainage area

The upland regions surrounding the Pantanal are occupied by a mosaic of different kinds of terrestrial biotic community. These are generally more important for the ecology of the Pantanal than they would be for wetlands with more seasonally stable hydrological conditions. During the dry seasons, many of the terrestrial animals are able to migrate from the upland regions covered by forest and savanna into the desiccated areas of the floodplain. Furthermore, the typical vegetation of the terrestrial regions penetrates into the Pantanal on somewhat elevated formations, such as the so-called *cordilheiras*, which can support different kinds of forest or savanna flora or have been modified into pasture land by local cattle ranchers.

Unfortunately, the terminology used by different authors for the various kinds of vegetation is not yet well defined, mainly because the dominant plant species in each floral aggregation are only beginning to be reported (Prance and Schaller, 1982). Various names for common plant communities are available in Portuguese, and only rather imprecise equivalent words are available in English. It is therefore necessary to provide a set of definitions, which will have to be modified in the future to accommodate the many plant communities that have not yet been incorporated in any classification system.

Figure 2.15: Typical hummocks, called *capões de mata*, that are scattered throughout the floodplain. These remain slightly above the high water level throughout the year. Most of the ranch houses have been built on cleared hummocks.

While the terrestrial vegetation in the Pantanal is confined to a rather small number of biotic communities (Section 3.1.2), that in the upland parts of the watershed shows an extraordinary variety. The most important of the terrestrial plant aggregations in the wetland region include the annuals on the floodplain during the dry season, the perennial plants on the hummocks (Fig. 2.15) or temporary islands that remain above the annual high water level, the gallery forests along the river banks (Fig. 2.16), and the artificial plant aggregations on and near the ranch houses. The upland regions, in contrast, are covered by more communities than can be defined and discussed here. However, it is necessary to describe only a few of the most important ones, especially those that cover vast expanses of the region and directly interact with the wetland biota. It is important to note that ecotones often support a much richer biota than either of the two kinds of biotic communities that they separate (Odum, 1971). Thus, the border region between the Pantanal wetland from the dry forests surrounding it is especially important to investigate (Veloso, 1947; Prado *et al.*, 1992).

Figure 2.16: Gallery forests line the rivers and streams in many places throughout the Pantanal. Their plant communities are generally the richest in terrestrial flora, not only in the Pantanal, but also in the catchment area.

One of the typical kinds of vegetation encountered in southern Mato Grosso is the savanna woodland or woodland savanna, called *cerrado* in Portuguese, and named and defined differently by various authors, depending upon whether they take the term in a strict or broad sense. As the following discussion shows, the present vocabulary used to describe the terrestrial communities and habitats surrounding the Pantanal is confusing, to say the least. The general term, *cerrado* is subject to a more differentiated definition, whereby several recognizable plant communities often encountered within the general "*cerrado*" areas marked on the maps are given names of their own. The term *cerrado* in the strict sense, limited to the savanna woodland but not used for the woodland savanna, may be a fire climax, that is, a kind of dry woodland dominated by species known to be resistant to fire (Fig. 2.17). The frequent burning of large areas during the dry season is a major determinant of the kind of vegetation present (Askew *et al.*, 1970). The resultant *cerrado* can also be called a shrub savanna, according to Por (1995), but Sarmiento (1983) reserved this term for the "*campo sujo*". The actual climax in the region has been called "*mata xeromorpha*" by Rizzini and Heringer (1962) and "sclerophyllous woodland" by Sarmiento (1983); it corresponds to what is locally called *cerradão*, which is a true savanna forest with a thin scattering of tall trees (Fig. 2.18).

The *cerrado* in the broad sense is extremely widely distributed in Brazil south and east of the Amazon forests. Colli (1991) estimated that it covers roughly 1,500,000 km^2 of Brazil. It occurs on all sides of the Pantanal in Brazil (Goodland and Ferri, 1979), and its typical flora and fauna have a great influence on the wetland, penetrating into it anywhere the terrain elevation is high enough to afford them some protection against the annual flooding. From the geological point of view, the alluvium rich in silica and poor in plant nutrients and humic matter provides a soil easily subject to erosion and through which water can drain rather rapidly. It therefore has only a limited ability to retain the enormous amounts of rainwater that fall in relatively short periods of time. As a result, the *cerrado* has the appearance of a semi-arid shrubland throughout much of the year. Eiten (1972) provided a general definition of it according to its typical vegetation.

Figure 2.17: A scene on the *alto plano* on which the streams arise that flow into the Pantanal. In the foreground is a pasture from which almost all of the natural vegetation has been removed. Behind it is a band of typical *cerrado* landscape dominated by woody species not only resistant to seasonal desiccation of the soil but also to frequent fires. Nearly all of the trees are deciduous. Some evergreen species are found among the line of trees separating the *cerrado* from the cultivated fields far in the background. These belong to a gallery forest bordering one of the streams.

To the north of the savanna forest, a vast belt of dry forest, or *mata seca*, extends far into the Amazon Basin (Askew *et al.*, 1970). In this kind of forest, the trees form a closed canopy, and some evergreen species are present (Ratter *et al.*, 1978). In southern Mato Grosso, this kind of forest is encountered only sporadically.

Other kinds of habitat located in and around the Pantanal are located on the large areas of well-drained, wooded *campos cerrados*, on which many species of woody plant typical of the widespread *cerrado* and *cerradão* adjacent to the Pantanal are able to develop. However, these woody plants are sparser, resulting in the name "wooded savanna" being applied in a strict sense only to this kind of community (Sarmiento, 1983). Where the trees on such fields have been cut or destroyed by fire, a community of insects develops, dominated by termites and ants, which produce the so-called *campo murundu* landscape (Fig. 2.19). These fields are usually unflooded or covered by only thin layers of water, permitting the termites to construct mounds on top of the charred tree stumps. Large ant colonies often occupy parts of the nests built by the termites, maintaining what has become one of the typical landscape forms in the Pantanal (Sections 3.1.2, 8.12.4). Areas located at higher elevations along the border of the Pantanal are usually occupied by *cerrado* with characteristics produced by the fires set to provide extensive cattle pastures, or by *cerradão*, which is less affected by fire. A few of the dry open woodlands have been cut to make way for the cultivation of sugar cane or corn. The biotic communities that develop on these sites will be discussed in Section 9.1.8.

Sarmiento (1983) provided a set of definitions for the main kinds of community encountered in the upland parts of the Pantanal watershed according to the percentage of land covered by woody plants. In sclerophyllous woodland or *cerradão*, more than 40% of the area is covered by such plants. Savanna woodland or *cerrado, sensu stricto*, has a 15% to 40% cover. If the cover is 2% to 15%, the community is wooded savanna or *campo cerrado*. If some woody plants are present but the cover is less than 2%, it is called tree or scrub savanna, depending upon how tall the woody plants are. This is the *campo sujo* in Portuguese. If only grasses and herbs are present, it is called savanna grassland or *campo limpo*. These definitions are easy to apply, but other authors have used the nomenclature differently, as mentioned above.

The greatest floral and faunal variety in southern Mato Grosso is found in the diverse small biotic communities confined to areas with relatively unusual conditions. Among these are the lines of gallery forest bordering upland streams, called *mata ciliar*; the many different kinds of plant community on the sides of hills and low mountains, including those dominated by different species of palm; and the vegetation on the rocky outcroppings near the borders of the lowland plain and the high plateau. The zonation of the plants on the outcroppings is particularly interesting, since xerophytic plants generally are predominant on the north sides, which are exposed to the most sunlight, while shade and moisture-loving species appear on the well shaded south side. A different kind of flora, rich in bryophytes and ferns, is found on the rocky walls continually subjected to a fine spray from waterfalls.

Figure 2:18: A thin forest with scattered tall trees that develops in areas not often subject to fires is called *cerradão*, although in much of the literature, it is not distinguished from the *cerrado*.

Figure 2.19: The landscape called mound savanna or *campo murundu* is marked by mounds where tree stumps were covered by large termite mounds, which gradually become overgrown with various plants and inhabited by many animals other than the termites that built them.

An interesting community is that inhabiting ground kept moist by seeping ground water. This so-called *campo úmido* occurs more frequently in the Pantanal than in the surrounding catchment area so it will be treated in Section 3.1.2 and subsequent sections.

The importance of the upland regions of the watershed to the equilibrium in the Pantanal will become increasingly apparent as more of these areas are cleared for farming, greatly increasing leaching and erosion. This discussion will be continued in subsequent sections.

2.2. Biogeographical features

Much information will be provided on the distribution patterns of the individual species in subsequent sections of this book. To avoid duplication of the information, only a brief outline of the biogeographical peculiarities of the Pantanal will be provided here. The structure of the flora and fauna in the wetland has been determined by the geological development of the region (Section 3), the present climate (Section 4), the hydrological system (Section 5), and to a lesser extent by the water chemistry (Section 6).

All of South America is often assigned to the Neotropical Region for want of a better way to characterize the biota of the southern part of the continent (Banarescu, 1995). However, a growing number of studies have established the existence of a common southern fauna which can best be described as circum-Antarctic. Various insect genera are found in the cooler parts of South America, Australia, and New Zealand. The taxa of this cool climate fauna of the southern hemisphere are usually best represented in South America, where it is distributed in southern Argentina and Chile and northward along the mountain ranges of the Andes well into the tropics (Brundin, 1966, 1967; Illies, 1965, 1966, 1969).

Among the southern fauna of temperate South America and the high elevations in the Andes, there are a considerable number of endemic microfauna species, including rotifers, cladocerans, and copepods (Sendacz, 1993). Among the truly Neotropical species, however, only a small proportion of the rotifers and cladocerans are not distributed on other continents (Koste, 1978; Sendacz, 1993), while the copepods, especially the harpacticoid species, are overwhelmingly endemic to South America (Reid, 1993). The tropical macrofauna, on the other hand, has many endemic South American taxa and only a few affinities to the fauna of Africa, exemplified by the fish families Cichlidae and Lepidosirenidae (Banarescu, 1990), and tropical Asia, where the snail family Pilidae, often called Ampullariidae, is represented in the same kinds of habitats as in South America and Africa. These affinities are almost exclusively at the generic or familial rather than the specific level. In contrast to the fishes and molluscs, many of the mammals are clearly immigrants from the Nearctic Region. The deer, puma, and raccoons are examples of such species. Moreover, some present day Nearctic faunal elements apparently originated in South America, crossing the Central American land bridge soon after it formed. The nine-banded armadillo is one example. The migratory bird fauna hardly need be mentioned, there are so many examples of species that make long north-south migrations and visit both the Nearctic and Neotropical regions each year (Sick, 1988). From the findings of the floristic and faunistic studies conducted during the research project, especially those presented in Sections 7 and 8, the biota of the Pantanal can be classified as predominantly Neotropical with many cosmopolitan and circumtropical species of microorganism present and relatively little representation by southern, circum-Antarctic plants and animals.

The phytogeographical peculiarities of the Pantanal are difficult to recognize because of the large number of macrophyte species that have been introduced during the past five centuries. Similarly, many South American species have been introduced to other parts of the tropics, where some, such as the water hyacinth (Gopal, 1987), have created serious economic problems (Ashton and Mitchell, 1989). There is hardly any trace of a flora endemic to the Pantanal. Nearly all of the macrophyte species encountered also occur in other parts of South America (Conceição and Paula, 1986), and the great majority of the terrestrial plants that are not known to have been introduced by man and most of the aquatic species, as well, are confined to the Neotropical Region. In contrast, the algae are nearly all known to be widely distributed, usually circumtropical or cosmopolitan (De-Lamonica-Freire and Heckman, 1996), but there is little documentation of introductions of the algal species from other parts of the world. Many of the species may well have already existed before the continents drifted apart, but they may also have been introduced from one continent to all of the others in recent historical times through shipping since vast amounts of unsterilized fresh water must have been transported as drinking water for the crews and ballast between South America and all of the other continents during the past 500 years. It will take a considerable amount of paleontological research to determine which of these two hypotheses is correct for each of the algal taxa.

Like the algal flora, the protozoan fauna of South America encompasses almost exclusively species found throughout the rest of the world (Cunha, 1916; Bovee, 1957; Green, 1975; Vucetich, 1970, 1972, 1976, 1980, 1989; Walker, 1982; Rolla *et al.*, 1990; Lopes, 1993). The sarcodine species found in the Pantanal (Hardoim and Heckman, 1996) have also been found in many other parts of the world (Deflandre, 1929a, 1929b,

1937, 1953; Gauthier-Lièvre and Thomas, 1958; Grospietsch, 1965, 1982; Decloitre, 1981). The ciliate species present are known from much of the rest of the world (Kahl, 1935), as well, with the exception of *Dileptus bivacuolatus*, which seems to be an endemic South American species (Hardoim and Heckman, 1996).

The geographical distributions of the main plant and animal taxa, including the metazoans, are summarized in Table 2.1.

Table 2.1: Generalized world distribution patterns of the species belonging to various higher taxa encountered in the Pantanal. Some revision in the future can be expected as the taxonomy of the South American biota is improved.

Taxon	General distribution of most species found in the Pantanal of Poconé and its surroundings
Cyanobacteria	Cosmopolitan or circumtropical
Algae	Cosmopolitan or circumtropical
Bryophyta	Cosmopolitan
Tracheophyta	Endemic Neotropical or introduced
Sarcodina	Cosmopolitan or circumtropical
Ciliophora	Cosmopolitan
Porifera	Endemic Neotropical with few exceptions
Platyhelminthes	Cosmopolitan
Rotifera	Cosmopolitan or circumtropical
Gastrotricha	Cosmopolitan
Ectoprocta	Endemic Neotropical, as currently classified
Oligochaeta	Cosmopolitan or circumtropical, few Neotropical
Hirudinea	Cosmopolitan and Neotropical
Mollusca	Endemic Neotropical
Arthropoda	
Acari	Endemic Neotropical
Araneae	Endemic Neotropical
Conchostraca	Circumtropical
Cladocera	Circumtropical and endemic Neotropical
Copepoda	Endemic Neotropical, few cosmopolitan species
Decapoda	Endemic Neotropical
Insecta	Endemic Neotropical with few exceptions
Vertebrata	Endemic Neotropical or migratory

The Rio Paraná System, the second largest river system in South America, generally flows north-south and includes the Rio Paraguay and its tributaries, along the upper reaches of which the Pantanal is located. Although the headwaters of rivers in the Amazon and Tocantins Systems are located on a plateau close enough to the rivers flowing into the Pantanal for captures to have occurred rather frequently, the faunas are not too similar. Banarescu (1995) reported that many important fish families in the Amazon region are missing from the Rio Paraná. These include the Osteoglossidae, Helegeneidae, Nandidae, Ctenoluciidae, and Crenachidae. Other families that are richly represented in Amazonian tributaries are scarce in the Paraná system. The ichthyofauna is actually a mixture of species that crossed over from the Amazon system and those that originated in the temperate zone of South America. According to the fish fauna, Géry (1969) placed the entire drainage area of the Rios Paraguay and Paraná in what he called the "Paranean" Region. Earlier, Pearson (1937) had estimated that only about 40% of the fish species in this region were also found in the Amazon Basin. Although much information is still required to determine the geographical origins of the Pantanal fish fauna, studies by Ringuelet *et al.* (1967) in the Argentinean sections of the Paraná River System revealed that among the 346 species known to occur in the river and neighboring water bodies, 131 were also found in the Amazon region, 54 were species also known from southeastern Brazil, and 161 were endemic to the system. It was surmised that there were more Amazonian elements found only in the upper Paraguay River system because of its proximity to the headwaters of Amazonian tributaries (Banarescu, 1995).

The exchange of fish fauna between the Amazon and Paraná River Systems is attested to by a large number of genera that are found nowhere else but in those two systems. These include *Lepidosiren, Prionoabrama, Astyanicinis, Macropsobrycon, Scides, Paulicea, Hemiodontichthys, Peckoltia, Pseudohemiodon, Trachydoras, Rhinodoras, Pterodoras, Brochis,* and *Pseudocetops* (Banarescu, 1990). Most of these genera are not represented in the Pantanal but occur in the large rivers to the south (Ringuelet *et al.*, 1967), the Rio Paraná System upstream from its confluence with the Rio Paraguay (Galetti *et al.*, 1990), or, as in the case of *Lepidosiren*, in the drier chaco areas, especially in Bolivia.

An interesting observation by Cunha Vieira (1951) of the freshwater porpoise, *Inia geoffroyensis* Blainville, 1817, in Mato Grosso, far upstream in the Rio das Mortes, provides some insight into the way in which even relatively large aquatic animals would have been able to cross from the Amazon Basin to the Rio Paraguay. The upstream section of the Rio das Mortes flows a rather long distance separated from several rivers flowing to the Rio Paraguay by low, rolling hills. Headwater captures due to blockage of the rivers can be assumed to have occurred rather frequently during the geological history of the plateau northeast of the Pantanal, and simple flooding would be sufficient to explain the movement of small fishes and amphibious species from one watershed into the other.

The aquatic snail fauna of the Pantanal was included by Jaeckel (1969) in the "Platensis Region", which is generally equivalent to the "Paranean" Region of Géry (1969). However, the faunal lists he provided were incomplete, and the species observed most frequently during his study are actually widespread beyond the drainage area of the Paraguay and Paraná Rivers (Section 8).

There are also great similarities between the microscopic aquatic flora and fauna in the Pantanal and the Amazon Basin. Cadima (1990) studied the algal flora in upland tributaries of the Rio Mamore in Bolivia, which flows into the Rio Madeira, a tributary of the Amazon. The chemical conditions in the relatively small rivers he studied, which are not far from the headwaters, are quite different from those in the Pantanal. Their water is much richer in calcium and in phosphate from mineral sources, and it is poor in iron. The headwaters are also well separated by mountains from the tributaries of the Rio Paraguay. Nevertheless, more than half of the euglenophyte species Cadima (1990) identified, 12 out of 23, were also encountered in the Pantanal (De-Lamonica-Freire and Heckman, 1996). That much smaller percentages of the other algal taxa were common to both river systems may be due in part to recent taxonomic revisions of some of the taxa (Section 7), especially among the diatoms, making it difficult to know exactly which species his names refer to.

In the middle section of the Amazon and its tributaries near Manaus, a much richer algal flora than that in the Pantanal of Poconé was encountered by Förster (1969, 1974), Uherkovich (1976, 1981), Uherkovich and Franken (1980), Uherkovich and Rai (1979), and Uherkovich and Schmidt (1974). A great many of the species encountered in the Pantanal have also been reported from the Amazon by these authors. There are also many similarities between the Pantanal flora that that in freshwater bodies of French Guiana (Bourrelly and Couté, 1982), although they seem somewhat less pronounced than those between the Amazonian Basin and both of these neighboring regions.

As in the case of the algae, a considerable number of species belonging to several microinvertebrate groups have been found both in the Pantanal and in the Amazon Basin. Almost all of the rotifer species found by Kretzschmar *et al.* (1993) and Heckman (1998) were also reported from the Amazon region by Koste (1972, 1974), Koste and Robertson (1983), and Koste *et al.* (1984). Again, the Amazon Basin was much richer in species, but most of those found abundantly in the Pantanal are also common in the Amazon System.

Although little detailed information is available about the macrophytes in the Bolivian part of the Pantanal, a cursory study by Frey (1995) revealed a flora encompassing 67 tracheophyte species, which is similar in structure to that in the Brazilian part of the wetland.

In conclusion, the Pantanal and its surroundings are not notable for an endemic aquatic flora and fauna. Adámoli (1992) designated the Pantanal as a "biogeographical crossroad", colonized jointly by biota from the chaco, *cerrado*, and Amazonian regions. Many cosmopolitan species are present, and almost all of the aquatic plants and animals that are limited in distribution to South America also occur in the Amazon Region, the lower reaches of the Rio Paraná System, or both. Of those few species limited to southern Mato Grosso, most of them occur mainly in the upland streams rather than in the Pantanal. Detailed information about the biogeography of the organisms encountered will be presented in Sections 7 and 8.

3. Geology

3.1. Physiognomy and soils

3.1.1. Recent formation of the Pantanal

The origin of the Pantanal has only recently received the attention of geologists, and there is not yet a broad consensus regarding the geological history of the region. In considering its most recent history, which is most important for its colonization by aquatic flora and fauna, the uppermost sedimentary layers provide some clues to recent climatic events. These upper layers obviously consist of alluvium, but Klammer (1982) found that beneath them, there are dunes apparently formed by wind. The subsurface layers must therefore be classified as loess, which indicates that the entire region was a desert until the most recent times on a geological scale, and the present climate resulted from a major change in the rainfall pattern. This hypothetical change could well have been brought about by the retreat of the Pleistocene glaciers, which would have been responsible for a greater summertime warming of the areas of South America situated in the south temperate zone and consequent changes in the wind patterns.

A striking feature of the entire plain on which the Pantanal, Cuiabá, and several smaller cities are situated is the almost semi-circular system of cliffs that separate the whole region from a high plateau to the north and east. From these, insights can be gained on the more ancient history of the wetland. To the east, there are hardly any recognizable mountains but only a low-lying plain surrounded by a plain at a higher elevation. This gives the impression that the bedrock on which the Pantanal lies simply caved in, causing the whole region to drop down to a lower level than the surrounding land. It seems to lie at the bottom of an enormous natural bowl well below the surrounding uplands.

This interpretation is disputed by some geologists, who maintain that the surrounding land was forced upward, and that the Pantanal and its surrounding plains somehow escaped the uplift. This interpretation can be supported by the fact that to the west and northwest, the region is bordered by ranges of low, ancient mountains and also by the presence of rock chimneys near Chapada dos Guimarães, giving evidence of volcanic activity a long time ago.

The upland region to the north and east is located on the Central Brazilian Shield (Sioli, 1984), which is part of one of the most ancient continental areas on earth. Certainly, it may well be that the rim of the bowl, running from the São Vincente upland region northward and westward toward Chapada dos Guimarães (Plate 1a) was at one time part of a range of mountains, including volcanoes, which completely eroded away during the past several hundred million years, leaving only a few chimneys of the long extinct volcanoes (Plate 1b) and a great deal of sand, which was distributed in rather even layers over the upland and lowland plains. Supporting this hypothesis is the fact that a large number of small rivers flowing into the Pantanal and several that flow toward the Amazon Basin have their headwaters at springs welling up at the same rate through the soil on the plateau throughout the entire year (Section 5), giving evidence of a complex subterranean hydrological system beneath the upper soil and lower sandstone layers.

3.1.2. Physiognomy and recent geological developments

The general landscape features of the Pantanal can only be described in general terms because the area is so vast, and exceptional formations are encountered throughout the region. Confining the discussion to the Pantanal of Poconé makes the situation somewhat more simple, as this entire region is a low plain without appreciable changes in the relief. The situation is different in the adjacent Pantanal of Cáceres to the west, where the wetland is crossed by low ridges, called *cordilheiras*, which run generally parallel to the course of the Rio Paraguay, as well as several ridges of mountains separating large areas of the wetland from one another. To the east in the Pantanal of Barão de Melgaço, there are isolated low hills at the borders of the wetland, and there are deep depressions which contain water at a depth of at least a few meters for the entire year. During the high water period, these special kinds of lakes, locally called *baias*, are connected with the Rio Cuiabá, but some of them may be cut off from the river during the dry season. In Mato Grosso do Sul, the state in which about 2/3 of the Pantanal is located, an even greater variety of landscapes and lake types are encountered.

Generally speaking, small differences in the elevation of the terrain result in the development of the distinct kinds of landscape encountered in the Pantanal. To avoid the risk of creating an oversimplified classification system for the kinds of habitat that exist in the region, it must be stressed here that the following descriptions are limited to some of the most abundant kinds of environment, A discussion of myriad smaller biotopes must be omitted. Nevertheless, it must not be forgotten that the exceptional regional habitats that occupy negligibly small areas of the Pantanal are important for preserving the species diversity of the wetland.

Among the lowest lying sites in the Pantanal of Poconé are those at the bottom of the main rivers, over which flowing water passes throughout the year. Somewhat higher are the beds of the smaller rivers crossing the wetland, over which flowing water passes during the rainy season and in which standing water is present during the rainless seasons. There are also some deep locations in the *baias* or large lakes located in various parts of the Pantanal. A large number of such lakes are encountered in the Barão de Melgaço Subregion, but a few are also found in the Pantanal of Poconé, including the Baia Piuval along the Rio Bento Gomes. The so-called Rio Pixaim is not actually a river but rather is a geological formation that consists of a deep trench permanently filled with water. In addition to the *baias*, trenches and other permanently flooded depressions that were apparently formed through various geological and hydrological processes as well as a large number of oxbows and backwaters are encountered along the courses of the main rivers.

The lowest-lying areas of the floodplain that are usually dry for at least a short period of the year form distinct kinds of landscape, the flooded meadow and inundated pasture, locally called *areas alagarveis*. These flooded meadows are those areas of the flood-plain covered by vegetation not modified by cattle grazing. Inundated pastures, on the other hand, develop under the influence of cattle herds, which graze away many of the common native plant species, making way for grasses, mainly introduced species, which are unpalatable to the cattle and therefore given a selective advantage over plants that the grazers prefer. These areas are generally free from woody plants, although a few trees and shrubs are tolerant enough to survive the flooding if it does not last too long.

The areas at elevations high enough to keep them above the high water level for almost all of the year are often occupied by woody vegetation (Ratter *et al.*, 1988). The most obvious of these are the gallery forests occupying the elevated banks bordering the larger rivers and the *capões de mata* or wooded islands, similar to the hummocks in North America. Also above the high water level of the *cheia* are the *cordilleiras*, or elongated elevated areas crossing the floodplain. These areas usually become islands bordered on all sides by inundated areas during the high water periods.

In addition to the shrub savanna, there are somewhat similar plant communities characterized by a greater number of grasses, which are known as wooded savanna or *campo cerrado*. Apparently, many of these communities have originated from attempts by the ranchers to produce more pasture land for their cattle. After a short period of time, the scrub vegetation partially regenerates, producing this kind of community.

Within the *cerrado* zone is a special kind of landscape called the earthmound fields or *campo murundu*. These mounds have apparently been produced by the nest-building work of termites over a long period of time. While Por (1995) described some of the fields that occur on the floodplain as "*murundu* islands" during the high water period, most of the *murundu* fields seem to be at elevations high enough to escape even partial inundation (Oliveira Filho & Martins, 1991).

In contrast to the *campo cerrado*, the *campo úmido*, a wet or marshy field, is characterized by waterlogged soil, generally kept continually moist by seepage of ground water (Dubs, 1994). The kind of vegetation it supports is quite distinctive and characterized by various annuals and only a few small woody plants. This kind of habitat is greatly limited in distribution. Most *campos úmidos* were encountered near upland springs, but a few were located at sites low enough to be inundated for short periods of time during the high water period (Section 2.1.6.).

Almost all of the terrestrial communities within the wetland are almost identical with those occupying vast areas surrounding the Pantanal. These were previously defined in Section 2.1.6. Only the hummocks support terrestrial communities confined mainly to the Pantanal.

In moist areas along streams and at partially shaded sites beside hills, a richer kind of forest sometimes develops. It is usually characterized by various deciduous tree species that grow among evergreens. These semi-deciduous forests are common in the upland regions surrounding the Pantanal, especially in hilly regions, but in the Pantanal itself, they are usually limited to a few of the gallery forests. The streams that provide such forests with a continuous water supply are fed by ground water emerging at a relative constant rate throughout the entire year (Heckman, 1995).

The presence of hills in some of the other subregions of the Pantanal increases the variety of the terrestrial plant communities and the diversity of the landscape. To complete a satisfactory study of the terrestrial communities in the northern part of the Pantanal should be a goal of subsequent research. Studies of the vegetation in the South with some reference to the Pantanal of Poconé, as well, are available for comparison (Pott and Pott, 1994). In addition, Dubs (1994) provided a map showing the distribution of the vegetation communities in the Pantanal north of the Rio Negro. The kinds of habitat encountered correspond closely with those in the Pantanal of Poconé (Scheßl, 1997), but considerable floral differences are apparent.

The physiognomy of the surface soil layers is important for determining the geological processes presently at work in the Pantanal. The wetland itself is still a region of alluvial sedimentation. This process is, in turn, governed largely by events in the upland parts of the watershed. The vegetation of the woodlands generally stabilizes the soils and covers them with a thin humic layer produced by fallen leaves. Where these woodlands are burned, however, a temporary erosion and leaching is promoted until a pasture vegetation is able to establish itself, beginning a seral regeneration process that leads to the restoration of a shrub savanna and then a dry woodland. When the vegetation over large areas is cleared, as it has been on the plateau to the east of the Pantanal for the culture of soybeans and sugar cane, the erosion process is enormously accelerated, promoting the formation of deep *voça rochas* or erosion canyons cutting deeply into the sandstone layers beneath the surface soils. This, in turn, can be expected to greatly increase the processes of siltation in the Pantanal.

3.1.3. Soils

The analysis of the ground water, soils, and sediments in the Pantanal and in the regions of the plateau to the north and east show that the substrate in the entire northeastern catchment area is mainly silica with scattered deposits of iron and associated metals, such as manganese (Table 3.1). The entire set of data shown gives the impression that the region has been leached for an extremely long period of time, and no soluble components of the upper rock and soil layers remain. The gold sand deposits of the Pantanal support this concept, since their presence suggests that a hypothetical ancient gold-bearing quartz has been completely eroded away, and the sand and gold have been roughly separated by wind and water, leaving the gold most concentrated in certain substrata where it is found today. The gold deposits, however, are confined mainly to subsurface sedimentary layers and not generally distributed, as the absence of detectable values for gold among the data presented in Table 3.1 indicates.

Table 3.1: The metal and metalloid content of the surface sediments in the Pantanal of Poconé. All values are shown in mg kg^{-1} dry weight. The analyses were conducted using atom absorption spectrometry, courtesy of Dipl. Biol. J. Kaiser. The sampling sites, as shown in Figure 2.6, are abbreviated as follows: PE = Rio Bento Gomes at the Poconé exposition park; BP = Baia Piuval, a large lake along the Rio Bento Gomes at the border of the wetland; BI = Birici, the channel followed by the water draining from the floodplain into the Rio Bento Gomes on the Fazenda Ypiranga; TP = the Rio Bento Gomes where it passes beneath the Transpantaneira Highway; RN = the Rio Novo where it passes beneath the Transpantaneira; RF = the Rio Frio where it passes beneath the same road; RP = the Rio Pixaim beside the Transpantaneira; PP = shallow pond adjacent to the Rio Pixaim along the highway.

Site of sample	PE	BP	BI	TP	RN	RF	RP	PP
Metal								
Ag	<0.05	-	<0.05	<0.05	<0.05	-	-	-
As	-	-	-	-	-	-	1.4	-
Au	<0.25	-	<0.25	<0.25	<0.25	-	-	-
Ca	117.2	-	40.4	429.2	350.4	-	-	-
Cd	<0.01	<0.2	<0.01	<0.01	<0.01	<0.2	<0.2	<0.2
Co	1.0	-	0.2	2.9	0.6	-	-	-
Cr	22.6	19.8	1.7	25.2	17.8	72.3	14.4	16.9
Cu	3.2	44.5	0.3	6.7	2.4	13.2	9.2	10.1
Fe	7821.8	-	841.8	6394.6	5528.9	-	-	-
Hg	<0.01	-	<0.01	<0.01	<0.01	-	-	-
Mg	143.4	-	42.9	235.4	102.0	-	-	-
Mn	52.8	-	10.0	214.8	18.1	-	-	-
Ni	2.5	16.0	0.6	2.9	1.4	7.3	7.4	8.0
Pb	4.5	12.2	1.0	6.8	2.3	12.9	15.2	14.4
Zn	2.6	33.3	2.2	6.6	3.2	8.0	18.2	37.8

Unfortunately, data on the geology of the subsurface soil layers and bedrock have never been systematically obtained in the northern Pantanal. It can only be reported that the richest deposits of gold dust lie in layers from just below the surface to about 50 m deep in the districts of Cangas and Poconé, where most of the active gold mines are located. Formerly, gold was obtained from other locations farther from the Pantanal, including the area near the town of Nostra Senhora de Livramento, located about 50 km north of Cangas and founded in 1731, 50 years earlier than Poconé, as a gold mining settlement. In these mines (Fig. 3.1), there is no evidence of bedrock layers, indicating that the deposits of fine loess and alluvium are relatively thick in the region, and the bedrock apparently lies much deeper.

A somewhat similar situation is encountered on the plateau to the east of the wetland, where the stripping of the dry forest vegetation to prepare the land for agriculture permits a rapid erosion of the soil (Table 3.2). Some of the small canyons or *voça rochas* formed are quite deep and deepen more each year. The sides of these erosion gullies consist of layers of sand and silt of various consistencies and colors, usually shades of red or orange, that show evidence of a long geological history of continuous erosion and deposition of sands. The lower layers of this material have already become sandstone, which is not compact and erodes rapidly as enormous quantities of water dropped by heavy rainstorms flow by. Outcroppings of the bedrock are few and generally confined to the regions near the cliffs separating the upland plateau from the plain in which Cuiabá and the Pantanal lie. This supports the general assumption that the region is geologically an ancient continental one, in which the prehistoric mountain ranges have been almost completely leveled by water and wind. It also explains the absence of calcareous rocks throughout nearly all of the region as well as the great scarcity of other soluble substances in the soils and sediments. Aluminum is also lacking in many sediment samples, which consist almost entirely of silicates.

Figure 3.1: Aerial view of one of the deep excavations at the Dois Paderes gold mine in Poconé, Mato Grosso.

The composition of some of the sediment samples taken from various water bodies in the region of the Pantanal (Table 3.2) indicated no natural contamination by metals that are potentially toxic to most animal and plant groups. Significantly large natural deposits of such metals and metalloids as lead, mercury, cadmium, copper, or arsenic are unknown, and their natural occurrence in the region is highly unlikely. The metals known to be present include gold, iron, and manganese, none of which seem to have acute toxic effects on animals or plants. Soluble ferrous ions in large quantities, however, may cause physiological problems for aquatic organisms that breathe through gills or their integuments because there is a rapid deposition of insoluble ferric compounds whenever the ferrous ions are oxidized, and such deposition on respiratory surfaces may hinder gas exchange. The iron and manganese seem to occur together, with the manganese being less abundant. Other metals present in relatively small amounts, such as calcium and magnesium, are generally encountered in the earth's crust, and their detection in the samples is not surprising.

Some data on the soils in the southern part of the Pantanal in the state of Mato Grosso do Sul were obtained by Amaral Filho (1986) and Dubs (1994). These authors presented information obtained by analysis of series of soil samples from depths of 20 cm and 40 cm near the Rio Negro in the Municipal District of Aquidauana. They found that all samples from the floodplain consisted of a fine sandy material with no hydromorphic features at all in the horizons near the surface. According to the classification system of Orioli *et al.* (1982), all of the samples from a depth of 20 cm were dystrophic soils because their base saturation was low. The base saturation is a measure of the number of exchangeable cations other than H^+ and Al^{3+} expressed as a percentage of the exchange capacity. The base saturation of the soils in the semi-deciduous forests was much higher, however, bringing these into the mesotropic category. These findings indicate that the soils of the northern and southern Pantanal floodplains are quite uniform, consisting almost exclusively of quartz sand. However, the upland areas may have more heterogeneous features, probably related to their vegetation cover.

Table 3.2: The metal and metalloid content of the surface sediments in streams and rivers in the catchment area of the Pantanal. All values are shown in mg kg^{-1} dry weight. The analyses were conducted using atomic absorption spectrometry, courtesy of Dipl. Biol. J. Kaiser. The sampling sites, as shown in Figure 2.6, are abbreviated as follows: CS = Corrego Salgadeira, a stream near Chapada dos Guimarães just downstream from a waterfall; RC = the Rio Coxipó, upstream from a high waterfall near site CS; CC = Rio Cuiabá in the city of Cuiabá; CV = Rio Cuiabá downstream from Varzea Grande; AL = a sandbank in the Rio Cuiabá at Santo Antonio de Leverger; PC = Rio Paraguay at Cáceres; PS = Rio Paraguay several kilometers upstream from its confluence with the Rio Cuiabá on the Fazenda Santo Antonio das Lendas.

Site of sample	CS	RC	CC	CV	AL	PC	PS
Metal							
Ag	<0.05	<0.05	<0.1	<0.05	<0.05	<0.05	<0.05
As	-	-	-	-	-	-	0.2
Au	<0.25	<0.25	<0.25	<0.25	<0.25	<0.25	<0.25
Ca	13.0	72.0	897.6	146.8	34.5	39.6	1336.86
Cd	<0.01	<0.01	<0.03	<0.01	<0.01	<0.01	<0.01
Co	0.7	3.7	5.8	1.0	1.5	0.6	3.7
Cr	2.5	6.9	9.7	1.8	5.4	2.4	12.9
Cu	2.4	3.1	6.1	1.4	0.6	1.0	16.8
Fe	4516.3	11189.9	14615.1	1741.3	4594.0	2640.5	10110.3
Hg	<0.01	0.02*	<0.02	<0.01	<0.01	<0.01	<0.01
Mg	24.8	127.5	1173.7	113.5	76.3	113.6	1746.6
Mn	10.7	45.2	293.5	36.0	46.2	20.2	49.3
Ni	1.2	5.3	7.8	1.4	1.0	1.0	13.0
Pb	2.4	4.2	8.7	1.2	1.7	2.0	9.8
Zn	4.6	11.6	11.6	2.8	1.3	2.0	24.7

---

* Due to a typographical error by the author, this value was previously incorrectly reported as 1.019 mg kg^{-1} (Heckman, 1995).

In the moist tropical climate, detritus breaks down quickly. Under these circumstances, the topsoil layers are thin, even beneath the stands of trees in the forested areas. On the floodplain itself, detritus is observed only seasonally, and there is therefore no loam formed at all. The turnover of the inorganic plant nutrients is therefore rapid and nearly complete during each season of transition. Apparently, a general lack of mineral nutrients for plants in the upper layers and a lack of material to bind such nutrients are typical of South American soils. Summarizing their findings, Beek and Bramao (1968) reported that a characteristic of the soils on the South American continent is "low natural fertility." Sanchez (1989) illustrated this convincingly through a comparison of the soil distribution in tropical America with those in the warm parts of Africa and Asia. Only 38% of the tropical land in Asia and the Pacific islands is covered by infertile soils, while the soil is similarly poor in 56% of the African tropics. The corresponding figure for the infertile soils in the Neotropics is 81%.

Information on the soil types found in the Pantanal was provided by Scheßl (1997), who found that there was considerable mixing of the types in the horizons he sampled. This would be expected in view of the past activity of wind and water on the soft upper sediment layers. Gleysols were found to be the dominant soils in most regions subjected to the influences of ground water and flooding. Anthrosols, which are soils showing anthropogenic influences on their formation, were also localized in the *campos murundus* and along the Transpantaneira Highway. In addition, Scheßl (1997) also reported the presence of fluvisols, arenosols, cambisols, solonetzes, planosols, alisols, acrisols, and vertisols, defined according to the FAO classification system.

3.2. Ancient geological history

In his review of the geological history of the Amazon Region, Fittkau (1974) provided some information about the uplands of the Central Brazilian Shield. He mentioned the great paucity of specific geological data on this region. From the north side of the plateau atop this ancient formation, the rivers flow toward the great Amazon Basin, while the south side is drained by rivers of several systems, including the Paraná, the system including the Rio Paraguay. According to Putzer (1968), the Paraná Basin encompasses about 1,300,000 km^2 and is filled with sediment 2000 m deep overlying a 1500 m layer of basaltic lava. The oldest sediment is the Eastern Paraguay sandstone, about 1000 m thick, which originated during the Ordovician (Putzer, 1962). The entire upper Rio Paraguay region, including the Pantanal, is located on the Central Brazilian Shield, which has a tongue-like extension southward into Paraguay. The river departs from this ancient kraton when it passes the Central Paraguayan Sill (Putzer, 1968).

The foothills of the Andes rise to the west of the Pantanal. This mountain chain is much younger than the Central Brazilian Shield, and its rocks are richer in minerals. However, because relatively little of the Pantanal wetland is located to the west of the Rio Paraguay, the water contributes little to the chemistry of the wetland.

It may be premature to present a geological history of the Pantanal region, even one presented in general terms. The fragmentary results of the geological sampling already completed is sufficient only for a greatly simplified outline of the events that resulted in the formation of the Pantanal and its watershed (Putzer, 1968). It is not unlikely that even some of the main events still remain undiscovered. Just as in the case of biological and chemical research in South America, basic geological studies have not been supported in a coherent or systematic way, so the following description of events must remain tentative until such studies are completed. However, an outline of the events generally thought to have occurred can be given as a framework around which to develop future studies.

3.2.1. Precambrian events

The ancient kraton of the Central Brazilian Shield was subjected to much tectonic activity during the Precambrian period. There was much breaking and folding of the stone during several cycles of erosion and sediment deposition accompanied by a number of transgressions of the ocean. Both basaltic and metamorphic rock originated before the Cambrian. At the end of the period, about 500 myrs ago, the tectonic activity ceased, leaving a massive block which remains today as two great land masses, the Guyanan and Brazilian Shields (Fittkau, 1974).

The southern part of the ancient massive kraton that became the Central Brazilian Shield is much younger than the northern part, which is now the Guyanan Shield. The oldest known South American stone formations, from 2500 to 3000 myrs old, are encountered on the northern edge of this shield (Posadas, 1966). Orogenesis on much of the rest of the Guyana Shield ceased about 2000 myrs ago, but granites and metasedaments only 1400 to 1800 myrs old can be found on the southern edge.

On the southern side of the Amazon Basin, the Central Brazilian Shield is divided into a northwestern and southeastern part by the Paraguay-Tocantins Geosyncline. The maximum age of the formations on the northwestern part of this shield is about 2000 myrs. In the region between the upper Rio São Francisco and the coast, the southeastern part also approaches 2000 myrs in age, but the rest of this part is only 500 to 1300 myrs old, at least at the sites already investigated (Fittkau, 1974).

Loczy (1969, 1973) believed that the Amazon Basin formed as a small rift valley as the Guyanan and Central Brazilian Shields began to separate. The northern slope of the Central Brazilian Shield therefore came to be drained by the Amazon System, while water from its southern slope flowed generally southward through several different river systems.

Fossils of a worm-like *Cloudina* species have been found by Fairchild (1978) in the Serra do Urucum south of Corumbá. These date from the late Precambrian and Cambrian. Such fossils have not yet been found in the northern part of the Pantanal.

3.2.2. The Palaeozoic

By the beginning of the Palaeozoic, tectonic events affecting the Central Brazilian Shield had almost completely ceased, and a long period of erosion and sediment deposition began. This eventually led to the leveling of the mountains and gradual formation of a plain. The substrate of the plateau region in the drainage area of the Rio Paraguay today consists largely of Palaeozoic sands (Fittkau, 1974), which were produced as the quartz formations were subjected to several hundred million years of weathering. Important on what is now the southern part of the kraton, presently constituting the Central Brazilian Shield, was the effect of ice cover, to which the region was periodically subjected. During the Permian, the climate gradually warmed sufficiently to eliminate the chance of ice formation, and the kraton has apparently remained free of ice since that time, except perhaps briefly during the Pennsylvanian.

In contrast to what is now the plateau to the northeast, the region presently occupied by the Pantanal must have been an extensive coastal plain near the edge of the vast land mass of Gondwanaland. It must have been at a high enough elevation, however, to keep it above sea level even when the level of the oceans became rather high. Putzer (1968) reported that during the Ordovician, the entire Andes region was at the bottom of the sea, and the seacoast was located along what is now the Bolivian and Paraguayan chaco. During the Silurian, the Amazon Basin was open to the sea on what is now the Pacific side through the present-day territory occupied by Peru and Ecuador. During the Devonian, the chaco basin sank, and in the Pennsylvanian, it passed through an ice age.

A mountain range, probably including numerous volcanoes, must have separated the lowland coastal plain from the upland region of the Central Brazilian Shield during much of the Palaeozoic. This range has nearly completely eroded away, and the remains of the volcanoes are evident only as some of the weathered stone chimneys and rocky outcroppings near the cliffs separating the Pantanal Basin from the *alto plano*, the upland plateau to the northeast and east of Cuiabá. However, some of these undoubtedly date from the Mesozoic. The climate apparently changed gradually through the Paleozoic from frigid or cold temperate to warm temperate or subtropical.

3.2.3. The Mesozoic

The period of relative inactivity during the Palaeozoic continued during the early Mesozoic, but by the Cretaceous, the South American continent began to separate and drift away from Africa as Gondwanaland progressively broke up. This movement began the upheavals that would produce the Andes during the Tertiary. Prior to the

Cretaceous, the Pantanal plain is thought to still have been located along or near the shores of the continent Gondwanaland. Grabert (1964) speculated on the existence of a brackish water lagoon, probably connected to the Pacific Ocean, which lost its contact with the ocean during the uplifting of the Andes. This hypothesis was accepted by Putzer (1984). However, no marine or freshwater fossil organisms have yet been found to support this (Por, 1995). If we assume, however, that the Pantanal was deeply eroded and then filled with sediments in relatively recent geological times, it may well be that any fossils present are buried deep beneath the recent fine sediment layers.

It is sometimes speculated that some of the ancient mountains near the borders of the Pantanal may have once been on a chain of islands called the "Archiplata" (Fittkau, 1969), just off the Pacific Coast of Gondwanaland. The lack of marine fossils, however, weakens this hypothesis, as well.

The warming trend begun in the Permian continued during the Mesozoic, probably related to a northward drift of the South American plate out of the frigid zone toward the equator. During this period, the Central Brazilian Shield rose to an elevation at which it would no longer be influenced by marine transgressions (Fittkau, 1974), which seemed to have come from the Atlantic side through the Amazon Basin, although there are still far too few geological data available to establish this with certainty. During the Triassic, the region of the present Paraná Basin was the largest desert in the world (Putzer, 1968). Climatic changes probably occurred rather frequently thereafter as tectonic activity increased, associated with movement of the land mass.

The Mesozoic seems to have been the period during which South America became fully isolated from all of the other present continents, with the possible exception of Antarctica. This can be assumed from the great increase in volcanic activity during the Jurassic. The volcanism reached a maximum during the Cretaceous, when much sandstone was spread around the Paraná Basin by the volcanoes (Putzer, 1968).

Its temperate climate during the late Mesozoic seems to have favored the development of a mammalian and avian rather than reptilian fauna, and South America seems to have been a haven for the development of early placental mammals, including the edentates, primates, and chiropterans. The edentates apparently remained among the prodominant mammalian fauna until the Central American land bridge formed, permitting the more advanced Laurasian fauna to invade the continent.

3.2.4. The Tertiary

The western limits of the upper Rio Paraguay watershed were formed during the Tertiary by the upheaval of the Andes. Petri and Fulfaro (1988) supported the hypothesis that the Pantanal sank below the level of the rest of the Central Brazilian Shield to the east at the end of the Eocene. The sediments in the Pantanal were deposited mainly during this period. This hypothesis is supported by the presence of travertines of Pliocene and Pleistocene origin, part of the Xaraes Formation, in the rock layer below the sediment deposits in southern Mato Grosso (Godoi Filho, 1986). The Xaraes Formation consists of chalks formed in freshwater lakes, and fossilized pilid snails seem to be present in the chalks.

Fittkau (1974) reported that the Amazon Basin had already taken on its present general form by the Pliocene, but the Andes continued to rise rapidly for the rest of the era. The Pantanal seems to have been a bowl-like depression between the rising Andes foothills and the Central Brazilian Shield into which large amounts of fine sediments were transported through the air or with runoff, depending upon the prevailing rainfall patterns. The amount of annual rainfall seems to have fluctuated between much and nearly nothing, producing the temporary Pantanal wetlands at times and a desert during other periods (Klammer, 1982; Ab'Saber, 1988). Speculation about the Pantanal being joined to a vast inland sea, perhaps connected to the Pacific Ocean, was discussed by Por (1995).

Events on the plateau to the north and northeast of the Pantanal during the Tertiary were of great importance for determining the distribution of the freshwater fauna of

South America. The headwaters of rivers flowing into the Amazon Basin are located on this *alto plano* in close proximity to those flowing southward into the Rio Paraná system. Hence, during periods of extensive flooding on this plateau, fishes and invertebrates were able to cross over from one watershed into the other. For insects with the ability to fly as adults as well as for amphibians and reptiles, the migration from one system to the other was even easier. It can be assumed, however, that epochs of severe drought in the Pantanal depression brought about periodic extinctions of many of the immigrant species. The present biota is thus a mixture of species that have recently immigrated across the plateau from the Amazon and Tocantins Basins or from the Atlantic coastal and Patagonian subregions via the Paraná River System.

3.2.5. The Quarternary

The recent history of the Pantanal is marked by events resulting from or influenced by considerable changes in the annual rainfall. These include the deposition of loess during a dry epoch, reported by Klammer (1982), and the subsequent sedimentation of alluvium after the Pantanal had developed in its present form. Thus, the Pantanal is covered by a rather thin layer of alluvium overlying deposits of loess, which may, in turn, overly deeper layers of alluvium. The present wet-and-dry period is thought to have begun about 13,000 years ago, while the desert seems to have existed between about 13,000 and 23,000 years ago (Ab'Saber, 1988). The Pantanal may be undergoing a process of becoming drier (Veloso, 1972) after a period of maximum humidity 5000 to 7000 years ago, during which large lakes formed, including perhaps the "Sea of the Xaraes", preserved in the oral histories of the native American populations (Por, 1995).

According to Ab'Saber (1988), the bottom of the trench in which the sedimentary layers are deposited is presently over 400 m below the current level of the sea. In the middle of the Pantanal, drillings by the Petrobras National Oil Company revealed that this sediment is about 500 m thick (Por, 1995). The deposition of these sediments has apparently occurred during fluctuating wet and dry periods related to glaciation episodes taking place on the colder parts of the continent from the Tertiary through the Quarternary (Por, 1995). With the fragmentary data presently available, many of the general events described by the authors cited are speculative, and no details of these events can yet be provided. Without doubt, the geological history of the entire Pantanal region is far more complex and varied then presently realized.

3.3. Conclusions

The geological region in which the Pantanal is located can be classified as a basin at the edge of the Central Brazilian Shield, an ancient continental land area, which was joined at one time to the ancient West African land mass, according to presently accepted theories of continental drift. This shield clearly shows the signs of having been leached and eroded for an extremely long period of time without any intervening period of cover by the sea. Marine fossils are not known from the area, giving further support to the assumption that the shield has not been at the floor of a sea, at least for most of its long geological history. The western boundary of the Pantanal basin is bordered by the foothills of the Andes, a geologically young mountain range.

Its location and relatively low elevation above sea level make it seem likely that during times long past, at least until the late Mesozoic or early Tertiary, the Pantanal was a coastal lowland, perhaps beside a large bay, along what would presently be the Pacific Coast of South America. Later, the bay may have been an inland sea with a permanent or temporary connection with the Pacific Ocean through a gap in the rising Andes. Since the tectonic uplifting of these young mountains which now run parallel to the West Coast of South America, the location of the Pantanal has moved close to the geographic center of South America.

The underlying rock strata in the Pantanal are located below deep layers of recent sand deposits. Near the center of the Pantanal, these are about 500 m deep. The material constituting the sedimentary deposits has been eroded from the upland regions surrounding the Pantanal, from which so much of the rock has been eroded away that former hills are no longer recognizable in many places and only the weathered chimneys of ancient volcanoes still mark their locations.

For part of the Quarternary, and perhaps for long periods of time at the end of the Tertiary, the Pantanal seems to have been a desert, and it has only recently come under the influence of its present wet-and-dry climate (Klammer, 1982). Unlike the Amazon region, its watershed is barely influenced chemically by the relatively young Andes range, and its input of minerals, except for iron and silicates, comes mainly out of the air with the falling rain (Section 5). Furthermore, the sea level changes that have apparently influenced the geological features of much of the Amazon Basin (Klammer, 1984) would have had little or no influence on the Pantanal and its upland watershed because they have remained entirely above the highest level of the sea since the Permian. The effects of the sea level variations and resulting air temperature changes would have had profound effects on the Pantanal because of the general climatic changes they provoked, however. Generally, because of the slow northward drift of the South American land mass out of the south frigid and temperate zones into the torrid zone, the air temperature has steadily been rising for a long period of time.

The geological history has greatly influenced the natural history of the Pantanal by limiting its aquatic fauna to those species that were able to successfully immigrate from the Amazonian or Patagonian regions in relatively recent times and adapt to the wet-and-dry climatic conditions. Extremely dry epochs have apparently brought about mass local extinctions of the aquatic species, and the present wetland fauna therefore consists nearly exclusively of immigrants that arrived since the most recent major climatic changes. This explains the absence of an endemic aquatic flora and fauna and rarity of any endemic species at all, as well as the general poverty of species, compared with the adjacent regions, especially the Amazon Basin. It also postulates periods during which the two watersheds had common sources in upland lakes or wetlands or when the capture of sections of upland rivers occurred, as these rivers shifted their courses from one watershed to the other during fairly recent times.

The Pantanal is a region still poorly known geologically, and the information presently available is fragmentary. There are enormous gaps in the knowledge of the local geological features and history of the subregions of the wetland. Intensive geological studies of the separate parts of the Pantanal and its watershed are urgently needed.

4. Climate

The Pantanal is located in the south torrid zone, far enough south of the equator to fall outside of the equatorial climatic belt, and it is subject to pronounced seasonal changes. During most years, there is little rainfall in its watershed from about the middle of April until mid-November. This period is characterized by clear skies and usually by hot weather. However, during the southern winter, periodic incursions of air from the South bring clouds and cool weather, but these generally result in reduced temperatures and light, fine precipitation for no more than a few days at a time. As the end of the dry season approaches, the weather remains uncomfortably hot with afternoon temperatures often reaching the mid-40s. The highest temperature personally observed in the center of Cuiabá was 47°C in early November. Figure 4.1 shows the weekly maximum and minimum air temperatures at the weather station in Varzea Grande during the study. The temperature was recorded only at six hour intervals, however. Therefore the highest temperature was usually reached two to three hours after the noon reading and three to four hours before the 6 p.m reading, and the maxima in the figure are somewhat below the actual highest temperature.

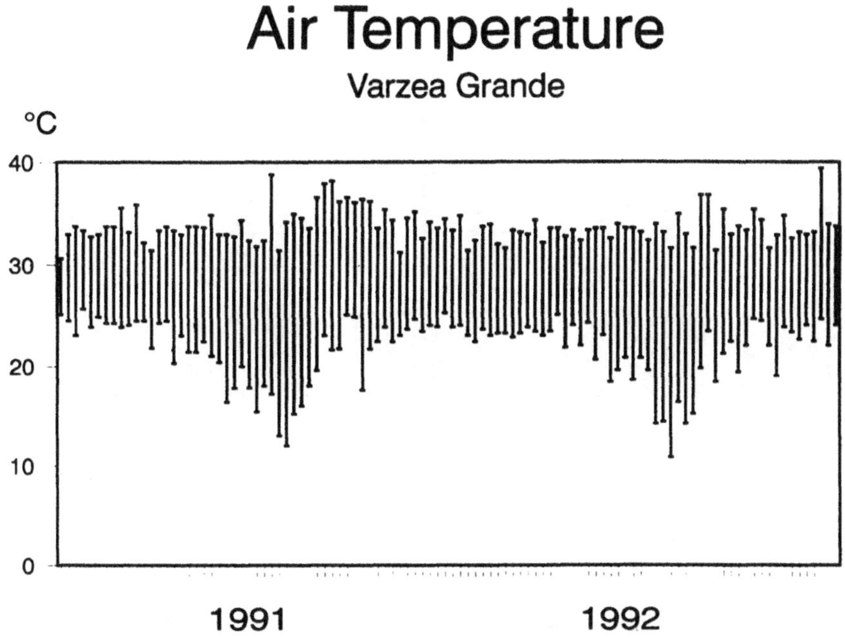

Figure 4.1: Weekly ranges of maximum and minimum temperature recorded at the meteorological station in Varzea Grande from January 1, 1991, to December 30, 1992. The graph was replotted from data previously shown by Heckman (1994a).

In mid-November, the rainy season usually begins, although pronounced year to year differences were observed during the study (Fig. 4.2). These probably reflect the relatively small size of the continental area responsible for the formation of the seasonal high and low pressure systems, which influence the direction of the prevailing winds during each of the seasons. In 1992, for example, a period of heavy rainfall followed by periodic thunderstorms began on September 2, and the northern Pantanal began to flood about two months later. Incidentally, during that same period, the ranchers in the southern part of the Pantanal were complaining of severe drought (Section 5).

Because of the year to year differences in the weather, it is difficult to classify the region of southern Mato Grosso precisely. It is therefore necessary to analyze local weather information over at least a number of years to determine how the general climate should be described. As yet, no regular collection of meteorological data has been initiated in the Pantanal itself. The nearest weather station to the Pantanal of Poconé is locate at Varzea Grande, the twin city of Cuiabá. During the study, data on the temperature, humidity, and rainfall were collected on the Fazenda Ypiranga about 15 km south of Poconé in the Pantanal, but this activity ceased as support for the limnological project there expired at the end of 1994. The data collected over a period of nearly three years at that site are insufficient to adequately estimate the year to year variations. Tarifa (1986) summarized the present knowledge of the climate in the Pantanal, stressing the need for systematic recording of meteorological data.

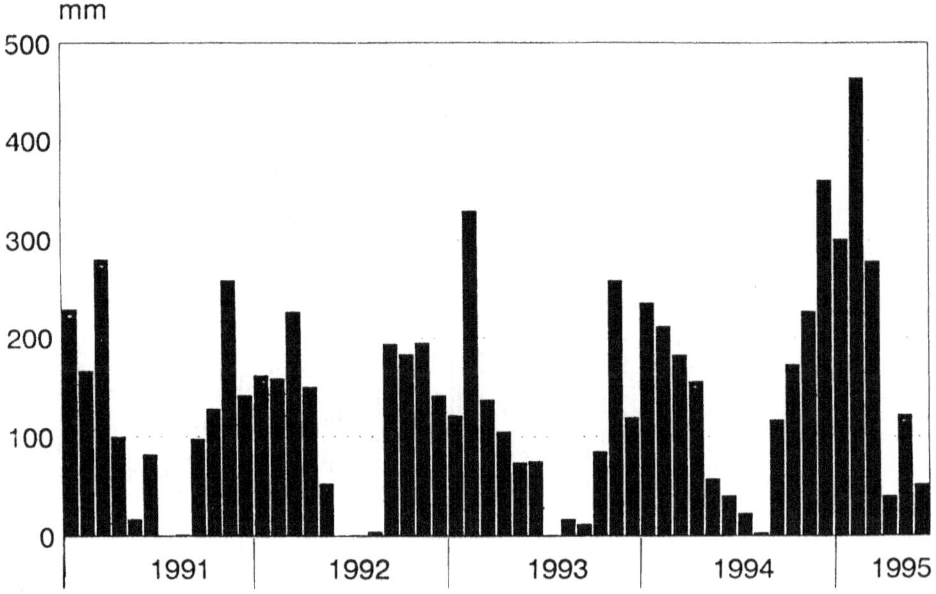

Figure 4.2: The monthly rainfall recorded at the meteorological station in Varzea Grande from January 1991 through June 1995.

Most of the weather data presented here were obtained at the meteorological station in Varzea Grande, which is located about 100 km north by northeast of Poconé. To determine the applicability of these data to the weather in the Pantanal itself, they were compared with the temperatures and amounts of rainfall recorded at the small weather station set up on the Fazenda Ypiranga. General patterns were thereby established to estimate how representative the Varzea Grande data are for the northern part of the Pantanal (Fig. 4.3).

As in the tropics around the world, precipitation is produced by small weather cells, which can often produce intense rainfall and strong winds. The vast stratus systems that bring steady rain for several days, which are typical for much of the temperate and frigid zones, do not form, except when large cold air masses move northward across South America from the Antarctic occasionally during the dry season, which coincides with the South American winter. This phenomenon, called a *friagem*, brings the lowest temperatures to the Pantanal that are normally recorded during the course of the year. During the rest of the year, intense rain may fall on one site, while only a few kilometers away, the weather remains hot and rather dry. This means that the weekly rainfall, as well as the minimum temperature and humidity during a given week, may be quite different at locations relatively close to one another. The comparison of the data obtained in the Pantanal with those recorded at the weather station in Varzea Grande, illustrated in Figure 4.3, shows that local differences can be pronounced, as in the case of the rainfall during September 1992. However, the total amounts of rainfall for that year at both locations were nearly equal.

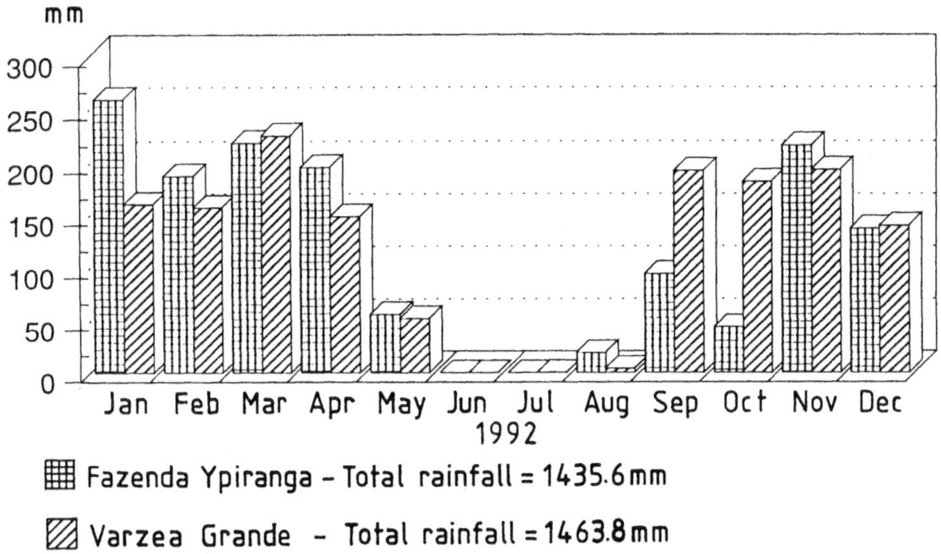

Figure 4.3: The rainfall at the meteorological station in Varzea Grande compared to that recorded by a rain gauge on the Fazenda Ypiranga in the Pantanal, about 15 km south of Poconé.

Difficulty in finding information in the literature on details of the climate is apparent from published climatological maps, which show the various zones in general ways without accounting for local rainfall and temperature patterns as affected by mountain ranges, wetlands, and other geographical features, which have a great effect on the local weather conditions. Obviously, the basic data on the weather are lacking, preventing any more detailed cartography of the climatic zones. For example, the discussion of the climate by Por (1995) was apparently based on information obtained near the southern borders of the Pantanal, and the details are not applicable to the situation in the North. Schnell (1987) reported an average rainfall for the entire wetland of 1200 to 1400 mm per annum. This range is obviously too low for Cuiabá, but it seems to be much more than falls on the central part of the Pantanal. A recent general discussion of the climate was provided by Alfonsi and Paes de Camargo (1986), and this report makes it clear how much more climatological research is needed to properly define the climate in the different subregions of the Pantanal. Especially needed is information on the quantitative differences in the weather from year to year.

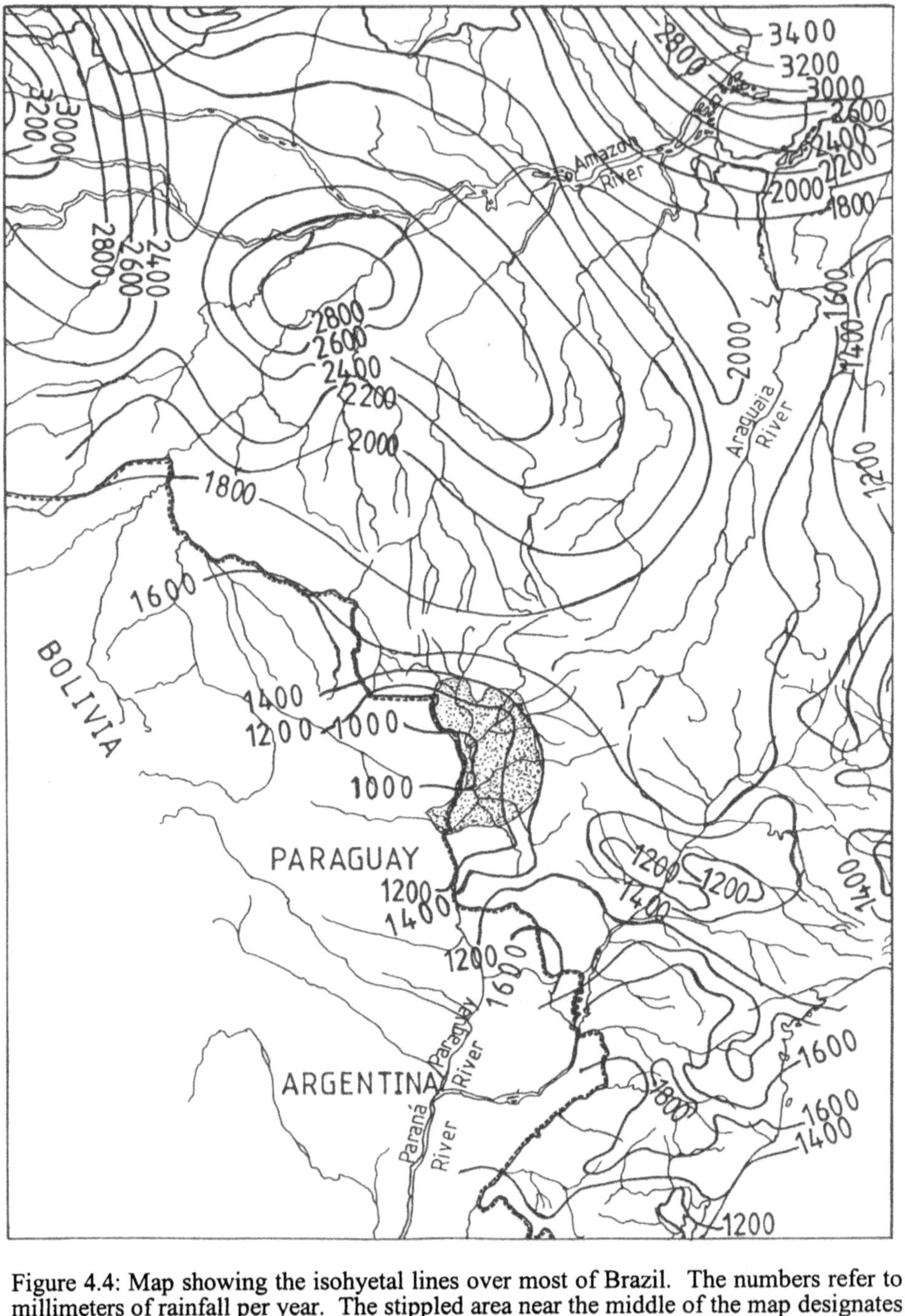

Figure 4.4: Map showing the isohyetal lines over most of Brazil. The numbers refer to millimeters of rainfall per year. The stippled area near the middle of the map designates the approximate extent of the Pantanal. The information was taken from a map prepared by the Coordinação Geral de Recursos Hídricos of the Secretaria de Energia, prepared from data recorded between 1931 and 1988.

Eidt (1968) referred to the climate in this entire region as a savanna climate. According to the maps of Landsberg *et al.* (1965), the Pantanal of Mato Grosso is located in the zone of the tropical humid summer climate with humid winters. The tropical wet-and-dry climate is shown to influence only the central and southern parts of the Pantanal, mainly in the state of Mato Grosso do Sul (Troll, 1965). In reality, however, the winters are quite dry, as shown by the rainfall data obtained in Varzea Grande during the years from 1990 through 1995 (Fig. 4.2). Some years display rainfall patterns typical of the wet-and-dry climatic zone, as evident for the year 1991. During other years, periods of rather heavy rainfall during one of the early or late dry months result in the dry period being too short for the climate to be classified as truly wet-and-dry. The months of July and August seem to be more consistently dry at Varzea Grande than the others, but sometimes there is rainfall during these months, as well. In spite of the fact that the dry season during some years may be too short for a typical wet-and-dry climate, the aquatic biota must be adapted to it because it prevails during other years, even at the northern borders of the wetland. Moving southward into the Pantanal, however, the dry period becomes decidedly longer as the total amount of annual rainfall steadily decreases (Fig. 4.4).

According to charts from 1993 published by the Coordinação Geral de Recursos Hídricos (CGRH), there is a center of maximum annual rainfall in the western Amazon Valley at the Peruvian border, where more than 3400 mm rainfall was measured. The site of a minor peak was located over the confluence of the Rios Aripuana and Madeira roughly 500 km east by northeast from Porto Velho. The rainfall at this site amounted to more than 2800 mm. There is a decrease in the amount of rainfall moving south southeast from this site toward the divide between the watersheds of the Amazon and upper Rio Paraguay System, where 1800 mm was measured for the year. At Cuiabá, about 1600 mm fell, and the northern part of the Pantanal is crossed by the 1400 mm isoieta. A regional minimum of less than 1000 mm was recorded near the confluence of the Rios São Lourenço and Paraguay near the borders between Mato Grosso, Mato Grosso do Sul, and Bolivia.

Even well to the north of the Pantanal in the Amazon catchment area, a well-defined wet-and-dry climate has been found to prevail, notwithstanding the map of Landsberg *et al.* (1965) showing this region to be well outside of this climatic zone. Members of a geographical expedition undertaken from 1967 to 1969 to a region about 260 km north of Xavantina in northeastern Mato Grosso recorded 1372 mm of rainfall during one year, virtually all of which fell during a 148 day period between November and April (Brown *et al.*, 1970).

The area under the influence of the continental wet-and-dry climate in which the Pantanal is located extends southward into Paraguay and crosses the Tropic of Capricorn into subtropical Argentina. Not far to the south of this area is the site of the summertime low and wintertime high pressure system that determines the seasonal wind direction and rainfall pattern (Prohaska, 1976). According to Eidt (1968), this pressure center is part of a continental pressure area extending from Northeast Brazil to the chaco in Argentina. A second pressure area is centered over austral South America. Because the continental land mass of South America is considerably smaller than that of Asia, it is not surprising that there is much more year to year variability in the wind and rainfall patterns. The still smaller continental area of Australia apparently makes the rainfall pattern even less predictable (Taylor and Tullock, 1985).

Due to several factors, the rainfall in the Pantanal can be assumed to have originated almost exclusively in the Atlantic Ocean, through which the warm Brazil Current flows off the East Coast of South America. In contrast, the Pacific shores of the continent are influenced by the cold Humboldt Current, and the water that evaporates is almost entirely deposited on the west slope of the Andes. Thus, a number of deserts have developed to the east of the Andes (Eidt, 1968).

Although the seasonal changes in rainfall are obviously produced by semi-annual reversals of the major wind patterns over South America, these movements of large air masses have little influence on the surface winds affecting the Pantanal. Because this wetland is located in a large basin surrounded by highlands, it is effectively sheltered from strong winds associated with the seasonal continental wind patterns. The winds in the Pantanal of Poconé are characteristically light and variable, which is responsible for making the hot, humid weather that prevails during much of the year even more uncomfortable. However, strong winds occasionally develop in association with thunderstorms, especially those at the beginning of the rainy season. These winds were strong enough in the town of Poconé on one occasion to uproot many of the trees and break large branches off others. These storms are of short duration and only affect a small area through which they pass.

Allanson (1979) and Rolla *et al.* (1994) provided detailed descriptions of some effects winds have on the physical and chemical conditions in tropical water bodies. Although their calculations generally seem to hold true, they are of little relevance for the Pantanal, where submerged, emergent, and floating macrophytes interfere with the wind action and water circulation patterns to a large and locally highly variable extent.

In summary, the climate of the Pantanal can be described as tropical wet-and-dry, while that to the north transitions into the humid tropical zone. The wet-and-dry climate is a primary determinant of the development of the flora and fauna in several particularly interesting regions of the world, the best known of which is monsoonal South and Southeast Asia (Heckman, 1974, 1979, 1997a). Similar regions, either smaller in size or with less regular periods of wet and dry, are encountered in Africa between the equatorial zone and the arid or semi-arid zones located approximately at the Tropics of Cancer and Capricorn and in a rather mountainous region of East Africa crossing the equator, which is noted for its large lakes. A small region subjected to this kind of climate is also found in Northern Australia and a small part of New Guinea. Floral and faunal similarities among these regions were discussed by Heckman (1997a), and intercontinental introductions of plant and animal species within this zone have often proven especially easy, creating remarkable amounts of economic loss when the undesirable features of the introduced species manifest themselves (Mitchell, 1985; Gopal, 1987; Halwart, 1994).

4.1 Seasons

As in the temperate zones, the tropical wet-and-dry zone has four distinct seasons. These greatly influence the habitat conditions and thereby also the activity of the biota. For many of the aquatic species, a period of dormancy is required during the dry season. Terrestrial species must develop rapidly and produced seeds before the next inundation (Section 10). The seasonal changes therefore determine to a great extent which species are able to colonize the Pantanal and which are not (Section 2.2).

In discussing the seasons, it is necessary to distinguish between the actual conditions that directly influence the biota and the seasons themselves as defined by the calendar. Generally, the *enchente*, as defined below, corresponds with the South American spring. The *cheia* coincides with summer. The *vazante* occurs during the autumn, and the *seca*, during winter. The actual seasonal changes as they influence the biota must be discussed individually because they do not always strictly follow the calendar. However, in the temperate zones, either "summer" or "winter" weather may occur in autumn, and the spring may be "late" or "early", just as the rainy season may begin late or early in the Pantanal. The seasons will be referred to by their descriptive Portuguese names and described individually below.

4.1.1. *Enchente*

The beginning of the rainy season may occur at any time between September and November, with rather great differences from year to year. The period of flooding associated with the early rainy season is referred to as the *enchente*, the Portuguese word for flood or flood tide. During this period, the drying pools of water remaining after the dry season and other low-lying areas become flooded with rainwater at various times, but they may dry up again after brief rainless periods. The long-term flooding of the Pantanal begins when the rivers, swollen from the rains on the upland catchment area, overflow their banks.

The transition period from the end of the *seca* to the early *enchente* is the season of the hottest air temperatures, which frequently exceed 40°C and occasionally rise above 45°C in Cuiabá. These temperatures are generally accompanied by a high relative humidity, making the region notorious among Brazilians for its uncomfortably hot and humid climate. The period from mid-October through November is most uncomfortable, with the heat and humidity being accompanied by long windless periods which produce poor visibility by allowing smoke and dust to accumulate in the air. While the daytime humidity usually falls somewhat below 100%, it generally does not sink below 70% to 80%. It seems to be a general characteristic of the tropical wet-and-dry climatic zone that the hottest season immediately follows the coolest season, and that a windless period during which the visibility remains poor precedes the onset of the rainy season (Heckman, 1979). Since the *enchente* corresponds to the spring in the temperate zones, its general features are caused by the springtime warming of the temperate continental areas until the wintertime high pressure area has been reduced to the approximate pressure of the marine areas off the continent. When these two areas are at nearly equal pressure, the vast cyclonic or anticyclonic wind systems disappear, and a period of calm sets in. As the continental land mass continues to heat up, a low pressure area forms, around which a wind pattern develops opposite in direction to that prevailing during the dry winter months. This is a vast anticyclone in South America, which bring moist sea air over the Pantanal. As this air moves across the hot tropical savanna, violent thunderstorms develop, and the rainy season has arrived.

The *enchente* is a period of rapid alterations in physical and chemical conditions. At the beginning of this season, the Pantanal resembles a semi-arid plain. This changes into a verdant terrestrial system immediately after the freshet, and a short time thereafter, within a period of only a few weeks, it becomes covered with a steadily deepening layer of flood waters from the rivers. The detritus left by the aquatic plants and animals that had died when the flood plain dried out together with the short-lived terrestrial plants decay quickly, raising the concentrations of dissolved inorganic nutrients. The humic matter changes the color of the water to brown, far above the 150 Hazen limit of the water color scale available during the study (Section 5.3.1). The concentration of dissolved oxygen is typically low at this time. Fishes, amphibians, and aquatic invertebrates depart from their dry season refuges and spread themselves throughout the large, newly flooded areas. At the beginning of this season, there is a short bloom of algae, and planktonic and micronektonic invertebrates may also be encountered. These do not persist for long after huge numbers of young fishes and tadpoles are produced during the intensive reproductive activity that typically occurs immediately after the freshet.

The date the *enchente* ends also varies from year to year, but within somewhat less of a time span than its beginning. Usually during the month of January, the water level in the northern Pantanal has come to approach its maximum, and the *enchente* can be said to give way to the *cheia*. By the end of the *enchente*, the maximum daytime temperatures are no longer so high, and the rainfall usually occurs more frequently but is somewhat less heavy. The water is generally rather clear, having lost its humic matter and the resulting brown color. There is generally a strong water flow in all of the water courses in the northern Pantanal, including the main tributaries of the major rivers. The volume of water carried by the rivers changes from day to day, sometimes considerably (Section 5).

4.1.2. *Cheia*

The *cheia*, which takes its name from the Portuguese word for "full" and is applied to high water periods and to the high tide, is the period during which nearly all of the Pantanal is under water. It begins about three to four months after the freshet, usually in January, and therefore corresponds to summer in the temperate zone. The northern Pantanal is flooded over large areas to depths from ½ to 2 m at this time. The only unflooded places are the hummocks covered by stands of woods, locally called *capões de mata*. These are usually elliptical or round and develop because the elevation is a few centimeters above the high water level reached during most rainy seasons. The ranch houses in the Pantanal have generally been built on sites formerly occupied by such woods, so they also remain above the water level during the season of flooding. In effect, the entire Pantanal becomes a shallow lake at this time with many small islands just above the level of the water.

Daily rainfall throughout the watershed maintains the high water level during this season. By this time, as already mentioned, the water has lost nearly all of its suspended matter and the typical brownish coloration of the early *enchente*. It remains extremely clear, permitting an uninhibited view of the sediment on the bottom. Submerged macrophytes form dense beds at moderately deep locations during this season, while emergent rooted vegetation is predominant in shallow waters, and floating plants develop rapidly in oxbows and backwaters of the permanent streams, in which the water is usually quite deep. By the end of the *enchente*, almost all of the free mineral nutrients have been incorporated in the tissues of living plants and animals, and no more than scarcely detectable concentrations of most ions, including those from phosphates and nitrogen compounds, were detected in nearly all water sampled in the wetland during the *cheia* (Section 6).

The *cheia* is characterized by stable weather conditions. Rainfall is frequent but only moderately heavy in comparison to the early *enchente*. The daily variations in the temperature of the air and water are minimal during this season, and the humidity remains high. Temperatures above 40°C are generally not encountered, and the maximum temperature during the daytime reaches only approximately 35°C (Fig. 4.1). Thus, the occurrence of especially high temperatures during the early summer, as reported by Por (1995), is actually not typical for the northern part of the Pantanal. Throughout this period, the water in the rivers and smaller water courses flows rapidly, but the water covering the floodplains is lentic. Many of the fishes produced during the *enchente* have already grown considerably and can be observed seeking the food items that have become scarce during the *cheia*. In the northern part of the Pantanal, this season generally lasts from January until early April, although in 1995, it had already begun before the end of December. During some years, the high water levels may not be reached until March. These irregularities are manifestations of the highly variable rainfall patterns in the catchment area.

The relative regularity of the rainfall during the *cheia* is reflected by the conditions in the water. The stability in physical and chemical conditions (Sections 5 and 6) is not only promoted by the relatively uniform daily weather conditions but also by the large volume of water covering the flooded wetlands. This is a great heat sink, which tends to heat and cool slowly, making temperature differences between day and night minimal.

It was noted that heavy rainfalls on the wetlands during this period have almost no effect whatever on the water level. This is rather determined by the levels in the main rivers, which are fed by rainwater on the more elevated sections of the watershed as well as in the wetland itself. In the absence of strong prevailing winds, the rainstorms over the central part of the wetland can be viewed as simply a return of evaporated water to the floodplain. The surface winds are light and variable during the late *cheia*, and the storms cells generally develop in the late afternoon after intensively sunny mornings, during which strong thermal updrafts of heavily moisture-laden air are produced.

4.1.3. *Vazante*

The *vazante*, taken from the Portuguese word for ebb tide, is the season during which the water level in the Pantanal falls rapidly. During this season, the rainfall diminishes greatly but does not cease altogether.

It should be mentioned that the word *vazante* is also applied to a kind of shallow water body on the floodplain, through which the floodwaters slowly drain as the water level sinks. This will be described in more detain in Section 9. To avoid confusion, the word *vazante* is used to refer exclusively to the season throughout the rest of the text.

The slow drainage of the water through the Paraguay River System begins to have an effect on the water level, which initially decreases rapidly and then more slowly as the water is no longer able to overcome the many low barriers to its flow. At this time, the aquatic macrophytes begin to die off and are replaced by terrestrial species, a process hastened by the extremely strong insolation from usually cloudless skies through the air cleared of suspended matter during the rainy season. Although the relative humidity rises to nearly 100% when the temperature falls during the nights, the rise in daytime temperatures lowers the humidity to below 60% during most days, promoting evaporation. Apparently because of its location at a higher elevation than the Pantanal, this pattern could not be well documented from the weather data at Varzea Grande (Fig. 4.5). Nevertheless, the maximum relative humidity recorded there during the day was seldom less than 80%, and the values during the late afternoon were lowest, but they almost always exceeded 50%.

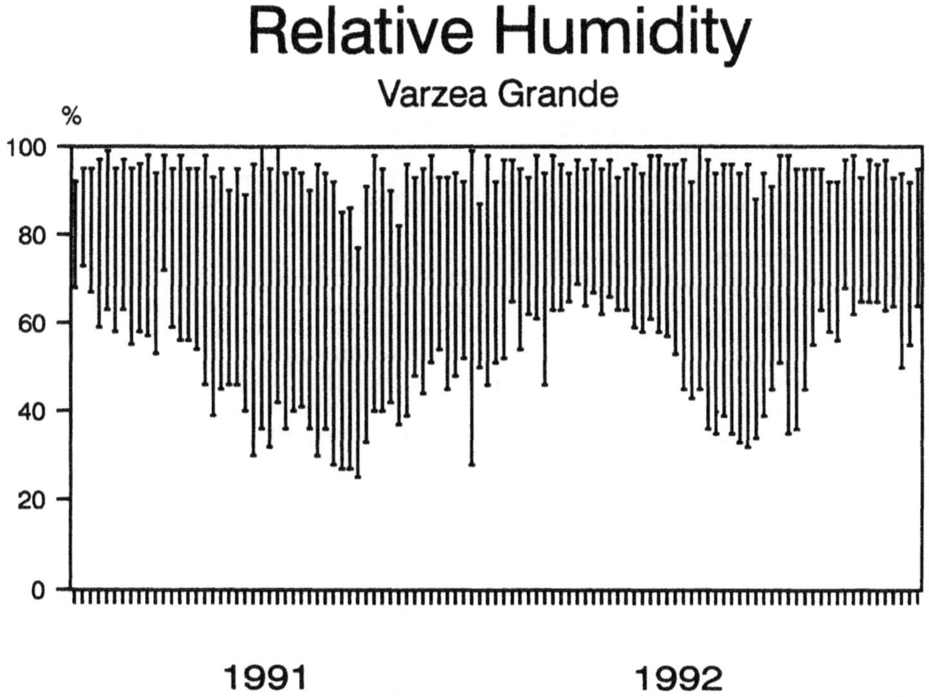

Figure 4.5: The weekly maximum and minimum relative humidity recorded at the meteorological station in Varzea Grande from January 1, 1991, to January 6, 1993. The actual ranges of humidity are greater than those shown because the data were recorded only three times each day, at noon, 6 p.m., and midnight.

The aquatic organisms can generally adapt to the water level decreases for about a month, and runoff continues to feed the water bodies in the Pantanal from its large drainage area. However, the water level falls gradually but steadily from the beginning of the *vazante* until the end of the subsequent *seca* (see below), and the transition of the conditions typical for one season to those characteristic of the next occurs at slightly elevated regions of the floodplain months before it does at the low-lying ones. It is thus difficult to define these two seasons using the calendar. In any case, about three months after the rains have markedly decreased in intensity, usually beginning in late April or May, much of the Pantanal has dried up, and the transition to the *seca* can be considered to take place in late July or early August at an "average" location in the northern Pantanal.

The *vazante* shows many similarities with its temperate zone equivalent, the autumn, although there is only a minimal shortening in the length of the day at the latitudes of the Pantanal. Just as most organisms in temperate zones must begin a period of dormancy or produce resting propagules in response to low temperatures, those in the tropical wet-and-dry zone must prepare for a period of inactivity induced by dryness. From the conditions in the water bodies, the *vazante* can be characterized as a season of rapid changes, with extremely large amounts of inorganic plant nutrients being released from decomposing aquatic plant and animal material. It is the season of accelerated saprobic activity as great masses of plant material produced during the rainy season die and decompose at extremely high rates through the action of microorganisms adapted to relatively high water temperatures. The areas at the uppermost levels of the floodplain become dry at the beginning of this season, and the rest of the floodplain is typically covered by a thin layer of water containing many intensively blooming aquatic plants that die off after the sites have nearly dried up.

4.1.4. *Seca*

The *seca* is the season during which most of the area of the Pantanal falls completely dry. Water remains at a few low-lying sites and in the rivers and streams, which become standing water bodies during this season. The transition period from the *vazante* to the early *seca* is the period of coolest temperatures in the Pantanal. It coincides with winter in the Southern Hemisphere and is characterized by regular northward movements of cold air masses from the Antarctic, which influence the weather as far north as the equator (Sioli, 1984). Such cold waves, called *friagems*, bring cool air and large areas of overcast to the Pantanal. They are sometimes accompanied by a fine, steady, mist-like precipitation, which scarcely influences the rainfall figures reported by the weather stations because the amount of water falling is scarcely measurable. Each *friagem* is generally followed by several days of rapid warming, which raise the daytime temperatures from slightly more than 10°C to about 35°C within about four or five days after the arrival of the cold air mass. During the period from late June through mid-August, the *friagems* occur at intervals of about five to eight days, although there are considerable year to year differences in their frequency and in the temperature decreases they bring about. During the warm periods between the *friagems*, the daytime humidity drops quite low, especially during the late part of the season. However, even on days when the afternoon relative humidity drops below 30%, the falling temperatures during the night almost always raise the relative humidity to 100%. This was observed week for week during the course of the studies at the small weather station set up on the Fazenda Ypiranga about 15 km south of Poconé.

The transition to the terrestrial phase begins early at higher elevations, and at the lowest lying sites, the aquatic habitat never completely disappears. During this period, the displacement of aquatic habitats by terrestrial ones is completed in many places, and profound changes occur even in those low-lying water courses that fail to dry up at all. Small and medium-sized rivers, in which strong water currents can be observed throughout the *cheia* and *vazante*, are transformed to lentic water bodies during the *seca*. The water quality also changes profoundly. The amounts of organic and inorganic substances in the water increase tremendously shortly before the shallow water bodies completely dry up. Sometimes, already high concentrations are increased even more by the release of materials from the bodies of fishes that suffer massive mortalities or from the excreta that enters the water wherever large flocks of wading birds congregate at the remaining standing water bodies on the floodplain during the late dry season. During some years, the freshet may occur late in the dry period, but subsequent rains fail to fall for a period of time long enough for the rainwater pools produced to completely dry out again. Blooms of algae or emergence of aestivating aquatic animals may take place during the short periods that these pools are flooded. In many low-lying parts of the Pantanal, a fast-growing community of short-lived annual, terrestrial herbs dominates the desiccated beds of shallow water bodies. These form an ecotone that moves across the drying areas on the floodplain (Heckman, 1997b). Generally, the dry period can be considered to occur roughly from late July or August through late October or early November, with great local and year to year variations in the exact dates.

The late *seca* and the transition period to the *enchente* are characterized by the hottest temperatures of the year. Evaporation of the shallow pools of water left on the floodplain is rapid, and the water temperatures greatly increases, with temperatures over $40^{\circ}C$ being encountered during the daytime (Section 5). Mass mortalities of varies aquatic plants and animals occur, although many of the species are adapted to aestivate in the sediment until the flood waters return.

Generally, the *seca* can be characterized as a season of extreme conditions, with the highest and lowest water temperatures, pH values, and oxygen concentrations being encountered, as discussed below. The dates on which the typical conditions of this season begin and end are also highly variable from year to year.

4.2. Physical changes in the surface water

During the course of the year, water temperature, pH, electrical conductivity, and chemical variables were determined at all of the sampling sites (Heckman, 1994a). These variables were also recorded routinely at other sites visited to compare regional conditions. The results of the chemical analyses will be reported in Section 6, while some of the data on the physical conditions are reported and discussed here. These data were determined using instruments manufactured by WTW. Temperature and electrical conductivity were recorded with the WTW LF 196 electronic sensor, and pH, using the WTW pH 196 electronic sensor.

In general, both water temperature, pH, and oxygen concentration are particularly dynamic factors, so continuous measurements were made for periods of at least 24 hours during different seasons. Each of these variables is greatly influenced by the daily weather, with the water temperature changing according to the air temperature, amount of direct insolation, and rate of evaporation at the surface, which is a function of the relative humidity. The pH changes result from gas exchange at the surface and between the water and the aquatic organisms. These processes are influenced, in turn, by the illumination, wind-driven water circulation, and rainfall. The oxygen exchange at the surface of the water is influenced by temperature, wind, and rainfall, while the release of oxygen through photosynthesis by the submerged water plants depends largely upon the amount and thickness of the cloud cover, inputs of nutrients with rainfall, and the wind-driven circulation of the water, which influences the amounts of inorganic nutrients reaching the plants. The data recorded about once each hour were used to plot the curves showing the diurnal cycles of these factors at different sites in Figures 4.6 and 4.7.

During routine sampling, the data recorded for these variables had to be evaluated on the basis of the seasonal peculiarities and daily weather. For example, during the high water period, diurnal changes are not considerable, and the temperature and pH recorded any time during the day are not far from the daily maximum and minimum values.

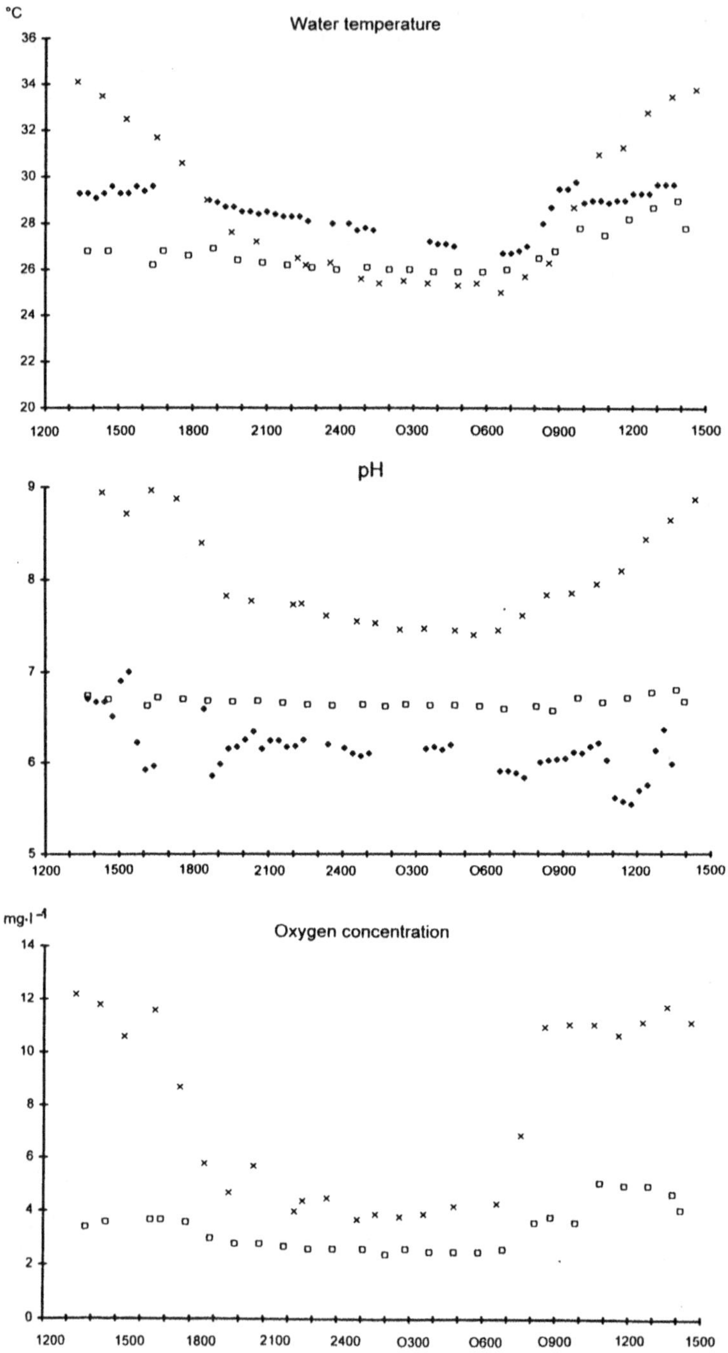

Figure 4.6: The water temperature and pH in the Rio Bento Gomes at the Poconé exposition park near the northern edge of the Pantanal recorded during 24 hour periods on February 4 and 5 (♦), March 24 and 25 (□), and August 25 and 26, 1992 (**x**), and the oxygen concentration recorded during the last two of these periods. These dates were chosen during the *cheia*, beginning of the *vazante*, and *seca*, respectively.

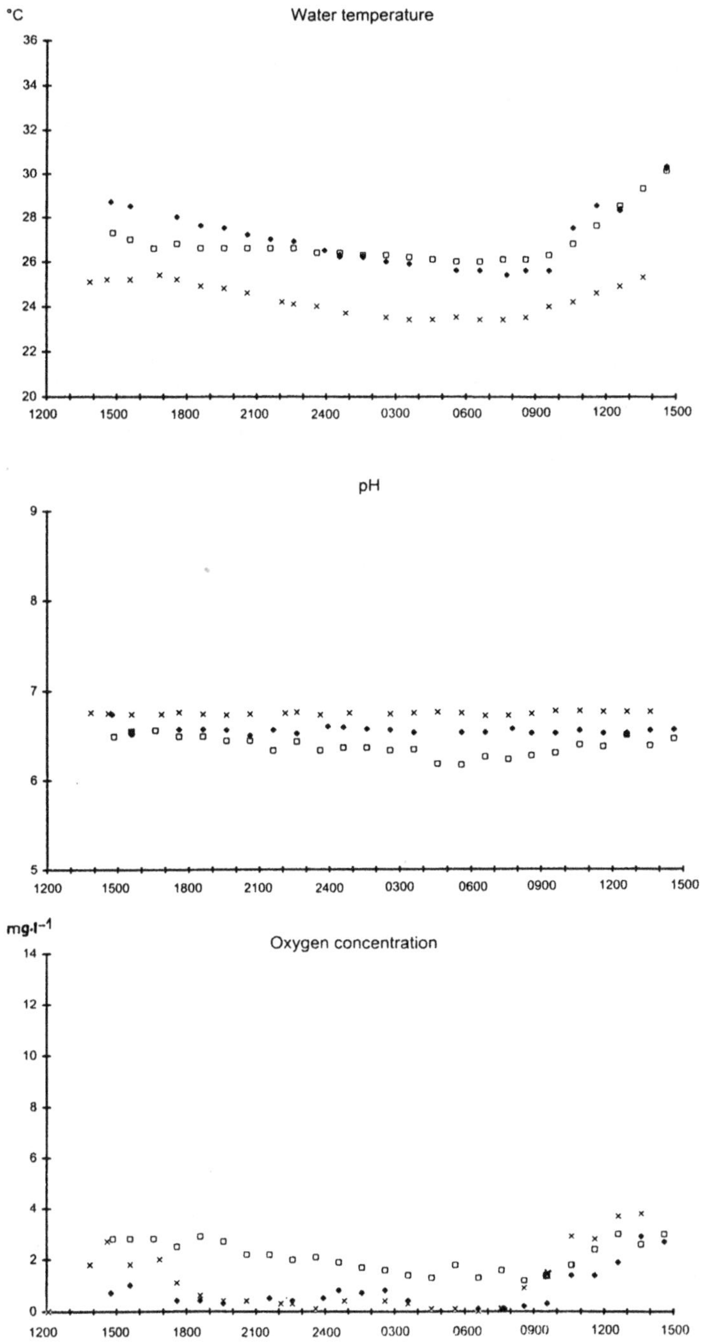

Figure 4.7: The water temperature, pH, and oxygen concentration during periods of about 24 hour in the Rio Bento Gomes during the *cheia* on February 18 and 19 (♦) and during the *seca* on September 22 and 23 (□), and the Rio Novo during the *cheia* on March 3 and 4, 1992 (**x**), where they cross beneath the Transpantaneira Highway in the Pantanal.

During the dry season, fluctuations are sometimes extreme, and no single water temperature value can be taken as representative for the day. Generally, the maximum temperature and pH values can be recorded during the midday hours, and the minimum values just before sunrise. However, the arrival of a *friagem*, that is, a cold air mass, can change both variables radically. The electrical conductivity (Section 5) and the concentrations of various ions derived from dissolved minerals (Section 6) proved much more stable over a 24 hour period than temperature, pH, and the oxygen concentration; however, the seasonal changes in these factors are considerable, and notable changes can occur as a result of a heavy rainstorm.

A notable feature of the pH curves is their relationship with the photosynthesis and decomposition processes taking place in the water body. This will be discussed in Section 6. It is sufficient to mention here that relatively large fluctuations in the pH during the course of a single day is usually an indication that the water body is eutrophic and rich in microorganisms, with the breakdown of large amounts of organic substances during the night lowering the pH and the considerable amount of photosynthesis during daylight hours raising it again. Compared to most water bodies on plains of the temperate zones (Heckman, 1986; 1990), those in the Pantanal show relatively little pH change over the course of a single day except for brief periods during the dry season. However, it can be assumed from the generally low electrical conductivity values that the water is poorly buffered, and the changes in pH shown in the curves are therefore much greater than they would be in more ion-rich water bodies in which the same amounts of decomposition and photosynthesis were taking place.

5. Hydrology

5.1. Water movements through the Pantanal

5.1.1. The annual cycle

The Pantanal is fed by small amounts of ground water throughout the year and large quantities of rainwater seasonally. Much of the water enters through a number of large rivers which converge on it. Most of these arise from the high plateau that surrounds its watershed. The main rivers flowing into the northernmost part of the Pantanal are the upper Rio Paraguay, Rio Cuiabá, and their many smaller tributaries (Fig. 2.3). South of the Rio Cuiabá, several other river systems flow toward the Rio Paraguay from the east, and a smaller number drain the watersheds in Bolivia and Paraguay and join it from the west. The Pantanal is then drained almost entirely through the Lower Rio Paraguay, a tributary of the Paraná River. The water eventually reaches the Atlantic Ocean through the Rio de la Plata between Argentina and Uruguay (Fig. 2.2).

Unfortunately, few data are available on the water movements in the Pantanal itself. There have been, however, fairly regular readings of the water level at several locations along the Rios Paraguay and Cuiabá since the early 1970s (Figs. 5.1 to 5.3). These provide much seasonal information on the flow of water into and out of the Pantanal.

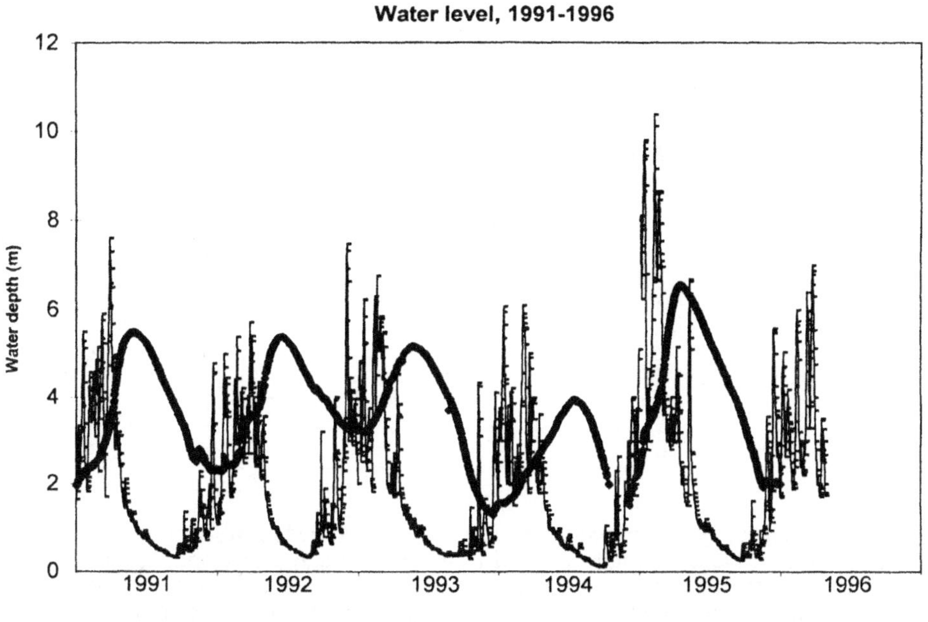

Figure 5.1: The daily water level in the Rio Cuiabá recorded upstream from the Pantanal at the city of Cuiabá from 1991 through early 1996 (■) and the level in the Rio Paraguay at Corumbá, downstream from the Pantanal, from 1991 through 1995 (♦), running together to form a thick line). A few gaps occur in the curves where data are lacking. From this graph, it is clear that the high water peak at Cuiabá occurs several months earlier than that at Corumbá and that the water level changes from day to day are often more than two meters upstream from the Pantanal, while they are seldom more than a few centimeters downstream from the wetland. This shows the flood retention effect of the Pantanal on the river system. The heavy rains of the freshet are evident from the sharp peaks in the curve for Cuiabá and as a small irregularity near the low point of the curve for Corumbá.

Nevertheless, far too few data are presently available to calculate the hydrological budget of the Pantanal. Some insights can be gained by comparing the inflow and outflow in the rivers, but this does not permit any sort of definitive quantitative calculations. Whether estimates of the water volume passing through the wetland during a relatively short period of time would provide better insights into the dynamics of the system is, in any case, doubtful, because the relatively little information already available amply supports the conclusion that the magnitude of the year to year differences is sufficient to render averages based on data collected over a period of only a few years statistically insignificant. To adequately characterize the hydrology of the Pantanal, even in a cursory manner, a program of monitoring for several decades will be necessary. No such program is presently under consideration, although the findings of this kind of research are vitally needed to estimate the extent of the freshwater resources of this vast wetland. Long-term data on the water level, recorded since 1940, are available for only one point at Porto Murtinho to the south of the main area of the Pantanal (Adámoli, 1986b).

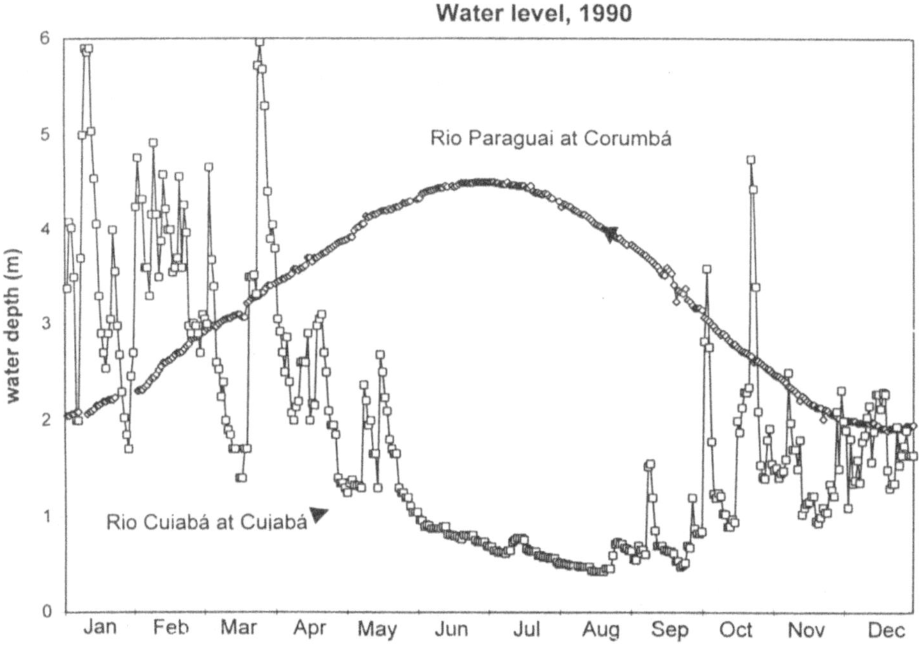

Figure 5.2: The daily water level in the Rio Cuiabá recorded upstream from the Pantanal at the city of Cuiabá (□) and the level in the Rio Paraguay at Corumbá, downstream from the Pantanal (◊) during the year 1990. The daily changes in the water level at Corumbá are so slight that the symbols run together in the figure to form a thick line. This figure shows a better resolution of the data points than do the curves in Figures 5.1 and 5.3.

The first problem is calculating the rainfall on the region (Section 4). There is no way of determining how much rainwater falls on any given area of the Pantanal itself due to a lack of any systematic collection of meteorological data (Section 4.1). Similarly, there is still no way of calculating the amount of evaporation. Estimates based on other regions in the tropics would be nothing more than guesses because the humidity of the region varies too greatly, as shown by the data in Figure 4.5 as well as those collected personally on the Fazenda Ypiranga for over two years during the project. Typically, the humidity during the nights throughout nearly the entire year is almost 100% on the Pantanal floodplain and over 90% for much of the year at Varzea Grande. During the day, it decreases as the temperature rises and sometimes becomes relatively low: often less than 50%, especially during the *seca*. A particular problem related to determining the amount of water loss due to evaporation is the fact that some of the water that evaporates during the late rainy season is carried into the upper atmosphere by strong daytime thermal currents and falls to earth as rainfall again, usually during the afternoons or evenings, undoubtedly not far from the site of its evaporation.

Finally, the amount of ground water seeping into the Pantanal and the amount of the water that escapes by seepage into deeper layers remains unknown.

Looking at the hydrological data for the Rio Cuiabá at Cuiabá and the Rio Paraguay at Corumbá during the year 1990 (Figs. 5.1), we see that the level of the water in the rivers upstream from the Pantanal at Cuiabá changes rapidly according to the local rainfall. The highest water levels are generally encountered near the end of the rainy season, usually in February, and after the rains have ceased, the level falls rather rapidly. In 1990, absolute peaks were reached in January and March, coincidental with periods of particularly heavy rainfall on the catchment area. During other years, peaks were recorded at other times of year (Fig. 5.2). Periods of heavy rainfall manifest themselves in the water level changes relatively quickly. At Cuiabá, this is due exclusively to rainfall on the upland parts of the drainage area, particularly the plateau to the north and east, which generally seems to be subject to more rainfall than the Pantanal itself due to the higher elevation above sea level and the resulting adiabatic effects on the windborne moisture in the air. Changes in the water level of the river between 1 and 2 meters within 24 hours are not at all infrequent. Since the wetland is located well downstream from Cuiabá, the rainfall there cannot have affected the water level shown for the Rio Cuiabá in Figures 5.1 and 5.2 at all. In contrast, the level of the Rio Paraguay at Corumbá rarely changes more than a few centimeters in a single day.

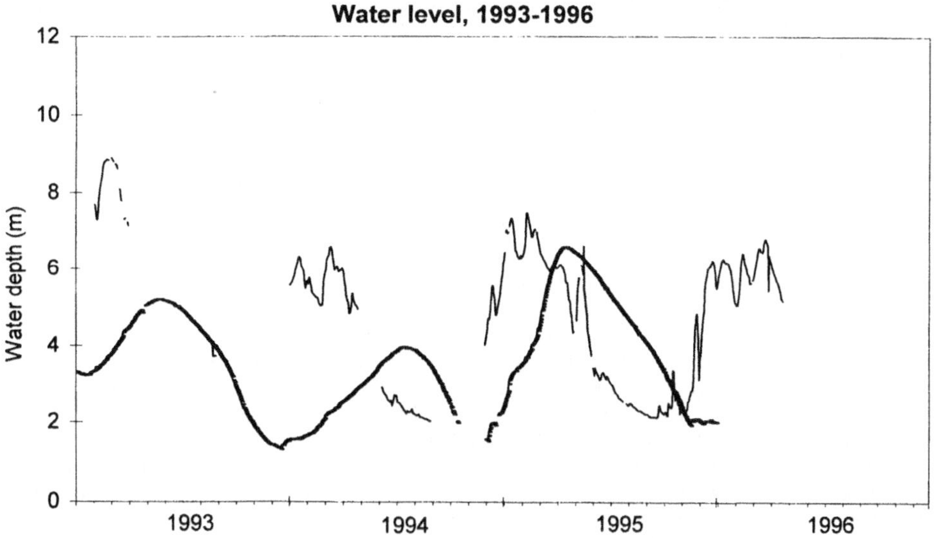

Figure 5.3: The daily water level in the Rio Cuiabá recorded upstream from the Pantanal at the town of Barão de Melgaço from 1993 through early 1996 (thin line) and the level in the Rio Paraguay at Corumbá, downstream from the Pantanal, from 1991 through 1995 (▬). The data for 1993 and 1994 from Barão de Melgaço are fragmentary but clearly show the high water periods. Because the river is wider there than upstream at Cuiabá, the dynamics of the water level fluctuations are different, the changes being somewhat less rapid. The year to year variability is obvious.

Figure 5.3 compares the changes in the water level at Barão de Melgaço at the northern edge of the Pantanal with those in the Rio Paraguay at Corumbá. Again, the changes in the Rio Cuiabá are much more rapid and of greater amplitude than those in the Rio Paraguay at Corumbá but somewhat less extreme than those at Cuiabá shown in Figure 5.2. A ground for this is apparently the connection of the river with several large lakes, the Baias Recreio, Porta de Fora, and Acorizal, which are available to act as immense natural flood retention basins which dampen the surges in the water flow at Barão de Melgaço.

The outflow from the Pantanal occurs through the Rio Paraguay system (Fig. 2.2). Downstream from the Pantanal, it must carry water at an increasing rate as the water level in the southern Pantanal increases. Because of the slight declination of the wetland and blockage by Quarternary sediments, the flow from north to south is relatively slow (Valverde, 1972; Alvarenga *et al.*, 1984). The result is a much later high water period at Corumbá than at Cuiabá and a continual rise in the water level throughout the early dry season in this section of the Paraguay River at Corumbá (Figs. 5.1 & 5.2). The four month displacement of the high water period from the season of peak rainfall is an important feature of the Pantanal of Mato Grosso do Sul, causing its hydrobiological dynamics to differ from those in the wetlands located in the State of Mato Grosso, where the peak periods of rainfall and the highest water levels more or less coincide.

Thus, the Pantanal can be considered to be a vast natural flood retention basin, which accumulates the pulses of water from the rivers feeding it during the rainy season and feeds the middle and lower sections of the Rio Paraguay at a rate that changes much more gradually during the course of the year. This is clear to see in Figures 5.1 to 5.3, in which the water level at the southern end of the Pantanal shows gradual increases and decreases with almost no sharp day to day or week to week fluctuations at all. The delay of about four months between the time the high water level has been reached in the northern part of the Pantanal at the peak of the rainy season and the time it occurs in the southern part, which is well after the beginning of the rainless seasons, is explained by the extremely slight declination of the vast Pantanal Plain, which reportedly averages only about 2 cm per km. The result of this is a slow passage of the water through the Pantanal itself. The high water peak can be visualized as a kind of wave that peaks at the northern end of the Pantanal at Poconé, Cáceres, and Barão de Melgaço at various times from late January to early March, depending on the distribution of the rainfall in the watershed during the particular year. The many rivers swell and overflow their banks, spilling water over the plain and forming the wave that moves slowly southward. It takes several weeks to reach the north central part of the Pantanal of Poconé drained by the Rios Novo, Frio, and Claro. At the same time, the water in the Rios Cuiabá and Paraguay rise and force water back into the many small water courses along their valleys until the whole northern Pantanal is under water. By this time, the water level in the northernmost part of the Pantanal is already falling slightly, although considerable fluctuations occur due to irregular periods of rainfall and sunny periods on the catchment area at higher elevations than the wetland. It is noteworthy that reports of drought by the ranchers in the southern Pantanal are often heard at this time of year, in spite of the sporadic thunderstorms, as mentioned in Section 4.

The high water wave finally reaches the central and southern parts of the Pantanal after the rainy season has ended, and the high water levels are measured at Corumbá well into the dry season in May or even June, forming the peaks shown in Figure 5.3. Veloso (1972) was one of the first to report that about 120 days are required for the high water crest to move from the northern to the southern borders of the Pantanal. This process repeats itself every year, but with considerably different high and low water levels and dates on which they are reach, reflecting the year to year differences in the weather.

The precise functions of the Pantanal in the global system are still largely unknown, although the role of the wetland in the hydrological dynamics of South America can be assumed to be profound. The water can be seen to flow over a geological sill located in the bed of the Rio Paraguay, which controls the outflow rate precisely. Thus, the Pantanal acts not only as a gigantic flood retention basin but also as an evaporation pan. These functions are certainly determinants of the entire hydrological and climatic system of central South America and most certainly affect the neighboring regions, as well.

An interesting corollary of these hydrological processes is the need to study the ecosystem dynamics in the northern and southern parts of the Pantanal separately. The conditions affecting the biota in these two regions are considerably different when the high water period reaches one section in the middle of the rainy season and the other during the rainless part of the year. The severity of the desiccation process on the aquatic biota during the low water period in the southern Pantanal should be tempered somewhat by the onset of the rainy season, which may prevent many of the small water bodies from completely drying out. With less severe conditions, the region should therefore support a larger number of aquatic and wetland species than the northern Pantanal. However, the feeding conditions for the piscivorous caimans and large water birds may not be as good. It must be left to future researchers to determine how much greater the species diversity is in the southern Pantanal than in the northern part of the region and how closely this can be correlated with the hydrological conditions.

The lower part of the Rio Paraguay-Paraná system in Argentina has been investigated by Bonetto (1975), who noted less regularity in the seasonal periods of high and low water than those prevailing in the northern part of the Pantanal. This reflects the great complexity of this section of the river system, which is influenced by the seasonal changes in discharge from both the upper Rio Paraguay and the upstream sections of the Rio Paraná and its tributaries, as well. Newly constructed dams along rivers in the Rio Paraná System have certainly brought about additional changes in the hydrodynamics of the entire system, which still must be quantitatively determined. The water level changes have considerable effects on the biotic communities along the rivers (Bonetto and Ferrato, 1969a), and the periodic flooding of the shallow areas of the floodplain seem to be important for the growth of fishes, which forage along the shores (Bonetto and Ferrato, 1969b). The water level changes in the Pantanal are thus manifested later in the year downstream throughout the entire river system. Together with the seasonal hydrological changes to the east, which influence the water level changes in the upper Rio Paraná, these fluctuations in the levels of the rivers are of great importance for the fish production throughout the entire system (Section 11.2). Natural floodplains beside water bodies and their biota are immensely important for the functioning of the entire fluvial ecosystem (Heckman, 1984), and the hydrodynamics of the Pantanal therefore assure that it plays a vital role for the fish and wildlife populations throughout a vast area of South America.

5.1.2. Year to year differences

Subsequent sections will refer to the considerable year to year differences in the hydrological cycle of the Pantanal. These can be considerable, with particularly dry or rainy periods lasting for several years. Extended dry or rainy periods generally result in corresponding expansions or contractions in the local distribution of the terrestrial and the aquatic biota. Reversals of these expansions and contractions occur when the long-term weather conditions again reverse themselves.

Such long-term changes, which still have shown no clear cyclical pattern since they have been recorded, may cause considerable losses among the affected species. Based on a compilation of data from various publications, Por (1995) reported that peak floods during this century occurred in the years 1905, 1920, 1932, 1954, 1959, 1974, 1979, and 1982. While virtually nothing definite is known about the losses of wildlife from excessive drought or flooding, cattle losses as high as 800,000 head were reported to have occurred during a flood in 1974 (Sucksdorf, 1989).

A problem in estimating the extent of year to year differences in the flooding results from the local differences in the time of year the high water level is reached in the individual rivers. This irregularity is a consequence of the local rainfall patterns, which show the considerable differences from year to year that were discussed above. For example, the high water in the Cuiabá and Paraguay Rivers upstream from their confluence may occur nearly simultaneously during some years and at different times during others. This has an effect on the depth and duration of the flooding in the southern part of the Pantanal as well as in the drainage areas of the two rivers in the northern part. Little can be said about this phenomenon at the present time because of the total lack of meteorological data from the valley of the Rio Paraguay from its source to the southern part of the Pantanal near Corumbá.

Even during the short period of the study, an impact on the biota was detected from year to year due to variations in the water depth (Fig. 5.4). These can be best illustrated through the variations in the local distribution patterns of the macrophytes (Section 10), although they affect most other groups of plants and animals on the floodplain, as well.

A corollary of the information presented concerns the effect that changing the flow system would have on the entire region of the Pantanal. The proposed Hidrovia project will be discussed later (Section 13.4, 13.5), but it should be mentioned here that extensive deepening of the Rio Paraguay to make it more navigable might well have an effect opposite to that planned during those years when the general water level becomes particularly low. If the flow in the river were not inhibited by natural shallow stretches, the Pantanal might be largely drained, and the water would flow past Corumbá at a rapid rate during the late rainy season and become too shallow for ship traffic thereafter. The water level fluctuations so familiar at Cuiabá would come to have a strong impact on the level of the waterways throughout much of the Paraguay-Paraná system, and during dry years, even small riverboats would not be able to pass at all during much of the late dry season. The economic losses from such a fiasco could be enormous.

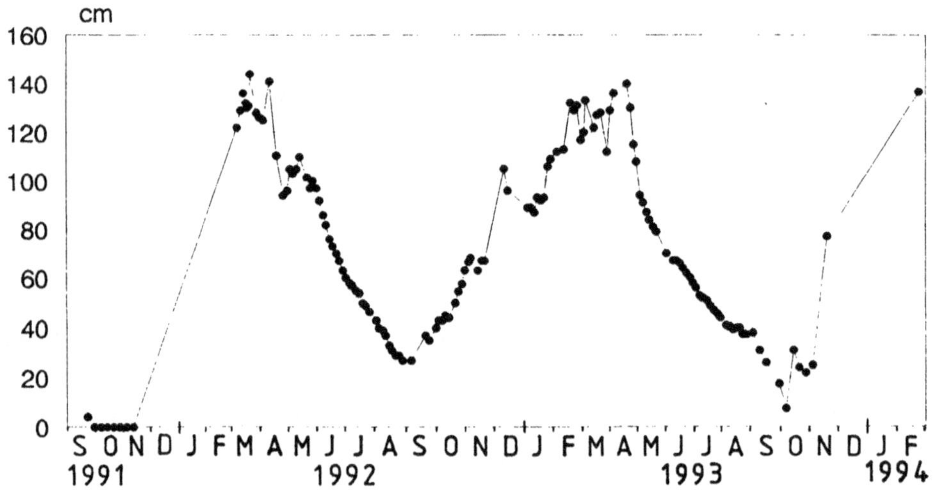

Figure 5.4: The water level on the flood plain read from a gauge at Birici Stream from September 1991 through February 1994. The site of the gauge was completely dry for several weeks during 1991, but it maintained at least a shallow layer of water for the rest of the study and through 1995.

5.2. Sources of water in the Pantanal

As stated above, the primary source of water in the northern part of the Pantanal is rainwater falling on the watersheds of the Rio Cuiabá and the Upper Rio Paraguay, which enters the Pantanal through many tributaries of these rivers. In addition to rainwater, however, some of the water in these streams comes from upwelling ground water on the highlands surrounding the Pantanal floodplain. Frequent observations of the springs feeding these rivers indicate that this water flows into the Pantanal at about the same rate the year round, but it is strongly augmented by the rainwater entering these streams during the rainy season. This natural ground water input from the neighboring highlands can obviously be drained from the region through the streams and rivers without causing any floods. This can be concluded from the fact that without rainfall, the floodplains nearly completely dry up.

There are still two sources of water that need some discussion. The first of these is rainfall directly on the Pantanal floodplain, and the second is ground water emerging from beneath the Pantanal.

Local rainfall begins on the wetland at about the same time after the long dry period as it does in other parts of the northern watershed. The thunderstorms of the freshet are usually intense and produce much local flooding. This rainwater, however, often seeps into the dry ground or evaporates within a few day. The freshet arrives at different times each year. During the four years of the study, the dates of the first rains heavy enough to be considered a freshet fell between September 2 and November 3. The freshet may be followed by a period of intensifying rainstorms, or it may precede a renewed dry period. The rainstorms prior to late December usually have surprisingly little effect on the water level in the tributaries passing through the northern Pantanal, which tend to rise slowly at first. Usually in late December, but occasionally much earlier, the water level suddenly begins to rise. This occurs even in years when the northern Pantanal has been having a rather long dry spell. Suddenly, generally within one or two days, water begins to feed backwards through the tributary systems and flood over the low-lying areas of the floodplain. This event, occurring as it sometimes does after several dry and sunny weeks, strongly demonstrates that the local rainfall has relatively little effect on the annual flooding of the Pantanal. It is the rain on the uplands that seems to be the decisive factor in feeding the flood.

Rains continue in the Pantanal throughout the rainy season, but it remains to be investigated how much of this rainwater is merely evaporated moisture from the region on which it again falls. The input of water originating from the oceans into the system depends on the winds, the circulation patterns of which over South America are still relatively poorly defined (Schwerdtfeger, 1976; Ratisbona, 1976). Local rains produced after local evaporation add nothing to the water reserves and cannot replace the large amounts of water being carried away through the Paraguay-Paraná River System. Furthermore, moving south and west from Poconé across the Pantanal, the amount of annual rainfall shows a steady decrease to the middle of the wetland. Figure 4.4 shows that the minimum average rainfall is encountered in the Bolivian part of the Pantanal just southwest of the confluence of the Rios Cuiabá and Paraguay. The decrease in the average annual rainfall results not from a reduction in the amount of rainwater that falls from the individual storm clouds but rather from a longer annual rainless period.

The final source of water to consider is the ground water beneath the Pantanal. Because the chemistry of this water is quite distinct (Section 6), its presence may be detectable. Apparently, the pressure from the surface water during the high water season in the Pantanal suppresses any seepage of ground water during the annual period of flooding. However, during the dry season, there are indications that water from the uppermost aquifer seeps onto the low-lying sections of the Pantanal in a few places, producing the *campo úmido* conditions.

An obvious seepage of water onto dry areas of the Pantanal was seldom observed. It did occur at two sites, however, one of them at the foot of a low hill adjacent to the Baia Porta de Fora in the Pantanal of Barão de Melgaço, and the other, also near the foot of low hills on the west side of the Rio Paraguay in the Municipal District of Cáceres. A third point, near the Poconé exposition park beside the Rio Bento Gomes, was observed for two years, in 1991 and 1992, but the site remained dry during subsequent years. This water may have been drainage water from the park, which was later rerouted, but its exact source was never determined.

All of these points of ground water seepage are marked by a peculiar kind of dry season vegetation consisting of wetland species that occur primarily, or perhaps exclusively in such habitats (Section 10). While this water has almost no significance for the hydrological balance in the Pantanal, it produces a profound effect on the biota wherever it appears. It supports a wetland aggregation that must otherwise die out in the region during the dry season. Not only do true wetland species occur at such sites, but special morphological forms of aquatic plants also develop (Section 7). Ground water seepage therefore provides a refuge for species that require a year-round water supply and are otherwise rare in the region.

5.3. Classification of South American surface waters

5.3.1. Water quality

Three major classes of surface water in tropical South America have been referred to as white, black, and clear water by Sioli (1984). The white water rivers generally have their sources on the slopes of the geologically young mountains in the Andes Range, including the foothills. This water carries fine suspended matter and a moderate load of inorganic nutrients, which give it a generally turbid white appearance. The largest of the white water rivers is the Rio Solimões, the main branch of the Amazon. The black water rivers originate in the forested mountains of northern South America. They are poor in nutrients and carry little suspended matter, but the color of the water is dark due to the presence of humic materials. The Rio Negro is the largest river of black water in Brazil. Where the white water of the Rio Solimões and black water of the Rio Negro meet near Manaus, mixing is gradual, and the sharp contrast between these two kinds of water is evident along the line where the two bodies of water meet. The mineral nutrient content of the water was regarded as such an important factor that Fittkau (1971) used it as a basis for dividing the Amazon Region into ecological zones.

The water of the rivers flowing into the Pantanal belong to the clear water class. The reason it has been given this name is obvious. The water from the sources of these rivers on the plateau of the Central Brazilian Shield to the north and east of the Pantanal contains little suspended matter and often has a Hazen value of zero. In contrast to water bodies in the white water system, where the light attenuation from suspended matter can limit the growth of algae (Schmidt, 1973), visible turbidity exists in the clear water streams only where fine sediment is stirred up in pools near the sources. The sediment is slow to resettle on the bottom because of the extremely low electrical conductivity of the water (Section 5).

As the ground water flows toward the Pantanal during the rainy season, it is increasingly diluted by rainwater, which considerably increases its electrical conductivity and content of various ions. During the rainy season, the water on the Pantanal floodplain is extremely clear due to its combined origin as clear ground water and rain. Only after a great deal of evaporation has taken place during the dry season does the appearance of the water show a significant change. This is a result of great turbidity increases related to the development of massive numbers of bacteria and algae. Another change takes place shortly after the flooding of the desiccated floodplain, when the water becomes coffee brown from the humic matter released through the decomposition of the terrestrial plant material. Interestingly, this water strongly resembles the black water in the Rio Negro System of Amazonia, not only in its appearance but also in many of its chemical features. It has a relatively low conductivity, generally well below 100 $\mu S \cdot cm^{-1}$, as well as low oxygen saturation, a relatively low pH, and a presumed high content of humic substances, attested to by its

Hazen value considerably exceeding 150, which was the maximum measurable value using the test kit available in the field (Table 5.1). It differs, however, in its initially high concentrations of dissolved inorganic nutrients, including sources of phosphorus and nitrogen (Section 6). As the aquatic plants begin to grow, however, this nutrient content decreases rapidly, and values reminiscent of the Rio Negro in the Amazon System are reached. However, there is a simultaneous increase in the dissolved oxygen concentration and a loss of the dark brown coloration. Thus, the "black water" that appears seasonally on the Pantanal floodplain does not fully correspond to the permanent black water of Amazonia because of the temporary nature of its characteristics.

The physical characteristics of the water in the streams feeding the Pantanal and at the sampling sites in the wetland itself were determined in the field using portable electronic equipment produced by WTW. The WTW LF 196 was utilized to obtain the water temperature and electrical conductivity values presented here.

The temperature of the ground water emerging from springs at the headwaters of the streams and rivers flowing toward the Pantanal is generally fairly constant throughout the year. The water from most of them is close to 20°C, but a few hot springs are also present in the catchment area (Table 5.2). Downstream from the springs, the temperature becomes increasingly influenced by the weather, particularly heavy rainstorms.

Table 5.1: The seasonal ranges of water color recorded on the Hazen scale at various sampling sites during the *seca* and *enchente* of 1994 using a color determination kit provided by Merck with distilled water as the standard. The values reported are true Hazen values determined on filtered samples. Only during the *seca* were differences between the true and apparent Hazen values occasionally observed.

	Seca	(n)	*Enchente*	(n)
Upland streams				
Rio das Mortes (not in the Pantanal catchment area)	0 - 10	(2)	30 - 40	(2)
Hot spring at Aguas Quentes	0	(3)	--	
Rio Arica Mirim	20 - 25	(2)	100	(1)
Sites in the Pantanal				
Rio Bento Gomes				
Poconé exposition park	55 - 65	(2)	>150	(2)
Fazenda Nova Estancia	50	(1)	>150	(1)
Baia Piuval	90	(1)	>150	(1)
Transpantaneira	40 - >150	(3)	100 - >150	(3)
Birici Stream on Faz. Ypiranga	70 - >150	(2)	>150	(2)
Rio Novo at Transpantaneira	>150	(1)	>150	(1)
Rio Frio at Transpantaneira	80	(1)	>150	(1)
Rio Pixaim at Transpantaneira	40	(1)	>150	(1)
Pond beside Rio Pixaim	60	(1)	>150	(1)
Rio Paraguay				
Faz. Sto. Antônio das Lendas	50	(1)	--	
Cáceres	50	(1)	--	
Cordraza	70	(1)	--	
Rio Miranda at Passo da Lontra	38	(1)	--	

Table 5.2 also shows the electrical conductivity of the water at the springs and in the upland streams that flow toward the Pantanal. In most cases, the conductivity values are extremely low initially, but they progressively increase as the water moves downstream toward the Pantanal. Rainstorms can radically increase the conductivity (Heckman, 1995). The cumulative concentrations of ions in the water are the determinants of electrical conductivity, which is therefore a collective variable indicating how many ions of all compounds are present in solution. The individual substances present will be discussed in Section 6.

Table 5.2: The daytime temperature (Temp., °C) and electrical conductivity (Cond., $\mu S \cdot cm^{-1}$) ranges recorded in upland springs and streams in the catchment area of the Pantanal and in the Rio Cuiabá at Porto Jofre and Porto Cercado during the *vazante, seca*, and early *enchente*. The first six streams arise on the alto plano to the east of the Pantanal. The Corrego Guanandi is located near Cangas, the closest town to Poconé, and the Corrego Salgadeira and headwaters of the Rio Coxipó are located near Chapada dos Guimarães (Heckman, 1995).

	Temp.	(n)	Cond.	(n)
Aguas Quentes	36.8 - 40.9	(4)	44.6 - 54.0	(4)
Corrego Filippa	23.9 - 24.3	(2)	6.43 - 8.54	(2)
Corrego Ajudicaba	20.0	(1)	5.16	(1)
Corrego Correia	22.6	(1)	2.49	(1)
Corrego Formoso	23.9 - 24.9	(2)	2.93 - 5.15	(3)
Rio Arica Marim	23.6 - 30.1	(2)	11.9 - 130.9	(2)
Corrego Guanandi				
(headwaters)	18.1 - 28.5	(18)	23.9 - 56.9	(19)
(impoundment)	20.5 - 31.1	(11)	38.7 - 97.6	(11)
Corrego Pintado	18.4 - 25.9	(6)	31.5 - 98.7	(6)
Corrego Salgadeira	26.0 - 26.2	(2)	2.03 - 2.29	(2)
Rio Coxipó				
(headwaters)	21.4 - 21.7	(2)	7.54 - 8.79	(2)
Rio Cuiabá	23.6 - 28.9	(2)	34.1 - 68.5	(2)

The seasonal differences in daytime temperature depicted in Figures 5.5 and 5.6 illustrate some of the important influences on the water temperature in different kinds of habitats in the Pantanal of Poconé. In Figures 4.6 and 4.7, the reason these data are less conclusive than those for electrical conductivity was made evident from the 24 hour curves of the water temperature, which show how strongly it can be influenced by the weather conditions, especially during the dry season.

The Corregos Guanandi and Pintado are located near the northern boundary of the Pantanal about 20 km northwest of Poconé. The Corrego Guanandi arises from springs in a marshy area densely covered by grasses and sedges and with much humic matter in the soil. These surroundings of the springs, which are apparently old *campos úmidos*, are atypical for the region of the Pantanal. All of the other streams referred to in Table 5.2 arise from springs in sandy soils on the plateau to the north and east of the wetland. There are hot springs at Aguas Quentes, which account for the high temperatures and distinctive chemistry of the water.

In the Pantanal itself, the physical variables of the water become subject to the daily weather. The temperature of shallow water may change radically during single 24 hour cycles, while greater volumes of water may show considerable temperature stability in spite of great fluctuations in air temperature. The 24 hour changes in the water temperature during the different seasons will be compared in Section 6. In contrast to the temperature, the electrical conductivity generally changes little during the course of a day, except during heavy rainstorms. However, seasonal changes are enormous.

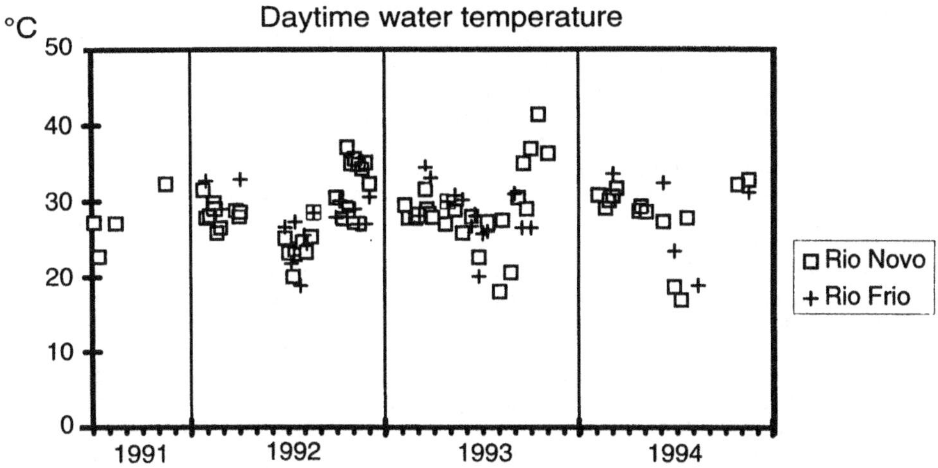

Figure 5.5: The daytime water temperature recorded 6 cm below the surface in the Rios Novo and Frio where they cross beneath the Transpantaneira Highway during the period of the study. The temperature was generally recorded between 11 a.m. and 4 p.m. and can be considered to be near the maximum for the 24 hour period.

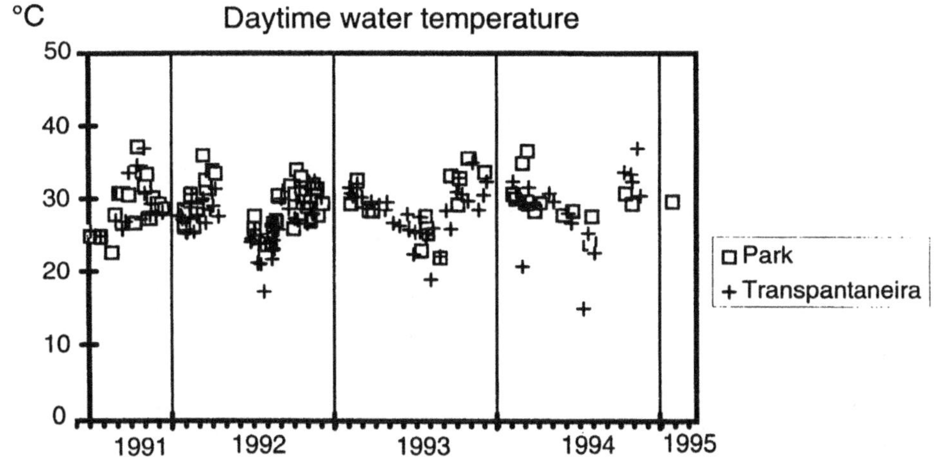

Figure 5.6: The daytime water temperature recorded 6 cm below the surface in the Rio Bento Gomes at the Poconé exposition park and at the point where it crosses beneath the Transpantaneira Highway during the period of the study. The temperature was generally recorded between 11 a.m. and 4 p.m. and can be considered to be near the maximum for the 24 hour period.

In Figure 5.5, great fluctuations in the water temperature are apparent during the rainless seasons in the Rios Novo and Frio, where they cross the Transpantaneira Highway. These two water bodies are lentic during the late *vazante* and *seca* and dry up completely or to a large extent before the freshet. The Rio Frio generally dries up before the Rio Novo, accounting for the great dry season gaps in the data. During the late *enchente* and *cheia*, there is often a rapid water flow through these water courses, and the temperature is considerably more stable. Because the water volume in these streams is small in comparison to the surface area, they cool rapidly during the *friagems* and heat up considerably during the late dry seasons, when there is little wind and intense insolation.

Unlike the streams that sometimes completely dry up, the Rio Bento Gomes remains moderately deep throughout the year (Fig. 5.6). This has an obvious moderating effect on the water temperature, which is sometimes considerably reduced during *friagems* in July or August near the Transpantaneira Highway, but less so upstream from the Baia Piuval. During these months, there is still some inflow of water from the more elevated parts of the catchment area that reaches the exposition park, and probably because this originated mainly as emerging ground water, its temperature is more stable.

The conditions in the large rivers are even more stable than those in the moderate-sized ones. In the Rio Paraguay within the Municipal District of Cáceres, the water temperature (Fig. 5.7) and electrical conductivity (Fig. 5.8) both seem to vary within relatively narrow ranges, although more data are required to better establish the seasonal patterns.

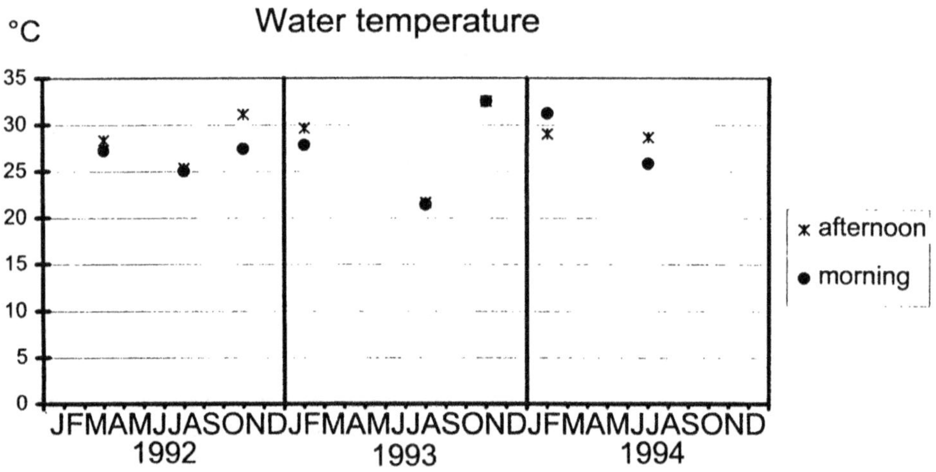

Figure 5.7: The water temperature recorded 6 cm below the surface in the Rio Paraguay. The temperatures of each vertical pair were determined during a single 24 hour period, one in the late afternoon and the other in the early morning, either at the Fazenda Santo Antônio das Lendas or in the city of Cáceres.

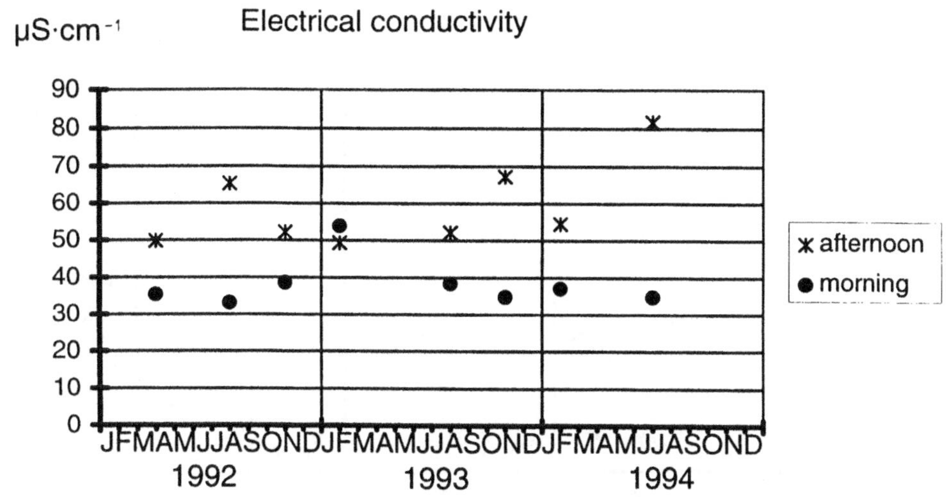

Figure 5.8: The electrical conductivity recorded 6 cm below the surface in the Rio Paraguay. The values of each vertical pair were determined during a single 24 hour period, one in the late afternoon and the other in the early morning, either at the Fazenda Santo Antônio das Lendas or in the city of Cáceres.

The seasonal changes in electrical conductivity depicted in Figures 5.9 and 5.10 show patterns similar to those of the temperature, but they can be considered much more reliable because electrical conductivity was shown not to fluctuate much at all during single 24 hour periods (Fig. 5.11) unless there is heavy rainfall. Hence, the values are more stable and less likely to be influenced by the time of day at which they were recorded. While there are obvious peaks at the end of the *seca* at all locations, the amplitude of these peaks in the water bodies that nearly dry out is considerably greater than it is in the Rio Bento Gomes, which contains a great water volume throughout the year. A more detailed discussion of the individual chemical factors responsible for the changes in electrical conductivity will be presented in Section 6.

In addition to the water bodies investigated during this study, several atypical kinds for the region are encountered in and near the Pantanal. Apparently, there is more of a variety of water sources in the southern part of the region since Wilhelmy (1958) reported the ground water to be well buffered and alkaline or neutral in pH, which the water feeding the springs to the north and east of the wetland certainly is not. Unusually mineral rich ground water is responsible for the presence of a few distinctive water bodies in the Pantanal. A few of these are saline lakes, several of which are known in the Nhecolândia Subregion (Cunha, 1943; Brum and Sousa, 1985; Mourão *et al.*, 1988). They are actually soda lakes characterized by high concentrations of sodium carbonate and with ammonium as the main nitrogen source. Their pH may reach about 10, and their electrical conductivity, 4,200 $\mu S \cdot cm^{-1}$ (Por, 1995).

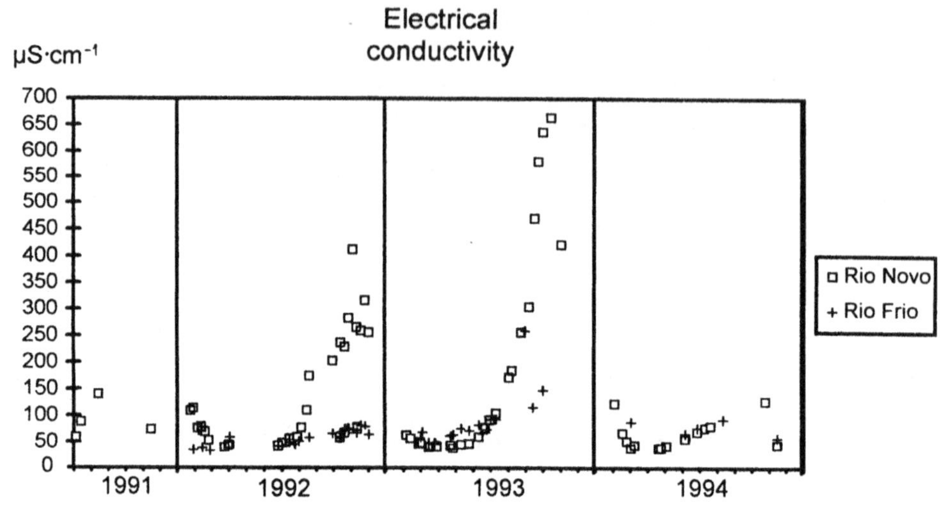

Figure 5.9: The electrical conductivity recorded 6 cm below the surface in the Rios Novo and Frio where they cross beneath the Transpantaneira Highway during the period of the study. The temperature was generally recorded between 11 a.m. and 4 p.m. and can be considered to be at or near the maximum for the 24 hour period.

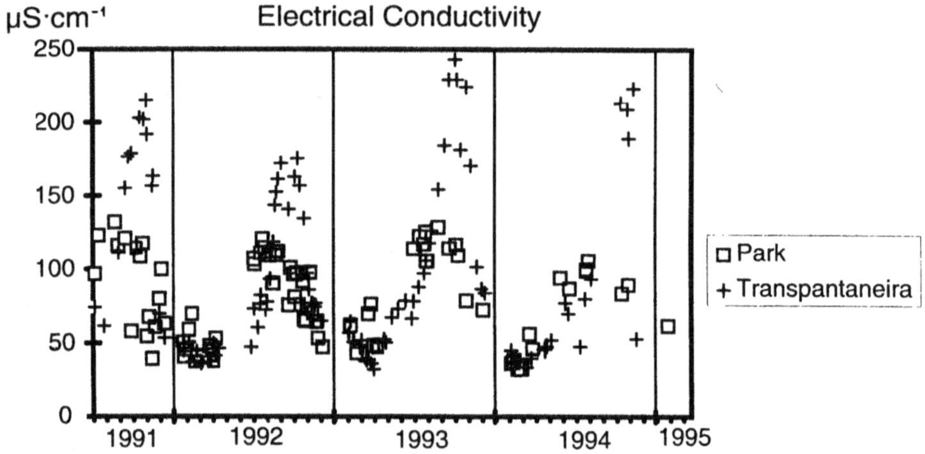

Figure 5.10: The electrical conductivity recorded 6 cm below the surface in the Rio Bento Gomes at the Poconé exposition park and at the point where it crosses beneath the Transpantaneira Highway during the period of the study. The temperature was generally recorded between 11 a.m. and 4 p.m. and can be considered to be at or near the maximum for the 24 hour period.

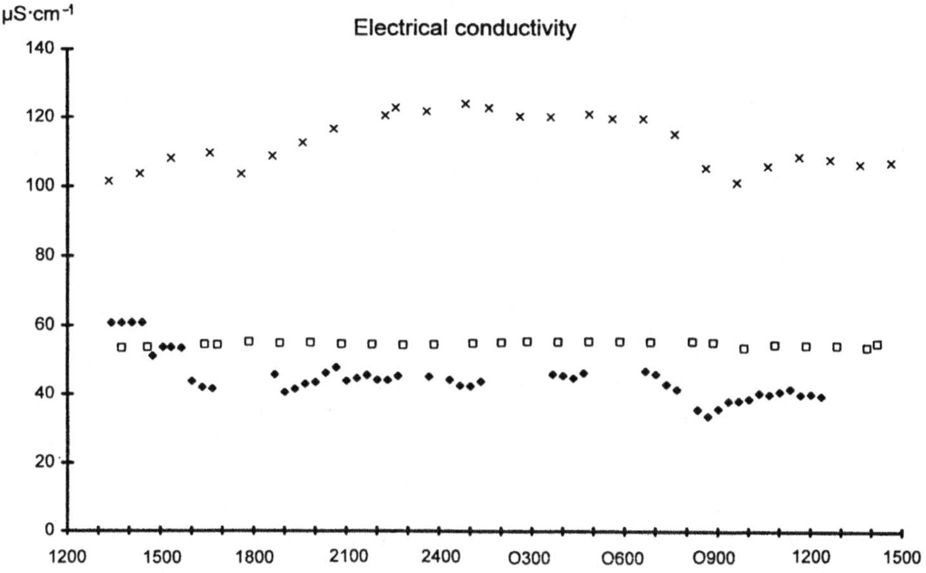

Figure 5.11: Electrical conductivity recorded 6 cm below the surface in the Rio Bento Gomes at the Poconé exposition park during 24 hour periods on February 4 and 5, March 24 and 25, and August 25 and 26, 1992. These dates were chosen during the *cheia*, beginning of the *vazante*, and *seca*, respectively.

Another atypical kind of water body is the calcium carbonate-rich river. These have been reported to exist near the headwaters of the Rio Paraguay and in the Rio Formoso, a tributary in the Rio Miranda System. These rivers flow from karst hills, which contain unusually large amounts of calcium salts, which are otherwise rare in the watershed. The Rio Formoso is reportedly so rich in calcium that it deposits travertine (Por, 1995). The effects of the calcium input from this river can still be detected in the downstream section of the Rio Miranda at the Passo da Lontra, where a total hardness value of 0.75 $mmol \cdot l^{-1}$ and carbonate hardness of 1.5 $mmol \cdot l^{-1}$ were recorded on August 15, 1994, about three months after the date on which the local water level reached its annual peak. The water is softer than that in most rivers of the world, but it is considerably harder than that in the nearby stretch of the Rio Paraguay at Cordraza, where values of 0.25 $mmol \cdot l^{-1}$ total and 0.5 $mmol \cdot l^{-1}$ carbonate hardness were determined on August 16, 1994. The electrical conductivity recorded in the Rio Miranda at the same time the hardness was determined proved to be 172.6 $\mu S \cdot cm^{-1}$, considerably higher than the 128 $\mu S \cdot cm^{-1}$ recorded for the same river by Mitamura *et al.* (1985). The conductivity encountered the next day in the Rio Paraguay at Cordraza was 50.8 $\mu S \cdot cm^{-1}$, compared with 40 $\mu S \cdot cm^{-1}$, recorded by Mitamura *et al.* (1985). The data presently available are insufficient for characterizing these river sections, as the seasonal fluctuations demonstrated in Section 6 clearly indicate.

For obvious reasons, the atypical water bodies in the Pantanal will not be dealt with here. None of them were encountered in the Pantanal of Poconé, and each is deserving of a detailed special study in the future.

5.3.2. River system morphology

The Rio Paraguay System is regarded as peculiar and has many atypical features, as discussed by Bonetto and Wais (1990). One of the main reasons for this is the location of the Pantanal wetland far upstream rather than along the middle or lower sections of the system, where most wetlands are encountered along other South American river systems. The wetland stores rainwater and releases it at a steady rate, accounting for a rather constant flow of about 4500 $m^2 \cdot sec^{-1}$ in the lower Rio Paraguay throughout the year. The north-south declivity is only about 37 cm.km-1, and this decreases considerably toward the south (Por, 1995).

Bonetto and Wais (1990) also mentioned that the separation of the Rio Paraguay and Amazon Watersheds is characterized by weak barriers to the movement of aquatic species from one to the other. They regarded the Pantanal as much richer in species than the lower regions of the Paraguay and Paraná River Systems, a statement that cannot be accepted without more thorough comparative studies, however. Although many Amazon elements have certainly crossed the level plains of the high plateau on which the headwaters of both Paraguay and Amazon tributaries arise almost side by side, the subsequent sections of this book will show that the wetland is characterized more by large populations of a few species rather than a great species diversity.

Other contributions of the Pantanal to the hydrology of the Rio Paraguay System as a whole include variable amounts of organic material in the form of whole, living water plants that are carried downstream from the wetland. This living vegetation and the detritus it later produces have a profound effect on the oxygen concentration, which is low in the sections of the river system downstream from the Pantanal (Bonetto *et al.*, 1981). This and the formation of great hydrodynamic backwaters at the confluence of the Rios Paraguay and Paraná make the conditions in the lower sections of the Paraguay River difficult to compare with those in other rivers (Bonetto and Wais, 1990).

Generally, the upper Rio Paraguay system can be regarded as distinct. The watershed had originally been dominated by dry forest and shrub savanna, which has now been largely replaced by massive soybean, corn, and sugar cane fields. Typical humid forest conditions exist almost exclusively in the gallery forests that form narrow bands along the banks of the streams. By the time the Rios Paraguay and Cuiabá have reached the Pantanal, both are lined extensively by sandy shores devoid of forests and frequently modified by the presence of small villages and ranches. The erosional shores, often lined by gallery forests and marked by cliffs that can be rather high, are locally called *diques*, while the alluvial sides are usually bordered by large areas temporarily flooded during the high water periods and overgrown by herbaceous aquatic plants with a few stands of trees on slightly higher ground. Frequently, there are also backwaters, minor channels, and entrances to *baias* on the alluvial side. After the main rivers enter the Pantanal, vast flood plains open up along both shores. The rivers remain distinct and separate from the floodplain lakes during the dry season, but the entire area becomes a vast shallow lake during the high water period.

5.4. The hydrological cycle and the biota

Due to the nature of the climatological and hydrological conditions in the Pantanal, the biota must adapt to periods of severe water shortage alternating with periods of flood. While the large vertebrates and some of the fish species can adapt to the changing conditions by migration, the plants and most of the smaller animals must develop survival strategies to maintain their populations through unfavorable seasons. These strategies will be discussed in detail in subsequent sections, but it is appropriate here to mention certain adaptations that concern the water supply.

The hydrological cycle discussed here is that observed in the Pantanal of Poconé and the other northernmost subregions of the wetland. In this part of the Pantanal, the lowest water levels are encountered at the end of the dry season, and the highest, near the end of the rainy season. As previously mentioned, the hydrological seasons in the southern part of the Pantanal are quite distinct from those in the North, producing considerable differences in the seasonal succession and strongly influencing the structure of the biotic communities.

In general, the plants on the flood plain are relatively short-lived annuals (Prado *et al.*, 1994). A few are perennials, such as *Ipomoea carnea fistulosa* and *Aeschynomene fluminensis*, but such species, which may well have been widespread in prehistoric times, are presently not too common because of heavy grazing by cattle during the dry season.

Like the terrestrial plants of the floodplain, the aquatic species are also mainly annuals, but a few are perennial, at least if local conditions permit their survival through the dry season.

In the Pantanal, a few plants have adapted to water shortages by storing it in their tissues. Most prominent among these are the cacti, *Cereus kroenleinii and C.* cf. *bicolor*, which are frequently encountered at the edges of the wooded islands that stand just above the high water level of the rainy season. Other species with succulent parts include *Bromelia balansae*, which can store water in the leaves, and *Nymphoides grayana*, an aquatic species which develops highly thickened leaves during the dry season and can apparently store enough water to keep the plants green until the next period of flooding. Aquatic and wetland herbs without means of water storage simply die off during the dry season and grow anew during the next period of flooding, usually from seeds but sometimes also from vegetative propagules. They are replaced by a sparse vegetation of terrestrial plants, mainly grasses, that are adapted to semi-arid conditions (Section 7).

The pronounced seasonal change in the flora that takes place twice a year on large areas of the Pantanal has given the landscape its characteristic appearance. A major part of the floodplain is free of woody perennials and is occupied instead by annual herbs or species that survive part of the year by producing vegetative propagules. During the season of flooding, these species are overwhelmingly aquatic, either emergent or submersed, while during the dry season, they are short-lived terrestrial annuals. The lack of plants that persist from year to year gives large regions of the Pantanal the appearance of expansive plains, green during the rainy periods, and largely brown during the dry season.

Scattered over this open plain are the temporary islands of trees and woody shrubs which form at sites elevated slightly above the high water level as well as long expanses of gallery forest along the courses of the rivers and streams. Unlike the aquatic and semi-aquatic plants, the terrestrial species of the permanently dry areas are predominantly perennials. Forests form on the alluvial deposits along the banks of the streams, which also remain somewhat above the level of the flood waters. Because tree species that are sensitive to inundation of the roots are eliminated, these wooded areas of the Pantanal are not particularly rich in species. There is no "terra firma" in the Pantanal proper as there is in Amazonia. The species diversity in the *cerrado* beyond the limits of the Pantanal is also somewhat reduced because of the dryness that prevails for much of the year and the inability of the porous soil to store water. An important effect of these plants on the hydrology of the region is their extreme reduction of transpiration during the dry period in order to maximize the storage capacity for the available water in the system. Most of the trees and shrubs are deciduous, losing all of their leaves during the dry season and growing new foliage as soon as the rains begin to fall. Many tree and shrub species, however, bloom intensely at specific times during the dry period, while they are still bare of foliage. All trees or shrubs of a given species produce flowers at the same time, maximizing the chances of successful pollination by providing hummingbirds, bees, and other potential pollen carriers with a rich nectar supply at times that food is otherwise scarce because of the dryness. Details on the seasonal cycles of the individual species will be provided in the subsequent sections.

Thus, as mentioned earlier, the pronounced hydrological cycle is a prime determinant of which species are able to survive in the Pantanal and thereby also a determinant of the species diversity. The fauna, like the flora, is forced to adapt an amphibious way of life and possess adaptations for surviving periods of high temperatures and desiccation, as well. As discussed in detail in subsequent sections, the majority of the invertebrate and fish fauna on the flood plain must compensate for high seasonal rates of mortality by reproducing at enormous rates. The population dynamics of these species are therefore overwhelmingly determined by the r-strategy. A parallel phenomenon is the autumn and winter mortality of herbal plants and invertebrates in the temperate zones. In both cases, the reduced numbers of reproductive individuals that have survived usually begin intensive reproduction during the spring, when the rains in the wet-and-dry zone and the increasing temperature in the temperate zone effect an improvement in the living conditions. These cyclical phenomena are lacking in the equatorial zone, except along rivers in which water level changes influenced by rainfall in adjacent wet-and-dry zones subject the littoral biota to some of the same stresses. In the Pantanal, the sparse terrestrial community that occupies the desiccated parts of the floodplain during the *seca* also has a counterpart in the temperate zones, namely, the community of cold stenothermal species, including mosses, lichens, collembolans, and even winter midges that fly at temperatures below freezing (Heckman, 1982a). Therefore, the climatic effects on seasonal activity in the temperate zones are closely paralleled by hydrological ones in the wet-and-dry climatic zones.

The significance of the adaptations of the biota and particularly the flora to the regional climatic and hydrological conditions can be appreciated when viewing the Pantanal from the air. The physical features of the wetland are produced to a large degree by the vegetation. During the high water period, vast areas are inundated, but movement of the water is obviously limited mainly to well defined water courses, which are bordered by bands of woody vegetation. The gallery forests are obviously great stabilizers of the river banks, which could well be eroded away within a few years if the woody vegetation were removed. Similarly, the dense meadows of emergent aquatic plants stabilize the hydrological conditions for the animals that require lentic conditions for reproduction and growth. Hence, there is not only an effect of the hydrology on the biota, the biotic systems determine the local hydrological conditions to a large degree, as well. Engineers and farmers who forget this can cause grave damage to the entire system.

6. Water chemistry

6.1. Methods

During a period of over 3½ years, the water at the network of sites described in Section 2.1 was analyzed for a series of chemical substances of importance for the biotic community. In addition, samples were taken at many other locations in the Pantanal and its watershed to provide a standard of comparison. The equipment used to determine the physical properties of the ambient water has already been discussed in Section 4.2.

The chemical analyses were completed in order to determine the actual conditions to which the organisms are exposed in the water bodies. For this reason, value was placed on rapid analyses in the field and repeated analyses at individual sites, both during 24 hour cycles and throughout the course of the year.

The analytical methods used varied in accuracy to some degree, and while most determinations were performed in the field with good results, several had to be completed in the laboratory for lack of simpler testing procedures. Field analyses were preferred to minimize errors arising due to transport of the samples over long distances on rough roads at high daytime temperatures. Because the organisms in the water are adapted to relatively high water temperatures (Section 4), cooling the water for transport could be expected to have disastrous results. The chemical variables most likely to be affected by transport include the concentrations of oxygen, nitrogen compounds, and free phosphate. Of these, only the nitrate concentration had to be determined in the laboratory, and the samples were always analyzed within a few hours after they were collected. The other analyses that had to be performed in the laboratory were those for chemical oxygen demand (COD), dissolved fluoride, and sulfate, as described below.

The methods selected for the determinations are described individually below.

pH

Both chemical indicators and electrical sensors react according to the ion balance between hydrogen and hydroxide ions, or ions with similar effects. These reactions occur best in well buffered solutions containing many ions. If no ions are present, these reactions cannot take place, and if there are few ions in the water, the reactions occur slowly, and unreliable results are obtained. In the Pantanal during the high water period, the electrical conductivity of the water is similar to that or rainwater in parts of the world where there is little air pollution. These conductivity values are low, usually little more than 30 $\mu S \cdot cm^{-1}$ (Figs. 5.8 & 5.9), but pH can still be recorded, although much time has to be allowed for the indicator dyes to change color or for the electronic sensors to stabilize. In the streams flowing into the Pantanal, it must be assumed that a measurable pH value does not exist. With electrical conductivity values between 2 and 5 $\mu S \cdot cm^{-1}$, it must be assumed that there are simply not enough ions present to permit a reliable determination of the pH, even when special electrodes are used. Indeed, slight movements of the electrodes cause large changes in the pH value to be registered, and it was usually necessary to leave the electrodes in the water for at least half an hour to achieve an apparent stability in the reading. Indicator paper scarcely changes color at all. Thus, the pH values shown for such streams are of little significance, as noted where they are presented. The significance of these values for the Pantanal increase in direct proportion to the electrical conductivity in the water, and therefore, the values recorded during the dry season must be regarded as quite satisfactory, while the data for the high water period must be interpreted with some caution. The pH detection strips with various indicators were supplied by Merck, and the WTW pH 196 was used for the electronic determination. Experience showed that the lower the electrical conductivity was at a given site, the longer it took for the electronic pH indication to stabilize. Similarly, during the *cheia*, it required well over 40 minutes for the pH indicator to complete its change to a stable color, and during this time, it often bleached, making it difficult to compare the color with the standards. Thus, dry season pH values can be considered representative of the water bodies at the time they were recorded, but the values determined in the wetland during the *cheia* and in the headwater streams throughout the year had to be correlated with the prevailing electrical conductivity to determine their reliability.

Dissolved oxygen

The concentration and saturation of dissolved oxygen was determined in the field using the WTW OXY 196. The electrodes were calibrated using the cylinder provided for that purpose by WTW, and the accuracy of the instrument during use in the field was frequently monitored using the same device. Thoroughly aerated tap water, assumed to be at 100% saturation, was used for additional accuracy checks in the laboratory.

COD

Chemical oxygen demand (COD) was determined photometrically after oxidation of the samples with chromate sulfuric acid in the presence of silver sulfate for two hours at a temperature of 140°C. These tests were prepared by WTW for use in their photometer. The samples were collected in the field 5 to 12 hours before analysis. To prevent loss of fine particulate matter and colloids, the samples were not filtered. They were visually inspected in the pipettes to assure that no visible particles of detritus were included. Duplicate samples were analyzed from time to time, and no significant deviations in the findings were ever determined, except in the case of one package of out-of-date reagent, which was discarded. The reagents, photometer, and light frequency filters were supplied by WTW.

Total hardness

Total hardness was analyzed in the field by a complexometric titration against Merck Titriplex III® using a mixed indicator. The maximum error expected is -0 and +0.1 $mmol{\cdot}l^{-1}$. This is due mainly to overshooting the end point of the titration by almost as much as the volume of the drop of the titrating solution coming out of the mini-burette. The error could be minimized by careful handling of the burette and allowing smaller quantities than one drop to run down the side of the test tube while vigorously swirling the mixture. Because the water was usually extremely soft, it was desirable to perform the titration as accurately as possible. The seasonal constancy of the results and comparisons with the data obtained by related analyses for carbonate hardness and calcium concentration, also determined by titration, indicated that the results obtained were quite accurate. Comparisons with hardness values reported in the literature are made difficult by the use of so many different units. The units used here are $mmol{\cdot}l^{-1}$ of alkali earth elements. The relationships among the most commonly used units for recording water hardness are 1 $mmol{\cdot}l^{-1}$ = 2 $meq{\cdot}l^{-1}$ = 5.6 German degrees of hardness = 100 ppm $CaCO_3$ = 7.02 English degrees of hardness = 10 French degrees of hardness.

Carbonate hardness

Carbonate hardness was analyzed by acidometric titration using a mixed indicator provided by Merck. The maximum error expected was -0 and +0.2 $mmol{\cdot}ml^{-1}$. The main source of error was the same as that in the analysis of total hardness, and the same remedies were used to minimize it.

Calcium

Two tests were used for calcium in the field. That found most reliable was titration utilizing the disodium salt of ethylene-dinitrilotetraacetic acid to chelate the calcium and calconcarboxylic acid as the calcium-specific indicator, sold as Titriplex® III by Merck. This test was used in 1991 and again in 1994 on a regular basis. The results always agreed well with those expected from the water hardness determinations.

The chromatic comparison of the red-violet complex in an aqueous medium after the reaction of Merck Calcospectral® was also employed. Its maximum error was calculated to be $\pm 1.5\ mg \cdot l^{-1}$. However, the shade of color was rather difficult to estimate during the day, and the test proved difficult to complete at night because of the weakness of the color reaction. Generally, the data obtained by this test at a given site during the same season showed considerably greater standard deviations than those obtained by the titration method. Furthermore, concentrations were occasionally determined that were much higher than expected when the water hardness values were extremely low, especially in the headwaters of streams where the conductivity remained less than $10\ \mu S \cdot cm^{-1}$. The values recorded during the years 1992 and 1993 are therefore considered less reliable than those obtained in 1994, when this test was abandoned in favor of the titration method. The reagents and color standards for both analytical methods were supplied by Merck.

Soluble iron

Dissolved iron was determined by chromatic comparison of the violet complex produced by the reaction of ferrous ions with a buffer of thioglycolate. The maximum error expected in using this method was ±30%, but interpolation of the color was possible and reduced the estimated error considerably. Laboratory tests in Germany, discussed below, included analyses of the total iron in water and sediment samples and ion balances to account for the electrical conductivity of the water. These confirmed that the results of the iron analyses in the field were accurate to a degree well within the calculated maximum of 30%. The reagents and color standards were supplied by Merck.

Soluble aluminum salts

The soluble aluminum concentration was determined in the field by matching the colors formed after the reaction of Chromazurol S® and aluminum in an acetate buffer with a standard color scale. The estimated error would have been $\pm 0.04\ mg \cdot l^{-1}$ if the concentration had ever been above the minimum detectable value. However, no water samples in the Pantanal or its surroundings ever contained a trace. The reagents and color standards were supplied by Merck.

Chromium ions

The analysis of chromium employed a colorimetric comparison of the red-violet chromous diphenylcarbazone complex formed by the simultaneous oxidation of diphenylcarbazide and the reduction of the chromic ions. The test is therefore not sensitive to all species of chromium. If any chromium ions had been detected, the maximum error would have been ±0.005 $mg \cdot l^{-1}$, but no trace of any detectable forms of the metal were ever encounterd in the water samples tested. The reagents and color standards were supplied by Merck.

Chlorosity

Chlorosity was determined by chromatic comparison of a standard color scale with the ferric thiocyanate produced by the reaction of chloride and mercuric thiocyanate in the presence of ferric ions; the maximum error is ±50%, but in practice, an average error of about 23% was attained with a maximum of 32%, as determined by comparisons with results obtained in a German laboratory by titration against silver nitrate (J. Kaiser, pers. comm.). This test used for analysis in the field was therefore more than adequate to demonstrate the large seasonal changes in chlorosity. The reagents and color standards were supplied by Merck.

Fluoride

The fluoride concentration in the samples were determined photometrically after decoloration of Spadns reagent. The analyses were conducted in the laboratory on the same day the samples were collected. The reagents and spectrophotometer were supplied by Hach.

Ammonium

The ammonium concentration was determined in the field immediately after the collection of the water sample by chromatic comparison after the Berthelot Reaction. The maximum error at concentrations up to 0.4 $mg \cdot l^{-1}$ after August 1992 was ±0.015. Prior to that month as well as thereafter whenever the concentration exceeded 0.4 $mg \cdot l^{-1}$, a less accurate test employing the same reactions was used, for which the maximum error reached ±0.15 $mg \cdot l^{-1}$. The reagents and color standards were supplied by Merck.

Nitrite

The nitrite concentration was determined in the field using a chromatic comparison after the Griess reaction. A maximum error of ±0.01 was assumed after August 1992 whenever the concentration was below 0.14 $mg \cdot l^{-1}$, but at the low concentrations usually encountered in the Pantanal, a greater, though unspecified accuracy can be assumed. Prior to that month and whenever the concentration exceeded 0.14 $mg \cdot l^{-1}$, a maximum error of ±50% is assumed. The minimum detectable concentration when employing the more accurate of the tests was 0.005 $mg \cdot l^{-1}$. The reagents and color standards were supplied in a test kit by Merck.

Nitrate

The nitrate concentration was determined in the laboratory on the same day the water samples were collected, and usually within 12 hours. The nitrate in the sample was first reduced to nitrite with metallic cadmium, and after the Griess reaction, the the total nitrite-N concentration was determined photometrically. The concentration of nitrite-N determined in the field was then subtracted from the result, and the remainder is assumed to have been in the form of nitrate-N in the sample prior to reduction. The results are reported as nitrate, calculated from the nitrate-N value. The reagents and spectrophotometer were supplied by Hach. The potential error of the determination was calculated to be less than 10%, and little change in the NO_3 concentration during transport was expected.

Phosphate

To determine the amount of free phosphate in the water, the method of chromatic comparison of the phosphomolybdenum blue produced by the reaction of the free phosphate with molybdate ions in a sulfuric acid solution and the reduction of the molybdosulfuric acid produced by ascorbic acid was selected. The maximum error of this test at the low concentrations of phosphate-P usually encountered in the Pantanal is ± 0.01 $mg \cdot l^{-1}$. However, during the dry season, concentrations higher than 0.15 $mg \cdot l^{-1}$ may be encountered, and at such times, the maximum error of the determination may be as high as 35%. The reagents and color standards were supplied by Merck.

Sulfate

The sulfate determination by the chromatic comparison of the brownish red complex produced by the reaction of iodate with tannin in a weak acid medium, provided by Merck for use in the field, proved too inaccurate and unreliable to use. The concentrations determined were suspected of being far too high, and laboratory tests employing a photometric analysis of the turbidity produced by the reaction of sulfate with barium ions in a sol confirmed this (Heckman, 1994a). Data are therefore available for only a limited number of samples collected in 1994 which were reliably tested in the laboratory. The results from 1994 seem quite accurate based on ion balances. The reagents were prepared in Cuiabá using analytical grade Merck chemicals with the kind assistance of Dipl. Biol. Joachim Kaiser.

Hydrogen sulfide

The amount of hydrogen sulfide in the water samples was determined in the field by chromatic comparison after the reaction of this compound with N,N'-dimethyl-1,4-phenylene-diamine dihydrochloride and oxidation with ferric ions to methylene blue. The maximum error of the test was ± 0.02 $mg \cdot l^{-1}$. The reagents and color standards were supplied by Merck.

Silicon compounds

The concentrations of silicon compounds in the water were determined colorimetrically in the field after the production of ß-silicomolybdic acid. The maximum error was ± 0.25 $mg \cdot l^{-1}$, and the maximum determinable concentration was 10 $mg \cdot l^{-1}$, which was not infrequently exceeded in some of the samples. The reagents and color standards were supplied by Merck.

Quality control

To check the accuracy of the results, water samples were sent to a private laboratory in Germany and analyzed with atomic absorption specrometry and other standard methods, through the courtesy of Dipl. Biol. Joachim Kaiser. The concentrations of ions that do not change readily were determined, and the results compared with the field data. For all tests checked in this way except that for sulfate, a maximum error of 32% was determined. The average error was 23% without any trend toward the results being either too high or too low. The sulfate test used in the field was rated by the manufacturer as having a great percentage of error under ideal conditions, but even so, the concentrations determined were far too high, and the error was estimated to be much greater than that reported by Merck. The data therefore had to be considered unreliable (Heckman, 1994a). As a substitute, the laboratory test described above was substituted for the field test.

Because or their relatively rapid changes due to microbial activity, the concentrations of nitrogen compounds and free phosphate could not be double checked in this way, although in the case of the phosphate, a rough estimate could be obtained from total phosphorus analyses, subtracting a quantity of organic phosphorus calculated from the COD value. Nevertheless, from ion balances made to account for the electrical conductivity values directly recorded in the field, the values determined for these substances seemed to be quite accurate. In any case, to determine the enormous seasonal differences in the amounts of dissolved substances in the water, the accuracy obtained was better than sufficient.

Ion balances were calculated to ascertain whether the values determined for all of the ions accounted for the approximate electrical conductivity of the water. These were performed by adding the expected concentrations of the ions not determined by analysis to the values obtained in the field. More exact calculations were conducted on samples analyzed for all common ions in Germany using Zeeman flame atom absorption spectroscopy or other methods, as noted in Section 6.3 for each ion.

6.2. Results

The results of water analyses conducted in all parts of the northern Pantanal and most of the lotic water bodies in the watershed revealed that there is an extreme shortage of soluble compounds of phosphorus and nitrogen as well as most other minerals in the watershed (Table 6.1), at least in its northernmost parts. Most of the streams and rivers feeding the Pantanal have been placed in the general class of "clear water" rivers (Sioli, 1984). This class, however, has been shown to be quite variable in its dissolved mineral content, in contrast with the other classes, the "black" and "white" water rivers. The general poverty in dissolved minerals is reflected by the conductivity and water hardness values, as well as by the concentrations of specific ions recorded. These are typically low during both the rainy and the dry seasons. Unlike nearly all other substances analyzed, silica is present nearly everywhere in moderately to very high concentrations (Table 6.1). There are also often moderate iron concentrations in the streams and rivers, especially near the springs feeding them.

The reason for the lack of soluble salts in the streams flowing from the plateau to the northeast of Cuiabá into the northern Pantanal is probably the great age of the substrate from which the ground water emerges (Fittkau, 1974). It has been suggested that nearly all soluble substances have been leached away during past hundreds of millennia, leaving only quartz and sandstone deposits and sometimes iron bearing rocks. It was noted with interest that during heavy rainfalls, the electrical conductivity of the water in the headwater streams increases about five-fold.

Thus, rainwater rather than ground water is the major source of mineral substances entering the Pantanal. This is quite a different situation than in most other parts of the world. Indeed during the high water periods, the physical and chemical properties of the

Table 6.1: Water hardness, chemical oxygen demand (COD), and the concentrations of various inorganic substances in water from streams in the catchment area of the northern Pantanal. The sites are the same as those shown in Figure 5.2. All hardness values are shown as $mmol \cdot l^{-1}$, COD as mg $O_2 \cdot l^{-1}$, and the concentrations as $mg \cdot l^{-1}$. The water samples were taken from the hot springs at Aguas Quentes (AQ), the Corrego Filippa (CF), Corrego Ajudicaba (CA), Corrego Correia (CC), Corrego Formoso (FO), Rio Arica Marim (AM), headwaters of the Corrego Guanandi (CGH), and Corrego Guanandi downstream from an impoundment (CGI). For CGI, n = 14, except for COD, NO_3, and Si, n = 4; CGH, n = 10; AQ, n = 4; CF and AM, n = 3; FO, n = 3; others, n = 1; N.D. = no data. Some of the data were taken from Heckman (1995).

	AQ	CF	CA	CC	FO	AM	CGH	CGI
Hardness total	0.10-0.15	N.D.	N.D.	N.D.	N.D.	0.025-0.5	0.05-0.3	0.1-1.5
CO_3	0.40-0.85	0.15	<0.1	<0.1	0.15-0.20	0.1-0.3	0.3-0.8	0.1-3.5
COD	c.12-31	N.D.	N.D.	N.D.	N.D.	<15-34	N.D.	21-37
Ca	<1-2	<1-6	<1	<1	5	<1-1	5-11	4-35
Fe	<0.1	1.0-1.3	0.3	0.5	0.2-0.3	<0.1-0.4	1.0-3.5	0.9-4.0
NH_4	0.01-0.08	0.02-0.04	0.02	<0.012	<0.012-0.03	<0.012-0.04	<0.012-0.1	<0.012-0.70
O_2	5.2-6.8	6.0-6.1	3.7	7.6	5.0-7.6	7.0-8.9	0.1-7.2	2.2-8.2
Cl	<3	<3	<3	<3	<3	<3	<3	<3-4
NO_2	<0.005-0.019	0.005-0.012	<0.005	<0.005	<0.005	<0.005-0.014	<0.005-0.016	<0.005-0.05
NO_3	0.18-0.42	N.D.	N.D.	N.D.	N.D.	<0.01-0.2	N.D.	<0.01-0.24
PO_4-P	0.02-0.05	<0.015	<0.015	<0.015	<0.015	<0.015	<0.015-0.020	<0.015-0.15

Pantanal water closely resemble those of rainwater (Section 4.2), with the exception of the high silica content. There are a few spring-fed springs and small rivers, however, that contain somewhat greater amounts of dissolved substances. For example, Neto *et al.* (1993) found that the water in some of the headwater tributaries of the Rio Cuiabá, which flow to the north of the mineral-poor laterite formations, has a higher than usual electrical conductivity, although it still cannot be considered high by world standards (Petr and Irion, 1983), the highest value being 344 $\mu S \cdot cm^{-1}$ in the Rio Cuiabazinho. However, most of the values were considerably lower, although there appears to be a discrepancy in one of the tables in this publication, where the values are reported in $mS \cdot cm^{-1}$, apparently contradicting the other table (Neto *et al.*, 1993).

The only time of year that the water in the Pantanal displays some interesting changes is during the annual dry season. At this time, another source of water seems to play a part. That is the ground water in the upper aquifers, which seeps out of the ground in a few places when the layer of flood waters has flowed out of the area, leaving behind nearly dry ground.

The ground water that was analyzed in and near the Pantanal displays chemical features that are quite distinct from both the stream and rain water flowing into the Pantanal. This water was taken from wells, including many in the city of Poconé, founded in 1781, so it is still not possible to say with certainty whether the inorganic substances in the water originated from natural deposits or whether they are breakdown products of long-buried sewage.

Some of the analytical results for the ground water in the Municipal District of Poconé are reported in Table 6.2. These data show that the deep ground water about 100 m or deeper below the surface is remarkably poor in mineral nutrients and displays many of the characteristics of the water from the clear water streams. This is also true for the near surface aquifers supplying wells on the west side of the Transpantaneira Highway. In contrast, extremely high concentrations of nitrate, nitrite, and chloride are encountered in water from wells fed by the uppermost aquifer in some parts of the city of Poconé. Other wells in the city are less polluted by these substances but nevertheless contain amounts considerably greater than those encountered in the surface water in the Pantanal during the rainy season.

The values determined might arouse a strong suspicion of anthropogenic contamination, except for the fact that wells on the Fazenda Ypiranga also contained extremely high concentrations of the nitrogen compounds and chloride, even though several of them were relatively far from the ranch house. This tends to leave the situation somewhat uncertain until more information about the upper ground water layer in other parts of the Pantanal is obtained. The public health problems associated with the use of water contaminated with large amounts of nitrite or nitrate for drinking was discussed in detail by Heckman *et al.* (1997) and will be examined in detail in Section 13.

In the Pantanal proper, the chemistry of the surface water undergoes enormous seasonal changes (Heckman, 1994a). During the high water period, the physical and chemical variables, except temperature, resemble those of the local rainwater. The generalized variables, such as conductivity and water hardness, remain low during the *cheia* but increase considerably in the shallow water bodies that rapidly dry up late in the dry season, as shown for conductivity in Figures 5.9 and 5.10. This conductivity sometimes increases to especially high values, obviously due to the release of various mineral salts, CO_2, and polar organic compounds from the decomposing vegetation.

Dissolved or colloidal compounds of silicon present a notable exception to the poverty in those inorganic and organic substances required by living organisms in the Pantanal waters. Values much above the analytical limits of the tests used, 10 $mg \cdot l^{-1}$ were often recorded in water samples. There is a general richness of siliceous minerals in the watershed, but this alone does not explain why the values observed were so high. In most parts of the world, the low solubility of the silicon compounds rather than their absence from the sediments is responsible for limiting the amount of dissolved silicates in the water. It was suspected that the presence of dissolved or colloidal compounds of silicon might be due to the presence of fluoride. For this reason, analytical tests for fluoride were initiated. These tests revealed that fluoride is indeed present at some locations; however, its occurrence shows no unequivocal relationship to the amount of silicon compounds in the water. It is most frequently encountered in the Pantanal in low concentrations in water samples characterized by unusually high chlorosity. Such water is often collected during the dry season, either due to the presence of excreta from many birds and caimans feeding on trapped fishes or the seepage of saline ground water (Section 6.2.5).

Table 6.2. Ranges of temperature, electrical conductivity, and concentrations of various ions in ground water samples from open wells in the vicinity of Poconé and in the northernmost part of the wetland, as well as from deep drilled wells used for the drinking water supply in Poconé, and from springs near small settlements in the northern Pantanal. The data are taken in part from Heckman *et al.* (1997). Unlike the surface water, the ground water in wells maintains an almost constant temperature.

	Private wells in Poconé n=27	Wells on the Fazenda Ypiranga n=7	Deep drilled wells in Poconé n=6	Springs, northern Pantanal n=6
Temp. (°C)	27.1-33.3	24.9-33.1	29.0-31.7	28.0-30.2
Cond. ($\mu S \cdot cm^{-1}$)	151-421	186-1325	87-238	12-76
pH	5.2-7.2	4.2-7.5	6.7-7.6	5.7-5.9
Hardness ($mmol \cdot l^{-1}$)				
total	0.4-1.3	0.4-2.0	0.3-1.0	0.0-0.3
CO_3	0.25-2.85	0.1-10.5	0.9-2.8	0.2-0.5
Conc. ($mg \cdot l^{-1}$)				
O_2	1.2-4.7	0.7-5.1	3.6-9.4	1.9-7.8
Cl^{1-}	6-80	6-40	1.5-4	<1-6
NO_2^{1-}	<0.005-0.4	0.005-0.2	<0.005-0.007	<0.005-0.008
NO_3^{1-}	16.3-75.7	No data	24.6	No data
PO_4^{3-}	<0.05-9.3	<0.05-3.7	<0.05-0.3	<0.05-4.7
Fe^{2+}	<0.1-2.8	<0.1-2.8	<0.1-0.8	<0.1-0.8
NH_4^{1+}	<0.01-8.0	<0.01-4.5	<0.01-0.04	<0.01-0.17

It should be noted that the high silicate values do not seem to have any effect at all on the conductivity of the water. Some of the clear water streams which are characterized by electrical conductivities less than 10 $\mu S \cdot cm^{-1}$ contain considerable amounts of silicon compounds (Section 6.2.10). These are likely to be present in colloidal form since they do not seem to ionize. As discussed further in Section 6.2.7, the dissolution of oxygen in the water often results a decrease in the electrical conductivity because it promotes the oxidation of the soluble ferrous ions to ferric, which precipitates as insoluble hydroxides and oxides. This phenomenon is most frequently observed near iron-rich springs.

6.2.1. Physical factors

The chemistry of the water is intimately related to physical factors, such as temperature, illumination, and electrical conductivity. Temperature affects the rate of both chemical and biological processes, and it can be the cause of major changes in biochemically mediated reactions when it becomes unusually high or low, or when its diurnal fluctuations become excessive. Changes in the illumination due to the length of the day, declination of the sun, or cloud cover, have few notable effects because light can be assumed to be available throughout the year in excess, permitting photosynthesis and other processes to occur without delays caused by insufficient light energy. Furthermore, the differences in the photoperiod between the longest and shortest day of the year do not amount to more than about two hours throughout the northern part of the Pantanal. This would require physiological adaptations to photoperiodism to be precise, which may well be the reason that many tropical plants exhibit no photoperiodic effects (Kaul & Bakaya, 1976). The water in the Pantanal is overwhelmingly clear, except for a short period at the end of the *seca* and beginning of the *enchente*, when much humic matter is released by the decomposing vegetation on the newly flooded parts of the floodplain (Section 5.3.1). Thus, the considerable differences in the light penetration encountered in the different classes of water in Amazonia (Furch *et al.*, 1985) are seldom encountered in the Pantanal throughout most of the year.

Electrical conductivity is not a cause but rather an effect of chemical processes. The more free ions are present in the water, the greater will be the conductivity. This variable is analogous to a sum of the effects of all ionized substances in the aqueous system, although this might be a rather oversimplified way of stating the situation. The data in Table 5.2 show some surprisingly low electrical conductivity values in some of the streams arising from the sandstone formations on the plateau to the north and east of the Pantanal. These are even lower than the reputedly low conductivity values in many streams feeding the Rio Negro System of the Amazon Basin, often being more than one order of magnitude below the values reported from the black water rivers of Amazonia. Even in the clear water rivers along the defunct Transamazonica Highway, the concentrations of most mineral ions contributing to the electrical conductivity were generally well below the world average (Furch, 1984a). Similarly low values in other water bodies of the world are rare (Green, 1970; Petr and Irion, 1983), but Petr (1983) found streams on New Guinea with a comparable electrical conductivity.

As the water in the upland streams flows toward the wetland, its electrical conductivity increases, especially during rainy periods. Furthermore, as it passes human settlements, it often receives polluted wastewater. As a result, in the neighboring Amazonian Drainage Area, particularly in the black water rivers (Goulding *et al.*, 1988), the electrical conductivity is generally even lower than it is in the surface waters of the Pantanal in spite of the mineral poverty of its headwater streams. Gibbs (1967) reviewed the geochemical factors that determine the features of the water chemistry in the Amazon Basin. In comparison, the geological contribution to the water chemistry in the upper Rio Paraguay System seems relatively simple, and a greater attention must be given to seasonal factors to account for the phenomena observed.

During the dry season, the pools of water on the floodplain evaporate, concentrating the ions and thereby elevating such collective variables as electrical conductivity and water hardness. In the Pantanal, locations at which the conductivity of the water was approximately that of rainwater during the *cheia* (Fig. 5.9) were found to have electrical conductivity values as high as 664 $\mu S \cdot cm^{-1}$ during the late *seca*. This is a general phenomenon in the wet-and-dry climatic zone, and McLachlan (1979) described events in Lake Chilwa, Malawi, that are qualitatively almost identical to the events in the Pantanal. However, African water bodies are much richer in mineral salts, and dry season conductivity values as high as 12,000 $\mu S \cdot cm^{-1}$ were recorded.

The seasonal conditions have a profound effect not only on the water temperature but also on the diurnal changes that the temperature undergoes. Standing water can be considered part of a large puffer system for the air temperature near the ground. Because its specific heat is high, it stores much heat, maintaining its temperature even when that of the air drops suddenly. Conversely, it heats up more slowly during sunny days than do the dry ground and air, and it cools more slowly at night. A corollary of this is that the temperature will remain more stable when the volume of water in a pond or lake is greater. This is demonstrated for water bodies in the Pantanal by Figures 4.6 and 4.7, which show 24 hour curves in water temperature, pH, and oxygen concentration in the upper 6 cm of various water bodies at different times of the year.

From these curves, it is apparent that the aquatic organisms live under constant temperature conditions during the *cheia* and unstable ones during the *seca*. Furthermore, the organisms in the shallow dry season water bodies must survive at temperatures that exceed 40°C at times (Heckman *et al.*, 1993; Kretzschmar *et al.*, 1993; Heckman *et al.*, 1996). Thus, a seasonal succession of biotic communities was observed, just as expected (Section 10), and the resulting biotic differences obviously contributed to the seasonal variations in the chemical properties of the water.

6.2.2. pH

The pH is a general factor reflecting a variety of physical and chemical properties that tend to make the water react more as an acid or as an alkaline. There is, however, a basic methodological problem in recording the pH in the northern part of the Pantanal watershed, as discussed above. The reliability and significance of the results therefore vary according to the electrical conductivity of the water. During the dry season, the water is rich in ions, and the pH can be recorded with little problem. During the high water period, however, the pH is rather difficult to determine, and both electrical probes and indicator dyes must be left in the water for a considerable length of time before they stabilize. The pH in the headwater streams of the Pantanal watershed, which often have an electrical conductivity below 5 $\mu S \cdot cm^{-1}$, was recorded after a long stabilization period for the electrodes, but the values are nevertheless of little significance because it is debatable whether such water has a measurable hydrogen ion concentration at all.

Generally, the pH remains near neutral. Typical of rainwater in regions with relatively little air pollution, it is generally slightly on the acid side of pH 7. Its diurnal fluctuations are generally not very great, except during the dry season when there is much decomposition and photosynthesis occurring in the water (Figs. 6.1 through 6.4). The saprobiotic processes and respiration of aquatic animals during the night typically reduce the pH to a minimum just at sunrise, and the photosynthetic processes during the day cause a peak in the pH to be reached in the early afternoon (Figs. 4.6 & 4.7). The elevation of the pH that occurs while photosynthesis is taking place apparently results from the absorption of the dissolved CO_2 from the water for carbon fixation through the dark reaction following the completion of the light reaction in the chloroplasts.

In contrast to the surface waters of the Pantanal, the ground water in the uppermost layer near Poconé is characterized by a low pH, usually between 5 and 6, but sometimes below 5 (Table 6.2). This may be a significant public health factor due to the high nitrite concentrations in many of the wells (Section 13).

Relatively high pH values are sometimes recorded during the day as a result of intensive photosynthesis by submerged plants. Such values are not encountered during periods when the water level is high, obviously because the mineral nutrient concentrations are not high enough to support intensive photosynthesis. The highest pH values are recorded in the shallow waters of drying pools on the floodplain during the *seca*. At the same sites, the values usually become quite low during the night, reflecting the intensive saprobic activity by the saprophytes and the animals that consume them. Thus, intense processes of decomposition and photosynthesis alternately decrease and increase the pH greatly only during the *seca*. The daytime pH values recorded during the study at the most frequently visited sampling sites in the Pantanal are depicted in Figures 6.1 through 6.4.

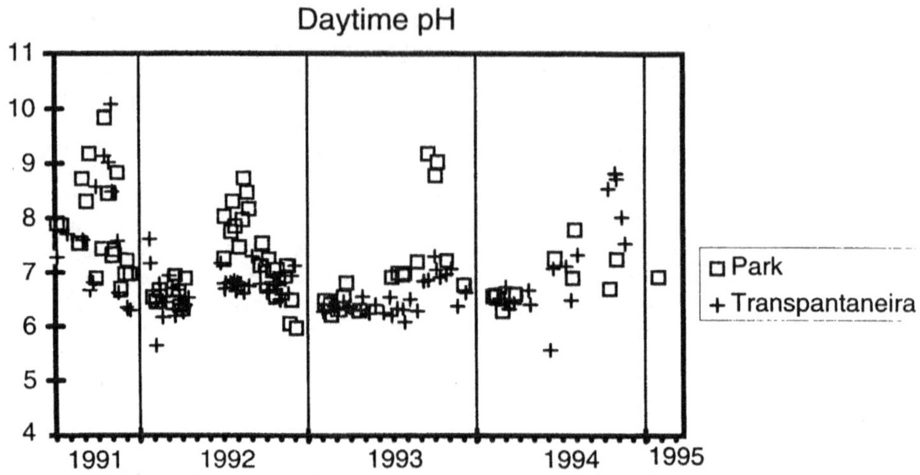

Figure 6.1: The pH recorded from late morning to late afternoon 6 cm below the surface of the Rio Bento Gomes at the Poconé exposition park and at the Transpantaneira Highway from July 1991 through January 1995.

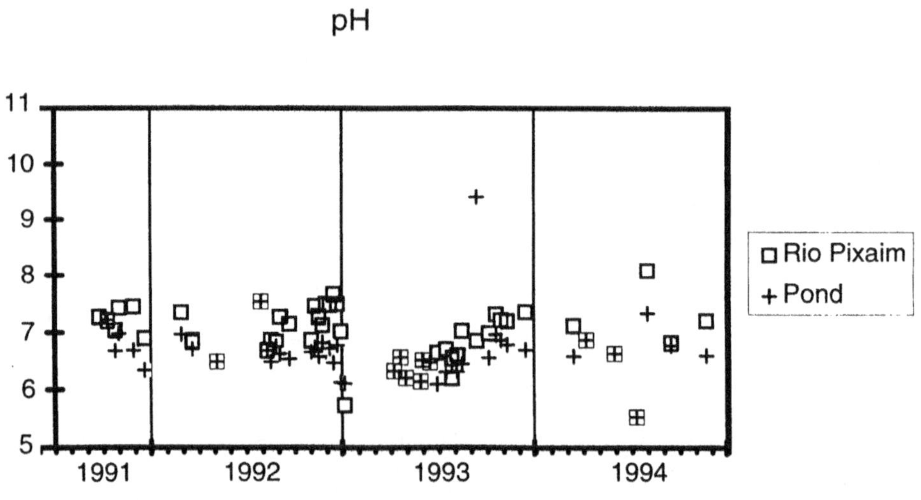

Figure 6.2: The pH recorded 6 cm below the surface during the late morning or afternoon in the Rio Pixaim and the adjacent shallow pond beside the Transpantaneira Highway from July 1991 through 1994. During the high water period, the river overflows into the pond.

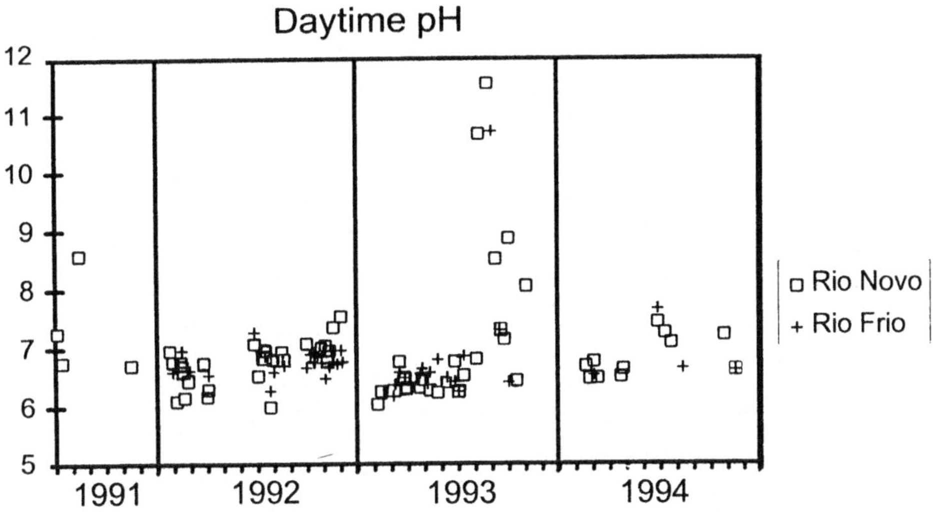

Figure 6.3: The pH recorded from late morning to late afternoon 6 cm below the surface of the Rios Novo and Frio at the Transpantaneira Highway from July 1991 through 1994. When the water was less than 6 cm deep, the electrode was held midway between the surface and the sediment.

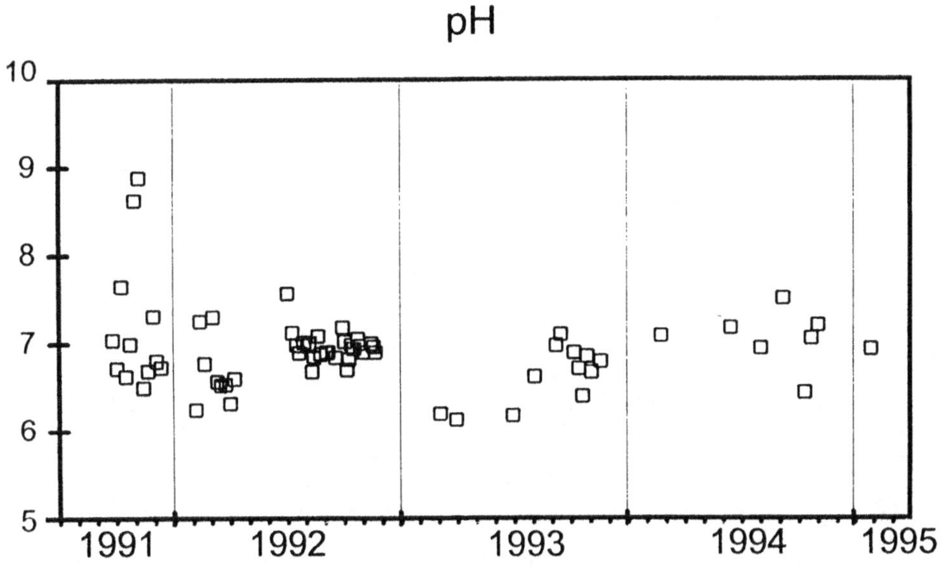

Figure 6.4: The pH recorded from late morning to late afternoon 6 cm below the surface of a low-lying drainage channel called Birici on the Fazenda Ypiranga about 15 km south of Poconé from July 1991 through January 1995. When the water was less than 6 cm deep, the electrode was held midway between the surface and the sediment.

The daytime pH in the Rio Bento Gomes typically increases during the dry season due to the development of rapid growth of algae promoted by the release of inorganic nutrients as the aquatic macrophytes and many of the fishes die off (Fig. 6.1). The excrement of piscivorous birds also promotes this growth. Each year, a pH between 8 and 9 was reached at the time of the freshet, when still more inorganic nutrients are leached into the river from the surrounding flood plain. Occasionally, the pH exceeds 10. In contrast, during the high water period, there is little plant growth due to a shortage of nutrients, and the pH remains slightly below neutral, like the local rain water. Occasionally, the pH may drop below 6 at times when much organic material is decomposing in the water.

The Rio Pixaim and its adjacent shallow pond are actually depressions that remain permanently filled with water throughout the year. Apparently, there is little input of inorganic nutrients from any inflowing water source, and the pH in these water bodies seldom departs from the neutral range (Fig. 6.2). For short periods of time, however, unusually low or high pH values can be recorded, generally associated with the development of large numbers of saprophytes or autotrophic algae, respectively. During the *cheia*, the pond and "river" are contiguous, and the physical and chemical variables of the water mass remain generally stable.

The water at the other sites usually remained in the neutral pH range. However, in 1993, enormous numbers of algae appeared in the Rios Novo and Frio, causing the pH to become unusually high (Fig. 6.3). The same phenomenon was observed in the low-lying site called Birici on the Fazenda Ypiranga in 1991 (Fig. 6.4), as described in detail by Heckman *et al.* (1993) and Kretzschmar *et al.* (1993).

6.2.3. Dissolved oxygen

The determination of dissolved oxygen is particularly important for understanding the biotic processes that influence the activities of the biota. Unfortunately, data on the oxygen concentration in water bodies of the Pantanal in the literature have almost always been reported without any information on the time of day at which they were determined. Dissolved oxygen is one of the variables subject to considerable diurnal variations. To gain information on the magnitude of these variations, curves of the changes during 24 hour periods were prepared at several sites during each of the seasons (Fig. 4.6 & 4.7).

The curves can be interpreted to estimate the amounts of photosynthesis and decomposition occurring beneath the surface of the water relative to one another. In tropical water bodies, such as those in the Pantanal, however, the information on the amounts of dissolved oxygen present at various times of day can only be used to calculate a portion of the total photosynthesis and decomposition in the aquatic milieu. The gas exchange by some aquatic plants occurs beneath the surface of the water, while the emergent plants can exchange carbon dioxide for oxygen directly with the atmosphere. However, some of the oxygen in their aerenchyme tissues are passed to the root systems, from which it can be released into the water (Jedecke *et al.*, 1989). Thus, while the oxygen concentration in the ambient water of the Rio Bento Gomes remained only 30% saturated throughout the day in April, due to the breakdown of the increasing amounts of detritus that enter the water during the *vazante*, the saturation of the water among the rhizoids of *Salvinia auriculata* reached 101% (Heckman, 1994b). This diffusion of oxygen from the roots and rhizoids of floating plants manifests itself only when the oxygen concentration in the water becomes low, but it is during such periods that the organisms of the interrhizon are most dependent upon this source of oxygen.

Direct exchange with the atmosphere is a second important source of oxygen, which depends upon the temperature, atmospheric pressure, and the prevailing oxygen concentration in the water. The submerged aerobic saprophytes require a supply of dissolved oxygen to fully break down the detritus. However, many species of heterotrophic microorganisms move to the surface of the water at times that the concentration is low and meet their needs by taking oxygen directly from the oxygen-rich surface tension layer. For this reason, oxygen-poor water in the Pantanal could almost always be recognized even before analysis by the thin film of bacteria and protozoans on the surface.

The 24 hour curves showing the dynamics of the oxygen concentration therefore can be taken as a measure of the seasonal metabolic activity in the water. Where little photosynthesis and decomposition is taking place, the curves reveal little difference between night and day. This is often the case during the high water season in the Pantanal, during which the photosynthesis is occurring at a low rate, almost exclusively in the leaves of the emergent macrophytes. If there is much photosynthesis and decomposition occurring beneath the surface of the water, the differences between the daytime and nighttime oxygen concentrations are considerable, as during the dry season (Fig. 4.6). Intermediate variations are observed during the seasons of rising and falling water levels.

The maximum and minimum oxygen concentrations during the course of the day reveal the nature of the dominant processes taking place in the water. When photosynthesis is occurring rapidly, the maximum value during the day is high. When the rate of this process greatly exceeds the saprobic activity, the minimum value for the entire 24 hour period remains relatively high (Fig. 4.6). When the heterotrophic processes are occurring at a rate considerably higher than the autotrophic ones, then the maximum daytime and minimum nighttime concentration will remain low. When the rates of photosynthesis and decomposition are both high, then the diurnal curve will include especially high maximum and very low minimum values. The curves in Figure 4.7 show that the decomposition processes were predominant in the Rios Novo and Frio at the times the data were recorded, but especially in September, at the end of the dry season. While seasonal trends are apparent from the curves, consistent seasonal patterns were observed at none of the sampling sites. Both low and high oxygen concentrations may be observed at any site during the same season. Within a few weeks and sometimes within a few hours after a particularly heavy rainfall, an oxygen-poor water body may become either supersaturated each day or develop stable conditions with the oxygen concentration remaining fairly high but below 100% saturation throughout the entire diurnal cycle, as in the Rio Bento Gomes at the Poconé exposition park during the *cheia* (Fig. 4.6). At that time, the oxygen saturations varied only between 30% and 70% of saturation.

In addition to the data from the 24 hour curves, single determinations can reveal much about the dynamics of the aquatic communities, provided the time of day they were recorded is noted. Those values recorded just before sunrise can generally be considered the minima for the entire 24 hour period, except during or just after periods of heavy rainfall. The values recorded in the early afternoon are generally at or near the maxima, unless heavy cloud cover reduced the insolation during the morning. Thus, interpretations of data obtained during these times of day can help to elucidate the relationships between the autotrophic and heterotrophic processes (Figs. 6.5 through 6.8). The data recorded reveal considerable differences among the water bodies investigated on the same dates and large short-term fluctuations in the conditions at any given site, especially during the seasons of rising and falling water levels. The daytime oxygen concentration in the Rio Bento Gomes reveals not only that the greatest peaks of supersaturation are almost always encountered during the *seca* or early *enchente* but also that the variability during all seasons is great (Fig. 6.5). Usually, but not always, higher concentrations were reached in the upstream section of the river where it entered the Pantanal rather than at the Transpantaneira Highway. This is apparently due to the greater load of detritus produced by the animals inhabiting the section of the river that passes through the wetland. During 1992, the maximum daily oxygen concentration remained particularly low near the Transpantaneira due to the large number of fishes that died and decomposed during that dry season.

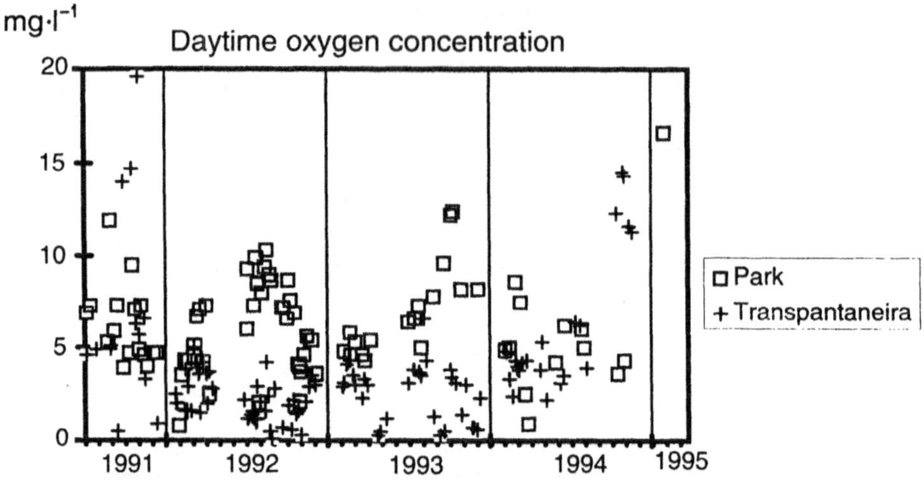

Figure 6.5: The oxygen concentration recorded 6 cm below the surface in the Rio Bento Gomes at the Poconé exposition park and at the Transpantaneira Highway from July 1991 through January 1995. The water was lentic during the dry season and became lotic during the rainy season.

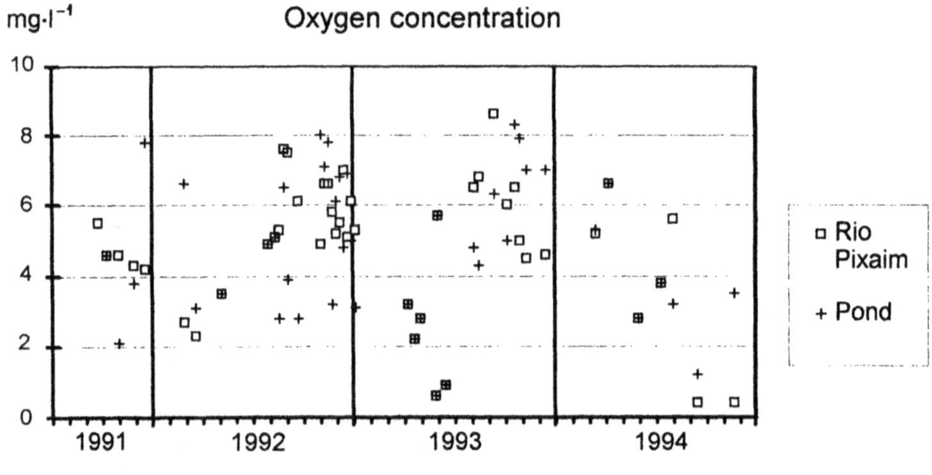

Figure 6.6: The oxygen concentration recorded 6 cm below the surface in the Rio Pixaim and the adjacent shallow pond beside the Transpantaneira Highway from July 1991 through 1994. During the high water period, the river overflows into the pond, and the two become a single water body (Fig. 2.13).

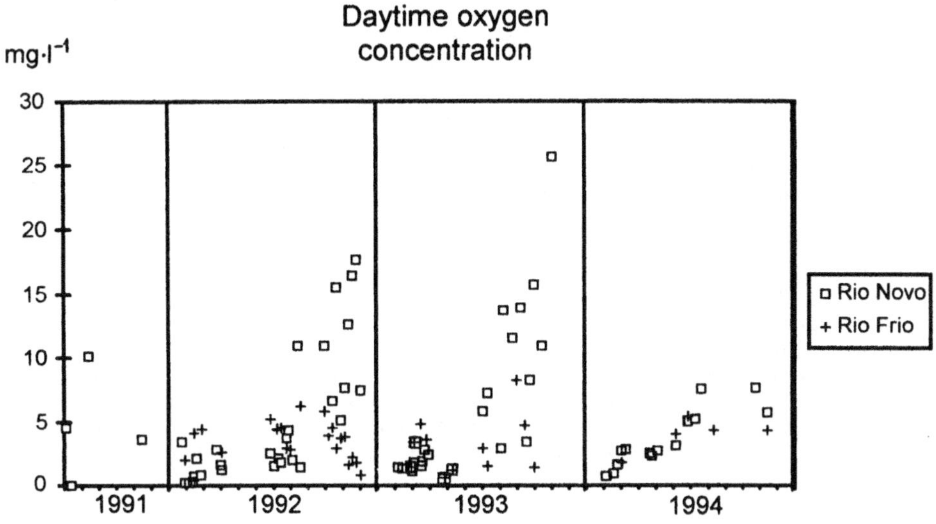

Figure 6.7: The daytime oxygen concentration 6 cm below the surface in the Rios Novo and Frio at the Transpantaneira Highway from July 1991 through December 1994, recorded during the late morning or early afternoon. When the water was less than 6 cm deep, the electrode was placed midway between the surface and the sediment.

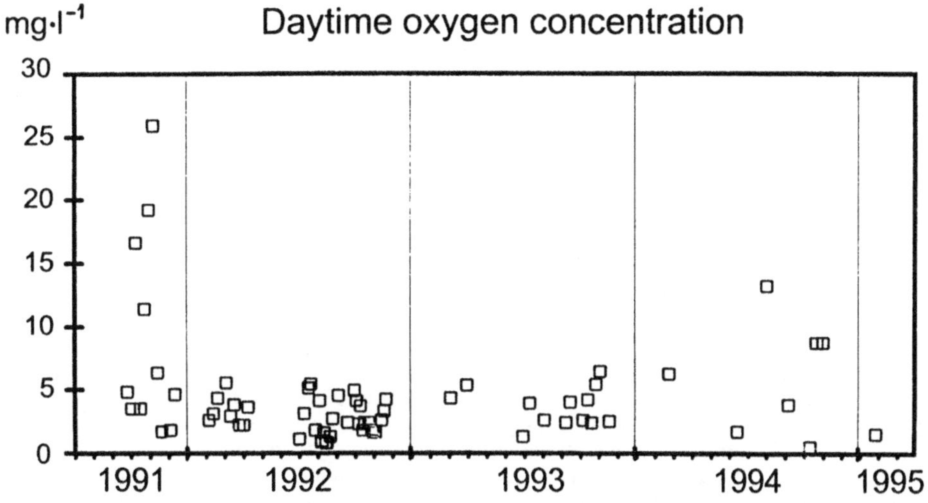

Figure 6.8: The oxygen concentration recorded from late morning to late afternoon 6 cm below the surface of a low-lying drainage channel called Birici on the Fazenda Ypiranga about 15 km south of Poconé from July 1991 through January 1995. When the water was less than 6 cm deep, the electrode was held midway between the surface and the sediment.

The least extreme variations in the maximum daytime oxygen concentration throughout the year were observed in the Rio Pixaim and an adjacent pond, which was dominated by emergent macrophytes (Fig. 6.6). The highest dry season concentrations were recorded in the Rio Novo (Fig. 6.7). These peaks generally correspond with the peak pH values, which were also recorded at this site during algal blooms. A similar peak at the Birici stream was recorded in 1991 (Fig. 6.8). The algae that produced the bloom were investigated by Heckman *et al.* (1993).

The oxygen concentration is an important variable for determining the suitability of the environment for the various species of aquatic organism, many of which depend partially or completely on dissolved oxygen for respiration (Section 10.4.2). The various processes affecting respiration and the oxygen supply in the waters of the Pantanal are depicted in Figure 6.9. This figure makes it evident that the processes by which oxygen reaches the water and is utilized by the organisms are simple and direct. The use of nitrate as an oxygen source apparently occurs rarely, and no evidence of sulfate reduction was ever found (see below). During periods of oxygen shortage, the fishes simply gulp air at the surface, while many of the microorganisms ascend to the surface tension layer. As discussed above, the floating macrophytes supply their own roots as well as the interrhizon community with air through their aerenchyme tissue.

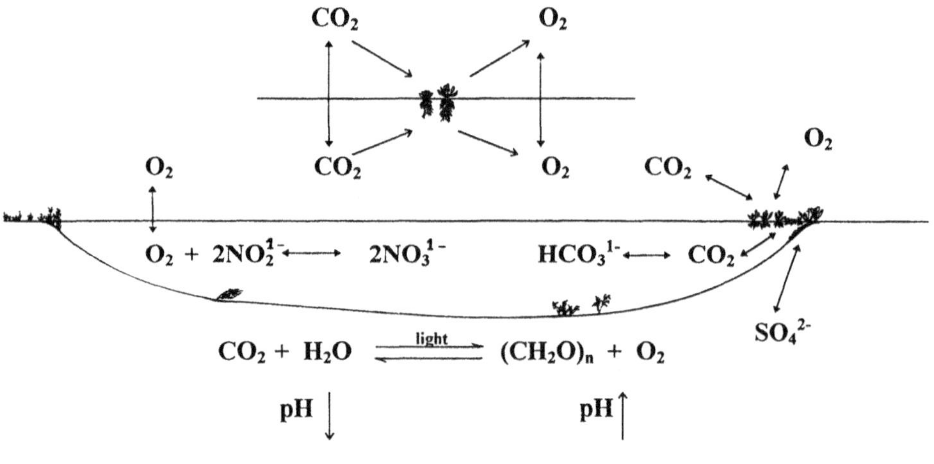

Figure 6.9: A simplified diagram of the processes affecting the oxygen concentration and pH of the water. Oxygen enters the water through diffusion at the surface, diffusion through the aerenchyme in the roots of emergent plants, and as a by-product of photosynthesis. It is also probably supplied from nitrate on rare occasions during the dry season. No evidence was found that sulfate is used as a source of respiratory oxygen. Oxygen leaves the water by diffusion when it is supersaturated and is consumed when organic substances are broken down. No evidence was ever found for activity by sulfur bacteria that generally substitute sulfur for oxygen or for production of hydrogen sulfide in the sediment, and detectable concentrations in the water were found only after mass mortalities of fishes. Emergent plants can exchange gas with the atmosphere, while submersed ones rely mainly on CO_2 from the water. When CO_2 is released during decomposition processes, the pH decreases, and when it is absorbed for photosynthesis, the pH increases, explaining the course of the 24 hour curves presented in Section 4.

From the data on oxygen assembled during the investigation, clear seasonal trends emerge, which would be completely masked by the many day to day differences due to the weather if a large number of samples had not been analyzed. During the *enchente*, the early rain storms increase the water depth and bring a slight increase in stability to the rapidly fluctuating oxygen concentrations in the remaining water bodies. The differences between day and night are usually not quite so great after the freshet. As the water in the rivers and streams begins to spill over onto the floodplain, it quickly loses most of its dissolved oxygen due to the rapid microbial decomposition of the detritus left over from the preceding flood phase or produced by the terrestrial plants during the dry phase. The rapidly rising water is nearly totally depleted of oxygen at times and usually is given a deep brown color by the humic matter present (Table 5.1). The substances producing this color, usually far above the 150 Hazen limit of the water color test kit employed, are generally thought to be the products of anaerobic processes. While the water in the rivers contains more oxygen than that on the floodplain, the daytime concentrations are low compared to those usually recorded during the other seasons. Even after the submerged macrophytes begin their rapid growth, the oxygen concentration remains low and only begins to increase after these plants have developed dense beds near the beginning of the *cheia*. In general, the oxygen concentration during this season can be characterized as low and stable during the diurnal cycle. However, short periods during which it rises considerably are sometimes encountered after several days of strong insolation and little rain.

During the *cheia*, the oxygen concentration remains rather high and stable throughout the diurnal cycle. It does not reach 100% saturation but typically remains at about 80% or somewhat lower where the beds of plants become particularly dense. Because epibenthic diatoms are sometimes present in shallow water during this season and because at deeper locations on the floodplain, macrophytes sprout from the sediment and first begin to develop at the bottom of the water column, the oxygen concentration was occasionally higher at the bottom than at the surface of the water for short periods of time during the late *enchente* and *cheia*. For example, at Birici stream during March, a concentration of 2.7 $mg{\cdot}l^{-1}$ of oxygen was encountered at the bottom of an 81 cm water column among sprouting *Cabomba furcata* and *Utricularia foliosa*, while the concentration was only 2.4 $mg{\cdot}l^{-1}$ in the upper 5 cm of the water column. This is an unusual situation in freshwater bodies. The surface water in the Pantanal, as elsewhere, generally contains more oxygen than the deep water, but the difference is generally less during the *cheia* than during other times of year even though the water is deepest. Because the water flows in many locations and because it is clear and free from organic substances, stratification remains minimal. Thus, in January, 1991, a 2.0 $mg{\cdot}l^{-1}$ concentration of oxygen was recorded at the surface of the Rio Bento Gomes at the Transpantaneira Highway, while the value at a depth of about 2 m was 1.9 $mg{\cdot}l^{-1}$. Similarly, values of 3.7 and 3.6 $mg{\cdot}l^{-1}$, respectively, were recorded in water 17 cm deep at the same location in March, while corresponding concentrations of 2.0 and 2.1 $mg{\cdot}l^{-1}$ were determined on April 1, 1992. The deepest site at which samples were analyzed during the study was located in the Rio Paraguay at Cáceres, where an oxygen concentration of 3.1 $mg{\cdot}l^{-1}$ was recorded at the surface in April, while 2.7 $mg{\cdot}l^{-1}$ was recorded at a depth of 4.25 m. Usually beneath rather dense beds of emergent plants, however, the O_2 saturation did sink to values considerably below those at the surface. For example, near the shores of the Rio Paraguay on the Fazenda Santo Antônio das Lendas, the oxygen concentration decreased from 2.7 to 2.0 $mg{\cdot}l^{-1}$ in only about 80 cm of water in April, and in the Rio Bento Gomes, a decrease from 3.9 to 2.8 $mg{\cdot}l^{-1}$ was recorded in a water column only about 2 m deep in early March. In standing water, the decrease with depth is even more rapid, especially near the sediment. In early March, at one location along Birici stream that was well shaded by emergent plants, the oxygen concentration of 2.9 $mg{\cdot}l^{-1}$ at the surface decreased to 2.6 at a depth of 50 cm, 2.1 at 1 m, and only 0.7 $mg{\cdot}l^{-1}$ at the sediment surface, 1.38 m deep. During the other seasons, the water at all sites, except in the main rivers and a few deep *baias*, is too shallow to permit an investigation of stratification. However, even in shallow waters on the floodplain, a microstratification is apparent, especially during the early morning, due to a rapid but uneven heating of the water bodies by the sun.

The *vazante* is a season during which the oxygen concentration gradually sinks and become increasingly unstable. This results from the increasing oxygen consumption by microorganisms responsible for breaking down the detritus left by the dying plants and animals. As the inorganic nutrients are released, however, algae may suddenly appear in enormous numbers, greatly increasing the concentration of dissolved oxygen during the day. This trend persists into the *seca*, when algae begin a general resurgence thanks to the ever increasing supplies of inorganic nutrients released from the detritus by the saprobic microorganisms. The dry season is characterized by maximum fluctuations in the oxygen concentration during the diurnal cycle, as Figures 6.5 through 6.8 amply illustrate. The concentration also fluctuates locally over periods of several days, as influenced by the activity of the fauna. The arrival of large flocks of birds to feed on the trapped fishes as well as mass deaths of aquatic organisms produce temporary decreases in the amount of oxygen available in the water. Such decreases are often followed by algal blooms, which may increase the O_2 saturation to over 400% during daylight hours. At night, the concentration at the same sites may decrease considerably, and in some cases, the water was found to become nearly anaerobic shortly before sunrise (Fig. 4.7).

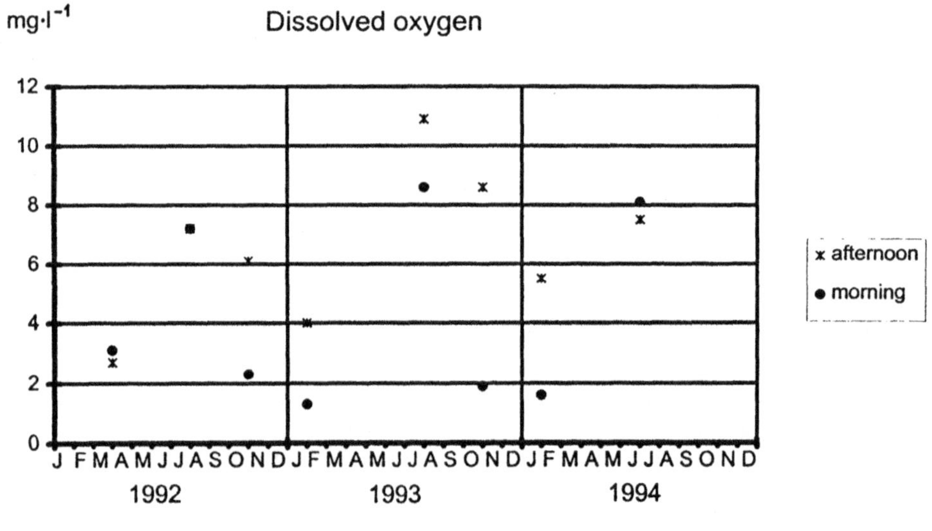

Figure 6.10: The oxygen concentration in the upper 6 cm of the Rio Paraguay at various times of year. The concentrations in each vertical pair were determined during a single 24 hour period, one usually in the late afternoon and the other in the early morning, either at the Fazenda Santo Antônio das Lendas or in the city of Cáceres. The data show that supersaturation can be encountered even in large rivers during the dry season, when inorganic nutrients become available.

The accumulation of some detritus on the sediment and the congregation of the algae in the upper water layers combine at the end of the *seca* and beginning of the *enchente* to produce the greatest decreases in the oxygen concentration with depth that are observed during the entire year. In the Baia Piuval during mid-November, for example, the concentration of oxygen decreased from 5.1 to 0.9 mg·l^{-1} from the surface to a depth of 1.2 m near sudd, a floating island of macrophytes. Elsewhere, from the surface to the sediment, 80 cm deep, the decrease was from 7.1 to 0.5 mg·l^{-1} on November 4 and 5.3 to 1.7 mg·l^{-1} on November 18, 1991. Close to the shore, the concentration decreased from 6.3 to 0.4 mg·l^{-1} over a depth of only 30 cm. It was noted by recording the concentrations at intermediate depths that a sharp decrease usually occurred when the water layers 1 to 2 cm above the sediment were reached.

The processes affecting the oxygen concentration in the Pantanal of Poconé are apparently similar to those affecting the water bodies in other parts of the northern Pantanal. Figure 6.10 shows some data for the Rio Paraguay, which, in spite of the volume of its water mass, is also subject to supersaturation with oxygen under certain conditions. The figure also shows that the greatest differences between the daily maximal and minimal oxygen concentrations probably occur during the *seca* and early *enchente*, as they do in the Poconé district. To the east of Poconé, in the district of Barão de Melgaço, the oxygen concentration in the *baias* seemed to remain somewhat lower than those recorded to the west, with 2.0 mg·l^{-1} being determined in the Baias Recreio and Porta de Fora near the end of the *vazante*, and a maximal concentration of 5.9 mg·l^{-1} being recorded for the Baia Recreio during the *cheia*. During the same season, oxygen concentrations of 2.9 and 3.5 mg·l^{-1} were encountered in the Baias Acorizal and Porta de Fora, respectively.

6.2.4. Alkali earth elements

The water hardness is a general measure of the amounts of alkali earth elements present in the water. Total and carbonate hardness were formulated based on relationships that are normally encountered in natural water bodies of Europe and North America, in which calcium is usually present in moderate to large amounts, and the proportions of the alkali earth elements in the sediment theoretically remain generally similar. In other parts of the world, the assumptions upon which the concept of water hardness is based do not necessary hold true.

Because the field test for magnesium proved unsuitable for analyzing this metal in the field, calcium was the only alkali earth element analyzed separately on a regular basis during the study. The values could generally be correlated with the total hardness, and both variables demonstrated that the water in the northern Pantanal remains in the "very soft" category throughout the year. Although the dry season values are clearly higher than those recorded during other times of year, even these are low in comparison with values recorded in many other parts of the world (Table 6.3).

The values recorded for the water hardness at the various sampling sites are presented in Figures 6.11 through 6.18. These show that the carbonate hardness in mmol·l^{-1} is almost always much higher than the total hardness. Obviously, the maxima of both hardness values are reached during the *seca* throughout the northern Pantanal, while during the *cheia*, the hardness remains low and rather stable. Considerable small scale fluctuations occur, especially during from the late *vazante* to the early *enchente*, and notable elevations in the hardness were observed only late in the *seca*, the most notable of these being in the Rio Novo during 1993 (Figs. 6.13 & 6.17). The concentrations of dissolved calcium, shown at selected locations, follow the general pattern of the total hardness (Figs. 6.19 through 6.21). However, the methodological problems of one of the tests used, discussed in Section 6.1 above, caused the data points for 1992 and 1993 to show considerable scattering. The results for 1994 are believed to be much more accurate.

The hardness of the water near the headwaters of streams in the catchment area deserves special mention (Table 6.1). In some of the springs, a total hardness of 0 was determined. The indicator dye used to determine the endpoint consistently changed color before any of the titrating solution was added. Although the carbonate hardness indicator did not do this, the slightest trace of titrating fluid caused the color to change, indicating that the end point had been reached. Generally, the electrical conductivity of the water tested was below 5 $\mu S \cdot cm^{-1}$ at such times.

Table 6.3. Values for total water hardness, converted to $mmol \cdot l^{-1}$, recorded by different authors in various parts of the world.

Kind of water body	Location	Total hardness ($mmol \cdot l^{-1}$)	Author
Rice field	Udon Thani, Thailand	0.28 - 0.46	Heckman (1979)
Rainwater	New Guinea	0.03 - 0.07	Petr (1983)
Glacial lakes	New Guinea	0.05 - 0.20	Löffler (1973), Petr (1983)
Mountain streams	New Guinea	0.05 - 0.91	Petr (1983)
Foothill lakes	New Guinea	0.48 - 0.94	Petr (1983)
Foothill rivers	Purari System, New Guinea	0.40 - 1.52	Petr (1983)
Lowland Purari River	New Guinea	0.50 - 0.75	Petr (1983)
Drainage ditches, Elbe floodplain	Northern Germany	0.7 - 5.0	Caspers & Heckman (1981)
Elbe Estuary and marshes	Northern Germany	2.0 - 5.9	Heckman (1986)
Headwaters of eight streams	Southern Mato Grosso, Brazil	<0.05	Heckman (1995)
Hot spring	Southern Mato Grosso, Brazil	0.15	Heckman (1995)
Pantanal of Poconé	Mato Grosso, Brazil		Original data
cheia		0.08 - 0.35	
seca-enchente		0.09 - 1.5	

The water in the Rio Paraguay in the District of Cáceres and the *baias* at Barão de Melgaço were perhaps even slightly softer than those in the Pantanal of Poconé. The total hardness in the Rio Paraguay in the northern part of the Pantanal varied from 0.15 $mmol \cdot l^{-1}$ during the *vazante* to 0.35 $mmol \cdot l^{-1}$ early in the *seca*. The carbonate hardness ranged from 0.45 $mmol \cdot l^{-1}$, recorded during the *vazante* and again during the *seca*, to 0.95 $mmol \cdot l^{-1}$ during the early *enchente*. In the Baias Acorizal, Porta de Fora, and Recreio, total hardness values between 0.1 and 0.2 $mmol \cdot l^{-1}$ were recorded, while the carbonate hardness varies from 0.2 $mmol \cdot l^{-1}$ in the Baia Porta de Fora to 0.9 $mmol \cdot l^{-1}$ in the Baia Recreio.

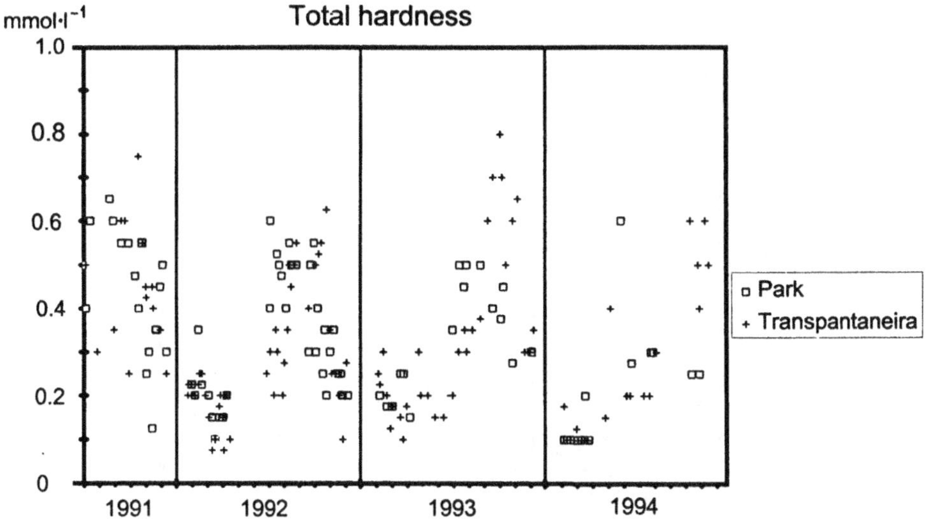

Figure 6.11: The total water hardness determined in the upper 6 cm of the water in the Rio Bento Gomes at the Poconé exposition park and at the Transpantaneira Highway from July 1991 through 1994. The water is lotic during the high water period and lentic or nearly so during the dry season.

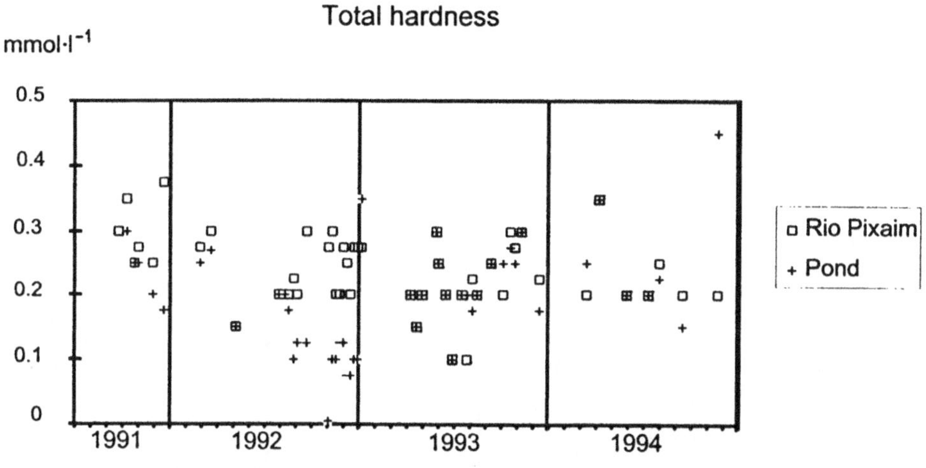

Figure 6.12: The total water hardness determined in the upper 6 cm of the Rio Pixaim and the adjacent shallow pond beside the Transpantaneira Highway from July 1991 through 1994. During the high water period, the river overflows into the pond (Fig. 2.13).

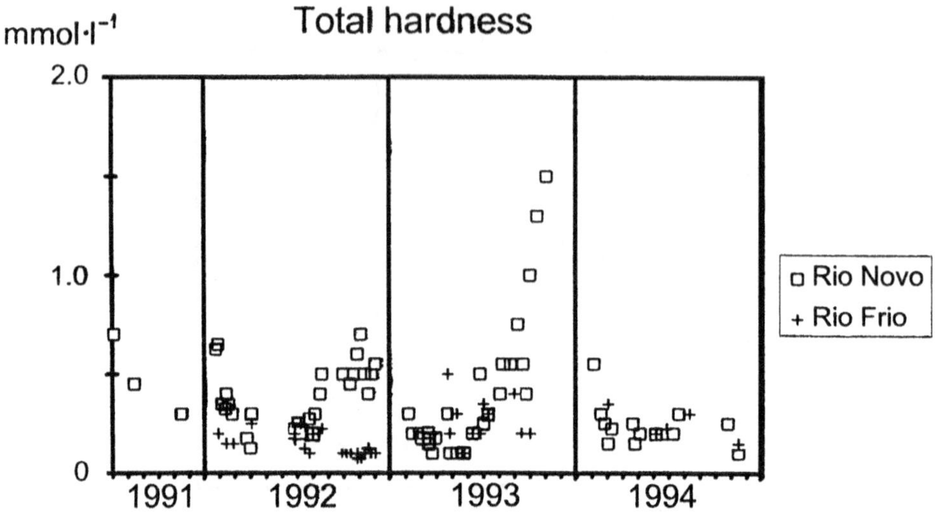

Figure 6.13: The total water hardness determined in the upper 6 cm of the water in the Rios Novo and Frio at the Transpantaneira Highway from July 1991 through 1994. When the water was less than 6 cm deep, the sample was taken from the upper half of the water column.

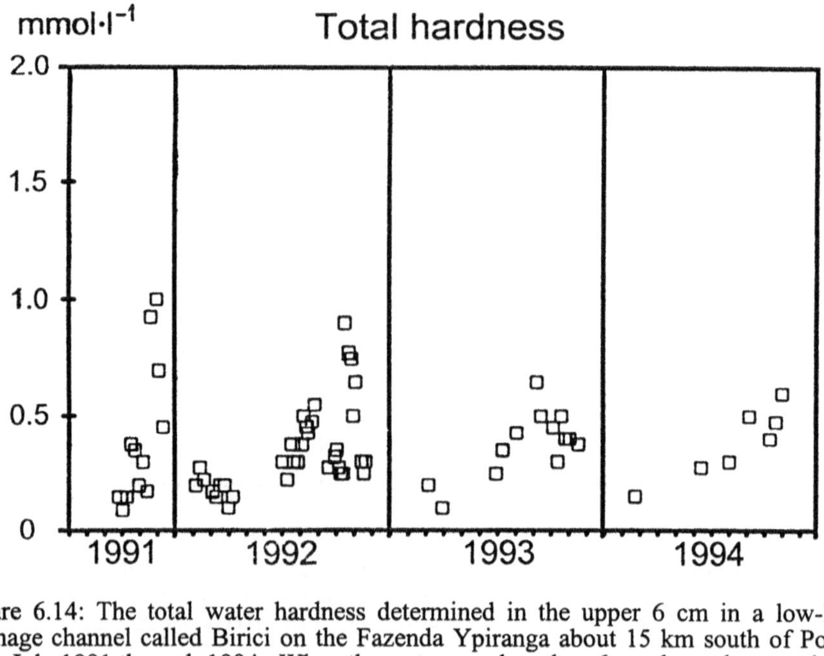

Figure 6.14: The total water hardness determined in the upper 6 cm in a low-lying drainage channel called Birici on the Fazenda Ypiranga about 15 km south of Poconé from July 1991 through 1994. When the water was less than 6 cm deep, the sample was taken from the upper half of the water column.

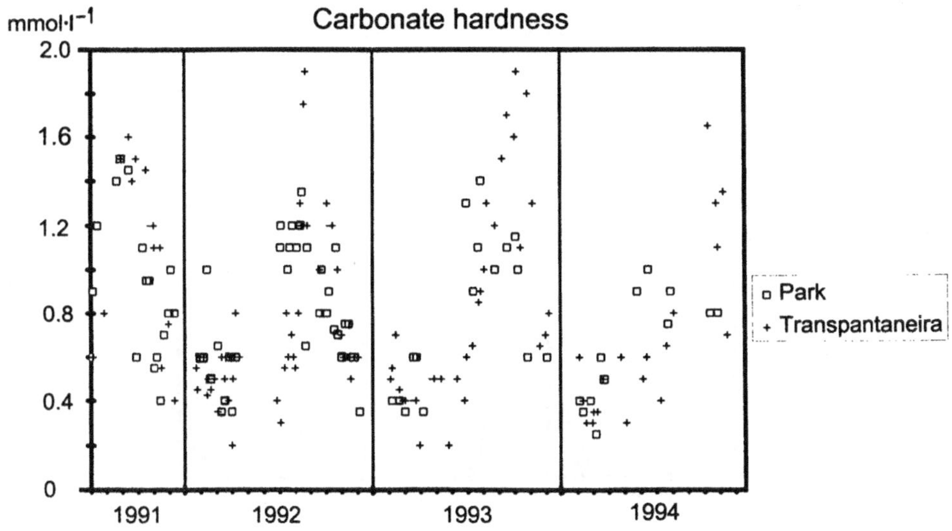

Figure 6.15: The carbonate hardness determined in the upper 6 cm of the Rio Bento Gomes at the Poconé exposition park and at the Transpantaneira Highway from July 1991 through 1994.

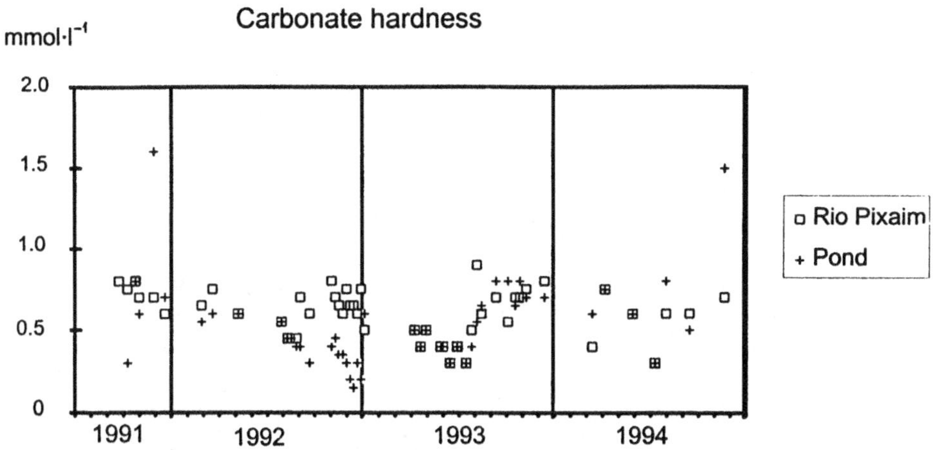

Figure 6.16: The carbonate hardness recorded in the upper 6 cm of water in the Rio Pixaim and the adjacent shallow pond beside the Transpantaneira Highway from July 1991 through 1994.

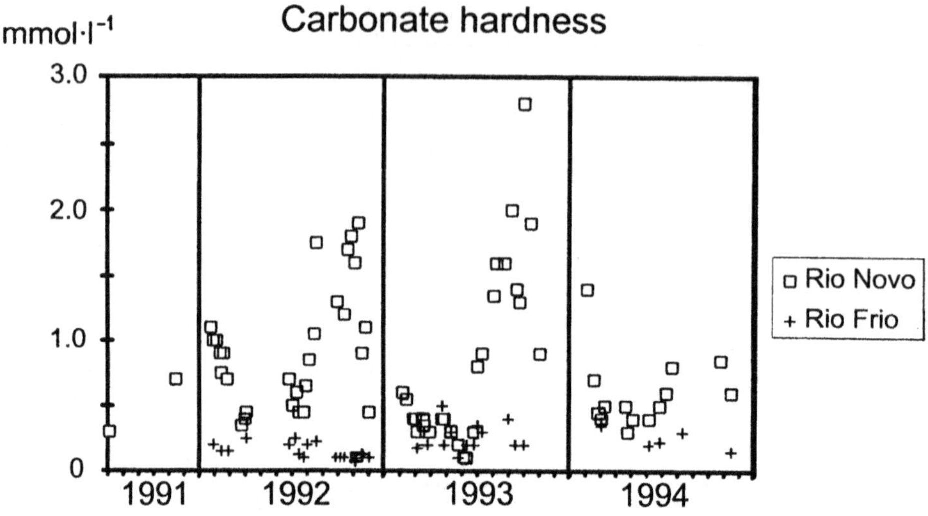

Figure 6.17: The carbonate hardness recorded in the upper 6 cm of water in the Rios Novo and Frio at the Transpantaneira Highway from July 1991 through 1994. When the water was less than 6 cm deep, the sample was taken from the upper half of the water column. The highest values recorded at any of the frequently visited sampling sites in the Pantanal were encountered in the Rio Novo during the dry season.

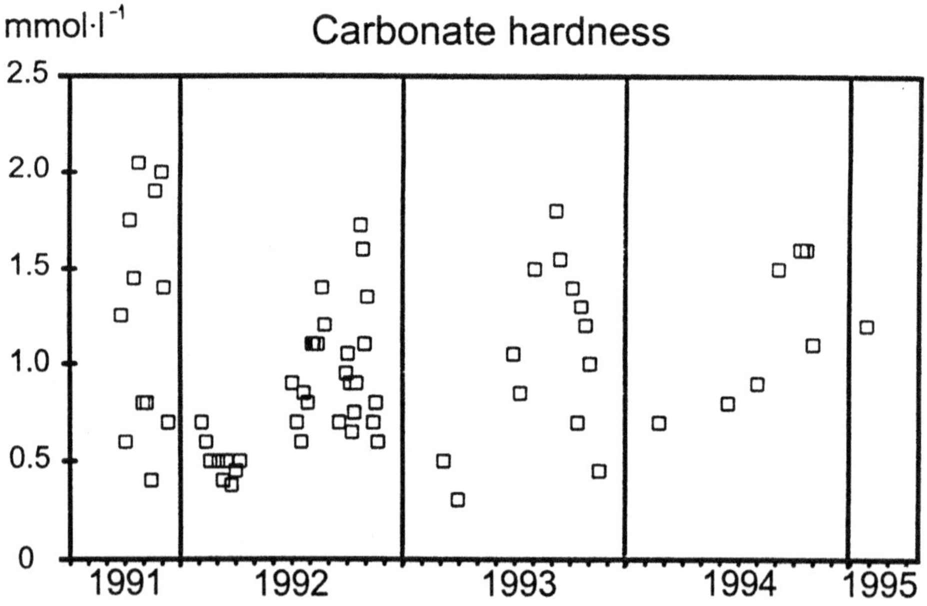

Figure 6.18: The carbonate hardness determined in the upper 6 cm of a low-lying drainage channel called Birici on the Fazenda Ypiranga about 15 km south of Poconé from July 1991 through January 1995. When the water was less than 6 cm deep, the sample was taken from the upper half of the water column.

According to Por (1995), the water hardness in the Rio Miranda should be relatively high for the Pantanal because a few of its tributaries, including the Rio Formoso, drain calcium-rich karst formations in the Serra da Bodoquena and are hypercalcic. Only one analysis was completed in the Rio Miranda at the Possa da Lontra on August 15, 1994, when a total hardness of only 0.75 $mmol \cdot l^{-1}$ and a carbonate hardness of 1.5 $mmol \cdot l^{-1}$ were recorded. These analyses was completed late in the rainless season, but in the southern part of the Pantanal, it was hydrologically only the early part of the *vazante* rather than the *seca*, as in the northern part (Section 5). These values were certainly higher than those usually recorded in the rivers of the northern part of the Pantanal, although a few of the dry season values were higher due to evaporation. A lower total hardness of 0.25 $mmol \cdot l^{-1}$ and a carbonate hardness of 0.5 $mmol \cdot l^{-1}$ were recorded the following day in the Rio Paraguay at Cordraza.

Although sources of calcium-rich water in the upper Rio Paraguay are few, Tricart (1982) reported that there are many karst springs near Fonte Progresso. The values he recorded near those springs were higher than those recorded anywhere in the upland parts of the watershed during the study. Even the water in the Rios Miranda and Paraguay downstream from inflows of relatively calcium rich the water must still be classified as soft, and the hardness values recorded are only slightly above the category "very soft", according to Höll (1968). Apparently, considerable dilution by the calcium-poor clear water streams and rainfall, as well as a probable absorption of the free alkali earth ions by the aquatic organisms, preclude a measurable elevation of the water hardness due to an input of the calcium from the few water sources that contain more than a trace of it.

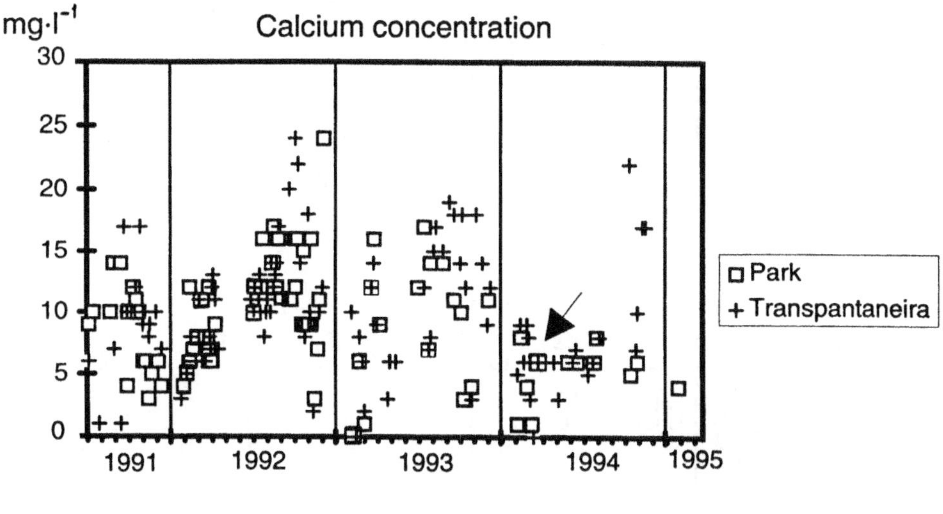

Figure 6.19: The calcium concentration determined in water samples from the upper 6 cm of the Rio Bento Gomes at the Poconé exposition park and at the Transpantaneira Highway from July 1991 through January 1995. The first data in 1991 were obtained using the titration method, but thereafter a color comparison method was used, which proved to have a much greater margin of error. The arrow indicates the time in 1994 when the color comparison method was abandoned in favor of the titration method. After this date, the data are considerably more accurate and show seasonal but little day to day variation.

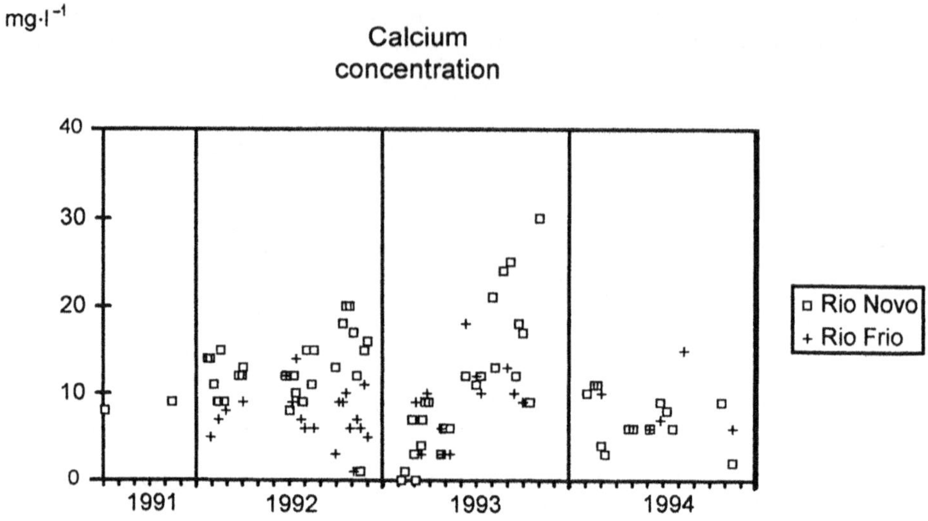

Figure 6.20: The calcium concentration determined in samples from the upper 6 cm of the Rios Novo and Frio beside the Transpantaneira Highway from July 1991 through 1994. The improved accuracy of the test used in 1994 was explained in the caption to Figure 6.18.

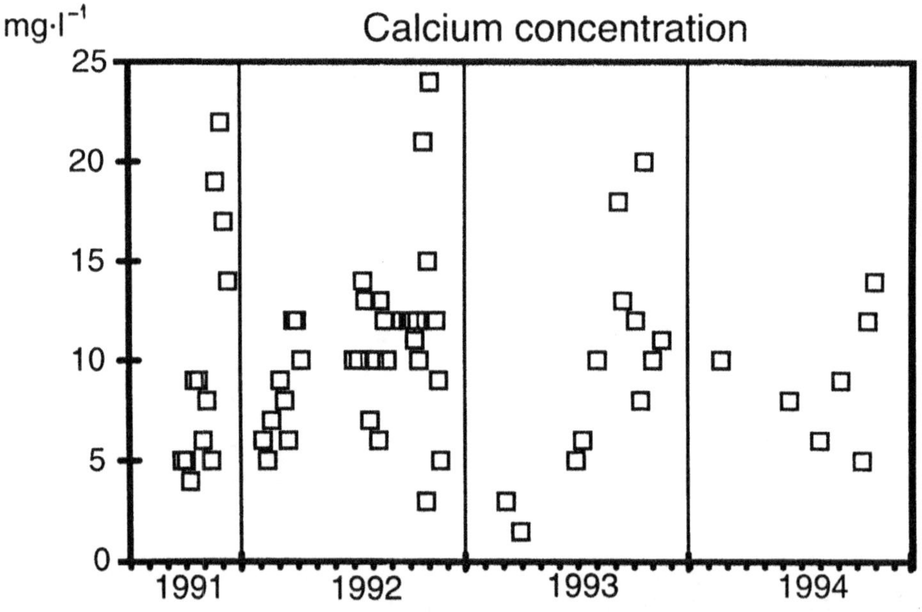

Figure 6.21: The calcium concentration determined in water samples taken from the upper 6 cm of the drainage channel called Birici on the Fazenda Ypiranga about 15 km south of Poconé from July 1991 through January 1995. The improved accuracy of the test used in 1994 was explained in the caption to Figure 6.18.

The effect of calcium on the toxicity of metal pollutants and ammonium in the water will be discussed in Section 13.2.1. It should be stressed, however, that only calcium among the alkali earth metals is thought to influence the toxicity of heavy metals on a variety of aquatic organisms (Jeffree *et al.*, 1995). This could well have an impact on public health as well as on the animals inhabiting the wetland, particularly those that feed on fishes near the sites of the gold mines.

6.2.5. Halides

The information on the halides is limited to chlorosity and the concentration of fluoride. The results of the substrate and stream water analyses from the upland areas of the watershed (Table 6.1) strongly suggest that the halides have been nearly completely leached from the upland parts of the northern Pantanal watershed, and no natural geological deposits of these substances remain. There are, however, analytical results for the uppermost layers of ground water that feed the wells within the town of Poconé and ranches to the south, which revealed relatively high concentrations of chloride, reaching about 80 $mg \cdot l^{-1}$ in Poconé and about 40 $mg \cdot l^{-1}$ in three wells on one ranch. During most of the year, the chlorosity of the surface water in the Pantanal was too low to be recorded accurately by the methods used in the field. It was generally considered to be about 1 $mg \cdot l^{-1}$ according to the trace of color produced during the analysis, and it was almost always below 3 $mg \cdot l^{-1}$, the minimum measurable value (Figs. 6.22 through 6.24). Analyses of deep wells used to obtain drinking water in Poconé also yielded values at or below the detection limits for the test (Table 6.2). According to the drinking water authorities in Poconé (SANEMAT), the water pumped from these wells comes from depths greater than 100 m (Heckman *et al.*, 1997). It should be noted that low chlorosity is not a characteristic of the water in all parts of the Pantanal, although it seems to be typical for most of the wetland. There are a few saline water bodies, mainly in the southern part of the Pantanal (Section 5.3).

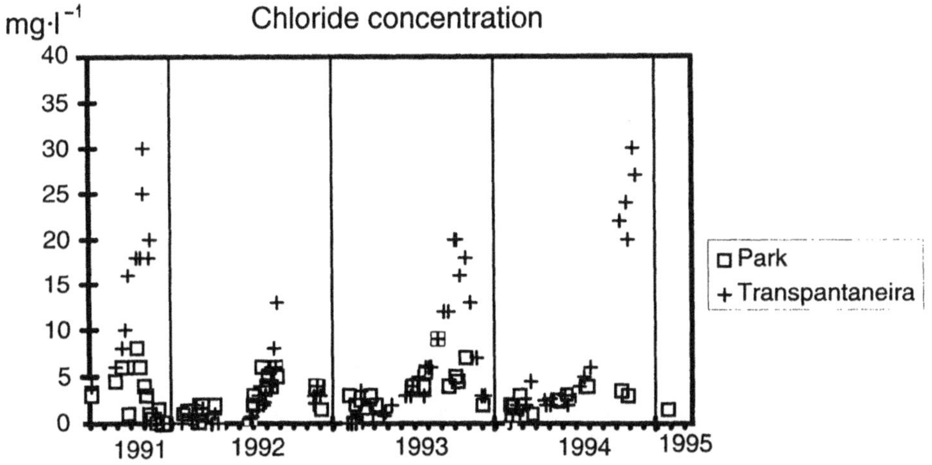

Figure 6.22: The chloride concentration determined in water samples from the upper 6 cm of the Rio Bento Gomes at the Poconé exposition park and at the Transpantaneira Highway from July 1991 through January 1995.

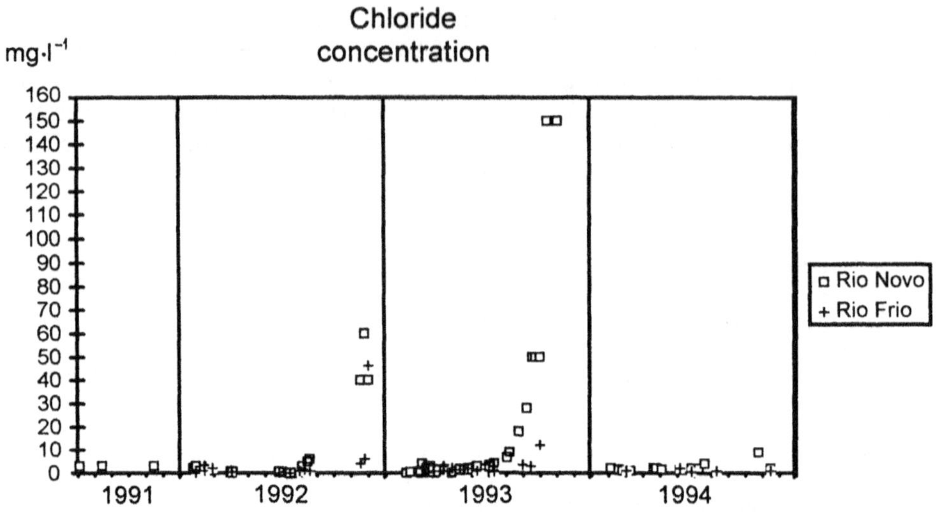

Figure 6.23: The chloride concentration determined in water samples from the upper 6 cm of the Rios Novo and Frio at the Transpantaneira Highway from July 1991 through 1994. The maximum dry season concentration in the Rio Novo was higher than those at the other sites regularly samples in the northern Pantanal, but it was in the same order of magnitude as those determined in some wells in and near Poconé.

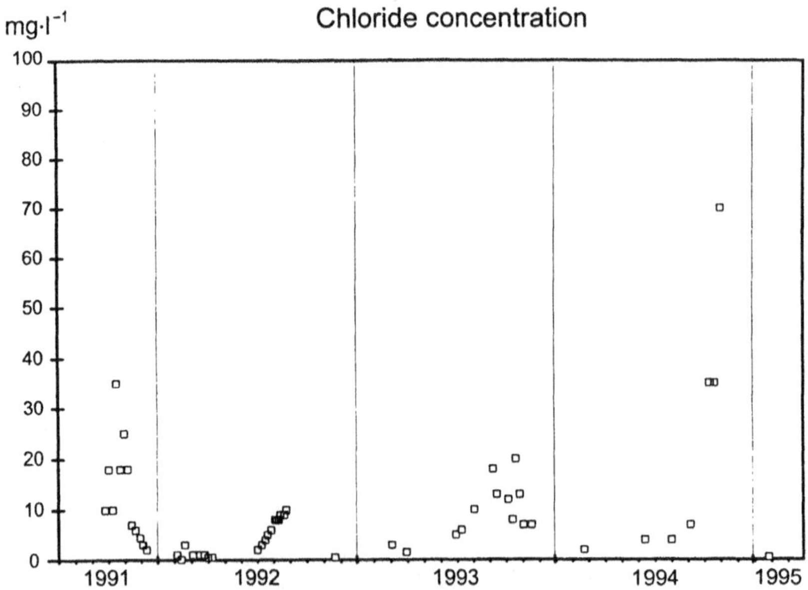

Figure 6.24: The chloride concentration determined in the upper 6 cm of the water in a low-lying drainage channel called Birici on the Fazenda Ypiranga about 15 km south of Poconé from July 1991 through January 1995.

The source of the high chloride concentrations in the upper ground water layer is therefore difficult to determine. The presence of high concentrations of nitrate and nitrite in these waters (Heckman *et al.*, 1997), however, strongly suggests that this chloride is a contaminant originating from old subsurface deposits of sewage. The town of Poconé was founded in 1781, and such contamination near the center of the town could originate from old deposits of wastes that have decomposed and slowly seeped into the uppermost ground water layer. This does not explain the high chlorosity of several wells on a ranch 15 km south of Poconé, however. In addition, a chlorosity of 150 $mg \cdot l^{-1}$ was recorded near the Rio Novo and the Transpantaneira Highway for short periods during the dry season. At these times, a seepage of ground water that kept the standing water in the stream from completely drying up was assumed. This water body was at least 1 km from the nearest ranch house and over 30 km from Poconé. Therefore, it is unlikely that the chloride source for this water body is domestic wastes. The matching cation in this case would presumably be sodium, which is assumed to be the main cation in solution in the absence of considerable quantities of calcium, as discussed above. However, high chloride concentrations were also encountered in the Rio Novo during the dry season, where some ground water seepage may have occurred. It is therefore not unlikely that there are some natural saline ground water sources in the northern Pantanal.

Among all of the sampling sites, the Rio Pixaim and its adjacent shallow pond showed the least variation in their chlorosity. The chloride concentration of all but 3 of the 65 samples analyzed at these sites was between 0 and 3 $mg \cdot l^{-1}$, that is, lower than the analytical method can accurately determine. The maximal chloride concentration was about 9 $mg \cdot l^{-1}$ recorded in the pond at the end of the *seca*. The other two values above 3, about 4 and 6.5 $mg \cdot l^{-1}$, were also recorded for water from the pond at the end of the *vazante* and beginning of the *seca*, respectively. These water bodies, being located near a restaurant and new tourist hostels, were not often visited by large flocks of water birds seeking fishes during the *seca*. Visits by such large numbers of birds have a temporary impact on the chlorosity of the water whenever they occur.

As already mentioned, because of the high concentrations of silicates in the water, analyses of fluoride were initiated to determine whether deposits of this halide might contribute to the dissolution of the silicates in the sediment. Fluoride was not abundant in the samples, but detectable amounts were discovered at a few sites in the Pantanal when the chloride concentration was particularly high, mainly during the *seca*.

During 1994, the only year in which the samples were tested for fluoride, only 18 out of 43 samples tested were found to contain any trace of fluoride at all. Five of these were streams fed by springs on the plateau to the east of Cuiabá, and two were in the Pantanal of Mato Grosso do Sul. The highest concentration in the Pantanal was 2.45 $mg \cdot l^{-1}$, encountered in the Rio Bento Gomes near the Transpantaneira Highway, where the chloride concentration was temporarily about 22 $mg \cdot l^{-1}$. The second highest concentration of 0.80 $mg \cdot l^{-1}$ was determined at Birici Stream, where the a chloride concentration of 70 $mg \cdot l^{-1}$ was simultaneously recorded. These relatively high concentrations were encountered in October, when the water level was near the minimum for the year. The other positive values were recorded only at these two sites and in the Rio Novo near the Transpantaneira Highway.

The chlorosity of two rivers was determined in the Pantanal of Mato Grosso do Sul. On August 15 and 16, 1994, chloride concentrations of about 3 $mg \cdot l^{-1}$ and below 1 $mg \cdot l^{-1}$ were recorded in the Rio Miranda and the Rio Paraguay, respectively. These are quite similar to those recorded in the surface waters of the Pantanal of Poconé during the *vazante*.

6.2.6. Aluminum

The tests for aluminum in the water invariably failed to detect any. On the one hand, this could be due to the lack of considerable amounts of the element in the sediment and subsoil. However, it could also be due to a lack of a sufficient acidity to ionize the aluminum compounds and thereby liberate them into the overlying water. To determine which explanation is most probable, an extensive series of sediment analyses is necessary. Almost no conclusive information is available on this subject in the literature, but from the consistency of the sediments examined as well as from the analytical results, it seems probable that aluminum will prove to be surprisingly rare in some parts of the northern Pantanal and its catchment area. It should be stressed that the situation in the southern parts of the wetland is probably different as surmised by personal examination of the soils near Corumbá.

6.2.7. Iron

Except for the silicates, the only minerals still present in considerable quantities in the underlying rock layers of the Pantanal watershed are those rich in compounds of iron and manganese. The color or many rock outcroppings and of the sediments in the Pantanal suggests the presence of these metals (Plate 1b). The abundance of iron was also confirmed by analyses of the water and sediments (Tables 3.1, 3-2, & 6.1). Although routine analyses of water samples for manganese were not conducted, its presence together with iron in emerging ground water at some sites is safe to assume because of its presence in the sediments (Tables 3.1 & 3.2). Because these two metals seem to possess no acutely toxic properties for any common forms of life, concern about their presence is generally focused upon the unpleasant taste they confer to drinking water when their concentrations are unusually high. The only other well documented ill effect of iron-rich water is the precipitation of the insoluble ferric compounds on the respiratory surfaces of aquatic organisms after the ferrous ions have been oxidized. In the Pantanal watershed, accumulations of ferric oxides and hydroxides are encountered almost exclusively near upland springs fed by ground water.

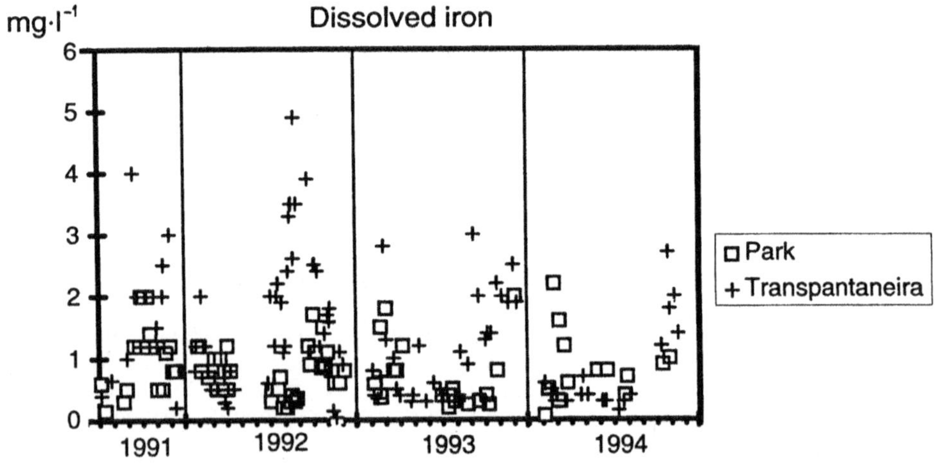

Figure 6:25: The dissolved iron concentration recorded 6 cm below the surface of the Rio Bento Gomes at the Poconé exposition park and at the Transpantaneira Highway from July 1991 through 1994. The water was lentic during the dry season and became lotic during the rainy season at both locations.

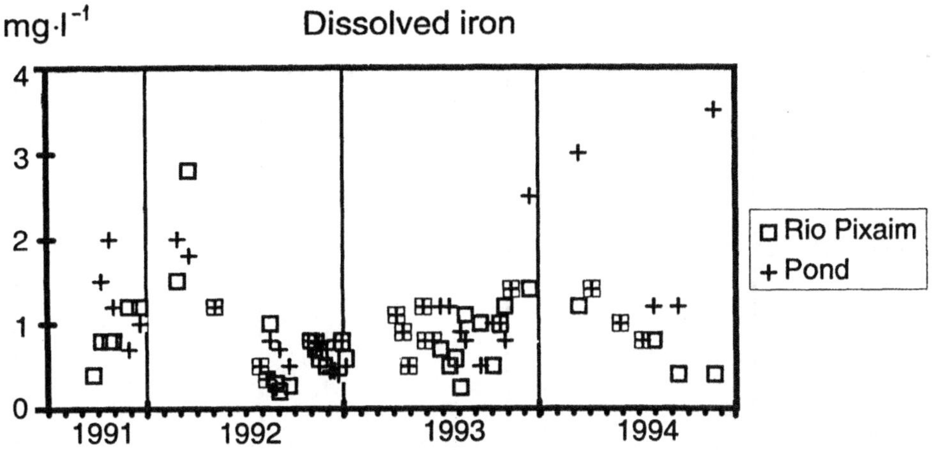

Figure 6.26: The dissolved iron oxygen concentration 6 cm below the surface in the Rios Novo and Frio at the Transpantaneira Highway from July 1991 through December 1994, recorded during the late morning or early afternoon. When the water was less than 6 cm deep, the electrode was placed midway between the surface and the sediment.

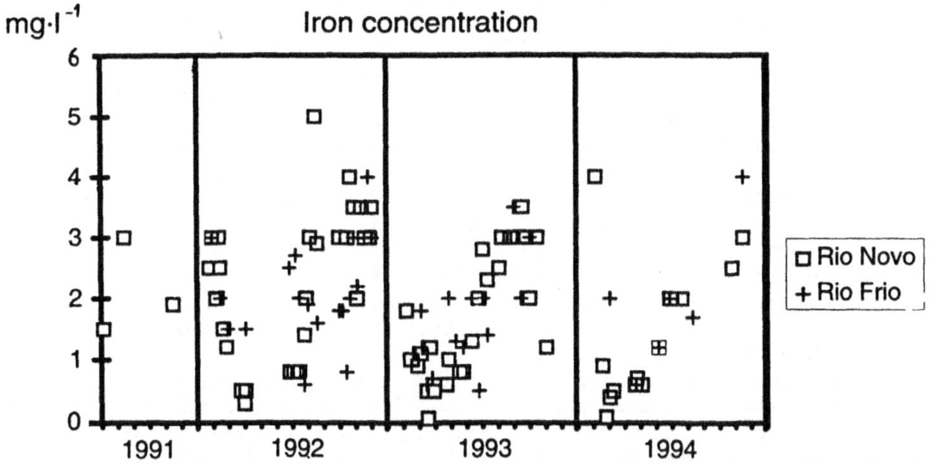

Figure 6.27: The dissolved iron concentration recorded 6 cm below the surface in the Rio Pixaim and the adjacent shallow pond beside the Transpantaneira Highway from July 1991 through 1994. During the high water period, the river overflows into the pond.

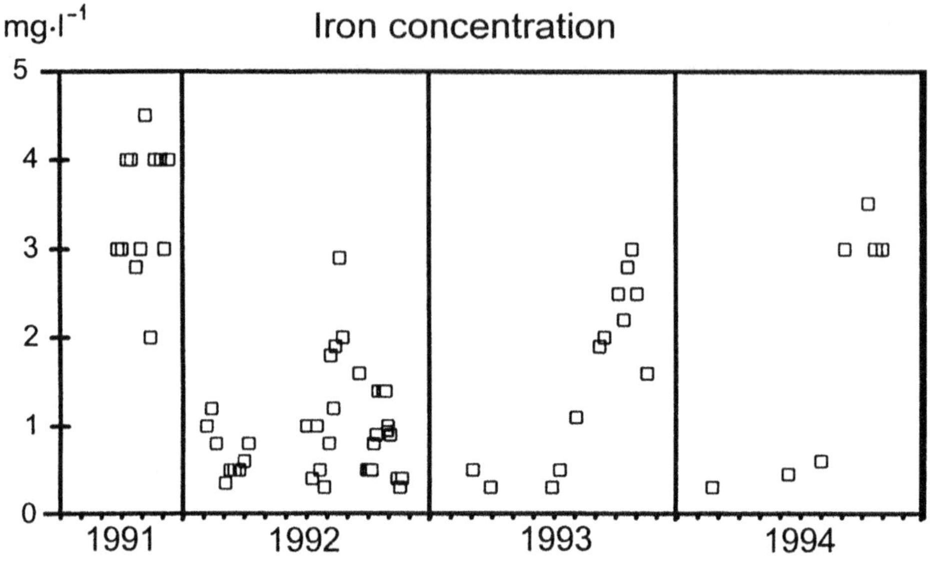

Figure 6.28: The dissolved iron concentration determined in water samples taken from the upper 6 cm of a low-lying drainage channel called Birici on the Fazenda Ypiranga about 15 km south of Poconé from July 1991 through 1994.

Generally, large amounts of ferrous ions in streams and water bodies in the Pantanal watershed suggest the emergence of ground water near the sites at which they are encountered. In upland streams, which flow mainly toward the Pantanal, some high iron concentrations are occasionally detected near the sources, where ground water emerges from springs. After it emerges, the iron is quickly oxidized from the ferrous to the ferric state, accompanied by a rapid decrease in the electrical conductivity of the water (Heckman, 1995). At such sites, objects in the water are generally covered by gelatinous iron hydroxides and oxides heavily colonized by iron bacteria (Section 7.1.2). Often, the surface of the water is covered by an oily-looking slick containing bacteria and iron compounds, as well (Section 7.1.2).

Such elevated concentrations of ferrous ions are also encountered at some sites in the Pantanal during the dry season (Figs. 6.25 through 6.29). At the sampling site near the bridge over the Rio Bento Gomes along the Transpantaneira Highway, for example, iron concentrations as high as 5 $mg \cdot l^{-1}$ were recorded during the dry season (Fig. 6.25). A similarly high concentration was recorded in the Rio Novo (Fig. 6.26). However, the general chemistry of the water in the Rio Novo suggests that during the *seca,* there is some emergence of ground water at or near this site (Section 6.2.5). Elsewhere, seasonal differences were obvious, but other than generally higher iron concentrations during the *seca*, the differences seem to have more to do with the daily weather than to seasonal factors (Figs. 6.26 - 6-28). As expected, when low oxygen concentrations prevailed for a long period of time, the soluble iron concentration tended to increase. On the one hand, without oxygen or another oxidizing agent in the water, the ferrous ions cannot be oxidized to the insoluble ferric state, and, on the other, emerging ground water rich in ferrous ions generally eliminates much of the free oxygen in the surface water with which it mixes as the ferrous ions are rapidly oxidized, usually mediated by the many iron bacteria.

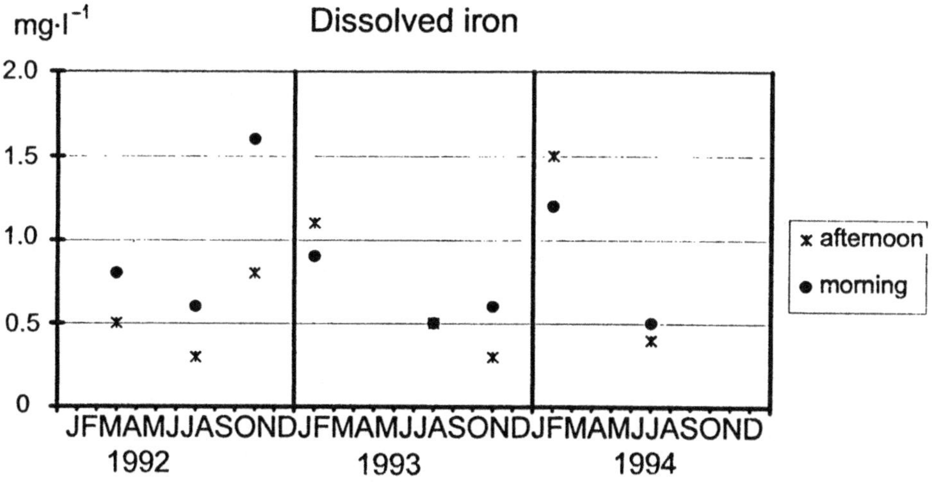

Figure 6.29: The dissolved iron concentration in the upper 6 cm of the Rio Paraguay at various times of year. The concentrations of each vertical pair were determined during a single 24 hour period, one in the late afternoon and the other in the early morning, either at the Fazenda Santo Antônio das Lendas or in the city of Cáceres.

In the large rivers and *baias*, the iron concentration was sometimes found to be moderately high (Fig. 6.29), but it never reached the concentrations encountered in the emerging ground water and on the flood plain during the *seca*. Because these large water bodies are generally well supplied with oxygen, it can be assumed that the ferrous ions entering them are rapidly oxidized to the ferric state and precipitate as hydroxides and oxides.

6.2.8. Compounds of nitrogen

The nitrogen cycle in the Pantanal is highly dependent on the seasonal climatic cycle. Analyses during the entire year revealed that free nitrate is seldom present in detectable concentrations in the surface water, and when it is present, the concentration is invariably low. The water samples were analyzed for nitrate considerably less often than for ammonium and nitrite because of the need to conduct this test in the laboratory and the lack of suitable facilities and reagents for such analyses during the early years of the project. Nevertheless, enough data are available to evaluate the seasonal processes promoting differential rates of plant growth. In the Pantanal of Poconé and in the Rio Paraguay at Cáceres, a total of 59 water samples were analyzed for NO_3 during the year 1994. In 41 of these, no trace whatsoever of nitrate was found. Most of the detectable concentrations were encountered during the *vazante*, when 6 out or 10 samples contained detectable quantities. The highest nitrate concentration recorded in the Pantanal, 0.44 $mg{\cdot}l^{-1}$, was detected at the beginning of the *seca* in the Baia Piuval. No nitrate was detected in two additional samples taken there later in the *seca* and in the early *enchente*. The small pond adjacent to the Rio Pixaim also contained detectable concentrations of nitrate during various seasons, with a concentration of 0.21 $mg{\cdot}l^{-1}$ being recorded on June 27, during the transition period from the *vazante* to the *seca*. Later in the *seca*, as well as during the *enchente* and *cheia*, no detectable amount of nitrate was found in the pond, but a concentration of 0.1 $mg{\cdot}l^{-1}$ determined in water from the Rio Pixaim on August 11, during the *seca*. Two samples from the Rio Frio were analyzed for nitrate, and only one contained a detectable amount of only 0.04 $mg \cdot l^{-1}$. All six samples from the Rio Novo contained no detectable amount. A total of 25 samples from the Rio Bento Gomes were analyzed, and nitrate was detected in only 11 of them, with 0.11 $mg \cdot l^{-1}$ being the greatest amount determined. This was recorded on June 13, during the late *vazante*, near the Transpantaneira Highway. The other amounts determined for the water there and at the Poconé exposition park varied from a trace to 0.08 $mg \cdot l^{-1}$. Two of the six samples taken at Birici stream on the Fazenda Ypiranga contained detectable amounts of nitrate. During the late *cheia*, a concentration of 0.2 $mg \cdot l^{-1}$ was recorded, while 0.1 $mg \cdot l^{-1}$ was determined during the late *vazante*. Five water samples from the Rio Paraguay were analyzed, and only one contained a determinable amount of nitrate: 0.04 $mg \cdot l^{-1}$. In contrast, water from the Rio Miranda in Mato Grosso do Sul was found to have a nitrate concentration of 0.15 $mg \cdot l^{-1}$.

In the upland parts of the watershed, a different situation was encountered (Table 6.1). A few analyses of the water from streams and the Rio Arica Mirim were almost always positive for nitrate. Two samples taken from the Rio Arica Mirim contained 0.18 and 0.20 $mg \cdot l^{-1}$. Traces of nitrate were even found in the extremely mineral-poor water of the streams near Chapada dos Guimarães and in the Rio das Mortes (Heckman, 1995). In addition, low concentrations from a trace to 0.09 $mg \cdot l^{-1}$ were detected in the Baias Acorizal, Recreio, and Porta de Fora near Barão de Melgaço. In contrast, the well water from the vicinity of the town of Poconé contains high concentrations of nitrate, as shown in Table 6.2 and discussed in detail by Heckman *et al.* (1997).

From the concentrations of nitrate, the data presented in Figures 6.30 through 6.37, and observations of the seasonal development of the vegetation in the field, it is concluded that the macrophytes dominate the available nutrient supply throughout the Pantanal for most of the year. The aquatic species absorb and use the rich supply of ammonium that is released into the water during the *enchente* as the drowned terrestrial plants and desiccated detritus still remaining from the previous rainy season decompose. Little of this ammonium can therefore be oxidized to nitrate in the free water. The growth of the plants is amazingly rapid as long as nutrients continue to be added to the water while its level is rising and flooding greater and greater areas of the plain. When the *cheia* is reached, the input of free nitrogen compounds ceases, and plant growth slows down. This is not apparent in the rivers of the Pantanal (Figs. 6.30 & 6.34), but it is well illustrated for the Rio Pixaim and its adjacent pond in Figure 6.31 for ammonium and Figure 6.35 for nitrite. Because the Rio Pixaim is more of a lake than a river, the seasonal input of nitrogen compounds resulting from the flooding is apparent. The ammonium and nitrite concentrations in the flowing water of the rivers are influenced by remote sources, sometimes outside of the wetlands. The ammonium concentration on the flood plain is reflected by the data for Birici stream. At this location as well as in the Rio Frio, which usually completely dries up during the *seca*, enormously high ammonium concentrations are occasionally encountered (Figs. 6.32 & 6.33). The presence of such high concentrations of inorganic nutrients usually provokes blooms of heat-tolerant algae (Heckman et al., 1993), which then deplete the nutrient supply somewhat. During other times of the year and during moister years when these shallow water bodies do not completely dry up, the ammonium concentration usually remains well below 1 $mg \cdot l^{-1}$, but this is enough to promote extremely rapid growth of the macrophyte vegetation. Thus, while the increases in the ammonium concentration during the last months of the year at Birici stream in Figure 6.33 may seem slight compared to the high dry season concentrations, they are quite sufficient to meet the needs of the autotrophs. The resolution of these early rainy season peaks are much better illustrated in Figure 6.31 because no excessively high dry season concentrations were encountered, and the scale of the graph reaches a maximum of only 1.0 $mg \cdot l^{-1}$. In contrast, the concentrations of ammonium and nitrite during the *cheia* were almost always minimal and often remained below the lowest detectable concentrations of the analyses employed. The plants, which have developed luxuriantly by the time the water level has reached its maximum, obviously monopolize the available nitrogen and ammonium, nitrite, and nitrate usually remain near or below the detection limit.

Because factors such as the oxygen concentration and presence of certain microorganisms influence the nitrite concentration, it is less reliable than the concentration of ammonium as a variable to estimate the amount of inorganic enrichment is entering the water. Nevertheless, its concentration is often higher than that of nitrate: a situation not often encountered in the well-enriched waters of the temperate zones. Because of the general shortage of inorganic nitrogen sources in the Pantanal throughout much of the year and the great affinity of plants for nitrate, it is assumed that it is absorbed by a plant almost as soon as it is produced.

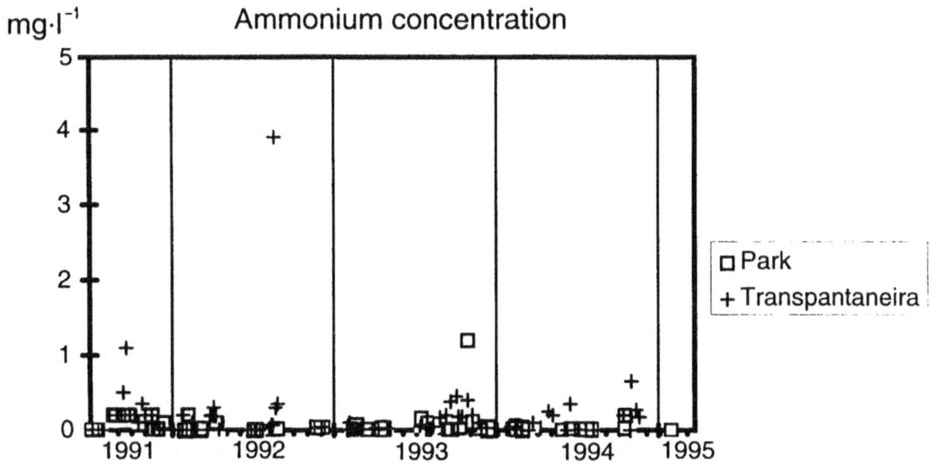

Figure 6.30: The ammonium concentration determined in samples from the upper 6 cm of the Rio Bento Gomes at the Poconé exposition park and at the Transpantaneira Highway from July 1991 through January 1995. The peak value in 1992 near the Transpantaneira Highway was recorded after a mass mortality of fishes at that site.

The levels of free nitrogen compounds only begin to rise again well into the *vazante*, when water plants and some aquatic animals begin to die off, or when large numbers of caimans and water birds congregate in one of the shrinking water bodies to feast on fishes, which have ever decreasing chances to escape. This rise in the amounts of free nitrogenous substances is reflected particularly well by the irregular and sporadic increases of ammonium detected ever more frequently as this season progresses. It is also during the *vazante* that determinable amounts of nitrate were encountered most frequently in the water samples from the wetland, as discussed above.

During the dry season, the process of death and decomposition within the aquatic community results in more ammonium being released. A great structural change in the algal community and periods of massive reproduction by various algal species (Section 7) are responsible for removing much of this inorganic enrichment, generally before the ammonium has had the chance to be oxidized to nitrite. It is also postulated that the plants have such a great affinity for free nitrate that any NO_3 ions formed at this time are immediately absorbed, explaining the absence nitrate from the water during the *seca* just as it was during the *cheia*. However, during the high water period, the macrophytes are apparently the main beneficiaries of the nitrate, while algae are more important consumers of this substance during the dry season. The algal community is only eliminated by the complete desiccation of the water bodies. At this time, a terrestrial macrophyte community consisting mainly of fast-growing annuals takes up some of the nitrogenous substances from the soil and holds them until the next period of flooding.

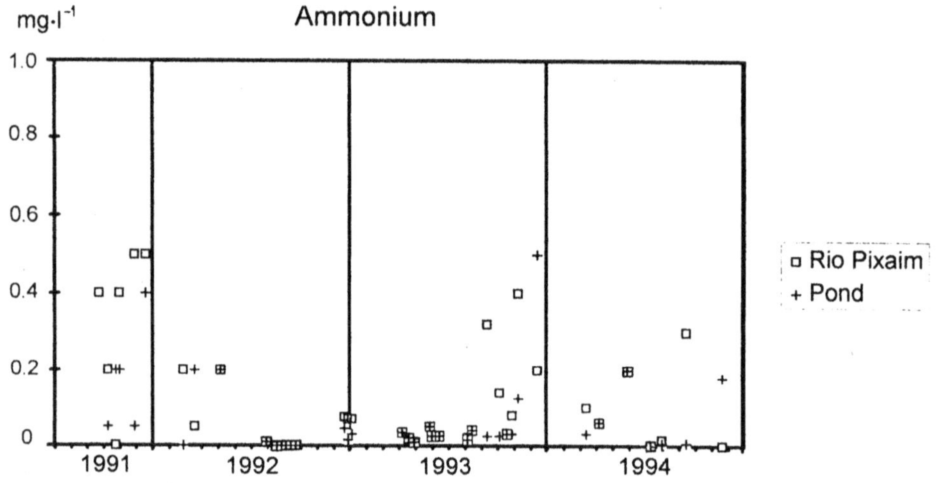

Figure 6.31: The ammonium concentration determined in water samples from the upper 6 cm of the Rio Pixaim and the adjacent pond beside the Transpantaneira Highway from July 1991 through 1994.

Nitrate losses due to denitrification processes are likely at the times that the water in various parts of the Pantanal becomes anaerobic during the *vazante* or *seca*. These losses can be replaced by nitrogen fixation, which occurs during thunderstorms (Jones and Bromfield, 1970) and within various prokaryotic species, including nodule bacteria found in many legumes, usually in the roots but also in the stems of the partially submerged species (Loureiro *et al.*, 1995), and the symbiotic cyanobacterium, *Anabaena azollae*, which was found and identified in the Pantanal during the investigation (Section 11.1.3). The relationship of this microorganism and one of its hosts, *Azolla mexicana*, was investigated by Holst and Yopp (1979).

The results of the chemical analyses demonstrate that nitrogen is often a limiting substance for primary production. This was generally determined by comparing the results of analyses for free phosphate with those of the nitrogen compounds. At different times at the individual locations, free phosphate was available in considerable concentrations, while there was no trace of ammonium or nitrate in the water. Similar results were obtained by Vassallo and Kieffer (1984) during studies conducted in the middle section of the Rio Paraná. They found that nitrogen compounds limited plant growth except during occasional periods when the NO_3 concentration exceeded that of NH_4, which, during their study, never reached the high values in the river that it did in the isolated water bodies in the Pantanal of Poconé during the dry season. The role of the nitrogen cycle in the ecosystem dynamics will be discussed further in Section 11.1.3)

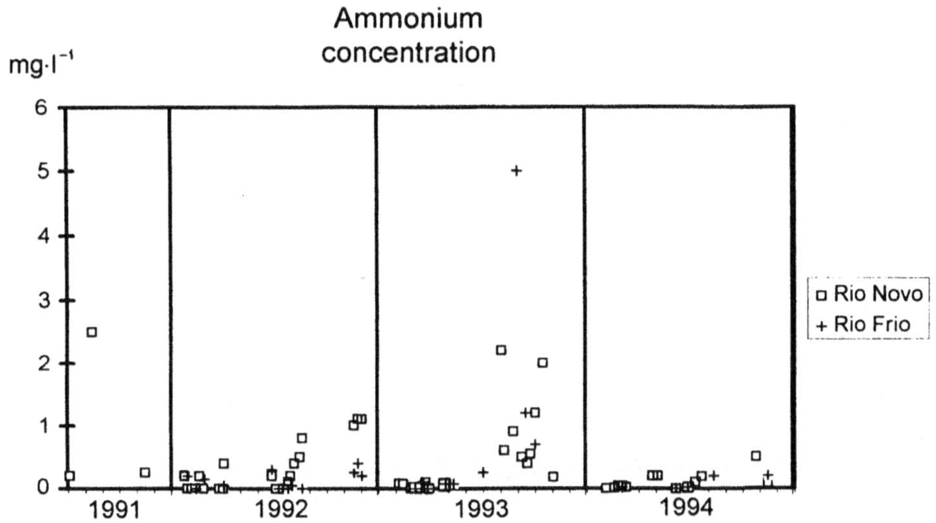

Figure 6.32: The ammonium concentration determined in water samples from the upper 6 cm of the Rios Novo and Frio at the Transpantaneira Highway from July 1991 through 1994. The peak concentrations in 1993 were associated with an algal bloom.

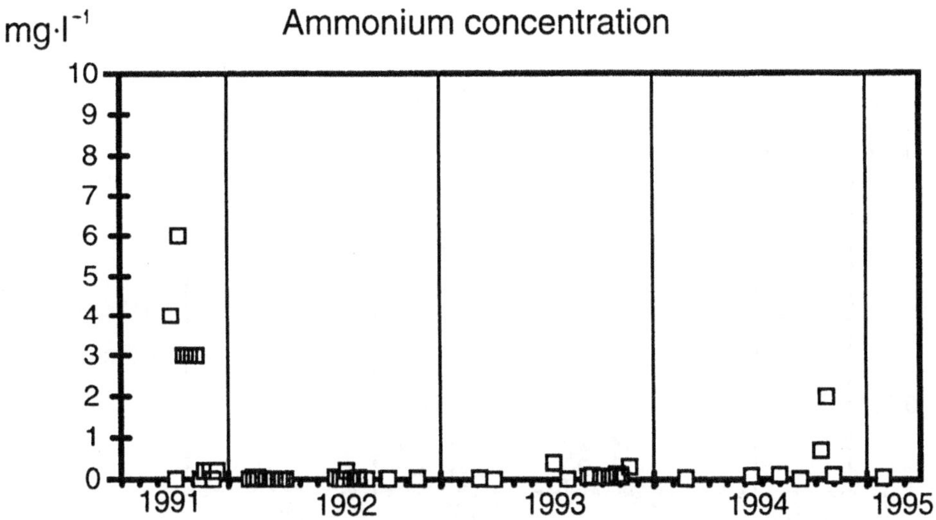

Figure 6.33: The ammonium concentration determined in water samples taken from the upper 6 cm of a low-lying drainage channel called Birici on the Fazenda Ypiranga about 15 km south of Poconé from July 1991 through January 1995. The peak values recorded in 1991 were associated with an algal bloom (Heckman *et al.*, 1993).

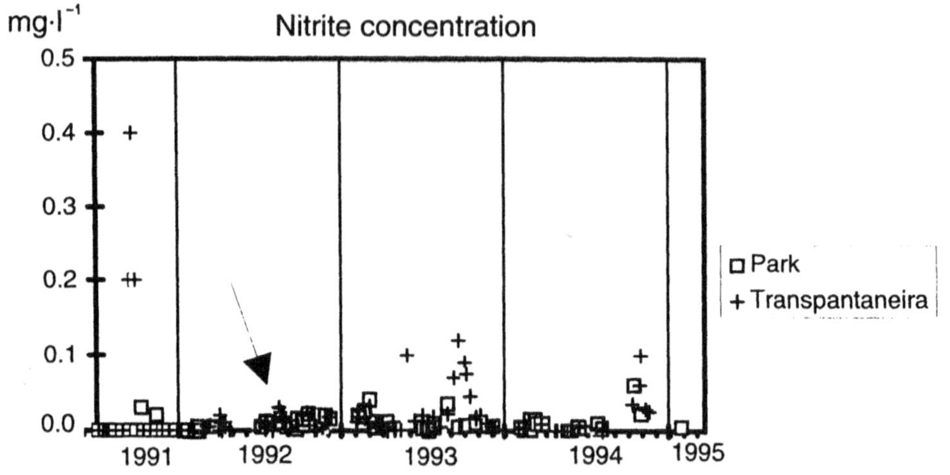

Figure 6.34: The nitrite concentration determined in samples from the upper 6 cm in the Rio Bento Gomes at the Poconé exposition park and at the Transpantaneira Highway from July 1991 through January 1995. The arrow marks the date on which a more sensitive analysis was introduced, which provided a better resolution and a detection limit of 0.005 $mg \cdot l^{-1}$ compared to the 0.1 $mg \cdot l^{-1}$ of the analysis used prior to that date.

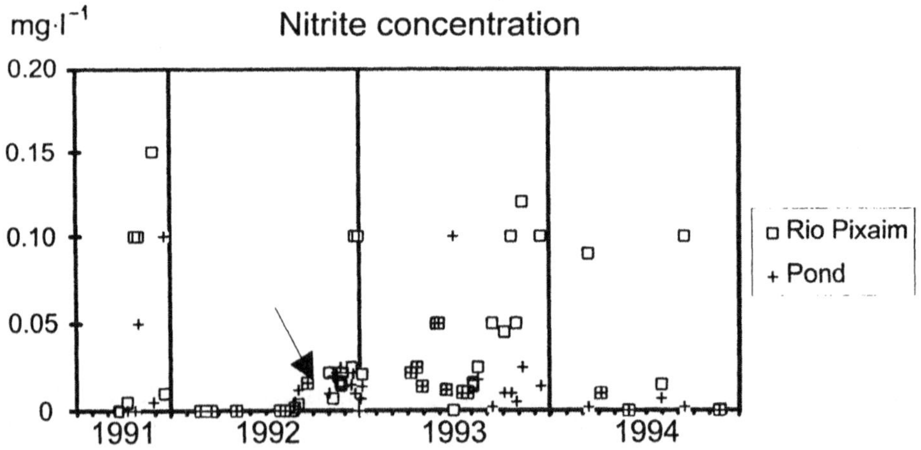

Figure 6.35: The nitrite concentration determined in water samples from the upper 6 cm of the Rio Pixaim and the adjacent pond beside the Transpantaneira Highway from July 1991 through 1994. The arrow has the same significance as that in Figure 34.

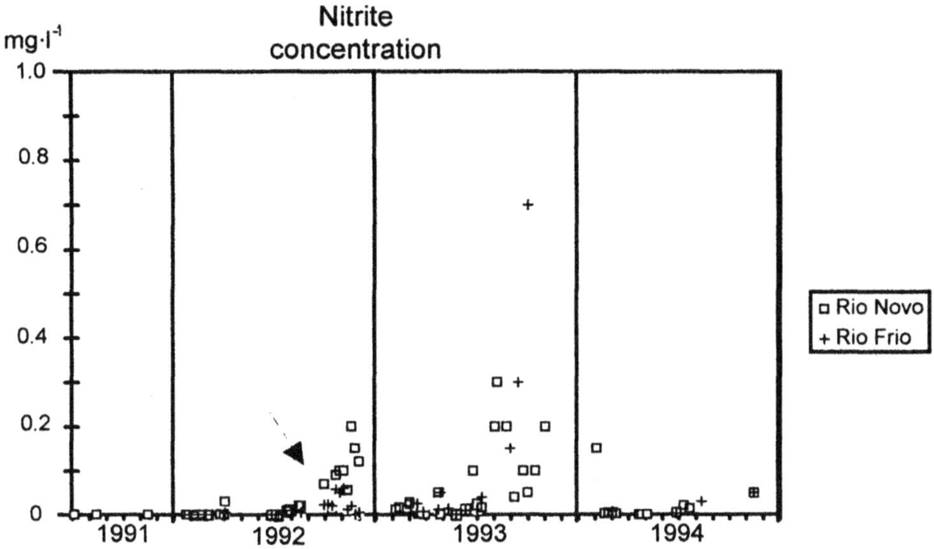

Figure 6.36: The nitrite concentration determined in water samples from the Rios Novo and Frio at the Transpantaneira Highway from July 1991 through 1994. The arrow has the same significance as that in Figure 34.

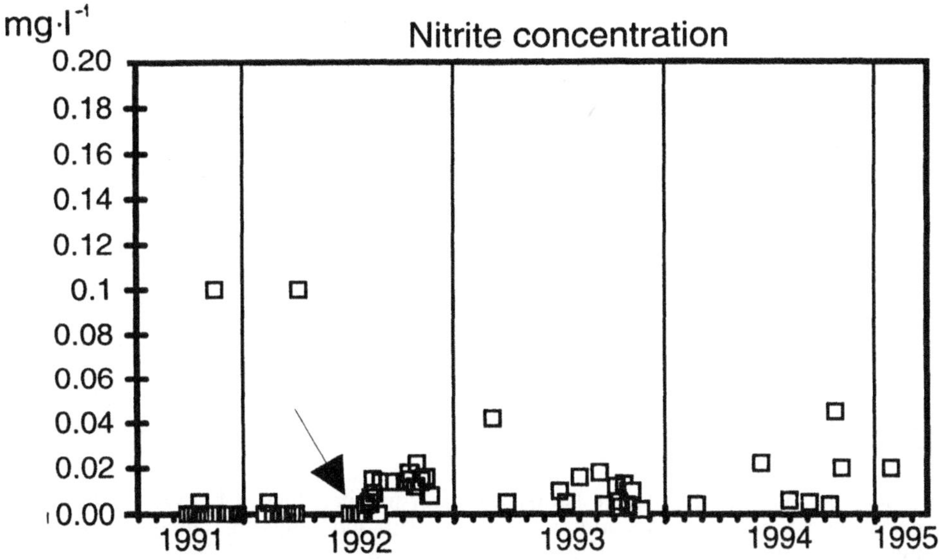

Figure 6.37: The nitrite concentration determined in water samples taken from the upper 6 cm of a low-lying drainage channel called Birici on the Fazenda Ypiranga about 15 km south of Poconé from July 1991 through January 1995. The arrow has the same significance as that in Figure 34.

6.2.9. Compounds of phosphorus

As in the case of the processes affecting the availability of nitrogen, those involved in the phosphorus cycle are critical for the biotic communities, because substances containing nitrogen and phosphorus form two groups of inorganic nutrients, the scarcities of which most frequently seem to limit primary production. The regional growth and survival of photosynthetic organisms depends upon the availability of phosphates or other phosphorus compounds, which are necessary for various universal biological processes, including energy storage and utilization within the cells. Unlike the case of the nitrogen compounds, however, there is an absolute limit on the supply of substances containing phosphorus since they cannot be obtained in any way other than tapping the reserves in the substrate, water, or living organisms. When leaching or erosional processes remove phosphates from the system, the potential absolute amount of plant and animal growth diminishes and cannot be restored without transporting more of such substances as solids or liquids into the system.

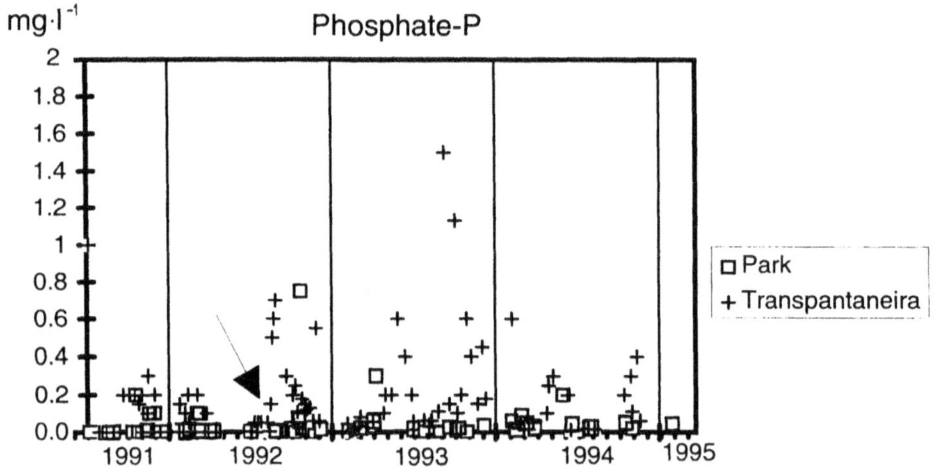

Figure 6.38: The concentration of phosphate-P determined in water samples from the upper 6 cm of the Rio Bento Gomes at the Poconé exposition park and at the Transpantaneira Highway from July 1991 through January 1995. The arrow marks the date on which a more sensitive analysis was introduced, which provided a better resolution and a detection limit of 0.005 $mg{\cdot}l^{-1}$ compared to the 0.1 $mg{\cdot}l^{-1}$ of the analysis used prior to that date.

The phosphorus cycle in the Pantanal closely parallels that of nitrogen. During the *cheia*, the amount of phosphorus in the living plant and animal tissues is maximal, and the amount in the free water is minimal. There is a recycling of the phosphorus twice each year as the aquatic community loses it to the short-lived terrestrial plants and these, in turn, release it again at the time of the next annual flood. The algae are able to utilize large amounts of phosphates only for a short period during the dry season (Section 11.1.3).

This seasonal availability of free inorganic phosphorus compounds is reflected by the analytical results depicted in Figures 6.38 through 6.41. The free phosphates present in the water during the *cheia* are usually near or below the detection limits of the tests used, while the maximum values are usually encountered during the *seca* (Fig. 6.38) Some unusually high phosphate concentrations may be encountered for short periods of time together with high ammonium concentrations during this season (Heckman *et al.*, 1993). The *enchente* and *vazante* are transition periods. However, large, irregular fluctuations in the phosphate concentration may be encountered at almost any time of year. For example, high concentrations in the Rio Pixaim and the adjacent pond were encountered during the *cheia* of 1994 (Fig. 6.39). A concentration nearly as high was determined in the Rio Frio during the *cheia* of 1992 (Fig. 6.40). Little more than a trace of ammonium had been present in the water at these sites during the periods the phosphate concentration rose. It is therefore seems likely that the growth of the dense emergent vegetation covering these two water bodies at the time had been retarded by a severe lack of nitrogen compounds. At the time, no nitrogen-fixing organisms had been observed. Therefore, the free phosphorus compounds were free to accumulate in the water without being absorbed by the plants, which were already adequately supplied with phosphorus but were severely lacking nitrogen sources. In contrast, large increases in the phosphate concentration during the dry season are usually accompanied by similar increases in the amounts of available ammonium, as at Birici stream in 1991 (Fig. 6.41), when there was a general release of inorganic nutrients from the remains of dead water plants and fishes in the drying pools of water on the flood plain.

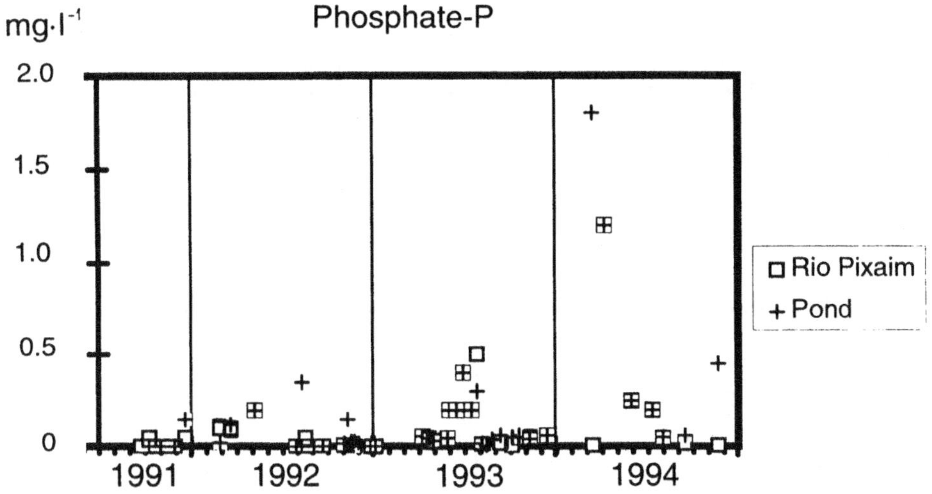

Figure 6.39: The concentration of phosphate-P determined in water samples from the upper 6 cm of the Rio Pixaim and the adjacent pond beside the Transpantaneira Highway from July 1991 through 1994.

Phosphorus inputs from ground water and erosion of the underlying rock layers are apparently at an absolute minimum in comparison with nearly every other region of the world (Heckman 1995). Some phosphates may well enter the system with dust washed down by the heavy rainstorms. This input may have been increased in recent times by the growing practice of burning large areas of woodland and shrub savanna to obtain open pasture. Relatively small amounts may enter the system through the upstream migration of large fishes via the rivers into the Pantanal and the arrival of birds from other regions, which add phosphate-rich excrements to the surface waters. Other sources are not apparent.

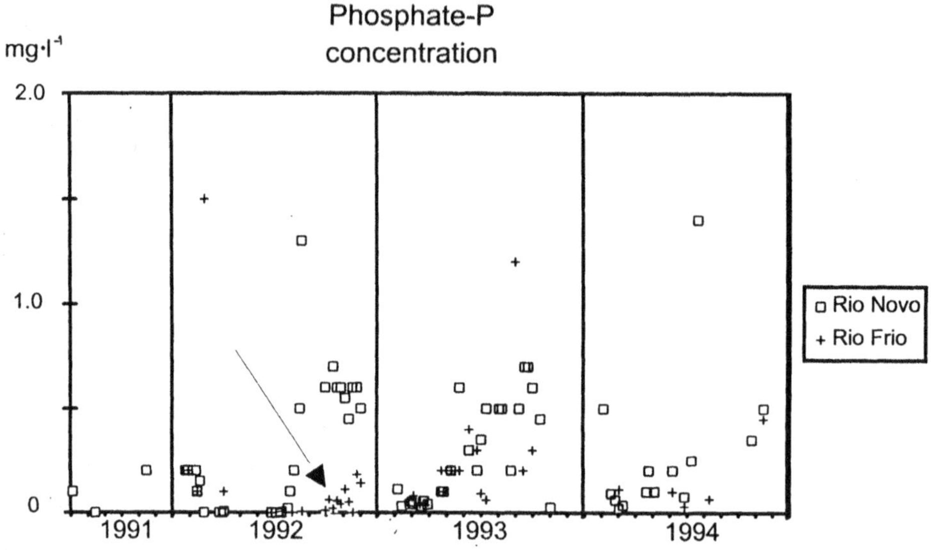

Figure 6.40: The concentration of phosphate-P determined in water samples from the upper 6 cm of the Rios Novo and Frio at the Transpantaneira Highway from July 1991 through 1994. The significance of the arrow was explained in Figure 6.38.

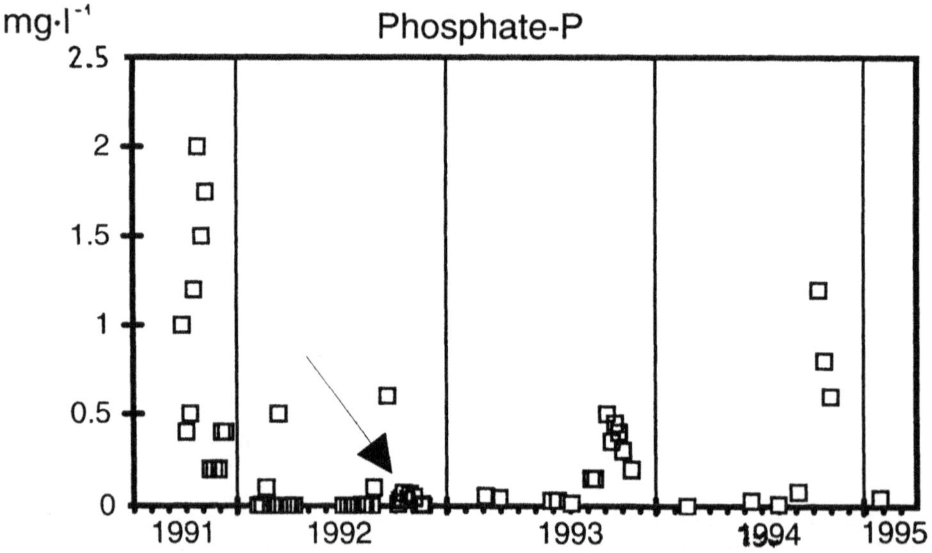

Figure 6.41: The concentration determined of phosphate-P in water samples taken from the upper 6 cm of a low-lying drainage channel called Birici on the Fazenda Ypiranga about 15 km south of Poconé from July 1991 through January 1995. The peak values recorded in 1991 were associated with an algal bloom (Heckman *et al.*, 1993). The significance of the arrow was explained in Figure 6.38.

Losses of phosphates from the Pantanal can occur through the downstream flow of the water in the rivers, which carry phosphorus either in the form of free phosphates or with living organisms. It is postulated that enormous amounts of this vital nutrient would have been lost to the Pantanal over the past thousands of years were it not for the conservative action of the macrophytes, which quickly bind the free material and anchor it in place until the passage of water through the Pantanal has reached its annual minimum (Section 11.1.5). The process of phosphorus loss and impoverishment of the Pantanal biota can be greatly accelerated through destruction of the vegetation by burning, which not only eliminates much phosphate-rich particulate matter through the air but also promotes leaching until the macrophytes are able to recover sufficiently to bind the phosphorus anew.

As mentioned in the discussion of nitrogen compounds, the analytical results indicate that phosphorus is the limiting inorganic nutrient for plant growth at times, while nitrogen limits it at others. Even at a single site, a phosphate limitation may occur during one season and a nitrogen limitation during another (Section 11.1.3).

6.2.10. Silicates

The only substances found in relatively large amounts in the upland sections of streams flowing into the Pantanal were compounds of silicon (Table 6.1). Because they had no notable effect on the conductivity of the water, they were assumed to be colloidal forms of silica or silicates. From the apparent geological history of the region (Section 3), it can be assumed that the processes of leaching and erosion have continued for an extremely long period of time, eliminating nearly all of the soluble substances and leaving mainly silicates, which have been eroded into fine silt and redeposited in sedimentary layers long enough to form sandstones. The sandstone layers are evident on the alto plano, where recent human activity has stripped away the vegetation, permitting the erosion of deep canyons, called *voça rochas*. Sand and silt deposits are also evident in the gold mines at the edge of the Pantanal.

The reasons the silica is found in the water as well as in the sediment were sought by testing the water for fluoride, which can dissolve silicates in an acid medium. The results, discussed above with the halides, revealed no particularly large amounts of fluoride in the Pantanal itself or in its watershed, although small but detectable amounts were encountered occasionally, especially in waters with elevated chloride concentrations due to the rapid drying of shallow pools on the flood plain during the *seca*. The springs on the *alto plano* to the northeast and east of the Pantanal, which are extremely poor in nearly all dissolved substances (Heckman, 1995), have detectable amounts of silicon compounds, but the concentrations are not nearly as high as in the Pantanal itself. The most probable forms in which the silicates are present in the water are colloids, which do not contribute to the electrical conductivity of the water. Even when the conductivity was extremely low, high concentrations of silicon compounds were detected (Table 6.4).

Although silicates might be expected to promote the development of large populations of diatoms, these algae were not as abundant as the green algae in the Pantanal during most of the year. However, some of the streams at higher elevations in the watershed contained enormous numbers of diatoms, as determined by examination of glass slides exposed for short periods of time on the sediment. Possible disadvantages faced by diatoms in the lowland water bodies of the Pantanal include very high seasonal water temperatures and a lack of rocks and coarse substrate particles on which to attach. These probably offset the favorable conditions produced by an excellent supply of silicates. This will be discussed in Sections 7 and 10.

Table 6.4: Results of the analyses for silicon compounds in surface waters of the Pantanal. The values are shown in mg·l^{-1}. The analyses were conducted only during 1994.

Site	(n)	Concentration of silicates (mg·l^{-1}) minimum	maximum
Enchente			
Rio Bento Gomes	4	1.6	9.5
Birici Stream	1	>10	>10
Rio Frio	1	4	4
Rio Novo	1	3.2	3.2
Cheia			
Rio Bento Gomes	12	4	8
Rio Novo	4	4	10
Rio Pixaim	1	4	4
Pond beside Rio Pixaim	1	3	3
Rio Paraguay	2	5	6
Vazante			
Rio Bento Gomes	9	4.5	8.5
Birici Stream	1	8	8
Rio Novo	3	5	6.5
Rio Frio	1	4.5	4.5
Rio Cuiabá	1	5	5
Rio Pixaim	1	3.5	3.5
Pond beside Rio Pixaim	1	3.5	3.5
Rio Paraguay	2	5.5	5.5
Seca			
Rio Bento Gomes	7	1.5	8
Birici Stream	3	6.5	>10
Rio Novo	1	2	2
Rio Frio	1	3.5	3.5
Rio Pixaim	2	3.0	3.2
Pond beside Rio Pixaim	2	3	6
Rio Paraguay	1	4	4
Rio Miranda	1	>10	>10

Compounds containing silicon were generally detected in large amounts when the content of inorganic ions corresponded roughly to those in rainwater. They were present in far greater concentrations than the maximum of 4.3 mg Si·l^{-1} reported for the Amazon region by Furch (1984b) and often exceeded the maximum value for the analytical methods used: 10 mg Si·l^{-1}. As mentioned above, the values recorded for the headwaters of the rivers on the plateau to the east of Cuiabá were only low to moderate, as were those determined in the pools of water with high concentrations of inorganic ions that were present during the dry season. Furch (1984b) also found lower concentrations in water poor in electrolytes. Nevertheless, a regression of the silicate values against the conductivity, however, revealed no sign of any statistical relationship between them, regardless of whether the data for streams and wells in the catchment area were included or whether only the values for the surface waters in the Pantanal were considered.

6.2.11. Compounds of sulfur

The sulfate test conducted in the field proved inadequate (Heckman, 1994a), not only because the test was not sensitive enough but also because the concentrations detected invariably proved much too high in comparison with the results of other kinds of test and from ion balance calculations. Through analyses carried out in the laboratory, it is known that sulfates are present in the water of the Pantanal in concentrations up to nearly 25 $mg \cdot l^{-1}$. In the headwaters of streams flowing into the Pantanal, sulfate concentrations of no more than about 2 $mg \cdot l^{-1}$ were detected. High sulfate concentrations in surface waters are often assumed to be typical of regions affected by industrial air pollution (Gambell and Fischer, 1964). However, concentrations in excess of 150 $mg \cdot l^{-1}$ were reported from southern Malaysia (Johnson, 1967), and even in a small mountain river in Selangor, West Malaysia, flowing into the Sungai Gombok, Bishop (1973) recorded an unusually high concentration amounting to a questionable 155 ppm, although the great majority of the other samples he analyzed contained no more than 1 ppm, and many were below the limit of detection for his analysis. There are thus some differences to expect when comparing the role of sulfate in tropical water bodies with those in waters of the temperate zones. However, in the Pantanal, only moderate concentrations seem to be encountered, although the results of the analyses conducted during the first years of the project proved to be inaccurate, leaving relatively few reliable data to judge from.

The possible use of sulfate as an oxygen source for aquatic organisms, particularly bacteria, has not been adequately investigated in tropical waters. The ensuing reduction of the sulfate might be expected to produce elemental sulfur or hydrogen sulfide. The scarcity of H_2S in water samples from the Pantanal, as discussed below, suggests that sulfate reduction by microorganisms does not occur to any great extent, however.

Hydrogen sulfide is apparently a substance that almost never accumulates under normal conditions in the Pantanal. All analyses for the presence this substance proved negative, except for a few conducted shortly after a mass death of thousands of fishes in the Rio Bento Gomes near the Transpantaneira Highway during the dry season in 1992. Even under these circumstances, where the water beneath a solid cover of dead fishes remained anaerobic for at least two weeks, the concentration of hydrogen sulfide remained low, barely reaching a value of 0.05 $mg \cdot l^{-1}$ at that time.

Analyses of foul-smelling mud from beneath old stands of emergent plants along the shores of the Baia Piuval also proved negative for hydrogen sulfide. However, a gas was produced, which most probably was methane, perhaps mixed with nitrogen released by denitrification processes.

From the results available, it can be postulated that the primary components of the sulfur cycle in the Pantanal are organic substances containing this element and free sulfate. An oxidation of any hydrogen sulfide produced to sulfite or sulfate at a rate far too rapid to allow any to accumulate can be assumed.

6.2.12. Organic substances

As a measure of the oxidizable organic substances in the water, the chemical oxygen demand (COD) was determined. The values obtained were seldom high, but during the dry season or *seca*, occasional values as high as 111 mg $O_2 \cdot l^{-1}$ were recorded in shallow pools of water during blooms of algae. During the seasons of initial flooding, the *enchente*, and the high water period, the *cheia*, high values were associated with inputs of decaying organic material from the terrestrial communities, which were released after inundation or washed into the water bodies with rainwater. Only during the period of decreasing water levels and flow through the water courses did the COD remain relatively low.

Table 6.5: COD determined on samples of surface water from the Pantanal. The values are shown in mg $O_2 \cdot l^{-1}$. The analyses were conducted only during 1994.

Site	(n)	COD minimum	maximum
Enchente			
Rio Bento Gomes	4	35	77
Birici Stream	1	82	82
Rio Frio	1	46	46
Rio Novo	1	84	84
Cheia			
Rio Bento Gomes	10	25	77
Rio Novo	3	30	96
Rio Pixaim	1	38	38
Pond beside Rio Pixaim	1	63	63
Rio Paraguay	2	36	37
Vazante			
Rio Bento Gomes	9	18	79
Birici Stream	1	31	31
Rio Novo	3	40	51
Rio Frio	1	32	32
Rio Cuiabá	1	31	31
Rio Pixaim	1	33	33
Pond beside Rio Pixaim	1	39	39
Rio Paraguay	2	<15	<15
Seca			
Rio Bento Gomes	7	32	99
Birici Stream	3	36	111
Rio Novo	1	89	89
Rio Frio	1	36	36
Rio Pixaim	2	36	41
Pond beside Rio Pixaim	2	33	40
Rio Paraguay	1	32	32
Rio Miranda	1	23	23

The COD in the water of streams in the upland parts of the watershed was generally low, often below the minimum reliable value for the test of 15 mg $O_2 \cdot l^{-1}$ (Table 6.1). In the Pantanal itself, however, the chemical oxygen demand was always above the minimal detectable limit, but it was often low, especially during the *vazante* and early part of the *seca* (Table 6.5).

The amounts of free organic substances in the water can change rapidly because of the relatively high water temperatures. This can be attested to by simple observation of the water bodies on a day to day basis. Detritus decomposes very rapidly, and even when relatively large amounts are produced during the periods that the water level is rapidly falling, dead plant material simply does not accumulate. Even after mass deaths of fishes, the carcasses do not take more than a few days to decompose. The extremely serious mass mortality in the Rio Bento Gomes in 1992 did cause the water to become nearly oxygen-free for over a week and produced a powerful stench, but the dead fishes had all disintegrated within about a week, and the foul mixture of organic substrate and the bacteria consuming it remained for only about a week longer. In cooler waters of the temperate zones, such a volume of dead fishes would have caused major pollution for a much longer period of time.

The organic material contributing to the values determined for the chemical oxygen demand includes those in the microorganisms in the water samples, tiny detritus particles, and colloid materials, as well as the humic material that gives the water its deep brown coloration during the *enchente*.

Somewhat comparable data from other parts of the Rio Paraná System were provided by Marglianesi (1973), who determined the permanganate oxidizability of the substances in the river water. These values ranged from 2.59 mg $O_2 \cdot l^{-1}$ in the upper Rio Paraná System to 9.24 mg $O_2 \cdot l^{-1}$ in the lower Rio Paraguay. Differences between the procedures and sensitivity of the tests can account for the much lower values detected downstream from the Pantanal.

6.3. Ion balances

To check the accuracy of the results and to provide some information on abundant ions that could not be analyzed in the field, supplemental analyses of water samples were completed in Germany in order to calculate ion balances (Figs. 6.6 to 6.9). From the results shown in Section 6.2., it is obvious that the isolated data on the alkali metal ions are not representative because considerable seasonal changes can be expected. However, these data do indicate that no surprises should be expected in the concentrations of the ions that generally account for the bulk of the electrical conductivity in the water bodies investigated in the northern Pantanal.

Table 6.6: Ion balance in the Rio Bento Gomes near the Transpantaneira Highway at 10:40 a.m. on July 10, 1994. The sodium, potassium, calcium, and magnesium concentrations were determined by AAS about one week later; the other values were determined in the field as described in Section 6.1.

Cations	mval·l⁻	Anions	$mval \cdot l^{-1}$
Na^{1+}	0.191	NO_3^{1-}	0.001
K^{1+}	0.080	S^{2-}	0.000
Ca^{2+}	0.184	SO_4^{2-}	0.000
Mg^{2+}	0.201	Cl^{1-}	0.113
Fe^{2+}	0.005	PO_4^{3-}	0.000
Mn^{2+}	0.000	BO_3^{1+}	0.000
NH_4^{1+}	0.000	Acidic capacity	0.400
Total	0.661		0.514

Table 6.7: Ion balance in the Rio Novo near the Transpantaneira Highway at 12:20 p.m. on July 10, 1994. The analyses were completed as noted in Section 6.1.

Cations	$mval \cdot l^{-1}$	Anions	$mval \cdot l^{-1}$
Na^{1+}	0.127	NO_3^{1-}	0.000
K^{1+}	0.093	S^{2-}	0.000
Ca^{2+}	0.271	SO_4^{2-}	0.000
Mg^{2+}	0.222	Cl^{1-}	0.056
Fe^{2+}	0.072	PO_4^{3-}	0.008
Mn^{2+}	0.000	BO_3^{1+}	0.000
NH_4^{1+}	0.004	Acidic capacity	0.600
Total	0.789		0.664

Table 6.8: Ion balance in the Rio Pixaim near the Transpantaneira Highway at 12:40 p.m. on June 27, 1994. The analyses were completed as noted in Section 6.1.

Cations	$mval \cdot l^{-1}$	Anions	$mval \cdot l^{-1}$
Na^{1+}	0.161	NO_3^{1-}	0.000
K^{1+}	0.060	S^{2-}	0.000
Ca^{2+}	0.279	SO_4^{2-}	0.000
Mg^{2+}	0.235	Cl^{1-}	0.028
Fe^{2+}	0.029	PO_4^{3-}	0.001
Mn^{2+}	0.000	BO_3^{1+}	0.000
NH_4^{1+}	0.002	Acidic capacity	0.600
Total	0.765		0.630

Table 6.6 shows the situation in the Rio Bento Gomes at the Transpantaneira Highway on July 10, 1994. The sample was collected at 10:40 a.m. on a clear and sunny day. The ratio of cations to anions equals 1.287 without organic ions being considered. Unfortunately, similar balances during the hours of darkness could not be calculated. It is likely, however, that the ratio would decrease considerably during the night when photosynthesis ceases but decomposition processes continue. During the late *vazante* and early *seca*, the release of inorganic nutrients through decomposition processes and photosynthesis intensify considerably, which causes the differences between the daytime and nighttime values of various chemical substances to become more pronounced (Figs. 4.6 & 4.7).

Similar ionic ratios in favor of the cations were calculated for the Rio Novo at the Transpantaneira Highway on the same day and the Rio Pixaim on June 27, 1994, as shown in Tables 6.7 and 6.8, respectively. However, in the shallow pond beside the Rio Pixaim, which was covered by floating and emergent rooted macrophytes, the cation:anion ratio was only 0.806 (Table 6.9). The minor imbalances can be compensated for by changes in organic compounds and by the well-known equilibrium shifts involving the relative amounts of carbonates, bicarbonates, and CO_2 or various other dissolved substances.

Table 6.9: Ion balance in the shallow pond on the north bank of the Rio Pixaim near the Transpantaneira Highway at 12:40 p.m. on June 27, 1994. The analyses were completed as noted in Section 6.1.

Cations	$mval \cdot l^{-1}$	Anions	$l \cdot l^{-1}$
Na^{1+}	0.110	NO_3^{1-}	0.000
K^{1+}	0.059	S^{2-}	0.000
Ca^{2+}	0.244	SO_4^{2-}	0.000
Mg^{2+}	0.235	Cl^{1-}	0.056
Fe^{2+}	0.043	PO_4^{3-}	0.005
Mn^{2+}	0.000	BO_3^{1+}	0.000
NH_4^{1+}	0.004	Acidic capacity	0.800
Total	0.695		0.861

6.4. Other chemical substances

Although the amount of chlorophyll a in the water can be an important factor in some aquatic systems (Nusch and Palme, 1975), analyses of this substance were avoided during this study for several reasons. The most important of these was the great volume of the aquatic plant leaves that extended above the surface of the water, thereby being capable of photosynthesis which is not detected by such analyses of the water. Moreover, during the dry season, the algae that were briefly abundant were often distributed more on the sediment and at the surface than in the water column, making a quantitative determination of the amount present in the water column impossible without a sampling program on a massive scale. The results of analyses by other authors seem to confirm this, with values from 0.6 to 2297 $mg \cdot l^{-1}$ being recorded from *baias* and *salinas* in the Pantanal of Nhecolândia (Mourão, unpublished, cited by Por, 1995), 0.74 to 58 $\mu g \cdot l^{-1}$ in the Rio Cuiabá (da Silva and Pinto-Silva, 1989), and 1.82 to 8.66 $mg \cdot l^{-1}$ in small lakes near Porto Jofre (Saijo *et al.*, 1987). Because these studies did not include intensive sampling programs during all seasons, it can be assumed that the ranges of the chlorophyll a concentration in the water is actually much greater than those reported.

Oils and greases were never observed in the waters of the Pantanal, and a contamination by these substances was not suspected. However, monitoring pollution of the water with such organic compounds should be undertaken in the Rio Cuiabá downstream from the cities of Cuiabá and Varzea Grande. In Brazil, urban pollution of the water courses has been rated as a problem of primary importance (Homewood, 1991), and while the Pantanal would be capable of absorbing considerable amounts of ordinary domestic wastewater, the industrial discharges and household chemicals that untreated wastewater invariably contains can create major hazards for man and wildlife in the Pantanal. The construction of sewage treatment plants and careful policing of factories are essential to protect the freshwater supply. Unfortunately, such measures are only being adopted with great delay, even in the more populous eastern part of Brazil, and it is considered likely that great damage will first have to become apparent before remedial action is taken.

Similarly, monitoring to detect various kinds of pesticide is urgently needed because there is still a lack of a competent authority to control the sale and application of these substances. This is a result of the general ignorance of the effects some of these chemicals can have on man and on the biotic environment (Heckman, 1993). Considerable lack of understanding was manifested during the last two years of the investigation by fogging programs in Mato Grosso using DDT. The fogging was initiated in an attempt to control the mosquito vectors of yellow and dengue fevers, diseases which had recently broken out for the first time in the State of Mato Grosso. Unfortunately, DDT has already become useless against the common mosquito species associated with human populations, including the vectors of these diseases. All such mosquito species can be assumed to have acquired resistance to this and other commonly used insecticides (Heckman, 1993), but this very persistent substance and its breakdown products have frequently caused severe environmental damage by endangering raptor populations and eliminating other important predators. Because the application of this chlorinated hydrocarbon was just being initiated as the investigation was in its concluding phases, nothing could be done to determine the entry of these chemicals into the Pantanal. It is assumed, however, that during subsequent years, there will be a slow but measurable increase of amount of DDT in the water courses as this substance is leached from the urban locations to which it was applied and drifts downstream into the Pantanal.

Other potentially dangerous substances that are coming into increased use include antibiotics and hormones, which have become commonplace in intensive cattle production. The effects of these substances are only beginning to receive attention, although they have become locally important environmental pollutants in water bodies receiving wastewater from various kinds of intensive animal culture. While the antibiotics can have extremely harmful effects on the microbiota and may interfere with the normal breakdown of organic detritus, the hormones seem to be having very hard to determine effects on vertebrates, including sex changes and anomalies in the reproductive behavior. Although the ranchers in the Pantanal have still done little to intensify their cattle production, the danger exists that the use of antibiotics and hormones will become more widespread as their availability increases.

7. Flora

7.1. Prokaryota

7.1.1. Cyanobacteria

Cyanobacteria are not among the dominant alga-like organisms in the Pantanal, except in isolated water bodies during the dry season, in which blooms of *Microcystis aeruginosa* (Kützing) Kützing sometimes develop in nutrient-rich waters of warm, shallow pools after large numbers of trapped fishes have died. Species of *Oscillatoria* may also develop large dark clumps, usually in partially shaded places. Otherwise, the species are represented chiefly by relatively isolated colonies among large numbers of eukaryotic algae.

Cyanobacteria in the Pantanal have been largely neglected by researchers. Drouet (1938) provided a list of species in collections from various parts of Brazil, including three species from Cáceres and two others from undisclosed parts of Mato Grosso, which at the time encompassed the present states of Mato Grosso do Sul and Rondônia, as well. The most recent list of species from the Pantanal was provided by De-Lamonica-Freire and Heckman (1996).

Chroococcaceae

The species in this family among the flora of the Pantanal include *Agmenellum quadruplicatum* (Meneghini) Brébisson, also called *Merismopedia glauca* (Ehrenberg) Nägler, which was typically found among the dense aggregations of green algae present in drying pools during the late part of the *seca* (Heckman *et al.*, 1993). The species with which it was associated, mainly species of Chlorococcales, including many species of *Scenedesmus*, were those with which it is often associated in phytoplankton of temperate zone water bodies, such as the Elbe River (Heckman, 1986, 1990; Section 9.1.1.). It shows no tendency to form blooms and is generally only a minor component of large, diverse communities.

As mentioned above, another species in this family, *Microcystis aeruginosa*, is seasonally abundant in the Pantanal, occasionally appearing in enormous numbers in small water bodies, nearly always as one of a few or nearly the only species of photosynthetic organism. *Microcystis flos-aquae* (Wittrock) Kirchner also seems to be present in such water bodies at times, tending to suggest that these species may actually be conspecific developmental stages, as long believed by many taxonomists before the two were recently recognized again as distinct (Komárková-Legnerová & Cronberg, 1994), based on the allozyme studies of Kato *et al.* (1991). Sometimes these species are associated with *Euglena sanguinea*, which also forms blooms but congregates on the surface tension layer of the water. In such cases, the water has a deep red film on the top, which, when broken, reveals bright green water beneath (Plate 2a). At such times, the water supports few organisms of other species. *Microcystis aeruginosa* is a wide-spread planktonic species, known in Brazil from the Rio Negro System in Amazonas (Uherkovich, 1976) to Rio Grande do Sul (Torgan *et al.*, 1995). It seems to be most abundant in relatively new impoundments and highly polluted lakes, and it often appears on the lists of phytoplankton species from such water bodies (Giani & Leonardo, 1988; Altafin *et al.*, 1995).

Oscillatoriaceae

This is the best represented Cyanobacteria family in the Pantanal. At times, some of the species are abundant locally, but generally their filaments of cells are observed among large masses of green algae, which are often the dominant primary producers among the microorganisms.

One of the species, *Oscillatoria agardhii* Gomont, is typically planktonic and is usually found together with the aggregation rich in Chlorococcales that develops in the drying pools during the late dry season and persists into the early rainy season. *O. agardhii* becomes abundant only after the rains have begun to increase the water depth in the pools and reduced the extreme diurnal fluctuations in water temperature (Heckman *et al.*, 1993). Like the green algae with which it is associated, its presence in the thick green alga mass in the shallow pools does not correspond to its typical planktonic form of life in the temperate zone.

The other four species of *Oscillatoria* have generally the same habits. They develop among other algae, including filamentous green species, for much of the year. When more nutrients than usual are available and the water is shallow, they form benthic carpets on the sediment, which coalesce through the active movement of the trichomes into dark clumps. During the day, these trap bubbles of released gases and float to the surface of the water, often carrying considerable amounts of sediment with them.

The largest of these species, and one of the most widespread in the Pantanal, is *Oscillatoria princeps* Vaucher ex Gomont. At some locations, the dominant benthic species were *Oscillatoria lacustris* (Klebahn) Geitler and *Oscillatoria limosa* C. A. Agardh. *Oscillatoria tenuis* C. A. Agardh was not found in such great masses, but considerable numbers of trichomes were found among green algae and submerged macrophytes. Generally, however, these species were not nearly as abundant as filamentous eukaryotic algae. These three species were reported from other parts of Brazil by Drouet (1938), who also found *Oscillatoria formosa* Bory ex Gomont in samples from Cáceres.

In addition to the *Oscillatoria* species, *Crinalium magnum* Fritsch was encountered once in pools of shallow water left after the first heavy rains of the *enchente*.

An unidentified and apparently still undescribed species of *Hormoscilla* was encountered in shallow pools of water in a branch of the Rio Novo near the Transpantaneira Highway during the *vazante*. The genus was first proposed for a marine species, but later, another one was discovered in fresh water. In light of the known habitats of the species already described, the discovery of this genus in the Pantanal was surprising. Interestingly, the dry season salinity at the site where it was discovered is sometimes extremely high for waters in the Pantanal (Section 6.2.5.).

Phormidiaceae

Two species of *Spirulina* were encountered in limited numbers in some of the water samples taken during the seasons of little rainfall. *Spirulina jenneri* (Stizenberger) Geitler was moderately abundant in some of the water samples taken from shallow water bodies during the *seca*, but it was never a species that produced blooms. Isolated filaments of *Spirulina subsalsa* Oersted occurred rarely during the *vazante*.

A species encountered on moist sediment near pools of water during the early rainy season was identified as *Phormidium* sp. Apparently, the long periods of dryness greatly limit the development of cyanobacteria in the soil, except in shaded places near permanent water bodies.

Nostocaceae

Nostocaceae was represented by one species that occasionally produces massive populations at the beginning of the rainy season (Fig. 7.1). Large numbers of *Anabaena spiroides* Klebahn filaments were observed in the shallow water bodies that had just been reflooded after a period of near dryness. Apparently, the species does not develop well at unusually high temperatures because it did not appear among the massive blooms of green algae that dominated the water before the first heavy rains. However, it quickly established itself a short time after heavy rainstorms early in the *enchente* (Heckman *et al.*, 1993). The species persists only for a relatively short period, not longer than three or four weeks, and disappears after the water becomes relatively deep. The species is apparently one of the commonest of its family in Brazil (Drouet, 1938).

A rare species in the Pantanal, *Anabaena sphaerica* Bornet & Flahault, was encountered once during the *vazante*.

The symbiotic species, *Anabaena azollae* Strasburger, develops in modified fronds of the *Azolla* species in the Pantanal. This cyanobacterium, like various others in this family, is capable of fixing nitrogen efficiently, thereby enhancing the productivity of the water bodies where it occurs. A detailed study of this symbiosis in *Azolla mexicana* was conducted by Holst and Yopp (1979).

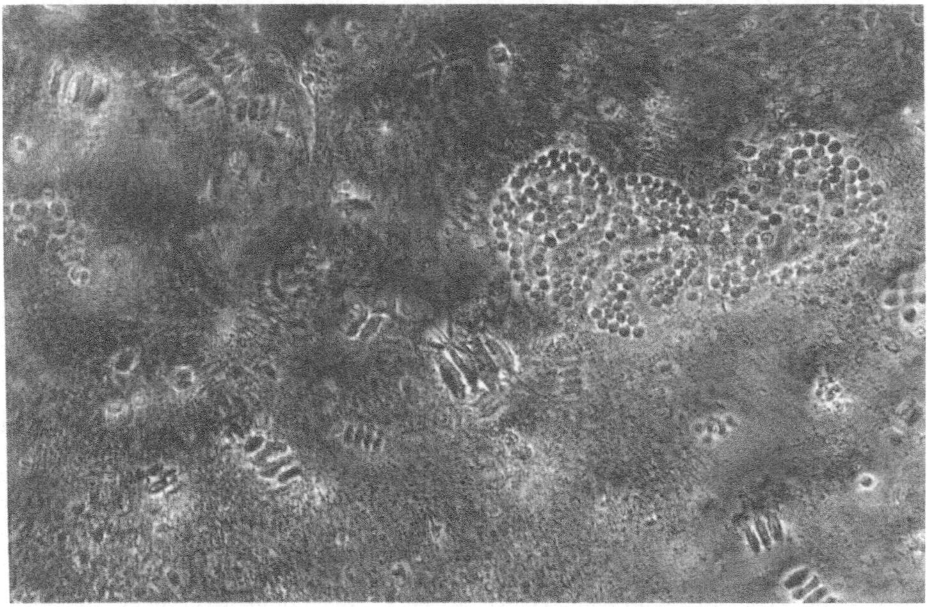

Figure 7.1: A bloom of *Anabaena spiroides* occurred at the beginning of the rainy season on the Fazenda Ypiranga in water already dominated by large numbers of chlorococcal green algae, including many species of *Scenedesmus*.

Pseudanabaenaceae

A more persistent but less numerous member of this family is *Pseudanabaena* cf. *constricta* (Szafer) Lauterborn, which was often observed in samples taken after the freshet from water bodies that did had not dried up completely during the dry season.

7.1.2. Schizomycetes

The free-living bacterial fauna of the Pantanal is rich and diverse. Although few bacteriological methods could be employed during the study, a number of species were identified according to their morphological peculiarities. An exhaustive survey would undoubtedly reveal most of the species present in freshwater bodies throughout the world. However, because of the local physical and chemical conditions prevailing during each of the seasons, the prerequisites for the development of all species are not met. Among the species conspicuous for their absence are the green and purple sulfur bacteria, for which the necessary hydrogen sulfide is apparently never present in sufficient quantities.

The abundant bacilli observed in water samples and assumed to be active in the decomposition of plant and animal material, including dead fishes, are assumed to belong to the cosmopolitan species encountered in freshwater bodies whenever suitable substrates are present. In addition, coliform bacteria are encountered in the rivers due to contamination with urban sewage. Recently, *Vibrio cholerae* Neisser has appeared in Mato Grosso for the first time, within only a few years of its arrival in South American from Asia.

The abundance of legumes in the Pantanal suggests that the nodule bacteria are of considerable importance for providing their hosts with nutrient material in the mineral-poor soil. The formation of the nodules containing the nitrogen-fixing bacteria was recently studied by Faria *et al.* (1989).

Unclassified species

According to Häusler (1982), there is still not enough information to group several of the common species into families, and these must be treated separately. Among these is the ubiquitous *Sphaerotilus natans* Kützing, which is a general indicator of domestic wastewater and other aquatic media which are heavily contaminated with organic material. This species was observed in the cities of southern Mato Grosso, where raw sewage empties into the rivers. The Rio Coxipó, for example, receives large amounts of domestic wastes from the houses near its banks, and this species is a characteristic organisms at the outlets of the sewage pipes. However, this bacteria was never observed in the Pantanal, where such concentrated wastewater is still not encountered. It could be recommended as an ideal indicator species for the presence of excessive sewage inputs at hotels and tourist centers in the wetland.

One of the bacteria with uncertain phylogenetic relationships is the filamentous iron bacteria, *Gallionella minor* Cholodnyi, which is widespread in the region of the Pantanal and occurs wherever there is an upwelling of iron-rich ground water. The reddish-yellow deposits of ferric hydroxide and ferric oxide marking such sites generally contain filaments of this species, which is apparently chemolithotrophic or in some cases chemoorganotrophic (Häusler, 1982). According to Perfil'ev and Gabe (1961) and Balousova (1968), by changing the conditions in the culture medium, bacteria with the characteristics of *Gallionella minor* can be produced from cultures of *Gallionella ferruginea* Ehrenberg. Thus, *G. minor* would seem to be a junior synonym of *G. ferruginea*, which has been reported to prefer water containing oxygen in concentrations less than 1 $mg \cdot l^{-1}$ and to be psychrophilic (Häusler, 1982). However, Rodina (1953) found the species at temperatures from 20.0°C to 22.7°C and at reduced abundance at 31.5°C. The iron-rich ground water that supports these iron bacteria in the region of the Pantanal generally contains little oxygen, often approximately 1 mg/l, and maintains constant temperatures close to 20°C. Hence, no ecological evidence can be provided that would support the premise that *G. minor* is a valid species.

Spirillaceae

The largest and one of the easiest to recognize aquatic bacteria under the microscope, *Spirillum volutans* Ehrenberg, was usually observed in aerobic water samples from the Pantanal which were particularly rich in decomposing material. These were most frequently collected during the *vazante*, and none were ever observed during the *cheia*.

This species reportedly prefers oxygen saturations from 1% to 9%, and its optimal temperature is 30°C (Häusler, 1982). Apparently, the relatively high concentrations of various salts inhibit the activity of this species during the *seca*. It seems to be inhibited by small amounts of sodium and by phosphate at a concentration of only 0.003 m (Cole & Rittenberg, 1971).

Methylomonadaceae

The sediment beneath the floating islands of sudd and along the shores of the Baia Piuval along the Rio Bento Gomes in the northern Pantanal produced some methane at times, allowing the presence of *Methylomonas* sp. to be presumed. The importance of methane production in the Pantanal of Mato Grosso do Sul has already been demonstrated by Hamilton *et al.* (1995).

Siderocapsaceae

Some of the most prominent species in the Pantanal include the iron bacterium, *Siderocapsa anulata* Kalbe, Keil, & Thiele. This species is particularly abundant in streams flowing into the Pantanal because these are frequently poor in organic nutrients but rich in ferrous ions that seep from subterranean deposits (Section 6.2.7.). As a result, there is a possibility for such microorganisms to gain energy from the oxidation of ferrous to ferric ions. The sites of upwelling waters rich in soluble iron are invariably marked by gelatinous masses of ferric hydroxide and ferric oxide, in which particularly large numbers of iron bacteria are encountered. In the headwaters of a stream to the northwest of Poconé, the Corrego Guanandi, this species was one of the only microorganisms to colonize glass slides exposed for several days to the flowing water. Usually, it was quite abundant and accompanied only by a few colonies of the Chrysophyte, *Anthophysis vegetans*.

Another species in this family, *Sideromonas* cf. *vulgaris* (Naumann) Beger, was found within yellowish deposits of insoluble ferric salts and in the surface film at sites where iron-rich ground water emerged and flowed into streams, including the Corrego Guanandi, to the northwest of Poconé. It is likely that the same species occurred in similar surface films observed in the Pantanal near Barão de Melgaço and Cáceres. Also in the iron-rich surface films and sediments were bacteria tentatively identified as *Ferribacterium* cf. *duplex* Brussoff, although its appearance did not fully fit the description of Häusler (1982). These bacteria are widespread in the Pantanal and are good indicators of the sites at which iron-rich ground water wells up.

7.2. Eukaryotic Algae

The density and species composition of the eukaryotic algal flora changes with great rapidity at times. Generally, the low concentrations of nutrients greatly limit the development of the algal populations, but at those times and locations at which nutrient substances are released, a great many algae appear in a short period of time. For example, when the decrease in the water level late in the dry season isolates many of the aquatic macrophytes in shallow pools of water, these plants begin to die and decompose, releasing amounts of nutrient elements sufficient to support large numbers of algae, Within a day, this can change the transparency of the water noticeably.

Among the algae, several groups are abundant and display considerable species diversity at times. Phytoflagellates and desmids are predominant among the green algae, and euglenophytes are also abundant. The diatom flora is also rich. A review of the seasonal activity patterns of the species observed is provided in Section 11.1.

The algae in the Pantanal belong overwhelmingly to cosmopolitan and circumtropical species (De-Lamonica-Freire & Heckman, 1996). Therefore, they can be identified with little difficulty using general literature prepared in other parts of the world. For this reason, the presence of many species in the Pantanal and other water bodies of southern Mato Grosso had been well documented before studies of other groups of biota had progressed beyond the exploratory phase. Braun (1882) provided an early report on the charophytes from the region, which had been studied by Dr. Otto Nordstedt. The publications by Bohlin (1897) and Borge (1903a, b) provided lists of algal species from the region, mainly those collected during the Regnell Expedition. Information on the Oedogoniaceae was provided by Hirn (1900), and other Brazilian algae, including some from Mato Grosso, were identified by Schmidle (1901). Later, Borge (1925) identified some of the algae collected during the Roosevelt-Rondon Expedition, which, however, seem to have been collected mainly from the part of Mato Grosso in the Amazon and Tocantins drainage areas.

These pioneer studies were followed by a period of neglect, during which little additional information on the algal flora was provided. Interest reawakened in recent years, when floristic surveys of the algae that had first been undertaken in the eastern parts of Brazil (R. Bicudo, 1968; Bicudo & Ventrice, 1968; Bicudo & Azevedo, 1977; Bicudo & Sormus, 1982; Bicudo & Samanez, 1984) and in the Amazon watershed (Förster, 1969, 1974; Thérézien, 1989a, b; Sophia & da Silva, 1989) were geographically extended into the Central Western section of the country (De-Lamonica-Freire, 1989, 1992; De-Lamonica-Freire *et al.*, 1992a; De-Lamonica-Freire & Sant'Anna, 1993; De-Lamonica-Freire & Heckman, 1996).

7.2.1. Chrysophyta

This group is represented by large numbers of cryptomonads, which appear chiefly during the dry season. Species in other families are also common but usually not not among the dominant algae in the water samples. This group has frequently been overlooked, and there are few references to the Brazilian flora in the literature. The species were identified mainly according to the descriptions provided by Starch (1985).

Chrysococcaceae

A fairly abundant alga during the dry season, *Chrysococcus rufescens* Klebs was fairly numerous in a few of the water samples, mainly those in low-lying areas of the flood plain, where the remains of the moribund macrophytes were decomposing in shallow water that did not dry up completely before the freshet. Elsewhere, the species was often present but never abundant.

Ochromonadaceae

One of the most widespread flagellates without chloroplasts was *Anthophysis vegetans* O. F. Müller, which occurred in slow-flowing streams in the upland parts of the Pantanal watershed as well as in the Pantanal itself. Cover glasses used as artificial substrates for microorganisms invariably were settled by a few colonies of this species within a few days.

Species of *Dinobryon* were not at all abundant in the Pantanal. The only species observed, *Dinobryon behningii* Swirenko, was encountered in water samples only during the dry season at sites that did not completely dry up.

Cryptomonadaceae

The taxonomy of this family is apparently in need of revision, since transitional forms between the most abundant species are often observed. According to the descriptions provided by Huber-Pestalozzi (1968), three cosmopolitan species are abundant in the Pantanal during the dry season, and a few of these flagellates persist into the early rainy season. *Cryptomonas erosa* Ehrenberg and *Cryptomonas ovata* Ehrenberg were amply represented by forms or varieties of both, as well as various transition forms, which probably lack taxonomic significance. *Cryptomonas reflexa* Skuja was also abundant among the cryptomonads observed. No algae in this family were ever observed during the high water period, and their abundance generally varies according to the amount of detritus that is present in the water.

Synuraceae

At least one species of *Synura* was observed in water samples from many sites in the Pantanal throughout the entire year. It was most abundant during the periods the water level was rapidly rising or falling.

Mallomonadaceae

A species of *Mallomonas* was observed only during the dry season in samples rich in other flagellates. It was never abundant.

7.2.2. Pyrrhophyta

Although this group is not represented by many species, several of those present can become numerically important among the free-swimming algae at times, especially during the dry season. They are seldom abundant during the blooms of green algae and euglenophytes, but they sometimes form their own blooms in the deeper water of small rivers in the Pantanal when other algae are not greatly abundant. They are also common just after the freshet has caused the water level to rise slightly but before the rivers begin to overflow and flood the Pantanal. The species were identified according to the descriptions provided by Popovsky and Pfiester (1990).

Gymnodiniaceae

One species in this family, *Gymnodinium cneoides* Harris, was identified in a single sample taken during the dry season from the Rio Bento Gomes near the Transpantaneira Highway. This sample also contained many other algae, including species of Euglenophyta and Chlorococcales, as well as large numbers of *Sphaerodinium cinctum* (Ehrenberg) Woloszynska.

Glenodiniopsidaceae

One species in this family formed blooms in small rivers in the Pantanal during the dry season. *Sphaerodinium cinctum* (Ehrenberg) Woloszynska was by far the most widespread and abundant dinoflagellate in the Pantanal of Poconé during the three and a half years of the study. It became rare during the *enchente* and was never observed during the *cheia* or *vazante*.

7.2.3. Bacillariophyta

The diatoms form an important part of the algal flora in the Pantanal at times. Usually, many individuals belonging to a large variety of species are present in the water samples. Only when the temperature rises and massive numbers of other algae appear in the water during the dry season do the diatoms decrease drastically in abundance. However, individuals of some species are able to persist under conditions that have not become too extreme.

Many of the diatoms develop well as epibenthic organisms on large and small stones in water courses. Large numbers of these algae were encountered in several streams near Cangas, along which impoundments containing much dead vegetation had increased the amounts of dissolved inorganic nutrients. In the Pantanal proper, however, the benthic substrate consists invariably of fine sand or silt. Substrates for settlement by benthic diatoms usually become available only after artificial modifications have been made. For example, small stones were dumped into the Rio Bento Gomes at the end of a road near the Poconé exposition park to provide a firm ground for launching small boats. The surfaces of these stones are often densely covered by diatoms, especially during the dry season when the water is shallow and the nutrient concentrations are high.

At most locations in the Pantanal where natural conditions prevail, diatoms are generally found as aufwuchs on plants or among the interrhizon community. The diatoms preferring the interrhizon are mainly vagile or filamentous species, while those on submerged plants are sessile, remaining either solitary or in small groups. Many species occur in both communities.

The taxonomy used here generally conforms with that of Krammer and Lange-Bertalot (1986, 1988, 1991a, 1991b), who placed a number of common nominal taxa in synonymy. Other standard works, including Patrick and Reimer (1966, 1977), were also consulted.

Melosiraceae

This family is represented by filamentous species that are generally associated with the interrhizon community but sometimes occur together with filamentous green algae at the edge of water courses, usually where nutrient-rich runoff is entering. *Aulacoseira granulata* (Ehrenberg) Ralfs is usually the most abundant species. It may appear in the free water during the dry season, but it is usually found in the interrhizon with various other algae. *Melosira varians* Agardh is most often associated with other plants, either in the interrhizon or on submerged macrophytes or filamentous algae. Both species can become abundant when the dissolved nutrient concentrations increase, but they disappear from the water when blooms of other algae occur. *Aulacoseira herzogii* (Lemmermann) Simonsen and *Aulacoseira italica* (Ehrenberg) Simonsen were also found in many samples during the *vazante* and *seca*; the former persisted into the *enchente*.

Fragilariaceae

This family is represented by extremely widespread species that can be found in freshwater bodies of all kinds almost anywhere in the world. *Fragilaria ulna* (Nitzsch)

Lange-Bertalot is one of the most widely distributed species in the Pantanal. A recent revision of the family (Krammer & Lange-Bertalot, 1991a) grouped two common morphological varieties, formerly called *Synedra ulna* and *Synedra acus*, in this single species of *Fragilaria*. The cells are typically found in groups loosely attached at one pole to plants or other objects. The slightest movements can cause them to break free and float in the water or sink to the sediment. They thus appear in samples of plankton and benthos as well as aufwuchs.

The other species in this family form long bands of cells attached along the long axis. Two of these species, *Fragilaria capucina* Desmazieres and *Diatoma mesodon* (Ehrenberg) Kützing, were seen on many occasions in samples from the Pantanal. The long filaments of cells were frequently found enmeshed among the submerged roots of floating macrophytes. They are thus adapted to life in the interrhizon as well as in the katharobic community.

Eunotiaceae

Representatives of this family include four species of *Eunotia* that are commonly encountered among samples of epibenthos and epibenthic detritus. They may also occasionally be encountered on submerged plants. The most common of these is *Eunotia flexuosa* (Brébisson) Kützing, which is quite variable in form. Considerably less abundant are *Eunotia serra* Ehrenberg, formerly known by its synonym, *E. robusta* Ralfs; *Eunotia bilunaris* (Ehrenberg) Mills; and *Eunotia camelus* Ehrenberg.

Naviculaceae

This family is not nearly as well represented by species in the Pantanal as it is in streams at higher elevations in the watershed. It is assumed that the fine substrate in the Pantanal is a deterrent to many of these species, which are vagile benthic algae.

The interrhizon community often supports large species of this family, including *Pinnularia major* (Kützing) Rabenhorst, *Pinnularia viridis* (Nitzsch) Ehrenberg, *Navicula cuspidata* (Kützing) Kützing, and *Stauroneis phoenicenteron* (Nitzsch) Ehrenberg. These species are often encountered on the artificial pebble substrate in the Rio Bento Gomes near the exposition park of Poconé and in streams just outside of the Pantanal, as well. According to Patrick and Reimer (1966), *P. major* is a common species in water with a low mineral content, while *S. phoenicenteron* is oligohalobic and indifferent to pH. This characterization of the autecology of these species indicates their suitability for life in the waters of the Pantanal.

A common species among the aufwuchs in the Pantanal is *Gomphonema augur* Ehrenberg, which occurs in moderate abundance during all seasons. The algae are sessile on various surfaces, including submerged macrophytes, small stones, and glass slides set out in the water to investigate the aufwuchs.

Several other species in the Naviculaceae were encountered on the surfaces of various objects in the Pantanal during the dry season. These include *Cymbela silesiaca* Bleisch, which was rather abundant at times in the littoral zone of the Rio Bento Gomes; *Navicula menisculus* Schumann, which was found mainly among benthic detritus in the same river; *Navicula radiosa* Kützing, which was widespread in the Pantanal; *Rhoicosphenia abbreviata* (C. Agardh) Lange-Bertelot, an epiphytic species frequently found on filaments of green algae, particularly species of *Oedogonium*; *Gomphonema angustatum* (Kützing) Rabenhorst, another sessile species observed only once during the dry season; and *Gomphonema clavatum* Ehrenberg, which appeared in a few samples during the *vazante* and *seca*.

Caloneis amphisbaena (Bory) Cleve was never observed in the Pantanal itself, but it was fairly abundant in streams and rivers at higher elevations flowing toward the Pantanal, such as the Arica Mirim, especially during the dry season. Less common in these electrolyte-poor waters was a large benthic species, *Pinnularia dactylus* Ehrenberg.

Bacillariaceae

This family was not well represented in the Pantanal. The species were generally observed sporadically and only in moderate numbers. The cosmopolitan *Hantzschia amphioxys* (Ehrenberg) Grunow was the most abundant species, occurring in moderate numbers, but only during the rainy season. It was one of the few common diatom species during the high water season. In contrast, *Hantzschia* cf. *sigmoidea* Hustedt was present only during the *vazante* and *seca.* This species, which greatly resembles *Nitzschia sigmoidea* (Ehrenberg) W. Smith, has appeared rarely in the literature and may be confined to the Neotropical Region.

Nitzschia palea (Kützing) W. Smith is also present in the region of the Pantanal; however, it is more abundant on the sediment of water courses upstream from the Pantanal.

Surirellaceae

Two cosmopolitan species in this family were regularly encountered in detritus samples from the Pantanal. *Surirella robusta* Ehrenberg was widespread and moderately abundant among the algae samples. Both *Surirella tenera* Gregory and *Surirella biseriata* Brébisson were observed only rarely, the former only during the dry season, and the latter, during the high water period. These species are usually epibenthic, but they are also encountered among the interrhizon.

7.2.4. Chlorophyta

The dominant groups among the microphytic primary producers were usually green algae. However, their role in the total primary production within the ecosystem was usually minor, except in the shallow water bodies remaining on the flood plain near the end of the dry season and immediately after the first few showers following the freshet. Green algae were also evident at a few sites along the edges of streams and rivers, where massive populations of zygnemataceans, most commonly *Spirogyra* spp., developed at different times of the year. At such sites, there was generally an inflow of water containing organic pollutants from ranches or cattle pens.

There were two distinct algal communities dominated by chlorophytes that developed seasonally. The first of these is typical of the nutrient-poor waters during the periods of high water and is characterized by few individuals but an enormous number of species, overwhelmingly desmids. De-Lamonica-Freire and Heckman (1996) identified 118 species belonging to the Desmidiales in samples from the Pantanal taken during the *cheia*. The second is the dry season aggregation that develops in drying pools and produces massive numbers of various species, the most abundant of which are in the family Scenedesmaceae, which was represented by 14 species during the *seca* (De-Lamonica-Freire & Heckman, 1996). This community persists into the early rainy season, when it shows a considerable increase in diversity, due mainly to the reappearance of desmids. However, it disappears after the macrophytes have grown and removed all of the free inorganic nutrients from the aquatic system (Heckman, 1994a; Section 6).

The ecology of the species in the various families present must be discussed individually because of the great differences among them. Similarly, the most recent taxonomic literature includes many extensive works treating only individual orders, families, or genera. The cosmopolitan distribution of most species permits the use of worldwide publications to identify the species. The publications used to identify members of the individual taxa are mentioned below.

Volvocaceae

The most common members of this family, *Pandorina morum* (O.F. Müller) Bory and *Eudorina elegans* Ehrenberg were both species that produced peak populations during the transition from the dry to the early rainy season. They were never observed during the high water season, and only *Pandorina morum* was observed during the *vazante*. The species are thus typical of nutrient-rich waters, and they become most numerous relative to other species of green alga when the highly enriched dry season pools become somewhat diluted and cooled by the first heavy rainfalls. *Gonium pectorale* O. F. Müller shows a similar pattern of development, being absent during the *cheia*, appearing rarely during the *vazante*, and becoming more abundant during the *seca* and early *enchente*.

Spondylomorum quarternarium Ehrenberg was not nearly as abundant as the other members of its family, and it was encountered only in relatively deep standing water near the Rio Bento Gomes during the dry season.

Unlike the other species of its family, *Volvox aureus* Ehrenberg was not observed during the dry season, apparently because of the extreme conditions of daytime heat, extremely high nutrient concentrations, and the predominance of other algal species. After the freshet, however, it became quite abundant at some of the sampling sites, especially those inundated most deeply by rainwater.

The taxonomy of the species has proven stable for a long period of time, and identifications were possible using the descriptions of Huber-Pestalozzi (1961).

Palmellaceae

Two of the shallow-water plankton species, abundant at times during the *seca*, were *Sphaerocystis schroeteri* (Lemmermann) Chodat and *Sphaerocystis planctonica* (Korschikoff) Bourrelly. Both of these might have been considered to have bloomed at some sites during this season if they had not been somewhat crowded out by other green algal species that were even more abundant. Both of these species rapidly disappeared during the *enchente*, indicating their need of abundant supplies of inorganic nutrients. The comprehensive work of Komárek and Fott (1983) was consulted for the identification of the species.

Scenedesmaceae

This is a dominant family among the algae in the Pantanal during the dry season (Fig. 7.2), and it is represented in the microflora throughout the year. The role played in the Pantanal by these species, which are considered to be planktonic in many parts of the world and have been frequently reported from other parts of Brazil (Uherkovich, 1976, 1981; Uherkovich & Schmidt, 1974; Uherkovich & Rai, 1979; Uherkovich & Franken, 1980), is discussed separately in Section 9, and their adaptation to high water temperatures, is treated in Section 10.

The seasonal activity cycle of these species was described by De-Lamonica-Freire and Heckman (1996). Of all of the species in this family, only *Scenedesmus quadricauda* (Turpin) Brébisson was observed during the *cheia*, and only a few specimens were found at that time. The *vazante* was also a season during which this family was poorly represented, and only by *Scenedesmus acuminatus* (Lagerheim) Chodat and *S. quadricauda*. In the dry season, however, all 14 species in this family that were identified during the study and various varieties of several of them occurred in the water samples. Among the species that formed true blooms at various locations during the *seca* were *Scenedesmus armatus* Chodat, *Scenedesmus intermedius* Chodat, and the two species that had already appeared during the *vazante*.

The dry season species encountered in the shallow water covering low-lying areas of the floodplain, generally following the taxonomy of Uherkovich (1966), include *Actinastrum hantzschii* Lagerheim, *Crucigenia fenestrata* (Schizuolle) Scmidle, *Crucigeniella rectangularis* (Nägeli) Komarek, *Dicloster arcuatus* Jao, Wei & Hu, *Scenedesmus armatus* Chodat var. *boglariensis* Hortobagyi, *Scenedesmus acuminatus* (Lagerheim) Chodat, *Scenedesmus acuminatus* var. *elongatus* G. M. Smith, *Scenedesmus acuminatus* var. *maximus* Uherkovich, *Scenedesmus acuminatus* var. *tortuosus* (Skuja) Uherkovich, *Scenedesmus bicaudatus* Hortobagyi, *Scenedesmus brasiliensis* Bohlin, *Scenedesmus carinatus* (Lemmermann) Chodat, *Scenedesmus denticulatus* Lagerheim var. *polydenticulatus* Hortobagyi, *Scenedesmus granulatus* var. *granulatus* Hortobagyi, *Scenedesmus i.* var. *intermedius*, *Scenedesmus intermedius* var. *bicaudatus* Hortobagyi, *Scenedesmus opoliensis* P. Richt, and *Scenedesmus quadricauda* var. *biornatus* Kiss. Most of these species persist in the water on the floodplain after the freshet (Fig. 7.2) but become progressively scarcer as the water level rises, first because their population densities per unit volume of water are greatly diminished as enormous amounts of flood water spread across the floodplain and later because of the rapid depletion of dissolved nutrients by the growing aquatic macrophytes (Heckman, 1994a; De-Lamonica-Freire & Heckman, 1996).

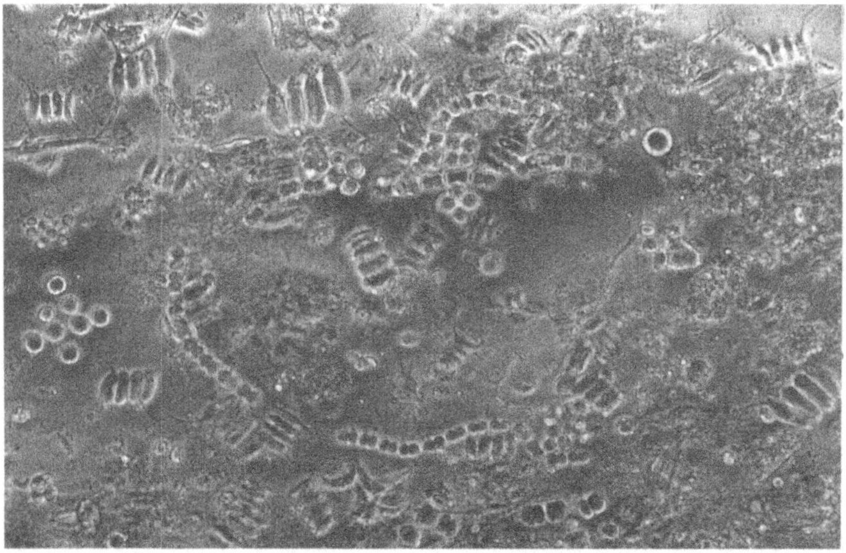

Figure 7.2: The shallow water on the floodplain before the freshet, which may reach temperatures above 40°C, contains enormous numbers of chlorococcal algae together with a few cyanobacteria, sometimes making the sediment resemble a green mud. Most of the species are commonly encountered in plankton in other parts of the world.

Oocystaceae

Among the species in this family are two that contribute to the massive blooms of algae in shallow pools of water during the late dry and early rainy season. In response to the great increases in the concentrations of dissolved mineral nutrients shortly before the water bodies completely dry up, such species as *Selenastrum rinoi* Komarek &

Comas Gonzales and *Kirchneriella lunaris* (Kirchner) Moebius form massive populations together with many species of the Scenedesmaceae (Heckman *et al.*, 1993). Also common but less abundant in the samples are *Ankistrodesmus bernardii* Komárek, *Ankistrodesmus falcatus* (Corda) Ralfs, *Ankistrodesmus spiralis* (Turner) Lemmermann, *Oonephris obesa* (W. West) Fott, *Tetraedron trigonum* (Nägeli) Hansgirg and *Closteriopsis acicularis* (G. M. Smith) Belcher & Swale. All of these species show a pronounced preference for eutrophic water, and none of them have been observed during the *cheia*, when the concentrations of mineral nutrients is lowest, or the *vazante*. The only one of the species not identified during the *seca* was *Tetraedron trigonum*, which occurred exclusively during the *enchente* (De-Lamonica-Freire & Heckman, 1996). Additional information on the South American distribution of the species in this family has been provided by Komárek & Comas González (1982), Sant'Anna (1984), and De-Lamonica-Freire & Sant'Anna (1993).

Coelastraceae

The species in this family are typical of the dry season aggregation that develops in the drying pools of water in which the concentrations of mineral nutrients become very high. None of them were observed during the *cheia* or *vazante*, and only one of them, *Coelastrum cambricum* Archer, persisted into the *enchente*. Of the five species identified, *Coelastrum intermedium* Korschikoff and *C. cambricum* were moderately abundant in the samples containing many other species of green alga, while *Coelastrum asteroideum* De Notaris, *Coelastrum microporum* Korschikoff, and *Coelastrum pseudomicroporum* Korschikoff were considerably less numerous. Descriptions of the species are found in Komárek and Fott (1983).

Characiaceae

Only a single species in this family was observed in the waters of the Pantanal. *Dictyosphaerium pulchellum* Wood occurred in moderate numbers during the dry season.

Golenkiniaceae

This family includes two species that were occasionally observed in massive numbers during the dry season (De-Lamonica-Freire & Heckman, 1996). Of the two, *Golenkinia radiata* Chodat was clearly more abundant, but *Acanthosphaera zackariasi* Lemmermann was also numerous, especially in water from the Rio Bento Gomes near the Transpantaneira Highway, where it occurred together with *G. radiata*.

Hydrodictyaceae

The peak season for the members of this family was the *seca*, during which large numbers of three taxa, *Pediastrum duplex duplex* Meyen, *Pediastrum duplex gracillimum* Desmazieres, and *Pediastrum tetras* (Ehrenberg) Ralfs var. *tetraodon* (Corda) Hansgirg, occurred in the water samples together with large numbers of other green algae, generally considered to be planktonic in other parts of the world (Heckman *et al.*, 1993). *P. d. gracillimum* appeared during the *vazante* and was not seen after the beginning of the rainy season, while the other two species appeared during the *seca* and persisted through the *enchente*. No species in this family were observed during the *cheia*, presumably because of the lack of dissolved minerals in the water. Both of these species are known from the Amazon region (Uherkovich & Schmidt, 1974).

Radiococcaceae

In the deeply green shallow water of drying pools during the *seca*, *Eutetramorus fottii* (Hindák) Komarek and *Eutetramorus planctonicus* (Korschikoff) Bourrelly occur among the large numbers of other green algal species (De-Lamonica-Freire & Heckman, 1996). They were not observed during the other seasons. Like various other green algae, they presumably require larger amounts of mineral nutrients than are present in the Pantanal during most of the year.

Zygnemataceae

As in most parts of the world, the systematic and ecological study of the filamentous algae in this family is hampered by the lack of reliable ways to identify the species of the vegetative cells. Only when zygotes are available can they be identified (Kadlubowska, 1984).

During the study of the algae in the Pantanal, only once were mature zygotes observed. The species was *Mougeotia scalaris* Hassall, probably the most widespread species of the genus.

Among the unidentified species, *Spirogyra* spp. were the best represented in the region of the Pantanal. At least four species were abundant, judging from the vegetative forms, which were of different dimensions and contained different numbers of chloroplasts.

Another common species in this family is *Zygnema* sp., which occurred in many water samples from various parts of the wetland. Somewhat less abundant was *Sirogonium* sp., which was generally most abundant in the Rio Pixaim, especially during the *cheia*. Special studies of this family were conducted in upland streams and water bodies near Chapada dos Guimarães (Dias, 1986) and Cáceres (Dias, 1989), but many more field studies will be necessary before a comprehensive species list can be prepared for the entire Pantanal.

Generally, the filamentous algae of this family develop during the rainy season. They are most abundant near the shores of water courses at sites where there is seepage of nutrient-rich wastewater, either from human settlements or from cattle ranches. They are also present on the floodplain, usually where there is enrichment of the water by cattle (Section 6), the wastes of which provide nutrients that are otherwise unavailable during the high water season.

Mesotaeniaceae

The mesotaeniaceae was represented by three species. *Mesotaenium mirificum* Archer and *Spirotaenia condensata* Brébisson ex Ralfs occurred only during the high water period. *Netrium digitus* (Ehrenberg ex Ralfs) Itzigsohn was observed during all seasons except for the *seca* and was represented by two varieties: *N. d.* var. *digitus* and *N. d.* var. *lamellosum* (Brébisson) Grönblad. According to Prescott *et al.* (1972), *M. mirificum* is a soft water species that is known from Europe and North America as well as from Brazil. The other two species are cosmopolitan or nearly so and are usually encountered in soft water.

Closteriaceae

Within the Closteriaceae, there are clear specific differences in the habitat preference. This is manifested by the seasonal occurrences of the individual species as well as by the microhabitat in the water bodies in which the specimens are usually collected. While most of the species prefer the nutrient-poor waters of the *cheia*, *Closterium ehrenbergii* Meneghini ex Ralfs and *Closterium moniliferum* (Bory) Ehrenberg are clearly algae that prefer eutrophic waters. *C. ehrenbergii* occurs throughout the year but appears much more frequently among the dense algal masses during the *vazante* and *seca* than any of the other species in this family. It is represented by two varieties: *C. ehrenbergii* var. *atumidum* Grönblad f. *atumidum*, and *C. ehrenbergii* var. *malinvernianum* (De Noteris) Rabenhorst. *C. moniliferum* was observed only during the *seca* and *enchente*. No other members of this family were more abundant during the *seca* than during the *cheia*, and all but four of the most common species could not be found during the *seca* at all (De-Lamonica-Freire & Heckman, 1996).

In addition to these two, 24 other species of *Closterium* were identified, 19 of which were seen during the *cheia*. Only a few of these were considerably abundant among the typical algal flora of the flooded plain. The other species were encountered seasonally in various microhabitats that were uncommon in the wetland. Species that remained mainly among submerged plants or in the interrhizon and were present throughout all seasons include *Closterium kuetzingii* Brébisson, *Closterium setaceum* Ehrenberg, *Closterium lineatum* Ehrenberg ex Ralfs, and *Closterium turgidum* Ehrenberg. *C. kuetzingii* was represented by two varieties; *C. kuetzingii* var. *kuetzingii* and *C. kuetzingii* var. *vittatum* Nordstedt, both throughout the year. Of the two varieties of *C. turgidum*, *C. t.* var. *turgidum* was encountered throughout the year, while *Closterium turgidum* var. *giganteum* (Nordstedt) De Toni was found only during the *enchente* and *cheia* in the region of Cáceres.

Those that seemed less tolerant of the dry season conditions and were absent only during the *seca* include *Closterium lunula* (O. F. Müller) Nitzsch, and *Closterium ralfsii* Brébisson ex Ralfs var. *hybridum* Rabenhorst. A group of species developed during the *enchente* and persisted through the early *cheia* but disappeared before the beginning of the *vazante*. These include *Closterium angustatum* Kützing var. *angustatum* Kützing, *Closterium malmei* Borge, *Closterium nasutum* Nordstedt in Krieger, *Closterium praelongum* Brébisson var. *praelongum*, *Closterium pritchardianum* Archer var. *pritchardianum*, and *Closterium tortum* Griffiths. None of these species were widespread or abundant at any location, although specimens of several were regularly encountered in samples from isolated sites during the rainy seasons. *Closterium archerianum* Cleve var. *archerianum* f. *compressum* Klebs, *Closterium closterioides* (Ralfs) Louis & Peeters, *Closterium cynthia* De-Notaris var. *cynthia*, *Closterium lagoense* Nordstedt, and *Closterium porrectum* Nordstedt var. *angustatum* W. & G. S. West were rare in the Pantanal and encountered only during the *enchente*, the early rainy season. All of the other species and varieties in this family were found only during the *cheia*: *Closterium abruptum* W. West var. *brevius* W. & G. S. West, *Closterium dianae* Ehrenberg var. *arcuatum* (Brébisson) Rabenhorst, *Closterium dianae* var. *pseudodianae* (Roy) Krieger, *Closterium limneticum* Lemmermann, *Closterium nematodes* Joshua, *Closterium parvulum* Nägeli, *Closterium tumidum* Johnson, and *Closterium venus* Kützing. Descriptions of the species were provided by Prescott *et al.* (1975) and Růžička (1977), and additional information about the distribution of the *Closterium* species in Brazil was provided by Bicudo and de Castro (1994).

Gonatozygaceae

Like most algae belonging to the Desmidiales, the species in this family occur chiefly during the *cheia*. Three species were identified during the study, *Gonatozygon monotaenium* De Bary, *Gonatozygon pilosum* Wolle, and *Gonatozygon aculeatum* Hastings. All of *vazante*. them was observed at other times as well: during the *enchente* and were moderately abundant during the high water period, and only *G. aculeatum* None were observed during the *seca*. The specimens conform to the descriptions of Růžička (1977), who reported all to be cosmopolitan in distribution, although *G. aculeatum* and *G. pilosum* are rare in the temperate zones.

Desmidiaceae

The desmids are the predominant algae present in the Pantanal during the high water period (Fig. 7.3). Most of these species are known throughout the world for their preference for water extremely poor in dissolved nutrients, including European moors (Krieger, 1937, 1939). This led many botanists early in this century to associate the presence of a large desmid species aggregation with acid water (Brook, 1965, 1981). Generally, in the temperate zones, the water bodies poorest in inorganic nutrients are bogs and moors, which also happen to be acidic and rich in humic substances (Burmeister *et al.*, 1980). As the studies by Heckman (1994a) and De-Lamonica-Freire and Heckman (1996) show, however, the richest desmid species aggregation, that might well be associated statistically with acid water in the temperate zones, develops in the Pantanal during the *cheia* in water with a pH near 7. Brook (1981) amply documented the occurrence of desmid species in water of greatly varying salinity, alkalinity, and hardness. However, he confirmed that the occurrence of the greatest number of species is generally associated with bog communities. The findings just discussed strongly suggest that this species aggregation develops preferentially in waters poor in dissolved inorganic nutrients, including calcium ions, and that the association with low pH is coincidental. Possible selective advantages for the desmid species in such waters include the scarcity of algal and macrophyte competitors, clear water with excellent light penetration, and a dearth of small herbivores that can graze on them. Furthermore, the rough and spiny surfaces of the desmids make them difficult for many tiny herbivores to ingest.

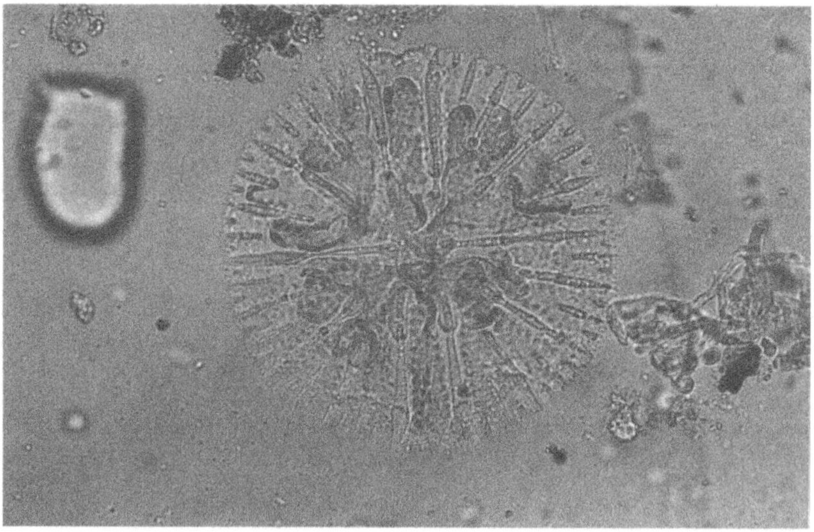

Figure 7.3: During the *cheia*, desmids become the dominant algae in the waters of the Pantanal. These algae, such as the *Micrasterias radiosa* depicted here, invariably occur isolated in the samples, displaying a characteristic diversity of species, each of which represented by very few individuals in any given sample.

There have been a number of recent taxonomic revisions of the desmids (Teiling, 1967; Bicudo, 1969; Bicudo & Azevedo, 1977; Bicudo & Sormus, 1982; Bicudo & Samanez, 1984; De-Lamonica-Freire, 1992a, b; Förster, 1969, 1974; Prescott *et al.*, 1975, 1977, 1981, 1982; Růžička, 1977, 1981; Tell *et al.*, 1994), unfortunately, without a consensus on the classification in all cases. A number of taxonomic works were consulted, and the nomenclature adopted here conforms with that used by De-Lamonica-Freire and Heckman (1996) for the reasons provided in that publication.

Of the 139 desmid species identified during the study and listed by De-Lamonica-Freire and Heckman (1996), only 10 were found during all seasons of the year. These species were *Cosmarium candianum* Delponte var. *candianum* f. *candianum, Cosmarium ginzbergeri* Grünblad, *Cosmarium margaritatum* (Lundell) Roy & Bisset, *Cosmarium moniliforme* (Turpin) Ralfs, *Euastrum ansatum* Ehrenberg ex Ralfs, *Micrasterias arcuata* Bailey var. *subpinnatifida* W. & G. S. West, *Micrasterias furcata* C. Agardh ex Ralfs, *Micrasterias truncata* (Corda) Brébisson var. *pusilla* G. S. West, *Pleurotaenium trabecula* Ehrenberg ex Nägeli, and *Staurodesmus convergens* (Ehrenberg) Telling var. *tellingi* De-Lamonica-Freire.

Only four species were observed only during the dry season: *Actinotaenium turgidum* (Brébisson) Teiling ex Růžička & Pouzar, *Cosmarium paraguayense* Borge, *Cosmarium subtumidum* Nordstedt var. *subellipticum* (Playfair) Krieger & Gerloff, and *Onychonema filiformis* (Ehrenberg) Roy & Bisset. Of these, only *Actinotaenium turgidum* was moderately abundant, while the rest were not encountered in more than one or two water samples.

A relatively large number of species, 20, were present during all seasons except the *seca.* The vegetative stages of many of these species appeared late in the *enchente* and disappeared again early in the *vazante.* They included *Cosmarium moerlianum* Lutkmuller var. *brasiliense* Borge, *Cosmarium pseudoconnatum* Nordstedt, *Cosmarium pseudopyramidatum* Lundell var. *pseudopyramidatum*, *Cosmarium vitiosum* Scott & Grünblad, *Desmidium baileyi* (Ralfs) Nordstedt var. *baileyi, Docidium baculum* Brébisson ex Ralfs var. *baculum*, *Micrasterias abrupta* W. & G.S. West, *Micrasterias borgei* Krieger var. *borgei, Micrastarias laticeps* Nordstedt var. *laticeps* Nordstedt, *Micrastarias laticeps* Nordstedt var. *acuminata* Krieger, *Micrasterias mahabuleshwarensis* Hobson, *Micrasterias radians* Turner var. *bogoriensis* (Bernard) G. S. West, *Micrasterias torreyi* Bailey ex Ralfs var. *nordstedtiana* (Hieronymus) Schmidle, *Onychonema laeve* Nordstedt, *Pleurotaenium minutum* (Ralfs) Delponte, *Staurastrum paradoxum* Meyen, *Staurodesmus dickiei* (Ralfs) Lillieroth, *Staurodesmus subulatus* (Kützing) Croasdale, *Staurodesmus validus* (W. & G. S. West) Thomasson, *Xanthidium convergens* (Ehrenberg) Teiling, and *Xanthidium mucronulatum* (Nordstedt) Förster.

A total of 12 species were present during the *enchente* and *cheia* but not found during the rest of the year: *Bambusina borreri* (Ralfs) Cleve, *Cosmarium bireme* Nordstedt var. *barbadense* G. S. West, *Cosmarium commissurale* Brébisson ex Ralfs var. *crassum* Nordstedt, *Cosmarium depressum* (Nägeli) Lundell var. *achondrum* (Boldt) W. & G. S. West, *Cosmarium exiguum* Archer var. *exiguum* f. *exiguum*, *Cosmarium pseudoprotuberans* Kirchner var. *pseudoprotuberans*, *Desmidium aptogonum* Brébisson ex Kützing var. *aptogonum, Euastrum elegans* (Brébisson) Ralfs, *Euastrum evolutum* (Nordstedt) W. & G. S. West, *Euastrum gemmatum* (Brébisson) Brébisson ex Ralfs, *Euastrum subintegrum* Nordstedt var. *brasiliense* Grönblad, and *Staurodesmus mamillatus* (Nordstedt) Teiling.

Only five species were observed exclusively during the *cheia* and *vazante*: *Cosmarium punctulatum* Brébisson, *Desmidium quadratum* Nordstedt, *Pleurotaenium nodosum* (Bailey) Lundell, *Staurastrum nudibrachiatum* Borge, and two varieties of *Xanthidium antilopaeum* (Brébisson) Kützing: *Xanthidium antilopaeum* var. *antilopaeum* and *X. a.* var. *canadense* (Joshua) Förster. In contrast, 39 species were encountered only during the cheia: *Cosmarium decedens* (Reinsch) Racborski var. *borgei* Krieger, *Cosmarium galiciense* (Gutwinski) De-Lamonica-Freire, *Cosmarium hexagonum* Nordstedt, *Cosmarium ornatum* Ralfs var. *pseudolagoense* Förster & Eckert in Förster, *Cosmarium orthostichum* Lundell var. *pumilum* Lundell, *Cosmarium refringens* Taylor var.

refringens, Cosmarium subspeciosum Nordstedt var. *subspeciosum* f. *brasiliense* Förster, *Cosmarium supraspeciosum* Wolle, *Cosmarium tinctum* Ralfs, *Cosmarium triangulare* Borge, *Cosmarium trilobatum* Reinsch var. *abscissum* (Schmidle) Krieger & Gerloff, *Desmidium aequale* W. & G. S. West var. *aequale* f. *aequale, Desmidium graciliceps* (Nordstedt) Lagerheim, *Desmidium laticeps* Nordstedt, *Euastrum fissum* W. & G.S. West var. *angustum* Scott & Grönblad, *Euastrum pectinatum* Brébisson ex Ralfs, *Groenbladia undulata* Nordstedt, *Hyalotheca dissiliens* (Smith) Brébisson ex Ralfs, *Micrasterias papillifera* Brébisson ex Ralfs, *Micrasterias ringens* Bailey, *Micrasterias thomasiana* Archer, *Pleurotaenium nodulosum* (Brébisson) De Bary, *Pleurotaenium ovatum* (Nordstedt) Nordstedt, *Spondylosium ellipticum* W. & G. S. West, *Staurastrum erinacoideum* Scott & Croasdale, *Staurastrum gladiosum* Turner var. *gladiosum* f. *ornatum* Laporte, *Staurastrum leptocladum* Nordstedt, *Staurastrum muticum* Brébisson ex Ralfs, *Staurastrum quadrangulare* Brébisson ex Ralfs, *Staurastrum rotula* Nordstedt, *Staurastrum sebaldi* Reinsch var. *ornatum* Nordstedt, *Staurastrum setigerum* Cleve var. *subvillosum* Grönblad, *Staurastrum striolatum* (Nägeli) Archer, *Staurastrum tohopekaligense* Wolle var. *trifurcatum* W. & G. S. West, *Staurastrum trachytithophorum* W. & G. S. West, *Staurastrum trifidum* Nordstedt var. *inflexum* W. & G. S. West, *Staurodesmus maximus* (Borge) Teiling, *Xanthidium mamillosum* (Grönblad) Förster var. *longispinum* (Borge) Förster, and *Xanthidium trilobum* Nordstedt. In addition, *Staurastrum leptocladum* Nordstedt is represented by two varieties: *S. l.* var. *cornutum* Wolle and *S. l.* var. *smithii* Grönblad.

Cosmarium contractum Kirchner, represented by *C. c.* var. *maximum* W. & G. S. West and *C. c.* var. *contractum*, as well as *Cosmarium impressulum* Elfving var. *impressulum, Cosmarium norimbergense* Reinsch var. *depressum* (W. & G. S. West) Krieger & Gerloff, *Cosmarium rectangulare* Gruner var. *hexagonum* (Elfving) W. & G. S. West, *Cosmarium trilobulatum* Reinsch var. *trilobulatum* f. *trilobulatum, Cosmarium urnigerum* Nordstedt, *Euastrum abruptum* Nordstedt var. *abruptum, Euastrum attenuatum* Wolle, *Euastrum rectangulare* Fritsch & Rich, *Micrasterias incisa* Brébisson ex Ralfs var. *excisa* Nordstedt, *Spondylosium planum* (Wolle) W. & G. S. West, *Staurastrum arctiscon* (Ehrenberg) Lundell var. *glabrum* W. & G. S. West, *Staurastrum asteroideum* W. & G. S. West var. *nanum* (Wolle) Grönblad, *Staurastrum claviferum* W. & G. S. West, *Staurastrum margaritaceum* Ehrenberg ex Ralfs *var. margaritaceum, Staurastrum minnesotense* Wolle, *Staurastrum orbiculare* (Ehrenberg) Ralfs, *Staurastrum wolleanum* Butler var. *heptacanthum* Irenée-Marie, *Staurodesmus cuspidatus* (Turner) Thomasson, and *Staurodesmus dejectus* (Brébisson) Telling, are the 20 species found exclusively during the *enchente.* Only eight species, *Cosmarium conspersum* Ralfs var. *rotundatum* Wittrock, *Cosmarium quadrum* Lundell var. *quadrum, Euastrum platycerum* Reinsch, *Micrasterias pinnatifida* (Kützing) Ralfs, *Pleurotaenium cylindricum* (Turner) Schmidle var. *stuhlmannii* (Hieronymus) Krieger, *Pleurotaenium subalternans* Borge var. *subalternans, Spondylosium pulchrum* (Bailey) Archer, and *Staurastrum leptacanthum* Nordstedt, were encountered exclusively during the *vazante*, while a total of only six were found during both the *enchente* and *vazante* but not during the other two seasons. These six were *Cosmarium galeritum* Nordstedt, *Euastrum didelta* (Turpin) Ralfs var. *quadriceps* (Nordstedt) Krieger, *Euastrum verrucosum* Ehrenberg ex Ralfs, *Pleurotaenium baculoides* (Roy & Bisset) Playfair var. *baculoides, Teilingia granulata* (Roy & Bisset) Bourrelly, and two varieties of *Staurodesmus crispus*: *S. c.* var. *borgei* (Scott & Prescott) Compère and *S. c.* var. *kalimantanus* (Scott & Prescott) Compère.

Four species were present during all seasons except for the *vazante*: *Cosmarium denticulatum* Borge, *Cosmarium pyramidatum* Brébisson ex Ralfs, *Micrasterias radiosa* Ralfs, and *Micrasterias tropica* Nordstedt. Similarly, four were found throughout the year except during the *cheia*: *Cosmarium pachydermum* Lundell var. *pachydermum, Pleurotaenium maskellii* Suxena & Venkateswarlu, *Staurastrum novae-caesareae* Wolle, and *Staurodesmus lobatus* (Borgesen) Bourrelly.

Three species, *Cosmarium obtusatum* Schmidle, *Cosmarium speciosum* Lundell, and *Staurastrum disputatum* W. & G. S. West var. *extensum* (Borge) W. & G. S. West, were encountered only during the *seca* and *enchente*, and two species were found only during the *seca* and *vazante*: *Pleurotaenium coroniferum* (Borge) Krieger, and *Pleurotaenium eugeneum* (Turner) W. & G. S. West var. *undulatum* (Borge) Krieger.

The most hard to explain combination of seasons is the *cheia* and *seca*. Only a single species, *Pleurotaenium ehrenbergii* (Brébisson) De Bary, was encountered exclusively during these two seasons, but it was rare in the samples, so the data probably do not provide sufficient information on its seasonal occurrence.

Peniaceae

Two species in this family were among the members of the Desmidiales found only when the concentrations of nutrients in the water were low. *Penium spirostriolatum* Barker, a cosmopolitan species (Růžička, 1977), was found only during the *cheia*, while *Penium subspirostriolatiforme* Grönblad occurred during the *enchente*. Both of these algae are rarely found but contribute to the high species diversity during the rainy season.

Oedigoniaceae

Filamentous algae in this family were widespread in the Pantanal but seldom abundant in the individual samples. Most of the specimens of algae in this family were collected in the Rio Bento Gomes. *Bulbochaete* sp. was observed only during the *enchente* and *vazante*, while one species of *Oedogonium* was collected during the *enchente*, and another, during the vazante. As in the case of the Zygnemataceae, the absence of zygotes makes specific identification impossible.

Chaetophoraceae

Unlike most of the algae in the Pantanal, *Chaetophora elegans* Agardh is a species that develops in fast flowing rather than standing water. Its occurrence is thus generally limited to the *cheia*, when the water is moving rapidly through the streams in the Pantanal. The species forms gelatinous green colonies on objects in the water, generally the twigs of woody plants. Where the current is strong, such as near the bridges along the Transpantaneira, these colonies can become quite numerous. Most of them have a spherical form, but larger ones can become elongated. This species is well known in streams of the Atlantic forest region in eastern Brazil, as well (Branco and Necchi, 1996). Much less common are species of *Draparnaldia*, which occur in the Rio Bento Gomes during the rainy season.

7.2.5. Euglenophyta

Among the commonest of the algae during the dry season, the euglenophyte species show a pronounced preference for water with high concentrations of mineral nutrients and much organic enrichment, as well. Most of the species seem to be at the opposite end of the trophic spectrum from the desmids, which are overwhelmingly represented by species typical of nutrient poor water. The species in this family that have been studied are physiologically capable of surviving either through photosynthesis or through heterotrophic utilization of organic compounds in the water (Kudo, 1966).

Although most euglenoids are encountered mainly or exclusively during the dry season (De-Lamonica-Freire & Heckman, 1996), several of them survive throughout the year, even during periods of extreme nutrient shortage. Several members of this group can be found swimming among the roots of the floating plants during the high water period, while the algal community is dominated by desmids, and few other species survive in active stages. The species can be grouped according to various criteria, including their seasons of occurrence, the microhabitat, and the degree of tolerance to environmental conditions (Heckman *et al.*, 1996). As in the cases of the other algae, they must be discussed on an individual basis to do justice to the complexities of their autecology.

Euglenaceae

The species in this family encountered in the Pantanal are overwhelmingly cosmopolitan or circumtropical (Huber-Pestalozzi, 1955). Their morphological characteristics and known distribution in South America have been reported by Tell and Conforti (1986). A detailed study of the tropical varieties of *Trachelomonas armata* (Ehrenberg) Stein was made by Bicudo and De-Lamonica-Freire (1993), and lists of the species known from the Pantanal of Poconé were provided by De-Lamonica-Freire *et al.* (1992) and De-Lamonica-Freire and Heckman (1996).

Euglena acus Ehrenberg, *Euglena spirogyra* Ehrenberg, *Phacus pleuronectes* (O.F. Müller) Dujardin, *Phacus longicauda* (Ehrenberg) Dujardin, *Phacus tortus* (Lemmermann) Skvortzow, *Trachelomonas armata* var. *armata*, *Trachelomonas hispida* (Perty) Stein em. Deflandre, *Trachelomonas volvocina* Ehrenberg, and *Trachelomonas volvocinopsis* Swirenko were the species in this family found to be most tolerant of environmental conditions, occurring during all seasons and in many of the microhabitats. With the exception of *T. volvocina*, they were seldom present in more than moderate numbers. All of them occurred in a great many water samples from all parts of the Pantanal. The species were especially abundant in samples taken from among submerged macrophytes, but they also appeared in the samples with many other euglenophyte species taken during the dry season (Fig. 7.4). Interestingly enough, these were the only species encountered during the *cheia*. Therefore, all members of this family that was found during the high water period were also present during all of the other seasons. They were also less abundant during the *cheia* than they were during the rainless periods of the year, and only *T. armata armata* and *T. volvocina* were moderately abundant during the high water period. All other species in this family were rare at this time.

Three species were found during all seasons of the year except the *cheia*: *Euglena tripteris* (Dujardin) Klebs, *Phacus suecicus* Lemmermann, and *Trachelomonas megalacantha* var. *megalacantha* Da Cunha.

Many of the species were found only during the dry season. Several of them are dominant in shallow, nutrient-rich pools of water that remain on the floodplain shortly before their complete evaporation. The temperature in these pools was found to reach at least 41.6°C (Heckman *et al.*, 1996). *Euglena sanguinea* Ehrenberg forms deep red coatings on the surface tension layer of these pools (Plate 2a), while *Euglena gracilis* Klebs is responsible for the similar bright green coatings at this time of year. *Euglena sanguinea* persisted into the *enchente* but disappeared completely from the samples as the water level began to rapidly rise. It was never observed in the deeper permanent water courses that drain the Pantanal. *E. gracilis* was never observed during the *enchente*, but it was moderately abundant in the water column during the *vazante* and appeared occasionally in samples from the interrhizon along the shores of the permanent water courses. Xavier *et al.* (1991) described similar blooms of *E. sanguinea* on the surface of fish ponds in the state of São Paulo, while Zakrys and Walne (1994) described the development of such blooms in North America.

Below the surface of the water in the shallow, evaporating pools of water on the flood plain during the *seca* were massive populations of *Trachelomonas molesta* Deflandre, *Trachelomonas armata* var. *longispina* Playfair emend Deflandre, and the ubiquitous *T. volvocina*, *T. hispida*, and *E. acus*. These species are accompanied by a variety of chlorococcal green algae and the somewhat less abundant *Euglena deses* Ehrenberg, *Euglena fusca* (Klebs) Lemmermann, *Euglena oxyuris* Schmarda, *Lepocinclis ovum*

(Ehrenberg) Lemmermann, *Phacus brachykentron* Pochmann, *Phacus parvulus* Klebs, *Phacus platalea* Drezepolsky, *Trachelomonas abrupta* Swirenko em. Deflandre, *Trachelomonas crispa* Balech., *Trachelomonas cervicula* Stokes, *Trachelomonas dubia* Swirenko em. Deflandre f. *acuminata* Deflandre, *Trachelomonas elliptica* (Playfair) Deflandre, *Trachelomonas lemmermannii* Woloszynska, *Trachelomonas robusta* Swirenko em. Deflandre, *Trachelomonas varians* Deflandre var. *varians* f. *globosa* Deflandre, *Strombomonas chodati* (Skvortzov) Deflandre, *Strombomonas encifera* (Daday) Deflandre, *Strombomonas gibberosa* (Playfair) Deflandre, and *Strombomonas scabra* (Playfair) Tell & Conforti.

Most of these floodplain species together with *Euglena limnophila* Lemmermann, *Phacus curvicauda* Swirenko, *Phacus mariae* Deflandre, *Phacus orbicularis* Hübner, *Trachelomonas acanthophora* Stokes, *Trachelomonas armata* var. *heterospina* Swirenko, *Trachelomonas lacustris* Drezepolski, *Trachelomonas superba* Swirenko em. Deflandre, and *Trachelomonas sydneyensis* Playfair occur in the interrhizon along the shores of permanent water bodies and appear in large numbers on the floodplain during the *enchente*. In addition, *Lepocinclis fusiformis* (Carter) Lemmermann, *Phacus pyrum* (Ehrenberg) Stein, *Trachelomonas bacillifera* Playfair, *Trachelomonas planctonica* Swirenko, and *Strombomonas australica* (Playfair) Deflandre var. *rectangularis* Deflandre were found among this aggregation only during the *enchente*.

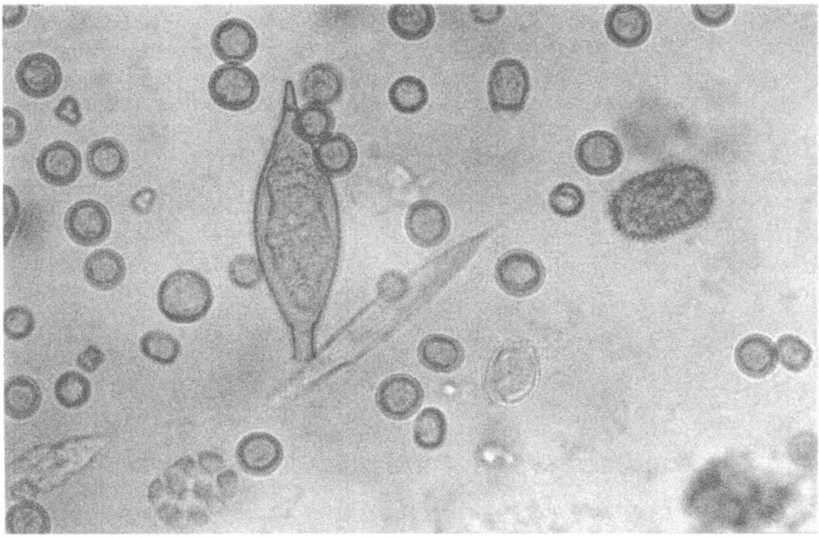

Figure 7.4: Many *Trachelomonas* species occur together in nutrient-rich water bodies during the *seca*, as in this micrograph of an unfiltered water sample, showing at least six species of *Trachelomonas*, a *Euglena*, and a green alga, *Pandorina morum*.

Two species. *Euglena ehrenbergii* Klebs and *Euglena intermedia* (Klebs) Schmitz, were encountered only during the *vazante*, the season during which saprobic species were predominant in most communities of microorganisms.

Although extensive, the floral list is not complete for southern Mato Grosso. During a study of this family in the Municipal District of Cáceres, Menezes and Fernandes (1987) compiled a list of pigmented Euglenaceae species, which includes several taxa not found during the study of the Pantanal (Table 7.1). Menezes and Fernandes (1989) expanded their studies to include other municipal districts near Cáceres. However, much of the area studied by these authors is not in the Pantanal but rather in upland ponds, lakes, and streams. To the north and west of the city of Cáceres, there are a good

Table 7.1: A list of the pigmented species of Euglenaceae found by Menezes and Fernandes (1987) in the Municipal District of Cáceres and adjacent municipal districts. Whether the presence of the species was detected in the Pantanal of Poconé during the study is reported at right. The presence is indicated only for the species and not for varieties.

Species	Variety, form, or (synonym)	Species seen in the Pantanal of Poconé
Euglena acus Ehrenberg		Yes
Euglena limnophila Lemmermann		Yes
Euglena oxyuris Schmarda		Yes
Euglena spirogyra Ehrenberg		Yes
Euglena fusca (Klebs) Lemmermann	(Syn: *Euglena spirogyra* var. *fusca*)	Yes
Lepocinclis autumnalis Chu		No
Lepocinclis fusiformis (H. J. Carter) Lemmermann emend. Conrad		Yes
Lepocinclis ovum (Ehr.) Lemmermann	var. *ovum*	Yes
	var. *dimidio-minor* Deflandre	-
	var. *ecaudata* Deflandre	-
Lepocinclis salina Fritsch	var. *salina*	No
Lepocinclis texta (Duj.) Conrad	var. *texta*	No
	var. *mammillata* (Cunha) Conrad	-
Lepocinclis turbiniformis Deflandre		No
Phacus contortus Bourelly		No
Phacus ephippion Pochmann		No
Phacus hamatus Pochmann		No
Phacus horridus Pochmann		No
Phacus longicauda (Ehrenberg) Dujardin		Yes
Phacus onyx Pochmann		No
Phacus raciborskii Drezepolski		No
Phacus tortus (Lemmermann) Skvortzow		Yes
Strombomonas encifera (Daday) Deflandre		Yes
Strombomonas maxima (Skvortzow) Deflandre		No
Strombomonas scabra	var. *scabra* (Playfair) Tell & Confor	Yes
	var. *longicollis*(Playfair)Tell & Conforti	-

Species	Variety, form, or (synonym)	Species seen in the Pantanal of Poconé
Strombomonas tetraptera Balech & Dastugue		No
Strombomonas urceolata (Stokes) Deflandre		No
Strombomonas verrucosa (Daday) Deflandre		No
Trachelomonas acanthophora Stokes	var. *acanthophora*	Yes
Trachelomonas alia Drezepolski emend. Middelhoek		No
Trachelomonas allorgei Deflandre		No
Trachelomonas armata (E.) Stein	var. *armata*	Yes
	var. *armata* f. *inevoluta* Deflandre	-
	var. *steini* Lemmermann emend Deflandre	-
	var. *litoralensis* Tell & Zalocar	-
Trachelomonas bulla Stein emend. Deflandre		No
Trachelomonas dastuguei Balech		No
Trachelomonas hispida	var. *hispida* (Perty) Stein emend. Deflandre	Yes
	var. *duplex* Deflandre	-
	var. *punctata* Lemmermann	-
	var. *erenulatocollis* (Maskell) Lemmermann	-
Trachelomonas lemmermannii Wol.	var. *acuminata*	Yes
Trachelomonas naviculiformis Deflandre		No
Trachelomonas parvicollis Deflandre		No
Trachelomonas pulcherrima Playfair	var. *latior* Playfair	No
Trachelomonas sydneyensis Playfair	var. *sydneyensis*	Yes
Trachelomonas similis Stokes	var. *spinosa* Huber-Pest.	No
Trachelomonas spinosa Stokes		No
Trachelomonas volvocina Ehrenberg	var. *volvocina*	Yes
Trachelomonas volvocinopsis Swirenko	var. *punctata* (Skv.) Bourrelly	Yes

many ponds which are much more eutrophic throughout most of the year than the waters of the Pantanal. These support a different macrophyte vegetation, providing a possible explanation for the local richness of the algal flora.

The fact that not even half of the species were reported for both the Pantanal of Poconé (De-Lamonica-Freire & Heckman, 1996) and the water bodies near Cáceres indicates that the southern Mato Grosso flora of Euglenaceae will be found to be even richer in future studies of other kinds of habitat. It is also a significant finding of all authors who have worked in Mato Grosso that there seem to be no species endemic to the region. All of the forms found fit the descriptions of cosmopolitan or circumtropical species well and undoubtedly are correctly assigned to those species.

In the upland streams surrounding the Pantanal, algae have also been collected. These waters are noted for their extreme lack of nearly all minerals (Section 6). A study by Menezes (1986) revealed only six species, one represented by three varieties, that were present in the waters near Chapada dos Guimarães. Of these, only two species, *Lepocinclis ovum* and *Phacus orbicularis*, were also found in the Pantanal (De-Lamonica-Freire & Heckman, 1996), and three, *Lepocinclis ovum*, *Phacus hamatus*, and *Phacus raciborskii*, also occurred in the municipalities near Cáceres (Menezes & Fernandes, 1987). *Phacus anomalus* Fritsch & Rich and *Phacus pusillus* Lemmermann, as well as one variety of *Lepocinclis ovum*, were found at neither of the other two sites.

Anisonemataceae

A ubiquitous and abundant species in the Pantanal is *Pseudoperanema trichophorum* (Ehrenberg) Larsen, which was never numerous in any water sample but occurred in a greater number of samples than any other Euglenophyte. It is a colorless, heterotrophic flagellate that consumes species of *Euglena*. Other species in this family were rarely found. *Anisonema acinus* Dujardin was extremely rare in the Pantanal, occurring only in clear water samples containing few other algae. However, it was observed regularly in samples taken from streams in the upland parts of the watershed, including the nutrient-poor headwaters of rivers on the plateau to the east of Cuiabá. *Tropidoscyphus octocostatus* Stein and *Tropidoscyphus decemlineata* Seckt were rare. They were found only in the Pantanal during the *vazante*, the season during which saprobic organisms were dominant in most communities of microorganisms.

7.2.6. Charophyta

Charophyte species were not found often in the Pantanal, but they were reported at various locations from time to time (Scheßl, 1997). These algae usually do not produce populations that appear at the same location from year to year. One stand of *Nitella acuminata* A. Br. ex Wallm. emend R. D. W., probably the most abundant charophyte species in the Pantanal, was observed in a shallow pool of rainwater covering dying terrestrial plants near the Transpantaneira Highway and the Rio Frio, shortly after an unseasonal rainfall during the *seca*. This species is widespread in Brazil and previously reported from the states of São Paulo, Minas Gerais, and Rio de Janeiro (R. Bicudo, 1968).

7.3. Fungi

7.3.1. Saprolegniaceae

Aquatic fungi of the family Saprolegniaceae are commonly observed on dead animals in water samples taken in the Pantanal. These are common aquatic saprophytes throughout the world (Seymour, 1970). Among the species observed, *Achlya racemosa* Hildebrand was the most abundant, appearing frequently on dead insect larvae and small crustaceans. The members of *Saprolegnia* present have not yet been identified to species.

7.3.2. Basidiomycetes

Terrestrial species of mushroom appear on various substrates in the Pantanal. The greatest variety of species is encountered on wood, including dead tree branches on the hummocks and in the gallery forests and on the planks used to construct the bridges along the Transpantaneira Highway. Other species grow on the manure of cows and other domestic animals. The mushrooms are not well represented in the Pantanal, most probably because of the dryness during much of the year and the flooding when moisture is present. However, there is a fairly diverse flora in the hummocks, where the wood of various tree species decomposes. Among the species observed during the study, most were fairly common at times, with the greatest variety being encountered during the *enchente* before local flooding had occurred.

A variety of species in the family Polyporaceae specialize in contributing to the decomposition of various kinds of wood. These shelf fungi could not be positively identified, and it is probable that several undescribed species are present in the Pantanal. Because the wood for the bridges is brought in from remote forests, it is safe to assume that the fungi growing on it are also not native to the Pantanal. Apparently, only the polypores that grow in the hummocks are members of the native flora.

Also represented in the Pantanal were species of Tricholomataceae. A fairy ring mushroom closely resembling *Marasmius oreades* (Bolt. ex Fr.) Fr. was encountered on the floodplain near the borders of a hummock during the dry season. In addition, the family Cortinariaceae is represented by a mushroom which grows solitarily at the edges of hummocks and closely resembled *Inocybe geophylla* (Sow. ex Fr.) Kum., a species widespread in North America.

The commonest species of mushroom that grows on the cattle dung during the early rainy season is probably *Psilocybe coprophila* (Bull. ex Fr.) Kum., a member of the Strophariaceae that is typically associated with cow and horse manure, as its English name, dung-loving psilocybe, implies.

7.4. Lichenes

A few lichens are abundant on tree trunks as well as on fence posts along the Transpantaneira, however, relatively few species occur in the Pantanal. In contrast, the lichen flora in the upland regions of the catchment area is quite rich. According to the information provided by the United States National Museum Herbarium on the Internet, among 27 type specimens from southern Mato Grosso represented in the collection, only four come from the lowlands around the Pantanal: three in or near Cuiabá and one, *Callopisma cuyabense* G. O. Malme, from Santo Antonio, apparently the town of that name along the Rio Cuiabá not far from the border of the wetland.

Two species are particularly prominent on the fence posts along the Transpantaneira Highway. One of these, probably a species of *Cladonia*, remains stiff and hard during the dry seasons but becomes soft when it resumes growth during the rainy season. The other species is an encrusting lichen which apparently belongs to the genus *Parmelia*. In addition to the species observed in the northern part of the Pantanal, one species is known from the Pantanal in Mato Grosso do Sul at Corumbá: *Porina melanops* G. O. Malme.

The lichen flora known from the upland parts of the Rio Paraguay drainage area includes three species known from the city of Cuiabá: *Buellia umbrina* G. O. Malme, *Callopisma fulvellum* G. O. Malme, and *Parmelia merrillii* B. A. Lynge. All of the other species have been found in the more elevated parts of the catchment area, mainly in the uplands near Chapada dos Guimarães. These include *Callopisma byrsonimae* G. O. Malme, *Callopisma chapadense* G. O. Malme, *Callopisma melanocheilum* G. O. Malme, *Laurera sanguinaria* O. Malme, *Parmelia acariospora* Zahlbruckner, *Parmelia annae* B. A. Lynge, *Parmelia cornuta* B. A. Lynge, *Parmelia digitata* B. A. Lynge, *Parmelia fatiscens* B. A. Lynge, *Parmelia laongii* B. A. Lynge, *Parmelia latissima* f. *microspora* B. A. Lynge, *Parmelia melanochaeta* S. Kurokawa, *Parmelia minima* B. A. Lynge, *Parmelia regnellii* B. A. Lynge, *Parmelia setigera* S. Kurokawa, *Porina elegans* G. O. A. Malme, *Pseudoparmelia cyphellata* B. A. Lynge, *Relicina incongrua M.* E. Hale, *Trypethelium subcatervarium* G. O. A. Malme, and *Lecidea stylophora* G. O. Malme.

7.5. Bryophyta

Bryophyte species are extremely rare in the Pantanal, apparently due to the high temperature and low humidity during the dry season. The only species that is fairly common and occasionally abundant is *Ricciocarpus natans* (L.) Corda, a small liverwort that floats on the surface of the water and is occasionally stranded on moist sediments. It is generally found at the edge of the Pantanal and in flooded areas along rivers with fine, organically enriched sediment rather than in the middle of the floodplain or in large, fast flowing rivers. A massive population completely covered the surface of a small roadside pond to the northwest of Cáceres, at a slightly higher elevation than the Pantanal. Other specimens were taken at the edge of the Rio Bento Gomes near the northern border of the Pantanal and along the marshes bordering the Rio Paraguay at Cáceres. The species is nearly cosmopolitan in distribution (Cook, 1974).

Mosses and other liverworts are quite common in the hilly regions of the Pantanal watershed, especially in the zone beside waterfalls constantly exposed to a fine spray, but apparently they do not occur at all in flatland habitats of southern Mato Grosso.

7.6. Pteridophyta

Although fairly widespread in hilly places and along streams outside of the Pantanal, terrestrial ferns and club mosses are uncommon in the Pantanal. Aquatic ferns, on the other hand, are among the dominant macrophytes throughout the floodplain. Unfortunately, the taxonomy of the South American pteridophytes requires much revision, and many ways have been proposed to classify the species during recent years. The assignment to higher taxa employed here follows the systematic organization of Kramer and Green (1990).

Azollaceae

Two species of *Azolla* occur in the waters of the northern Pantanal. *Azolla filiculoides* Lamarck and *Azolla mexicana* Presl. There are still many open taxonomic questions about these species, and Svenson (1944), in his concise revision of the genus in the New World, may not provide the final answers on their identification. Both species were observed at the same site along the Rio Bento Gomes during different times of year, and there have been numerous verbal reports of other species occurring in the Pantanal, but no specimens of other species were ever identified during the study. Under the influence of the strong sunlight in the Pantanal, these ferns typically develop a deep red color (Plate 2b) unless shaded by other vegetation, in which case they maintain the bright green of the newly sprouting fronds. The important nitrogen-fixing cyanobacteria, *Anabaena azollae*, which lives in symbiosis with these ferns, was investigated in *Azolla mexicana* by Holst and Yopp (1979), as mentioned above (Section 7.1.1.).

An interesting terrestrial growth form of *Azolla mexicana* was observed during the dry season along the Rio Bento Gomes at a site where a constant outflow of ground water kept the soil moist. The stranded aquatic ferns continued to grow and reproduce vegetatively until the water returned during the next *enchente*. They formed spreading rosettes on the wet ground with a strong contrast between the deep red of the older plants and the bright green of the newly growing fronds.

The significance of these plants as hosts for the nitrogen-fixing cyanobacterium, *Anabaena azollae*, is discussed further in Sections 7.1.1 and 11.1.3.

Dennstaedtiaceae

The large terrestrial fern, *Pteridium aquilinum* (L.) Kuhn var. *arachnoideum* (Kaulf.) Brade, was encountered near wooded areas on more elevated parts of the wetland. It was much more abundant in the upland regions of the watershed than in areas subject to temporary flooding.

Lycopodiaceae

One club moss species was observed in the watershed of the Pantanal, but it was never found in the wetland itself. *Lycopodium* cf. *reflexum* Lam. was observed along the headwater streams on the plateau to the east of Cuiabá. Specimens were observed on fairly bare areas of sand near the edge of water coming from springs.

Marsiliaceae

After the field studies in the Pantanal had been concluded, the strong impression was conveyed from the data that the Pantanal is generally too poor in nutrients to support species of *Marsilia*. During the visits to the wetland, the local species of the genus, *Marsilia polycarpa* Hooker et Grev., was only encountered on the rarest occasions, and always where wastewater rich in minerals was covering the ground. The largest stand grew on the floodplain at the gate to the Fazenda Ypiranga, where water flowing from the ranch bore the wastes of domestic animals directly onto to surface of the sediment. The chemical analyses demonstrated that this location was usually rich in compounds of phosphorus and nitrogen (Section 6). There was also a stand of *M. polycarpa* near Porto Cercado, where similar conditions prevailed. It was completely absent from other locations in the Pantanal.

The taxonomy of the genus *Marsilia* has frequently been revised, but the delimitation of the species is still problematic. The range of the form or forms presently considered to belong to the species *Marsilia polycarpa* extends over a vast area from Mexico to southern Brazil, and it is also reported from the Society Islands and Tahiti (Mickel and Beitel, 1988).

Pteridaceae

The aquatic fern, *Ceratopteris pteridoides* (Hooker) Hieron. (Syn: *Bertholyceratopteris pteridoides*), was sometimes encountered floating on the rivers that cross the Pantanal, buoyed up by its aerenchyme tissue (Fig. 7.5). The species is generally recognized by its swollen petioles and two strikingly different kinds of fronds, one broad and the other with filiform lobes. It usually reproduces by means of latent adventitious, bud-like propagules formed in the leaf axils (Albuquerque, 1981). The species is found only in the New World (Tryon *et al.*, 1990).

Along upland streams in the watershed of the northern Pantanal, *Pityrogramma calomelanos* (L.) Link is the most abundant littoral fern. The species is not at all abundant in the Pantanal and was found only along some of the water courses at the margins of the wetland. However, Scheßl (1997) encountered it on sudd. The species is known throughout the Neotropical Region from Florida to Argentina (Lellinger, 1989).

Figure 7.5: The floating fern, *Ceratopteris pteridoides*, shown on the surface of the Rio Bento Gomes, is remarkable for the distinctly different kinds of fronds borne on each plant.

Salviniaceae

One of the ferns, *Salvinia auriculata* Aublet, has to be rated as a macrophyte that determines the character of large areas of the Pantanal. It is the substrate for important biotic communities, and it affects the structure of adjacent communities, as well. By blocking off large areas of the water's surface from light, it effectively prevents all development of submerged macrophytes where it achieves dominance. It then provides a substrate for the establishment of a pseudoterrestrium, a community of aquatic species, usually belonging to predominantly terrestrial invertebrate families, that live atop the floating plant mats (Heckman, 1994b). These species and their habits will be discussed in detail in Section 9.1.6.

Among the rhizoids of this submerged water fern, the interrhizon community of small invertebrates and fishes develops particularly richly. This is because the rhizoids are long, finely branched, and soft, allowing the invertebrates to get deep enough inside to be completely concealed from the many predators in the water below (Plate 3a).

The characteristics of the species employed for identification were described by Sota (1962). This aquatic fern is native only to the American tropics and subtropics (Lichtenstein, 1939). However, congeneric species occupying comparable ecological niches are found throughout the tropics and are among the dominant aquatic plant species in many parts of the tropical wet-and-dry climatic zone (Heckman, 1997a). Although Schneller (1990) reported that they usually occur in nutrient-rich water, *Salvinia auriculata* is able to develop massive populations in the nutrient-poor water of the Pantanal.

7.7. Dicotyledonae

Acanthaceae

One species of herb in this family, *Justicia laevilinguis* (Nees) Lindau, is a widespread component of the floodplain community during the period of flooding. It is an emergent species of the rainy season that is generally absent from the deeper areas that remain submerged for most of the year. It produces a dehiscent fruit that bursts to scatter the seeds on the surface of the water. Two additional species of *Justicia* were reported from the Pantanal by Scheßl (1997): *J. anagallis* Lindau and. *J. lavandulaefolia* (Nees) Lindau.

This family is also represented in the region of the Pantanal by several terrestrial plants. Species of *Ruellia* are commonly encountered in wooded areas of the Pantanal and in the dry forested regions around its borders. Among the most common of these are *Ruellia gemminiflora* H. B. K. and *Ruellia tweediana* Gris. An unidentified species of this genus was reported by Dubs (1994) from a moist grassland, the habitat called *campo úmido*, near the Rio Pixaim. Dubs (1993) also listed *Ruellia erythropus* Lindau among the species from Mato Grosso and Mato Grosso do Sul, based on its occurrence in European herbaria. Another member of this family typically encountered in the *cerrado* rather than in the wetland is *Stenandrium pohlii* Nees. Pott and Pott (1994) reviewed the reports of acanthacean species in the region of the Pantanal.

Aizoaceae

Glinus radiatus L. is a widespread species that has been classified as a pioneer around the edges of drying lakes and ponds (Pott and Pott, 1994). It occurs occasionally in the northern part of the Pantanal.

Amaranthaceae

A circumtropical wetland species with its origin in South America, *Alternanthera philoxeroides* (Martius) Grisebach, is widespread in the wetlands of Mato Grosso and ranges southward on the floodplains along the river system through Mato Grosso do Sul and into Paraguay (Wolf, 1990). Especially well adapted to the conditions in the tropical wet-and-dry climatic zone, it has developed a life cycle timed to the annual cycle in the Pantanal (Prado *et al.*, 1994). It has been introduced to similar wetlands of North America, Australia, and Asia (Cook, 1974), where it quickly became an important component of the local flora, as in the Mekong Valley of Thailand (Heckman, 1979).

A less abundant species that was observed in the roadside ditches along the Transpantaneira Highway is *Amaranthus lividis* L., a small annual herb. Its relative scarcity may be due to its quality as forage for cattle and other herbaceous mammals. Scheßl (1997) also found *Amaranthus spinosus* L. and *A. viridis* L. in the Pantanal.

Gomphrena elegans Martius is a littoral species that occurs sporadically along the Rio Bento Gomes and beside roadside water bodies in the Pantanal of Barão de Melgaço. Also reported for the Pantanal by Scheßl (1997) are *Gomphrena celosioides* Martius and *G. vaga* Martius.

Several terrestrial members of this family are also encountered along roads in the Pantanal. *Pfaffia glomerata* (Spreng.) Pedersen was commonly observed at the edges of the Transpantaneira Highway and near the high-water line along the Rio Bento Gomes. It is found throughout the warm parts of South America, and Siqueira (1989) reported that it occurs mainly on river banks in the *cerrado* region of Goiás. Information on some of the substances produced by this plant was provided by Shiobara *et al.* (1993). Less common was the *pombeiro-branco*, *Iresine macrophyla* R. E. Fries, a tree found at isolated locations at relatively high elevations near the Transpantaneira Highway. Additional species in this family reported by Scheßl (1997) from the Pantanal include *Alternanthera kurtzii* Schinz ex Pedersen, *A. paronichyoides* A. St. Hilaire, and *Pfaffia iresinoides* (H.B.K.) Spreng.

Anacardiaceae

Among the deciduous species of the shrub savanna that is occasionally encountered at drier locations in the Pantanal is the *conçaleiro*, *Astronium fraxinifolium* Schott. This tree, which reaches 25 m in height, was observed near the Transpantaneira Highway as well as on some of the ranches in the Pantanal. It was also moderately abundant at higher elevations in the catchment area. The wood of this tree is valuable (Guarim Neto, 1991).

Another tree that produces valuable lumber grows at elevations above the floodplain: *Myracrodruon urundeuva* (Engl.) Fr. All. It is also known by its synonym, *Astronium urundeuva* (Fr. All.) Engl. and is locally called the *aroeira*. This species was encountered much more often in the *cerrado* than at the borders of the Pantanal. Guarim Neto (1991) and Dubs (1994) reported the species to be abundant in semi-deciduous as well as dry forests.

The cashew, *Anacardium occidentale* L., is a native tree of the northern Brazilian coastal region and is particularly abundant in Piauí and Mananhão (Lorenzi, 1992). From its original range, it has been transported to many parts of Brazil. At the borders of the Pantanal, these trees are not uncommon on high ground near human habitations, and several of them have been planted on ranches and near tourist facilities in the wetland itself. Another cashew species, *Anacardium humile* Saint Hilaire, is native to the *cerrado* and dry forests surrounding the Pantanal. Its aromatic fruit is not often eaten, although it has potential for cultivation (Pott and Pott, 1994).

Other species of this family that can be encountered along roads in the Pantanal or around the borders of the wetland are *Schinopsis brasiliensis* Engl. and *Tapirira guinensis* Auld.

Spondias lutea L., the *cajá*, is a native species of Africa that has been introduced to South America and may be encountered sporadically at sites in the Pantanal that are not flooded during the high water period. Guarim Neto (1991) found this species to be abundant near the Taiamã Ecological Research Station in the District of Cáceres. Pott and Pott (1994) reported that it requires some calcium in the soil. Its fruit is edible.

The mango, *Mangifera indica* L., has been introduced to all parts of the tropics from its homeland in the mountains of Burma. Mango trees are grown on many of the ranches in the Pantanal, and they are also found growing along roadsides, apparently from discarded seeds.

Annonaceae

Annona phaeoclados Martius is a shrub that grows at various sites near the edges of the hummocks that usually remain above the high water level during the *cheia*.

The congeneric *Annona coriacea* Martius, a small tree reaching a maximum height of about 12 m, grows along the roads at the northern edge of the Pantanal. The tree is also common in the upland *cerrado* and dry forests. Its fleshy flowers are eaten by the insects that pollinate them, and its fruit is edible.

Xylopia aromatica (Lam.) Martius is a tree that reaches about 10 m, which was observed along the roadside near Barão de Melgaço.

Unonopsis lindmanii Fries is a species of dry and gallery forests that occurs in all parts of the Pantanal (Pott and Pott, 1994). It was observed only along the northern section of the Transpantaneira Highway, but it was somewhat more abundant at higher elevations in the Pantanal watershed.

Annona dioica Saint Hilaire, a shrub that reaches a height of about 1 m, is widespread in the dry forest areas near the borders of the Pantanal as well as in upland areas of the watershed, where it is one of the small woody plants that sometimes colonizes pastures. It was rare along the roads in the Pantanal. Its flowers are pollinated by beetles that feed on its fleshy petals (Pott and Pott, 1994).

Duguetia furfuracea (Saint Hilaire) B. and H. was observed only in a few upland areas at the edges of patches of dry forest and along roads. It is a shrub that reaches a height of two to three meters and is sometimes eaten by cattle (Pott and Pott, 1994).

Asimina pygmaea Dun. is another small shrub that is found in the *cerrado* at the northern edge of the Pantanal. It was not observed in the wetland.

Apocynaceae

Rhabdadenia pohlii M. Arg. is a moderately common vine that grows on shrubs in patches of forest along roadsides at the edges of woodlands near the Pantanal. *Fosteronia pubescens* A. P. de Candolle, a large shrub that can reach about 6 m (Pott and Pott, 1994), grows along roads and in gallery forests in the Pantanal but is more abundant in the *cerrado* of the watershed. *Prestonia coalita* (Vell.) Woods is a vine that was observed occasionally at the edge of dense forests along small rivers in the Pantanal.

Dubs (1994) reported the presence of *Secondaria densiflora* A. L. de Candolle in the gallery forest along the Rio Pixaim, and *Tabernaemontana siphilitica* (L. f.) Leeuwenb. from the banks of the Rio Jauru in the Municipal District of Cáceres. *T. siphilitica* was also reported from the Taiamã Ecological Research Station along the Rio Paraguay under its synonym, *Bonafousia juruana* (K. Schumacher) Mgf. (Guarim Neto, 1991).

Several other species in this family are more common at elevations above the wetland and rarely, if ever, observed in the Pantanal itself. *Aspidosperma tomentosum* Martius is a small tree, reaching about 8 m, which is most common in upland dry forest areas, especially where fire had previously been used to clear the vegetation. *Himatanthus obovatus* (M. Arg.) Woods can be observed at many locations along the highway from Cuiabá to Poconé. The yellow oleander, *Thevetia bicornuta* Muell. Arg., is a shrub encountered mainly on the plateau to the east of Cuiabá but also rarely at lower elevations. It was not observed in the Pantanal. In addition to these, *Mandevilla illustris* (Vell.) Woodson has been reported from dry grassland, *campo cerrado*, and *Tabernaemontana flavicans* Willdenow from semi-deciduous forests near Chapada dos Guimarães (Dubs, 1994).

Guarim Neto (1991) reported two additional species to be present along the shores of the Rio Paraguay at the Taiamã Ecological Research Station in the District of Cáceres: *Rhabdadenia macrostoma* (Bentham) M. Arg. and *Thevetia peruviana* (Pers.) K.

Schumacher. Also observed in the Pantanal by Scheßl (1997) were *Catharanthus roseus* (L.) Don, *Mesechites trifida* (Jacquin) M. Arg., and *Rauwolfia mollis* S. Moore.

Aristolochiaceae

Aristolochia cf. *esperanzae* Kuntze is a woody vine that was observed at one location near the edge of the Pantanal, not far from the exposition park of Poconé. The large flowers roughly resemble the head of a bird and contain large numbers of beetles, which presumably pollinate them.

Asclepiadaceae

This family was represented by *Funastrum clausum* (Jacquin) Schl., a perennial vine that occurs at the edge of dense gallery forests along the Rio Bento Gomes and near Birici Stream on the Fazenda Ypiranga 15 km south of Poconé. A second species in this family, *Asclepias mellodora* A. Saint Hilaire, was reported by Dubs (1994) from a *campo úmido* near the Rio Pixaim. Another woody vine in this family observed in various places along the Transpantaneira Highway was tentatively identified as *Schubertia grandiflora* Martius and Zucc. In addition, an unidentified species of *Metastelma* was encountered along roads near the borders of the Pantanal. Two species of this genus are already known from the region: *Metastelma stenolobum* Decaisne in A. P. de Candolle, reported by Dubs (1993), and *Metastelma berterianum* (Spreng.) Decaisne, also referred to as *Tassadia berterianum* (Spreng.) W. D. Stevens, a vine with limited distribution in the wetland (Pott and Pott, 1994).

Balanophoraceae

A species in this family that inhabits the Pantanal is *Helosis brasiliensis* Schott and Endl., a root parasite of angiosperms that prefers moist, shaded locations. Guarim Neto (1991) reported it to be abundant in a gallery forest along the Taiamã River in the District of Cáceres. Dubs (1993) also listed a second root parasite, *Langsdorffia hypogaea* Martius, among the flora in the general region of the Pantanal.

Bignoniaceae

Two species of *Tabebuia* are extremely widespread trees in the Pantanal wetland, and several others are typical species in the woodlands of southern Mato Grosso. *Tabebuia heptaphylla* (Vel.) Tol., locally called the *piúva-do-pantanal* and sometimes referred to as *Tabebuia ipe* (Pott and Pott, 1994), is a species characteristic of the Pantanal floodplain, since it is more tolerant of temporary inundation than most other trees. The leaves are shed during the dry season, and a period of intensive blooming occurs several weeks before the beginning of the rainy season while the trees are still leafless. New foliage begins to appear after the trees have finished blooming. They are prominent in the region of Poconé during the dry season when they are covered with pinkish violet flowers (Plate 3b). While these trees are in flower, their great abundance and wide distribution in the Pantanal is evident. They are common throughout the entire wetland (Pott and Pott, 1994).

The other common species of this genus in the Pantanal, *Tabebuia ochracea* (Chamisso) Standl., has yellow rather than pink flowers. Like *T. impetiginosa*, it loses its leaves and produces flowers during the dry season, but always at a different time, generally later, than the congeneric species.

Tabebuia impetiginosa (Martius) Standl., called the *ipê-roxo* or *piúva-da-mata*, is widespread in the *cerrado* or dry forest vegetation, but it also occurs in sparse gallery forests as well as on relatively high areas within the Pantanal that do not become inundated. It was found to be dominant, accounting for 31% of all trees in the semi-deciduous forests of Mato Grosso do Sul (Dubs, 1994).

In addition to the abundant species, several others that are typical of the dry forests grow at the edge of the Pantanal, and isolated trees are encountered along the Transpantaneira and other roads. *Tabebuia caraiba* (Martius) Bur. and *Tabebuia dura* (Bour. and K. Schumacher) Spreng and Sandl. are both abundant in the dry upland forests but uncommon in the Pantanal. *Tabebuia vellosoi* Tol. occurs sporadically in the Pantanal, generally in or at the edge of gallery forests. In addition, the presence of *Tabebuia aurea* (Manso) Bonpland and Humboldt in woodlands subject to occasional, short-term flooding was reported by Joly (1980) and Pott and Pott (1994).

Along the roads through the Pantanal as well as along the rivers, two other trees in this family are encountered: *Jacaranda cuspidifolia* Martius and *Jacaranda ovalifolia* R. Brown (Syn: *J. mimosaefolia* D. Don). Both have probably been introduced from the upland regions, where they are fairly common. *J. ovalifolia* is sometimes planted as an ornamental tree in Cuiabá.

Anemopaegma arvense (Vell.) Stellfeld ex de Souza is a rather small tree encountered mainly in the drier upland *cerrado*. It is not uncommon in southern Mato Grosso.

An African species in this family has been planted as an ornamental tree in various parts of southern Mato Grosso, especially in the cities and towns. *Spathodea campanulata* Beauv. was frequently observed in Cuiabá and Poconé, and its cultivation on ranches in the Pantanal is therefore to be expected.

In addition to the trees, the family is represented by several species of *Arrabidaea*, which are woody vines that spread over the trees and shrubs in many parts of the wetland. The large flowers of one prominent species are yellow, while those of another are bluish violet. The genus is in urgent need of revision, and until this is undertaken, the species cannot be identified with certainty.

Bombacaceae

This family is represented by a woody species, *Pseudobombax tomentosum* Martius ex Zucc., which is a common tree in the upland dry forest areas that rarely occurs along roads near the northern edge of the Pantanal.

Boraginaceae

Among the littoral species along the Transpantaneira Highway and on river banks is *Heliotropum indicum* L. These plants are generally found isolated or in small groups in the Pantanal, typically among supralittoral vegetation and not on temporarily flooded areas. The congeneric species, *Heliotropium filiforme* H. B. K., occurs sporadically along roadsides, mainly at locations partially shaded by trees or hills. In addition to these, Scheßl (1997) reported the presence of *Heliotropium lagoense* (Warm.) Gürke and *Heliotropium procumbens* Mill. on the floodplain.

One species of tree in this family, *Cordia glabrata* (Martius) A. L. de Candolle, is a component of the semi-deciduous forests around the Pantanal. Small trees of this species were also observed to occur along the roads in the Pantanal itself. Other members of this family, including *Cordia sellowiana* Chamisso, were observed only in the dry upland forests but not in the Pantanal itself.

Burseraceae

One of the widespread trees in the region of the Pantanal is *Protium heptaphyllum* (Aubl.) March., which was observed at various locations at higher elevations throughout the northern Pantanal. According to Pott and Pott (1994), this aromatic tree, which reaches about 10 m in height, is browsed on by cattle. It is widespread in tropical and subtropical South America.

Cabombaceae

A species belonging to a genus of popular aquarium plants, which, according to the last revision by Brako and Zarucchi (1993), should be called *Cabomba furcata* Schultes and Schultes filius, is a major component of the floodplain flora of the Pantanal. After the first flooding of the dry areas during the *enchente*, this species begins to grow rapidly and has produced dense underwater beds by the *cheia*. At this time, its pinkish purple flowers begin to appear above the surface of the water for pollination (Fig. 7.6), after which the stems curve downward to permit the development of the fruit to be completed underwater. There has been much controversy over the taxonomy of this species, and in most recent publications, it has been called *Cabomba piauhyensis* Gardner in Hook. The taxonomic revisions have been made mainly or exclusively from material collected in the northern part of South America and Central America. In spite of its abundance, its distribution south of the Amazon region went unnoticed by botanists until recently. Fassett (1953), for example, knew of no reports of this species in the watershed of the Rio Paraguay.

Figure 7.6: The flowers of *Cabomba furcata* extend above the surface of the water on the floodplain during the *vazante*.

Cactaceae

Cacti were fairly abundant in the region of the Pantanal, occurring at the edges of the hummocks between the woody vegetation and the floodplain, that is, in the supralittoral zone during the *cheia*. The species were reported by Scheßl (1997) to be *Cereus kroenleinii* (Kiesling) N. P. Taylor and *C.* cf. *bicolor* Rizzini and Mattos Filho. In addition, one large cactus, probably *Cereus peruvianus* Mill., had been planted in front of the ranch house on the Fazenda Ypiranga.

Capparideae

An unidentified species, *Capparis* sp., was observed along the roads at the northern edge of the Pantanal. Its occurrence was sporadic.

Caricaceae

The papaya, *Carica papaya* L., has been cultivated in Central and South America since Precolumbian times. The species has been transported around the world and has become one of the most widely cultivated tropical fruits. In the Pantanal and adjacent regions, it is found on many of the ranches but not on temporarily inundated areas (Fig. 7.7)..

Figure 7.7: The papaya, *Carica papaya*, is one of the introduced plants that is most frequently cultivated near human habitations in the Pantanal.

Cecropiaceae

Cecropia pachystachya Trec. is an abundant tree along the roads of southern Mato Grosso (Fig. 7.8). It is generally considered a species that rapidly colonizes localities that have been disturbed by human activities: a pioneer tree species noted for its rapid growth. It is rare in the Pantanal, occurring only at the edges of the wetland, generally near roads.

Combretaceae

This family includes some woody vines that are widespread and abundant at the water's edge in many parts of the Pantanal. *Combretum laxum* Jacquin, often referred to as *Combretum jacquini* Grisebach, is a species that forms dense thickets in many parts of the Pantanal. It is common along the Transpantaneira Highway, where it is rooted just at the high water level on the embankment. Blooming is most intensive during the *vazante* and *seca*, so that the seeds can be dispersed by the wind before the next *enchente*. In the *cerrado* of Goiás, Silva and Valente (1996) found that this species can grow to a height of 7 m along streams and take on the form of a tree as it becomes older.

Another member of this family is a small tree that is common along the embankment of the Transpantaneira Highway. *Combretum lanceolatum* Pohl was observed mainly near bridges under which small rivers flowed. In addition, *Combretum leprosum* Martius was reported by Dubs (1994) from semi-deciduous and deciduous forests near the Rio Juaru in the Municipal District of Cáceres, and *Combretum duarteanum* Cambess was reported to occur together with *C. laxum* in Mato Grosso do Sul.

Figure 7.8: *Cecropia pachystachya*, recognized by its large compound leaves, is considered an indicator of areas disturbed by human activity. It is common along the roads and on ranches in the Pantanal.

A native of the region from Madagascar through India to the Ryukyu Islands, *Terminalia catappa* L. has been planted in most cities and towns of southern Mato Grosso. This tree is already found at various locations around the outskirts of the Pantanal.

Compositae

Although there are no fully aquatic members of the Compositae in the Pantanal, a number of littoral species are quite abundant at the water's edge, especially where human activities have modified the native vegetation. In addition, several composites are pioneer plants on areas affected by man, particularly roadsides and near human habitations. Although they are not wetland species, they penetrate into the Pantanal along the roadways and on the ranches. Their seeds are probably dispersed mainly by the vehicles of ranchers, fishermen, and tourists.

Vernonia scabra Pers. was an extremely abundant shrub along the Transpantaneira Highway from the region of the Rio Bento Gomes to the Rio Pixaim. Elsewhere, it occurred sporadically. According to Pott and Pott (1994), it is found in all parts of the Pantanal. This plant grows mainly during the rainy season and blooms after the rains have almost ceased in the *vazante* and early *seca*. Its wind-borne seeds are dispersed during the late *seca*. The species is robust and resistant to adverse conditions, including fire, sprouting again from underground stems after the emergent parts of the plants have been burned. Typically, the leaves of plants along the roadsides are covered by a heavy coat of dust during the dry season, but they are able to survive this until the subsequent rainy season. The hardiness and growth characteristics of the species have made it a troublesome weed on the pastures used for cattle. These plants are eaten by capybaras, and their seeds, by birds (Pott and Pott, 1994). Guarim Neto (1984) reported the presence of the congeneric *Vernonia ferruginea* Less. in the upland regions of southern Mato Grosso, but the potentially troublesome *Vernonia rubricaulis* H. B. K., reportedly responsible for poisoning cattle in Mato Grosso (Tokarnia and Döbereiner (1982), was not observed during the study.

Along the Transpantaneira Highway, *Bidens gardneri* Bak. was observed regularly near the water's edge. In similar locations along the roadsides of the Pantanal is the less abundant vine, *Mikania capricorni* Rob., and the perennial herb, *Eupatorium macrocephalum* Less. In addition, Dubs (1994) reported the presence of *Wulffia baccata* (L.) Kuntze on the banks of the Rio Pixaim.

Four additional roadside plants, which were encountered in many locations in southern Mato Grosso, *Ageratum conyzoides* L., *Eupatorium candolleanum* Hook. and Arn., *Eupatorium asperulaceum* (DC) Baker, and *Aspilia latissima* Malme grow rapidly during the rainy season. These species are apparently being introduced to the Pantanal by the vehicle traffic on the Transpantaneira Highway and other roads through the wetland.

Pacourina edulis Aubl. was encountered only at the lower elevations of the Pantanal near Campo Jofre. It was observed blooming on areas recently left dry by the receding waters during the *vazante.*

Tridax procumbens L., is a common plant of waste places in and near the city of Cuiabá. In the Pantanal, it is found almost exclusively along the embankment of the Transpantaneira Highway, where it is not abundant. Its life span is only 80 to 100 days, and it produces a large number of seeds (Lorenzi, 1991), making it an ideal pioneer species under the conditions in southern Mato Grosso.

According to Dubs (1994), *Urolepis hecatantha* (D.C.) R. M. King and H. Rob., also called *Eupatorium hecatanthum* (DC) Baker, is common in the Pantanal of Mato Grosso do Sul. Its presence in the northern Pantanal was noted by Scheßl (1997).

Convolvolaceae

The most abundant genus of this family in the Pantanal is *Ipomoea*, which is represented by both aquatic and terrestrial species. *Ipomoea carnea fistulosa* (Martius ex Choisy) D. Austin is one of the dominant wetland species, which typically grows on areas that are dry for rather long periods during the rainless seasons and under water during the *cheia*. It has a stiff but flexible woody stem, which grows upright and generally remains thin, having nearly the same diameter at the base and near the tip.

Another abundant member of the genus is *Ipomoea sagittata* Poiret, which is a thick, fleshy vine that creeps on the surface of the floodplain. The plants bloom intensively during the dry season shortly after the flood waters have receded. Blooming continues throughout the dry season. The leaves are frequently eaten by chrysomelid beetles.

Merremia umbellata (L.) Hall. is a common species of herbaceous vine along roads in the northern Pantanal. It is often encountered in the literature under its synonym, *Ipomoea umbellata* (L.) Hall., as discussed by Pott and Pott (1990). *Aniseia cernua* Moric. is another vine which climbs over the roadside plants and also over emergent aquatic species in the temporarily flooded roadside marshes, such as *Thalia geniculata*. Its white flowers are most abundant during the *vazante*.

The large genus *Merremia* is also represented in the Pantanal by *Merremia* cf. *capitata*, which is a vine seen along the roads in various places, both in the wetland and in the surrounding upland areas. The genus, which seems to be very rich in species in Brazil, is in need of taxonomic revision.

Also fairly common in the region of the Pantanal is *Ipomoea quamoclit* L., which generally grows in dry open fields and waste places, but not on the floodplain of the Pantanal. It is a common roadside plant in the city of Cuiabá.

Costaceae

At the edges of densely wooded areas in the upland parts of the Pantanal watershed, an unidentified species of *Costus* was observed occasionally. The presence in similar locations of a second species in this genus, *Costus* cf. *arabicus* L., was reported by Pott and Pott (1994) and Dubs (1994) from semi-deciduous forests. This species seem to have been introduced as an ornamental plant from Asia.

Dilleniaceae

One of the species able to tolerate waterlogged soils for part of the year is *Curatella americana* L. It is therefore one of the few that is observed near the edges of permanent water bodies that flood over their banks during the *cheia* and in sparse scrubland on the uppermost areas of the floodplain. This species, locally called the *lixeira*, seems to be more abundant in the southern part of the wetland (Dubs, 1994) and in the upland areas to the north of the Pantanal. Oliveira-Filho and Martins (1991) found it to be the predominant species on the *murundu cerrado*, the dry forests that develop on earthmounds, where it far outnumbered all other species of woody plant. It was roughly second in the dominance ranking in the pebble *cerrado* and was also relatively abundant in the phyllite *cerrado*. However, it was completely absent from the interfluvial and sandstone *cerrados*. Oliveira-Filho and Martins (1991) further noted that this plant is more often encountered growing as a shrub with several small trunks than as a small tree with a single trunk.

This family is also represented in southern Mato Grosso by *Davilla elliptica* Saint Hilaire and *Davilla grandiflora* Saint Hilaire & Tulasne, two woody shrubs that were found by Oliveira-Filho and Martins (1991) in the dry forest areas near Cuiabá.

Droseraceae

The insectivorous *Drosera sessilifolia* Saint Hilaire was not personally observed in the Poconé District, but it was rather abundant on some small *campos úmidos* fed by seeping ground water near the Paraguay River in Cáceres. Silva (1996) reported the species to be widely distributed throughout Brazil, Venezuela, and Guyana.

Erythrolaceae

Two species in this family, *Erythroxylon anguifugum* Martius and *Erythroxylon* cf. *suberosum* St. Hilaire, were found in the mound savanna by Scheßl (1997).

Euphorbiaceae

One aquatic species of this family, *Caperonia castaneifolia* (L.) Saint Hilaire, is a characteristic species of the flooded areas of the Pantanal. Although few vegetative plants survive the dry season, it appears quickly after the first flooding, and its stem continues to elongate to keep the upper part of the plant above the surface of the water. Blooming begins during the high water period and intensified greatly during the *vazante*. The species is characterized by two distinct leaf forms. One is broadly elliptical, and the other, long and narrow with the edges serrated like a chestnut leaf, a characteristic that gives the species its Latin name (Fig. 7.9).

Another aquatic species found in isolated groups at the edges of some of the streams and small rivers draining the Pantanal is *Phyllanthus lindbergi* Muell. Arg. The ecology of this species is poorly known. *Alchornea castaneifolia* (Willd.) A. Juss. is a species found on sudd (Scheßl, 1997).

Figure 7.9: Emergent stems of *Caperonia castaneifolia* extend above the shallow water of the floodplain during the *cheia*. The species was never dominant, but it is extremely widespread. The shape of the leaves resemble that of the chestnut.

Euphorbia thymifolia L. is a European species that was introduced to South America and has found a niche as a prostrate herb in the pioneer community on bare ground. In the Pantanal, it was abundant on the access roads to the ranches, especially where these roads were reinforced by gravel, such as on the Fazenda Ypiranga.

One of the less common herbs encountered in the upland regions surrounding the Pantanal is *Dalechampia* cf. *weddelliana* Baillon. It was not observed in the wetland itself during the study, but it was encountered occasionally along roads to the northeast of Cuiabá.

This family is also represented by woody plants in southern Mato Grosso. *Jatropha vitifolia* Mill. and *Jatropha elliptica* (Pohl) Bail. were observed along some of the roads at the northern edge of the Pantanal and along the Rio Bento Gomes upstream from the wetland.

Sebastiana schottiana Muell. Arg. is encountered in the *cerrado* around the borders of the Pantanal as well as in hilly regions to the north of the wetland. It sometimes appears in the literature under its synonym, *S. brasiliensis* Spreng.

Mabea paniculata Bruce ex Benth. was observed in the northern part of the wetland, mainly along the roads. These were tall shrubs or small trees that were most abundant along the edges of upland evergreen forests, generally along streams and small rivers.

Croton spp. are other widespread trees that was most common in the dry upland forests. *Croton cuyabensis* Pilg. occurred in the northern Pantanal in places subjected to flooding (Scheßl, 1997).

Flacourtiaceae

This family is represented in the dry forests surrounding the Pantanal by *Casearis sylvestris* Sw., a large shrub that sometimes reaches the size of a small tree. Although nowhere abundant, Oliveira-Filho and Martins (1991) found it on every kind of dry forest near Chapada dos Guimarães, except for the sandsone *cerrado*.

Haloragaceae

Myriophyllum mattogrossense Hoehne is one of the fine aquarium plants that is native to the Pantanal. It is the only species of the family with deep red leaves (Plate 4a), although it may lose this color when grown in the shade. Its taxonomic status has long been in doubt, but (Orchard and Kasselmann, 1992) thoroughly investigated its biology and confirmed the validity of the species. Difficulties in recognizing the species, which is ordinarily characterized by densely clustered leaves and its deep red color, have frequently arisen because these characteristics disappear when the plants are grown in subdued light. They not only lose the color, but the stems become more elongated, making the leaves lose the densely clustered appearance.

The species appears shortly after the dry areas become flooded during the *enchente*, and luxuriant submerged beds have developed by the *cheia*. Intensive blooming begins at this time, and by the *vazante*, the vegetative parts of the plants are rapidly dying off. By the *seca*, the plants have disappeared from the floodplains.

Hydrophyllaceae

A species well adapted to the transitions from wet to dry and dry to wet is *Hydrolea spinosa* L., which is extremely variable in its morphology throughout the Pantanal (F. Martins, pers. comm.), raising the suspicion among some botanists that more than one species might be referred to under the name. Although the plant has thorns, it is a favorite food of the cattle, which frequently feed on the newly growing stems before the thorns have become hard. The older stems are somewhat woody. The species generally begins to develop rapidly with the onset of the rainy season. The plants continue to grow vegetatively after the fields become flooded. The upright stems elongate as the level of the flood water rises, keeping the upper parts of the plant well above the surface of the water. Some of the plants begin to bloom during the *cheia*. As the water level begins to fall, blooming intensifies, and a great many flowers are produced as the water recedes and exposes the waterlogged sediment. The upright stems persist during the dry period, even on some areas that remain dry for a few months, but the species is generally absent from the highest levels of the floodplain.

The genus *Hydrolea* is well represented by species in similar tropical floodplain habitats throughout much of the world (Davenport, 1988).

Labiatae

Among the emergent aquatic herbs common during the period of flooding are species of *Hyptis*. One of the commonest species on the moderately shallow flooded areas of the northern Pantanal is *Hyptis lorentziana* O. Hoffmann, which produces many new shoots during the early rainy season and begins to bloom at about the time the water level is highest. Blooming intensifies during the months of little rainfall and continues until the areas completely dry out. The plant is moderately resistant to drying, and if the period of desiccation does not last too long, the plants are able to resume vegetative growth after the freshet. Thus, the plants are generally much more abundant in the lower lying parts of the temporarily flooded areas on the floodplain.

Hyptis suaveolens (L.) Briq. and *Hyptis crenata* Pohl are common roadside species in the upland dry forest region in the catchment area of the Upper Paraguay River. They are sometimes observed along roads at the borders of the Pantanal.

Other species of the genus also occur in the Pantanal, including *Hyptis lappacea* Bentham, *Hyptis hygrobia* Briquet, *Hyptis mutabilis* (Rich.) Briquet, *Hyptis hirsuta* H.B.K., *Hyptis brevipes* Poit., *Hyptis recurvata* Poir, *Hyptis velutina* Pohl ex Bentham, *Hyptis microphylla* Pohl ex Bentham, and *Hyptis lutescens* Pohl ex Bentham (Scheßl, 1997).

A herbaceous species in this family, *Leonotis nepetifolia* (L.) R. Br., is a common plant found most frequently in partially shaded locations at the edges of woods, more in the upland parts of the watershed than in the wetland itself. It was found to be widespread but never abundant.

Plectranthus amboinicus (Lour.) Spreng., also known by its synonym, *Coleus amboinicus* Lour., is a plant that was introduced to Brazil and has spread to various parts of the country. It has become one of the roadside plants in the region of the Pantanal.

Lecythidaceae

This family is represented by a tree species that is not uncommon in the gallery forests in the region of the Pantanal: *Cariniana* sp. It is a large tree found along upland streams as well as along rivers in the wetland.

Leguminosae *sensu lato*

This family is treated here in its broader sense, although various recent taxonomic works have broken it up into several families.

Legumes of many species are widespread in the Pantanal, but few of them are aquatic. One of the most fully aquatic species of the family is *Aeschynomene fluminensis* Vell., sometimes called *A. flumensis*, which is one of the only woody plant species that grows on the deeper parts of the floodplain. It is a low shrub, usually with a thickened base and an erect central stem, which tapers considerably just above the base. It is distinguished readily from the congeneric *Aeschynomene sensitiva* Sw., which is widespread in generally drier locations throughout southern Mato Grosso, by its green rather than blackish fruit. Where cattle graze in the Pantanal, the *A. fluminensis* plants are characteristically chewed down to their woody bases, in spite of their thorns. This limits the local abundance of the species considerably. Dubs (1994), however, reported large, nearly monospecific stands of this species in Mato Grosso do Sul. Whether cattle grazing in the region is light was not reported.

The aggregation of rooted emergent and floating plants on the broad, temporarily flooded littoral area along the Paraguay River on the Fazenda Santo Antonio das Lendas included several aquatic legumes that were not at all abundant in the Poconé District. Among these were *Sesbania exasperata* H. B. K., an erect emergent plant, and *Neptunia prostrata* Baillon (Syn. *N. oleracea* Lour.), a prostrate plant with horizontal stems and modified roots inflated with air that float on the surface of the water (Fig. 7.10). Another less abundant species in this river bank aggregation was *Mimosa pellita* H. and B. These species grow and flower during the rainy season. Their vegetative parts die off during the *seca*, when the riverbank dries out. *M. pellita* is circumtropical and is well known from Thailand and Australia, and it has sometimes been referred to as *M. pigra* L. in the literature.

Littoral species that can grow on areas flooded for short periods of time include *Discolobium pulchellum* Bentham and *Discolobium psoraliaefolium* Bentham. *Calliandra parviflora* Bentham is a common species in forested areas of the Pantanal, including some temporarily flooded areas near Poconé. It also occurs in dry upland forests.

Chamaecrista serpens (L.) Greene is a slender herb that grows along roadsides and on river banks in the northern part of the Pantanal, where it is represented by the variety *C. s.* var. *grandiflora* (Bentham) Irwin and Barneby. The variety *serpens* also occurs in Brazil, but the ranges of the two varieties do not overlap, and *grandiflora* is the only one known from the region of the Pantanal (Irwin and Barneby, 1982).

Indigofera suffruticosa Mill., called *mata-pasto-preto*, was encountered occasionally along roadsides near Poconé. Plants of this genus are the sources of a vivid indigo dye. This species, sometimes known by its synonym, *I. anil* L., is typical of the secondary vegetation at disturbed sites (Lewis, 1987).

Figure 7.10: The floating stems of *Neptunia prostrata* on the surface of the water near the shores of the Rio Paraguay.

Other roadside plants include legumes introduced to Brazil and spread to different parts of the country by man. These include species observed in and on the outskirts of the Pantanal, such as *Macroptilium* cf. *lathyroides* Urb., *Pithecolobium tortum* Martius, *Pithecolobium saman* Bentham, and *Dioclea floribunda* Bentham. *Pithecolobium* is alternatively spelled *Pithecellobium*.

Relatively uncommon terrestrial, herbaceous legumes are also encountered along the Transpantaneira Highway, including *Centrosema* cf. *pubescens* Benth., a few of which were observed blooming at isolated locations. The genus is also represented in the Pantanal by two species not observed during the study. In the southern parts of the wetland in Mato Grosso do Sul, Pott and Pott (1994) observed *Centrosema vexillatum* Bentham growing on temporarily flooded areas and *Centrosema brasilianum* (L.) Bentham among secondary vegetation along the edges of hummocks and in unflooded upland areas.

Among the terrestrial species, many of the trees as well as herbs in the Pantanal belong to this family. *Senna occidentalis* (L.) Irwin & Barneby, *Senna aculeata* (Bentham) Irwin & Barneby, and *Senna alata* (L.) Irwin & Barneby are widespread woody shrubs on areas of the floodplain that remain under water for only short periods of time during the rainy season. *S. occidentalis*, which reaches a maximum height of about 2 m, is widespread and abundant in the northern Pantanal, while the other two species, shrubs reaching 3 m, are considerably rarer, although *S. aculeata* forms prickly thickets at a few locations at the edges of hummocks on the floodplain. At the present time, these species may also be encountered in the literature under the names *Cassia (Senna) occidentalis* L., *Cassia (Senna) aculeata* (Bentham) Irwin and Barneby, and *Cassia (Senna) alata* L., as cited by Irwin and Barneby (1981). It is not unlikely that *Senna* will become generally accepted as a genus rather than a subgenus, as various authors have already adopted the change as proposed by Irwin and Barneby (1982).

Dipteryx alata Vog. is a common species of the gallery forests and woodlands around the margins of the Pantanal. It is a hardwood tree valued because of the natural resistance of its wood to decay and attacks by insects (Lorenzi, 1992). It also produces edible seeds. The quality of its wood has resulted in a great demand for these trees, which apparently accounts for the disappearance of the species throughout much of its former range. However, Dubs (1994) found it to be still rather abundant in the semi-deciduous forests of the Pantanal in Mato Grosso do Sul, and it was found to be the second most dominant species of the *murundu cerrado*, a dry woodland that develops on earthmounds from 0.2 to 20 m in diameter and 0.1 to 2 m above the level of the surrounding ground. This kind of *cerrado* is most common in the upland regions of southern Mato Grosso (Oliveira-Filho and Martins, 1991).

A smaller tree in the gallery forests of the northern Pantanal is *Andira inermis* H. B. K., which was observed along the Rio Bento Gomes upstream from the Pantanal. Another familiar tree in the gallery forests of the northern Pantanal is *Pterocarpus rohrii* Vahl.

Dimorphandra mollis Bentham is observed at various locations in the Pantanal where the terrain is high enough to keep the roots above the flood waters during the *cheia*. Oliveira-Filho and Martins (1991) found the species in pebble, phyllite, and interfluvial *cerrado* in the upland areas of the watershed, but the species seemed to be absent from the *murundu* and sandstone *cerrado* areas that they investigated. Somewhat more common in the upland dry forests is *Hymenaea stigonocarpa* Martius, which Oliveira-Filho and Martins (1991) reported to be fairly common on the *murundu*, pebble, phyllite, and interfluvial *cerrado*. However, they did not find it on the sandstone *cerrado*, which covers rocky outcroppings, sometimes covered by thin layers of sandy soil, at the upper edge of the escarpments delimiting the plateau of Chapada dos Guimarães. *Vatairea macrocarpa* Aublet is another tree that they found to be rather abundant throughout all kinds of dry forests, except the sandstone *cerrado*. It also penetrates into the upland areas separating parts of the Pantanal floodplain.

Other woody legumes of the upland areas in the watershed include two species of *Sclerolobium*, which seem to display a habitat selectivity that minimizes their occurrence together in the same areas. While *Sclerolobium aureum* Bentham was found to be rather abundant in areas of *murundu*, pebble, and phyllite *cerrado*, only *Sclerolobium paniculatum* Vog. was encountered in interfluvial and sandstone *cerrado* (Oliveira-Filho and Martins, 1991). Furthermore, *S. paniculatum* was represented in the interfluvial *cerrado* only by the variety *S. p.* var. *rubiginosum* (Tulasne) Bentham and on the sandstone *cerrado* only by *S. p.* var. *subvelutinum* Dwyer.

Among the trees encountered along the roads in the Pantanal are *Caesalpinia peltophoroides* Bentham, *Acacia* sp, *Cassia grandis* L.f., *Cassia ferruginea* (Schrader) Schrader ex A. P. de Candolle, *Anadenanthera falcata* (Bentham) Spreg., *Hymenaea courbaril* L., *Amburana cearensis* (Fr. All.) A. C. Smith, *Enterolobium contortisiliquum* (Vel.) Morong, *Bowdichia virgilioides* Kunth., *Desmodium canum* (Gmel.) Schinz. and Thell., *Desmodium incanum* A. P. de Candolle, and *Desmodium adscendens* A. P. de Candolle. All of these species are found throughout southern Mato Grosso and penetrate into the wetlands mainly along the roads and banks of the main rivers but also occasionally on the hummocks. Some of these species are planted as ornamental trees (Lorenzi, 1992).

In addition to the trees, many small woody legumes are encountered along the roads in the Pantanal. *Chamaecrista nictitans patellaria* (A. P. de Candolle) Greene, often referred as *Cassia patellaria* A. P. de Candolle by authors who regard *Chamaecrista* as a subgenus of *Cassia*, was observed at many sites along the Transpantaneira Highway near Poconé and occasionally along the same road in the wetland itself. The conspecific *Chamaecrista desvauxii* Collad. was also seen occasionally on the outskirts of the wetland. *Cassia tora* L., also referred to as *C. obtusifolia* L. and *Senna obtusifolia* (L.) Irwin and Barneby, is commonly encountered along the roads in the Pantanal. Various mimosas also grow in similar locations in the Pantanal and a few meters above the high water level in the wetland. *Mimosa obtusifolia* Willdenow and *Mimosa adenocarpa* Bentham were observed at various times along the Transpantaneira near Poconé. Other species of *Mimosa* are also present in the Pantanal, especially along the roads, but taxonomic complexities within this genus make identification difficult. A taxonomic revision of the genus is urgently needed.

The genus *Bauhinia* is well represented in the region of the Pantanal, and the number of species is apparently being increased by introductions from other parts of South America. *Bauhinia mollis* (Bong.) Dietr. and *Bauhinia pentandra* (Bong.) Vog. are abundant in the dry forests surrounding the Pantanal. They reach heights of only 4 m and 5 m, respectively (Lorenzi, 1992), and are often pioneer woody species that invade pasture land. *Bauhinia forficata* Link is a typical tree of the Atlantic forest (Lorenzi, 1992), which has been introduced in limited numbers to the cities of Mato Grosso, apparently as an ornamental plant. *Bauhinia glabra* Jacquin is a common woody vine along the Transpantaneira Highway, especially along the stretch near Poconé. *Bauhinia longifolia* Link is apparently the smallest member of the genus encountered in southern Mato Grosso. It is a creeping woody plant that is encountered in the *cerrado*, along roadsides, and at the edge of cattle pastures. It is a species introduced to Brazil that has spread particularly rapidly in regions of cattle ranching.

One small creeping species that forms dense "lawns" is *Arachis pintoi* Krap. and Greg., a species that does not grow on areas that are temporarily flooded. It is sometimes planted around public buildings in Cuiabá.

Pterodon emarginatus Vog., sometimes known by its synonym, *Pterodon pubescens* Bentham, is a common species of tree that occurs sporadically in interfluvial *cerrado*, in which the soil is a fine, sandy alluvium (Oliveira-Filho and Martins, 1991). The species is present but not abundant in the Pantanal.

A small tree reaching a height of only about 6 m, *Stryphnodendron obovatum* Bentham, formerly referred to by its synonym, *Stryphnodendron adstringens* (Martius) Coville, is abundant in various kinds of dry forest at the borders of the Pantanal as well as along roads and at the edge of small forests in the wetland. Pott and Pott (1994) designated it as a typical species of the upland dry forests.

A somewhat similar distribution has been reported for *Plathymenia reticulata* Bentham, a tree that reaches a height of about 10 m (Pott and Pott, 1994). It was observed in many locations on the plateau to the east of the Pantanal.

Trees belonging to the genus *Inga*, presently considered to be the species *Inga affinis* A. P. de Candolle and *Inga fagifolia* Willdenow, are encountered among mixed stands of trees in southern Mato Grosso. *Inga uruguensis* Hooker and Arnotti is a small tree reaching a maximum height of about 10 m. It is found sporadically in gallery forests and among stands of trees in and near the Pantanal. Less abundant are *Inga heterophylla* Willdenow and *Copaifera langsdorffi* Desv., which are relatively rare in the region of the Pantanal. Oliveira-Filho and Martins (1991) observed these trees in the *cerrado* near Chapada dos Guimarães. Lewis (1987) stressed the need for a revision of this genus, and the taxonomy of the local *Inga* trees may eventually be changed. The flowers produced are noted to be especially good sources of nectar for bees and apparently also for birds, as well.

Senna splendida (Vog.) Irwin and Barneby and *Senna georgica* Irwin and Barneby are both shrubs or arborets with long branches that can entangle among other plants somewhat like vines. *S. georgica*, represented by the variety *georgica* that reaches a height of 5 to 6 m, is obviously an upland species that grows on areas above the elevation of the wetland, while *S. splendida* var. *splendida*, which grows to a height of 6 or rarely 10 m, is not a native to southern Mato Grosso at all, but has been planted, apparently for ornamental purposes, in various locations near the Pantanal, as it has in other parts of Brazil, probably Paraguay, and even Africa and the Malay Peninsula (Irwin and Barneby, 1982). A considerable number of ornamental trees belonging to the Leguminosae are encountered in southern Mato Grosso. The roadside species, *Bergeronia sericia* Micheli, ranges southward into Paraguay and Argentina (Burkart, 1952). It is seldom encountered in the open *cerrado*, perhaps because it is not resistant to fire (Pott and Pott, 1994).

The dry and semi-deciduous forests surrounding the Pantanal support the *angico*, *Anadenanthera peregrina* (L.) Spegazzini, a tree commonly cut for lumber in Mato Grosso (Guarim Neto, 1991). *Andira cuyabensis* Bentham also occurs in the dry forest areas of southern Mato Grosso. Oliveira-Filho and Martins (1991) reported it to be one of the most abundant trees on the *murundu cerrado*, or earthmound forest, which they selected for study in the region between Cuiabá and Chapada dos Guimarães. It also occurred in the pebble and interfluvial kinds of *cerrado*, distinguished most readily by the upper soil layer, which consists of quarzite pebbles intercalated with phyllite on the pebble *cerrado* and fine alluvial sand on the interfluvial *cerrado*. Much less abundant on the pebble *cerrado* was another legume, *Diptychandra aurantiaca* Tulasne, which is also encountered occasionally by these authors in natural forests, called *cerradão* (Section 2), on the phyllite *cerrado*.

Another common species in the dry forests surrounding the Pantanal is *Machaerium acutifolium* Vog., a tree that reaches a height of about 14 m. Less abundant is the conspecific *Machaerium aculeatum* Rodd., which is most abundant in the towns and near ranches. It may have been introduced to the region, as Lorenzi (1992) did not report the species to be native to Mato Grosso. The tree, *Clitoria fairchildiana* Howard, is present along roads in the dry forests near the border of the wetland. It produces woody beans that explode with a loud bang when ripe, widely scattering the seeds. Lorenzi (1992) reported the species to be a rainforest tree from the Amazon region and northeastern Brazil, but it also appears to be abundant in southern Mato Grosso.

Among the small, creeping plants that grow at the edge of roads and on bare areas near human habitations is *Indigofera sabulicola* Bentham. It was observed along the Transpantaneira a few kilometers south of Poconé and near the Rio Bento Gomes.

Erythrina fusca Lour. is a tree that reaches a maximum height of about 15 m and occurs in moist forests along the road north of the Fazenda Santo Antonio das Lendas near the Rio Paraguay. Pott and Pott (1994) reported its occurrence in temporarily flooded gallery forests along that river.

Another abundant tree in the region surrounding the Pantanal as well as in wooded areas of the wetland itself, especially along roads, is *Anadenanthera colubrina* (Vel.) Brenan, which reaches a maximum height of about 20 m. According to Pott and Pott (1994), the species is represented in southern Mato Grosso by the variety *cebil* Grisebach, distinguished from the variety *colubrina* by its yellow rather than white flowers. It has appeared in the literature under the synonyms *Anadenanthera macrocarpa* (Bentham) Brenan and *Piptadenia macrocarpa* Bentham.

Machaerium aculeatum Raddi is a somewhat rare species that inhabits woodland with sandy soil. It is a tree that reaches a maximum height of about 15 m (Pott and Pott, 1994).

The Leguminosae is also represented in the Pantanal by the flame tree, *Delonix regia* (Boj. ex Hooker) Raf., which received its name because it appears to be in flames while it is blooming, generally during the late dry season. The species is native to Madagascar. It has been introduced to all parts of the tropics as an ornamental tree. Flame trees are not only present in the cities but also on ranches and along the roads in the Pantanal.

In addition to these species, Dubs (1994) reported the presence of *Swartzia jorori* Harms among secondary vegetation near the Rio Pixaim and *Indigofera lespedezoides* H.B.K. and *Vigna lasiocarpa* (Bentham) Verdc. from secondary forests and along the banks of the Rio Jauru near Cáceres, respectively. Pott and Pott (1994) reported *Swartzia jorori* to be abundant in various kinds of forests in Mato Grosso.

Lentibulariaceae

Certainly, the poverty of nutrients in the Pantanal during the high water periods contributes to the competitive success of the *Utricularia* species there. Their ability to trap small animals and utilize their nitrogen and phosphorus compounds for their own growth is a factor that promotes their development during the *cheia*, when they become dominant components of the submerged flora. Two species are particularly common on the floodplains of the northern Pantanal, *Utricularia foliosa* L., characterized by its branching stems and relatively large emergent yellow flowers, and *Utricularia poconensis* Fromm-Trinta, with long, nearly unbranched stems and emergent white flowers with brownish purple or pinkish markings. Both species share the same habitat preferences and grow vegetatively at the same time of year (Fig. 7.11).

Figure 7.11: The shallow, clear waters of the floodplain are typically colonized by two common species of plant that capture small animals in bladder-like traps formed on the leaves to supplement their nutrient supplies. Both *Utricularia foliosa* with regularly branching stems and *Utricularia poconensis* with its irregular, almost unbranched stems, are shown together here.

Considerably less common in the northern Pantanal is *Utricularia gibba* L., which produces small yellow flowers. A dense stand of this species was noted in a roadside ditch along the Transpantaneira Highway between the Rios Frio and Pixaim. It was not seen anywhere else during the study. This species seems to be considerably more abundant in eastern Brazil, where it is the predominant species in some water bodies (Pompêo and Moschini-Carlos, 1995).

Utricularia inflata Walker, characterized by yellowish flowers, is reported for the northern Pantanal, but its presence was not confirmed during the study.

There has been considerable confusion in the past about the identity of the *Utricularia* species in the Pantanal. The comprehensive monograph by P. Taylor (1989) has eliminated the confusion and provided information on the worldwide ranges of the species. Two of them, *U. foliosa* and *U. gibba*, are found in both the Old and New Worlds, as well as in temperate and tropical waters. *U. inflata* is confined to the New World, occurring from the colder regions of the north temperate zone through equatorial South America to Argentina (Dawson, 1973), while *U. poconensis* may be limited to South America, although it has not been recognized as a species long enough to permit a compilation of definitive range information.

Loganiaceae

Two species in this family have been reported from the Pantanal: *Strychnos nigricans* Progel, occurs in secondary forest areas near Porto Cercado and the Rio Pixaim (Dubs, 1994), while *Strychnos pseudoquina* Saint Hilaire, a small tree reaching a height of 6 m, was reported to be present but fairly rare by Pott and Pott (1994).

Loranthaceae

The trees of the Pantanal are frequently parasitized by two species of *Psittacanthus,* formerly referred to by the synonym, *Loranthus*, both of which have roots modified to penetrate deeply into the wood of the host to obtain nutrients. *Psittacanthus warmingii* Eichl. is widespread, attacking a variety of trees. The seeds are apparently spread by birds. *Psittacanthus cordatus* Hoffmannsegg, is somewhat less abundant but is nevertheless widespread in the Pantanal. Large numbers of this species parasitized the trees near the Baia Piuval. Occasionally, the seeds of these species sprout on the metal wire of fences, on which the young plants secrete a sticky substance to firmly bind themselves to the wire. These form buds and survive until the small shoots start to grow. They then dry up because the metal wire fails to provide nutrients. On trees, it can be assumed that the formation of shoots from the buds is accompanied by the invasion of the host tissues by the modified root system of the parasite.

Lythraceae

Two of the common roadside plants along the Transpantaneira Highway and in moist places along other roads in southern Mato Grosso were *Cuphea melvilla* Lindl. and *Cuphea speciosa* (Anders.) O. Kuntze, tall herbaceous plants that generally grow in the upper part of the sublittoral zone or in partially shaded places with some fallen leaves covering the soil. *C. speciosa* was particularly common between the Rio Novo and Rio Frio, while *C. melvilla* occurred most abundantly along the stretch of the highway several kilometers south of Poconé. Both species are widely distributed but only locally abundant. A third species, *Cuphea odonellii* Lourteig, was identified by Dubs (1994) on wet savanna, *campo úmido*, near the Rio Pixaim.

In addition to the *Cuphea* species, the family Lythraceae is represented in the region of the Pantanal by a tree that reaches 18 m in height: *Lafoensia pacari* Saint Hilaire. Pott and Pott (1994) reported that it reaches a maximum height of only 10 m in the region of the Pantanal. It is primarily a species of the dry upland forests, occurring from Minas Gerais to Santa Catarina and westward to Mato Grosso do Sul (Lorenzi, 1992) and Paraguay. It is not a common species in the wetland, but it has occasionally been introduced along the roads.

Physocalymma scaberrimum Pohl was reported by Dubs (1994) from a semi-deciduous forest near the Rio Jauru in the Municipal District of Cáceres.

Malpighiaceae

A common species of the dry woodland, *Byrsonema orbignyana* A. Jussieu, occurs in the Pantanal, mainly as a roadside tree. It is found only on sandy soils and frequently along the Transpantaneira Highway. Its fruit is edible, and it is used for making beverages. It is also an important food for various species of bird.

At similar locations but somewhat less abundant is *Banisteriopsis pubipetala* (A. Jussieu) Cuatr., a shrub with vine-like branches that tends to form small thickets.

Malvaceae

This species includes several roadside plants, which are somewhat more abundant at the outskirts of the Pantanal than in the wetland itself. *Urena lobata* L. is an annual with woody stems that occurs along the upper shoulders of the Transpantaneira Highway near Poconé and along other roads in the upland parts of the watershed. It has a growth cycle lasting about 150 days, flowers during the late *cheia* and early *vazante*, and reproduces exclusively by seeds (Aranha *et al.*, 1988). *Pavonia* sp. is another small plant found in somewhat similar places along the roads throughout much of southern Mato Grosso.

Also observed in various places along the roads of southern Mato Grosso is the introduced species, *Sida* cf. *santaremensis* H. Monteiro, which is among the species spread far beyond its native range by man.

An additional species, *Hibiscus striatus* Cav. was found on the banks of the Rio Jauru near Cáceres by Dubs (1994).

Melastomataceae

One herbaceous or weakly lignified species in this family, *Rhynchanthera novemnervia* A. P. de Candolle, is encountered throughout the Pantanal, especially in the supralittoral zone around water bodies and along the roads. It reaches a height of about 1 m. A second species, *Mouriri guianensis* Aubl., was found in the gallery forest along the Rio Pixaim and in semi-deciduous woodland near the Rio Jauru in the Municipal District of Cáceres by Dubs (1994).

Menyanthaceae

This family includes a characteristic plant species of the Pantanal that is particularly well adapted to the wet-and-dry climate. *Nymphoides grayana* (Grisebach) Kuntze proved to be a particularly problematic species to identify. It was illustrated as *Nymphoides kirkii* N. E. Br. by Riehl and Baensch (1995) and has been known as *Villarsia kirkii* (N. E. Br.) De Wit as well, although the structure of the ovary resembles

Nymphoides rather than *Villarsia* (M. Scheßl, pers. comm.). Prado *et al.* (1994) called the species *Villarsia kirkii*, and there have apparently also been references to it as *Limnanthemum kirkii*. In the past, it has frequently been confused with *Nymphoides indica* Thw., which has flowers somewhat similar in structure, but they are white rather than yellow. Many references to *Nymphoides indica* in the Pantanal probably refer to *Nymphoides grayana*, although *N. indica* may well occur along some of the large rivers. The true identity of the species was revealed by R. Ornduff, who examined material supplied by M. Scheßl (A. Pott, pers. comm.).

Figure 7.12: The leaves and flowers of *Nymphoides grayana* during the *vazante*.

Nymphoides grayana is a characteristic species of the Pantanal floodplain because it develops both aquatic and terrestrial forms, and its vegetative parts often survive weeks of dryness on the areas of the Pantanal that become terrestrial during the dry season. The aquatic growth form is somewhat similar to those of other *Nymphoides* species (Fig. 7.12), but the development of drought resistant leaves seems to distinguish this species from others in the genus. During the periods of dryness, the species develops thick leaves, somewhat resembling succulents but with a white spongy underside. These leaves persist for long periods of time on dried sediments, and if the dryness does not last too long, the plants survive vegetatively until the next period of flooding.

Nymphoides indica has frequently been reported from the northern part of the Pantanal, but it was never seen during the study. The white flowering species in the region may be the form known as *Nymphoides humboldtiana* (Kunth) O. Kuntze, which, however, may well prove to be a synonym of *N. indica* when more information about its morphology becomes available (Scheßl, 1997). In any case, white flowered species of *Nymphoides* are exceedingly rare in the northern Pantanal, while the yellow-flowering *N. grayana* is ubiquitous.

Moraceae

The palm trees in the Pantanal, most commonly *Scheelea phalerata*, are frequently parasitized by *Ficus pertusa* L. f., which eventually completely overgrows the host and becomes a tree. The seeds are usually transported by birds or fructivorous mammals and begin to grow in the upper parts of the palm. The palm grows taller while the fig engulfs its trunk and sends roots downward to the ground. A less abundant "strangler fig" is *Ficus luschnathiana* (Miq.) Miq., which has a similar life history.

Another species in this family, *Sorocea klortzschiana* Baillon, has been reported from Juscimeira in southern Mato Grosso by Dubs (1994), and Neves *et al.* (1993) reported *Ficus elliotiana* Moore to be one of the few species endemic to the Pantanal.

Musaceae

This family is represented by a rather uncommon species, *Heliconia marginata* (Griggs) Pittier, which was observed on the banks of the Rio Jauru near Cáceres by Dubs (1994).

Myrtaceae

A common hardwood tree in the region of the Pantanal is *Eugenia brasiliensis* Lam., which occurs in gallery forests as well as in upland woodlands. In the Pantanal, its abundance has apparently been reduced by the demand for its wood. The congeneric *Eugenia florida* D.C. was reported by Dubs (1994) from a secondary forest near Porto Cercado on the Rio Cuiabá.

Calyptranthes eugenioides Cambess. in Saint Hilaire is a common woody shrub in the dry forest areas surrounding the Pantanal, which penetrates into the wetland along roads and ridges that remain above the water level during the *cheia*.

The guava, *Psidium guajava* L., is a tree that is abundant along the Transpantaneira Highway. Many small guava trees have most certainly grown from seeds spread by the visitors to the Pantanal. The species is also common on the ranches but is not encountered at all in the patches of woodland distributed throughout the wetland. The fruit has considerable local economic importance, being used in various food products of tropical America (Zayas, 1968). The congeneric *Psidium guineense* Sw. was observed by Dubs (1994) along the Transpantaneira Highway near the Rio Pixaim.

Nyctaginaceae

This family is represented by a tree found at scattered locations in the Pantanal. One tall tree belonging to this species, *Pisonia zapallo* Grisebach, was observed on high ground near the shore of the Baia Piuval. Its fruit, which is usually ripe by the start of the rainy season, is armed with minute spines, making it stick stubbornly to clothing or the skin of passing animals.

Among the species introduced by man to the Pantanal is *Bougainvillea glabra* Choisy, which is planted on some of the ranches. It has not shown any tendency to spread to uncultivated areas.

Nymphaeaceae

A common species at the relatively low-lying locations on the floodplain and in shallow backwaters of the streams is *Nymphaea amazonum* Martius and Zuccarini (Fig. 7.13). The adaptation of its life cycle to the wet-and-dry climate was discussed by Prado *et al.* (1994). Prance (1980) provided information on its blooming, which occurs strictly at night. The flowers usually close within 15 to 30 minutes after sunrise but may open again on a second night. Their strong odor attracts nocturnal beetles, which pollinate them. Scarab beetles of the genus *Cyclocephala* are apparently important for the pollination of these water lilies. A second species, *Nymphaea gardneriana* Planchon was found by Scheßl (1997) to be common on the floodplain.

Figure 7.13: Thick beds of *Nymphaea amazonum* form on the surface of shallow permanent water courses sheltered from strong currents as well as on the temporarily flooded areas of the floodplain. The flowers open at night and usually remain open for no more than about half an hour after sunrise.

Figure 7.14: *Victoria cruziana* on the surface of floodplains along the Rio Paraguay at Cáceres.

One small population of *Victoria cruziana* Orbign. was encountered in shallow backwaters along the western shores of the Rio Paraguay near the bridge at Cáceres in 1992 (Fig. 7.14). It persisted until the dry season 1993, after which it disappeared while most of the backwater was drying up. The species was not observed again. Its taxonomy is somewhat confusing at the present time. Por (1995) referred to it as *Victoria cruziana* Orbign., a name sometimes regarded as a synonym of *Victoria amazonica* Sowerby, which has sometimes been referred to as a junior synonym of *V. regia* Lindl. Until additional studies of this spectacular water plant are undertaken, the name *V. cruziana* will be maintained for the local Pantanal species.

Olacaceae

Dulacia candida (Poeppig) O. Kuntze has been reported from the gallery forests along the Rio Pixaim (Dubs, 1994). Its presence was not noted during the study.

Onagraceae

The widespread aquatic and wetland species of *Ludwigia* are important components of the macrophyte communities in the Pantanal throughout the entire year. Most of these are known to be widely distributed in South America, and some have ranges including other continents (Munz, 1947). One of the most fully adapted species of this genus to the aquatic environment is *Ludwigia sedoides* (Humboldt and Bonpland) Hara, which is abundant throughout the Pantanal, especially in the low-lying parts of the flood plain (Plate 4b). It is generally absent from the permanent water bodies, but it is most common in habitats that dry up for only short periods during the dry season or remain somewhat moist throughout the entire year. The plant forms a rosette of leaves on the surface of the water, from which emerge the stems with the flower buds, most frequently during the *vazante* and early part of the seca. The plants are able to survive in nearly dry soil after the flood waters have receded, and the vegetative plants can persist into the following year if the rains come soon enough to prevent their complete desiccation. Otherwise, new plants growing from seed replace the old ones.

Ludwigia helminthorrhiza (Martius) Hara is another aquatic species that dies off if drought persists. Unlike *L. sedoides*, it commonly grows along the edges of permanent water bodies. It is not too abundant in the Pantanal, but one stand was observed each year in a backwater of the Rio Paraguay on the Fazenda Santo Antonio das Lendas, along the same river near the former research station of EMPA in Cáceres, and in small rivers near Porto Cercado. This plant produces long horizontal stems that are buoyed up by swollen modified roots filled with gas. The stems remain just below the surface of the water, and the emergent leaves extend well above the surface. Flowers of the typical *Ludwigia* form and yellow in color appear at the leaf axils. The species is also known locally by its synonym, *Ludwigia natans* Humboldt & Bonpland This synonymy has been discussed by Ramamoorthy and Zardini (1987).

Even better adapted to the conditions in the Pantanal is *Ludwigia inclinata* (L. f.) Gomez, which develops both a submerged aquatic stage and a terrestrial wetland form according to the prevailing conditions (Fig. 7.15). These two forms developed by the same plant at different times are so different that casual observers cannot believe that both are the same species. The submerged form, which rapidly develops immediately after the flood waters cover the plain, has long, band-like red leaves. This form is a common, distinctive floral element over large areas of the floodplain. During the *cheia*, flowers begin to appear with the typical form and yellow color of most aquatic species of the genus. These appear singularly on emergent flower stalks and disappear under water again after pollination. The number of these flowers increases steadily as the water level decreases during the *vazante*. When the water level has fallen so low that the aquatic plants begin to die off, the terrestrial form of *L. inclinata* appears above the surface of the water. Its leaves are short, semi-elliptical, and green, and flowers are frequently borne from the leaf axils.

Ludwigia octovalvis (Jacquin) Raven is a typical littoral species of the genus, which is uncommon in the Pantanal but occurs at many locations along the borders of the wetland, such as beside the Rio Bento Gomes near the Poconé exposition park and the Fazenda Nova Estancia, downstream from the park. The species now has a circumtropical distribution.

Less frequently observed in the region are *Ludwigia pilosa* H. B. K., *Ludwigia tomentosa* (Camb.) Hara, and *Ludwigia nervosa* (Poiret) Hara, three littoral species encountered along the edges of the river and streams and occasionally along roads crossing the floodplain during periods of flooding. They seem to be somewhat more abundant on marshy near the Rio Paraguay in the Pantanal of Cáceres. Scheßl (1997) provided collected numerous specimens of *Ludwigia* and provided a list of 16 identified species.

Figure 7.15: *Ludwigia inclinata* develops as a submerged plant, typically with red, band-like leaves. During the *vazante*, it develops emergent shoots bearing quite different leaves and yellow flowers, as shown here.

Ludwigia rigida (A. P. de Candolle) Hara is a species found in somewhat drier locations in the Pantanal, where it forms small stands at relatively elevated sites on the floodplain, especially along roads. Nevertheless, it was never observed far from water.

Ludwigia anastomosans (A. P. de Candolle) Hara is the only species of tree in the genus. It is a littoral species with a range limited to the general region of the Pantanal. Normally, these plants appear as shrubs, but a few manage to reach a height of about 8 m. During the study, this tree was observed only along the shores of the Piuval Lake on the Fazenda Ypiranga near Poconé.

Passifloraceae

Two species of this family have established themselves in the Pantanal. They are particularly abundant along the Transpantaneira Highway, where their seeds have probably been introduced by travelers from other parts of Brazil. *Passiflora cincinnata* Mast is by far the most abundant species in the wetland, especially along the northernmost part of the Transpantaneira (Plate 5a). It is a vine that climbs over other plants between the road and the edge of the water. It is cultivated for its fruit in the eastern provinces of Bolivia (Brücher, 1977). The second species, *Passiflora giberti* N. E. Brown, is not nearly as common, although it can be found at various sites along the northern part of the Transpantaneira as far south as the point at which it crosses the Rio Frio.

That these species seen to be restricted to roadsides suggests that they are not native to the Pantanal and are probably distributed by seeds transported by visitors to the wetland.

Piperaceae

This family is represented in the Pantanal wetland by *Piper tuberculatum* Jacquin, which is found throughout the Neotropical Region in secondary forests and on temporarily flooded areas. Another species or the family, *Peperonia circinnata* Link, is known from southern Mato Grosso. Dubs (1994) reported its presence in semi-deciduous forests near Chapada dos Guimarães.

Polygalaceae

One species in this family, *Polygala* sp., was found on waterlogged sandy soil near the Rio Paraguay in the District of Cáceres as well as near the springs forming the headwaters of the rivers on the plateau to the east and northeast of Cuiabá. It seemed to be rare in or absent from the Poconé District. Scheßl (1997) found a total of eight species in this genus in the Pantanal, including *Polygala longicaulis* H.B.K., *Polygala timoutoides* Chodat, and six species not positively identified.

Polygonaceae

This family is represented by littoral species, which often form dense beds along the shores of the streams and rivers, especially near the borders of the Pantanal. Because of the variability of some of the species and their extensive ranges, there has been a good deal of confusion about the identity of South American species. Some of the taxonomic problems were clarified by Duke (1960), whose delimitations of some of the Neotropical species are accepted herein.

Polygonum hydropiperoides Michaux was particularly abundant at littoral sites cleared by man along the Rio Bento Gomes, such as near the exposition park of Poconé. *Polygonum acuminatum* H. B. K. is an abundant species that forms dense stands along the shores of the Rio Paraguay and its branches near Cáceres. In the Pantanal of Poconé, it was observed in more isolated groups, a few of which were observed along the Rio Bento Gomes. It seems to be strictly a littoral plant along large rivers.

In addition of the two littoral species, the genus is represented by *Polygonum punctatum* Elliott, a member of the wetland plant aggregation that dominates low-lying, temporarily terrestrial areas of the Pantanal during the dry season. The plants continue blooming after the area becomes flooded again and only die off after the water increases to nearly a meter in depth (Prado *et al.*, 1994). Many emergent stems of the plants bearing developing fruit are observed among the rapidly growing aquatic species in water that can be quite deep.

Polygonum persicaria L. is a terrestrial species that occurs sporadically in the region of the Pantanal. Its frequent occurrence among the roadside vegetation gives the impression that it is continually introduced from other parts of Brazil but fails to permanently colonize the Pantanal or its surroundings.

One atypical member of this family is a terrestrial tree species, which often grows at the edge of the woods on hummocks and other slightly elevated areas that remain above the high water level. Its taxonomy is somewhat confusing, and it has been reported under various specific names but is generally referred to as *Triplaris americana* L. at the present time. Each tree produces only male or female flowers, and some of them host aggressive ants in the genus *Crematogaster*, which make their nests in the hollow stems of the tree (Pott and Pott, 1994). This seems to be a form of symbiosis, since the presence of the ants, which have a formidable stinging bite, certainly protects the tree from breakage by large animals. Formerly, two similar species of *Triplaris* were recognized, one with ants and the other without. However, it is generally agreed now that both belong to a common species. These trees were fairly common near the Baia Piuval and along the Transpantaneira to the south of the Rio Pixaim. In addition to the Pantanal, it is known from the Amazon and Orinoco Valleys (Pott and Pott, 1994). Benson (1985) provided additional information about the symbiosis of plant and ant species in tropical South America.

Portulacaceae

A small, prostrate perennial herb with short upright shoots, *Portulaca fluvialis* Legr., grew in a few clearings near the high water level of the *cheia*. It is apparently a pioneer species that does not survive long in competition with the other floodplain herbs.

A species introduced from Europe, *Portulaca oleracea* L., is now a common roadside plant along the northernmost section of the Transpantaneira Highway and other roads near Poconé. Since its introduction, it has become a widespread species in cultivated fields, along roads, and in places cleared of native vegetation throughout Brazil. It is a herbaceous annual that reaches a maximum height of only 40 cm, but nevertheless, a single plant can produce about 10,000 seeds, which remain viable in the soil for as long as 19 years (Lorenzi, 1992).

Rhamnaceae

This family is represented in southern Mato Grosso by a medium-sized tree that reaches about 15 m: *Rhamnidium elaeocarpum* Reiss. It blooms at the end of the *seca* and produces its small, sweet fruit during the *cheia*. When the fruit is ripe, the tree is visited by various birds, including large flocks of parakeets. These noisy birds were observed each year in the branches of trees of this species on the campus of the Federal University of Mato Grosso. Other representatives of this species were found at scattered locations around the edges of the Pantanal and near some of the roads leading to ranch houses.

Rubiaceae

This is a family well represented in the Pantanal by several species of small, low herbs. One of these, *Diodia kuntzei* Schumacher, is a dominant species covering large areas of the floodplain during the transition from aquatic to terrestrial conditions and for a period of time after the flood waters have returned during the *enchente* (Prado *et al.*, 1994).

In addition to the herbs, this family includes a tree and several shrubs that are common in the Pantanal. *Genipa americana* L. is a tree that has been observed along the Transpantaneira Highway south of the Rio Pixaim. *Tocoyena formosa* (C. and S.) Schumacher is a tall shrub reaching 4 m in height that is more abundant in the upland regions of the watershed but is also observed at a few locations along the

Transpantaneira Highway, most frequently south of the Rio Pixaim. The congeneric species, *Genipa spruceana* Steperm., has been reported from the Pantanal of Mato Grosso do Sul by Dubs (1994). *Chomelia obtusa* Chamisso and Schlechter is a miniature tree, reaching only 5 m in height, that is rather abundant along the northernmost section of the Transpantaneira Highway as well as on some of the hummocks that are rather heavily forested.

Palicourea rigida H.B.K. is known from the upland regions to the northeast of the Pantanal. Dubs (1994) reported its presence in a dry savanna region, the *campo cerrado*, near Chapada dos Guimarães. Dubs (1994) also confirmed the presence of *Borreria capitata* (R. and P.) A. P. de Candolle and *Sipanea pratensis* Aubl. in the Pantanal of Mato Grosso do Sul. *Tocoyena foetida* Poepp. & Enal. is a tall shrub more common in the dry forest regions at higher elevations than the Pantanal wetland. *Randia armata* (Sw.) is a Neotropical xerophyte that grows in the sandy soil of semi-deciduous and gallery forests. A plant tentatively identified as *Spermacoceodes glabrum* (Michx.) O. Kuntze was encountered along the roadsides around the outskirts of the Pantanal.

Rutaceae

Fagara hassleriana Chod., a small tree with a maximum height of about 10 m, characterized by a thorny trunk, was observed at many locations in the Pantanal, mainly near human settlements, roads, and areas elevated above the floodplain. It flowers during the *vazante*, and its fruit ripens during the *seca*. Pott and Pott (1994) reported its presence in the low-lying region of Campo Jofre near the southern end of the Transpantaneira to the north of the Rio Cuiabá.

A congeneric tree species reaching about 16 m in height, characterized by a deeply furrowed bark, *Fagara chiloperone* (Martius) Engl., occurs in the upland regions surrounding the Pantanal. It begins to bloom at the end of the *vazante* and continues through the early *seca*. Its fruit ripens during the *enchente*.

Two other tree species in this family are found in the semi-deciduous upland forests: *Zanthoxylum rhoifolium* Lam. and *Zanthoxylum riedelianum* Engl., neither of which were abundant.

Salicaceae

Salix humboldtiana Willdenow is an abundant littoral tree along the rivers in the Atlantic forest. It occurs along rivers in the Pantanal, as well, although it is not very abundant. This tree is a pioneer species in humid, deforested areas (Lorenzi, 1992).

Samydaceae

Casearia parvifolia Willdenow, sometimes reported under its synonym, *Casearia decandra* Jacquin, is a small tree that commonly occurs in the gallery forests along rivers and in narrow strips of woodland along small streams in the wetland. It seems to be even more common along upland streams in the dry forest area.

Sapindaceae

This family includes a species of woody shrub common on the hummocks that remain above the water level during the *cheia*: *Paullinia pinnata* L. The seeds of this species frequently sprout on the floodplain and are observed growing during the dry seasons, but the seedlings drown during the flood (Prado *et al.*, 1994). This assures that the trees remain only on relatively high ground. As long as the upper parts of the plants can remain above the water level, the species can survive with its roots submerged in waterlogged soil for long periods of time. It is therefore well adapted to conditions in the Pantanal, explaining why it is one of the most abundant tree species on the hummocks scattered across the floodplain. It has also been reported from gallery forests along the Rios Pixaim and Cuiabá (Dubs, 1994).

A smaller plant in the same genus is the *guaraná*, *Paullinia cupana* HBK, a species with seeds that contain 5% caffeine, making it by far the most prolific alkaloid producer of any tropical plant (Brücher, 1977). Extracts of its seeds are used for making a popular beverage in Brazil. *Guaraná* bushes are found in the dry forest region in the Pantanal watershed and at a few sites along the Transpantaneira Highway in the northern part of the wetland.

A species of woody shrub that produces medicinal substances used locally (Guarim Neto, 1991) is the *cinco-folhas*, *Serjania erecta* Radlk., which sometimes appears on the enclaves of dry forest vegetation in the Pantanal.

Magonia pubescens St. Hill is a medium-sized tree that reaches about 15 m in height (Pott and Pott, 1994). It was not observed at many locations in or near the Pantanal, but a row of three trees of this species was located along the road at the entrance to the Fazenda Ypiranga. The trees were rooted at a site roughly 20 cm above the average high water level during the *cheia*.

A rather small evergreen tree, *Sapindus saponaria* L., is encountered on the wooded islands and along roads in the wetland as well as in semi-deciduous forests in the upland regions surrounding the Pantanal.

This family is also represented by tree species that are encountered mainly in the dry woodlands and *cerrado* surrounding the Pantanal and sometimes in the gallery forests. *Dilodendron bipinnatum* Radkl. was uncommon in the region of the northern Pantanal, occurring chiefly near human settlements.

Allophyllus edulis (Saint Hilaire) Radlk., the *pitombeira*, is a rare tree in the region, encountered only in upland semi-deciduous forests. It produces edible fruit (Guarim Neto, 1991).

Scrophulariaceae

This family is represented in the Pantanal by *Bacopa* sp., which is frequently overlooked because it resembles species of *Cabomba* and other submerged plants with fine leaves. *Bacopa myriophylloides* (Bentham) Wettstein was abundant on the floodplain during the rainy season. Dubs (1994) reported it to be a common species in the Pantanal of Mato Grosso do Sul.

Scoparia dulcis L., locally called the *vassourinha*, was found along the Transpantaneira Highway near the Rio Bento Gomes and at other locations around the outskirts of the wetland. It is a species introduced from Europe to Brazil (Barroso, 1952), and it is becoming widespread, especially in regions of cattle ranching.

Solanaceae

A variety of species in this family occur in various parts of the Pantanal. All of them are littoral or roadside species, usually with a wide distribution but isolated occurrence. *Solanum sisymbrifolium* Lamarck is one of the most widespread species in the Pantanal. It is most common between the roads and the edges of the water in the roadside ditches.

Solanum viarum Dum. is a supralittoral species that was observed at various locations along the northern part of the Transpantaneira Highway and other roads in the wetland. It is found throughout all parts of the Pantanal (Pott and Pott, 1994).

Solanum lycocarpum A. Saint Hilaire was found at many locations along the roads in the Pantanal.

Sterculiaceae

One of the most abundant roadside shrubs along the northernmost section of the Transpantaneira Highway is *Helicteres guazumaefolia* H. B. K., a plant commonly encountered in the dry forests and *cerrado*, as well. It becomes scarcer between the Rio Bento Gomes and Rio Novo and is rather rare in the wetland south of the Rio Pixaim. It is also locally abundant in the regions of Cáceres and Barão de Melgaço. Pott and Pott (1994) reported that it occurs in all subregions of the Pantanal, while Dubs (1994) mentioned that an unidentified species of *Helicteres*, most probably this one, was present in a secondary forest near the Rio Pixaim.

Several species of *Melochia* are found along roadsides and in other open areas of the Pantanal and adjacent regions. *Melochia villosa* (Mill.) Faw and R. was frequently observed at the edges of the flooded areas during the rainy season. It was especially abundant along highways. Also common in similar locations along the roads, including the Transpantaneira Highway, was *Melochia parvifolia* H. B. K., which penetrates well into the wetland along transportation routes. Pott and Pott (1994) reported that this species provides forage for cattle. *Melochia arenosa* Bentham in Hook. was observed in similar locations throughout most of the year, but it was not so abundant. *Melochia simplex* A. Saint Hilaire occurred in upland regions near the Pantanal and was also noted in the northern part of the wetland.

Dubs (1994) reported the presence of *Guazuma ulmifolia* Lam. in semi-deciduous forests covering upland regions of southern Mato Grosso.

Ulmaceae

Trema micrantha (L.) Engler is a small tree that reaches 6 m. Originally a species of the Atlantic forest, it has been introduced throughout South and Central America to reforest degraded areas (Pott and Pott, 1994). It is found in upland deciduous forests (Dubs, 1994), but its presence was not noted in the wetland during the study, although it was not uncommon in the *cerrado* surrounding the Pantanal.

Umbelliferae

Hydrocotyle ranunculoides L. is abundant along the Rio Bento Gomes. Its horizontal stems run just below the surface of the water, supporting the leaves which emerge at the end of long petioles. This seems to be one of the species that survives only along permanent water courses and is apparently not adapted to grow on the temporarily flooded areas of the Pantanal.

Dubs (1994) reported the presence of *Hydrocotyle verticillata* in the Pantanal of Mato Grosso do Sul. The identity of this species is not clear since the name is presently applied to a valid Australian species apparently not known in South America, while the name *sensu* Thunberg is a synonym for *H. vulgaris* L., a European species.

Verbenaceae

This family is well represented in the region of the Pantanal. *Vitex cymosa* Bert. is a common tree in the wetland with moderate resistance to flooding, and as a full grown tree, to fire as well. It reaches a height of 20 m and is encountered on relatively high ground throughout the Pantanal. It loses its leaves during the dry season, flowers from

September to November, and produces fruit about two to three months later (Pott and Pott, 1994). It often invades cleared land (Dubs, 1994).

Aloysia virgata (Ruiz et Pav.) A. L. Jussieu is a small tree that reaches about 5 m in the Pantanal but can reportedly grow to 10 m in Argentina (Pott and Pott, 1994). It is found at isolated locations in the wetland and its surroundings and is most abundant along the roads.

Stachytarpheta elatior Schrader is a herb reaching about 80 cm that grows at moderately high elevations on the floodplain during the dry season, usually near the edges of the hummocks. The congeneric *Stachytarpheta cayennensis* (L. C. Rich) Vahl is a herb that is widespread along the roads in and near the Pantanal.

Phyla reptans (H. B. K.) Greene is a small prostrate herb that grows on open areas of the floodplain, at the edges of woods, and even on the surface of little-used roads.

Lippia alba (Mill.) N. E. Brown is a small shrub that is found in many places on the floodplain, at the edges of woods, along roads, and on areas recently burned. Pott and Pott (1994) reported that the species is tolerant of fire, regrowing rapidly after being damaged by the flames. *Lantana trifolia* L., a tall shrub, grows at similar locations but is somewhat less abundant. An unidentified species of *Lantana* was also observed in the *cerrado* around the outskirts of the Pantanal.

Vitaceae

Among the commonest woody vines along the edges of the flooded areas are the species of *Cissus*. *Cissus gonoides* L. formed dense thickets atop species of shrub near the sampling site at Birici. The fruit of this species is eaten by birds, which are apparently the chief agents for dispersing the seeds.

A common congeneric species, *Cissus spinosa* Camb., is a dominant plant of the thickets that sometimes form along the edges of the wooded islands at elevations above the high water level in the Pantanal. It generally forms on the sides of the woodlands with the strong exposure to the sun. Seedlings of these species sometimes sprout on the floodplains during the dry season, but the plants drown during the flood (Prado *et al.*, 1994). A less abundant species, *C. erosa* L. C. Rich., was observed at various sites along the Transpantaneira Highway between the Rios Bento Gomes and Pixaim.

Other species of *Cissus* are common in the upland regions surrounding the Pantanal. *C. sicyoides* L. is a vine sometimes encountered at the edges of gallery forests, while *C. campestris* (Bak.) Camb. and *C. erosa* L. C. Rich. are encountered mainly in the dry forests and *cerrado*.

Vochysiaceae

The species of this family are most abundant in upland forests, and their occurrence in the Pantanal is limited mainly to the hummocks above the high water mark and the gallery forests. *Vochysia divergens* Pohl is a common woody pioneer species that appears on land used for pasture (Pott and Pott, 1994). It grows into a tree up to 18 m in height. Two other trees of this genus have been reported only from the southern part of the Pantanal. *Vochysia rufa* Martius has been found to be a dominant tree on hummocks in Mato Grosso do Sul (Dubs, 1994), while *Vochysia tucanorum* Martius is an evergreen tree known as a pioneer species on areas with poor soil (Lorenzi, 1992).

Callisthene fasciculata (Spr.) Martius is another widespread tree species in the region of Poconé, and it was reported by Dubs (1994) from Cáceres, as well. Its resistance to fire gives it an advantage over other trees on pastures that are periodically burned during the dry season (Pott and Pott, 1994). *Qualea grandiflora* Martius is another fire-resistant species that is abundant in the upland areas surrounding the Pantanal. It is more a species of the *cerrado* than the wetlands, however. Dubs (1994) reported the presence of *Qualea grandiflora* in the Pantanal of Cáceres. It is a tree of *cerrado* areas with sandy soils poor in calcium that reaches a height of 15 m (Pott and Pott, 1994). Its flowers are mainly yellow. The wood of this tree is of suitable quality for furniture (Hueck, 1966).

Qualea dichotoma (Warm.) Stafl. was often seen in the *cerrado* surrounding the Pantanal. It is a tree reaching a height of about 18 m (Lorenzi, 1992) that can be quickly distinguished from the congeneric species in the region by its mainly white flowers, which usually have faint violet markings on the petals. A third species of the genus, *Qualea parviflora* Martius, is known from the dry forests of Mato Grosso do Sul (Lorenzi, 1992). It reaches only about 10 m in height and can be distinguished by its light violet flowers with a white midline stripe on the corola. It was not observed in or near the Pantanal of Poconé.

7.8. Monocotyledonae

Alismataceae

Several members of this family are among the dominant plant species in the wetlands. Revisions of the South American species of *Sagittaria* and *Echinodorus* were provided by Rataj (1972, 1975) and Haynes and Holm-Nielsen (1994). *Echinodorus paniculatus* Micheli and *Echinodorus macrophyllus* (Kunth) Micheli are observed from the *enchente* to the *vazante* on the entire Pantanal floodplain. They are emergent rooted plants that grow each year on the flooded areas as well as in the shallow waters of small, slow-flowing streams. During periods of dryness, they rapidly die off. Although they seem to be somewhat resistant to dryness, they have little chance to survive because they are so preferred by herbivorous species, from insects to cattle. Their leaves are rapidly consumed as the water level decreases, permitting terrestrial herbivores to reach them. It is possible that the plants could survive for several years if protected from herbivores, but in the Pantanal, their life span does not generally extend beyond the early dry *vazante*. Much less frequent but found in the same kinds of habitat as the other two species is *Echinodorus glaucus* Rataj, which is believed by Haynes and Holm-Nielsen (1994) to be a form of *E. paniculatus*, making the name *E. glaucus* a junior synonym. The name *Echinodorus muricatus* Grisebach has been used in place of *E. macrophyllus*, and it is sometimes also regarded as a distinct variety of that species. However, Haynes and Holm-Nielsen (1994) reduced it to synonym of *E. grandiflorus*, mentioned below.

Figure 7.16: Flowers of *Echinodorus tenellus*, a short-lived annual inhabiting a moving ecotone that follows the receding water during the *vazante* (Heckman, 1997b).

The life cycle of the tiny species, *Echinodorus tenellus* (Martius) Buch., is quite unlike those of the other species in this family. The plants appear, apparently sprouting from seeds left from previous years in the sediment, as soon as the flood waters have receded from an area during the *vazante*. The plants grow rapidly and begin flowering. They form dense beds on the exposed sediment, where they persist as long as the soil has sufficient moisture (Fig. 7.16). They are essential components of a dynamic ecotone that moves across the floodplain during the *vazante* (Heckman, 1997b).

An additional species in the genus, *Echinodorus grandiflorus* (Chamisso and Schlechter), often referred to by the synonym, *Echinodorus floribundus* Seubert, has been reported from the Rio Jauru near Cáceres by Dubs (1994). Considerably less abundant than the *Echinodorus* species in the northern Pantanal are *Sagittaria guyanensis* H.B.K. and *Sagittaria rhombifolia* Chamisso, both of which grow at sites that are under water for only about half of the year. The leaves of *S. guyanensis* resemble small lily pads on the surface of the water (Fig. 7.17). Both species are distributed widely in the district of Poconé, but neither seems to become abundant.

The Pantanal flora also includes *Echinodorus longiscapus* Arechavaleta, which is present in the northern Pantanal but was not found at any of the sampling sites during the study.

Figure 7.17: *Sagittaria guyanensis* is a typical species on relatively shallow areas of the floodplain that are usually inundated to a depth of no more than 30 to 40 cm for three to four months each year. It is not very abundant.

Amaryllidaceae

A fairly common flower in the Pantanal is *Amaryllis miniata* Ruiz & Pavon, which is observed in small patches throughout the wetland (Fig. 7.18). Because these plants grow mainly from bulbs and are not well dispersed by seeds, they are vulnerable to decimation by flower pickers and elimination by crowding from the introduced grass species, which have spread rapidly throughout nearly the entire wetland.

Two additional rare species in this family have been reported to occur in the Pantanal of Poconé (Pott and Pott, 1994): *Alstroemeria* cf. *psittacina* Lehm. and *Hippeastrum belladonna* L. Neither was observed during the study; however, Dubs (1994) reported the presence of an unidentified species of *Hippeastrum* along the river banks near Porto Cercado. The species of this family are all especially subject to selective elimination by man, particularly through habitat alteration. Picking the flowers and uprooting the plants to sell or take to home gardens present additional danger to the survival of these species in the Pantanal. They will probably require some form of legal protection to survive.

Figure 7.18: A few beds of *Amaryllus miniata* like this one were observed at the edges of hummocks.

Araceae

One aquatic species of this family is one of the dominant species in the Pantanal. Water lettuce, locally called the *erva de Santa Luzia, Pistia stratiotes* L., is extremely widespread among the floating plants in the Pantanal, usually not in monospecific beds but intermixed with other floating species. The roots provide shelter for the members of the interrhizon community, although they do not offer as much concealment as the rhizoids of *Salvinia auriculata*, with which *Pistia stratiotes* is often found in the Pantanal. The influence of its roots on the pH of the microhabitat was shown by Prado and Moraes (1990) to depend upon the amounts of ammonium and nitrate in the water. Flowers of this species are generally considered to be rare. However, after a period of intensive vegetative growth in the shade during the late rainy season, many cultured specimens produced their tiny flowers at the same time (Fig. 7.19). However, most reproduction observed in the Pantanal was vegetative by means of underwater runners.

In addition to the extremely abundant aquatic species, the family is represented by an epiphytic vine, *Philodendron imbe* Schott, which is most typically observed growing on the palm, *Orbignya oleifera*, near the edges of hummocks just above the high water level throughout the Pantanal. It is an abundant species in these terrestrial communities bordering the floodplain as well as on trees along the inland borders of the gallery forests. Scheßl (1997 also reported the presence of *Taccarum weddellianum* Brongniart and *Xanthosoma striatipes* (Kunth) M. Madison, on termite nests in the Pantanal.

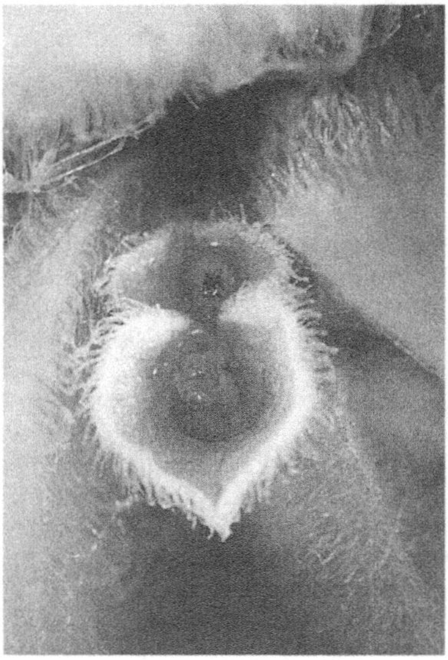

Figure 7.19: A flower of *Pistia stratiotes*, barely 5 mm long, that was produced by plants cultured in the shade.

Bromeliaceae

The bromelias were represented by relatively few species in the Pantanal because most ephphytic species seem to be excluded by the dryness during the *seca*. One common species, the *gravatá*, *Bromelia balansae* Mez, typically grows right on the border of the wooded areas at the high water line of the *cheia*. The plants flower during the late *seca* (Plate 5a) and produce fruit during the rainy season.

Pineapple plants, *Ananas sativus* (Lindl.) Schultes, were observed in several places along the Transpantaneira Highway about 20 km south of Poconé. It is assumed that these, like so many other plants at the edge of that highway, grow from seeds and other propagules thrown from the automobiles by visitors or that they fall from the mud sticking to the underside of the vehicles. The species is native to Brazil and has been cultivated since Precolumbian times. Through introductions by man, the species has attained a circumtropical distribution. A congeneric species that is not cultivated, *Ananas ananassoides* (Baker) L. B. Smith, occurs in the *cerrado* near the Pantanal. It has also been reported from the wetland in the state of Mato Grosso do Sul (Dubs, 1994).

In addition to the bromelias that grow on the ground, one epiphytic species, *Tillandsia loliacea* Martius, is encountered on trees in the Pantanal. It is not abundant for various reasons, including the frequent fires, excessive dryness during the *seca*, and high daytime air temperatures, but it occurs in gallery forests and other rather dense stands of trees along water courses in the wetland.

Cannaceae

Only one species in this family, *Canna glauca* L., is fairly well distributed along the water courses and in wet woodlands throughout the Pantanal. It is subject to competition from the introduced plants, and its small stands are often overgrown by fast-growing grasses. This is one of the species most vulnerable to habitat alteration, and it should be given legal protection in Brazil.

Commelinaceae

This family is represented by a tiny but abundant *Murdannia* species, which is a component of the dry season herb aggregation in the dynamic ecotone between the parts of the floodplain still under water and the fully terrestrial parts (Heckman, 1997b). It is an extremely small plant, rarely exceeding 5 cm in height, with small white flowers.

A less common member of the Pantanal flora is *Commelina nudiflora* L., characterized by its blue flowers. Isolated plants are generally encountered along the edges of the flooded areas, especially near the Transpantaneira Highway, and on *campos úmidos* permanently marshy from the emergence of ground water. It is widespread in the Pantanal as well as at the edges of semideciduous forests.

In addition to these two species, Dubs (1994) reported the presence of *Commelina erecta* L. at several locations in the Pantanal of Mato Grosso do Sul. The South American species of this family, especially the species of *Murdannia*, are in need of systematic study and taxonomic revision.

Cyperaceae

The sedge family is extremely rich in species in the Pantanal. At many locations, sedges are the dominant species in the macrophyte communities. Two species are typical of and predominant in special wetland habitats. One of these, *Cyperus giganteus* Vahl, forms dense stands reminiscent of papyrus marshes in the Old World (Fig. 7.20). These generally appear in moderately deep water that persists throughout nearly all of the dry season, although the water supply may be limited to the moisture content of damp mud during the driest times of the year. This species normally reaches a height of more than 2 m and is normally the tallest plant where it occurs.

The other species that dominates its habitat is *Oxycaryum cubense* (Poep. and Kunth) K. Lye, also referred to as *Scirpus cubensis*. This plant is rather small, seldom reaching 0.5 m in height, but it forms dense, nearly monospecific stands on shallow, waterlogged ground during the dry season (Fig. 7. 21), as well as on the *batume*, or floating islands of vegetation, throughout the year. This sedge was particularly abundant on floating islands in the Baia Piuval, where samples were frequently taken.

Additional sedge species are abundant in the Pantanal, mainly among mixed communities of species belonging to various other families. *Cyperus haspan* L. was particularly widespread among the typical vegetation on the shallower parts of the floodplain, as well as in various littoral habitats. *Cyperus brevifolius* (Rottb.) Hassk. and *Fimbristylis dichotoma* (L.) Vahl generally occur in similar habitats as *C. haspan* but do not survive in water as deep because of their smaller size. *Cyperus ferax* L. C. Richard seems to be a species preferring flooded wetlands to drier areas. It was found growing on the same areas as the three species just discussed, but it occurred there only during the *enchente* (Prado *et al.*, 1994). In contrast, young plants were observed on the desiccated floor of a drying pond during the dry season. They bloomed and went to seed before the end of the *seca* and died off after the pond was flooded again during the *enchente.*

Cyperus zollingeri Steud., often known by its synonym, *Cyperus bracteatus* Hort., is a species introduced from tropical Asia and Africa. It has colonized the floodplain, where it is often sighted together with the native species during the *vazante* and *seca*. *Cyperus luzulae* (L.) Retz and *Cyperus surinamensis* Rottb. are two of these abundant floodplain species that occupy the supralittoral zone during the rainy season and *vazante* and produce their seeds before dying off during the *seca*. The relatively tall *Cyperus odoratus* L. inhabits similar locations but is not nearly as abundant as these smaller sedges.

Eleocharis elegans (H. B. K.) Roemer and Schultes and *Eleocharis acutangula* (Roxburgh) Schultes are characteristic species of the rainy season. They emerge above the surface of the water to bloom on many of the temporarily flooded areas from the late *enchente* to the time that these areas dry up during the rainless months.

Cyperus esculentus L. was observed growing along roadsides and small rivers as well as in other isolated locations in and near the Pantanal. The largest stands of this sedge were observed within the city of Cáceres, where it was the dominant plant in several roadside ditches.

Figure 7.20: The largest sedge in the Pantanal, *Cyperus giganteus*, appears to fill an ecological niche somewhat similar to that of papyrus in Africa. However, the species forms only isolated stands and did not come to dominate large areas of the wetland.

Scleria flagellum-nigrorum Berg is a tall sedge with weak stems made rough by hooked spines, which extend onto the leaves. The plants can tear the skin when brushed against and are apparently avoided by the cattle. They typically entwine among woody plants at the edge of wooded areas to support themselves. They have almost developed into vines, but they seem to lean on the other plants rather than climb on them. They generally grow best on the shaded sides of the hummocks on the Pantanal floodplain.

Lipocarpha sphacelata (Vahl.) Kunth var. *gracilis* (Rich. ex Pers.) Boeck., sometimes called *Hypolytrum gracilis* Rich. ex Pers., is a plant typically found on marshy ground, either temporarily exposed to the air during the dry season or, much more typically, in places kept permanently marshy by seepage: the *campos úmidos*. The plants, which seldom reach a height of more than 5 cm, were found on ground pepetually moist from seepage water near the exposition park of Poconé and the Rio Paraguay at Cáceres. A few were also observed on wet ground near drying pools of water on the Fazenda Ypiranga. They grow almost exclusively during the dry season, producing flowers and seeds within a period of a few weeks. A much taller dry season species that typically grows on wet ground near drying water bodies as well as in soils permanently moist from seepage water on *campos úmidos* is *Lipocarpha humboldtiana* Nees, often called *L. sellowiana* Kunth, which is present throughout much of the Pantanal and surrounding wetlands.

Eleocharis minima Kunth developed on the Fazenda Ypiranga and at various other locations in the Pantanal. During the *vazante*, their thick, dark green, underwater shoots become evident at the bottom of the water bodies. As the water bodies dry up during the late *vazante* or *seca*, the typical terrestrial forms of the plant appear (Prado *et al.*, 1994). They are usually no more than three or four centimeters high and sometimes even shorter. These form dense mats in the dynamic ecotone of the desiccating floodplain. They produce flowers and fruit and die off in only a few weeks before the soil becomes completely dry (Heckman, 1997b).

A large stand and several smaller groups of *Eleocharis fistulosa* Schultes was found in the vicinity of the Rio Paraguay near Cáceres. It seems to occur mainly in waterlogged soils.

Figure 7.21: *Oxycaryum cubense*, also commonly known as *Scirpus cubensis*, is a very widespread species in the Pantanal, where it not only forms dense stands in shallow water or on moist soil on the floodplain but also grows on the sudd, the floating islands of vegetation on the Baia Piuval.

A sedge with a pronounced preference for drier terrestrial habitats in southern Mato Grosso is *Rhynchospora nervosa* (Vahl) Boeck. However, it also occasionally grows in the supralittoral zone along lakes, rivers, and ditches in the Pantanal of Poconé, , where it is considerably rarer than on the lawns and meadows in Cuiabá. In the wetland, it was most often seen along the edge of the Transpantaneira Highway, roads leading to ranches, and other elevated areas subject to frequent disturbance by human activity. In the literature, this species is often referred to as *Dichromena ciliata* Vahl. It is prominent wherever it grows due to the bleached appearance of the leaf bases surrounding its inflorescences (Lorenzi, 1991).

Observations of the Brazilian flora in many different kinds of habitat suggest that the family Cyperaceae was originally a dominant family in lowland South America with species that fill many of the ecological niches held by grasses on the Eurasian and North American continents. Unfortunately, there has been relatively little recent systematic, ecological, and zoogeographical research on this family, and the most comprehensive publication that includes many of the Pantanal species is almost 60 years old (Barros, 1938). Through human influence, the grasses, including many species introduced from other continents, are gradually displacing the sedges from the South American savannas and wetlands.

Eriocaulaceae

This family was represented by two species in the catchment area of the northern Pantanal. Neither was observed in the Pantanal of Poconé, but both were common species in areas kept continually moist and generally clear of woody vegetation by seepage water. These two species, *Philodice hoffmannseggii* Martius and *Syngonanthus gracilis* Martius, grew around the springs that fed the headwaters of the rivers on the plateau to the east of Cuiabá, in waterlogged, sandy soils near the Rio Paraguay in Cáceres, and in similar areas near Barão de Melgaço. *P. hoffmannseggii* was generally on the periphery of such areas, where the soil was dry, while *S. gracilis*, the more abundant of the two, preferred the wettest soils, often occurring at the edge of springs.

Gramineae

As expected, grasses were found to be among the most widespread and abundant plant species, both on the relatively dry areas at higher elevations and on the flooded meadows and islands of floating plants. Scheßl (1997) provided a list of about 100 species he found in the Pantanal. As mentioned above, many of these species are known to have been introduced from other continents by European immigrants and their descendants. Among these are species that have become dominant elements of the floodplain flora in the Pantanal. In assessing the grass flora in various parts of Bolivia, Killeen and Hinz (1992) concluded that the Pantanal was the richest of all habitats in species of this family.

The largest members of the Gramineae, the bamboos, are found along the roads in the Pantanal. The largest stands along the Transpantaneira Highway were encountered south of the Rio Pixaim. However, the species were somewhat more abundant in the districts of Cáceres and Barão de Melgaço than to the south of Poconé. Flowers were not observed on these plants during the study, so the species remain unidentified. The only species reported to occur in the *cerrado* of southern Mato Grosso is *Actinocladium verticillatum* (Nees) McClure ex Soderstrom, a species adapted to resist the effects of the frequent fires (Soderstrom, 1981). Specimens from Xavantina, well to the north of the Pantanal, were examined by Filgueiras (1995), but the species probably is one of those growing in or near the Pantanal. Other species probably belong to the genus *Bambusia*, which have been planted in gardens around the world.

A few aquatic grasses are abundant on the floodplain during and just after the flooding. *Paspalidium paludivagum* (Hitch and Chase) Parode and *Hymenachne amplexicaulis* (Rudge) Nees were the only species that occurred in moderately deep water during the high water periods. Both are native to tropical America (Burkart *et al.*, 1969; Allem and Valls, 1987). The other aquatic species of the floodplain were concentrated in the relatively shallow water up to about half a meter in depth. *Panicum laxum* Swartz and *Reimarochloa brasiliensis* (Spreng.) Hitchcock were the most abundant of these species in the Pantanal. *Echinochloa crus-galli* (L.) Beauv. and *Panicum colonum* L., often appearing in the literature under its synonym, *Echinochloa colona* Link, also grow in such habitats. However, in the Pantanal, these were uncommon species. Both of them were often observed in the shallow water of temporarily flooded fields during the rainy season.

Panicum repens L. and *Leersia hexandra* Swartz also grow on the floodplains, but they are present mainly while the water is only a few centimeters deep or while the fields are not flooded. As long as there is some moisture in the soil, these species thrive. They do not persist much longer than the water plants, however, after the formerly flooded areas have become baked dry and hard.

Setaria brachiata Kunth, often referred to by its synonyms, *Panicum chloroticum* Nees and *Panicum dichotomiflora* Michaux, occurs in somewhat deeper water than *P. repens*. It is usually encountered along the shallow streams that drain the Pantanal. Its inflorescences are most abundant during the *vazante*.

On the low-lying areas that are flooded for more than half of the year, several species develop characteristically during the *vazante* and *seca*. *Setaria parviflora* (Poir.) Kerguélen develops enormous populations on the sediment of the drying shallow water bodies near the end of the *seca*. The grass produces its spikelets and goes to seed in a matter of a few weeks and is drowned by the flooding during the subsequent rainy season, making it a characteristic species of the dynamic ecotone (Heckman, 1997b). The name *Setaria geniculata* (Lam.) Beauv. was commonly used in the past to refer to this grass until it was placed in synonymy with *S. parviflora* by Judziewicz (1990).

Another small annual species with a short life span on the wet sediment exposed as the water on the floodplain dries up during the *seca* is *Eragrostis hypnoides* (Lam.) Britton, Stern & Poggenberg. It is a helophyte that reaches a height of only about 20 cm and is reportedly a selective hygrophyte that is considered rare in Brazil (Smith *et al.*, 1982).

One aquatic species, *Paspalum repens* Bergius, occurs less frequently on the floodplain of the northern Pantanal than in the littoral zone of water courses, on the surface of backwaters, and on floating islands along the margins of the few deep lakes, such as the Baia Piuval. Its considerably greater abundance in deeper water in the District of Poconé may reflect the preference of the cattle for this grass. The congeneric *Paspalum conspersum* Schrader occurs along streams and rivers, especially around the edges of the Pantanal, while *Paspalum conjugatum* Berg. occurs sporadically near the Pantanal, usually in roadside ditches.

Grasses common on the floodplain during the terrestrial phase and near the water's edge during the high water season include Neotropical species as well as various species introduced from other parts of the world. *Andropogon bicornis* L. has become a characteristic species on those areas of the floodplain that are flooded only briefly during the *cheia*. The seasonal growth cycle begins with the onset of the rains and continues until the wind-borne seeds are produced during the early dry season. *Andropogon hypogynus* Hackel in Martius, which is somewhat less abundant but nevertheless a fairly common Pantanal species, has a similar seasonal growth cycle. Allem and Valls (1987) discussed the importance of these grasses as forage for cattle on the floodplains in the southern part of the Pantanal, particularly the Nhêcolandia, Paiaguás, Abobral, Aquidauana, and Miranda Subregions.

By the later part of the *seca*, the grasses of the genus *Andropogon* have already turned brown and dried out. They are grasses typical of regions with poor but well-drained sandy soil, which are of poor quality for grazing animals (Lorenzi, 1991), giving them a selective advantage over more palatable grasses. They are thus able to become the dominant species in a relatively short period of time, which has given them a reputation as troublesome weeds among the ranchers of the Pantanal. A third species of the genus, *Andropogon sellowius* (Hackel) Hackel, occurs in isolated patches in well drained soil along the Transpantaneira Highway and other roads around the borders of the Pantanal. The 40 to 80 cm tall grass is known from open fields with sandy soil throughout Brazil (Smith *et al.*, 1982). The rapid spread of these species in the Pantanal has been facilitated by their wind-borne seeds.

Another floodplain species, *Panicum rudgei* Roemer and Schultes, does not produce structures to facilitate dispersal by the wind. Apparently, animals would be the main agents of dispersal. The species is locally abundant in the northern part of the Pantanal of Poconé but is not widespread.

One of the grass species introduced from Africa that is preferred by the cattle is the annual species, *Hyparrhenia rufa* (Nees) Stapf. It has spread throughout Brazil and is considered one of the desirable pasture species by cattle ranchers (Aranha *et al.*, 1988). Generally, this grass has become common where grazing is not too heavy. Because it is eagerly consumed by the cattle, it is often displaced by the less desirable *Andropogon* species when large herds are present. During the late rainy season, the alternating yellow and green sections of its tall stems are ubiquitous along most of the roads in the upland regions of southern Mato Grosso, and in the early part of the dry sesaon, it flowers and goes to seed. The desiccated stems and leaves of this grass are often consumed by the many grass and brush fires that break out, often due to arson, during the later part of the dry season. In the Pantanal, this grass has scarcely penetrated the wetlands and is confined to a few of the more elevated, well drained sites. Apparently, it is not capable of surviving in places subject even to short-term floodiing.

Among the supralittoral species that are found at scattered locations between the roads and the adjacent floodplains as well as near water bodies at elevations somewhat above the Pantanal are the widespread *Eleusine indica* (L.) Gaertner, *Dactyloctenium aegyptium* (L.) Beauv., *Axonopus leptostachyus* (Fluegge) Hitchcock, *Digitaria insularis* (L.) Maz and Ekman, *Axonopus purpusii* (Mez) Chase, and a species of *Paspalum* in the group *plicatulum* Michx., which was found along the Transpantaneira Highway near the Rio Bento Gomes. Usually, the young plants of the supralittoral zone sprout during the rainy season but do not begin to grow rapidly until the *vazante*. They flower, go to seed, and die off during the *seca*, and decompose beneath the flood waters during the *enchente*, making way for the aquatic plants of the new generation, which grow either from seeds of from underground propagules.

Grasses preferring dry, well drained soils, particularly at sites strongly influenced by man, such as roadsides, may occur sporadically in the Pantanal, especially along roads and on the fazendas. One such grass is *Cenchrus echinatus* L., which produces burr-like fruits that are dispersed by passing animals. These fruits frequently attach to the clothing of people walking in the Pantanal, especially during the dry season. *Rhynchelytrum repens* (Willd.) C. E. Hubbard is a common grass of the upland regions near the Pantanal, which penetrates into the wetland along the roads. It is an excellent forage grass that was introduced from southern Africa (Lorenzi, 1991). *Panicum maximum* Jacq. is a large grass, often exceeding 2 m in height, which was introduced from Africa and has become a common weed throughout much of Brazil (Lorenzi, 1991). It is common in the city of Cuiabá and near human habitations throughout southern Mato Grosso.

Other widespread species that occur in the Pantanal drainage area but were not encountered in the wetland itself during the study include *Paspalum notatum* Flügge, *Paspalum plicatulum* Michx., and *Chloris caribaea* Sprengel, grasses of roadsides and cleared fields; *Pennisetum setosum* (Swartz) L. C. Richard, a species of upland pastures; and *Imperata brasiliensis* Trinius, which formed dense stands in partially shaded, marshy places. Many of these species will certainly be introduced along the Transpantaneira Highway in the future due to the transport of the seeds by vehicles from other parts of Brazil, especially from adjacent parts of Mato Grosso.

Dubs (1994) reported the presence of *Schizachyrium condensatum* (Kunth) Nees at the edge of a pond in Mato Grosso do Sul. According to Por (1995), a species of wild rice, *Oryza perennis* Moench, was cultivated in the Pantanal by native tribes during Pre-Columbian times. The name of another wild rice species that has been reported from the Pantanal, *Oryza subulata* Nees, is probably a synonym of *Oryza sativa* L., the common cultivated species of rice. The Pantanal is a wetland that would seem ideally suited to the cultivation of rice. However, this grain is in oversupply in Brazil because much can be grown in the eastern states, and there is little demand because most of the Brazilian population is accustomed to bread made from terrestrial grains as the staple carbohydrate source.

In addition to the grasses observed in the wetland, several species that occur in the *cerrado* of Goiás were reported by Filgueiras (1995) to occur in the Pantanal or in the upland parts of its catchment area. *Panicum mertensi* Roth in Roemer was found along the shores of the Rio Taquari in Coxim, and specimens of the valuable forage grass for cattle, *Paspalum virgatum* L., were found in the southern districts of the Pantanal, including Miranda and Nabelique, as well as near Corumbá. Apparently less common are *Leptochloa virgata* (L.) P. Beauv., encountered in Cáceres, and *Loudetia flammida* (Trin.) C. E. Hubbard, found near Corumbá. Among the species found in the northern part of the catchment area of the Pantanal are *Trachypogon montufari* (Kunth) Nees, reported from Chapada dos Guimarães, and *Anthropogon villosum* Nees and *Echinolaena inflexa* (Poir.) Chase, good forage grasses for cattle that grow in the *cerrado*. *Hackelochloa granularis* (L.) Kuntze was found between Rondonopolis and São Lorenzo, where it is a pioneer species that sometimes invades cultivated fields. Due to difficulties in identification, the distribution of the grasses in southern Mato Grosso is still poorly known, and most or all of these species can be expected to occur in the upland regions of the Pantanal of Poconé, as well.

Heliconiaceae

One species in this family, *Heliconia marginata* (Griggs) Pitt., locally called *pacóva*, was reported by Guarim Neto (1991) to be locally abundant along the shores of the Rio Paraguay near the Taiamã Ecological Research Station in the Pantanal of Cáceres. He noted that the leaves of this plant have been commonly used by the poorer inhabitants of the Pantanal for roofing to cover their houses.

Hydrocharitaceae

Small beds of *Egeria najas* Planchon were encountered near the Rio Bento Gomes and the Transpantaneira Highway during the final year of the study. The species bloomed intensely during the *vazante* and died off when the water completely dried up during the *seca*.

Iridaceae

Herbertia sp. is among the species that flower during the rainy season, appearing as small rosettes of leaves with isolated flowers emerging from the center. Generally, these plants grow partially concealed by somewhat taller plants on the more elevated areas of the floodplain.

Lemnaceae

The duckweed species were not at all abundant in the Pantanal, most probably because of the scarcity of nutrients in the water. Where they did occur, the water bodies were typically receiving much organic enrichment from large herds of cattle or outflow of nutrient-rich waters from ranches. The largest concentration observed was found on the surface of a pond on the Fazenda Santo Antônio das Lendas near the Rio Paraguay, which was completely covered by *Eichhornia crassipes* and *Salvinia auriculata*, with *Wolffia brasiliensis* Weddell, *Wolffiella lingulata* (Hegelmaier) Hegelmaier, and *Lemna aequinoctialis* Welwitsch filling in the spaces between the larger floating plants and forming a dense cover on areas not occupied by the other species (Fig. 7.22).

Somewhat rarer in the area was *Spirodela intermedia* W. Koch, which occurred occasionally in flooded roadside ditches near ranches. In addition to these species, several others have been found in the Pantanal of Mato Grosso do Sul by Pott and Cervi (in press): *Lemna valdiviana* Phil., *Wolffiella welwitschii* (Hegelmaier) Monod, *Wolffiella oblonga* (Phil.) Hegelmaier, and *Wolffia columbiana* Karsten. In the entire Brazilian part of the Pantanal, Pott and Cervi (in press) found the Nabileque Subregion to be richest in species of this family and attributed this to the greater fertility of the soils. Observations in the Poconé Subregion and the data on its water chemistry (Section 6) suggest that this part of the Pantanal is probably the poorest in these species, while Cáceres is probably the richest district to the north of the Cuiabá River.

Landolt (1986) revised the taxonomy of the species in this family and provided updated information on their distribution. *Lemna valdiviana, Wolffiella oblonga*, and *Wolffiella welwitschii* range throughout the tropics of both the Old and New Worlds, while *Spirodela intermedia* seems to be an endemic Neotropical species, and *Lemna valdiviana, Wolffia brasiliensis, Wolffia columbiana*, and *Wolffiella lingulata* are found in both the torrid and temperate zones of the New World, at least in regions that do not become too cold during the winter.

Figure 7.22: Apparently because of its poverty in nutrients, the water of the Pantanal is usually free of species belonging to the family Lemnaceae. Large numbers of *Wolffiella lingulata* together with *Lemna aequinoctialis* were interspersed among *Salvinia auriculata*, which covered a shallow pond in the middle of a large holding pen for cattle. Manure from the animals is responsible for the enrichment of the water.

Liliaceae

Species of the lily family are found singly or in small groups on moist and partially shaded areas in the Pantanal. *Heleotropium* sp. is a common species that usually blooms among short-lived annuals on the temporarily flooded areas shortly after the flood waters have receded during the *vazante*. Is is widespread but was never found to be abundant.

Limnocharitaceae

This family includes at least two common species in the Pantanal. *Hydrocleys nymphoides* (H. and B. ex Willd.) Buchenau, sometimes referred to by its synonym, *Hydrocleis commersonii* L. C. Richard, the most abundant of the two, appears as the flood waters arrive during the *enchente*, forming leaves of a typical submerged form. After these leaves have begun to develop, the long emergent stalks of the flowers begin to appear above the surface of the water. A congeneric species may also be present, but its identity was not confirmed. Scheßl (1997) thinks it may be *Hydrocleys parviflora* (L.) Buch.

Limnocharis flava (L.) Buch. seems to prefer shallower areas than *Hydrocleys commersonii*, at locations where the flood waters do not persist as long (Fig. 7.23). It is often seen with its emergent leaves and flowers in water only a few centimeters deep. Both of these species are widely distributed in the Pantanal but are usually not abundant. They are typical species of the temporarily flooded areas.

Figure 7.23: An uncommon species on shallow areas of the floodplain near Poconé, *Limnocharis flava* is a short-lived water plant well adapted to the tropical wet-and-dry climate.

Marantaceae

One of the species that dominates large areas of the wetland along the northern section of the Transpantaneira Highway is *Thalia geniculata* L. It has emergent leaves on long petioles that can extend more than a meter above the surface of the water (Fig. 7.24).

The plants form sparce stands covering flooded areas that do not dry up for more than a few weeks during the year. When dryness persists too long, the leaves turn brown and dry up, and the stands progressively die off. After several wet years, the stands have generally spread considerably, but after a particularly dry *seca*, they die off again over relatively large areas. Flowers appear from the late rainy season through the early part of the *seca*. The emergent leaves on long stems permit these plants to utilize the direct sunlight and shade the rest of the aquatic vegetation. Near the Rio Bento Gomes, they almost always shared the habitat with floating plant species, especially *Salvinia auriculata*. While physically sharing the same areas, competition is avoided, since *T. geniculata* forms an elevated canopy above the floating species and takes its nutrients from deep in the sediment, while *S. auriculata* can receive direct sunlight through the thin canopy of emergent leaves and absorbs its nutrients from the water rather than the sediment.

In southern Mato Grosso, this family is also represented by *Monotagma plurispicatum* (Koern.) K. Schum., which was also reported by Dubs (1994) to occur near Chapada dos Guimarães.

Figure 7.24: Very large areas of the flooded ditches bordering the Transpantaneira Highway are dominated by *Thalia geniculata*. The plants survive as long as the soil remains moist during the *seca*. During particularly dry years, some of the stands are killed off.

Orchideae

The aquatic orchid species, *Habenaria aricaensis* Hoehne, has been reported from the Pantanal of Mato Grosso by Guarim Neto (1991) and from Mato Grosso do Sul by Por

(1995). It is a tall, attractive plant that is usually found rooted on the floodplain. None were observed during the study.

The epiphytic vanilla orchid, *Vanilla palmarum* Lindl., was observed in the region of the Pantanal. Guarim Neto (1991) reported that it is most frequently encountered in the wetland on the *acuri* palm, *Scheelea phalerata*.

Palmae

A variety of palms grow on the plain and in the foothill regions of southern Mato Grosso. The distribution of the species is interesting, since each seems to have a preferred elevation for growth. Only two of the species are typical Pantanal palms: *Scheelea phalerata* (Martius) Burret, locally called the *acuri* or *bacuri* and sometimes referred to by its synonym *Attalea phalerata* Martius ex Spreng., and *Orbignya oleifera* Bur. the *aguaçu*, which is often the dominant species on the hummocks slightly above the high water level of the floodplain, nearly all of which support some of these palms. They occasionally form nearly monospecific stands in the northern part of the Pantanal (Fig. 7.25). They are frequently overgrown by parasitic members of the fig family, *Ficus* spp., which eventually completely cover the host and form large independent trees. Apparently, the rough surface of the upper parts of the trunks, typically still bearing the lower parts of the dead leaf stems, is particularly prone to trapping and holding the seeds of the strangler figs, which are often deposited in the droppings of visiting birds. Other epiphytes transported to the palm trees by animals that do not kill the host include *Philodendron imbe* and *Vanilla palmarum* (see above). The stands of these palms on the hummocks form the foundation for an extremely important habitat frequented by a variety of vertebrates in the Pantanal, including large lizards, monkeys, and a great many different bird species, as discussed in Section 8.

Figure 7.25: *Orbignya oleifera* is the most characteristic palm species in the Pantanal of Poconé. It is often the predominant tree on the hummocks. The tree shown here, visited by hyacinthine macaws, is playing host to common local epiphytes, including *Philodendron imbe* and parasitic figs.

Also found at a few sites along the Transpantaneira Highway is the *macaúba*, *Acrocomia aculeata* (Jacq.) Lodd., which is abundant in the upland parts of the Pantanal watershed but not in the wetland. The fruit of this palm has a taste somewhat like cheese and is highly prized by some of the people in Mato Grosso. Those trees along the highway have evidently been accidently or intentionally planted by visitors to the Pantanal.

One member of the palm family seems to be endemic to Mato Grosso. The *urumbamba*, *Desmoncus cuyabensis* Barb. Rodr., is a vine-like palm, apparently named after the Cuiabá River. Pott and Pott (1994) tentatively identified an abundant vine in gallery forests along the rivers of the northern Pantanal as belonging to this species.

In the floodplain along the Rio Paraguay, several stands of the *carandá*, *Copernicia alba* Morong, were encountered on the Fazenda Santo Antônio das Lendas. The trees were visited by flocks of hyacinthine macaws while their fruit was available. However, Por (1995) reported the fruit of *Acrocomia aculeata*, referred to by its synonym, *A. sclerocarpa* Martius, to produce the fruit preferred by this rare parrot species (Section 8).

The buriti palm, *Mauritia vinifera* Martius, apparently occurs only along the eastern edges of the Pantanal, including the region of Barão de Melgaço. Pott and Pott (1994) also reported the species from the Aquidauana District of the Pantanal.

The other species of palm in southern Mato Grosso do not usually occur in the Pantanal, although they may be encountered in upland regions nearby. *Bactris gasipaes* Kunth is a species known from the Amazon region, and *Polyandroeocos caudescens* (Martius) Barb. Rodr. is a palm of the Atlantic forest. Both are present in southern Mato Grosso, perhaps planted intentionally. *Bactris glaucescens* Drude also occurs in the region, where it is probably a native tree. *Lepidocaryum guainiense* (Spruce) Spruce, sometimes referred to by its synonym *Mauritia guainiensis* Spruce, is present in upland regions of the watershed, where it is abundant at a few locations, especially near human habitations. This specific name appears alternately in species lists as *guainiensis* and *guianensis*, apparently due to an old typographical error.

Pontederiaceae

Among the most characteristic species of the shallow Pantanal wetlands are *Eichhornia azurea* (Swartz) Kunth. and *Pontederia lanceolata* Nuttal. These two species both remain rooted in the sediment, with *E. azurea* being dominant in relatively deep water owing to its potential for great elongation of its main stem, and *P. lanceolata* covering large areas in which the water is shallower, preferably less than half a meter, except for periods of a few weeks when the flood waters are at their peak (Fig. 7.26). The relationship between the water depth and the relative dominance of these two species was outlined by Prado *et al.* (1994). Both are among the most abundant species in the entire Pantanal and can be considered among the most characteristic species in the entire wetland. They illustrate for the macrophytes a typical feature of the Pantanal: an enormous dominance of a few species over a biotic community not notably diverse. Similarly low species diversities among various groups of animals will be given close attention in Section 8 and subsequent parts of this book.

A second *Pontederia* species, *P. cordata* L., characterized by blue rather than white flowers, is rare in the Pantanal. Several small stands, however, were observed near the Transpantaneira Highway to the south of the Rio Pixaim and north of Campo Jofre. Recent proposed revisions of the family have reduced *P. lanceolata* to a subspecies called *P. cordata lancifolia* (Muhl.) Torrey. However, these revisions have been rejected by most experts on the family, and the classification system of Gopal (1987), which generally conforms to the earlier systematic work on the family by Castellanos (1959), is maintained here. For practical purposes, this greatly simplifies studies of the genus in the Pantanal, since the color of the flowers permits the two species to be instantly distinguished.

Other rooted emergent species of this family include *Reussia subovata* (Seubert) Solms-Laubach, which is widespread on the temporarily flooded areas during most of the year, blooming intensively during the *vazente* and disappearing during the *seca*. A somewhat larger species, *Reussia rotundifolia* (L.) Cast., is commoner along the edges of deeper water bodies than on the flood plain (Fig. 7.27). It is therefore more a littoral than a wetland species. Some authors include the species of *Reussia* in the genus *Pontederia*.

Figure 7.26: Vast areas of the floodplain are dominated by *Pontederia lanceolata* during the period of inundation. It is perhaps the most abundant aquatic plant in the Pantanal of Poconé. A pink clutch of pilid snail eggs is evident below the flower. The holes eaten in some of the leaves are typical of those made by the common grasshopper, *Cornops aquaticum*.

Unique among the Pontederiaceae in its free-floating form of growth is *Eichhornia crassipes* (Martius) Solms, which is generally confined to streams and rivers, from which it is frequently carried by the wind into backwaters and oxbows, where it forms unbroken mats on the surface of the water. The species also participates in the formation of floating islands or sudd, locally called *batumi*, as described in Section 9. Where the species can be found at any one time generally depends upon the direction of the prevailing winds as well as on the water currents. In the Rio Bento Gomes, large stands of this plant developed in the upstream section near the exposition park of Poconé from 1988 through 1991, then moved several kilometers downstream, where dense woody vegetation blocks the entrance into the Baia Piuval. They remained there until 1993, when some of these plants were carried back to the region of the exposition park by the wind. However, by the beginning of 1995, they had still failed to form more than small clusters along the banks of the river, where they were loosely anchored by emergent rooted plants.

Eichhornia crassipes is generally excluded from the floodplain by the emergent aquatic plants, especially *E. azurea* and *Pontederia lanceolata*, which block its drift. Apparently due to the general poverty in inorganic nutrients during much of the year, the individual plants remain much smaller than in Asia and other continents to which the species has been introduced. They seldom extend more than about 35 cm above the surface of the water.

Figure 7.27: *Reussia rotundifolia* is a characteristic littoral species at the edge shallow drainage channels in low-lying parts of the floodplain.

Smilacaceae

A common vine in the dry forest areas, *Smilax benthamiana* A. DC., was reported to occur on small enclaves of dry forest vegetation in the Pantanal (Guarim Neto, 1991). The species was never observed near the floodplain, but is was seen along the margins of woods in the upland *cerradão*.

Typhaceae

Typha domingensis Pers. is the only species of its genus that was observed among the wetland flora near the Pantanal of Poconé. It is not a typical member of the regional wetland flora but grew only in dense but isolated stands around the edges of the wetland and in highly enriched water bodies surrounded by cattle pastures. One of these was in a shallow pond heavily enriched by the wastes from large cattle herds fenced in on the Fazenda Santo Antônio das Lendas near the Rio Paraguay in the Municipal District of Cáceres (Fig. 7.28). The pond was eutrophic, and the surface of its water was covered by dense masses of floating plants, including *Eichhornia crassipes* and several species in the family Lemnaceae, which were also rare in unenriched waters in and near the Pantanal (see above). It was also seen in a few ponds at the margins of the Pantanal in the District of Barão de Melgaço.

Figure 7.28: *Typha domingensis* is a rare species in the northern Pantanal, although it is somewhat more common in upland ponds that receive some input of nutrients. This is the largest stand of the species observed during the study. The pond in which it developed is heavily enriched by cattle held in pens surrounding it.

Xyrideae

On sandy patches around the springs feeding the headwaters of the rivers on the plateau to the east of Cuiabá as well as on sandy ground kept moist by seepage water near Cáceres, *Xyris* cf. *savanensis* Miq. is a predominant floral element. The species is characteristic of bare sandy areas kept constantly damp by ground water. A second species of this family, *Xyris laxifolia* Martius, occurs on the eroded areas near springs at the headwaters of the streams flowing into the Pantanal. It was not observed in the wetland. Wanderley (1989) found these plants, which exceep 1 m in height, at scattered locations in Goiás similar to those at which it was observed in Mato Grosso.

The presence of *Xyris jupicai* L. C. Richard was reported at the edge of a pond in the Pantanal of Mato Grosso do Sul (Dubs, 1994). It was also observed in the upland *cerrado* regions of the watershed, where it grows at somewhat drier locations than the other species of the genus.

8. Fauna

8.1. Sarcodina

In general, the amoebas and their relatives play a much greater role in the Pantanal than do other free-living protozoans. This situation is somewhat reversed from many water bodies in the temperate zones, in which ciliates are far more abundant than sarcodines. Speculative reasons can be found for this, such as the great abundance of small predatory fishes that can feed on ciliates more easily than on shelled testaceans or spiny heliozoans, and the concentration of nutrients among the roots of floating plants, which gives microorganisms that creep on surfaces some advantage over free-swimming forms. Temperature may also play a role, but more has to be discovered about the tolerances of protozoans adapted to tropical conditions.

The Pantanal has virtually no endemic sarcodine fauna and only one species apparently confined to the state of Mato Grosso. All species encountered fit the descriptions of cosmopolitan and circumtropical species extremely well (Decloitre, 1981; Deflandre, 1929a, b, 1937; Gauthier-Lièvre and Thomas, 1958; Grospietsch, 1965). Other researchers studying this fauna elsewhere in Brazil (Cunha, 1916; Green, 1975) and Argentina (Vucetich, 1970, 1972, 1976, 1980) have come to the same conclusion.

8.1.1. Lobosea

Amoebidae

The amoebas without shells or skeletal structures are not abundant in the Pantanal. They are observed from time to time in water samples containing submerged plants or detritus. The taxonomy of this family is confused, to say the least, but the species observed occasionally among detritus during the *vazante* showed no differences from the common free-living amoeba, *Amoeba proteus* (Pallas, 1766) in its various shapes and forms.

Pelomyxidae

The taxonomy of this family, which is often included within the Amoebidae, has been particularly confused because large numbers of names have been applied to various forms and different sizes that probably all belong to a single species. Using the descriptions of Page (1976), the one observed in the Pantanal among detritus during the dry season must be the cosmopolitan *Pelomyxa palustris* Greef, 1874, and the specimens are near the small end of the size range, perhaps reflecting nutritional shortages in the water bodies or temperature effects. The amoebas of this species were generally about 350 µm in diameter along the long axis. The form, content, and movement of the cells closely resembles those of the much larger individuals observed in the temperate zone of Europe. They were occasionally encountered during the dry season among detritus near the shores of permanent water bodies or trapped among the roots and rhizoids of floating plants.

Hartmannellidae

The species of this family have, like the pelomyxids, been given many different names without any real evidence being presented that specimens assuming different forms and shapes are really taxonomically distinct from one another. To this family belong the so-called *limax*-amoeba, which are small in comparison to other amoebae without shells. Using the key of Page (1976), the most abundant species in the Pantanal can be designated *Saccamoeba stagnicola* Page, 1974. It was abundant only during the initial flooding period of the early rainy season. One of the most abundant species of this group in Northern Germany resembles very closely the forms observed in the Pantanal, and both conform excellently to the description of Page (1976).

Arcellidae

In the freshwater bodies of tropical South America, the testaceans seem to play a more important role than the ciliates, a situation quite different from that in the waters of the North Temperate Zone (Hardoim and Heckman, 1996). These organisms hold various positions in the food web, but most seem to be omnivores, feeding on detritus, bacteria, algae, or small animals, depending upon what is available. Some of the testaceans, however, are apparently primarily predators. The autecological observations made on each of the species is provided separately.

The testaceans in this and the following families were characteristic species of the interrhizon and epibenthos of the Pantanal. Many species were present, and some of them were quite abundant seasonally. These constitute a major element in the Pantanal food web (Section 11).

The arcellid species are well represented both in the Pantanal proper and in the streams of the surrounding watershed. The most abundant species, seen the year round, are *Arcella discoides* Ehrenberg, 1871, which is encountered in almost every kind of habitat, and *Arcella hemispherica* Perty, 1862, represented by two subspecies, *A. h. hemispherica* and *A. h. undulata* Deflandre, 1928. Pinto (1925) had previously reported *A. discoides* from Mato Grosso without specifying a locality.

Arcella vulgaris Ehrenberg, 1832, was seen the year round in the Pantanal, but generally only isolated individuals were observed among submerged plants, and the species cannot be considered common. Green (1975) recorded two "forms" of *A. vulgaris* from northern Mato Grosso: *crenulata* and *undulata*, with *crenulata* being the more abundant of the two. Although *Arcella dentata* Ehrenberg, 1838, was also seen during all seasons, it seemed to be a species preferring the flowing water of streams in the watershed and was rare in the lentic waters of the Pantanal. Other species are less common, and most were not encountered during all seasons (Hardoim and Heckman, 1996). *Arcella megastoma* Penard, 1913, was sometimes abundant among submerged plants; it was observed only during the high water and dry seasons. Inexplicably, it was never encountered during the seasons of rapidly rising or falling water levels.

In contrast to the other arcellids, *Hyalosphaenia subflava* (Cash, 1909) was common during the high water period and persisted until the water level had fallen considerably after the end of the rainy season. It was absent during the dry and early rainy season, when most of the testacean species are easiest to find.

Also present among the arcellid fauna but not very common were *Arcella conica* (Playfair, 1918), *Arcella costata* Ehrenberg, 1847; *A. gibbosa* Pénard, 1890; *Arcella lobostoma* Deflandre, 1928; *Arcella rota* Daday, 1905; and *Arcella rotundata* Playfair, 1918, generally observed among submerged plants or in the detritus they produced when they died off during the *vazante* (Hardoim, pers. comm.), as well as *Pyxidicula cymbalum* Penard, 1901, and *Pyxidicula operculata* (Agardh, 1827). The two species of *Pyxidicula* both preferred inhabiting deposits of sapropel that form during the rainless season.

Besides the study of the relationship between the seasonal succession and occurrence of the testacean species in the Pantanal (Hardoim and Heckman, 1996), the only previous published study of this group in Mato Grosso was conducted by Green (1975) in oxbow lakes along the river systems in the northern part of the state, which flow into the Amazon and Tocantins Rivers. In these water bodies, he encountered all of the species of *Arcella* observed in and near the Pantanal except for *A. dentata*. He did not find any species of the genera *Hyalosphaenia* or *Pyxidicula*, either.

In the uppermost sections of the Rio Paraná System in the Brazilian states of Mato Grosso do Sul and Paraná, Velho *et al.* (1996) reported the presence of *Arcella brasiliensis* Cunha, 1913, and *Arcella norestina* Vucetich, 1973, two rare species. They also recorded the more common *Arcella mitrata* Leidy, 1879, for the first time from Brazil. In addition, they encountered *Arcella costata*, which was rare in the Pantanal of Poconé and Green (1975) had already reported from Mato Grosso. Since only five arcellid species found in the Pantanal were also found by Velho *et al.* (1996), and the commonest Pantanal species, *Arcella hemispherica*, was not among them, it can be assumed that there are significant differences between the conditions in the upper Rio Paraguay and those in the upper Rio Paraná. Intensified studies of the testacean fauna in both of these river systems together with the sections downstream from their confluence are needed to determine the reasons for the distribution patterns of these species.

Centropyxidae

A testacean species among the most numerous in the Pantanal and unquestionably the most widespread is *Centropyxis aculeata* (Ehrenberg, 1838), which occurred on every kind of surface during every season. During the *cheia*, it sometimes produced massive populations on the surfaces of submerged plants. Its presence in Mato Grosso had already been reported by Pinto (1925) and Green (1975). *Centropyxis discoides* Ehrenberg, 1843, was also abundant during the high water period, but it was not encountered during other times of the year. It was encountered creeping on submerged macrophytes. The other species in this family were seldom encountered. Most of them seemed to prefer relatively large, permanent water courses. *Centropyxis platystoma* (Penard, 1890) and *Centropyxis spinosa* Cash, 1905, were found on submerged plants during the dry season, while *Centropyxis marsupiformis* (Wallich, 1864) was encountered among the detritus produced by the dying macrophytes during the *vazante*, the season the water level rapidly decreases. *Centropyxis constricta* (Ehrenberg, 1838) was encountered only among filamentous algae in ground water seeping out onto a swampy area in the Pantanal of Barão de Melgaço. All of these species seem to be omnivores that subsist mainly on saprophytic organisms.

Somewhat rarer species have also recently been observed by E. Hardoim (pers. comm.) in the Pantanal of Poconé. These include *Centropyxis ecornis* (Ehrenberg, 1841), and *Centropyxis hirsuta* Deflandre, 1929, observed only in the Rio Bento Gomes at the edge of the Pantanal, including the Baia Piuval. Neither was ever observed during the *cheia*.

The *Centropyxis* fauna encountered in the northern drainage areas of Mato Grosso is apparently quite different from that in the Pantanal. While Green (1975) did not report the presence of *C. marsupiformis, C. platystoma*, or *C. spinosa,* he encountered four species that were never observed in the Pantanal: *Centropyxis cassis* (Wallich, 1864), *Centropyxis minuta* Deflandre, 1929, *Cyclopyxis kahli* Deflandre, 1929, and *Cyclopyxis impressa* (Daday, 1905). During a recent study of the uppermost parts of the Rio Paraná System in the Brazilian states of Paraná and Mato Grosso do Sul, Velho *et al.* (1996) encountered a centropyxid fauna including distinct elements encountered either in the Pantanal or in the Amazonian drainage system of Mato Grosso but not in both regions. For example, they found *C. marsupiformis*, which had not previously been reported from Brazil north of Rio Grande do Sul but which also occurs in the Pantanal, as well as several species reported by Green (1975) from the Amazon drainage area of Mato Grosso but not yet discovered in the northern part of the Pantanal, including *Cyclopyxis kahli* and *Cyclopyxis impressa*. The gaps in the ranges of these species, however, is most certainly due more to the dearth of studies of the protozoan fauna than to any real zoogeographical considerations.

Difflugiidae

The difflugiids are important members of the biotic community, mainly as primary consumers of algae as well as saprophytes, the food preferences apparently being species specific. The species also demonstrated distinct preferences for microhabitats (Hardoim and Heckman, 1996). *Difflugia corona* Wallich, 1864, which was common throughout the entire year, showed a clear preference for aquatic macrophytes, over which it creeps in search of its food (Fig. 8.1). Its diet seems to include many algae, which it captures on its pseudopodia (Fig. 8.2). Just as abundant throughout the year are *Difflugia acuminata* Ehrenberg, 1838, which is usually benthic or found creeping among filamentous algae. It probably subsists mainly on detritus and saprophytes. *Difflugia lobostoma* Leidy, 1879, which can be encountered on any kind of surface, is probably chiefly a predator. *Difflugia oblonga* Ehrenberg, 1838, was common throughout most of the year among benthic detritus, but it was not observed during the high water period. It apparently subsists on saprophytes and substances in the detritus. *Difflugia tuberculata* Wallich, 1864, is also abundant throughout most of the year, occurring among macrophytes and algae in the Rio Bento Gomes. Species less frequently encountered include *Cucurbitella mespiliformis* Pénard, 1901, and *Difflugia elegans* Penard, 1890, species generally found on submerged plants. Both of these species were encountered during the high water period.

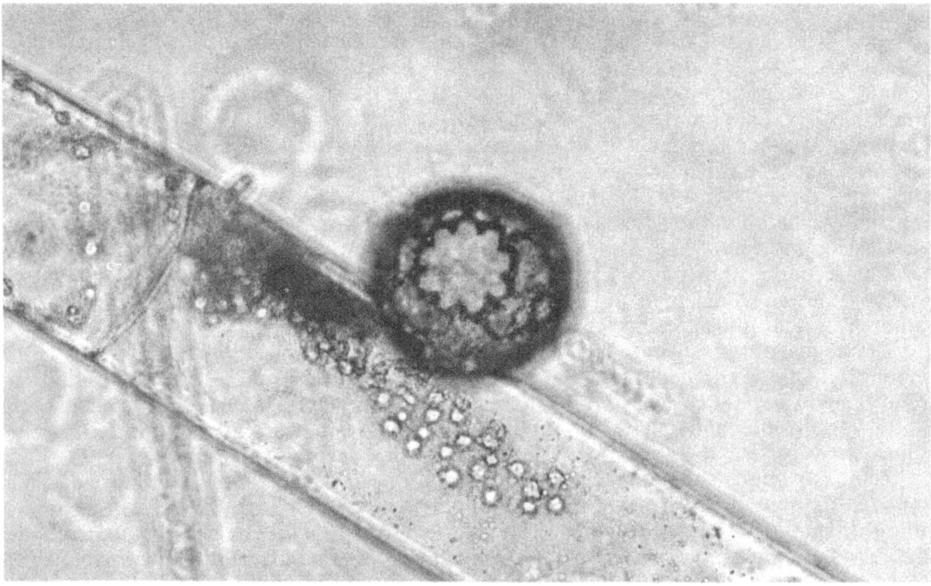

Figure 8.1: Testaceans are often the predominant protozoans in the water bodies of the Pantanal. A shell of *Difflugia corona* shows its characteristic aperture. It is a common species on actively growing submerged vegetation.

Among the rarer difflugiids are *Difflugia bacillifera* Pénard, 1901, *Difflugia ventricosa* Deflandre, 1926, and *Difflugia urceolata* Carter, 1864, all found in detritus deposits during the dry season, and *Pontigulasia compressa* (Carter, 1864), found in the same kind of microhabitat during the early rainy season. *Difflugia lanceolata* Penard, 1890, was observed only in one of the roadside ditches along the Transpantaneira Highway and in seepage water in a swampy area of the Pantanal near Barão de Melgaço.

Unlike the arcellids, the difflugiids incorporate foreign bodies composed of silica in their shells (Vucetich, 1989), making them somewhat dependent on contact with the sediment (Deflandre, 1953). At times they are among the dominant non-photosynthetic microorganisms in the community. Their grazing acts to regulate the development of algae and saprophytes on the surfaces of the water plants. The clean water in the Pantanal during the *cheia* with its rather constant supply of dissolved oxygen seems to promote the development of species such as *Difflugia corona* and *Cucurbitella mespiliformis*. Other difflugiids, such as *Difflugia oblonga*, however, do not seem to be adversely affected by low oxygen concentrations and benefit from the enormous supply of saprophytes during the seasons when much detritus is decomposing in the water.

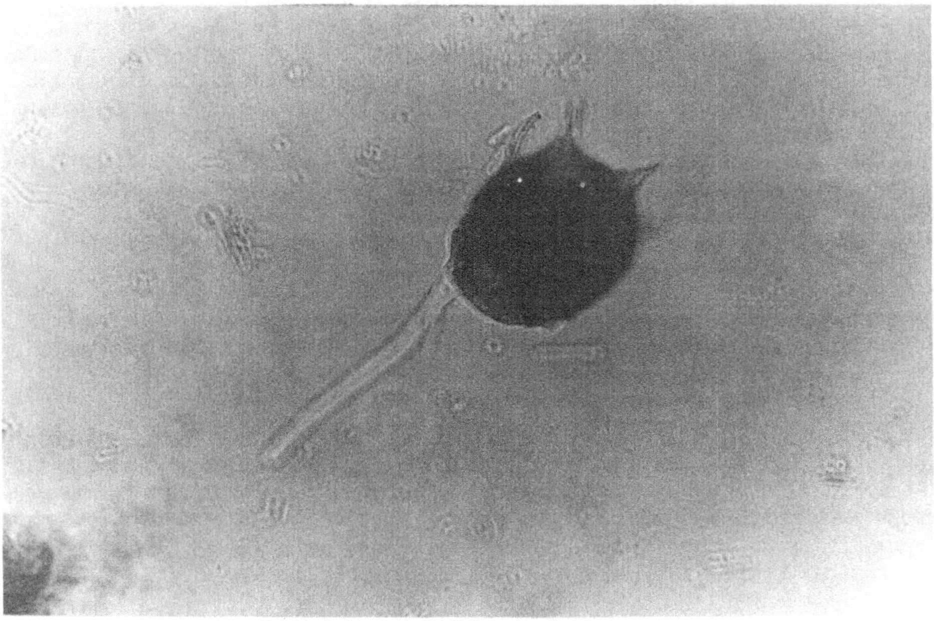

Figure 8.2: A living *Difflugia corona* showing a pseudopod extending far in front of the shell.

The presence of these species, except for *Cucurbitella mespiliformis*, *Difflugia ventricosa*, and *Pontigulasia compressa*, was previously reported for Mato Grosso by Green (1975). In addition to the species that occur in the Pantanal, he encountered *Difflugia brevicolla* Cash, 1909; *D. globularis* Dujardin, 1837; *D. gramen* Penard, 1902; *D. lismorensis* Playfair, 1917; *D. minuta* Rampi, 1950; *D. penardi* Hopkinson, 1909; *D. schurmanni* van Oye, 1932, *D. smilion* Thomas, 1953, and *Hoogenraadia cryptostoma* Gauthier-Lièvre and Thomas, 1958, in the northern part of the State of Mato Grosso in the drainage area of the Rios Tocantins and Amazon. He also described a new genus and species from his material: *Suiadifflugia multipora* Green, 1975, which is probably the only protozoan species known only from the State of Mato Grosso. It has recently been tentatively identified in the stomachs of fishes which prey heavily on testaceans in the Pantanal (Silva Neto *et al.*, in press).

Cryptodifflugiidae

Hardoim (pers. comm.) observed *Cryptodifflugia operculata* (Playfair, 1917) during the dry season among detritus. It is not one of the abundant testaceans in the Pantanal.

Nebelidae

This family was represented by two of the most common testacean species in the Pantanal and three uncommon to moderately abundant species. *Lesquereusia spiralis* (Ehrenberg, 1840) was encountered on every kind of surface during all seasons of the year. It was abundant on plants in the Rio Bento Gomes and appeared in most of the samples of submerged macrophytes that were taken there. Also abundant, but much less so than *L. spiralis,* was *Lesquereusia modesta* Rhumbler, 1895, which is distinguished mainly by the presence of foreign bodies in the shell. It was not observed during the *vazante* and was rare during the *seca,* but during the rest of the year it was quite common in detritus samples. Both of these species seem to be widespread circumtropical elements of the microfauna. They are reported from many warm regions that have been investigated and are rare or absent from the waters of the temperate zones. Pinto (1925) had previously reported *L. spiralis* to be present in Mato Grosso, and Green (1975) indicated that both of them are widespread in the northern part of Mato Grosso.

A moderately common species, *Nebela collaris* (Ehrenberg, 1871), was encountered during every season except the *cheia* at various locations throughout the northern Pantanal. *Awerintzewia cyclostoma* (Pénard, 1901) was abundant only during the dry season, when it was encountered in the deep water bodies among submerged plants.

8.1.2. Filosea

Vampyrellidae

An interesting sarcodine was observed on species of the filamentous alga, *Spirogyra* spp., in the standing waters near the Paraguay River at Cáceres. *Vampyrella lateritia* (Fresenius, 1856) is a herbivorous species that fastens itself to the alga filaments and penetrates into the individual cells with its pseudopods to consume the contents.

Euglyphidae

Euglyphids were extremely common in the Pantanal, occurring throughout the year among the submerged vegetation. The most abundant of these species was *Euglypha acanthophora* (Ehrenberg, 1841), which appeared in a great many water samples collected during all seasons. It was generally encountered in samples taken from among living submerged plants.

Somewhat less abundant were *Euglypha filifera* Penard, 1902, found among detritus during the early rainy season, and *Euglypha tuberculata* Dujardin, 1841, found throughout the year at all of the sampling sites in the wetland. The Pantanal testacean fauna in the Poconé District also includes *Euglypha laevis* (Ehrenberg, 1845), a species taken from among detritus and on artificial substrate during the *enchente,* as well as *Euglypha cristata* Leidy, 1874, and *Euglypha strigosa* (Ehrenberg, 1938), observed during the other seasons of the year by Hardoim (pers. comm.).

Specimens of *Trinema enchelys* (Ehrenberg, 1818) were collected from seepage water that kept various areas of *campo úmido* moist the year round. The species was observed during the *vazante* and *seca.* At other times of year, the sites at which the water seeped out on the surface were covered with flood waters.

All of the euglyphid species were previously reported from the northern part of the state of Mato Grosso by Green (1975).

Cyphoderiidae

The only species in this family encountered during the study, *Cyphoderia ampulla* (Ehrenberg, 1840), was a member of the seepage water micro-community. It was observed only during the *vazante*, when the flood waters receded from the area kept permanently moist by the seepage water.

8.1.3. Heliozoea

In the Pantanal, Actinophryidae was the only heliozoean family encountered. Its most common representative was *Actinosphaerium eichhorni* (Ehrenberg, 1840), a cosmopolitan species encountered in the water column, most frequently in rather sparse beds of submerged plants. Among the many specimens seen were a few green ones, which have been designated as *Actinosphaerium eichhorni viride* (Ehrenberg, 1833). They apparently contain zoochlorellae, but little is known about them with certainty. The size of the specimens encountered ranged from small to extremely large. The smallest individuals were generally semi-planktonic, suspended in the water but invariably occurring near submerged plants. The large specimens were usually encountered among deposits of detritus on the sediment or trapped among plant masses.

A slightly less abundant species is *Actinophrys sol* Ehrenberg, 1830, which also occurs in open water or among plant beds. It apparently has the same feeding habits as *A. eichhorni*, but is not capable of consuming food items quite as large.

Actinosphaerium eichhorni and *Actinophrys sol* are omnivores that feed on small particulate matter in the water and perhaps simple, dissolved organic substances. Bacteria, tiny algae, or small ciliates become immobilized on the spines and are then transported with the flowing cytoplasm toward the center of the animal and engulfed by small pseudopods (Kudo, 1966).

A third heliozoan was somewhat rarer than the two free-floating ones. This sessile species, *Clathrulina elegans* Cienkowski, 1867, usually occurred attached to small bits of detritus. It is apparently an epipelic omnivore.

No heliozoans were observed during the high water period or during the periods of high water temperatures in the dry season, but they were fairly common during the periods of rising and falling water levels. They are much more frequently encountered than the amoebae without hard structures, but it may well be that the naked amoebae are better able to conceal themselves within various objects and therefore escape attention.

8.2. Zooflagellida

The occurrence of zooflagellids was confined almost exclusively to the dry season, during which rich accumulations of detritus collected in the drying pools of water, and enormous numbers of saprophytic bacteria appeared to break the substances down. Only isolated individuals were observed, and none of them were abundant enough to systematically investigate.

8.3. Ciliophora

In comparison to the ciliate fauna in the temperate zones, that in the Pantanal has to be considered depauperate. The testaceans and rotifers have apparently taken over many of the niches in the micro-communities from the ciliates, and with the exception of a few species, ciliates have to generally be considered uncommon. Several of the few species that become extremely numerous at times are those that usually creep on surfaces, using their cirri similarly to legs. Others are free-swimming but generally remain near plants of other solid objects in the water.

Hardoim and Heckman (1996) provided some possible reasons for the scarcity of ciliates, and a list of factors that can be considered negative for many species in this group can be proposed to explain their scarcity:

1) They have more difficulty than other microorganisms surviving the periods of desiccation;
2) During the period that bacteria, algae, detritus, and other possible foods are abundant, they do not adapt well to the temperature, which may exceed 40°C during the day;
3) During the seasons that the water temperature is only moderately warm, bacteria and algae are extremely scarce, depriving ciliates of a rich food supply;
4) Various rotifer and gastrotrich species become abundant at times, perhaps eliminating ciliates by competition;
5) Predator pressure from the young of many fish species and various invertebrates is great.

The ciliates of the Pantanal all belong to widespread species, many of which are cosmopolitan or nearly so (Kahl, 1935). The absence of endemic species and the general scarcity of all ciliates in the water bodies of the Pantanal suggest that the wetland was recently colonized and that only the species with the widest environmental tolerances are able to survive there. Only one of the species seems to be endemic to South America: *Dileptus bivacuolatus* Da Cunha, 1915.

Various classification systems for the ciliates have been proposed. The species discussed below are generally arranged according to Foissner and Foissner (1988) as modified by Hardoim and Heckman (1996).

8.3.1. Kinetophragminophora

Loxodidae

One species in this family, *Loxodes rostrum* (O.F.Muller, 1786), was fairly abundant among detritus during the *seca*. In some samples, moderate numbers were observed, but it never formed notably large populations.

Holophryidae

The holophryids are seasonally important elements of the ciliate fauna, but they are never really dominant, and none were observed during all seasons. Two free-swimming species were common in water samples from the Pantanal during the *vazante* and *seca*. *Prorodon teres* Kahl, 1927, was fairly common during the *vazante*, when it could be observed among the detritus left by the moribund aquatic macrophytes. *Placus luciae* Kahl, 1926, occurred regularly among the finer detritus encountered in the permanent water bodies during the *seca*.

Colepidae

Species of *Coleps* were among the few ciliate species that can be considered fairly abundant members of the micro-communities in the Pantanal, at least seasonally. *Coleps hirtus* Nitzsch, 1817, was identified during all seasons among submerged plants and detritus. At times, the species was fairly abundant. It is a cosmopolitan species, which is generally associated with the katharobic community.

Coleps octospinus Noland, 1925, is a somewhat larger species that was observed among submerged plants and detritus only during the *vazante.* It was moderately abundant when present at all, but it was not often encountered.

Coleps nolandi Kahl, 1930, was observed in the same kind of micro-community as the other two species of the genus. It was found only during the *vazante* and *seca*.

Didiniidae

This family includes one of the most widespread and abundant ciliates in the Pantanal, *Mesodinium pulex* (Claparède and Lachmann,1858), which is usually abundant in water samples or not present at all. It seems to be mainly katharobic, swimming in the open water among masses of submerged vegetation, but it is also found in samples rich in detritus. In the open water among plankton, however, it is generally not encountered, even while it is abundant among nearby masses of plants. In the Pantanal, it was found both in the lentic waters of the flood plain and in the littoral zone of lotic waters, including the Rio Bento Gomes near the Poconé exposition park. In European waters, this species also seems to prefer beds of submerged vegetation as its habitat, but it also ventures into the open water, especially when much suspended matter is present (Heckman and Schade, 1989).

The family Didiniidae also includes two species that prey on relatively large organisms. *Didinium nasutum* (O. F. Müller, 1786) was observed regularly during all seasons and was particularly abundant during the dry season, when rather large populations were encountered. These ciliates specialize in feeding on species of *Paramecium*, which are generally ingested whole (Kudo, 1966). Large populations of *D. nasutum* were generally encountered after species of *Paramecium* had become rather abundant in the detritus and alga-rich waters left during the *seca. Mesodinium balbianii* Fabre-Domergue, 1888, was observed rarely among algae and detritus accumulating near the Rio Novo as the water rose during the *enchente*.

Trachelophyllidae

Chaenea teres (Dujardin, 1841) inhabited a markedly different kind of community than most of the other ciliates encountered in the Pantanal. It was observed rarely in the areas newly flooded with rainwater during the early *enchente.* In this community, the short-lived terrestrial macrophytes were drowning, thereby yielding detritus for a rapidly developing saprobic community to break down.

Lacrymariidae

Lacrymaria olor (O. F. Müller, 1776) was observed repeatedly during the *enchente* concealed within clumps of detritus, its long trunk probing the area for bacteria. The individuals were nearly always solitary in the samples. This species seems to be a rather abundant member of the shallow water saprobic community that is active in breaking down the detritus left by the drowned terrestrial vegetation during the *enchente.*

Tracheliidae

The tracheliid fauna of the Pantanal includes perhaps the only typical South American protozoan species encountered during the study, *Dileptus bivacuolatus* Da Cunha, 1915, which was discovered and described on the continent and does not seem abundant anywhere else. The species was not abundant, but it was encountered on rare occasions among detritus samples during the *vazante*. The other two species in the family were also encountered during periods of peak saprophyte growth during the *vazante*. *Paradileptus conicus* Wenrich, 1929, was sometimes present in moderate numbers, while *Trachelius ovum* Ehrenberg, 1831, was nearly always represented by isolated individuals, but usually in samples containing at least a few of other ciliate species.

Actinobolinidae

One rather rare species in this family was encountered among detritus in the Pantanal: *Actinobolina vorax* Wenrich, 1929, was observed only during the *enchente*.

Amphileptidae

This family is represented in the Pantanal by four cosmopolitan species, the most common of which is *Litonotus fasciola* (Ehrenberg, 1870). which was observed throughout the year, except during the high water period. It occurred in samples of plants, detritus, and sediments. Generally, the species is vagile on surfaces, including the underside of the surface tension layer of the water.

During the *vazante*, *Litonotus procerus* Penard, 1922, and *Litonotus* cf. *rotunda* (Kahl, 1926) were observed in samples of water containing detritus from the moribund water plants. These species were not encountered during the *seca*, perhaps because the temperature and concentrations of various inorganic substances were too high (Hardoim and Heckman, 1996).

Litonotus lamella (Ehrenberg, 1829) was observed only during the *enchente* in samples from newly flooded water bodies in which the remains of the drowned terrestrial plants were decomposing.

Microthoracidae

One member of this family, *Drepanomonas dentata* Fresenius, 1858, was observed in water samples taken for a short period of time during the *seca*. The samples were heavily enriched with decomposing organic matter.

Chlamydodontidae

Three species in this family were encountered during the study. Of these, only *Pseudochilodontopsis algivora* (Kahl, 1935) was moderately abundant. It occurred only during the *enchente*. The other species were rare. Both *Chilodonella uncinata* (Ehrenberg, 1838) and *Chilodonella acuta* Kahl, 1935, were encountered only during the high water period in small clumps of detritus on the surface of the water. *C. uncinata* typically creeps on the underside of the surface tension layer of the water, where surface-dwelling bacteria are abundant.

Lynchellidae

One species in this family, *Gastronauta membranaceus* Engelmann, 1875, was found among drowning terrestrial plants, mainly grasses, in shallow pools of rainwater on the flood plain shortly after the freshet.

8.3.2. Oligohymenophora

Ichthyophthiriidae

In addition to the purely free-living ciliate species, *Ichthyophthirius multifiliis* Fouqué, 1876, was observed in water samples. This is a notorious external parasite of fishes in aquaria. In the Pantanal, it attacks some of the same species that it might infest in aquaria. Infectious swimming forms of the parasite were encountered in the Rio Bento Gomes during the dry season.

Parameciidae

Only two species of *Paramecium* were found during the study. The nuclei of specimens of the genus without zoochlorellae were stained with methyl green, and all proved to be *Paramecium caudatum* Ehrenberg, 1838. These were fairly common during the *enchente* and *cheia*, especially in epibenthic detritus deposits. *Paramecium bursaria* (Ehrenberg, 1836), a species containing zoochlorellae, was encountered in several samples collected during the dry season.

Frontoniidae

This family was represented by two species of *Frontonia*, which feed on relatively large organisms, including various groups of algae. *Frontonia leucas* Ehrenberg, 1838, was fairly abundant throughout the entire year, mainly in water samples taken from among submerged plants. The species was most abundant in the permanent pond beside the Transpantaneira Highway near the Rio Pixaim. It is katharobic rather than planktonic, although it may be found in the free water near the submerged vegetation. The specimens collected often contained relatively large filaments of cyanobacteria and algae, including diatoms. *Frontonia acuminata* Ehrenberg, 1838, is a somewhat smaller species that was observed during the dry and early rainy seasons among submerged plants, in detritus, and among various species of algae. All species in this family recorded for the Pantanal are cosmopolitan in distribution and common throughout most of the world.

Urocentridae

One of the few ciliates that was often observed, sometimes in large numbers, was *Urocentrum turbo* (O. F. Müller, 1786). At times during the dry season, massive populations were present in a few samples, the individuals rotating continually at the end of their threads. They were generally encountered among masses of submerged plants and were especially abundant in the permanent pond along the Transpantaneira Highway near the Rio Pixaim. None were observed during the *enchente*, but they were first noticed during the late *cheia* and increased steadily in numbers until the *seca*.

Pleuronematidae

This family is represented by a widespread species, which was never found to be abundant in the Pantanal. *Pleuronema crassum* Dujardin, 1841, was observed among the newly developing submerged plants and algae during all seasons except the *enchente*.

Cyclidiidae

Cyclidium glaucoma O. F. Müller, 1786, was observed while many of the plants were dying off during the dry season. *Cyclidium flagellatum* Kahl, 1926, was fairly abundant in a few water samples containing detritus that were collected during the *vazante*. These are two rather small species which apparently feed mainly on bacteria and small particles of organic matter.

Vorticellidae

Surprisingly, the sessile protozoa in this family were sparsely represented in the Pantanal. They were seldom seen, and only isolated individuals were observed. The most abundant species was *Vorticella campanula* Ehrenberg, 1831, which was most frequently found on cover slips of microscope slides set out for several days as artificial substrate to collect aufwuchs. The species was observed in the Pantanal only during the *cheia*, but it was seen at other times in permanent streams in the catchment area.

Vorticella similis Stokes, 1887, was seen only rarely, attached to detritus during the *enchente*.

The absence of many of the cosmopolitan vorticellid species in the Pantanal can be explained by the lack of the mesotrophic and eutrophic habitats they prefer, except during seasons when the water temperature is often high.

Epistylidae

This family is not well represented in the Pantanal. An unidentified and probably undescribed species of *Epistylis* was observed only once during the dry season attached to the antenna of a cyclopoid copepod, and an unidentified species of *Pyxidium* was observed rarely on the shells of *Biomphalaria straminea*.

Operculariidae

The only species observed during the study was *Opercularia rugosa* Kellicott, 1884, which was among the sparse periphyton during the *enchente*.

Vaginicolidae

Only one species in this family was observed in the Pantanal. It was found attached to filamentous algae during the *vazante*. The individuals observed fit the description of *Pyxicola affinis* Kent, 1881, but a positive identification was not possible because of the paucity of individuals observed.

8.3.3. Polyhymenophora

Spirostomatidae

Water containing much detritus was the preferred habitat of two *Spirostomum* species: *Spirostomum teres* (Claparède and Lachmann, 1859) which was observed only rarely during the *vazante*, and *Spirostomum minus* Roux, 1901, which was more abundant during the *enchente*. Both species are typical members of the saprobic community, probably subsisting nearly entirely on bacteria.

Stentoridae

Only three *Stentor* species were encountered during the study. *Stentor roeseli* Ehrenberg, 1835, was by far the most abundant, inhabiting short tubes constructed in small masses of detritus. The species is a typical member of the interrhizon community. It was represented in many water samples, sometimes abundantly. *Stentor igneus* Ehrenberg, 1838, generally found swimming freely, was observed exclusively during the rainless *vazante* and *seca*. It generally appeared in samples containing submerged plants.

Another free-swimming species, *Stentor amethystinus* Leidy, 1880, was present but rarely seen during the study. It occurred mainly in samples taken from among submerged plants containing much detritus.

Halteriidae

A small, free swimming species, *Halteria grandinella* (O. F. Müller, 1786), was usually observed in the open water among masses of submerged plants. It is not planktonic and seldom strays far from the plants. Sometimes, relatively large numbers of this species can be found in a single water sample, and the species proved particularly abundant in the waters beside the Transpantaneira Highway near the Rio Novo during the late dry and early rainy seasons. Members of this species presumably feed nearly exclusively on bacteria and the simple organic compounds released from detritus by their action.

Strobilidiidae

Strobilidium gyrans (Stokes, 1887) was one of the more abundant ciliates in the Pantanal. It was observed rather often among submerged plants during the seasons of rising and falling water levels. It was rare during the dry season and not observed at all during the high water period. The habits, micro-habitat, and abundance of this species are similar to those of *Urocentrum turbo*.

Recently, the name of this species has been called into question. Foissner and Foissner (1988) resurrected the long unused name, *Strobilidium caudatum* (Fromentel, 1876) to replace its frequently cited name on the grounds of the priority rule. In addition to the practical reason that this change is causing much confusion in the literature, especially in view of the fact that Kahl (1935) applied *S. caudatum* to a completely different species, this change must be rejected according to the rule of zoological nomenclature suppressing long-unused names, even if these names prove to have priority to generally used names for the species. A decision on this matter is pending (Hardoim & Heckman, 1996).

Spirofilidae

Stichotricha secunda Perty, 1852, constructs tubes in detritus to conceal itself, extending only the proboscis-like oral end of its body out to filter its food, mainly bacteria. It was often observed when dead plant material was abundant in the water, providing this species with a food supply, apparently bacteria, and much substrate in which to construct its tube-like hiding places. It was most abundant while the aquatic vegetation was dying off during the *vazante* and somewhat less abundant among the decomposing terrestrial plants in the water during the *enchente*. It was seen most frequently in the waters near the Rios Novo and Bento Gomes.

Urostylidae

Uroleptus dispar Stokes, 1886, was fairly abundant among detritus during the dry season as well as the *enchente*. A species similar in appearance, *Paruroleptus piscis* (Kowalewski, 1882), was observed among submerged macrophytes at various times during the *enchente*.

Kahliellidae

Hypotrichidium conicum Ilowaisky, 1921, is katharobic, and it was quite widespread and moderately abundant shortly after the flooding during the *enchente*. It was found among water plants near the shores of the small rivers draining the wetland and on adjacent, newly flooded areas of the floodplain. Moderate numbers of this species were present whenever it was found.

Oxytrichidae

The species in this family generally creep along surfaces, including the surface tension layer, and rarely swim for short distances in the open water. Oxytrichids were well represented in the Pantanal during the *enchente*, when all three species identified during the study were encountered. The most abundant species was *Stylonychia mytilus* (O. F. Müller, 1773), which was observed creeping on all kinds of surfaces, including detritus, aquatic macrophytes, algae, and the floating layers of bacteria on the surface of the water. It occurred during all seasons. *Steinia inquieta* (Stokes 1887) occurred rarely during the *vazante* and more frequently during the *enchente*. The other species of this family, *Opisthotricha euglenivora* Kahl, 1935, was found only during the *enchente*. As its name implies, *Opisthotricha euglenivora* generally occurs in water rich in euglenoid species, on which it feeds.

Euplotidae

Interestingly enough, the only member of this family present, *Euplotes patella* (O.F.Müller, 1773), occurred in moderate numbers only during the *cheia*. The individuals generally creep using their cirri in a manner reminiscent of legs on all kinds of surfaces. Like most of the ciliates in the Pantanal, this is a cosmopolitan species that is extremely widespread in many kinds of water body.

Aspidiscidae

Probably the most frequently encountered species of ciliate in the Pantanal is *Aspidisca costata* (Dujardin, 1842), which typically creeps on the surface of detritus particles or, less frequently, on living plants. It was observed at one time or other in water samples from all sites and during all seasons. Somewhat less frequently seen but nevertheless abundant is *Aspidisca lynceus* Ehrenberg, 1838, which has similar habits and occurs in the same kind of habitats.

As in the case of *Strobilidium gyrans* and for the same general reason, Wu and Curds (1979) resurrected the long unused name, *Aspidisca cicada* (O. F. Müller, 1786) to replace *Aspidisca costata*. This change must also be rejected according to rule of zoological nomenclature suppressing "forgotten names", which may not be recognized as valid because of priority after they have not been used for more than 50 years.

8.4. Porifera

One species of freshwater sponge in the family Spongillidae, *Drulia browni* (Bowerbank, 1863), is very widespread and abundant in the Pantanal. The sponges are most numerous on the branches and roots of trees along the small rivers, where they can be found totally desiccated during the dry season (Fig. 8.3). In addition to their occurrence in the rivers, these sponges were also encountered on dead tree trunks and fence posts on the floodplain at sites deep enough to remain under water for at least five to six months each year.

Just as the spongillids produce gemmulae in response to the coming of winter in the temperate zones, the species in the tropical wet-and-dry region of South America produce them to survive the dry season (Penney and Racek, 1968). It therefore seems likely that the sponges with the ability to form gemmulae originally developed either in the temperate or the tropical wet-and-dry climatic zone and from there colonized regions with seasonal alternations between favorable and unfavorable conditions, caused either by low temperatures or dryness. The success of *Drulia browni* in the equatorial regions of Amazonia can be attributed to the extreme water level changes in the rivers, effected largely by the seasonal changes in rainfall feeding those tributaries located in tropical wet-and-dry zones.

Mothes de Moraes (1983) reported the presence of this species in the Rio Bento Gomes near Poconé as well as at sites as remote as Venezuela and the Amazon region. The species is also known from the extreme southeastern part of Brazil in oxbows along rivers in the Rio Uruguay System (Volkmer-Ribeiro *et al.*, 1983).

Figure 8.3: A desiccated skeleton of the freshwater sponge, *Drulia browni*, along the shores of the Rio Bento Gomes during the dry season. Because the gemmulae of this species remain enmeshed in the old skeleton, they are able to continue growth on its very tough framework as soon as the river rises high enough for the water to cover the skeleton again.

In the lower sections of the Rio Paraná System in Argentina, Ezcurra de Drago (1993) reported the presence of ten species of freshwater sponge. The relative poverty of the sponge fauna in the Pantanal reflects the harshness of the conditions, including the long periods of dryness and the lack of nutrients during the high water period. Ezcurra de Drago (1993) also noted the presence of a particularly large number of endemic species among the South American sponge fauna.

8.5 Platyhelminthes

Like the algae and protozoa, the free-living flatworms are represented in the Pantanal mainly by very widespread members of the Neorhabdocoela, although a number of species in the Rio Paraná and Paraguay System seem to be endemic to South America (Noreña-Janssen, 1995a, b). No turbellarians were ever encountered in the Pantanal, although a species of *Dugesia* was found to be abundant in an upland stream within the watershed. In contrast, parasitic trematodes are apparently widespread, and their cercarial stages are, at times, quite abundant.

Catenulidae

A relatively rare member of the interrhizon in the Pantanal is the cosmopolitan species, *Catenula lemnae* A. Dugès, 1832, which was observed only during the *vazante* among decaying leaves of aquatic plants. Apparently, this species prefers water containing many small saprobic organisms, and the conditions in the Pantanal are therefore seldom conducive to its development. Marcus (1945) described several new species of *Catenula* of dubious validity from material collected in the states of São Paulo and Paraná, but most of his specimens were assigned to *C. lemnae*.

Microstomidae

One of the ubiquitous species encountered in many water samples from the Pantanal was *Microstomum lineare* (O. F. Müller, 1774), which occurred occasionally in samples of detritus taken among submerged plants along the permanent small and medium-sized water courses that contain standing water for much of the year. The species has been found to be common in a lake in contact with the Rio Paraná System in Argentina (Noreña-Janssen, 1995a, b).

At least two additional microstomids, both species of *Gieysztoria*, were encountered in water samples from among the rhizoids of *Salvinia auriculata* in the Rio Bento Gomes near the Transpantaneira Highway during the late dry season, one near the Transpantaneira Highway and the other near with Poconé exposition park. The identification of the flatworms in this genus is problematic, and the majority of known species in the Paraná River System have only recently been described (Noreña-Janssen, 1995b).

Polycystidae

The cosmopolitan *Gyratrix hermaphroditus* Ehrenberg, 1831, was encountered in many samples taken from among the interrhizon community, mainly during the rainless seasons. The species is apparently not uncommon throughout the entire river system as it was found to be abundant in a lake along a tributary of the Rio Paraná in Argentina (Noreña-Janssen, 1995a, b).

Macrostomidae

Immature specimens of one and possibly two species in this family were observed on occasion in the Pantanal. They are probably undescribed species (Faubel, pers. comm.).

Dugesiidae

One unidentified species of *Dugesia*, most similar to *Dugesia arndti* Marcus, 1946, was considerably abundant on the underside of large rocks in the Corrego Guanandi just downstream from a small impoundment in the district of Cangas several kilometers north of the Pantanal. No turbellarians were ever found in the wetland or any of its water bodies. The taxonomy of the South American Dugesiidae has never been satisfactorily resolved. *Dugesia arndti* is considered to be a species with many synonyms, although it is still not certain which taxa are valid because of the great natural variability of the morphological features, even within a single population.

Cryptogonimidae

Two species of digenic trematode in this family have been found in caimans in the Pantanal, *Caiman yacare*. *Caimanicola marajoara* Freitas and Lent, 1938, and *Proctocaecum dorsale* Catto and Amato, 1993a, have both been found in the wetland by Catto and Amato (1993a), and *P. dorsale* in known only from the Pantanal.

Telorchidae

This family is represented by two species of digenic trematode known only from *Caiman yacare* in the Pantanal and only recently described (Catto & Amato, 1993b): *Pseudotelorchis yacarei* Catto and Amato, 1993b, from the intestine and *Pseudotelorchis caimanis* Catto and Amato, 1993b, from the uterus.

Unidentified Trematoda

Massive numbers of unidentified gymnocephalus cercariae with two eye spots and unforked tail pieces were observed in the flooded ditches along the Transpantaneira Highway south of the Rio Bento Gomes during the *seca* of 1991. Many were attached to moribund fishes the shallow water. They were not observed during subsequent years.

8.6. Nematoda

Nematode species were common within the interrhizon as well as among benthic detritus during the dry season. The species could not be identified with reasonable certainty, but one of the commonest within the root masses is thought to be *Tobrilus* sp. The nematode fauna is rich seasonally, being nearly completely absent during the *cheia* and becoming abundant as the macrophytes begin to die off during the *vazante*.

8.7. Rotifera

The rotifer fauna of the Pantanal was characterized by various species that developed massive populations seasonally for short periods of the year. Few individuals were encountered during the high water season, but as the water level decreased during the *vazante*, an increasing number of species were encountered. During the dry season, a few of them develop dense populations at some locations, and the maximal species diversity and number of individuals are observed during the early part of the rainy season (Kretzschmar *et al.*, 1993).

The population dynamics of the rotifer species is a vital factor for the reproductive cycle of certain fishes. The maximal rotifer abundance is usually reached at the time most fish species begin their spawning, and the importance of rotifers as a first food for the hatchlings of many fish species is well known and has been documented from all parts of the world (Auvinen, 1988; Bok, 1989; Treasurer, 1990; Nyonje and Radull, 1991; Opuszynski and Shireman, 1993; Ziliukiene, 1994). Primarily freshwater rotifer species have even been found suitable as the first food of marine fishes (Young, 1994). Generally, species with hatchlings that are relatively small may be expected to depend on the presence of rotifers, while species larger at the time exogenous feeding begins tend to rely on small crustaceans as the first food (Ludwig, 1994). Obviously, the success of the annual recruitment to replace the great losses suffered by many of the relatively small fishes in the Pantanal during the dry season is dependent upon the presence of large rotifer populations (Heckman, 1994b).

Another notable feature of the rotifer fauna in the northern Pantanal is the poverty of species. Aside from those that were extremely abundant, at least for a short period during the annual cycle, the species recorded from the water samples taken during the study tended to occur irregularly and without any notable seasonal pattern. Several species were encountered only once or twice during the entire study. Given the area of the wetland, almost all of the known rotifer species might be expected to occur there at one time or another. An exhaustive study concentrated only on these organisms would certainly yield a longer list than that recorded here. This is particularly true in view of the great abundance of migratory water bird species, on the legs and feathers of which rotifers can be carried from distant parts of the earth. However, most of these imported species obviously do not seem to establish successful populations in the Pantanal, and the rarer ones contribute little to the community dynamics of the established biotic elements. The extreme abundance of hungry small fishes as well as the seasonal scarcity of food particles suitable for most rotifers contribute to the faunistic poverty of the group (Section 11.2).

Brachionidae

The family Brachionidae accounts for a good number of the rotifer species in the Pantanal, including some of the most abundant ones that are active during the dry season. A study of very warm pools of drying water encountered on the flood plain by Kretzschmar *et al.* (1993) revealed the presence of *Brachionus urceolaris* (O. F. Müller, 1773), *Brachionus bidentatus bidentatus* Anderson, 1889, and *Brachionus b. inermis* (Rousselet, 1906) in a community characterized by the presence of large numbers of *Asplanchna sieboldi*, as discussed below. The temperature of the water in these pools, which were very rich in chlorococcal algae, exceeded 40°C during the day.

In addition to these species, *Brachionus quadridentatus* (Hermann, 1783) and *Keratella americana* Carlin, 1943, were found to be very abundant in the large permanent lakes along small rivers at the edge of the Pantanal, including the Baia Piuval, where *Brachionus ahlstromi* (Lindeman, 1939) was also observed during the early dry season. These species seem to be truly planktonic in these lakes, but they are common only during the late dry and early rainy seasons. Their scarcity or absence during the high water periods is apparently due to the foraging by many tiny fishes as well as water movements through the lakes, which overflow large areas of the adjacent floodplain. The reduction in the availability of bacteria and algae possibly also contributes to the seasonal decline of the planktonic species. *Keratella cochlearis* (Gosse, 1851) was also encountered in a few of the samples from the Baia Piuval as well as from some of the pools of shallow water on the floodplain during the dry season. However, it was generally much less abundant that *K. americana*.

Two brachionids were typical inhabitants of plant-filled shallow water bodies that seldom completely dry up. *Brachionus patulus* (O. F. Müller, 1786) was commonly observed in a pond near the Rio Pixaim during the late dry season. This shallow water body receives some wastewater from a small restaurant and hotel, making it slightly more eutrophic than other areas on the floodplain. However, any nutrients entering the water are rapidly incorporated by the many rooted emergent and floating plants, allowing their presence to escape detection by chemical analyses of the water alone. During the high *cheia*, water from the Rio Pixaim overflows into the pond, and the two water bodies remain contiguous for several months. At this time, *Platyias quadricornis* (Ehrenberg, 1832) was observed in the interrhizon among the rhizoids of *Salvinia auriculata*.

One additional brachionid, *Brachionus angularis* (Gosse, 1851) was occasionally observed in water samples taken during the *vazante*.

Euchlanidae

One of the fairly abundant dry season species among the interrhizon and filamentous algae in the small rivers draining the Pantanal was *Euchlanis dilatata* Ehrenberg, 1832. When it is present at all, it is usually moderately abundant.

Mytilinidae

Mytilina mucronata (O. F. Müller, 1773) was a fairly common species among water plants in shallow water bodies. It was most abundant during the dry season and was not observed at all during the high water period. Although the species is thought to prefer cool water, Koste and Böttger (1992) found the species in an Ecuadorian water body at a temperature of 25.9°C.

Trichotridae

The family Trichotridae includes two species encountered as isolated individuals in the plant-rich littoral zones of small rivers, lakes, and ponds. They were observed only during the dry season and never in the temporary water bodies of the flood plain. *Trichotria tetractis* (Ehrenberg, 1830) was collected in water samples from various locations in the Pantanal, including the Baia Piuval and other sites along the Rio Bento Gomes. *Macrochaetus sericus* (Thorpe, 1893) also inhabited the Rio Bento Gomes and was also found in the pond beside the Transpantaneira Highway near the Rio Pixaim.

Colurellidae

The Colurellidae was a well represented family in the Pantanal. Among the most abundant species observed swimming among water plant masses was *Lepadella patella* (O. F. Müller, 1786), which was apparently active throughout the entire year, but was encountered most frequently during the dry season. It is reported to be a cosmopolitan species (Koste, 1978).

Other dry season species, which were observed in many samples during the *vazante* and *seca* together with *L. patella*, include *Lepadella triptera* Ehrenberg, 1830, *Lepadella acuminata* (Ehrenberg, 1834), and *Lepadella cristata* (Rousselet, 1893).

Colurella uncinata (O. F. Müller, 1773) and *Colurella obtusa* (Gosse, 1886) were abundant and often observed creeping on the surfaces of submerged plants and detritus in water samples taken at many locations in the Pantanal. Like most other rotifer species, they were most frequently seen from the end of the *vazante* through the early *enchente*.

One species in this family, *Squatinella mutica* (Ehrenberg, 1832), is modified for swimming on the underside of the surface tension layer of the water. Specimens were observed during the early *seca* on the surface of the water among the floating vegetation in the Rio Bento Gomes.

Lecanidae

This species rich family was not well represented in the samples taken in the Pantanal of Poconé, apparently because of shortages of food organisms and high predator pressure. The species most often observed was *Lecane lunaris* Ehrenberg, 1832, which was abundantly represented by the subspecies *Lecane lunaris crenata* (Harring, 1913). However, *Lecane l. lunaris* was also found to be represented among the Pantanal fauna. The species was seen particularly often in water samples from the Rio Bento Gomes at the Transpantaneira Highway at the end of the *vazante* and beginning of the *seca*, but always in very limited numbers. It was rare or absent from samples taken from temporary water bodies on the floodplain.

The most common species in the subgenus *Monostyla* seems to be *Lecane bulla* (Gosse, 1886), which was observed rarely in samples taken from among water plants.

Other lecanids observed during the study included *Lecane ungulata* (Gosse, 1887), which is generally encountered in the same kind of macrophyte beds as its congeners seek out. Because Turner and da Silva (1992) provided no precise collecting data, the locations along the Transpantaneira Highway at which they claim to have found considerably more species of *Lecane* could not be visited to verify the presence of the populations or obtain notes on their seasonal occurrence.

Notommatidae

Scaridium longicaudum (O. F. Müller, 1786) is a rather frequently observed species during the dry season. It is generally collected with samples of water plants from the littoral zone or masses of plants floating in the middle of the rivers draining the Pantanal. It was never observed in the temporary water bodies on the floodplain; rather, it was collected from among the plant masses at the edges of the Rio Bento Gomes near the Transpantaneira Highway at the end of July and the permanent pond beside the Rio Pixaim in mid-September.

A second species of this genus, *Notommata glyphura* Wulfert, 1935, was observed once in the Rio Novo in late July. It was identified in a sample of water containing much moribund plant material from a shallow pool free from small fishes due to high daytime temperatures, low nighttime oxygen concentrations, and the presence of more than a dozen hatchling caimans.

Trichocercidae

The trichocercids most common in the Pantanal, *Trichocerca braziliensis* (Murray, 1913) and *Trichocerca* sp., were generally found among the interrhizon, but they were also very abundant in the *Asplanchna sieboldi* aggregation, which dominates the shallow pools of water on the floodplain during the *seca*. Their great abundance during the dry season attests to their resistance to water temperatures sometimes exceeding 40°C (Kretzschmar *et al.*, 1993).

Synchaetidae

The synchaetids in the Pantanal were the most abundant species encountered in the planktonic aggregations during the dry season, but they could also be collected at times with water samples from littoral plant masses throughout the entire year. Their rapid jumping movements may give them some extra degree of protection against the intensive predation by small fishes during the high water period. *Polyarthra vulgaris* Carlin, 1943, was the most abundant synchaetid in most of the permanent water bodies of the Pantanal and was among the most abundant of all rotifers throughout the year. *Polyarthra remata* (Skorikov, 1896) was also observed, but not as frequently.

The true identity of the Neotropical form designated "*Polyarthra* (*vulgaris* ?)" by Koste (1972) because of differences from the European form in the position of the lateral sensory setae and in habitat preferences deserves some attention. Koste (1978) finally accepted his Amazonian rotifer as a genuine form of *P. vulgaris*, and it has proven to be one of the numerically dominant species in the Brazilian limnofauna. Schaden (1978) found it to be one of the most common species in plankton samples from the Rio Negro System at locations where the electrical conductivity varies between 6.8 and 10.4 $\mu S \cdot cm^{-1}$, even lower than the values recorded for the water in the Pantanal, although higher than those encountered in some of the nearby upland streams (Section 5). In addition, it was found at all sampling points in two floodplain lakes along the lower Amazon River in the State of Pará by Bozelli (1992). Personal observations have also confirmed that it is abundant in tributaries of the Rio São Francisco, especially in impoundments. It therefore seems to be one of the predominant rotifer species in Brazil. The temperature of the water it inhabits in Brazil is considerably warmer than the cool water it is reported to prefer in Europe (Koste, 1978).

Asplanchnidae

A typical dry season rotifer in the very warm pools of shallow water on the Pantanal floodplain is *Asplanchna sieboldi* (Leydig, 1854), which was abundant enough to characterize the species aggregation discussed by Kretzschmar *et al.* (1993). This circumtropical rotifer has long been known as a warm stenothermal element, which occurs in water bodies with relatively low levels of saprobity (Koste, 1978). These rotifers are large and prominent in the shallow water bodies filled with chlorococcal algae, particularly species of the family Scenedesmaceae. They were most abundant at low-lying sites along the temporary stream called Birici on the Fazenda Ypiranga, but specimens were also found in similar shallow water bodies along the Rio Novo. *A. sieboldi* seems to be very widespread in tropical South America, occurring in the drainage areas of the Amazon and São Francisco Rivers as well as in the Pantanal (Dabés, 1995).

Dicranophoridae

This species-rich family was represented in the Pantanal by *Paradicranophorus* cf. *sudzukii* Donner, 1968, and *Dicranophorus forcipatus* (O. F. Müller, 1786), both of which appeared only occasionally in samples. Both species are reportedly carnivorous, although *P. sudzukii* may also consume diatoms. *D. forcipatus* is noted for attacking relatively large prey (Koste, 1978).

Although the single specimen observed closely matched the description of *P. sudzukii* provided by Koste (1978), the identification of this species remains questionable because it is apparently rare and was previously unreported from South America.

Testudinellidae

A common epibenthic and interrhizonic species encountered during the dry season in the Pantanal is *Testudinella patina* (Hermann, 1783), which is widespread but never abundant. Generally, only one individual was observed in a water sample, but rarely two were present. The reason for its solitary occurrence remains to be investigated.

Flosculariidae

During the rainless seasons, particularly near the end of the *vazante*, tubes containing *Floscularia melicerta* (Ehrenberg, 1832) were observed on dying plants and pieces of detritus. Generally, the dwellings of these rotifers remain well concealed in masses of soft material, apparently formed mainly by bacteria. This gives the rotifer a better chance to avoid the many predators that are still active during this time of year.

Filiniidae

The species of *Filinia* are generally encountered in very large numbers in the open water near masses of water plants. The species found to be most common in the Pantanal during the study was *Filinia longiseta* (Ehrenberg, 1834), which was represented by the subspecies *Filinia longiseta limnetica* (Zacharias, 1893) as well as by *Filinia l. longiseta*. Both subspecies were most abundant at the beginning of the rainy season, when they occurred among the short-lived plankton aggregations of the permanent lakes as well as in the areas of the floodplain adjacent to the streams. The species was usually absent from the warm water aggregations in shallow water bodies during the late dry season, but it became active very soon after the freshet.

Collothecidae

Among the few tube-dwelling sessile species found occasionally in the interrhizon community was *Collotheca campanulata* (Dobie, 1849), which was encountered during the late *vazante* in the Rio Novo near the Transpantaneira Highway.

Philodinidae

A diverse fauna of bdelloid rotifers was not encountered in the Pantanal. The ubiquitous *Rotaria rotatoria* (Pallas, 1766) was abundant on many occasions in water containing much decaying plant material during the dry season. It was collected in the littoral zone of the permanent water bodies as well as on the floodplain. This is a truly cosmopolitan species that is encountered in many different kinds of aquatic habitats (Donner, 1965).

In streams, springs, and marshy places in the upland parts of the watershed, an interesting bdelloid fauna exists, which should be studied more intensively in the future. Distinctive species, including *Habrotrocha angusticollis* (Murray, 1905), were found among mosses beside springs on the plateau to the east of the Pantanal, where the headwaters of many rivers flowing into the Pantanal were located. The water from the springs feeding the rivers were very poor in dissolved minerals, as attested to by the very low electrical conductivity reported in Section 5 and by Heckman (1995). The species is cosmopolitan in distribution (Donner, 1965).

Other rotifer populations in the Pantanal

In addition to the species encountered during the study, a large number of species have been encountered in the Baia Acorizal in the Pantanal of Barão de Melgaço and at other sites along the Transpantaneira Highway during the late *vazante* and early *seca* in 1988 (Turner and da Silva, 1992). The list of species encountered by these authors is presented in Table 8.1.

Baia Acorizal has considerably different limnological characteristics from the shallow water habitats encountered along the Transpantaneira Highway and therefore seems to have a particularly rich rotifer fauna (Table 8.1). That the list of rotifer species in the Pantanal is far complete is indicated by the fact that two very common dry season species encountered during the study, *Keratella americana* and *Trichocerca braziliensis*, were not reported at all by Turner and da Silva (1992), although their list indicates the presence of considerably more species than were encountered during the later study.

Table 8.1: Rotifers collected from the region of the Pantanal by Turner and da Silva (1992) during a study in 1988 and those found during the research project conducted from 1991 through 1994 (Heckman, 1998). This list includes only taxa presently regarded as species; those reported as subspecies, varieties, and forms of a single species are omitted. BA = Baia Acorizal in the district of Barão do Melgaço, SW = salgadeira swamp at an unspecified location, TP = five sites along the Transpantaneira Highway and the Rio Pixaim in the district of Poconé; PP = all sampling sites in the Pantanal of Poconé during the study from 1990 through 1994 (Heckman, 1998). Taxa not identified to species are omitted.

Species	BA	SW	TP	PP
Brachionidae				
Aneuropsis fissa (Gosse, 1851)	+	-	-	-
Brachionus ahlstromi (Lindeman, 1939)	-	-	-	+
Brachionus angularis (Gosse, 1851)	-	-	-	+
Brachionus budapestinensis Daday, 1885	+	-	+	-
Brachionus bidentatus Anderson, 1889	-	-	+	+
Brachionus caudatus Barrois & Daday, 1894	+	-	+	-
Brachionus dolabratus Harring, 1915	+	-	-	-
Brachionus donneri Brehm, 1950	+	-	-	-
Brachionus falcatus Zacharias, 1898	+	-	-	-
Brachionus mirus Daday, 1905	+	-	-	-
Brachionus patulus O. F. Müller, 1786	+	-	+	+
Brachionus quadridentatus (Hermann, 1793)	+	-	+	+
Brachionus urceolaris (O. F. Müller, 1773)	+	-	-	+
Keratella americana Carlin, 1943	+	-	+	+
Keratella cochlearis (Gosse, 1851)	+	-	+	+
Keratella lenzi Hauer, 1953	+	-	+	-
Platyias brevispina (Daday, 1905)	+	-	-	-
Platyias latiscapularis Koste, 1974	+	-	-	-
Platyias leloupi Gillard, 1957	+	-	+	-
Platyias quadricornis (Ehrenberg, 1832)	+	-	+	+
Euchlanidae				
Euchlanis dilatata Ehrenberg, 1832	+	-	-	+
Euchlanis incisa Carlin, 1939	+	-	-	-
Euchlanis meneta Myers, 1930	-	-	+	-
Euchlanis mucronata (Ahlstrom, 1934)	-	-	+	-

Species	BA	SW	TP	PP
Mytilinidae				
Mytilina acanthophora Hauer, 1938	+	-	-	-
Mytilina compressa (Gosse, 1851)	+	-	-	-
Mytilina mucronata (O. F. Müller, 1773)	+	-	-	+
Mytilina ventralis (Ehrenberg, 1832)	+	-	+	-
Trichotridae				
Macrochaetus collinsi (Gosse, 1867)	+	-	-	-
Macrochaetus longipes Myers, 1934	+	-	-	-
Macrochaetus sericus Thorpe, 1893	+	-	-	+
Trichotria tetractis (Ehrenberg, 1830)	-	-	-	+
Colurellidae				
Colurella bicuspidata (Ehrenberg, 1832)	+	-	+	-
Colurella obtusa (Gosse, 1886)	+	-	+	+
Colurella uncinata (O. F. Müller, 1773)	-	-	+	+
Lepadella acuminata (Ehrenberg, 1834)	+	-	-	+
Lepadella apsida Harring, 1916	+	-	-	-
Lepadella cristata (Rousselet, 1893)	-	-	-	+
Lepadella dactyliseta (Stenroos, 1898)	+	-	-	-
Lepadella donneri Koste, 1972	-	-	+	-
Lepadella imbricata Harring, 1914	+	-	-	-
Lepadella patella (O. F. Müller, 1786)	+	-	+	+
Lepadella quinqueconstata (Lucks, 1912)	+	-	+	-
Lepadella rhomboides (Gosse, 1886)	-	-	+	-
Lepadella triptera Ehrenberg, 1830	-	-	-	+
Paracolurella aemula (Myers, 1934)	-	+	+	-
Squatinella mutica (Ehrenberg, 1832)	-	-	-	+
Lecanidae				
Lecane (L.) curvicornis (Murray, 1913)	+	-	+	-
Lecane (L.) doryssa Harring, 1914	+	-	-	-
Lecane (L.) elongata Harring & Myers, 1926	+	-	-	-
Lecane (L.) haliclysta Harring & Myers, 1926	-	-	+	-
Lecane (L.) hastata (Murray, 1913)	-	-	+	-
Lecane (L.) leontina (Turner, 1892)	+	-	+	-
Lecane (L.) ludwigii (Eckstein, 1893)	-	-	+	-
Lecane (L.) luna (O. F. Müller, 1776)	+	-	-	-
Lecane (L.) papuana (Murray, 1913)	+	-	+	-
Lecane (L.) rhytida Harring & Myers, 1926	-	-	+	-
Lecane (L.) signifera (Jennings, 1896)	+	-	+	-
Lecane (L.) unguitata (Fadeev, 1925)	-	-	+	-
Lecane (L.) ungulata (Gosse, 1887)	-	-	-	+
Lecane (Monostyla) aguessi (De Ridder, 1960)	+	-	-	-
Lecane (Monostyla) arcuata (Bryce, 1891)	-	-	+	-
Lecane (Monostyla) bulla (Gosse, 1886)	+	-	+	+
Lecane (Monostyla) closterocerca (Schmarda, 1859)	+	-	+	-
Lecane (Monostyla) copeis Harring & Myers, 1926	-	-	+	-
Lecane (Monostyla) cornuta (O. F. Müller, 1786)	+	-	+	-
Lecane (Monostyla) elachis (Harring & Myers, 1926)	+	-	-	-
Lecane (Monostyla) furcata (Murray, 1913)	+	-	-	-
Lecane (Monostyla) hamata (Stokes, 1896)	+	-	+	-
Lecane (Monostyla) lunaris (Ehrenberg, 1832)	-	-	+	+

Species	BA	SW	TP	PP
Lecane (Monostyla) monostyla (Daday, 1897)	-	-	+	-
Lecane (Monostyla) obtusa (Murray, 1913)	+	-	-	-
Lecane (Monostyla) quadridentata (Ehrenberg, 1892)	-	-	+	-
Lecane (Monostyla) styrax (Harring & Myers, 1926)	+	-	-	-
Proalidae				
Proales decipiens (Ehrenberg, 1831)	-	-	+	-
Proalinopsis staurus Harring & Myers, 1924	-	-	+	-
Lindiidae				
Lindia pallida Harring & Myers, 1922	-	-	+	-
Lindia truncata (Jennings, 1894)	-	-	+	-
Notommatidae				
Cephalodella gibba (Ehrenberg, 1838)	-	-	+	-
Cephalodella mira Myers, 1924	-	-	+	-
Monommata astia Myers, 1930	+	-	-	-
Monommata maculata Harring & Myers, 1924	-	-	+	-
Monommata phoxa Myers, 1930	-	-	+	-
Notommata allantois Wulfert, 1935	+	-	-	-
Notommata copeus (Ehrenberg, 1834)	+	-	-	-
Notommata glyphura Wulfert, 1935	-	-	+	+
Scaridium longicaudum (O. F. Müller, 1786)	+	-	+	+
Taphrocampa annulosa Gosse, 1851	-	+	+	-
Taphrocampa selenura Gosse, 1887	-	-	+	-
Trichocercidae				
Trichocerca (Diurella) insulana (Hauer, 1938)	-	+	+	-
Trichocerca (Diurella) obtusidens (Olofsson, 1918)	+	-	-	-
Trichocerca (Diurella) similis (Wierzejski, 1893)	+	-	+	-
Trichocerca (Diurella) tenuior (Gosse, 1886)	+	-	+	-
Trichocerca (Diurella) tigris (O. F. Müller, 1786)	+	-	+	-
Trichocerca braziliensis (Murray, 1913)	-	-	-	+
Trichocerca (T.) chattoni (De Beauchamp, 1907)	+	-	-	-
Trichocerca (T.) elongata (Gosse, 1886)	+	-	-	-
Trichocerca (T.) gracilis (Tessin, 1890)	+	-	-	-
Trichocerca (T.) longiseta (Schrank, 1802)	+	-	-	-
Trichocerca (T.) pusilla (Lauterborn, 1898)	+	-	+	-
Trichocerca (T.) rosea (Stenroos, 1898)	+	-	-	-
Trichocerca (T.) simonei Smet, 1989	+	-	-	-
Trichocerca (T.) stylata (Gosse, 1851)	+	-	-	-
Gastropidae				
Ascomorpha ecaudis (Perty, 1850)	+	-	+	-
Ascomorpha ovalis (Bergendal, 1892)	-	-	+	-
Ascomorpha saltans Bartsch, 1870	-	-	+	-
Gastropus hyptopus (Ehrenberg, 1838)	+	-	+	-
Gastropus minor (Rousselet, 1892)	+	-	-	-
Gastropus stylifer Imhof, 1891	-	-	+	-
Synchaetidae				
Polyarthra dolichoptera Idelson, 1925	+	-	+	-
Polyarthra longiremis Carlin, 1943	+	-	-	-
Polyarthra remata Skorikov, 1896	+	-	-	+
Polyarthra vulgaris Carlin, 1943	+	-	+	+
Synchaeta pectinata Ehrenberg, 1832	-	+	+	-

Species	BA	SW	TP	PP
Asplanchnidae				
Asplanchna sieboldi (Leydig, 1854)	+	-	+	+
Dicranophoridae				
Dicranophorus forcipatus (O. F. Müller, 1786)	-	-	+	+
Dicranophorus hercules Wiszniewski, 1932	-	-	+	-
Dicranophorus luetkeni (Bergendal, 1892)	-	-	+	-
Paradicranophorus sudzukii Donner, 1968	-	-	-	+
Testudinellidae				
Testudinella ahlstromi Hauer, 1956	-	-	+	-
Testudinella amphora Hauer, 1938	-	-	+	-
Testudinella patina (Hermann, 1783)	+	-	+	+
Flosculariidae				
Beauchampia crucigera (Dutrochet, 1812)	-	-	+	-
Floscularia melicerta (Ehrenberg, 1832)	-	-	-	+
Floscularia ringens (Linnaeus, 1758)	-	-	+	-
Limnias ceratophylli Schrank, 1803	-	-	+	-
Limnias melicerta Weisse, 1848	-	-	+	-
Sinantherina spinosa (Thorpe, 1893)	+	-	-	-
Conochilidae				
Conochilus coenobasis Skorikov, 1914	+	-	-	-
Conochilus natans (Seligo, 1900)	+	-	-	-
Hexarthridae				
Hexarthra braziliensis Hauer, 1953	+	-	+	-
Hexarthra intermedia Wiszniewski, 1929	+	-	-	-
Hexarthra mira (Hudson, 1871)	+	-	-	-
Filiniidae				
Filinia longiseta (Ehrenberg, 1834)	+	-	+	+
Filinia opoliensis (Zacharias, 1898)	+	-	-	-
Filinia passa (O. F. Müller, 1786)	-	-	+	-
Filinia saltator (Gosse, 1886)	+	-	-	-
Filinia terminalis (Plate, 1886)	+	-	-	-
Trochosphaeridae				
Horaella thomassoni Koste, 1973	+	-	+	-
Collothecidae				
Collotheca campanulata (Dobie, 1849)	-	-	-	+
Collotheca natans Tschugunoff, 1921	-	-	+	-
Collotheca ornata Tschugunoff, 1921	-	-	+	+
Collotheca pelagica (Rousselet, 1893)	+	-	-	-
Collotheca tenuilobata (Anderson, 1889)	-	+	+	-
Atrochidae				
Cupelopagis vorax (Leidy, 1857)	-	-	+	-
Philodinidae				
Rotaria rotatoria (Pallas, 1766)	-	-	-	+

8.8. Gastrotricha

Chaetonotidae

The gastrotrich fauna of the Pantanal is comparatively rich. Many individuals belonging to a variety of species were encountered throughout the year, although the general abundance of the group was generally much greater during the dry season. The most common gastrotrich species belong to the genus *Chaetonotus*, with *Chaetonotus acanthodes* Stokes, 1887, *Chaetonotus murrayi* Remane, 1929, and *Chaetonotus larus* (O. F. Müller, 1784) being represented in water bodies along the Transpantaneira Highway during the early *seca*, and *Chaetonotus hystrix* Metschnikoff, 1865, occurring throughout much of the year among plants near the shore of the Rio Bento Gomes at the exposition park near the northern borders of the Pantanal.

C. acanthodes was found at various times in the Rio Bento Gomes, where it was seen most frequently near the Transpantaneira Highway. It is a cosmopolitan wetland species (Rudescu, 1967), which Schwank and Bartsch (1990) found to be an acidophilic species in Europe, where it has been found in moors at a pH between 4.3 and 5.9.

The permanent pond near the Rio Pixaim was particularly rich in gastrotrichs, which were often collected from among the interrhizon community. *C. murrayi* was abundant only in this habitat during the *seca*. In Europe, where the species is rare, it has been encountered in moors, where it was reported to occur at a pH of 5.4 (Schwank and Bartsch, 1990).

Considerably less abundant in the Pantanal than the two congeneric species, *C. larus* was confined to the waters of the deep roadside ditches adjacent to the Rio Novo. It is probably a cosmopolitan species, although Rudescu (1967) reported it only from the Old World. In Europe, it is known from a wide variety of freshwater habitats and is one of the commonest gastrotrich species (Schwank and Bartsch, 1990).

Lepidodermella squamata (Dujardin, 1841) was nearly as abundant as the *Chaetonotus* species and more widespread than any one of them. It occurred in most of the water bodies sampled, being most abundant during the dry season in water bodies without conspicuous algal blooms. The species is cosmopolitan and known from a very wide variety of freshwater habitats (Rudescu, 1967).

A particularly abundant species in small water bodies remaining at the end of the dry season is *Polymerurus callosus* Brunson, 1950, a gastrotrich reported by Schwank and Bartsch (1990) to be exclusively a New World species. During the year of the study, it was observed at Birici Stream, where it was abundant seasonally.

Dasydytidae

The only species in this family observed in the Pantanal was *Halidytes festinans* Voigt, 1909, which is a littoral species preferring to live in masses of macrophytes. In Europe, this species reportedly occurs on sediments containing much decaying matter more frequently than among water plants (Schwank and Bartsch, 1990). It may not have previously been found in South America, although a congeneric species, *Halidytes crassus* (Greuter, 1917), was reported by Schwank and Bartsch (1990) to occur in Argentina. It can be assumed that all species in this genus are much more widely distributed than presently reported because of their relative rarity.

8.9. Ectoprocta

Colonies of ectoprocts were not encountered during the study. However, floatoblasts were collected at times, usually at the edges of the rivers and streams flowing into the Pantanal. The floatoblasts were produced by the species *Plumatella* cf. *siolii* Wiebach, 1970. The general taxonomy of South American ectoprocts is presently very confused because the colonies have very seldom been found (Wiebach, 1970). Some of the older reports of *Plumatella fungosa* (Pallas, 1768) from South America may actually be this species because the shape of the floatoblasts is rather similar. The species was originally distinguished on the basis of the fine structure, as observed in specimens collected from various parts of the Amazon River System (Wiebach, 1970). *P. siolii* has also been found in Colorado, North America (Wiebach, 1974). However, the question remains whether the small differences in the fine structure of the floatoblasts are really sufficient grounds for distinguishing a separate species. The considerable differences in water chemistry and other environmental factors may well be responsible for these differences. Floatoblasts collected during earlier studies in South America were assigned to a number of *Plumatella* species that, like *P. fungosa*, have been found in many parts of the world (Thomasson, 1963). These species are apparently not at all rare in South America, but the colonies may develop, reproduce, and die off rapidly in the warm water and are therefore seldom observed.

8.10. Annelida

8.10.1. Oligochaeta

Naididae

In spite of the very intense predation by small fishes in the Pantanal, *Stylaria fossularis* (Leidy, 1852), is able to survive and occasionally produce notably large populations, especially in the shallow water left in the streams and small rivers during the dry season.

The only other abundant representatives of this family belong to the genus *Dero*. From the very limited information available on South American oligochaetes, it seems that this genus is probably the most widespread of the family in the inland water bodies of the Neotropical Region. *Dero (Aulophorus) borellii* Michaelsen, 1900, seems to be the most abundant species of the genus in the Pantanal, although a second, unidentified species was also encountered. Persia (1980) provided a short summary of the oligochaete fauna of Argentina and reported the presence of eight species of *Dero* in the subgenus *Dero* and five species in the subgenus *Aulophorus*, but he did not include *D. borellii* in his list. Varela (1990), however, reported it to be a common South American species that ranges from Brazil and the Paraguayan chaco southward into Argentina. There is also a doubtful report of ist occurrence in North America.

Aelosomatidae

Aelosoma cf. *evilinae* Marcus, 1944, is a common species of the interrhizon community. Unlike most species of this community, it prefers the sparce rhizoid and root system of small water plants, in the Pantanal, particulary *Azolla mexicana*. It was encountered mainly during the dry season. This species constructs small portable tubes of tiny bits of the *Azolla* fronds, which it pulls along with it. The flattened head emerges from the tube and sticks to the substrate. It then pulls the tube up to point of attachment and extends the head again to grab the substrate farther ahead. In this way, it move with great speed and ability through the rhizoid masses of the water ferns. Little has been reported on the interesting habits of this species because of the common practice of fixing the material collected before observing the organisms.

8.10.2. Hirudinea

All leeches observed in the Pantanal belong to the family Glossiphoniidae. A member of this family often observed in the Pantanal during the study was *Helobdella stagnalis* (Linnaeus, 1758), a species known under many synonyms in South America. This small leech feeds mainly on body fluids that it sucks from small organisms, both invertebrates and vertebrates (Elliott and Mann, 1979). Young and Procter (1986) the feeding habits of this leech in England by serological methods and concluded that it is an opportunistic predator.

Oligobdella cheloniae Ringuelet, 1978, was encountered on the caimans in the Pantanal. Leeches of this species are also very abundant on the sediment and objects in the water at sites frequented by many caimans. A group of more than 12 specimens were collected from the base of a caiman's back leg while it sunned itself on the shore. The digestive system of each was gorged with blood.

8.11. Mollusca

8.11.1. Gastropoda

Ancylidae

Hubendick (1964) resolved some of the taxonomic confusion about this family in the Neotropics by reducing the number of extant genera to one: *Gundlachia* Pfeiffer, 1849. In the Pantanal, at least one species is common, *Gundlachia moricandi* (d'Orbigny, 1837), which was found during the *cheia* and *vazante* on the submerged stems of emergent plants.

Planorbidae

The only common planorbid observed in the northern Pantanal was *Biomphalaria straminea* (Dunker, 1848), a species known to act as the intermediate host of schistosomes. The cercarial emergence of the most dangerous of these parasites for man, *Schistosoma mansoni* Sambon, 1907, was described in detail by Favre *et al.* (1995). Although this suitable intermediate host is present in the Pantanal, the parasite is presently not a public health problem in southern Mato Grosso.

Only small individuals of *Biomphalaria straminea* were sporadacally encountered during the study. Apparently, conditions are less than optimal for this species in the northern Pantanal.

Pilidae (= Ampulariidae)

Three pilid species are extremely abundant both in the permanent streams and small rivers and on the temporarily flooded areas of the Pantanal, namely *Pomacea linearis* (Spix, 1927), *Pomacea scalaris* (d'Orbigny, 1846), and *Marisa planogyra* Pilsbry, 1933. These species are able to survive the dry period by seeking the few permanent water bodies or, more frequently, by digging into the sediment and aestivating until the next rainy season (Kretzschmar and Heckman, 1995). After they are released from the hardened mud by the flood waters, they reappear at the surface and immediately begin producing eggs, which are attached by the *Pomacea* species in pink masses on the emergent parts of plants during the night or are laid underwater by *M. planogyra* in gelatinous masses.

Snails of this family are widespread in the tropics and are very prominent among the aquatic invertebrates of the tropical wet-and-dry climatic zone, especially in Southeast Asia (Brandt, 1974; Heckman, 1979) and South America (Pilsbry, 1933; Kretzschmar and Heckman, 1995; Heckman, 1998). Their success in shallow water bodies where the physical and chemical conditions greatly fluctuate is explained by various morphological and physiological modifications. The species inhabiting temporary water bodies are able to dig into the sediment before the water dries up and survive with the operculum tightly closed. Their possession of respiratory organs functioning both as gills and as lungs (Berry, 1974; Berthold, 1991) makes them almost independent of the oxygen, carbon dioxide, and ammonium content of the water. Because they can feed directly on living macrophytes, as observed both in Southeast Asia (Heckman, 1979) and in South America, they have a rich supply of food during the high water period, when most microorganisms are scarce in the water. Furthermore, relatively little calcium is deposited in the shell, permitting the species to thrive in very soft water, as in the Pantanal (Heckman, 1994a).

These three snails form the staple diet of several vertebrate species in the Pantanal and are therefore of immense importance for maintaining the diversity of the fauna. The large snails are consumed in large quantities by the snail kite, limpkin, and the caiman lizard. In addition, the juvenile snails are probably consumed by the larvae of the common aquatic lampyrid in the Pantanal, *Photinus* sp.

8.11.2. Bivalva

Mycetopodidae

Two species in this family are common in the Pantanal: *Anodontites trapesialis* (Lamarck, 1819) and *Anodontites trigonus* (Spix, 1827). These species are generally encountered along stretches of the shore with fine, hard-packed sediments, into which the bivalves burrow deeply. They were also observed in low-lying areas of the floodplain which dry up completely during short periods at the end of the dry season. The burrows of these bivalves along the banks of the small rivers and streams usually remain underwater, but during particularly dry years, they may become exposed to the air. At such times, the bivalves remain closed until the water level increases again. The mortality at such times is high, and many of the shells were found to contain larvae of diptera feeding on the remains of the molluscs. Nevertheless, some of them seem capable to surviving at least short periods of time in desiccated sediments.

Both of these species are distributed throughout large parts of South America. *Anodontites trapesialis* ranges throughout the drainage areas of the Amazon and Paraná Rivers (Haas, 1969). Simone (1994) demonstrated the great anatomical variability of this species, which accounts for the exceptionally long list of its synonyms. *Anodontites trigonus georginae* Gray, 1834, was found to be the largest and most abundant bivalve in a river draining a humid subtropical forest at an elevation of 500 m in Bolivia (Maldonaldo & Goitia, 1993).

Psammobidae

One psammobid species, *Tagelus* cf. *gibbus* (Spengler, 1794), was found buried in particularly fine sediment near the shores of the Rio Bento Gomes between the Poconé exposition park and the approaches to the Baia Piuval. This sediment remains very firm, giving the bivalve the chance to hold its position very tenaciously and use its muscular body, which is much too large to be completely contained in the shell, to resist attempts to pull or dig it out.

Hydriidae

The only hydriid encountered during the study was *Castalia ambigua* (Lamarck, 1819), which was extremely abundant in the sediment of the Rio Bento Gomes. Specimens were also found in the Rio Pixaim. It occurred mainly in relatively coarse sediments near the shore. Specimens were never found in the temporary water bodies of the floodplain.

This species is extremely widespread and abundant in the freshwaters of South America, being encountered in the Amazon region (Mansur and Valer, 1992) as well as far upstream in the system of the Rios Paraguay and Paraná.

8.12. Arthropoda

8.12.1. Acari

Most water mites fill quite different ecological niches as larvae and as adults. The larvae of most species are parasitic on adult aquatic insects, including many dipterans. The adults are small aquatic predators. Only a few species were observed as adults in the Pantanal, but they occurred in large numbers for short periods, indicating that considerably more could be found by exhaustive searching. This opinion was reinforced by the observation of larval mites on large numbers of different flying insects.

Eylaidae

Occasionally, *Eylais* sp. was encountered in very shallow water bodies in the process of drying up on the flood plain during the late dry season.

Arrenuridae

A number of species in this family occur in the Pantanal. Two species of *Arrenurus* were recognized by the form of the males, but females suspected to belong to one or two others were also observed.

8.12.2. Araneae

The aquatic and semi-aquatic spiders in the Pantanal are concentrated mainly in the pseudoterrestrium, the community that inhabits the emergent parts of the floating and long-stemmed anchored plants on the surface of the lakes and floodplains, and in the littoral zone. The main food of most of these species seems to be the adults of small aquatic insects that have just completed metamorphosis after emerging from the water.

Argiopidae

A characteristic species of the pseudoterrestrium is the orbweb spinner, *Actinosoma pentacanthum* (Walckenaer, 1841), which was observed almost exclusively on the beds of floating vegetation rather than on rooted emergent plants. In the Pantanal, it was generally found of *Eichhornia crassipes* rather than the smaller floating species, such as *Pistia stratiotes* and the water ferns. When disturbed, it may conceal itself underwater. This spider does not seem to construct elaborate webs but rather builds very small orbs among the leaves of the emergent plants. Levi (1995) reported that the species was also found on floating beds of *Paspalum*, assumed to be *Paspalum repens*, in the Amazon region near Manaus. He also reported its occurrence in Peru, Ecuador, Bolivia, Paraguay, Uruguay, Argentina, and the state of Espirito Santo along the Atlantic coast of Brazil. The type may have been collected in French Guiana. Apparently the species occurs almost anywhere in South America where its special habitat can be found. Observations of this species in the field suggest that it feeds mainly on very small insects, such as chironomids, which are often seen in the webs.

Tetragnathidae

Among the commonest spider species in the Pantanal and its vicinity is *Tetragnatha nitens* (Audouin in Savigny, 1825), which typically builds its webs among littoral water plants. However, occasionally individuals are observed on emergent rooted plants, often rolled in a narrow leaf. The species seems to be very common in Brazil and apparently has a wide range in the tropics and subtropics, having been described from specimens taken in Egypt. It has since been found at several sites in North America, as well as in the West Indies and Central America, Peru, and the Galápagos Islands. Levi (1981) suggested that this species may be native to South America, although there are relative few reports of its presence in the literature.

Like several congeneric species in Europe, *T. nitens* is a typical inhabitant of the littoral zone of slow and moderate flowing streams and standing water bodies. It generally conceals itself in a blade of grass or other narrow leaf and waits for its prey to become trapped in the web it constructs on littoral plants just above the surface of the water. Personal observations in Minas Gerais indicate that small aquatic insects undertaking their first flights after metamorphosis are the main prey of these spiders. Chironomids were particularly abundant in Minas Gerais, but in the Pantanal, this group is not as well represented, and members of other Diptera and Lepidoptera families as well as terrestrial homopterans are apparently more important food items in the Pantanal.

Lycosidae

One lycosid spider frequently seen hunting on the surface tension of the water concealed itself within emergent foliage of the pseudoterrestrium when approached.

8.12.3. Crustacea

The number of small crustacean species in the Pantanal is limited by the lack of nutrients and heavy predation by small fishes that virtually eliminates the zooplankton for much of the year. Planktonic species are encountered in considerable numbers only for a period of several weeks at the end of the dry season and beginning of the rainy season, when there is a large amount of decomposing detritus in the water accompanied by bacteria and algae and after the fish population has been decimated by the dryness, predation, and poor water quality of the late *seca*. During this period, a variety of cladocerans and copepods can be encountered. During the rest of the year, the active crustacean fauna is represented by relatively large decapods.

Taking the Rio Paraguay-Paraná System as a whole, a progressive enrichment of the microcrustacean fauna seems to occur moving downstream. For example, while macrocrustaceans were almost completely absent from the streams and impoundments upstream from the Pantanal, three species of copepod were observed in the Pantanal of Poconé. In the Brazilian section of the Rio Paraná System in the state of Mato Grosso do Sul, Lima *et al.* (1996) found 12 copepod species, and in the middle section of the Rio Paraná in Argentina, about 50 species were distinguished (Paggi and Paggi, 1990). The cladoceran fauna shows a similar enrichment moving downstream, with only 10 highly adaptable species being encountered in the Pantanal of Poconé, 38 in the

Brazilian section of the Rio Paraná System in Mato Grosso do Sul (Lima *et al.*, 1996), and about 90 in the middle section of the Rio Paraná System in Argentina (Paggi & Paggi, 1990). It is therefore considered probable that the microcrustacean fauna of the southern districts of the Pantanal will prove considerably richer than that encountered near Poconé.

Conchostraca

The only species in this group encountered in the Pantanal is *Cyclestheria hislopi* (Baird, 1859), a species in the family Cyclestheriidae with circumtropical distribution, which is very abundant in the Amazon region and widespread in other wet-and-dry climatic regions of the world, including monsoonal Southeast Asia (Heckman, 1979). This is the only conchostracan with a brood pouch, in which parthenogenetically produced offspring develop before release as miniature adults (Roessler, 1995). This kind of parental protection permits the species to successfully rear large numbers of offspring to a reproductive stage with a minimum number of losses.

In the Pantanal, this species was not particularly common, occurring in only a few shallow water bodies with a year-round water supply and a large number of floating and rooted emergent macrophytes. It joins the interrhizon community in the Pantanal, which is well protected from predation by the roots and rhizoids of the plants.

Cladocera

This group of small crustaceans is represented by several planktonic species, which have brief periods of abundance in larger water bodies of the Pantanal at the end of the dry season, as well as by various species that live in the sediment or the interrhizon community. Apparently the extremely great predator pressure from the many small fishes prevents large populations of cladocerans from developing, except in the baias near the end of the dry season.

Daphniidae

The daphniids are represented by a very abundant planktonic species, *Ceriodaphnia cornuta* Sars, 1885, which is very widely distributed in Brazil. Many were found in the Baia Piuval late in the *seca* and early in the *enchente*. None were encountered during the rest of the year.

A daphniid that prefers to live among submerged vegetation, *Simocephalus serrulatus* (Koch, 1841), was also fairly abundant in the Baia Piuval and in the Rio Bento Gomes near the Transpantaneira Highway, but only during the late *seca* and early *enchente*.

The third species in this family observed in the Pantanal generally swims inverted on the underside of the surface tension layer. *Megafenestra aurita* (Fischer, 1849) was found only in samples from the Baia Piuval near the end of the *seca*. A second scapholeberine species is known from Paraguay (Dumont & Pensaert, 1983) and probably also occurs in the region of the Pantanal. Although it was described under the name *Scapholeberis armata freyi* Dumont & Pensaert, 1983, Dumont (pers. comm.) considers it to be a separate species, *S. freyi*.

Other species were reported from the Rio Paraná System in the state of Mato Grosso do Sul by Lima *et al.* (1996), and they can be assumed to be present, at least in the southern part of the Pantanal. *Simocephalus vetulus* (Schodler, 1852, is a cosmopolitan species that can almost certainly be encountered throughout the Pantanal wherever sufficient food and shelter are present. *Daphnia gessneri* Herbst, 1967, on the other hand, probably does not occur in the northern part of the wetland because the conditions are not conducive for such a planktonic species.

Sididae

Among the species encountered in the plankton aggregation near the shores of the Baia Piuval during the dry season are *Diaphanosoma birgei* Korinek, 1981, and the nearly cosmopolitan *Sida crystallina* (O. F. Müller, 1776). They were always encountered together with other caldocerans, especially *Ceriodaphnia cornuta* and other planktonic species. Various authors have described several species of *Diaphanosoma* from tropical South America and Africa, but they very closely resemble other species and their validity remains to be demonstrated. Among the additional sidids found in the Brazilian part of the Rio Paraná System to the south of the Pantanal are *Sarsilatona serricauda* (Sars, 1901); *Diaphanosoma fluviatile* Hansen, 1899; *Diaphanosoma spinulosum* Herbst, 1975; and *Diaphanosoma brevireme* Sars, 1901(Lima *et al.*, 1996).

Bosminidae

Bosmina longirostris (O. F. Müller, 1785) is an extremely abundant planktonic species in the fairly deep permanent water bodies of the Pantanal during a short period at the end of the dry and beginning of the rainy season. Several congeneric species, including *Bosmina hagmanni* Stingelin, 1904, and *B. tubicen* Brehm, 1939, are also present in the Rio Paraná System (Lima *et al.*, 1996), but none were identified with certainty during the study. In addition, *Bosminopsis deitersi* Richard, 1895, was found by Lima *et al.* (1996) to be the most common species in the Brazilian section of the Rio Paraná System.

Moinidae

The late dry season plankton community included *Moina micrura* Kurz, 1874, which was less abundant in the open water than among the littoral vegetation. Most of the specimens collected came from the Baia Piuval. Elsewhere, the species was very rare.

To the south of the Pantanal in the Rio Paraná System, *Moina minuta* Hansen, 1899, has been reported to be abundant (Lima *et al.*, 1996).

Macrothricidae

Although no macrothricids were observed during the study, everal have been reported from the Rio Paraná System to the south of the Pantanal. These include *Macrothrix spinosa* King, 1853; *Macrothrix triserialis* Brady, 1886; *Grimaldina brazzai* Richard, 1892; and *Onchobunops tuberculatus* Fryer and Paggi, 1972 (Lima *et al.*, 1996). Their presence in the faunistically poorer Poconé District is probable, although these species apparently have little impact on the ecosystem.

Ilyocryptidae

The only Ilyocryptid species identified with certainty n the Pantanal is presently referred to as *Ilyocryptus spinifer* Herrick, 1884. It was present in the sediment below only a few centimeters of water along the shores of the Rio Bento Gomes near the exposition park of Poconé during the *seca*. Other species in the genus have recently been described from the Rio Paraná System in Argentina, somewhat complicating the identification of specimens. In the past, nearly all specimens of this genus from South America were listed as *Ilyocryptus sordidus* (Liéven, 1848), the presence of which has yet to be confirmed in South America since the newly distinguished taxa were first described as distinct species (Paggi, 1989). However, one micrograph of an *Ilyocryptus* specimen collected from the Rio Bento Gomes during the study clearly shows eight preanal spines on the postabdomen, which has generally been considered to be a diagnostic feature of *I. sordidus*.

Chydoridae

A member of the cosmopolitan genus, *Chydorus*, is an inhabitant of detritus aggregations as well as dense beds of submerged plants. It was the only cladoceran that was observed during all seasons, as well as at nearly all of the sampling sites in the wetland. However, its appearance in the samples was only sporadic.

The only other chydorid observed during the study was *Alona costata* Sars, 1862, which was found among epibenthic detritus in shallow water containing dying macrophytes. Because the fishes had already departed, it was fairly safe from predation. The species was never very numerous, but it was observed fairly often.

In the Rio Paraná System to the south of the Pantanal, Lima *et al.* (1996) found a rich chydorid fauna of 18 species, but the family is much less abundance in the Poconé region, apparently due to the lack of food and great predator pressure throughout much of the year.

Copepoda

Calanoida

The calanoid copepods show interesting distribution patterns, as there is an apparent absence of cosmopolitan species, and even circumtropical ones may not occur (Matsumura-Tundisi, 1986). Although it may presently be premature to designate certain species as endemiic to specific river systems in South America, such endemism seems to be much more common among calanoids than among cyclopoids and other groups of microscopic crustaceans, although this may not hold true after all of the groups have been more thoroughly investigated. All species observed during the study were planktonic members of the Diaptomidae, and only one was identified with certainty.

In the Baia Piuval, *Notodiaptomus coniferoides* (Wright, 1927) is the most abundant planktonic copepod. It was encountered in large numbers during the late dry and early rainy season. During the high water period, it was not observed. None were ever encountered on the floodplain, but a few were also observed in the Rio Bento Gomes near the Transpantaneira Highway, downstream from the Baia Piuval. This species was first reported from northern South America, including the Brazilian states of Rondonia, Amazonas, and Pará (Brandorff, 1976). It was later reported from the Itaipu Reservoir in the Rio Paraná System (Matsumura-Tundisi, 1986), and has since been found to be among the commonest copepod species in various parts of the Pantanal (Reid and Moreno, 1990) and one of the most widespread calanoids in Brazil (Matsumura-Tundisi, 1986).

In addition to *N. coniferoides*, some of the permanent lakes of the Pantanal are populated by *Notodiaptomus dietersi* (Poppe, 1891), and *N. venezolanus deevoyorum* Bowman, 1973. Matsumura-Tundisi (1986) determined the presence of these species in just a few samples, mainly from the region of Barão do Melgaço. From her reports, species of the genus *Notodiaptomus* seem to become more numerous moving downstream in the Paraguay-Paraná River System, and more species would be expected in that part of the Pantanal located in Mato Grosso do Sul. However, the strongly seasonal occurrence of the species observed in the Baia Piuval suggests that the appearance of large numbers of the different species is somewhat irregular from one year to the next, and the largest swarms are generally encountered during the dry and early rainy seasons, almost exclusively in the large water bodies where piranhas and other large piscivorous fishes keep the numbers of small predatory species low (Section 11.3). Thus, frequent sampling of the many *baias* during the seasons of low water levels will probably reveal the presence of considerably more species of diaptomids than are presently known from the region.

The diaptomid fauna in the Brazilian part of the Rio Paraná System seems quite distinct from that in the Pantanal. Lima *et al.* (1996) failed to find the commonest species known from the Pantanal, but found instead five species, three of which were identified as *Notodiaptomus* cf. *amazonicus* (Wright, 1935), *N. isabelae* (Wright, 1936), and *N. iheringi* (Wright, 1935). In addition, the presence of *Argyrodiaptomus furcatus* (Sars, 1901) was reported.

Cyclopoida

Copepods in the family Cyclopidae were not commonly encountered in the Pantanl, probably because of the extreme predator pressure by fishes. However, several species appear in water samples seasonally and are occasionally moderately abundant for short periods of time, generally not longer than about two weeks during the *seca*. *Thermocyclops minutus* (Lowndes, 1934) and *Mesocyclops meridianus* Kiefer, 1926, were recognized among the specimens collected, and a third species, probably *Ectocyclops* sp., was observed once at the water's edge in the Rio Bento Gomes near the Transpantaneira Highway. This species habitually pushed its way as far out of the water as possible on the shore, a rather unusual behavior for a cyclopoid.

In the Brazilian section of the Rio Paraná System to the south of the Pantanal, Lima *et al.* (1996) also found *Thermocyclops minutus* and *Mesocyclops meridianus* to be the most widespread and abundant cyclopoid species. In addition, they encountered *Thermocyclops decipiens* (Kiefer, 1929), *Microcyclops anceps* (Richard, 1897); *Paracyclops fimbriatus* Kiefer, 1936; and *Eucyclops ensifer* (Fischer, 1853).

Branchiura

One specimen of the argulid, *Dolops carvalhoi* Castro, 1949, was found in shallow water near the shore of the Rio Paraguay on the Fazenda Santo Antônio das Lendas. According to Castro (1986), the only host this species was found on is *Rhaphiodon vulpinus*, a species common throughout the Rio Paraná System. However, Thatcher (1991), in his study on Amazonian fish parasites, reported the species from *Pseudoplatystoma fasciatum*, *Pseudoplatystoma tigrinum*, *Phractocephalus hemiliopterus*, *Colossoma macropomum*, *Colossoma bidens*, and *Pellona castelnaeana*. None of these external parasites of fishes were ever observed in the region of Poconé, which is very surprising in view of the great abundance of suitable hosts.

Ostracoda

Ostracods were not often encountered in the surface waters of the Pantanal. However, several inhabited the open wells on some of the ranches. The only fairly common species in the wetland was encountered during the dry season. It is a member of the family Candonidae, *Candona* sp.

Decapoda

Palaemonidae

Two decapod species are very abundant throughout the Pantanal. One of these, a freshwater prawn, *Macrobrachium amazonicum* (Heller, 1862), is abundant among the littoral vegetation along small streams and rivers. It was commonly found in the Rio Bento Gomes where it crosses the Transpantaneira Highway, as well as in shallow water bodies near the Rio Pixaim. However, it could be encountered almost anywhere in the Pantanal during the high water period, except on areas that do not remain flooded for very long.

This species has eliminated the requirement of most decapods to return to seawater, or at least brackish estuaries, to reproduce. Its larval stages can complete their development in the fresh water of the Pantanal, in spite of its normally low chlorosity. This has made the species a resident freshwater prawn of the Pantanal, like its Southeast Asian counterpart, *Macrobrachium lanchesteri* (De Man, 1911). As discussed elsewhere (Heckman, 1974, 1979, 1997a), however, *M. lanchesteri* has developed a brood pouch in which the larval stages are passed. Other South American species of *Macrobrachium* are also able to reproduce in freshwater thanks to the possession of a brood pouch, and Margalhães (1985) reported that *M. amazonicum* is the only species known in the Amazon System with free-swimming zoea larvae. These larvae tend to congregate among the roots of floating plants (Collart and Moreira, 1993), where they can feed on microscopic members of the interrhizon community and escape predation by fishes. The larvae were found to undertake vertical migrations, with peak activity at about 10 p.m. (Moreira and Collart, 1993).

According to Kensley and Walker (1982), the congeneric Amazonian species have an even more abbreviated larval development than *Macrobrachium amazonicum*, permitting a relatively early release of the juveniles from the brood pouch. An interesting question is whether such modified larval development in tropical Asia and South America is a case of parallel evolution within the family Palaemonidae or whether these are geographically separated forms of a single phylogenetic complex that colonized the entire tropics in prehistoric times.

A second prawn species is known from the northern part of the Pantanal. *Palaemonetes ivonicus* (Holthuis, 1950) inhabits shallow water zones along the larger rivers.

Some economic importance has been attributed to this species by a few authors, as reviewed by Holthuis (1980), but studies have been confined to northern South America, and references to this species in the Pantanal are limited mainly to a few unpublished dissertations. All of the adult specimens collected from among the aquatic macrophytes in the shallow water of the floodplain during the study conformed to the description of *M. amazonicum* provided by Holthuis (1952). Apparently, *Palaemonetes ivonicus* is a species that has a strong affinity for the rivers, generally keeping its population separate from that of *M. amazonicum*.

Trichodactylidae

In addition to the prawns, the Decapoda is abundantly represented in the Pantanal by freshwater crabs in the genus *Trichodactylus*. In the Rio Paraguay and Paraná System, two species are well represented: *Trichodactylus parvus* Moreira, 1912, seems to be the predominant freshwater crab in the wetlands in the upstream sections of the system, including the Pantanal, while *Trichodactylus borellianus* Nobili, 1896, seems to be a species of the rivers, which is commoner to the south and is apparently the most widespread species of the genus inhabiting the freshwater sections of this river system in Argentina (Lopretto, 1995). With regard to its capability for reproduction without returning to the sea, *Trichodactylus parvus* has Southeast Asian counterparts displaying the same kind of brooding behavior, which permits the rapid passage of the larval stages beneath the abdomen of the female. The young are released as miniature crabs. In Southeast Asia, *Somanniathelphusa sinensis* (Stimpson, 1907) has developed an analagous brood pouch beneath the abdomen for this purpose (Heckman, 1979).

Because the crabs of the region have been so infrequently studied, many taxonomic problems still exist. A variety of synonyms seem to exist for the common species, although a revision of thee genus will be necessary to clarify the taxonomic status of some of the forms. Moreover, both of the common Pantanal species were recently placed in the new genus, *Mikrotrichodactylus*, by Rodriguez (1992), who made

Trichodactylus borellianus the genotype. Because the differences between these and the other species of *Trichodactylus* are so minor, however, Margalhães and Türkay (1996) suppressed this new genus.

Due to the taxonomic problems and paucity of studies on the ecology of the crab fauna, the exact delimitation of the ranges of the two common species and their ecological segregation within the aquatic system are still scarcely known.

8.12.4. Insecta

Collembola

The collembolan fauna of the Pantanal includes species that live of the surface tension film of the water, usually near shore or among the floating and emergent water plants. The aquatic species belong overwhelmingly to the Symphypleona, probably due to their more compact form, which seems to give them some advantage in escaping predators.

Hypogastruridae

Although not represented among the aquatic fauna, this family was represented by a terrestrial species, *Choreutinula travassosi* (Arlé, 1939), which was collected at the edges of hummocks on the Fazenda Ypiranga. It is also known from the upland parts of the watershed.

Neanuridae

Two neanurids are known from Mato Grosso: *Hylaenura infima* (Arlé, 1959) and *Kenyura porculus* (Arlé, 1959). Both of them are tiny springtails with a subterranean habitat, the former eyeless and the latter with a reduced number of ocelli. They are encountered only occasionally beneath pieces of wood or stones, and their habits are still unknown.

Entomobryidae

Entomobryids are known from the Pantanal, and a few individuals were collected at the edge of hummocks beside the sampling site at Birici on the Fazenda Ypiranga. *Lepidocyrtus atratus* Mari Mutt, 1983, and *Dicranocentrus silvestrii* Absolon, 1903, were collected during the dry season, but neither of them was abundant. These species had not previously been reported from Mato Grosso, and *Lepidocyrtus atratus* was previously known only from Venezuela.

Along streams in the upland parts of the catchment area, specimens of *Paronellides alticolus* (Arlé, 1939) were collected during the *cheia*. The species is widespread is Brazil, but like other South American collembolans, reports of its occurrence are relatively few.

In addition to the species observed during the study, three other entomobryids have been described from southern Mato Grosso: *Entomobrya (Entomobrya) aipatse* Arlé, 1959; *Entomobrya (Entomobrya) uambae* Arlé, 1959; and *Lepidosira villasboasi* Arlé and Guimarães, 1981. In addition, two other are known from Mato Grosso do Sul: *Seira atrolutea* (Arlé, 1939) and *Seira brasiliana* (Arlé, 1939). The extreme dearth of ecological and zoogeographical studies on this insect group in South America make it difficult to judge their role in the Pantanal. Although only relatively recently described, some of these species, such as *Seira brasiliana*, are known to be very widespread. Further research on these insects is urgently needed.

Bourletiellidae

Various species in this family live directly on the surface tension layer of the water, generally near the shore or among masses of floating macrophytes. Large numbers of *Deuterosminthurus richardsi* Arlé, 1971 were found on the surface of the small pond near the Rio Pixaim, and the species was enountered at many of the other sampling sites, as well. It is the only aquatic springtail that can be described as widespread and abundant on water bodies in the Pantanal of Poconé. The species was previously known from Pará.

A second species of the genus, *Deuterosminthurus aueti* Arlé, 1961, is known from Mato Grosso. It was not observed during the study.

Sminthurididae

Less abundant but often accompanying *Deuterosminthurus richardsi* on the surface tension layer of the water was a sminthuridid identified as *Sminthurides macroceros* Arlé, 1961, a species described from Mato Grosso. Identification is somewhat problematic because only a thorough description of male is available, and females were collected. The individuals seem to remain on or close to the water plants of the pseuroterrestrium, while *D. richardsi* is typically found resting on the surface tension layer of the water.

Ephemeroptera

Larval ephemoropterans were observed in only a few water samples in the Pantanal. The adults of several species, however, appeared frequently at light traps. Because these were most often collected at the edge of the Pantanal or a few kilometers outside of the floodplain, it is assumed that most of the species are inhabitants of permanent streams, ponds, and impoundments outside of the Pantanal. The floodplain and small rivers draining the Pantanal do not seem very conducive for the development of Mayfly populations, perhaps because of a very strong predator presssure and the failure of most species to adapt to the conditions in the interrhizon, the habitat in which most small aquatic invertebrates of the Pantanal are able to survive.

Polymitarcyidae

Immediately before the freshet, usually in September, large numbers of *Campsurus striatus* Needham & Murphy, 1924, were collected at nights around artificial lights near the Rio Bento Gomes and elsewhere near the city of Poconé. The larvae apparently are not abundant in the Pantanal itself but are extremely abundant in the water bodies at the edges of the wetland. This species was previously reported from the Pantanal near Corumbá in Mato Grosso do Sul, and no previous references to the species from the northern part of the Pantanal have been found in the literature.

A second, somewhat similar congeneric species is also known from the southern part of the Pantanal. *Campsurus notatus* Needham & Murphy, 1924, differs in lacking some of the dark markings of *C. striatus*. It was not observed during the study, nor was *Tortopus harrisi* Traver, 1950, another species described from Mato Grosso do Sul.

Baetidae

Larval specimens of the family baetidae were observed occasionally in the Pantanal. Their identity is not certain. Three species of *Callibaetis* were described from the region of the wetland in Mato Grosso do Sul: *Callibaetis viviparus* Needham &

Murphy, 1924; *Callibaetis pallens* Needham and Murphy, 1924; and *Callibaetis jocosa* Navás, 1912. Little is known about their ecology, and their larvae remain undescribed.

In addition, rather large swarms of adult *Baetodes* sp. were found flying around lights in the city of Cangas to the north of the Pantanal during the *cheia*. Unfortunately, most South American species in this genus were described only from larvae, and the species of the adults are still not distinguishable.

The larvae of a species from savanna woodlands in the region of the Pantanal, *Cloeodes hydation* McCafferty and Lugo-Ortiz, 1995, was reported by Nolte *et al.* (1996) to survive as long as nine hours exposed to the air. This ability to resist desiccation is an obvious asset for survival in the temporary water bodies of southern Mato Grosso.

Odonata

Calopterygidae

In the standing water of the Pantanal, this family was not represented. However, *Mnesarete guttifera* (Selys, 1873) proved to be quite common along water courses upstream from the Pantanal, where its larvae are apparently able to develop in water with an electrical conductivity of less than 3 µS·cm. It was observed regularly near Chapada dos Guimarães and along the Corrego Guanandi near Cangas not far from the northern border of the Pantanal.

Not seen during the study were species of *Hetaerina*, a number of which have been reported from Mato Grosso (Longfield, 1929). *Hetaerina laesa* Hagen in Selys, 1853, was encountered by Longfield (1929) in upland forests near Chapada dos Guimarães. The other species may be confined to the Amazon drainage area.

Coenagrionidae

The only abundant zygopteran species in and beside the waters of the Pantanal itself belong to this family. A consideraable number seem to live in the interrhizon as larvae and prey on the species in the pseudoterrestrium, in which the adults are very frequently seen. Many of the adults appear during the *vazante* and *seca* to exploit the emerging adults of small aquatic insects as a food source. The genus *Helveciagrion*, which seems very well adapted to the temporarily flooded areas, is represented in the Pantanal by *Helveciagrion simulacrum* (Calvert, 1909), the adults of which were found on the floodplains during the late dry season, and *Helveciagrion chirihuanum* (Calvert, 1909), which appeared along the edges of the Rio Bento Gomes at the northern edge of the Pantanal near Poconé during the same season. Both of these species have long been known from southern Mato Grosso (Machado, 1980). These are rather delicate damselflies, and they would probably have difficulty surviving as adults if frequently exposed to very heavy rainfall.

The adults of *Telebasis willinki* Fraser, 1948, were abundant beside the rivers and streams in the Pantanal during all seasons except the *vazante*.

Commonly found along the Rio Bento Gomes at the northern edge of the Pantanal is *Homeoura chelifera* (Selys, 1876), which is another species that lives among the emergent plants on flooded areas. The adults appear during the dry season and seem to spawn most intensively during the onset of the rainy season. The species is widespread in South America, occurring as far north as Venezuela (Rácenis, 1958) and as far south as Argentina (Ris, 1913).

Also represented among the odonatan fauna of the wetland is *Ischnura capreolus* (Hagen, 1861), a small species that was found at the northern edge of the Pantanal. The adults are also most common during the late dry season.

Species of the genus *Acanthagrion* are also known from the region of the Pantanal, although they were not identified during the study. These include *Acanthagrion temporale* Selys, 1876; *Acanthagrion chararum* Calvert, 1909; and *Acanthagrion truncatum* Selys, 1876. Apparently, these species are adapted better to upland water bodies than to the wetland. This is also the true of *Acanthallagma luteum* Williamson and Williamson, 1924 Two additional species, *Acanthagrion chacoense* Calvert, 1909, and *Acanthallagma strohmi* Williamson and Williamson, 1924, are reported only from the Pantanal of Mato Grosso do Sul. Also apparently confined to the streams of the Pantanal watershed is *Argia* sp., which was only found beside permanently flowing water.

Pseudostigmatidae

The presence of *Mecistogaster linearis* (Fabricius, 1776) in southern Mato Grosso is probable since Longfield (1929) reported it from an undisclosed location in the state. However, no specimens in this family were presonally observed during the study.

Protoneuridae

Although *Neoneura sylvatica* Hagen in Selys, 1886, was not encountered in the Pantanal itself, it was found to be rather abundant along streams in southern Mato Grosso, including the Corrego Guanandi, which flows through the district of Cangas toward the Rio Bento Gomes. Apparently, the larvae of this species are stream dwellers that do not often find conducive conditions in the predominantly standing waters of the Pantanal. This species is common and widely distributed in South America.

Perilestidae

This family is represented by at least four species in the upland regions of Mato Grosso, but only one is known from the catchment area of the Pantanal. None were observed in the wetland itself during the study. Among the species encountered by Longfield (1929), *Perilestes solutus* Williamson and Williamson, 1924, was fairly numerous near Chapada dos Guimarães. The remaining species may not be encountered south of the Amazon drainage area.

Aeschnidae

Most aeschnids are strong fliers which may be found far from their breeding places. None were observed in the Pantanal, although they were seen occasionally on the plateau to the east of the wetland. Longfield (1929) reported several species from southern Mato Grosso, the most abundant of which along the Rio Cuiabá was apparently *Coryphaeschna viriditas* Calvert, 1952, at the time not yet distinguished from *Coryphaeschna virens* Rambur, 1842. Other aeschnids that occur in the upland regions of the catchment area include *Coryphaeschna adnexa* (Hagen, 1861) and *Castoraeschna longfieldae* (Kimmins, 1929).

Libellulidae

From the late rainy period through the first dry months, large swarms of libellulids are observed flying over the Pantanal in search of insects. These dragonflies are very strong flyers and remain in the air for long periods of time during the day. They tend to follow large animals in order to capture the insects that they disturb or the mosquitos that they attract. They were often observed snatching the diurnal mosquito, *Psorophora varipes*, out of the air during the late rainy season (see below).

A very common species observed throughout the catchment area is *Orthemis ferruginea* Fabricius, 1775, which generally travels in swarms, seeking insects over terrestrial regions.

Among the species that are active as adults during the rainy season are *Brachymesia herbida* (Gundlachi, 1889) and *Erythemis attala* (Selys, 1857), which are common in the Pantanal but less likely to be seen in groups of more than two or three.

Among the species with very brightly colored wings, was *Diastatops intensa* Montgomery, 1940, the adults of which were observed each year during the high water period, occasionally in considerable numbers. They were seen most frequently along the Transpantaneira Highway and on the Fazenda Ypiranga. This species is not a strong flier and generally remains on emergent plants above the water. Adults are common only during the *cheia*, although a few may persist into the *vazante*.

One of the commonest of the species throughout the Pantanal during the rainy season is *Erythrodiplax castanea* (Burmeister, 1839), which is a very widespread dragonfly throughout much of tropical South America. These species were generally observed in small swarms patrolling along the edges of the water bodies in search of insects among the plants. The species is found from Trinidad to Ecuador, Peru, Bolivia, and Paraguay and throughout most of Brazil.

Also very widespread and abundant, especially during the rainy season, was *Erythrodiplax umbrata* (Linnaeus, 1758), a species that could be easily recognized from a distance from the dark bands it usually displayed across the middle of the wings, from the nodus to the stigma. Unlike *E. castanea,* however, this species did not form swarms. Individuals were generally observed alone or in small groups of two to four. This is one of the most widespread of the South American lowland dragonflies, occurring from North America throughout the West Indies and Central America to Argentina. It seems to occur in all parts of Brazil. The adults were most abundant in the Pantanal during the rainy season, from November through March.

Less abundant was *Erythrodiplax fusca* (Rambur, 1842), the adults of which were observed during all seasons in many different parts of the wetland.

Micrathyria hesperis Ris, 1911, was encountered rather frequently from the beginning of the *enchente* and through the *cheia* along the Transpantaneira Highway and on the nearby ranches. These dragonflies did not seem to undertake long flights, and individuals were commonly seen perched on emergent plants on the temporarily flooded areas. Adults were most abundant during March.

One of the smallest anisopterans encountered in the Pantanal is *Nephepeltia leonardina* Rácenis, 1953. The adults of this species were observed along the edges of hummocks during the dry season near Birici stream. They were regularly seen during their season of flight only at this sampling site. Their habits were more typical of zygopterans than most other anisopterans, since they normally waited on the leaves of woody plants for a chance to quickly grab their prey and return to their resting places. They are not strong fliers; however, they were difficult to capture because they flew into the thick foliage when persued.

A rather rare species in the Pantanal, *Podothemis nemesis* Ris, 1909, was encountered occasionally near the Rio Bento Gomes. The adults were observed during the early rainy season.

In addition to those identified, an interesting specimen of *Orthoclada* sp. was taken during the dry season at the Baia Piuval. It could not be readily assigned to any known species (A. Machado, pers. comm.).

The dragonfly fauna of the Pantanal and its catchment area includes a number of species not identified during the study. Some of these were catalogued by Longfield (1929) from the Rio Paraguay drainage area, mainly in the upland regrons near Chapada dos Guimarães and at the headwaters of the Rio Paraguay. These included *Diastatops pullata* (Burmeister, 1939); *Diastatops obscura* (Fabricius, 1775); *Perithemis cornelia* Ris, 1910; *Perithemis mooma* Kirby, 1889; *Uracis ovipositrix* Calvert, 1909; *Uracis imbuta* Burmeister, 1839; *Uracis imfamata* (Rambur, 1842); *Uracis fastigata* (Burmeister, 1839); *Rhodopygia hollandi* Calvert, 1911; *Miathyria marcella* Selys, 1857; *Erythrodiplax connata* (Burmeister, 1839); *Erythrodiplax attenuata* (Kirby,

1894); *Erythrodiplax latimaculata* Ris, 1911; *Erythrodiplax ochracea* (Burmeister, 1839); *Erythrodiplax venusta* (Kirby, 1897); and *Erythrodiplax paraguayensis* (Förster, 1904), which is probably the species she referred to as *E. minuscula* Rambur, 1942. Any of these species might be encountered in the Pantanal, at least in near the gallery forests, but they seem to be species more adapted to the *cerrado* and forested regions of the *alto plano*.

Gomphidae

The commonest species in this family, which was observed frequently seeking prey above the floodplain during the *enchente* and *cheia*, was *Aphylla theodorina* Navas, 1933. Some of the adults persisted through the *vazante*, but none were observed during the *seca*. Larvae of an *Aphylla* species, probably this one, were frequently encountered in the Rio Bento Gomes, particularly in the Baia Piuval. The exuviae were found on plants most often during the late *enchente*. Other gomphid species were not observed in the wetland, although several occurred at more elevated locations of the catchment area. Longfield (1929) reported the presence of *Progomphus recticarinatus* Calvert, 1909, and *Progomphus intricatus* Hagen, 1858, from Chapada dos Guimarães.

Epigomphus hylaeus Ris, 1918, was described from a Mato Grosso specimen without any more precise location. Two more recently described species from the State of Mato Grosso which apparently occur rarely or not at all in the Pantanal are *Archaeogomphus nanus* Needham, 1944, and *Archaeogomphus vanbrinki* Machado, 1994.

Orthoptera

Acrididae

The pseudoterrestrium has a rich orthopteran fauna. These include the grasshoppers *Cornops aquaticum* (Bruner, 1906), an insect that can feed on most aquatic plant species (Ferreira, pers. comm.) but seems to feed heavily on species of the Pontederiaceae in the field. An exhaustive study of this species by Ferreira (pers. comm.) revealed that these alert diurnal grasshoppers remain on the plants they feed upon unless disturbed by a potential predator. They first move to the opposite side of a vertically emergent leaf so that an approaching man or animal cannot seen them, and if the threatening creature comes closer, they usually hop and fly to a distant plant. Sometimes, however, they dive into the water and conceal themselves among the plant roots until it seems safe for them to emerge again. On occasion, these grasshoppers have been seen swimming from plant to plant at a depth of nearly one meter in the very clear water during the *cheia*. The eggs are deposited mainly beneath the leaf axils or in cavities in the stems of aquatic and littoral plants. Apparently, the adults die off during the dry season, and the eggs left by the previous generation hatch shortly after the freshet.

Pauliniidae

One of the species in the floating plant community is *Marellia remipes* Uvarov, 1929, a species generally associated with *Hydrocleys nymphoides* and apparently *Nymphaea* spp., as well as species of the Hydrocharitaceae not encountered in the Pantanal. Carbonell (1957) made a detailed study of its habits and reported that the species is noted for avoiding vertically emergent plants, remaining instead on those leaves that float horizontally on the surface of the water. The grasshoppers prefer to escape by diving underwater rather than flying away, and the eggs are deposited on the underside of floating leaves, where they remain submerged until hatching, when the newly emerged nymphs climb onto the upper surfaces of the leaves. They are strong swimmers, both on the surface and underwater, and fly more frequently at night than during the day. As in the case of *Cornops aquaticum*, a considerably larger variety of plants are consumed by individuals in captivity than are eaten normally by the grasshoppers under natural conditions.

Another grasshopper, *Paulinia acuminata* (De Geer, 1773), is associated with the floating fern, *Salvinia auriculata*, on which it feeds under natural conditions. Barthlott *et al.* (1994) found an extreme degree of adaptation by the herbivore to its host. The grasshopper not only mimics the water fern in its color but also in the physical patterns of its untrastructure, as revealed by the scanning electron microscope. Both the plant and the insect have a water repellent layer of epicuticular wax arranged in pentagons of similar size.

Gryllotalpidae

Two species of mole cricket, *Neocurtilla* spp., are frequently observed in subterranean tunnels beneath objects near the water's edge. Along the Rio Bento Gomes near the Poconé exposition park, the tunnels of the larger species were particularly abundant. On smooth sand surfaces along the shore during the rainless seasons, the courses of these tunnels are clearly marked by cracking of the sediment. The smaller of the two *Neocurtilla* species was found only along the shores of the Rio Paraguay near the Fazenda Santo Antônio das Lendas.

Gryllidae

The pseudoterrestrium community includes at least two species of cricket, *Argizala* sp. and an unidentified one. These are encountered throughout the year in various stages of development, indicating that they do not have pronounced seasonal adaptations to the wet-and-dry climate. For this reason, they seem confined to the plant aggregations dominated by *Eichhornia* spp., which generally remain in places flooded the year round. These species were very abundant on the islands of floating vegetation on the Baia Piuval but rare in the shallower streams crossing the floodplain.

Isoptera

Because most of the termites are not at all aquatic and most of the species are more typical of the upland savanna and forests than of the wetland, their ecology is not treated intensively here. However, several species are characteristic of the terrestrial areas of the Pantanal region and should be mentioned. One of the typical kinds of landscape in the Pananal is the earthmound savanna or *campo murundu*, which takes its name from the mounds constructed on tree stumps by large colonies of termites. However, these mounds do not always house the species that constructed them. After the builder has left, many other species as well as various ants may move in. Many of these termites are important food items for the widest variety of insectivorous animals and are apparently the main source of nourishment for the anteaters, worm lizards, and caecilians of the region. The winged adults swarm in enormous numbers and provide a seasonally important item in the diets of anurans, bats, and crepuscular birds. The knowledge of the species in southern Mato Grosso was enhanced considerably by the study of Mathews (1977), and some additional information on species important in the wetland was provided by Dubs (1994).

Kalotermitidae

This family is known only from the plateau region northeast of the Pantanal near Barra do Garças, where Mathews (1977) encountered *Neotermes monoculus* Snyder,

1926, previously known only from Bolivia, and *Neotermes acceptus* Mathews, 1977.

Rhinotermitidae

This family includes one confirmed wetland species, *Heterotermes tenuis* (Hagan, 1858), that Mathews (1977) encountered in an abandoned mound constructed by *Cornitermes snyderi* (see below) near Xaventina. The other species, representing three subfamilies, Heterotermitinae, Coptotermitinae, and Rhinotermitinae, are apparently confined mainly to the upland regions surrounding the Pantanal, although penetration by all of them into the *cerrado* and *cerradão* on the *cordilleiras* crossing the Pantanal is almost certain. None of them are known to build mounds themselves, but they are frequently encountered feeding on the leavings in old mounds of other species (Mathews, 1977). They include *Heterotermes longiceps* Snyder, 1924, and *Heterotermes sulcatus* Mathews, 1977, both in the subfamily Heterotermitinae, as well as *Coptotermes testaceus* (Linnaeus, 1758), the only member of the Coptotermitinae, and the members of the Rhinotermitinae, *Rhinotermes marginalis* (Linnaeus, 1758) and *Dolichorhinotermes longilabius* (Emerson, 1925).

Serritermitidae

Only a single species was reported by Mathews (1977) from southern Mato Grosso: *Serritermes serrifer* (Bates, 1858), known from the plateau region northeast of the Pantanal but never reported from the Pantanal itself.

Termitidae

This family is richly represented in the Pantanal and its immediate surroundings. Several of its species are familiar in the Pantanal and are apparently well adapted to the wetland. One of the most prominent is *Rotunditermes bragantinus* (Roonwal and Rathore, 1976), which is a species often encountered in the islands of woodland, the *capões de mata*, just above the high water level of the floodplain (Dubs, 1994). *Dentispicotermes pantanalis* Mathews, 1977, seems to be another species with excellent adaptations for inhabiting the hummocks on the Pantanal. Mathews (1977) found the species in an abandoned nest of *Armites euamignathus* Silvestri, 1901, in a grove of *Mauritia vinifera* on a hummock. He noted what seemed to be permanent underground waterways which the workers were apparently using. As a defense, they secreted a yellow fluid with a very strong smell. Both of these species seem to have developed a high degree of adaptation to wetlands.

Other common species in the Pantanal include the mound-builders themselves. These termites are responsible for the formation of a very typical kind of landscape in southern Mato Grosso (Ponce and da Cunha, 1993): the earthmound savanna (Section 2.1.6). *Armites euamignathus* is one species that constructs large nests on the dry pastures and natural savannas, generally on top of tree stumps and on pieces of wood. However, some of these nests are constructed quite near the high water level in the wetland, suggesting that they are adept at waterproofing some of their tunnels or "moving upstairs" during the *cheia*. Tall narrow nests in wooded areas are constructed by *Cornitermes snyderi* Emerson, 1952, which Mathews (1977) reported to inhabit the entire region between Goiás and Bolivia and found abundantly at the northern margins of the wetland. Dubs (1994) reported this species to be very common in the Pantanal of Mato Grosso do Sul, as well. Apparently, the most abundant mound builder in the North, including the drier parts of the wetland, is *Cornitermes bequaerti* Emerson, 1952, a species abundant both in the *cerrado* and in gallery forests. A third species in this genus, *Cornitermes cumulans* (Kollar, 1832), is known to inhabit southern Mato Grosso only from the presence of a queen found in the stomach of an armadillo by Mathews

(1977). A mound builder not reported from the open areas in the Pantanal but confined mainly to gallery forests is *Armitermes neotenicus* Holmgren, 1906, which constructs low, dome-shaped nests, always on the ground but sometimes against the sides of trees (Mathews, 1977). *Armitermes cerradoensis* Mathews, 1977, is one mound builder that is apparently native mainly to the upland dry *campo* and *cerrado* regions, while *Rotunditermes rotundiceps* (Holmgren, 1906) is a species with similar construction skills that is known with certainty only from the *alto plano* to the northeast of the Pantanal. It was reported by Mathews (1977) to inhabit forests rather than savannas.

Few if any of the species identified by Mathews (1977) are endemic to the region of the Pantanal. Although the species *Nasutitermes coxipoensis* (Holmgren, 1910) was discovered near and named for a district in Cuiabá near the confluence of the Coxipó and Cuiabá Rivers, it was later found in São Paulo (Araujo, 1958a) and Minas Gerais (Araujo, 1958b), demonstrating that it is a very widespread species in Brazil.

In addition to the builders of mounds on the ground, there are several species in this family that construct their mounds in trees. These species may be encountered almost anywhere in the wetland since they are safe from the floods. Such nests are extremely common, especially in large trees. The builders include *Microcerotermes strunckii* (Sörensen, 1884), *Nasutitermes surinamensis* (Holmgren, 1910), *Nasutitermes ephratae* (Holmgren, 1910), *Constrictotermes cyphergaster* (Silvestri, 1901), *Labiotermes labralis* (Holmgren, 1906), and *Armitermes holmgreni* Snyder, 1926.

Other species known from the wetlands of southern Mato Grosso inhabit the mounds built by the other species. According to Mathews (1977), they include *Spinitermes trispinosus* (Bates, 1858), which inhabits nests of *Cornitermes bequaerti* and *C. snyderi*; *Spinitermes nigrostomus* Holmgren, 1906, found in abandoned mounds of *Armites euamignathus*; *Cavitermes parmae* Mathews, 1977, in a nest of *C. snyderi*; *Armitermes peruanus* Holmgren, 1906, in a mound of *Armites euamignathus*; *Velocitermes heteropterus* (Silvestri, 1901) together with *Nasutitermes kemneri* Snyder and Emerson, 1949; *Triangularitermes triangulariceps* Mathews, 1977, in a mound of *C. snyderi*; and *Paracornitermes hirsutus* Araujo, 1954, in an abandoned mound of the same species. This leaves *Dentispicotermes pantanalis* in the Pantanal and several newly described species in the *cerrado* as candidates for the distinction of being endemic to Mato Grosso. However, from the present knowledge about the courtship migrations of many species and their food preferences, it is unlikely that any of them will prove to be endemic to the state.

In addition to the species known from the wetlands, quite a few species of termite are known only from the *cerrado* and upland gallery forests of southern Mato Grosso (Mathews, 1977). Although the regional termite fauna is still poorly known, the number in this family is already considerable. The great majority are species that inhabit the large nests of the mound builders, either still occupied or deserted, or simply live in wood that is in contact with the soil. These include *Anoplotermes turricola* Silvestri, 1901, and eleven unidentified species of the genus; *Grigiotermes metoecus* Mathews, 1977, and an unidentified species of the same genus; *Ruptitermes xanthochiton* Mathews, 1977; *Ruptitermes reconditus* Silvestri, 1901; *Cylindrotermes parvignathus* Emerson in Snyder, 1949; *Cylindrotermes flangiatus* Mathews, 1977; *Amitermes* sp.; *Microcerotermes exiguus* (Hagen, 1858); *Microcerotermes indistinctus* Mathews, 1977; *Spinitermes robustus* Snyder, 1926; *Spinitermes allognathus* Mathews, 1977; *Cavitermes parvicavus* Mathews, 1977; *Genuotermes spinifer* Emerson, 1950; *Orthognathotermes aduncus* Emerson in Snyder, 1949; *Orthognathotermes gibberorum* Mathews, 1977; *Termes medioculatus* Emerson in Snyder, 1949; *Termes bolivianus* (Snyder, 1926); *Inquilinitermes fur* (Silvestri, 1901); *Inquilinitermes microcerus* (Silvestri, 1901); *Crepititermes verruculosus* Emerson, 1925; *Neocapritermes opacus* (Hagen, 1858); *Planicapritermes planiceps* (Emerson, 1925); *Dihoplotermes inusitatus* Araujo, 1961; *Nasutitermes bolivari* (Snyder, 1959); *Nasutitermes callimorphus* Mathews, 1977; *Nasutitermes kemneri*; *Nasutitermes nigriceps* (Haldeoman, 1858); *Nasutitermes stricticeps* Mathews, 1977; *Nasutitermes tatarendae* (Holmgren, 1910); *Nasutitermes unduloceps* Mathews, 1977; *Nasutitermes wheeleri* Emerson, 1925; *Syntermes molestus* (Burmeister, 1839); *Diversitermes diversimiles* (Silvestri, 1901);

Diversitermes aporeticus Mathews, 1977; *Cortaritermes silvestrii* (Holmgren, 1910); *Labiotermes leptothrix* Mathews, 1977; *Armitermes festivellus* Silvestri, 1903; *Curvitermes angulariceps* Mathews, 1977; *Curvitermes strictinasus* Mathews, 1977; *Curvitermes odontognathus* (Silvestri, 1901); *Curvitermes planiculus* Mathews, 1977; *Curvitermes projectidens* Mathews, 1977; *Angularitermes clypeatus* Mathews, 1977; *Angularitermes orestes* (Araujo, 1970); *Cyranotermes timuassu* Araujo, 1970; *Convexitermes clevelandi* Snyder, 1926; and several probably undescribed species of *Subulitermes*.

Several species known from the *cerrado* are less likely than the others to occur in the wetland. *Rhinchotermes nasutissimus* Silvestri, 1901; *Rhinchotermes myctobius* Mathews, 1977; *Rhinchotermes diphyes* Mathews, 1977; *Velocitermes glabrinotus* Mathews, 1977; *Velocitermes paucipilis* Mathews, 1977; *Velocitermes velox* (Holmgren, 1906); and *Parvitermes bacchanalis* Mathews, 1977, are all litter feeders that build no mounds but rather live on the forest floor among soil and leaves. Such species would be more subject to displacement due to flooding in the Pantanal. Large termites that inhabit tropical savannas and rainforests throughout South America, including those in southern Mato Grosso (Mathews, 1977), include *Syntermes grandis* (Rambur, 1842) and *Syntermes snyderi* Emerson in Snyder, 1924, which feed on grasses and leaves without building elaborate nests.

Heteroptera

Hemiptera

The Pantanal has a particularly rich fauna of water bugs, probably because most of these species are immune from predation by fishes due to the repellent chemicals produced in their bodies. In fact, the only invertebrate species that can be found abundantly in the open water during the high water period are small notonectids and corixids, which no fishes were ever observed to consume. On many occasions, fishes were observed to rob the prey that notonectids or gerrids had seized on the water surface, but the same fishes never attempted to attack the insect itself, attesting to the effectiveness of its chemical repellents. In addition to defending themselves with chemicals, many water bugs can inflict painful jabs with their mouthparts, which sting for a long time, apparently due to injected chemicals.

Corixidae

This family is represented in the region of the Pantanal by members of the genus *Tenagobia*, the most abundant of which is *Tenagobia incerta* Lundblad, 1928, identified from numerous specimens from the Rio Bento Gomes near the exposition park of Poconé. Other congeneric species seem to be less abundant, although at least two, surmised from the distinctness of female specimens, seem to be present. The identification of the species is rather difficult, and the revision of Nieser (1977) had to be utilized because some of the morphological features reported by Hungerford (1948) to be characteristic of certain species are incorrect.

In addition to the Rio Bento Gomes, *Tenagobia incerta* occurs as one of the most numerous of all aquatic insects along some of the other small rivers flowing through the Pantanal. It is generally observed resting on the sediment in shallow waters, but it also occurs less frequently in the interrhizon of floating plants. The corixid fauna of the Pantanal also includes *Tenagobia hungerfordi* Deay, 1930, known only from the region of Bolivia, Paraguay, and Corumbá, presently in Mato Grosso do Sul, which became politically independent from Mato Grosso after the publication of Hungerford's monograph in 1948. Therefore, the presence of this species has been confirmed only in the southern Pantanal. More likely to be encountered in the region of Poconé is *Tenagobia schadei* Lundblad, 1928, known from Surinam to Paraguay. However, among the many specimens of its genus taken in the Pantanal, none within its reported size range were encountered.

In addition to the species of *Tenagobia*, two species of *Heterocorixa* are known from the catchment area of the northern Pantanal (Hungerford, 1948). *Heterocorixa brasiliensis* Hungerford, 1928, is known only from Mato Grosso, although it has been reported but not confirmed from other localities in Brazil. *Heterocorixa chapadiensis* Hungerford, 1928, was reported from "Chapada" without the authors mentioning which Chapada. There are many places with this name in Brazil. Due to its occurrence in Mato Grosso, however, it is suspected that Chapada dos Guimarães is meant. Both of these species can be expected in the upland areas to the north and east of the Pantanal, but corixids in their size ranges were never encountered in the wetland itself.

Gelastocoridae

The toad bugs were typical littoral species, found at the water's edge both along rivers and streams and along the margins of the flooded fields. *Gelastocoris flavus* (Guérin-Méneville, 1835) is a member of the dynamic ecotone that moved slowly across large areas of the Pantanal during the rainless seasons. Long known under its synonym, *G. nebulosus* (Guérin-Méneville, 1844), this species can be encountered both in shallow water of the sublittoral zone and on moist sediments at the water's edge. When it hops, it can easily be mistaken for one of the newly transformed toadlets that temporarily shares its habitat. There is some controversy over the taxonomy of this species. Todd (1955) considered this taxa to encompass a number of forms previously described by other authors as valid species. However, Estévez and Schnack (1980) attempted to revive some of the synonyms on the grounds of some minor morphological differences. The validity of the species in question must await more information on natural intraspecific variability.

Gelastocoris willneri (De Carlo, 1954) was the only other gelastocorid species identified in the region, but it was found only near the sources of the streams on the plateau east of Cuiabá and never in the Pantanal itself.

Notonectidae

The family Notonectidae is represented in the Pantanal by small species belonging to the genus *Buenoa*. *Buenoa salutis* Kirkaldy, 1904, was apparently the most abundant of these, but it was occasionally seen together with *Buenoa amnigenus* (White, 1879), which also was observed on the floodplain during all seasons except the *enchente*. *Martarega gonostyla* Truxal, 1949, was encountered on the floodplain, as well, but specimens were collected only from the shallow pools of water remaining during the *seca*.

In addition to the species in the Pantanal, *Buenoa fuscipennis* (Berg, 1879) inhabits southern Mato Grosso. Only one specimen was identified after being collected under a light at the time of the freshet.

Nepidae

This family is taken *sensu lato*, although there has been a tendency recently to break it up, thereby placing the species mentioned here in the family Ranatridae. Only the genus *Ranatra* was observed during the study; the nicht filled by *Nepa* and its relatives in other parts of the world is apparently occupied mainly by belostomatids in Brazil.

Adult *Ranatra wagneri* Hungerford, 1929, were collected in a shaded pool on the flood plain on the Fazenda Ypiranga. They were found only during the early *vazante*, when the water in the pools was still fairly deep. This species was previously reported from Misiones, Argentina, and Santa Catarina, Brazil (De Carlo, 1964). It is similar morphologically to *Ranatra brevicauda* Montandon, 1905, known from the same locations, and the two will almost certainly prove to be conspecific, making *R. wagneri* the junior synonym.

Also known from the Rio Paraguay System to the south of the Pantanal are *Ranatra macrophthalma* Herrich-Schaffer, 1853; *Ranatra signoreti* Montandon, 1907; and *Ranatra rabida* White, 1879. None of them were observed during the study.

Belostomatidae

This family is well represented in the northern Pantanal. Although the species were seldom encountered in the water and net samples, the presence of many species was confirmed from specimens taken around street lights in the towns of Poconé and Cangas, mainly during the early rainy season, when flights of massive numbers of insects occur, usually about one to two days before heavy thunderstorms. The species fly together with large numbers of aquatic beetles and other insects of numerous families and orders. The attraction of belostomatids to lights during their nocturnal flights has been noted by Link and de Souza (1983), who recommended using light traps to collect them. This family is particularly rich in Neotropical species. Among the works consulted to identify the specimens were those of De Carlo (1938) and Lanzer-de Souza (1991).

The species found most frequently in Poconé include *Lethocerus maximus* De Carlo, 1938, a species reaching 11 cm in length, and *Lethocerus grandis* (Linnaeus, 1758), which is nearly as long, sometimes exceeding 10 cm. Among the somewhat smaller species are *Lethocerus delpontei* De Carlo, 1933; *Lethocerus annulipes* (Herrick-Schäffer, 1846); *Lethocerus melloleitaoi* De Carlo, 1933; *Belostoma discretum* Montandon, 1903; *Belostoma gestroi* Montandon, 1903; *Belostoma bosqi* De Carlo, 1930; *Belostoma elongatum* Montandon, 1908; and *Belostoma micantulum* Stål, 1860.

B. micantulum is a species that has been particularly often studied, especially in Argentina. Consoli *et al.* (1989) noted that individuals kept in the laboratory fed on the larvae of *Aedes fluviatilis* Lutz, 1904, suggesting that it might be valuable for mosquito control. Studies on the population dynamics revealed that males and females are present at a ratio of about 1:1 and that the entire egg complement of a female can be attached to the back of a single male (Schnack *et al.*, 1986). Its rate of reproduction in Paraguay was determined from a study of 12 males with eggs attached to the hemielytra. The number carried varied from 30 to 95 and averaged 52.3 (Strickman, 1981).

In addition to the species found in and at the edge of the Pantanal, *Belostoma ribeiroi* De Carlo, 1933, was observed in a stream near one of the waterfalls at Chapada dos Guimarães. One specimen bearing eggs on its back was collected during July of 1991.

Belostoma dentatum (Mayr, 1863), found from Venezuela to Argentina, almost certainly occurs in the region of the Pantanal, as well, but it was not seen during the study.

Pleidae

The pleids encountered during the study were *Neoplea gauchita* Bachmann, 1968, observed in drying pools of water on the Fazenda Ypiranga during the late dry season and shortly after the freshet, and *Neoplea* cf. *argentina* (Drake and Chapman, 1953), which was observed during the same season along the edges of the Rio Bento Gomes near the exposition park of Poconé. Only adults were encountered. The difference between these two nominal species are relatively minor, arousing the suspicion that they may actually be conspecific.

Naucoridae

Naucorids are not commonly seen in the Pantanal. Apparently, the most abundant species encountered is *Pelocoris impicticollis* Stål, 1876, the adults of which were observed along the Rio Bento Gomes near Poconé at the beginning of the rainy season, when so many aquatic hemipterans take to the air. Other adults were observed along the Transpantaneira Highway near the Rio Bento Gomes. According to Nieser (1975), this species has been reported from Corumbá near the southern borders of the Pantanal in Mato Grosso do Sul under the synonym, *P. horvathi* Montandon, 1905. Lopez-Ruf (1991) reported it for the first time as far south as Argentina.

A second species identified with near certainty was *Pelocoris politus* Montandon, 1895. One male was found concealed in the interrhizon during the late *seca* bearing eggs attached to its dorsum. Apparently, it is a secretive species. Hence, it is seldom seen, although it may be fairly common in the Pantanal.

Larval specimens, most similar to the description of *Pelocoris poeyi* (Guérin-Méneville, 1835), were observed among the floating vegetation in the shallow pond beside the Rio Pixaim during the early rainy season. Unfortunately, the taxonomy of this genus in South America is in need of revision, and Nieser (1975) discussed the possible synonymy of several species known from Brazil. Thus, it is presently not possible to provide a name for the species with a fair degree of certainty that the name is valid. *P. poeyi* has been reported no farther south than Pará, but it may not be distinct from *Pelocoris binotulatus* Stål, 1862, which is known from Rio de Janeiro but has not yet been reported to the west of that state. The larval specimen collected near the Rio Pixaim, however, lacks any sign of the red patch on the dorsal side of the femur, which is thought to be a distinctive characteristic of *P. binotulatus*, according to Nieser (1975). However, it is not known whether this character is an adult trait which has not yet been developed by the larvae.

Hydrometridae

The only hydrometrid species identified in the Pantanal was *Hydrometra argentina* Berg, 1879, an abundant species on the surface tension layer among emergent plants during the rainy season. Adults were most abundant during the early *vazante* when the water level in the wetland was still rather high. These water measurers were almost never found on the rivers and streams of the Pantanal but congregated rather on the flooded plains that largely dried up during the *seca*.

These insects are predators on small arthropods with live in and among the pseudoterrestrium. Like other water bugs, they seem immune from all attacks by predatory fishes and hence can move freely over open areas of the water during daylight hours.

Veliidae

The most abundant species of this family in the Pantanal seems to be *Microvelia mimula* White, 1879, which is abundant on the surface of the water among emergent and floating plants during the periods of flooding. The species was observed in this kind of habitat both on the floodplain and along the water courses at locations where there was no water flow. It generally moved freely across the surface tension layer, taking the smallest insects as its prey and leaving the larger ones to the other heteropterans with which it shared the habitat. Like the other surface-dwelling members of its order, it seems immune from predation by fishes, apparently because of unpalatable substances it secretes.

The veliids are also well represented in the permanent streams flowing into the Pantanal as well as in the wetland itself. The most abundant species in this family belong to the genus *Rhagovelia*, characterized by a structure on the tarsi of the middle legs that permit these insects to move extremely rapidly on the surface tension layer of the water.

Most of the *Rhagovelia* species congregate on the surface of small, rather fast flowing streams. The standing water of the Pantanal is not at all attractive to them, and they are extremely rare in the wetland. In contrast, they are ubiquitous on the small water courses in the Pantanal drainage area. Among the most common local species are *Rhagovelia whitei* (Breddin, 1898), observed regularly along the streams near Chapada dos Guimarães, and *Rhagovelia amazonensis* Gould, 1931, which occurred together with the former species on the surface of the Corrego Guanandi near Cangas a few kilometers from the northern borders of the Pantanal.

Gerridae

The gerrid fauna of the Pantanal was relatively rich, but as in the case of the veliids, not nearly as rich as that in the upland streams. Among the gerrid fauna encountered in water courses upstream from the Pantanal is a species among the largest in the family, *Cylindrostethus palmaris* Drake and Harris, 1934. It was encountered frequently in streams near Chapada dos Guimarães and in the Corrego Guanandi. Rather large groups moved on the surface of the water above deep pools, where the current was not quite as fast as elsewhere. A few members of this species were also observed in the Pantanal itself during the rainy season, but it is likely that these isolated individuals had simply reached the standing water on the floodplain during nocturnal flights and would probably either return to the upland streams for reproduction or perish.

Among the most abundant gerrids inhabiting the Pantanal itself include species of *Limnogonus*, which were present in large swarms, including immature individuals, both on the floodplain and in the upland streams. *Limnogonus aduncus* Drake and Harris, 1933, was one of the most abundant species in the region, being encountered on the surface of all kinds of water bodies, both temporary and permanent. The species is one of the most widespread members of the genus in South America, occurring from Trinidad southward to Venezuela, Guyana, Surinam, Ecuador, Paraguay, and throughout Brazil (Drake and Harris, 1934), and recently being reported from Argentina, as well (Crespo, 1984).

Similarly widespread was *Limnogonus ignotus* Drake and Harris, 1934, which was observed both in the Pantanal and in the upland streams. The type specimen was obtained in Paraguay, and the species seems to have its center of distribution in the region of the Pantanal, as it also occurs in Bolivia and Argentina.

A smaller species, *Limnogonus* cf. *franciscanus* Stål, 1859, was also encountered in large numbers on the surface of various water bodies in the Pantanal. More commonly, these water striders congregated on the surface of streams and small rivers draining the Pantanal during the dry and very early rainy seasons, when the water was not at all flowing.

Limnogonus lotus White, 1879, which occurred in small groups on the surface of pools near the springs at the headwaters of rivers arising on the plateau to the east of the Pantanal as well as near the shore of the Rio Bento Gomes at the Poconé exposition park, was one of the least abundant species in the region.

One species of *Rheumatobates* was encountered on the floodplain in the northern part of the Pantanal. It generally formed rather large groups in shaded locations beside hummocks and at the edge of gallery forests.

Halobatopsis sp. was abundant at times, both on the surface of small upland streams and on pools of water on the floodplain during the *cheia* and through the rainless months until the pools began to dry up.

Homoptera

Cicadidae

The species in this family are overwhelmingly terrestrial and are therefore only mentioned here. A few of the species occur in the supralittoral zone of the Pantanal water bodies, where their favorite food plants apparently occur. One species was abundant on trees in the cities of Cuiabá and Cáceres but was observed in the Pantanal only in *cerrado* at the edges of the wetland.

Membracidae

This family was represented by at least two species in the Pantanal. One of them, *Membracis* sp., was abundant on the supralittoral woody vegetation along the shores of the Baia Piuval. Its body is brightly patterned with dark brown, yellow, and reddish markings, and it has a high, conspicuous crest on the thorax. The second species, probably a member of the genus *Centrotus*, was found only in near the Rio Paraguay in Cáceres.

Cicadellidae

In contrast to the cicadas, the cicadellids play an important role in the aquatic ecosystem. A variety of species inhabit the pseudoterrestrium, where they feed on various aquatic plants. Other species appear frequently in the littoral zone on various species of plant.

Aphididae

Like the cicadellids, aphids are among the prominent small herbivores of the pseudoterrestrium. Although often seen water plants collected in the wetland, they never become abundant, apparently because of the presence of two abundant coccinellid beetles in their biotic community (see below). The host of one of the species appeared to be *Azolla mexicana* because those aphids were often found on that species and were not observed on other plants. In addition, many of the emergent rooted species on the flood plain were frequent food sources of the aphids.

Neuroptera

The only aquatic member of this order is the corydalid species, *Corydalus primitivus* Weele, 1909, which was never encountered in the wetland but was moderately abundant in the upland streams on the plateau to the east of the Pantanal, in spite of the low ion content of the water (Sections 5 and 6). This is a large species, rarely found as an adult but with a larva that is a common faunal element of the streams near Chapada dos Guimarães and various tributaries of the Rio São Lourenço.

Trichoptera

Caddis flies were surprisingly rare in the Pantanal, apparently because of their inability to adapt to the periodic desiccation, high temperatures, and great seasonal predator pressure from the fishes. The adults collected were invariably encountered near permanent rivers. They seemed to be much more abundant in the small upland streams than in the vast wetland, where larvae were never observed.

Hydropsychidae

The only larvae observed in the region of the Pantanal were those of *Smicridea* sp., which were never seen in the wetland but were abundant in the Corrego Guanandi at the outlet of a small impoundment used as a source of cooling water for an alcohol refinery. Because of the presence of the impoundment, there was a continuously strong flow of water the year round. During the last year of the project, 1994, the impoundment was drained, and the habitat for this species dried up.

In addition to the Smicridea species, adult *Macronema* cf. *santaeritae* Ulmer, 1905, were collected in light traps during early evenings of the enchente. Larvae were never observed.

Psychomyiidae

Adults of one species in this family, *Austrotinodes* sp., were collected in light traps at night along the Transpantaneira Highway near Porto Jofre. The kind of habitats in which the larvae develop are unknown, but there are both lowland marshes and gallery forests in the region, and the Rio Cuiabá itself was not far from the site of collection.

Lepidoptera

Nocturnal collections of insects in and near the Pantanal invariably yielded large numbers of moths. Daytime observation also confirmed that a large variety of butterflies are present in the wetland. During the rainy season, vast numbers congregate along the Transpantaneira Highway seeking salt left with the excrement of the many cattle. A long list of species known from the region was provided by Brown (1986). This fauna, which includes many small nocturnal species, merits detailed investigation. However, because the vast majority of the species live on the terrestrial plants, they are not considered further here. However, mention must be made of the aquatic species present and their role in the wetland ecosystem, which has been largely overlooked in the past. Two exceptions, however, have been given attention worldwide as possible agents to control various water plants, as discussed below.

Various systems of classifications have been proposed for the genera in the family Pyralidae. Here, the system recently adopted by Bachmann (1995) is employed, maintaining the family intact and regarding the groups other authors have raised to the rank of family as subfamilies.

The family Pyralidae *sensu lato* has long been noted for including aquatic species, which are able to carry their air supply underwater to feed on aquatic macrophytes (Heckman, 1983a). Most of the species in the Pantanal, however, are inconspicuous members of the pseudoterrestrium, living on or mining the emergent rooted and floating plants.

Larvae tentatively identified as *Samea multiplicalis* Guenee, 1874, develop into a moth in the subfamily Pyraustinae which consumes much *Pistia stratiotes*. Consumption by the caterpillars, which usually conceal themselves near the bases of the leaves during the day (DeLoach *et al.*, 1978; DeLoach & DeLoach, 1979), seems to account for the relative scarcity of this species in the Pantanal. The grasshoppers that feed on *P. stratiotes* generally do not seem to consume it so ravenously as to eliminate the plants. *Sameodes albiguttalis* (Warren, 1889), is another pyraustine moth of the pseudoterrestrium that consumes the water hyacinth, *Eichhornia crassipes*. It occurs sporadically where this species is present. Its habits have been extensively studied because it has been considered as a candidate for introductions to other continents to control the water hyacinth (DeLoach and Cordo, 1978).

Coleoptera

Gyrinidae

The gyrinid fauna of the Pantanal consisted of enormous numbers of individuals belonging to relative few species. Problems in the specific identification were aggravated due to poor descriptions, primarily by two entomologists, who worked mainly during the first three quarters of this century. They saw no need to provide illustrations and seldom described essential morphological structures, such as the male genitalia. The fact that large numbers of the type specimens were apparently destroyed during the Second World War makes it likely that some of these species may never be recognizable with certainty again.

By far the most abundant species in the Pantanal could be identified with a reasonable amount of certainty as *Gyretes levis* Brullé, 1837, which formed dense swarms on the surface of the water in various parts of the Pantanal. Much less common was *Gyrinus ovatus* Aubé, 1838, which was observed at only one location near the Fazenda Ypiranga during the early dry season. In the Rio Bento Gomes near the exposition park of Poconé at the edge of the Pantanal, *Gyretes burmeisteri* Ochs, 1929, was also observed.

A particularly interesting species was encountered only on the surface of streams near the cliffs separating the Pantanal plain from the plateau to the northeast near Chapada dos Guimarães. It is among the largest of all gyrinids, *Enhydris tibialis* Régimbart, 1876, which was originally described from Chapada and has since been found in isolated locations near the Pantanal and as far east as Goiás. Three congeneric species of typically large size are found in geographically isolated locations in eastern and northern South America.

In addition to the species observed during the study, at least five other species occur in the region of the Pantanal. These are *Gyretes multisetosus* Ochs, 1929; *Gyretes dubia* Ochs, 1929; *Gyretes angustipes* Ochs, 1965; *Gyretes impiger* Ochs, 1965; and *Gyretes angustata* Régimbart, 1883.

Dytiscidae

Beetles regarded by some as belonging to a separate family, Notaridae, are included arbitrarily in this family. Generally, English and American authors treat notarine dytiscids as members of a separate family, Notaridae, while Continental authors, such as Guignot (1961) and Freude *et al.* (1971), have treated them as a subfamily of the Dytiscidae. In any event, they will be discussed separately from the other dytiscids.

Many Dytiscids are known from the southern parts of the Pantanal, since a large collection of specimens was obtained from the vicinity of Corumbá at the early part of the century. Many were described by Zimmermann (1921). Most of the others are species known to be widespread in South America east of the Andes.

Notarinae

The notarine species are numerous in lowland Mato Grosso, and during the study, many were encountered in the Pantanal, usually among the interrhizon community. Both adults and larvae were easily collected by removing the floating plants, especially *Salvinia auriculata*, and slapping the roots or rhizoids on the surface of a flat pan. The insects that were shaken out could be collected before they were able to retreat again into the rhizoid masses. The larvae still cannot be identified, but several species of adult can be recognized from the descriptions in the literature, especially those by Sharp (1882), which are still the best for several of the species. Among the extremely common species encountered are *Suphisellus grammicus* (Sharp, 1882), shown in Plate 6a; *Suphisellus rufipes* Sharp, 1882; *Suphisellus curtus* (Sharp, 1882); *Suphisellus rufulus* Zimmermann, 1921; *Suphisellus hieroglyphica* Zimmermann, 1921; *Suphisellus similis* Zimmermann, 1921; *Hydrocanthus debilis* Sharp, 1882; *Hydrocanthus laevigatus* Brullé, 1836; *Hydrocanthus socius* Sahlberg, 1844; *Suphis minutus* Régimbart, 1903; and *Canthydrus ovatus* Régimbart, 1903. All but three of these species were encountered abundantly in the interrhizon. *S. hieroglyphica* and *S. similis* were encountered in the sediment near the shores of small streams and rivers, while *S. curtus* was found only beneath street lights shortly before the freshet.

In addition to the species observed, *Notomicrus gracilipes* Sharp, 1882, *Notomicrus traili* Sharp, 1882, *Notomicrus malkini* Young, 1978, *Notomicrus brevicornis* Sharp, 1882, and *Mesonoterus crassicornis* Régimbart, 1889, are known from Mato Grosso, mainly from the region of Cuiabá, where many of the type specimens were obtained during the 19th century.

Laccophilinae

The genus *Laccophilus* is represented by several abundant species in the Pantanal, most of which are found concealed within the interrhizon community. When dislodged from their shelters in the roots or rhizoids, they swim rapidly to seek another hiding place in the root masses. This behavior pattern apparently helps them escape the intensive predation by the many small fishes in the water bodies. The species in this subfamily that were encountered most frequently in the interrhizon community during the study are *Laccodytes* cf. *phalacroides* Régimbart, 1895, and at least three species of *Laccophilus*. In addition, several species were observed in shallow pools of water on the flood plain during the dry season.

Hydroporinae

Among the species encountered in the interrhizon are a few tiny hydroporine species, which apparently occur in enormous numbers in the Pantanal. These include *Anodocheilus silvestrii* Régimbart, 1903, and *Anodocheilus sarae* Young, 1974, which can be found in almost every stand of floating plants during the rainy and early dry periods. The small size of these beetles, only slightly longer than two and one millimeter, respectively, permits them to effectively conceal themselves within the root and rhizoid masses. Individuals were also collected from soft sediment near the shores of water bodies during the dry season.

In addition to these species, a few larger hydroporines occur in the wetland but are rather uncommon. The adults of most appeared during the *enchente* soon after the first rains. *Pachydrus obesus* Sharp, 1882, was encountered in the Rio Novo near the end of the dry season as well as in rainwater pools on the floodplain just after periods of heavy rainfall. The sediment caked to their elytra was evidence that they had just emerged from their aestivation in the dried sediment.

The adults of several species were encountered at the Poconé exposition park at the onset of the rainy season. They were not found during other seasons of the year. These include *Celina intacta* Zimmermann, 1921; *Neobidessus bokermanni* Young, 1981; *Hemibidessus conicus* Zimmermann, 1921; *Hemibidessus* sp.; and *Macrovatellus* sp. In addition, other species were encountered during the dry season together with many hydrophilids beneath small stones that had been dumped at the site to make the sediment firmer for launching boats. These included *Bidessodes knischi* Zimmermann, 1921, and *Neobidessus corumbensis* Zimmermann, 1921.

Adult hydroporine dytiscids were frequently abundant in the pools of water drying on the floodplain during the seca. All of these species bury themselves deeply in the sediment, coming to the surface only to renew their air supplies and returning rapidly. Where small stones had been placed in the water near the exposition park, these beetles, together with several hydrophilids, moved deep into the interstitial zone for concealment after renewing their air supply. Among the species that inhabited the sediment on the floodplain were *Neobidessus alternatus* (Régimbart, 1889), *Neobidessus corumbensis*, *Bidessonotus* sp., and *Hemibidessus conicus*.

Just before the first heavy rainstorms at the beginning of the rainy season, several hydroporine species were found among the many other aquatic insects flying around the street lights in Poconé or caught in light traps at the northern borders of the wetland. These included *Celina reclusa* Guignot, 1955, and *Anodocheilus silvestrii*.

Colymbetinae

Aglymbus sp., which may still be undescribed, was found near the end of the dry season in a shallow pool of water along the Rio Novo near the Transpantaneira Highway. Apparently, the larvae of the known species of this small genus inhabit the water trapped in species of Bromeliaceae (Scott, 1912).

Dytiscinae

Megadytes laevigatus (Olivier, 1795) was encountered in a shallow-water interrhizon community during the *seca*. The smallest South American species of its genus, it is widely distributed, although there has been much confusion about its identity in the past (Mouchamps, 1957).

Hydrophilidae

The hydrophilids are well represented in the Pantanal, and several species are extremely abundant in the wetland. The most prominent member of this family is *Tropisternus collaris parananus* Sharp, 1883, which has at times been regarded as an independent species but was included by Blackwelder (1944) as a subspecies of the enormously widespread *Tropisternus collaris* (Fabricius, 1775). This beetle is a characteristic species of shallow pools of rainwater that were slightly above the level of the water on the floodplain but persisted throughout the rainy season. A particularly large population was also present along the banks of the Rio Bento Gomes near the Poconé exposition park, where the adults concealed themselves beneath stones and pieces of wood that were dumped at one site to reinforce the banks of the river for the launching of boats. These sites were enriched by large amounts of fish offal left by the fishermen, who cleaned their catch there. The beetles could be seen rapidly coming to the surface for air and returning to the soft sediment or spaces between the stones on the bottom, where they remain fully concealed for long periods of time. These habits suggest that they are endobenthic with the requirement of replenishing their air supplies at the surface, habits that are still unknown among beetles of the temperate zones. Accordingly, the species is encountered only in shallow water, generally with few submerged aquatic plants but often with drowning terrestrial species.

Nearly as abundant on the floodplain as *T. collaris* is *Tropisternus laevis mergus* Say, 1837. The South American representatives of this subspecies of the widespread *T. laevis* (Sturm, 1826) are usually referred to by the synonym, *T. nitens* LaPorte, 1840. To make the matter more confusing, this subspecies is more or less sympatric with *T. laevis laevis* in southern Brazil and Argentina, and the morphological differences between the two are minimal.

The species of *Tropisternus* are among the aquatic beetles that aestivate as adults in the dried mud during the annual *seca*. They appear in the water within hours after the first heavy rains, while their dorsal surfaces are still caked with the sediment in which they were buried during the dry season.

One of the most abundant species encountered on the temporarily flooded areas of the floodplain was *Berosus truncatipennis* LaPorte, 1840, which spends the dry periods as an adult in sediment and emerges in pools of rainwater after the first thunderstorms begin to flood the low-lying areas of the Pantanal. The larvae develop among the interrhizon during the rainy season, and metamorphosis occurs before the water level has decreased very much during the *vazante*. This is the largest of the abundant species of *Berosus* encountered in the Pantanal. It is widely distributed in South America, occurring from Guyana to Argentina (Oliva, 1989).

Several smaller species of the genus are common in the Pantanal, as well. Adult *Berosus fratellus* Knisch, 1921, appeared as soon as the first rainwater pools appeared on the floodplain. Specimens taken near the sampling site at Birici still had caked sediment adhering to the elytra. During the early *enchente*, this species was enormously abundant in many pools of turbid water. They were again abundant in the pools of water left on the floodplain during the *seca*. Between the *enchente* and *seca*, adults were occasionally encountered within the interrhizon together with larvae probably belonging to this species. Another abundant member of the shallow water community on the floodplain is *Anacaena corumbana* Orchymont, 1938, which was originally described from the southern part of the Pantanal, as the name implies. Somewhat less abundant was *Berosus palposus* Knisch, 1921, which was encountered on the floodplain of the Fazenda Ypiranga during the early *enchente*. Apparently, this species was previously known only from the region of Corumbá in Mato Grosso do Sul (Oliva, 1989). *Berosus chalcocephalus* Germain, 1865, and *Enochrus* sp. were also encountered in drying pools of water on the floodplain during the late dry season, but they were not as abundant as the other species of the family.

Other species of the genus were abundant along the shores of the Rio Bento Gomes, where small stones had been dumped to make the sediment firm for launching boats, as already described. Like the *Tropisternus* species, *Berosus subtilis* Knisch, 1921, and *Berosus pleuropunctatus* Mouchamps, 1963, lived in the interstices among the stones. They were able to penetrate much deeper into the crevices than the larger *Tropisternus* species, constituting an unusual kind of interstitial fauna consisting of species that were forced to return periodically to the surface to exchange their air supplies. Occasionally, the species were also encountered among *Salvinia auriculata* rhizoids. Their trips to the surface were made rapidly, and they were never encountered for long in the open water or above the sediment. Otherwise, they would have quickly been attacked by one of the many predatory fishes in the habitat.

Among the plants in the shallow water zone of the Rio Paraguay on the Fazenda Santo Antônio das Lendas, *Berosus masculinus* Knisch, 1921, was found to be fairly abundant during the early rainy season.

One of the species found almost exclusively in the interrhizon was *Derallus rudis* Sharp, 1887, which has a body form ideally suited for moving with the ventral side against the surface tension layer. The species was quite common anywhere sufficient vegetation was available on the surface of the water to provide these insects with possibilities for concealment.

The species *Helochares femoratus* (Brullé, 1838) was encountered in the Rio Novo near the Transpantaneira Highway in September, 1993. Adults were encountered at the surface of the water among dense beds of floating plants. The species has previously been reported from the states of Corrientes, Argentina, and Rio Grande do Sul, Brazil (Orchymont, 1939). Beetles in the genus *Helochares* differ in behavior from most other members of the family in carrying their eggs on their undersides (Richmond, 1920).

A species common on the flooded areas during the rainy season was *Paracymus* cf. *sedatus* Wooldridge, 1973, the adults of which appeared is shallow water during the *enchente* and were not encountered again until the late *vazante* or early *seca*. The species was formerly known only from Argentina (Wooldridge, 1973).

Enochrus lampros Knisch, 1924, was not found in the wetland, but it occurred in the upland streams of the catchment area.

The largest hydrophilid species found in the region of the Pantanal is *Hydrophilus ensifer* Brullé, 1838, often referred to by its synonym, *Stethoxus ater* (Oliver, 1792), which was most frequently found flying around street lamps at night during the mass flights of aquatic insects observed just before the freshet. This beetle was most abundant in the district of Barão de Melgaço and occurred in the suburbs of Cuiabá, as well. In contrast, it was not common in the region of Poconé, although a number of specimens were also found flying around the street lamps in that town, as well. The species is widespread, occurring from Central America and the West Indies to Argentina (Bachmann, 1969).

Coccinellidae

Two coccinellid species were abundant in the pseudoterrestrium community. The more abundant of the two, *Coleomegilla maculata* (De Geer, 1775), was particularly common on the emergent parts of the *Eichhornia* species. The other species probably belongs to the genus *Adelia*. Adults of both were found during all seasons.

Lampyridae

Aquatic lampyrids were especially prominent in the Pantanal at night during the dry season. Their identification is still problematic because their systematic position has not been satisfactorily determined. They are tentatively assigned by various authors (Trémouilles *et al.*, 1995) to the genus *Photinus*. The larvae live in the interrhizon community, generally beneath *Salvinia auriculata*, during the day. At night, they creep out onto the upper fronds and glow. During the *vazante* and early *seca*, they make a darkened mat of *Salvinia* appear like a reflection of the sky. There is also at least one species, but probably two, which have terrestrial larvae that glow on the soil. These can be distinguished by the intensity of their light as well as by their habitat.

The larvae are apparently predators that feed nocturnally on invertebrates at the surface of the water. The adults are briefly quite abundant during the dry season, when the males undertake their annual mating flights shortly before the first rains.

Chrysomelidae

Several aquatic plants of the Pantanal are typically consumed by chrysomelid beetles. The most common of these were observed regularly on species of *Ipomoea*. They belong to at least two species of the Cassidini, probably both to the genus *Coptocycla*. A specimen apparently belonging to a third species in this subfamily was also collected during the study. Large numbers of adults and larvae of these species were observed on the Fazenda Ypiranga about 15 km south of Poconé during the late *seca* and early *enchente*, but at least a few adults were encountered during each of the seasons.

Among the other chrysomelids observed in the Pantanal was a species in the subfamily Halticinae. It seems to live on littoral plants, on which it was encountered along the Rio Paraguay near Cáceres.

Curculionidae

As expected, this family was best represented by terrestrial species in the region of the Pantanal. However, there are also several aquatic weevils that are important components of the wetland communities, especially in the pseudoterrestrium community. One of the most prominent of these is *Ludovix* cf. *longirostris* Pascoe, 1870, a species with unusual habits for the family, consuming the eggs of other insects rather than plant juices (Trémouilles *et al.*, 1995).

Two less common species in the Pantanal have received considerable attention as possible agents to control *Eichhornia crassipes* around the world. Julien (1987) listed *Neochetina bruchi* Hustache, 1928, and *Neochetina eichhorniae* Warner, 1970, as insects that could be used to control this frequently introduced aquatic weed. They have already been imported into Florida for this purpose (Spangler, 1981), a practice that has proven extremely dangerous in the past. Both of these weevils have specialized in feeding on members of the Pontederiaceae.

Diptera

Tipulidae

The Pantanal supported surprisingly few tipulids. Those few encountered belonged exclusively to the subfamily Limoniinae. Specimens of *Teucholabis* cf. *jocosa* Alexander, 1913, and *Limonia (Dicranomyia)* cf. *distans* were collected in light traps set up by placing white trays of ethanol beneath the headlamps of the project car along the Transpantaneira Highway and at other sites throughout the northern Pantanal.

The tipulid fauna known from the Western Central part of Brazil numbered only 40 species when the list of Alexander and Alexander (1970) was compiled. Only 10 or these belong to the subfamily Tipulinae; the rest are limoniines. These species are listed in Table 8.2.

Chaoboridae

During the study, only one chaoborid species, *Chaoborus* sp., was encountered. Adults were collected among floating plants, mainly *Eichhornia crassipes*, along the shores of the Baia Recreio in Barão de Melgaço.

Culicidae

Among the enormous numbers of mosquitos that made themselves evident during nearly four years of excursions into the Pantanal, only three species were represented. This is considerably fewer than the number of species encountered in the city of Cuiabá and fewer than the number in the *cerrado* region, as well. It is noteworthy that none of these species seem to transmit diseases to man, although one of them, *Anopheles (Nyssorhynchus) albitarsis* Arribalzága, 1878, is often listed as a malaria vector (Pessôa and Martins, 1988). Since malaria seems never to have occurred in the Pantanal, it is evident that these mosquitos do not transmit it. Whether they can or not has not been determined, but lacking a pool of infected mammalian hosts, this question cannot be answered at the present time. According to Lane (1953), the subspecies *A. a. albitarsis* never enters houses and is not known to carry malaria, while the subspecies *A. a. domesticus* Galvão and Damasceno (1944), known to range from Pará to Rio de Janeiro, enters houses and acts as a malaria vector. Although mosquito taxonomy has tended to become overly complex since 1953, which may well be a result of a lack of clarity of what constitutes a species, the general concept of Lane (1953) to treat these two forms as subspecies seems reasonable.

Table 8.2. The tipulid species known from the Central Western region of Brazil (Alexander & Alexander, 1970).

Tipulinae
- *Brachypremna breviventris* (Wiedemann, 1821)
- *Tipula armatipennis* Alexander, 1912
- *Tipula decolorata* (Alexander, 1935)
- *Tipula eurymera* Alexander, 1945
- *Tipula guato* Alexander 1912
- *Tipula schwarzmaieri* Alexander, 1940
- *Tipula smithi* Alexander 1912
- *Zelandotipula laevis* (Alexander, 1914)
- *Zelandotipula ringens* (Alexander, 1935)
- *Zelandotipula tarda* (Alexander, 1935)

Limoniinae
- *Helius larotypa* Alexander, 1926
- *Helius pallidipes* Alexander, 1926
- *Helius rubicundus* (Alexander, 1922)
- *Limonia (Dicranomyia) distans* (Osten Sacken, 1859)
- *Limonia (Geranomyia) biargentata* Alexander, 1929
- *Limonia (Geranomyia) cinereinota* (Alexander, 1913)
- *Limonia (Geranomyia) immerita* Alexander, 1929
- *Limonia (Geranomyia) tumidibasis* Alexander, 1938
- *Limonia (Peripheroptera) imcommoda* (Osten Sacken, 1887)
- *Limonia (Rhipidia) annulicornis* (Enderlein, 1912)
- *Limonia (Rhipidia) bruchiana* Alexander, 1929
- *Limonia (Rhipidia) shadei* Alexander, 1929
- *Atarba punctiscuta* Alexander, 1922
- *Ctenolimnophila bivena* Alexander, 1921
- *Epiphragma imitans* Alexander, 1913
- *Epiphragma pupillata* Alexander, 1913
- *Hexatoma goyazensis* Alexander, 1940
- *Hexatoma lopesi* Alexander, 1942
- *Lecteria mattogrossae* Alexander, 1913
- *Polymera microstictula* Alexander, 1929
- *Polymera obscura* Macquart, 1838
- *Polymera regina* Alexander, 1926
- *Polymera superba* Alexander, 1913
- *Gnophomyia pallidapex* Alexander, 1929
- *Gonomyia schadeana* Alexander, 1935
- *Molophilus sabethoides* Edwards, 1927
- *Neognophomyia immaculipennis* (Alexander, 1926)
- *Sigmatomera rufithorax* (Wiedemann, 1828)
- *Sigmatomera occulta* Alexander, 1914
- *Teucholabis jocosa* Alexander, 1913

Anopheles albitarsis is a species active as adults in the Pantanal solely during the dry season. The females are crepuscular and seek blood meals most intensively at sundown. However, they will occasionally feed almost any time during the night, provided the air temperature is not too low. During the periodic *friagems*, or cold spells, these mosquitos are generally not encountered. Rarely, if the weather is not too hot and the sky overcast, these mosquitos may attack during the day. On hot days, however, none of these insects are seen in the open. Within a few days after the freshet, the adults disappear from the Pantanal.

One of the most aggressive mosquitos in the Central West region of Brazil is *Psorophora (Janthinosoma) varipes* (Coquillett, 1904). In the Pantanal, these mosquitos attack particularly savagely during the midday hours, when enormous numbers can emerge from woods or thickets and swarm over a person or other mammal at once. On occasion, they also feed at night, but with little vigor. The individuals are quite variable in size, and they are frequently thought to be members of more than one species by casual observers. However, on examination under the dissecting microscope, they invariably prove to belong to the same species.

Psorophora varipes is a species primarily of the rainy season. The adults prefer high air humidity and become rarer after the rains have ceased. When the *seca* begins, adults can no longer be found. There is only a short period of overlap between the occurrence of the adults of *Anopheles albitarsis* and *Psorophora varipes* at the end of the *vazante*, usually in May or June, so a person is usually bothered by mosquitos either during the day or early in the evening but seldom during both day and night. From shortly after the freshet until the rainy season begins in earnest, adults of both species may be absent from most of the Pantanal.

The third mosquito species present in the Pantanal, *Culex (Culex) maxi* Dyar, 1928, is observed in the adult stage only for a short period of time during the period immediately following the freshet. The adults are most common near streams and rivers, but their abundance never approaches that of either of the two other species. It is likely that this species of *Culex* seeks hosts other than man.

The Pantanal obviously supports few mosquito species. In contrast, the *cerrado* area supports a greater number of species, including those found in the Pantanal and, in addition, *Psorophora (Janthinosoma) ferox* (Humboldt, 1820) and *Sabethes (Sabethes) belisarioi* Neiva, 1908. The greatest species diversity of the Culicidae is found in the city of Cuiabá, where species generally associated with man in the tropical regions of the world are encountered, as evidenced by recent outbreaks of yellow and dengue fever. Fortunately, malaria is still not being transmitted in the cities of southern Mato Grosso.

Chironomidae

The chironomid fauna of the Pantanal is not overly rich, apparently due to a lack of an abundant food supply and to the intense predation by small fishes. However, several species are abundant enough to form prominent swarms of males along the Transpantaneira Highway shortly before sundown.

The relative scarcity of chironomid species in the Pantanal is probably due in part to the periodic desiccation of the shallow water habits in which the larvae would be able to develop without extremely high predator pressure from the numerous fishes in deeper waters. One species, however, *Fissimentum desiccatum* Cranston and Nolte, 1996, has developed a tolerance to desiccation and has thereby been able to find a niche in the wetland fauna. Larvae taken from the Rio Bento Gomes were reared in the laboratory, where they were found to live in the sediment and survive even after it had dried and started to crack (Cranston and Nolte, 1996). It is in the tribe Chironomini of the subfamily Chironominae.

Along the Rio Paraguay on the Fazenda Santo Antônio das Lendas, adult *Parachironomus supparilis* (Edwards, 1931) were collected in the evening during the dry season. The species is widespread in South America and has previously been reported from Paraguay Apparently, the larvae develop in shallow water zones of the large rivers.

A somewhat rarer species from the same kind of habitat along the Rio Paraguay is *Goeldichironomus* sp. Some members of this genus develop on the sediment of medium-sized rivers in eastern Brazil, becoming abundant in waters with moderate organic enrichment. In the oligotrophic rivers of the Pantanal, it is not likely that these midges would become too numerous.

The upland streams to the north and east of the Pantanal harbor a poorly known chironomid fauna. Among the few studies conducted on these insects is a report of the presence of *Apedilum elachistus* Townes, 1945, a species previously known only from North America, which can develop from an egg to an adult in less than seven days (Nolte, 1995). In addition, specimens belonging to the diverse genus *Polypedilum* were encountered. The adults were collected during the *seca*.

Ceratopogonidae

Although biting midges of this family are numerous and often annoying along fast-flowing upland streams, they can seldom be found in the waters of the Pantanal. Obviously, the lentic waters and periodic shortages of dissolved oxygen, as well as the abundance of predatory fishes, are limiting factors for this family. During the study, only *Atrichopogon flavipes* Lutz, 1914, a widespread species in South America, was found near the large rivers. The adults of the known species of this genus suck the body fluids of other insects (Wirth and Cavalieri, 1977).

Psychodidae

Psychodids in the Pantanal were associated mainly with human habitations and is mainly found in bathrooms and outhouses. The larvae apparently develop in sewage.

In addition, species of *Psychoda* were collected in light traps from time to time. Blood-sucking sand flies were never observed. Unfortunately, little is known about the South American psychodids that do not suck blood.

Stratiomyidae

Stratiomyids were not abundant in the Pantanal, but they were represented by a single species, found on only one occasion during the study.

Syrphidae

The syrphids were never abundant in the Pantanal, but the adults of a few species were observed occasionally on flowers. In addition to several species with terrestrial larvae, most in the subfamily Syrphinae, the eristaline, *Lycastrirhyncha nitens* Bigot, 1859, was encountered on flowers along the Rio Bento Gomes during the *seca*. The species is widespread in South America (Thompson, 1972).

Tabanidae

Tabanids were represented by a four common species in the Pantanal as several that are seen only occasionally. The one most frequently encountered was *Lepiselaga crassipes* (Fabricius, 1805), a species that attacks man and other mammals. The larvae have been found to develop in association with *Pistia stratiotes* by Dunn (1934) and Fairchild (1940). It is therefore a characteristic species of the interrhizon community.

The second most abundant horsefly that seeks human blood is *Chrysops* sp., the adults of which were found throughout the *seca* and early part of the *enchente*.

Less common but also regularly encountered was *Tabanus* sp., a species that does not seem to seek warm-blooded prey but rather specializes in feeding on the blood of the caimans while they sun themselves of the shore during the day. This horsefly alights on the eyelids of the reptiles and gorges itself on the blood after biting the relatively thin skin between the tiny scales. Apparently, the eyelids are the only vulnerable part of the body because elsewhere the scales are set too closely together. A dead caiman that was starting to decay was covered by a large number of these horseflies, which were obviously attracted to the bloody fluids starting to ooze from the head.

A large species of *Esenbeckia* was seen from time to time during all seasons except the *seca*. From the size of its mouthparts, it is assumed to be able to successfully attack cattle and other large mammals.

Sciomyzidae

Only one specimen of a sciomyzid was collected during the study. The larvae of these flies usually feed on snails.

Dolichopodidae

This large family is represented in South America by many species that frequent water bodies and wetlands. The adults are generally encountered in relatively small numbers among large numbers of other dipterans. Little is known about the larvae, which presumably develop in the sediment of water bodies, waterlogged soils, or plants that hold water at the bases of the leaves. *Diaphorus (Diaphorus) chamaeleon* Becker, 1922, was among the insects encountered in the low Pantanal near Campo Jofre during the late dry season. A second species of *Diaphorus* was found along the shores of the rivers among many ephydrids and ceratopogonids during the *seca*. *Paraclius* sp. was found in similar locations throughout the northern Pantanal during the same season.

Along the upland streams near the northern border of the Pantanal, adult *Pelastoneurus ochreifacies* Van Duzee, 1931, were found in large numbers, which is unusual for a dolichopodid, most other species being rather solitary. Of 15 specimens collected from swarms along the Corrego Guanandi near Cangas, only one was a male, which was remarkable for having a strongly modified antenna. In addition, *Condylostylus* sp. was encountered in moist locations near the Rio Cuiabá upstream from the Pantanal.

Sphaeroceridae

Adult sphaerocerids are common on the shores of the water bodies during the dry seasons. They are rare during other times of the year. The taxonomy of this family has been neglected for a long time. Since Duda (1925) provided his voluminous work on the non-European species, no major revisions of the South American species have been undertaken. The species in the Pantanal are apparently scavengers, which are attracted to the decaying organic material at the edge of the receding water during the *vazante* and *seca*. Most of the adults collected in the Pantanal were found among small flies of other families along the edges of the rivers draining the Pantanal. The most abundant species belong to Duda's subgenus *Collinella*, a taxa which has been subjected to a series of revisions. None of the proposed changes have been generally accepted. The genus *Limosina*, which was used by Duda (1925) as the generic name of the three species encountered in the Pantanal, is now often reduced to a subgenus of *Leptocera*. Using the most widespread taxonomy at the present time, the species encountered in large numbers during the study are *Leptocera paraguayensis* (Duda, 1925), *Leptocera aequipilosa* (Duda, 1925), and *Leptocera striata* (Duda, 1925). These species were found at the same locations along the Rio Bento Gomes, Rio Frio, and Birici Stream during the same time of year, the late *seca*. The larvae of these small flies are apparently important in breaking down the detritus that accumulates along the edges of the water bodies as the water level falls.

Ephydridae

Some ephydrid species are noted for their tolerance of extreme forms of pollution and to elevated water temperatures. Along the edges of the desiccating water bodies during the late *vazante* and *seca*, large numbers of adult ephydrids were collected. The adults of many of these species congregated along the edges of the drying pools of water on the floodplain, and numerous specimens were obtained along the banks of the Rio Bento Gomes. Most of the species belong to the subfamily Notiphilinae, including *Paralimna (Phaiosterna) obscura* Williston, 1896, and *Brachydeutera neotropica*, Wirth, 1964, but the subfamily Ephydrinae was also present, represented by *Scatella obscura* Williston, 1896. *Brachydeutera neotropica* occurs from the United States southward to Brazil. A great many adults were found resting on the surface tension layer of a pool of water in a dirt road in Cáceres, and the species is apparently one that has spread with human beings and benefits from the organic pollution they cause.

Along streams in the upland region near the edge of the Pantanal, *Notiphila (Notiphila) erythrocera* Loew, 1878, was collected on occasion. This is a widespread species that occurs from North America southward to Rio de Janeiro and has accumulated a number of common synonyms, possibly including *Notiphila (Notiphila) punctifera* Cresson, 1917, which is known from Paraguay and Mato Grosso do Sul and which Cresson (1917) believed to be a valid species.

Hymenoptera

Formicidae

The ant fauna in the Pantanal and adjacent regions of Mato Grosso is highly diverse. Because the species are mainly terrestrial, their ecology will not be considered in detail here. However, several species are particularly prominent in the wetland and should be mentioned.

One of the ants often encountered on moist pastures in the Pantanal is *Camponotus* sp., which builds large nests in the termite mounds on fields above the high water level of the floodplain. Another species commonly seen on areas somewhat above the high water level is the leaf-cutter ant, *Atta sexdens* (Linnaeus, 1758), which is frequently seen carrying pieces of leaves or petals to its nest. This material is used to culture the mushrooms on which the ants subsist. The species occurs along the Transpantaneira Highway but is seldom observed near the flooded areas. It is much more abundant in the *cerrado* and in the cities of southern Mato Grosso than in the wetlands.

Paratrechina fulva (Mayr, 1862) generally builds small mounds above their underground nests, which usually extend only a few centimeters above the level of the ground but may exceptionally reach a height of approximately 20 cm (Dubs, 1994). This protects the species from brief periods of flooding at locations near the high water level of the *cheia*. This is one of the ant species that often moves into termite mounds.

Other ants present in the Pantanal include the fire ants, *Solenopsis wagneri* Santschi, 1916, which have been referred to by their synonym, *Solenopsis invicta* Buren, 1972, by most recent authors. These ants were reported to attack hatchling caimans in their nests (Cintra, 1985).

Ichneumenoidea

This superfamily is represented by a number of aquatic species about which little is known. The larvae of many members of this group develop as parasites in other insects. Others develop inside of plants.

Other hymenopterans

The Pantanal supports a rich variety of bees in the family Apidae, and wasps of the family Vespidae, as well as many representatives of other families. The importance of these insects for maintaining the structure of the flora should not be underestimated. Besides the species that pollinate a wide variety of flowers, there are a number of specialized wasps that are apparently the only species capable of pollinating individual plants. The best known of these relationships involve the pollination of certain figs of the genus *Ficus* by wasps that develop enclosed within the flowers.

8.13. Chordata

8.13.1. Pisces

According to the preferred habitat of the species, the fish fauna can be divided into three general groups. The first are the species of the large rivers, which generally do not stray far from the deep channels except perhaps for spawning. Many of these fishes reach a rather large size and are often caught commercially or for private comsumption by the people living near the rivers. They are not typically fishes of the Pantanal but rather occur in those stretches of the rivers that pass through the wetland as well as in rivers elsewhere. Some of these species undertake long migrations (Bonetto and Pignalberi, 1964; Lowe-McConnell, 1987). An interesting phenomenon sometimes observed in the large rivers of the Pantanal is the mass migration of many species together, collectively called a *piracema*, as described by Por (1995). During these events, many large fishes can be observed curning up the surface of the water and occasionally leaping into the air. During the investigation, researchers riding in motorboats were occasionally hit by small fishes that made long leaps, apparently because they had been disturbed by the vibrations caused by the boat's propeller, and many fishes leaped out of the water in the wake of the boat. During the *paracemas*, similar leaps are made by large fishes, and persons have reportedly been hit by fishes weighing many kilograms (Por, 1995). To learn more about the fishes of the rivers, a study confined to the Pantanal would not be sufficient. Many of this group are known to travel both upstream and far downstream from the wetland during their seasonal migrations (Bayley, 1973).

The second group inhabits backwaters, small rivers, and streams draining the Pantanal. They are fishes that participate more directly in the food web of the wetland, often traveling short distances from the streams onto the floodplain to seek food or breed. These are among the species most important in the diet of the caimans, which generally do not encounter the large river species and seldom come into contact with the species of the third group, most of which are too small to contribute much to the diet of an adult caiman. The large piscivorous water birds also rely heavily on these species for food.

The third group encompasses floodplain fishes, generally small in size, which survive near the shores of the permanent water courses or, exceptionally, in the dried sediment during the rainless seasons and migrate throughout the shallow waters of the wetland while they are flooded. These species form an important part of the diet of the small piscivorous water birds as well as some of the larger fish species that live along the edges of the flood plain and occasionally leave the deeper waters to seek food.

Unfortunately, little ecological information on the extremely rich fish fauna of Brazil is available. Almost nothing has been reported on the biology of these fishes in their natural habitats, and the available publications concern mainly those species suitable for aquaria. Comprehensive works with information on all known species in the region are somewhat outdated, the most recent of these being the publications of Fouler (1948, 1959). More recently, useful publications on some of the individual families have appeared, such as the reviews of the characoids by Géry (1977) and the family Cichlidae by Goldstein (1973). Revisions of individual genera are few, but those available are cited in the discussions of the families below. Some meristic and morphometric information has been provided on the commercially important fishes, but this is far from sufficient to permit definitive decisions on their taxonomy, and complete descriptions of their ecological requirements have seldom been attempted. The following section provides some observations on the fish species in the Pantanal and seeks to provide a brief outline of their roles in the ecosystem. The discussion of the known habits of these fishes are provided on a species to species basis below.

The present state of the taxonomic work on the ichthyofauna of Brazil is too unsettled to provide a definitive list of the species that can be encountered in the Pantanal. When revisions are made, some of the names below will undoubtedly become junior synonyms of other species, while others will prove to have been applied to several distinct species that have not yet been distinguished. It is hoped, however, that the identities of the species observed will still be recognizable from their present appellations combined with their ranges within the Pantanal. Many of the smaller species have been collected and preserved and are presently in the collection of the author. In all, Marins *et al.* (1981) estimated the ichthyofauna of the Pantanal to encompass 405 species. Considerably fewer described species have been positively identified from its waters, however, and Ringuelet *et al.* (1967) reported that only 346 species were known from the entire Paraná River System and its neighboring water bodies.

Elasmobranchia

The elasmobranchs are represented in the Pantanal only by the family Potamotrygonidae. Of the three species of freshwater ray, only one was observed during the study. One large specimen of *Paratrygon motoro* (Müller & Henle, 1841), which usually hides in the sediment of the rivers, was found dead on the surface of the Rio Bento Gomes during the massive dry season fish kill in 1992. It measured about half a meter in length and width.

The other species of ray known from the rivers of the Pantanal are *Paratrygon brachyura* (Günther, 1880) and *Paratrygon falkneri* Castex and Maciel, 1963. Most of the published information about these species comes from the Amazon region.

Osteichthyes

Lepidosirenidae

One lungfish or *pirambóia*, *Lepidosiren paradoxa* Fitzinger, 1836, has been reported from the region of the Pantanal (Por, 1995) as well as from the Bolivian chaco. It is an anguilliform species that was not encountered at all in the wetland during the study, and it seems to prefer drier regions, where there is less competition from and predation by other fishes. A maximum length of 125 cm can be reached by this species (Sterba, 1987).

Pristigasteridae

One pristigasterid species, *Pellona flavipinnis* (Valenciennes, 1847), is part of the Pantanal ichthyofauna. It occurs in the large rivers of the Amazon System as well as in the Rio Paraguay and its tributaries draining the Pantanal. This fish is considered to be of low quality as food because of its many small, forked bones, dry flesh, and lack of flavor (Magalhães, 1931)

Characoidei

The superfamily Characoidea is the most species-rich taxon in the Pantanal, and the Characidae and Serrasalmidae are among the most species-rich families in the region. Unlike the cichlids, the majority of the large fishes in this family usually avoid the floodplains and remain in the large and medium-sized rivers that drain the Pantanal, although most make short forays over the flooded banks of the water courses during the high water period. The best known members of this superfamily range from the piranhas, which are predatory fishes that consume relatively large prey, to many small, attractive species that have become well established in aquaria kept by hobbyists in many parts of the world. Most of the species that are locally important food fishes also belong to this group.

Characidae

This family is delimited according to Géry (1977), who raised several of its former subfamilies to familial rank. Although not universally accepted, this classification system permits a better organization of the text, separating the predatory piranhas, for example, from omnivorous and herbivorous fishes.

The subfamily Characinae includes a rather small, predatory fish, *Charax gibbosus* Linnaeus, 1758, which inhabits the tributaries of the Rio Paraguay that drain the Pantanal as well as some of the lakes in and near the wetland. The species was observed in the Rio Bento Gomes downstream from the exposition park of Poconé. It is known to be widespread in lowland, tropical South America. Apparently, this predator has a preference for small fishes, as Sterba (1987) reported that the stomach contents of specimens caught in natural water bodies contained many partially digested scales of other fish species. Furthermore, Knöppel (1972) reported that the stomachs of specimens from the Rio Chapare in Bolivia and the Laguna los Guateques in Venezuela were filled with fishes. However, he also reported that the remains of dragonfly larvae were found in the stomach of one taken from the Rio Purús in Amazonas.

Five species assigned or formerly assigned to the genus *Roeboides* are suspected to belong to the Pantanal ichthyofauna. They are mainly small fishes with somewhat transparent bodies that prefer to live among water plants rather than in the open water of the rivers. They have been reported from the basin of the Paraguay-Paraná River System, but not all of them have been previously reported with certainty from the Pantanal. *Asiphonichthys condei* Gery and Kröppel, 1976, referred to by Sterba (1987) as *Roeboides microlepis* (Reinhardt, 1849), its synonym according to Riehl and Baensch (1991), has been reported from Venezuela and from Paraguay at Asuncion. The other species thought to occur in the Pantanal and water courses in its catchment area are *Roeboides bonariensis* Steindachner, 1879; *Roeboides prognathus* Boulenger, 1895; *Roeboides descalvadensis* Fowler, 1932; and *Roeboides paranensis* Pignalberi, 1975.

The subfamily Rhaphiodontinae is represented by one fish that is sometimes caught by local fishermen. The *dourado-cachorro, Rhaphiodon vulpinus* Agassiz, 1829, is a river fish found throughout the Rio Paraná System. In the region of the Itaipu Reservoir, this species is still abundant but is not at all prized by the fisherman (Agostinho and Zalewski, 1995). In the region of the Pantanal, the *dourado-cachorro* apparently remains in the large rivers and enters the adjacent lakes, but it avoids those areas of the temporarily flooded plains that are located far from the rivers.

The subfamily Triportheinae includes locally common but poorly studied fish species, better suited for aquaria than for the local food market. A common fish in the rivers of the Pantanal is the narrow hatchetfish, *Triportheus angulatus* (Spix, 1829), small schools of which were observed in the Rio Bento Gomes downstream from the Poconé exposition park. It was probably also present in the Rio Pixaim, but none of the specimens observed were collected for identification. This fish feeds primarily on insects, migrates almost continually through the river systems, and spawns for the first time at a length of about 16 cm during the first year of life (Dourado, 1981).

Triportheus sp. is a second, congeneric species found in the same water bodies. It is probably an undescribed species, but its identity will first become clear after a revision of the subfamily is undertaken.

The subfamily Bryconinae is represented by mainly herbivorous fishes that inhabit the rivers. The *piraputanga, Brycon hilarii* Valenciennes, 1849, occurs almost exclusively in the rivers in and near the Pantanal. It apparently feeds mainly in vegetable matter, including the seeds of plants that fall into the water. It grows to a large size and is a food fish with considerable economic importance in Mato Grosso. A second fructivorous species, *Brycon orbignyanus* (Valenciennes, 1849), is known to be an important species for the fishery downstream from the Pantanal, and it probably enters the Pantanal at times, as well. This species disappeared from the region of the Itaipu Reservoir after it was filled, apparently because its feeding grounds in the temporarily flooded areas along the river were permanently inundated after the construction of the dam (Agostinho and Zalewski, 1995). This indicates that the species requires extended floodplain systems for its development. It is widespread in South America and apparently ranges throughout the entire Amazon and Paraguay-Paraná River Systems (Fowler, 1948).

The subfamily Paragoniatinae includes many species suitable for aquaria. One of the commonest of the small fishes that forms large schools in the shallow water zones of the streams draining the Pantanal is *Prionoabrama paraguayensis* (Eigenmann, 1914), which is an omnivore with a preference for animal prey. These fishes give the impression of being extremely hungry as they rapidly dart around examining everything that falls on the surface of the water to see if it is edible. A piece of bread, meat, or any other food item is generally attacked by these fishes within seconds after it is dropped into the water. Small insects in the open water, which the exception of beetles and aquatic heteropterans, are instantly attacked and devoured. Although aquatic heteropterans are safe from attack, apparently due to their chemical repellents, the fishes will rob them of their prey. These and other small fish species with similar habits are responsible for keeping the shallow waters completely free of planktonic invertebrates for much of the year.

Apparently to avoid large fish predators, these fishes congregate in the shallow water and do not venture into deeper parts of the streams and rivers. They move onto the floodplains, as well, as soon as these become flooded in the early rainy season. The species is widespread in the systems of the Rios Paraguay and Uruguay, and a similar species, *Prionoabrama filigera* (Cope, 1870), inhabits the nearby upland rivers in the Amazon watershed (Eigenmann, 1918).

Other fishes suitable for aquaria belong to the subfamily Cheirodontinae. Among the species in this subfamily in the Pantanal is *Cheirodon* cf. *piaba* Lütken, 1874, a small fish that does not exceed 5 cm in length. This is an extremely abundant school fish in the Rio Bento Gomes System in the northern part of the Pantanal, and it was observed in the Rio Pixaim, as well. In aquaria, these fishes consumed small invertebrates and also bit off pieces of larger food objects, if somewhat reluctantly. Their life span seems quite short. The identity of the species is assumed on the basis of its range. Sterba (1987) discussed the difficulties of distinguishing this species from several congeneric ones and provided some notes on synonymy within the genus. *C. piaba* is apparently the only species of *Cheirodon* known from the region of the Pantanal; it also occurs in the Rio São Francisco (Sterba, 1987) and Rio Araguaia Systems (Lowe-McConnell, 1991).

Other small aquarium fishes that prefer water well overgrown with macrophytes along the flooded shores of small rivers in the Pantanal are *Aphyocharax paraguayensis* Eigenmann, 1915, and *Aphyocharax rathbuni* Eigenmann, 1907, both of which seem to subsist almost exclusively on small invertebrates hidden among submerged water plants and in the interrhizon community beneath masses of floating macrophytes. The habits and reproductive biology of *A. rathbuni* have been thoroughly reported by Franke (1980), and the frequent occurrence of testaceans among its stomach contents has been reported by Silva Neto *et al.* (pers. comm.).

The subfamily Stethaprioninae is represented by an abundant school fish of a size suitable for aquaria, *Poptella orbicularis paraguayensis* (Eigenmann, 1907), which reaches a maximum length of 11 cm. Sterba (1987) reported that *Poptella orbicularis* (Cuvier and Valenciennes, 1848) is widely distributed from the Guianas through the Amazon Basin to the system of the Rio Paraguay. In the Pantanal, large schools were encountered in the Rios Bento Gomes and Pixaim. A small school adapted well to a large aquarium in the laboratory, where the fishes fed on solid pieces of food, including fish flesh and boiled eggs, which they collectively tore apart. The maximum life span is assumed to be four to five years, as individuals more than three years old began to show morphological signs of senility. Sterba (1987) reported that 1000 to 2000 eggs are released during spawning, confirming that the characteristics of the species conform to the general life cycle of most Pantanal fish species: fast growth and rapid maturity, high reproductive rate, and a short life span.

The subfamily Tetragonopterinae is well represented in the Pantanal. The genera and species occur both in the river systems and on the floodplains.

The species of *Astyanax* are predominantly river fishes, which usually remain in the large and medium-sized water courses of the Pantanal. The larger individuals are generally observed in small groups rather than large schools. *Astyanax alleni* (Eigenmann and McAtee, 1907) inhabits the Rio Bento Gomes System, including the Baia Piuval. Adult *Astyanax bimaculatus* (Linnaeus, 1758) are found throughout the permanent water bodies of the Pantanal, and schools of juveniles frequent the floodplains during the *chiea* in search of food, which seems to consist mainly of invertebrates. The adults are also known to consume plants (Riehl and Baensch, 1991). Meschiatti (1995) studied the diet of this species in an oxbow lake in the State of São Paulo and found that vegetable matter, including filamentous algae, constitutes a large part of the diet during all seasons. Insects, including Mayflies and terrestrial species, as well as fish scales, were also found in the stomachs. The species is also abundantly represented in other lakes adjacent to the upper Rio Paraná (Galetti *et al.*, 1990), in the Rio Xingu and Araguaia Systems to the north (Lowe-McConnell, 1991), and in rivers of the temperate zone in Argentina (Bistoni *et al.*, 1992).

Among the rather small fishes in the Pantanal that are suitable for home aquaria are the yellow-banded Moenkhausia, *Moenkhausia sanctaefilomenae* (Steindachner, 1907), and three congeneric species: *Moenkhausia dichroura* (Kner, 1858), *Moenkhausia oligolepis* Günther, 1864, and *Moenkhausia intermedia* Eigenmann, 1908. These fishes are not uncommon in the small rivers and streams and sometimes in the areas flooded during the high water period. *M. sanctaefilomenae* is widely distributed throughout the lower Paraná System in Argentina (Cordiviola de Yuan and Oliveros, 1979) and has become an aquarium fish maintained by hobbyists throughout the world (Sterba, 1987). It is apparently the species of its genus most abundant in all parts of the Pantanal. *M. dichroura* is apparently almost as abundant, being known from many parts in the wetland. The reported range of *M. oligolepis* often excludes the Pantanal (Britski, 1969). According to Eigenmann and Myers (1929), *M. oligolepsis* is found in the northern part of South America, and they did not report its occurrence south of the

Amazon System. However, it has been found in the Rio Paraguay System in Paraguay as well as near Corumbá and Cáceres, attested to by specimens in the Field Museum of Natural History. Apparently, the least abundant of all of the *Moenkhausia* species is M. *intermedia,* which may be confined to the southern part of the Pantanal, having been collected and identified with certainty only in Paraguay. There have been a large number of reports of unidentified *Moenkhausia* species near Cáceres and elsewhere in the Pantanal. Knowledge of the distribution of these species within the wetland is therefore still unsatisfactory

Gymnocorymbus ternetzi (Boulenger, 1895) is another denizen of the small streams and rivers in the Pantanal that has become a popular aquarium fish. It is known to breed readily in captivity. Its eggs hatch within 24 hours, and the young fishes feed on rotifers and copepods (Sterba, 1987).

Like several other characid genera, the genus *Hyphessobrycon* is characterized by separate species confined to individual river systems in South America (Sterba, 1987), and *Hyphessobrycon callistus* (Boulenger, 1900) may be endemic to the Paraguay River System. However, there are questionable reports of its presence in the southern Amazon Basin. It is an attractive reddish species with black markings, known as the "Mato Grosso" in Brazil. It reaches a length of only about 4 cm, but most of the individuals observed in the Pantanal were somewhat smaller. They were caught regularly in seives and hand nets beneath plants in water from 10 to 40 cm deep along the Rio Bento Gomes near the Transpantaneira Highway. These species were not the most abundant in the region, but they were regularly encountered at various times during all seasons. Captured specimens were observed to consume small animals in the size range of rotifers and small cladocerans as well as hard-boiled egg yolk. They were reluctant to bite off pieces of fish flesh and other large pieces of food, and they did not seem to feed on plant material.

Another endemic species of a widespread genus is Ulrey's tetra, *Hemigrammus ulreyi* (Boulenger, 1895), which reaches a size of only 5 cm (Sterba, 1987). It is a rather delicate-looking fish that usually forms schools in shallow water during the dry season and moves short distances onto the floodplain during the flood period. A more common, congeneric species with an extensive range in South America is the bassam tetra, *Hemigrammus marginatus* Ellis, 1911, which was occasionally observed swimming in large schools in the Rio Bento Gomes near the exposition park at the edge of the Pantanal and occasionally in the same river near the Transpantaneira Highway. These fishes consume small animals and also tear pieces off large chunks of animal matter, such as fish flesh. They proved to be hardy in aquaria but show morphological signs of senility after only two to three years.

A somewhat larger species suitable for aquaria is *Markiana nigripinnis* (Perugia, 1891), which reaches a length of about 10 cm. It was reported by Sterba (1987) to be omnivorous. The fish apparently inhabits only the rivers, and it was not seen often during the field studies.

The silver tetra, *Tetragonopterus argenteus* Cuvier, 1819, is a widespread species found throughout tropical and subtropical river systems of South America. In the Pantanal, isolated schools of this species were present in the small to medium-sized rivers draining the wetland. It was observed most often in the Rio Bento Gomes downstream from the exposition park of Poconé.

Other species in this subfamily known from the Pantanal include *Piabucus melanostoma* Holmberg, 1891, which generally remains near the bottom of the rivers seeking benthic invertebrates. During the high water season, it leaves the rivers to undertake short raids in search of food on the floodplain. The confusing astyanax, *Astyanax abramis* (Jenyns, 1842), and a congeneric species, *Astyanax* sp. were found occasionally in the Rio Bento Gomes. In addition, *Bryconamericus exodon* Eigenmann, 1907; *Creatochanes melanurus* Eigenmann and Eigenmann, 1891; and *Psellogrammus kennedyi* (Eigenmann, 1903) inhabit the small rivers of the Pantanal. In addition, the presence of *Hyphessobrycon anisitsi* (Eigenmann, 1907) has been reported from the upper Rio Paraguay by Argentinian authors (Cordiviola de Yuan and Oliveros, 1979).

An additional species that has been added to the fauna of the Pantanal, or at least to its drainage area, is *Hyphessobrycon herbertaxelrodi* Géry, 1961, a small fish reaching only 3.5 cm in length. It is apparently known only from the Rio Taquari near Coxim not far from the border between Mato Grosso and Mato Grosso do Sul.

In addition to the characids discussed here, various species have been commonly found in the Rio Paraná System downstream from the Pantanal. Several of them probably occur in the Pantanal, as well. For example, according to Cordiviola de Yuan and Oliveros (1979), *Holoshesthes pequira* (Steindachner, 1882) is also a common species in the upper Rio Paraguay System, and it probably occurs near Cáceres.

Serrasalmidae

The best known serrasalmids in the freshwater bodies of Brazil are the piranhas of the subfamily Serrasalminae, which are represented by several species in the Pantanal. They invariably remain in the relatively deep water of the larger rivers and some of the moderately shallow tributaries, where they are general or specialized predators. From the rivers, the piranhas frequently enter the baias and may conceal themselves beneath the floating vegetation. The most common species include *Serrasalmus spilopleura* Kner, 1860; *Serrasalmus marginatus* Valenciennes, 1847; *Pygocentrus ternetzii* Steindachner, 1908; and *Serrasalmus brandti* (Reinhardt, 1874).

Generally, these fishes are encountered in small groups. Only *Pygocentrus nattereri* Kner, 1860, seems to form relatively large schools, making it a dangerous predator to relatively large animals (Por, 1995). However, this was not among the species observed during the study, although it is known from the region of the Pantanal (Sazima & Machado, 1990). The piranhas are easily captured on hook and line with small pieces of meat as bait, and they are a popular food fish for the poorer members of the rural society. Their teeth form incredibly sharp cutting edges, and they can easily remove a large slice of skin with a single bite. One of the botanists working in the research project was bitten on the finger while trying to remove a piranha from a net and bled profusely for a considerable length of time before he could be brought to a pharmacy in Poconé to obtain an astringent and bandages to stop the bleeding.

As mentioned above, *Pygocentrus nattereri* is a predator that hunts in schools, a habit that has given piranhas a fearsome reputation. In contrast, *Serrasalmus marginatus* generally remains alone, feeding on the fins and scales of other fishes, which it cleanly severs with its sharp teeth. *Serrasalmus spilopleura* seems to be a common species throughout the entire Paraná-Paraguay River System, as it was also found to be abundant in lakes adjacent to the upper Rio Paraná (Galetti *et al.*, 1990) and in rivers of the temperate zone in Argentina (Bistoni *et al.*, 1992). Meschiatti (1995) reported it to be mainly piscivorous.

The subfamily Mylinae includes several species that inhabit the rivers of the Pantanal but frequently forage on the floodplain during the high water period. Two species of *Mylossoma* are known from the region: *Mylossoma orbignyanum* Valenciennes, 1849, and the Paraguayan metynnis or *pacú*, *Mylossoma paraguayensis* Norman, 1928. The species of this genus that have been studied in the Amazon region are omnivores, which feed predominantly on seeds and fruit but also consume arthropods (Goulding, 1981). The diet of the two species in the Pantanal is probably similar.

Other species in this subfamily include *Metynnis maculatus* (Kner, 1859) and *Metynnis mola* Eigenmann and Kennedy, 1903, which are apparently exclusively vegetarian or nearly so (Sterba, 1987). They are river fishes, which apparently do not venture far onto the floodplain during the high water period. These species seem to be most abundant in the lakes along the courses of rivers in the Pantanal, such as the baias near Barão de Melgaço and the Baia Piuval along the Rio Bento Gomes.

The fructivorous *Colossoma (Piaractus) mesopotamicus* Holmberg, 1887, is a commercially important fish species in the Paraná River System. It is one of the species that is locally called *pacú*. Agostinho and Zalewski (1995) suggested that this fish requires temporarily flooded areas for feeding and development. After the flooding of the Itaipu Impoundment in southeastern Brazil, this species disappeared from the section of the river upstream from the dam. Thus, great annual changes in the water level, a characteristic of much of the middle and upper Paraná River System, may be essential for the survival of this species.

Another species also referred to by the name, *pacú*, is *Colossoma (Piaractus) brachypomum* (Cuvier, 1817). Some aspects of the population dynamics of this species in the Rio Cuiabá have been investigated by Ferraz-de-Lima *et al.* (1984a, b), who referred to the species by its synonym, *Colossoma mitrei* Berg, 1895. They found that these fishes swim upstream to spawn, mainly in November but occasionally in October or December, as well. This is a river fish with a relatively long life span for the region. It first matures at an age of three years and lives long enough to reach a size attractive for fishermen.

Myloplus levis (Eigenmann and McAtee, 1907) has also been reported from the region of the Pantanal. Its genus is considered by some authors, including Sterba (1987), to be a subgenus of *Myleus*.

Gasteropelecidae

This family is represented in the Pantanal by a single species, the silver hatchetfish, *Thoracocharax stellatus* (Kner, 1860), large schools of which were observed in the Rio Pixaim during the dry season. For unknown reasons, it was not encountered in the Rio Bento Gomes or other water bodies at the edge of the Pantanal. Like other fishes of this family, *T. stellatus* feeds mainly or exclusively on small invertebrates. The species was originally described from specimens taken from the Rio Cuiabá (Fowler, 1948).

Erythrinidae

Among the most abundant fishes on the floodplain are two species in this family: the golden trahira, *Hoplerythrinus unitaeniatus* (Spix, 1829), and the *traíra*, *Hoplias malabaricus* (Bloch, 1794). During the dry season each year during the study, enormous numbers of these non-migratory fishes were trapped in the shallow pools of water as the Pantanal gradually dried up, and frequent mass deaths were observed of those not first eaten by the numerous piscivorous birds that fatten themselves during their season of plenty. One *Hoplerythrinus unitaeniatus* kept in an aquarium with a small school fishes regularly bit these fishes in half, consuming one half and leaving the rest. Later, it was able to adapt to hard-boiled egg, on which it survived for about three years and continues to survive as this book is being written (Hardoim, pers. comm.). Lowe-McConnell (1991) observed that these fishes also feed on aquatic and terrestrial insects in streams. However, Dourado (1981) reported that insects are much less important in the diet of *Hoplias malabricus* than fishes and prawns, and also that the first spawning occurs at an age of about one year and a length of 22 cm. After this, spawning is continuous. Both species usually seek a hiding place on the bottom in which to rest during the day, coming out only when they are hungry. They are widespread in South America. *Hoplias malabaricus* has also been encountered in lakes adjacent to the upper Rio Paraná (Galetti *et al.*, 1990) and is apparently distributed throughout the entire Paraná-Paraguay River System (Meschiatti, 1995). It also inhabits rivers of the temperate zone in Argentina (Bistoni *et al.*, 1992). Both species are also common in the Xingu and Araguaia River Systems on the Amazonian side of the low hills dividing the watersheds (Lowe-McConnell, 1991).

The survival of *Hoplias malabaricus* on the floodplain during the dry season is apparently facilitated by its tolerance to low oxygen concentrations. Parma de Croux (1983) demonstrated that these fishes first begin to show a reaction to low oxygen concentrations below 2 ppm, and the lethal concentration is from 0.3 to 0.8 ppm at temperatures from 13°C to 28°c.

Characidiidae

Only one species in this family is known from the Pantanal. Unfortunately, the genus *Characidium* is in need of taxonomic revision, and the species present could only be tentatively assigned to *C. fasciatum* Reinhardt, 1866, which, according to Sterba (1987), is still known with certainty only from northern South America, including the Amazon Basin. Only a few individuals were captured, most in shallow water among plants along the Rio Bento Gomes near the Transpantaneira Highway. They reportedly feed on microscopic invertebrates and reach a maximum length of 10 cm (Sterba, 1987). Even when kept in a large aquarium, these fishes always attempted to escape during the night and were able to find the tiniest spaces between the frame and the glass cover of this aquarium to flee.

Lebiasinidae

The taxonomy of the species in this subfamily is somewhat unsettled. Only one species was found during the study: a member of the subfamily Pyrrhulininae. Originally described as *Pyrrhulina australis* Eigenmann and Kennedy, 1903, the taxon has now been demoted to a subspecific status as *Pyrrhulina brevis australe*, but there is still much confusion about the taxonomy of the *P. brevis* group (Géry, 1977), the fishes of which are known as short-lined pyrrhulinas. These fishes live among submerged plants along small upland streams flowing toward the Pantanal but were not observed in the wetland itself. They deposit their eggs on the leaves of plants and apparently defend them (Sterba, 1987). The species is also known from the Rio Xingu and Araguaia Systems (Lowe-McConnell, 1991).

Hemiodidae

This family is known to be represented in the Pantanal by two species. *Hemiodus orthonops* (Eigenmann and Kennedy, 1903) forms small schools in some of the rivers and streams that drain the wetland. These fishes seem to be omnivorous with a preference for invertebrates (Sterba, 1987), but no specific information is yet available. The Paraguay darter, *Parodon affinis* Steindachner, 1879, is a benthic fish that was found occasionally in the Rio Bento Gomes near the Transpantaneira Highway. It has sometimes been placed in its own family, Parodontidae.

Anostomidae

In the Pantanal, the species in this family belong to the subfamily Anostominae. *Leporinus lacustris* Campos, 1945, is a widespread species in the Rio Paraná-Paraguay System. Galetti *et al.* (1990) reported it to be a fish that prefers standing water, including lakes connected to but largely isolated from the rivers. *Leporinus striatus* Kner, 1858, inhabits the Pantanal, but it is not abundant, at least in the northern part. It is extremely widespread in tropical South America. Knöppel (1972) examined the stomach contents of specimens from Ecuador and Colombia and found only an oligochaete, ants, and a terrestrial weevil.

The presence of several other species in this family has been reported from the Pantanal but could not be confirmed during the study. The species, *Abramites hypsilonotus* (Günther, 1868), is probably represented in the water courses of the region by the subspecies, *Abramites hypsilonotus ternetzi* Norman, 1926. These fishes feed mainly on algae, soft sprouts of macrophytes, and small animals (Riehl and Baensch, 1991).

Frederic's leporinus, *Leporinus frederici* (Bloch, 1795), is said to be a sponge-eating fish that has been reported from the Pantanal (Carvalho, 1986) as well as from lakes adjacent to the upper Rio Paraná System (Galetti *et al.*, 1990). However, Sterba (1987) believes that this species is not found south of the Amazon Basin. Reports of its presence in the Pantanal may be due to confusion with *Leporinus tschenaulti* Cuvier and Valenciennes, 1848, a species found from the Guianas through the Rio Paraguay System that closely resembles *L. frederici*. However, the presence of *L. frederici* in the upstream sections of rivers in both the Xingu and Araguaia River Systems (Lowe-McConnell, 1991), where many other Pantanal species also occur, suggests that there is a strong possibility of its presence in the Pantanal. Neither of these species was captured during the study. *L. frederici* kept in artificial water bodies in northeastern Brazil were reported to be omnivorous fishes, which consumes algae, seeds, and molluscs, with no mention made of sponges. It is known to reproduce after reaching a length of 16 cm during its first year of life (Dourado, 1981). Specimens reported to be *L.* cf. *frederici* from Surinam were found to have fed on decapods, including prawns, and probably snails (Knöppel, 1972). Better diagnoses of this species will be required in the future to permit better determinations of its ecological characteristics and range, as well as those of the congeneric species with which it may be confused. These include *Leporinus macrocephalus* Valenciennes, 1849; *Leporinus silvestrii* Boulenger, 1902; and *Leporinus gilberti* (Valenciennes, 1847), all of which may be among the Pantanal ichthyofauna.

Reports of *Schizodon fasciatus* Agassiz, 1829, from the Pantanal probably resulted from its confusion with *Schizodon borelli* (Boulenger, 1900), which is one of the species known from the upper Rio Paraguay System. Another congeneric species, *Schizodon isognathus* Kner, 1859, is also found in rivers of the wetland. Both of these species are generally confined to the larger water courses, where they are sometimes caught by fishermen.

Curimatidae

Revisions of the subfamily Curimatinae (Vari, 1989a, b; 1992a, b), to which the species in the Pantanal belong, have resulted in considerable modifications of the faunal lists for the Pantanal since the publication of species lists by Britski (1969, 1972). The genus *Curimata* has disappeared from the region because all of the species formerly assigned to it have been transferred to other genera (Vari, 1989a). One or more species of fish formerly referred to as *Curimata gilbert* (Quoy and Gaimard, 1824), sometimes spelled *C. gilberti*, belong to other species, and this taxon probably does not belong on the list of the Pantanal ichthyofauna. According to Vari (1989a), the fishes in this complex should all be placed in the genus *Cyphocharax*, with the most typical species of this group in the Pantanal being *Cyphocharax gillii* (Eigenmann and Kennedy, 1903), a fish reaching a length of less than 10 cm. Sazama (1988) reported that the individuals of this species defend territories among the submerged parts of plants rather than forming schools, as other species in this family do. They ingest sediment containing various algae and other microorganisms (Sazama, 1986). Other species that have been mistakenly identified as *C. gilbert* include *Cyphocharax voga* (Hensel, 1869), a widely distributed species in southern Brazil, known from the Rio Paraguay System as far north as Paraguay. It probably ranges at least as far as the southern part of the Pantanal. *Cyphocharax modestus* (Fernández-Yépez, 1948), which reaches at least 12 cm and was referred to as *C. gilberti* by Britski (1972), inhabits the Rio Paraná System. There is one questionable report by Géry *et al.* (1987) of a *C.* cf. *modestus* in the Rio Paraguay. *Cyphocharax spilotus* (Vari, 1987) is also known from the Rio Uruguay and lower Rio Paraguay, but it probably does not range as far upstream as the Pantanal.

Curimatella dorsalis (Eigenmann and Eigenmann, 1889), also known by the synonym, *Curimatella australe* Eigenmann and Kennedy, 1903, is a wide-ranging species in South America, inhabiting the systems of the Orinoco, Amazon, lower Paraná, and Paraguay Rivers. Specimens with a standard length as long as 113.7 mm have been taken in the Rio de la Plata. The species of this genus usually form small schools in the rivers of the Pantanal. They are noted for taking shelter among dense vegetation.

Several species of *Steindachnerina* also inhabit the system of the Rio Paraguay. *Steindachnerina conspersa* (Holmberg, 1891) has been reported from Paraguay, and *Steindachnerina nigrotaenia* (Boulenger, 1902) is also known from the river systems to the south of the Pantanal. Their confirmed presence as far north as the Pantanal remains to be reported. Both were originally assigned to the genus *Curimata*. A third species in this genus is apparently present at least in the southern part of the Pantanal. Originally called *Curimata elegans paraguayensis* Eigenmann and Kennedy, 1903, it was placed in synonymy with *Curimata elegans nitens* (Holmberg, 1891). It reportedly ranged from Argentina through the Rio Paraguay system. Vari (1992a) examined the specimens assigned to this taxon and found that some of them actually belonged to the species *Curimatella dorsalis*, while one of the paratypes was actually a specimen of *Steindachnerina brevipinna* (Eigenmann and Eigenmann, 1889). Difficulties in the taxonomy of the group in the past have prevented the fauna in this subfamily from being properly catalogued in the Pantanal, and additional species, including one still undescribed, are suspected of inhabiting the Pantanal.

The feeding habits of two species assigned to *Curimata* were investigated by Sazama and Pelligrini Caramaschi (1989) in the Pantanal. Members of this genus seem to be generally omnivorous, but the taxonomic confusion has hindered the compilation of information on the diets of the individual species.

In addition to the species formerly considered to belong to *Curimata*, this subfamily is represented by *Psectrogaster curviventris* Eigenmann and Kennedy, 1903, in the Pantanal. These fishes form schools in the main tributaries draining the Pantanal, generally preferring deeper water to the shallow water of the floodplain. A type specimen of this species was collcted in the Baia Buritizal (Vari, 1989b).

Prochilodontidae

Species belonging to three subfamilies of the Prochilodontidae are known to be present in the catchment area of the Pantanal. Of these subfamilies, Prochilodontinae is the best represented.

The *curimbata, Prochilodus lineatus* (Valenciennes, 1847), is one of the migratory river fishes that has become important for the fishery in some streams of the Rio Paraná System. Its young utilize flooded areas along the river to grow and mature before returning to the river as adults (Agostinho and Zalewski, 1995). There is a corresponding transition in its diet from plankton to detritus, which accompanies this change of habitat (Bayo & Cordiviola de Yuan, 1992). Almeida *et al.* (1993) found algae to form the largest part of the diet, but some zooplankton and benthic animals were also consumed. Its presence has been documented from the middle sections of the Rio Paraguay (Carozza & Cordiviola de Yuan, 1991), and most of its population apparently remains downstream of the Pantanal, except during the spawning season, when the fishes migrate upstream, at least as far as the Pantanal of Mato Grosso do Sul (Almeida *et al.*, 1993). Resende (1992) reported that the gonads of the fishes in the Pantanal of Mato Grosso do Sul mature from August through February and that most of the reproduction in the uppermost sections of the Rios Miranda and Aquidauana apparently occurs from December through February. After spawning, the fishes return downstream.

A congeneric species also referred to by the common name, *curimbata*, is *Prochilodus platensis* Holmberg, 1889, a species known from the lower sections of the Rio Paraná System (Bayley, 1973), where it accounts for 60% of the "ichthyomass" in the floodplain pools (Cordiviola de Yuan & Hassan, 1989), as well as from river systems of the temperate zone in Argentina (Bistoni *et al.*, 1992). It is a migratory fish in the river system (Pignalberi & Cordiviola de Yuan, 1973) but apparently is not common as far upstream as the Pantanal.

The subfamily Cynopotaminae is represented by fishes that inhabit the rivers and adjacent lakes of the Pantanal. The species known from the region include *Cynopotamus kincaidi* (Schulz, 1950), *Galeocharax humeralis* (Valenciennes, 1847), and another, unidentified and perhaps undescribed species of *Cynopotamus*. Little has been reported about their habits, but it is likely that the species of *Cynopotamus* are predators, as Knöppel (1972) reported the presence of a fish in the stomach of *Cynopotamus essequibensis* Eigenmann, 1912, which inhabits French Guiana and Surinam.

The subfamily Salmininae is represented by only a single species in the rivers of the Pantanal, albeit a important one for the fishery. One of the most popular food fishes caught in the Rio Cuiabá is *Salminus maxillosus* Valenciennes, 1849, locally called the *dourado*, a name applied to other species of the genus in other parts of Brazil, as well. This species is a large river fish, which does not venture onto the floodplains but frequently enters the baias along the Rio Cuiabá near Barão de Melgaço. It enjoys protection during the late dry and early rainy season, which is its normal spawning season. However, the fishing laws are widely disobeyed due to sporadic enforcement. Whether it spawns in the region of the Pantanal is yet to be determined; however. Lowe-McConnell (1987) discussed upstream migrations of at least 400 km by this and other relatively large species used for food in the Rio Paraná System. However, her information came mainly from the studies of Bayley (1973), who concentrated his efforts chiefly on *Prochilodus platensis*, which is not a common fish in the Pantanal. The maximum average speed of migration upstream by a marked *dourado* in Argentina was reported by Bonetto *et al.* (1971) to be 0.9 km per day over a distance of 62 km. Fishes of this species are known to spawn in the upper sections of tributaries in the Andes (Lowe-McConnell, 1987). However, it may well be that breeding grounds of *Salminus maxillosus* are also located in the upper Rio Paraguay System, where the water quality is quite different from that in the Andes, since the species is also encountered in lakes adjacent to rivers in the upper Rio Paraná System (Galetti *et al.*, 1990). It also occurs in rivers in the temperate zone of Argentina (Bistoni *et al.*, 1992).

Marking studies of this and other species were conducted in the Rio Paraná System in Argentina by Bonetto and Pignalberi (1964), who noted, among other things, that a *Salminus maxillosus* migrated 1000 km downstream in 60 days. It was suggested that the large river fishes they studied migrated downstream to feed and upstream to spawn. Feeding studies of this species revealed it to be mainly predatory on a wide variety of invertebrates during its early stages of development, while older individuals include fishes in the diet (Rossi, 1989).

Gymnotoidei

Gymnotidae

The genus *Gymnotus* is apparently represented by two species in the Pantanal. Their identity is not yet certain, although one of them is probably *G. carapo* Linnaeus, 1758, a species also known from the upper Rio Paraná (Galetti *et al.*, 1990) and upper Araguaia (Lowe-McConnell, 1991) Systems. These species generally remain in the water courses and do not travel onto the shallower parts of the floodplain, although they may be found in the smaller streams at times. They seem to prefer rather slow-flowing water in backwaters and beneath plants near the shore of the rivers. The activity of these fishes is crepuscular or nocturnal, and they are not commonly seen, although they can be found in many parts of the Pantanal. Fishes of this genus are widespread in Brazil and neighboring countries.

Rhamphichthyidae

Members of the knife fish family are generally inhabitants of the littoral zones of the large rivers or the sections of the lakes and oxbows overgrown with plants. *Eigenmannia virescens* (Valenciennes, 1847) is apparently the most abundant species of this family in the Pantanal. The males reach a length of 45 cm, but the females are shorter (Sterba, 1987). Like many other fishes of the Pantanal, this knife fish is noted for a rapid rate of growth. In addition to the Rio Paraná system, it is also known from the Rio Dulce in the temperate zone of Argentina (Bistoni *et al.*, 1992).

Other knife fishes reported for the Pantanal include *Rhamphichthys marmoratus* Castelnau, 1855, mentioned by Britski (1969), and an additional species, probably *Rhamphichthys rostratus* (Linnaeus, 1766). No specimens of these fishes were collected during the study.

Siluroidei

Doradidae

Seven species of doradid catfishes constitute an important part of the fish fauna in the Pantanal. They possess strong, serrated pectoral spines, which are extended and locked in position when the fish is threatened. This makes it dangerous for a predator to swallow one. Piscivorous birds and caimans are adept at crushing the bones before these fishes are swallowed, but this procedure may take as long as five minutes. Cormorants are frequently seen working on this problem after one of these fishes has been captured. If the bird is still unsuccessful in swallowing the fish after about five minutes, it discards the dangerous prey.

Platydoras costatus (Linnaeus, 1766), a popular aquarium fish known as the chocolate catfish, was abundant in the Rio Bento Gomes during the dry season of 1992. It is a small fish that reportedly reaches a maximum length of 22 cm (Riehl and Baensch, 1991). Large numbers appeared in the shallow water near the Transpantaneira Highway in July and August of that year, and many of them died as the water reached its lowest level of the year. The mortality continued after the first rains. The species was not abundant in subsequent years. Apparently, these fishes move through the small river systems in large schools, so they are periodically either abundant or absent at a particular site. The species is also known from the Amazon System.

The Paraguay doradid, *Trachydoras paraguayensis* (Eigenmann and Ward, 1907) is a small fish, usually about 8 cm in length (Axelrod *et al.*, 1987). It seems to be distributed mainly in the small rivers crossing the Pantanal. Only isolated individuals apparently belonging to this species were observed during the study.

Another rather small fish with branching barbels called Eigenmann's doradid, *Doras eigenmanni* (Boulenger, 1895), is also reported from the Pantanal. Its length is approximately 10 cm (Axelrod *et al.*, 1987). This nocturnal, secretive fish was not observed during the study.

Other doradids that may be encountered in the rivers and streams draining the Pantanal are *Anadoras grypus* (Cope, 1872); *Oxydoras kneri* Bleeker, 1862; *Pterodoras granulosus* (Valenciennes, 1833); and *Rhinodoras dorbignyi* (Kröyer, 1855).

Auchenipteridae

This family was represented by two species in the Pantanal. *Auchenipterus nigripinnis* (Boulenger, 1895) and a species tentatively identified as *Trachycorystes porosus* Eigenmann and Eigenmann, 1888. In an aquarium, these species proved resistant to toxic substances in the water, surviving after the water was contaminated by a mixture of xylene and formaldehyde, which was used by botanists to protect wood samples in a closed room. The vapors, which entered the aquarium through the aeration system, killed all of the fishes except for the auchenipterids.

Pimelodidae

This family is represented by at least 17 catfish species in the region of the Pantanal, some of which are quite widespread. Several are important for the local fishery. They seem to be almost exclusively nocturnal, at least as far as their excursions from deep streams and lakes into shallow water are concerned. One of the species, *Paulicea luetkeni* (Steindachner, 1875), locally called the *jaú*, can attain a weight of about 100 kg (Ravazzani *et al.*, 1991).

Another species, *Pseudoplatystoma corruscans* (Agassiz, 1829), is an economically important food fish, locally known as the *pintado*. It is often caught in the large rivers in and near the Pantanal, but during the high water period, it forrages on the floodplains. This is one of the species that attracts many sport fishermen from all over Brazil to the Pantanal. The fishes caught usually have reached a weight of about 25 kg (Ravazzani *et al.*, 1991). In a study of the species in Argentina (Cordiviola, 1966), the maximum age of the specimens caught was 14 years, and the greatest fork length was 154 cm. The species was not found to exceed 53 kg in weight.

Other important fishes for the commercial and sport fishery are the *barbado*, *Pinirampus pirinampus* (Spix, 1829); the *cachara*, *Pseudoplatystoma fasciatum* (Linnaeus, 1766); the *jurupensum*, *Sorubim* cf. *lima* (Schneider, 1801); and the *jurupoca*, *Hemisorubim platyrhynchos* (Valenciennes, 1840). Most of these fishes are caught from boats in the Cuiabá and Paraguay Rivers. Like the other important food fishes in the Pantanal, they are endangered by the destructive, illegal fishing methods that can be practiced because of the lack of financial support for the forestry police (Section 12.3).

Goulding (1981) reported that *Pseudoplatystoma fasciatum* has been heavily exploited in the Amazon System since about 1970, and that by 1978, its population had already been considerably depleted. The predatory species, *Pinirampus pirinampus* and *Paulicea luetkeni*, are still important commercial species in the Itaipu Impoundment, although the relative representation of the individual fish species in the local fauna is still changing rapidly (Agostinho and Zalewski, 1995).

In addition to the large species important for the fishery, the family includes various small species suitable for aquaria. *Pimelodella gracilis* (Valenciennes, 1840) is somewhat large for aquaria, since it reaches a length exceeding 30 cm. In the Pantanal, it seems to be a river fish that seeks out temporarily flooded areas along the rivers only at night, if at all. It is also known from rivers of the temperate zone in Argentina (Bistoni *et al.*, 1992).

Pimelodus maculatus Lacépède, 1803, is an attractive spotted fish that may reach a length of about 30 cm. It seems to be a catfish that generally remains concealed alone or in small groups during the day in the streams and small rivers draining the Pantanal. Meschiatti (1995) found that this fish consumed much sediment in an oxbow lake along the Rio Mogi-Guaçu during the dry season, but he also summarized reports on its diets elsewhere, where various researchers reported it to be omnivorous (Arcifa et al., 1988), insectivorous (FUEM, 1987), and partially piscivorous (Basile-Martins, unpubl. diss. from 1978, cited by Meschiatti, 1995).

Other attractive species that are sometimes kept in aquaria are the golden mandi, *Pimelodus blochi* (Cuvier and Valenciennes, 1840), which has sometimes been referred to in the literature by its synonym, *Pimelodus clarias* (Bloch, 1795), and *Pimelodus ornatus* Kner, 1857, which are both mainly nocturnal, benthic catfishes (Axelrod *et al.*, 1987; Sterba, 1987). *P. blochi* has an extensive range east of the Andes: from Central America to Argentina.

In addition to the species mentioned here, Pinto (1974) reported the widespread *bagre urutu*, *Genidens genidens* (Valenciennes, 1839), as well as *Netuma barbus* (Lacépède, 1803) and *Rhamdia quelen* (Quoy and Gaimard, 1824) from the waters of Mato Grosso.

Additional species known or thought to occur in the region of the Pantanal include *Iheringichthys labrosus* Kröyer, 1874; *Pimelodella taenioptera* Ribeiro, 1914; *Pimelodella mucosa* Eigenmann and Ward, 1907; and *Megalonema platanum* (Günther, 1880).

Sometimes, the juveniles of the large pimelodid species are exported from South America for aquarium hobbyists in other parts of the world. These species are actually well suited for extremely large aquaria, but they quickly outgrow those of normal commercial sizes.

Callichthyidae

One of the most abundant benthic school fishes on the floodplain of the Pantanal is the aeneus catfish, *Corydoras aeneus* (Gill, 1858), which is particularly abundant on temporarily flooded areas. In addition, *Corydoras microps* (Eigenmann and Kennedy, 1903) is reported from the region, and sometimes schools of what appeared to be one of the two species were observed in the same habitats at different times. This brings up taxonomic problems, since Regan (1912) considered *C. microps* to be a synonym of *C. aeneus*. No ecological differences were noted between the *Corydoras* populations in the Pantanal, and the fishes seemed to display a considerable variability during the different stages of growth. The schools formed generally contain individuals about the same size: small during the period of initial flooding and rather large during the dry season. During the *vazante*, the schools venture into shallow water during the day, spreading out to consume the microorganisms that develop on the surface of the sediment and keeping out of reach of large predatory fishes. When approached from the land side, they form compact schools and flee toward deeper water. The fishes seem fast and circumspect enough to avoid being trapped in the small pools of water that dry up completely during the dry season, and they follow the receding water into the lowest-lying areas. During the dry season of 1991, however, the water level sank particularly low, and large numbers of this species were found dead in a small pool of water along the course of Birici Stream, a kind of shallow water course known locally as a *vazante*. With the onset of the rainy season, the survivors of the dry period begin reproduction, and schools of juveniles are observed during the later part of the *enchente*. Tiny fry have not been observed, suggesting that reproduction takes place in deeper waters of the permanent streams or rivers.

Growth is rapid, and the fishes seem to reach maturity at the age of about one year, but more accurate studies should be undertaken to confirm this. Another species, questionably assigned to *Corydoras hastatus* Eigenmann and Eigenmann, 1888, is also known to occur in the Pantanal.

Nijssen (1970) briefly reviewed the information on the distribution of the *Corydoras* species in South America. Because of the great difficulties in distinguishing the species, little exact information is available, except in small areas where special studies were conducted, such as Surinam. In any case, there was no information available at the time on the occurrence of the *C. aeneus* group, including *C. aeneus* and *C. microps*, if these both prove to be distinct species, in the region of the Pantanal.

Another callichthyid, *Hoplosternum littorale* (Hancock, 1828), inhabits the shallow waters along the edges of rivers. It is widespread in tropical South America but not abundant in the Pantanal.

Loricariidae

In the small streams and rivers of the Pantanal, the family Loricariidae contributes a number of species, that are often quite numerous, to the ichthyofauna. These fishes are not truly "bottom-feeders" as they are sometimes reported to be. Rather they are most frequently found in the interrhizon community. They are actually best described as "surface feeders" or "scrapers", preferring to move along objects near the surface on which tiny aufwuchs or vagile microorganisms are moving. They actually seem best adapted to moving along stones, but the normal sediment of the Pantanal includes no stones. It is rather a fine sand or silt. Therefore, they generally apply their scraping mouths to the floating plant root systems to remove the microorganisms that are enmeshed there.

Unfortunately, the taxonomy of this group is in great need of revision, and many species that are commercially sold to aquarium hobbyists cannot presently be identified (Sterba, 1987). It is still not possible to identify many species encountered in the field, as well.

The species of the subfamily Loricariinae most numerous in the Pantanal is *Rineloricaria parva* (Boulenger, 1895), which was captured in enormous numbers at various times among the roots of floating plants, most frequently on *Salvinia auriculata*. Apparently, the fishes are omnivores that scrape microscopic organisms from among the interrhizon community.

In addition to the small species, several large loricariines were observed in the Rio Bento Gomes and Rio Novo systems. Frequently during the dry season, caimans and piscivorous birds were observed feeding on them. One of these is *Loricaria* sp., the identity of which will only be revealed after a taxonomic revision of the genus or family has been completed.

In addition to the loricariines, the family is represented by at least one species in the subfamily Hypoptopomatinae. This small loricariid inhabits the shallow water along the edges of small rivers in the Pantanal. It resembles *Otocinclus vittatus* Regan, 1904, a species known to inhabit the region (Sterba, 1987). Only a few isolated individuals were observed in the Rio Bento Gomes during the study.

Belonidae

Small groups of the needlefish, *Potamorrhaphis eigenmanni* Ribeiro, 1915, are often observed on the surface of the water in or near the small rivers draining the Pantanal. This migratory species, which apparently feeds mainly on small fishes, is often seen in the water beneath the bridges along the Transpantaneira Highway. It is seldom seen on the temporarily flooded areas.

Cyprinodontidae

A poorly known annual has been described from the district of Cáceres. *Rivulichthys rondoni* (de Miranda Ribeiro, 1923) has been cultivated in aquaria, where it produces large eggs, 2 mm in diameter, which hatch after a period of dryness lasting as long as a year. The species may be endemic to the region of the northern Pantanal.

Synbranchiformes

This order is represented in the Pantanal by a single species of the family Synbranchidae, an eel-like fish with an exceptional adaptation to waters of the wet-and-dry climatic zone: *Synbranchus marmoratus* Bloch, 1795, sometimes referred to by its synonym, *Synbranchus vulgaris*. This is one of the only species that is able to survive the long dry periods underground and can therefore reduce its dry season mortality rate below those of most other floodplain fish species. Like other members of its family, it is able to breath air, which it holds in its balloon-like gill chambers modified into its main respiratory organ. Thus, these swamp eel or *mussum* can live like an amphibians, leaving the water and travelling rather long distances over land (Sterba, 1987). During the *vazante*, swamp eels are often found under pieces of wood, stones, and other objects at the edges of drying water bodies. As these become progressively drier, the eels disappear into tunnels constructed into the mud while it is still soft. Large wading birds, including the jabiru and great white heron, can often be observed pulling the eels out of their tunnels before the mud has become hard enough to give them protection from the probing beaks of these predatory birds (Fig. 8.4). During the late dry season, drying pools of water are encountered in the lowest-lying areas on the floodplain in which large numbers of trapped fishes are perishing and in which a great many simply dry up in spite of extremely intensive feeding by huge flocks of piscivorous birds (Section 10.4). By this time, however, the swamp eels have moved deep into the sediment to await the next rainy season, and they seem to be among the only common fishes that do this. Only the feral and domestic pigs seem able to dig these fishes and the aestivating crabs and snails out of the hard-baked sediment during this season. Nevertheless, the mortality rate among these favorite food fishes of wading birds seems to be high because most of the specimens observed are much smaller than the maximum length of 150 cm reported for the species by Sterba (1987). The species is widely distributed in South America from the Amazonian region to the temperate zone of Argentina (Bistoni *et al.*, 1992).

Figure 8.4: *Synbranchus marmoratus*, shown in the beak of a great egret, *Casmerodius albus*, is a favorite food of the egrets and stork jabirus because it lacks long spines.

Perciformes

Cichlidae

Several cichlid species are among the dominant fishes on the floodplain of the Pantanal. The family is represented predominantly by fishes that prefer shallow water. In the smaller rivers, the larger species of this family are invariably found near the shore. They typically hide themselves in the dense vegetation to breed. Most species give considerable protection to the young during the breeding season.

The *apaiari*, *Astronotus ocellatus* (Cuvier, 1829), is known among aquarium fish hobbyists as the oscar. It reaches a length of about 30 cm and is considered an excellent food fish in the Amazon System, where it is most frequently encountered on the floodplain (Goulding, 1981). Mago (1970) found this species in shallow floodplain waters near the Apure River in the Orinoco System of Venezuela, where its breeding normally coincides with the onset of the rainy season, just as it does in the Pantanal. However, this water body is in the northern hemisphere, and the rainy seasons at these two locations begin roughly half a year apart. Individuals kept in artificial water bodies in Brazil were reported to be mainly predatory, consuming insects, prawns, and small fishes.

At times, the species in the Pantanal has erroneously been called *Astronotus crassipinnis* (Cuvier, 1829), which might become the valid name if the species were distinct from the form present in the Amazon System. These fishes are frequently observed in the shallow waters along the rivers and streams crossing the floodplain. It is large enough to constitute a fair amount of the caiman's diet, particularly during the dry season, when the declining water levels and low oxygen concentrations force these fishes to the surface in the shallow water, making them vulnerable to the caimans. They seem preferable as food to the numerous catfishes because the body spines that the caimans must crush before swallowing the fishes are much softer in this cichlid than they are in most catfish species. *Astronotus ocellatus* is a widespread species, which seems to have colonized habitats similar to the Pantanal all over South America.

The breeding habits of the oscar in aquaria were described in detail by Paiva and Nepomuceno (1989), and its commercial culture in artificial ponds offers much promise. Even in culture, rainfall plays a great role in inducing the spawning of the species, although in Amazonas, it could breed the year round. Sexual maturity is reached in slightly less than one year, but if the fish survives the dry periods, it can continue to spawn throughout its nine to ten year life span (Paiva and Nepomuceno, 1989). In artificial water bodies of northeastern Brazil, the first spawning was observed to occur when the fishes had reached a total length of about 18 cm (Dourado, 1981). The parents provide some protection for the hatchlings, which characteristically seek protection by congregating on the lateral surfaces of the parents' bodies (Sterba, 1977).

A somewhat less abundant species in the small rivers of the Pantanal was *Cichla ocellaris* Bloch and Schneider, 1801, which was observed regularly in the Rio Bento Gomes near the Transpantaneira Highway, especially during the high water periods. It seek moderately deep, open water and avoids shallow places, leaving the small rivers before or during the *vazante*. According to Dourado (1981), its diet consists mainly of live fishes and prawns, but insects are consumed occasionally. It reproduced for the first time at an age of about one year after a length of about 26 cm has been attained. The presence of this species in southern Mato Grosso may well have resulted from introductions by man. According to Saint-Paul (1986), this species has been introduced into many water bodies since 1940. Information on the reproduction was provided by Braga (1952), Fontenele (1955), and de Godoy (1965). As long as the water temperature remains above 15.6^{o}C and below 37.9^{o}C, the species can reproduce rapidly, producing from 1500 to 8000 eggs at a time. Survival of the young is enhanced by the parents, which guard the hatchlings.

Among the smaller cichlids are three species of small mouth breeder: the gold-cheeked flag cichlid, *Aequidens paraguayensis* Eigenmann and Kennedy, 1903, sometimes known under the synonym, *Bujurquina vittata* (Heckel, 1840); *Cichlasoma dimersus* (Heckel, 1840); and *Gymnogeophagus balzani* (Perugia, 1891). These fishes spread over the floodplains with the first flood water and generally conceal themselves in the dense masses of drowning terrestrial plants. They begin to breed at this time, and schools of tiny fry guarded by the parents are already encountered before the flood waters have become deep. The adult *Aequidens paraguayensis* have the interesting habit of leaping out of the water when a wading person approaches and skimming on the side over the surface of the water toward the person. They are easily taken in the hand at this time. Apparently, this is a means of the fish to escape large underwater predators. If the fish jumps into a large leaf or other object on the surface of the water, it is no longer visible from below.

Another species of *Aequidens* that is not a mouth breeder also inhabits the region of the Pantanal. *A. dorsigerus* (Heckel, 1840) is observed occasionally among dense vegetation on the floodplain during the high water period.

Among the species encountered frequently in the shallow waters along the edges of streams and on the floodplains during the high water period is *Crenicichla lepidota* Heckel, 1840, a species with an extensive distribution in South America. The species is predatory, with small fishes and large insects accounting for a large part of the diet (Sterba, 1987).

Apistogramma corumbrae (Eigenmann in Regan, 1906) was long believed to be a species endemic to water bodies in the Rio Paraguay System (Ringuelet *et al.*, 1967). However, Cordiviola de Yuan and Oliveros (1979) reported that individual specimens were obtained in lakes as far downstream as the middle Rio Paraná. It is a small fish, reaching only 5.5 cm in total length, which was encountered rather seldom along the edges of streams and small rivers in the Pantanal, generally partially concealed in submerged macrophytes.

Chaetobranchopsis australis Eigenmann and Ward, 1907, is another species encountered in the shallow water zone of the small rivers draining the Pantanal.

Reports of *Cichlasoma festivum* (Heckel, 1840) in the Pantanal (Müller and Weimer, 1976) could not be confirmed. Sterba (1987) reported the range of this species to encompass the region of the Guiana and Amazon Basin. *Batrachops semifasciatus* Heckal, 1840, also reported for the Pantanal, was not seen during the investigation, either.

Other cichlids reported by Müller and Weimer (1976) or by Sterba (1987) to inhabit the Rio Paraguay System that were not observed in the northern part of the Pantanal during the study include *Aequidens tetramerus* (Heckel, 1840), *Apistogramma borelli* (Regan, 1906), *Apistogramma pleurotaenia* (Regan, 1909), *Apistogramma trifasciatum* (Eigenmann and Kennedy, 1903), *Cichlasoma facetum* (Jenyns, 1842), and *Crenicichla saxatilis* (Linnaeus, 1758). It is suspected that these species, known from the Pantanal of Mato Grosso do Sul, seldom range northward into Mato Grosso because of the differences in the dry season conditions (Section 5). A further problem is the considerable taxonomic confusion that still results in synonyms appearing in the literature, sometimes resulting from disagreement about whether a fish is a local race of a widespread species or a valid species endemic to the Paraná River System. Because of their great representation among the South American ichthyofauna, cichlids are particularly often the objects of such disputes.

Sciaenidae

Freshwater species of this primarily marine family occur in the rivers flowing throught the Pantanal. There are two native species: *Pachyurus bonariensis* Steindachner, 1879, and *Plagioscion ternetzi* Boulenger, 1895.

A third species of the family, *Plagioscion squamosissimus* (Heckel, 1840), was introduced to the lower Rio Paraná System in 1967 (Cruz *et al.*, 1990). It will therefore probably be found in the Pantanal at some time in the near future after it has colonized the entire river system. It feeds mainly on other fishes and prawns but also consumes a few insects and spawns for the first time at the age of about a year (Dourado, 1981).

Ageneiosidae

This family includes the *palmito*, *Ageneiosus brevifilis* Valenciennes, 1840, an important food fish of the rivers, sought by both sport and commercial fishermen.

8.13.2. Amphibia

The choruses of small anurans of various species are a characteristic feature of the shallow flooded meadows of the Pantanal at the beginning of the rainy season. The great numbers of frogs and diversity of species in these choruses has been described in a publication by Lutz (1946). The choruses continue for long periods of time, although they decrease considerably in intensity after the rainfall has begun to diminish in March or April.

Apoda

Caeciliidae

At different times during the early rainy season, two specimens of *Siphonops paulensis* Boettger, 1892, were found on the Transpantaneira Highway about 7 km south of Poconé. They had apparently been run over by cars and were injured, although the loose stones on the unpaved roads had apparently prevented them from being flattened. Both died soon thereafter. Apparently, they leave their underground abodes and seek higher ground when flood waters begin to cover the wetland. The caecilians of this species are terrestrial and seem to prefer dry forest areas, where they reportedly consume worms and both larval and adult arthropods (Cei, 1980). According to Taylor (1968), this species seems to be most abundant in the regions surrounding the Pantanal as it was found to be most abundant at locations in Bolivia, Paraguay, and Argentina adjacent to the Pantanal and at Corumbá in Mato Grosso do Sul. However, specimens have also been collected from Goiás, Rio Grande do Sul, and Rio de Janeiro. Because these animals inhabit subterranean burrows, their habits are poorly known. In contrast to the live-bearing species of the family Typhlonectidae, *S. paulensis* reproduces by laying eggs.

Typhlonectidae

Although never encountered in the Pantanal, the presence of the live-bearing swamp dweller, *Chthonerpeton indistinctum* (Reinhardt & Lütken, 1862), in the wetland at the present time is considered almost certain.

Salientia

Microhylidae

One microhylid species, *Elachistocleis bicolor* (Valenciennes, 1838), was frequently heard and observed along the dirt roads at elevations just above the high water level of the floodplain during the early rainy season. These small frogs congregate in the deep pools of water that form in depressions after the freshet and persist until the end of the rainy season. Very often, the males were observed calling under the headlights of the project bus stopped in front of rainwater pools on the road during the first few hours after sunset. This species seems to deliberately seek out such temporary pools and to avoid the deeper locations in which aquatic vegetation develops. This may well allow its larvae to grow without exposure to most predators of the wetlands, except for the dragonfly larvae, which also develop in such rainy season pools. Little is known about the habits of the adults, which are almost never seen, except during breeding in the early rainy season. Cei (1980) speculated that they live underground, feeding on termites.

This frog is sometimes treated as a subspecies of *Elachistocleis ovalis* (Schneider, 1799).

A second microhylid species with more of less the same breeding season, *Dermatonotus muelleri* (Boettger, 1885), was not seen as frequently, perhaps because it breeds more commonly in shallow water among aquatic plants than in the pools of water along the roads. The main breeding season for this species is also the early rainy season. Its habits are also poorly known. A specimen of *D. muelleri* was found at night on a street in the town of Poconé after a heavy rainfall. Both of these species are widely distributed in lowland South America from Bolivia to Argentina (Lutz, 1972).

Bufonidae

One of the most abundant large amphibians in the Pantanal is *Bufo paracnemis* Lutz, 1925, which forms breeding choruses at various times of the year but most often during the *vazante* and *seca*. This results in large numbers of larvae being concentrated in drying pools of water. It may be a means of protecting the tadpoles from predation by large fishes and other aquatic predators, which have already fled the locations to keep from being trapped as the water level rapidly falls. Whether or not such a seasonal breeding schedule is necessary as a survival strategy will require more research into the ecology of this species and its possible predators because the tadpoles do not seem palätable to most of the fishes that sometimes share deeper water with them. In any case, reproduction during this season produces some dramatic moments as the tadpoles attempt to complete metamorphosis before they are left to dry in the sun by the receeding water. Many times, large groups were observed to die while some of the siblings succeeded in transforming just enough to hop away from the water as tiny toadlets before their larval habitat dried up.

Once transformed, the toadlets, only about 7 mm long, leave the water and seem quite protected from desiccation by their tough skin. They move rapidly toward the permanent terrestrial vegetation that grows at higher elevations and take shelter among the plants.

The adults can attain a rather large size, commonly over 15 cm in snout-vent length, and often frequent well lighted places at night, where they fatten themselves on the diverse insects that fly into the lights and fall to the ground. Their preference for human habitations was already noted by Müller and Hellmich (1936) in Paraguay. The adults are relatively safe from predators because of the poisonous secretions they produce; however, they suffer considerable mortality on the highways.

Concerning the name of this species, it should be noted that Gallardo (1962) examined specimens of *Bufo paracnemis* and *B. schneideri* Werner, 1894, and expressed the opinion that they may belong to the same species. This would make the widely used name, *B. paracnemis*, a junior synonym of *B. schneideri*. However, because Gallardo (1962) did not state his opinion in definite terms, Frost (1985) has maintained both names in his world list of amphibians.

Other species in this family that occur in southern Mato Grosso include *Bufo arenarum* Hensel, 1867, and *Bufo granulosus* Boulenger, 1894, represented in the region of the Pantanal by *B. granulosus major* Müller and Hellmich, 1936. These species were not encountered in or near the wetland but are rather species of drier, forested upland regions of the watershed.

Hylidae

The hylids are represented in the Pantanal and the adjacent forested areas in its watershed by a number of species with various habits. One of these, *Hyla raniceps* (Cope, 1862), is a member of the pseudoterrestrium community, sleeping by day, usually curled up in a leaf of *Eichhornia crassipes* or *E. azurea*, which gives it maximum concealment because of its curved shape (Plate 6b). The species is active at night, hunting insects or forming choruses. During the dry season, this tree frog seeks cool and preferably somewhat moist places to wait for the next rains. They were frequently seen in the bathrooms of ranch houses, pumping stations for well water, and other buildings with a permanent water supply. Individuals kept in houses readily ate the large tropical cockroaches, *Periplaneta americana* Linnaeus, 1958, and other insects, which they made long leaps to seize. They are general predators on almost all arthropods small enough to swallow, including some cockroaches about 80% of their own length. This species has an extensive range in South America east of the Andes, occurring from the Guianas to Argentina.

Although Lutz and Lutz (1973) believed that only *H. raniceps* occurs in the Pantanal wetland, at least one other common species, *Hyla pulchella* Duméril and Bibron, 1841, was observed there. This species is represented in South America by many local races and subspecies. Like *H. raniceps*, these tree frogs, which probably concealed themselves in trees in earlier times, frequently find shelter in rather small objects constructed by man, especially those containing water, such as the cases of water pumps, bathrooms of the ranch houses, and even the small shelters for the rain gauges set out in the field. However, unlike *H. raniceps, H. pulchella* does not seem to be member of the pseudoterrestrium community but rather remains in or near the wooded areas. A study of a population in Argentina revealed that dragonfly larvae and water bugs, especially belostomatids and notonectids, influence the survival and age structure through predation on the tadpoles (Kehr & Basso, 1992).

Several other hylids not identified during the study are also known from the Pantanal (M. Carvalho, pers. comm.). These include *Hyla minuta* Peters, 1782; *Hyla nana* Boulenger, 1889; *Hyla punctata* (Schneider, 1799), and an unidentified species in the *Hyla geographica* group. Small treefrogs, probably belonging to these species were observed at times by researchers working in the Pantanal ecology project when they emptied the metal canister used to collect and record precipitation on the Fazenda Ypiranga. They could not be removed through the narrow opening and were not collected or identified. However, they belonged to species other than the common *H. raniceps* and *H. pulchella*, and they were not observed elsewhere in the Pantanal. *Scinax fuscomarginata* (A. Lutz, 1925), *Scinax* cf. *egleri*, and an unidentified species in the *S. fuscovaria* group can also be included in the list of Pantanal herpetofauna (M. Carvalho, pers. comm.).

In addition, two species adapted to living in human habitations, especially bathrooms, water tanks, and pumping stations, are encountered in the region of the Pantanal: *Phrynohyas venulosa* (Laurenti, 1768) and an unidentified species in the *Senax rubra* group.

Other species of southern Mato Grosso seem to prefer the upland areas, often remaining near human habitations. The species *Hyla tritaeniata* Bokermann, 1965, was described from specimens found in the region of Cuiabá and the upland forests near São Vincente. Also reported from these upland forests by Lutz and Lutz (1973) was *Hyla albopunctata* Spix, 1824, a species also observed during the study along the outskirts of the Pantanal. This species has an extensive range in the Central Western Region of Brazil and adjacent parts of Argentina, and Lutz and Lutz (1973) believe that it may be a subspecies of *Hyla daudini* B. Lutz, 1973, which is common throughout much of the county. In previous publications, *H. daudini* usually appears under the name *H. boans* Daudin, 1802: a name preoccupied by another species. Two additional tree frogs known from the region of the Pantanal but which might not frequent the wetland are *H. polytaenia* Cope, 1869, and *H. rubicundula* Reinhardt and Lütken, 1862.

The type specimen of *Hyla melanargyrea* Cope, 1887, a species ranging as far north as Surinam, was captured at Chapada dos Guimarães. This species has been treated as a subspecies of *H. senicula* Cope, 1868, by some authors, but others have maintained that it is a valid species (Frost, 1985).

In summary, while the hylids are not nearly as numerically abundant as the anurans that live on the ground, as judged by the choruses heard during the course of the year, at least one species, *Hyla raniceps*, is a widespread and common element of the Pantanal fauna. The family is by no means confined to the large wetland areas in southern Mato Grosso, and most of the local species seem to avoid the floodplain altogether.

Leptodactylidae

The leptodactylids are rather well represented in the region of the Pantanal. Although a few members of this family are present in southern Africa and Australia, the overwhelming majority of its species inhabit the Neotropical Region (Lynch, 1971). *Leptodactylus chaquensis* Cei, 1950, a species strongly resembling a spotted ranid, was frequently seen on the Transpantaneira Highway at night. These frogs make long hops and therefore can move rapidly on land. They are active during most of the year and are not seen at night only during the driest periods. None were ever found during the day, attesting to their considerable skill at concealment. In the northern part of the Pantanal, the range of this frog apparently overlaps that of *Leptodactylus ocellatus* (Linnaeus, 1758), a closely related species with a range throughout nearly all of South America east of the Andes. Both species have been assigned to the same "species group" (Frost, 1985), and it would therefore be desirable to investigate these frogs in the northern Pantanal to determine whether interbreeding occurs. Heyer *et al.* (1990) found *L. ocellatus* in clearings of a forested region in the State of São Paulo, which has a rich amphibian fauna that apparently includes few species also occurring in the Pantanal. A detailed study of the diet of *L. ocellatus* tadpoles by Lajmanovich (1994) revealed that they subsist mainly on a combination of green algae, diatoms, and the remains of macrophytes.

Two rather small species, *Leptodactylus podicipinus* (Cope, 1862) and *Leptodactylus fuscus* (Schneider, 1799) were found at various sites along the Transpantaneira Highway. Small leptodactylids were especially common near the Rio Novo and in the roadside ditches to the south of the Rio Bento Gomes. These frogs were generally observed aestivating beneath pieces of wood near bridges during the dry season. According to M. Carvalho (pers. comm.), they are quite abundant in the Pantanal.

Physalaemus nattereri (Steindachner, 1863) is another rather abundant leptodactylid in southern Mato Grosso. It is also among the frogs that form active, loud choruses during the early *enchente*, and individuals were frequently seen on the roads during rainstorms. Its type specimen was captured in Cuiabá. It has appeared in the literature under the name *Eupemphix* but was included in *Physalaemus* by Lynch (1970).

A large *Leptodactylus labyrinthicus* (Spix, 1824) was found on a street in Cuiabá after the freshet in 1991. The aquatic species, which is apparently secretive, was not observed again. It is large enough to feed on vertebrates, including small birds and snakes (Cei, 1980).

The occurrence of *Physalaemus biligonigerus* (Cope, 1861) is considered likely in the region of the Pantanal, although it was not reported from Mato Grosso by Frost (1985). The name was considered to be a synonym of *P. fuscomaculatus* (Steindachner, 1864) until Milstead (1963) justified the recognition of the species as valid. The presence of *Pseudopaludicola falcipes* (Hensel, 1867) is also considered probable, although specimens were not identified with certainty.

Generally, this family, like most of the Pantanal fauna, is represented by widespread species. Species with limited distribution, on the other hand, may be encountered in the upland region of the catchment area. For example, the type specimen of *Leptodactylus syphax* Bokermann, 1968, was found in the São Vincente mountains near Cuiabá. This species is in the *"pentadactylus"* group and is apparently confined to a few locations in eastern Mato Grosso and regions to the northeast (Bokermann, 1968).

Phyllomedusidae

Phyllomedusa hypocondrialis (Daudin, 1802) is abundant during the early rainy season in shallow water bodies throughout most of the Pantanal as well as in temporary upland ponds, sometimes together with such leptodactylids as *Adenomera* cf. *hylaedactyla* (Cope, 1868), *Pseudopaludicola ameghini* (Cope, 1887), *Physalaemus biligonigerus*, and *Physalaemus centralis* Bokermann, 1962, most of which were not seen in the Pantanal.

Pseudidae

Two species of pseudids are commonly encountered in the Pantanal. According to M. Carvalho (pers. comm.), they are *Pseudis paradora* (Linnaeus, 1758) and *Lysapsus limellus* (Cope, 1862).

8.13.3. Reptilia

Two reptile species are among the most prominent animals in the Pantanal and thereby responsible for attracting tourists: the caiman or *jacare-do-Pantanal*, *Caiman yacare* (Daudin, 1802), and the anaconda or *sucuri*, *Eunectes notaeus* Cope, 1862. In addition to these, the Pantanal offers an attractive habitat to a variety of other reptiles, some of which are frequently encountered but most of which are relatively rare. The drier upland regions surrounding the Pantanal and covered by the various kinds of *cerrado* vegetation support a more diverse reptile fauna.

Figure 8.5: This tortoise, *Testudo carbonaria*, was encountered crossing the desiccated floodplain from one hummock to another during the day.

Chelonia

Testudinidae

Only one terrestrial tortoise, *Testudo carbonaria* Spix, 1824, also known as *Geochelone carbonaria* and *Chelonoidis carbonaria*, was observed in the Pantanal during the study (Fig. 8.5). It has to be considered rare in the wetland, although it is a rather common species in dry forest regions and is said to be quite abundant in the moister hillside forests of Mato Grosso. The tortoises are tame and readily accept food from the hand after a short period of adjustment. They are mainly vegetarian, consuming herbs and fruits, preferably bananas and other soft pulpy ones. A large number of this species are kept in the zoological collection on the campus of the Federal University of Mato Grosso, unfortunately with little possibilities for reproduction.

The increase in the burning of woodlands will seriously endanger the tortoise populations, not only because some of them are burned to death but also because of the destruction of their preferred habitats, the stands of forest in which palms and other fruit trees grow. The grasslands offer the tortoises little food and no chance to find shelter.

Pelomedusidae

During the study, one aquatic side-necked turtle, *Acanthochelys macrocephala* (Gray, 1844), was observed in the Pantanal. It was an adult encountered while it was crossing the Transpantaneira Highway near midday while the water level was relatively high but decreasing (Fig. 8.6). This species is widespread in South America and has been reported from Venezuela, the Guianas, Bolivia, Paraguay, and many parts of Brazil (Wermuth and Mertens, 1961).

Figure 8.6: Water turtles are rare in the Pantanal. This side-necked turtle, *Acanthochelys macrocephala*, was observed crossing the Transpantaneira Highway during a late morning of the *vazante*.

The reason for the rarity of side-necked turtles is unknown. It is thought that the species would be well suited to life in the Pantanal. However, all of the pelomedusids in Brazil have to be considered endangered because in the regions where they are most abundant, located mainly in the Amazon drainage area, they are regularly captured by fishermen, and their eggs are often sought as food.

In summary, the turtles are among the most endangered reptile species due to increasing human activity and the lack of a well-organized program to protect living resources.

Squamata

Amphisbaenia

One of the most interesting groups encountered in the Pantanal are the amphisbaenids, also called worm lizards or *cobras-de-duas-cabeças*. Two species proved to be rather common in southern Mato Grosso: *Amphisbaena alba* Linnaeus, 1758, and *Amphisbaena leeseri* Gans, 1964. *A. alba* is widely distributed from northern Paraguay, Bolivia, and much of Brazil northward to Peru, Colombia, Venezuela, the Guianas, Trinidad, and possibly Panama, while *A. leeseri* seems to be limited to Central Brazil (Gans, 1967). In addition, *Amphisbaena silvestrii* Boulenger, 1902, a species limited to Central Brazil and adjacent Bolivia, is known from the wetland, as well (M. Carvalho, pers. comm.). These "worm lizards" reputedly live in the underground passages of termite nests and are infrequently encountered (Gans, 1969). In the Pantanal and surrounding parts of the watershed, specimens were not infrequently found on the roads during the *enchente*. Either the flood waters had driven them from their underground passages, or they normally undertake migrations at this time, when the high humidity protect them from desiccation. The number of body annuli, which are among the morphological features employed by various authors to distinguish the species, is influenced by the temperature (Vanzolini, 1968a). Thus, considerable more must be learned about this group, not only to gain sufficient ecological information to determine their roles in the community but also to better define the species. Attempts were made to keep worm lizards from the Pantanal alive in aquaria with large numbers of termites. They seemed to feed on the termites by squeezing the insides of the abdomen out with their jaws, but they apparently did not get enough to eat and died after several weeks.

In addition to the wetland species, *Amphisbaena neglecta* Dunn and Piatt, 1936, is known from southern Mato Grosso but only from the upland *cerrado* near Chapada dos Guimarães and eastward into central Goiás (Gans, 1967).

A fifth species in this genus, *Amphisbaena vermicularis* Wagler, 1824, has been reported from the region of the Pantanal (Gans & Amdur, 1966). It had also been reported from Cuiabá and Cáceres in the 19th century.

Two additional species of the family have ranges that probably include southern Mato Grosso: *Leposternon macrocephalum* Wagler, 1824, and *Leposternon infraorbitale* (Berthold, 1859).

Lacertilla

Anguidae

In the region of the Pantanal, one legless lizard, *Ophiodes yacupoi* Gallardo, 1966, is known to occur. It was not encountered in the wetland during the study and is apparently more at home in the *cerrado*.

Iguanidae

One of the three large lizard species in the Pantanal belongs to this family. The iguana or *camaleão*, *Iguana iguana* (Linnaeus, 1758), the type species of the family Iguanidae, is among most prominent terrestrial reptiles in the Pantanal (Fig. 8.7). Individuals over 2 m in length, including the tail, are occasionally seen, and smaller individuals are quite common. Their presence is usually revealed by the loud squawking of numerous birds of various species, which seem to accompany and seek to drive these lizards away wherever they happen to be. Small iguanas no more than about a year old are especially bright green and are not uncommon along the Transpantaneira Highway. These lizards are mainly fruit and leaf eaters, but they may include some animal protein in their diet. During the high water period, the iguanas often escape pursuit by diving into the water and swimming away. They are good underwater swimmers, holding their legs close against the body and swimming with their strong tails. Generally, the local residents of the Pantanal do not eat these lizards, but they are sometimes killed by the highway workers who clear the vegetation along the roads. Large ticks are sometimes observed attached to this lizard. This species is not presently endangered in the Pantanal.

Figure 8.7: The common iguana, *Iguana iguana*, is very common in the Pantanal. An unidentified species of tick is attached to his neck.

Studies of iguana behavior in Panama (Rand, 1968) revealed that they deposit eggs at collective nesting sites, the one studied being on an island. The nests are constructed, and the eggs laid during the early dry season so that the young hatch at the beginning of the rainy season. The appearance of young iguanas in the Pantanal during the early *enchente* suggests that the same seasonal pattern is followed, albeit with a sixth month displacement in accordance with the locations being in different hemispheres. The nests are quite deep, providing some protection from nest robbers. Rand (1968) also reported that the iguanas spend most of their time high in the tree canopy of the local rainforests. In contrast, they were only occasionally seen in trees in the Pantanal. Instead, they occurred mainly in low thickets and on the ground in dense brush or at the edge of the water on the floodplain during the rainy season. The aggressiveness of the birds and scarcity of woods with closed canopies apparently make arboreal life more difficult for these lizards.

The smaller iguanid species are much more abundant in the drier parts of southern Mato Grosso, particularly near outcroppings of rock, than they are in the Pantanal. Only *Tropidurus* cf. *torquatus* (Wied, 1815), was observed in the Pantanal, and its occurrence there is exceptional. It was found only in relatively dry places at the edge of the wetland, generally on concrete structures, such as the foundation of bridges. Near the city of Cuiabá and in other upland areas near human habitations, this species is extremely abundant. A large population inhabited the campus of the Federal University of Mato Grosso. The congeneric *Tropidurus spinulosus* (Cope, 1862) is also a member of the Pantanal herpetofauna (M. Carvalho, pers. comm.).

Other iguanids encountered in the Pantanal or at its outskirts include *Stenocerus caducus* (Cope, 1862) and *Polychrus acutirostris* Spix, 1825, neither of which seem to be common in the wetland.

Hoplocercus spinosus (Fitzinger, 1843) is a species typical of the *cerrado* and does not seem to occur in the Pantanal itself. It is nevertheless encountered in the drier uplands around the outskirts.

Teiidae

Two of the three abundant species of large lizard in the Pantanal, as well as numerous smaller ones, are teiids. A less common species of considerable size also belongs to this family. The caiman lizard or *víbora-do-Pantanal, Dracaena paraguayensis* Amaral, 1950, is the heaviest of the large lizard species in the Pantanal (Fig. 8.8). In contrast to the iguana, it is heavy-bodied and slow-moving, and individuals can be approached rather closely. Attempts to touch this animal are countered by hard slapping movements of the tail, with which the lizard defends itself against its enemies. The skin is covered by particularly hard, plate-like scales. The entire body of the animal closely resembles that of a caiman, hence its common name. Only the head with its long sensory tongue reveal at a glance the animal's membership in the Lacertilia. Several sightings of "caimans in trees" proved to be this species. The caiman lizard has particularly powerful jaws used to crush the shells of large snails, which are apparently one of the lizard's main sources of nourishment. Hence, the species is generally found near water, in which it escapes its enemies by diving beneath the surface and swimming away. On occasion, caiman lizards with amputated limbs or tails have been observed, probably attesting to attacks by fishermen or road workers. According to the revision of the genus by Vanzolini and Valença (1966), this is one of the few species endemic to the region of the Pantanal. A congeneric species is known from the eastern part of the Amazon Basin.

The Pantanal is also the home of the tegu lizard or *teiu*, *Tupinambis teguixin* (Linnaeus, 1758). Unlike the other large lizard species discussed above, it does not like to enter the water but rather hides on the hummocks above the high water level. During the late *vazante*, as the falling water level exposes increasingly large areas of the floodplain, these lizards are frequently seen wandering far from their woodland homes, possibly on courtship tours (Fig. 8.9). When disturbed, they flee extremely rapidly on their hind legs toward the nearest stand of trees. Like the iguana and caiman lizard, the tegu is not locally eaten but is nevertheless subject to attacks by man. However, it is more adapt at flight and escape than the caiman lizard and iguana. It is most endangered by cars on the roads during its traveling season, and many tegus were observed dead on the road. The species is known from the Guianas throughout Brazil to northern Argentina and Uruguay (Peters & Donoso-Barros, 1970).

A congeneric species of more limited range, *Tupinambis mariane* (Duméril & Bibron, 1839), is also present in the Pantanal (Araujo, 1994). It apparently occurs less frequently than the common tegu, but its distribution and habits still require much investigation. Peters and Donoso-Barros (1970) had considered *T. mariane* to me a junior synonym of *T. teguixin teguixin*.

Figure 8.8: The body of the caiman lizard, *Dracaena paraguayensis*, superficially resembles that of a caiman. It is a large, slow moving lizard that seems to feed mainly on snails.

The Pantanal is an attractive habitat for some smaller lizard species, including *Ameiva ameiva* Linnaeus, 1758, which is one of the most abundant terrestrial vertebrates along the Transpantaneira Highway. Small groups of these lizards can sometimes be observed moving along the margins of the road, where they looks for prey beneath fallen leaves and other objects. The species is somewhat variable in color, with the green in the pattern generally becoming more extensive as the individuals becomes older. According to Vitt & Colli (1994), only *Iguana iguana* has a greater range throughout the Neotropical Region than this teiid, but the behavior and ecology of *A. ameiva* is similar in whatever kind of habitat it is found in. In the wet-and-dry regions, its reproductive pattern is seasonal, with reproduction largely curtailed during the dry season (Colli, 1991).

Other teiids known from southern Mato Grosso and the region of the Pantanal include *Kentropyx viridistriga* (Boulenger, 1894), which has a range including Paraguay and Argentina as well as the Central Western Region of Brazil (Peters & Donoso-Barros, 1970). *Cnemidophorus ocellifer* (Spix, 1825) is also a common species, observed most often running along the edges of roads. *Pantodactylus schreibersi* Wiegmann, 1834; *Bachia dorbignyi* (Duméril & Bibron, 1839); *Bachia bresslaui* (Amaral, 1935); and *Micrablepharus maximiliani* (Reinhardt & Lütken, 1862) are also included among the regional herpetofauna. Generally, however, they are species encountered on the outskirts rather than in the wetland itself, although they may be encountered along the roads far into the wetland.

Teius tiyou (Daudin, 1802), sometimes reported for the Pantanal, is known to occur only in Mato Grosso do Sul and only reaches the extreme southern border areas of the Pantanal (M. Carvalho, pers. comm.).

Figure 8.9: The abundant tegu lizard, *Tupinambis teguixen*, is a very fast moving teiid that runs on its hind legs to escape predators.

Figure 8.10: Skinks are seldom seen in the Pantanal, and only *Mabuya frenata* was observed during the study, although other species are reported from the wetland, as well.

Scincidae

Only a single member of the family Scincidae, *Mabuya frenata* (Cope, 1862), was observed in the Pantanal (Fig. 8.10). This species was observed among the large cacti of the genus *Cereus*, probably *Cereus peruvianus*, which are abundant at the edges of the small wooded areas above the high water level of the flood season. The congeneric *Mabuya guaporicola* Dunn, 1935, *Mabuya dorsivittata* Cope, 1862, and *Mabuya bistriata* (Spix, 1825) are also found in the region of the Pantanal, but they were not observed during the study (M. Carvalho, pers. comm.). Peters and Donoso-Barros (1970) considered *M. bistriata* to be a junior synonym of *M. mabouya mabouya* (Lacépède, 1788).

Geckonidae

The most common gecko in the region of the Pantanal is *Phyllopezus pollicaris* Spix, 1825, represented by the subspecies, *P. p. przewalskii* Koslowsky, 1895, which is found in most houses in southern Mato Grosso, including ranch houses in the Pantanal. The species is familiar on walls under lights at night, where it captures insects flying to the light.

An unidentified species of the genus *Lygodactylus*, probably introduced from tropical Africa or Madagascar, is also known from Mato Grosso and Bahia (Vanzolini, 1968b). Much is still to be learned about the distribution and spread of native and introduced gecko species in South America, and some of the available information was summarized by Vanzolini (1968c).

Serpentes

In contrast to that in the adjacent *cerrado* and pasture regions, the snake fauna of the Pantanal is not rich. During the study, only three species with semi-aquatic habits were abundant, and two terrestrial species that live in the drier areas of the wetland were observed. However, many snakes may be encountered near the borders of the Pantanal that are not essentially wetland species, especially during the dry season. Furthermore, snakes are abundant on the high ground of the *cordilleiras* on which many of the cattle ranches are located.

The natural snake populations in the Pantanal are not presently endangered, but they will require protection in the form of public education programs should the number of visitors to the Pantanal increase (Section 13). Although none of the species of the wetland are poisonous, they are often found dead or scarred due to human aggression. Snakes near the highway are sometimes needlessly killed by fishermen, and a large number are killed on the highways by automobiles near Poconé at night.

Boidae

The Boidae is fairly well represented in Mato Grosso. As mentioned above, the largest snake in the Pantanal is the anaconda or *sucuri*, *Eunectes notaeus* Cope, 1862, which is much smaller than the Amazonian anaconda but nevertheless reaches a length of more than 4 m (Fig. 8.11). Amaral (1977) reported the average size of adults to be only 3.3 meters, while adults of the congeneric anaconda in the Amazonian region, *Eunectes murinus* (Linnaeus, 1758), average more then 8 m. *E. notaeus* is a general predator, feeding on mammals, large birds, and also on the caiman (Strussmann & Sazima, 1993). One of these snakes was observed consuming a great white heron, and one individual in captivity at the university animal park escaped from its enclosure and ate a *lobinho*, *Speothos venaticus*, in a neighboring enclosure.

Normally, the anaconda remains concealed among plants underwater, where it is in an ideal position to suddenly seize passing prey. During the dry season, they are forced to leave the drying water bodies and undertake migrations overland to deeper lying water bodies. The species is also common in water bodies of the chaco region to the west of the Pantanal in Bolivia (Gallardo, 1979).

Other boids are encountered in the upland regions surrounding the Pantanal. These include the *jibóia, Boa constrictor* Linnaeus, 1758, and *Epicrates cenchria crassus* (Cope, 1862), which were fairly frequently observed along the highways, *B. constrictor* during the day and *E. cenchria crassus* at night. Both species seem to feed mainly on mammals and birds, their habits being fairly well known through observations in zoological parks.

Figure 8.11: The most frequently observed snake in the Pantanal is the relatively small anaconda species, *Eunectes notaeus*. This individual was leaving a drying pool of water during the *seca* to seek a deeper one. It was closely observed by two smooth-billed anis, *Crotophaga ani*, which followed the snake as it moved, always keeping its head in sight and making bowing movements with their wings partially open as it moved along.

Typhlopidae

The presence of *Typhlops brongersmai* Vanzolini, 1972, in the Pantanal of Poconé was reported by Strussmann and Sazima (1993). The species is insectivorous and was encountered by these authors in a decaying palm trunk. None of these secretive, fossorial reptiles were personally encountered during the study.

Leptotyphlopidae

The occurrence of *Leptotyphlops albifrons* (Wagler, 1824) and *L. koppesi* Amaral, 1955, in southern Mato Grosso is probable because they are generally distributed in the western Brazil. Because of their secretiveness, however, more precise range information and reports of confirmed occurrences in the wetland are still unavailable. Much must still be learned about the habits of these interesting reptiles.

Anomalepedidae

The only member of this family known from the region of the Pantanal is *Liotyphlops ternetzii* (Boulenger, 1896). This secretive snake belongs to a family believed to include typically myrmecophilous species. The habits of most are poorly known. This species was reported by Amaral (1977) to occur from Paraguay eastward to the state of São Paulo, apparently with a population center in Mato Grosso and Mato Grosso do Sul.

Colubridae

The taxonomy of the species included in many colubrid genera in South America is not satisfactory, and the nomenclature of several species will probably be revised in the years to come. It will nevertheless be possible to determine which species have been referred to from the names used below, which are mainly taken from the works of Peters and Orejas-Miranda (1970) and Amaral (1977), with amendments as cited in the text.

A large colubrid snake that is common in the Pantanal is the *jaracuçu-malha-de-piava, Hydrodynastes gigas* (Duméril, Bibron, & Duméril, 1854), which is diurnal and frequently seen moving across the floodplain during the dry season. These snakes are also abundant in and near water during the rainy season, but they usually remain concealed. Adults may exceed 2.5 m in length. When approached, these snakes usually flatten their necks vertically to reveal a pattern of yellow, vibrate their tails, and display a threatening behavior while fleeing in the direction of water or woodland. Strussmann and Sazima (1993) reported that this species together with the anaconda accounted for 34% of all snakes they encountered during their survey in the Pantanal. They reported that the diet changes seasonally. During the period of flooding, it consists mainly of frogs and fishes, while in the dry season, mammals and dead fishes and frogs may be consumed. It is also suspected that birds and their eggs may also be eaten by this general predator (Strussmann & Sazima, 1990).

The *caninana*, *Spilotes pullatus pullatus* (Linnaeus 1788), is a somewhat less abundant, large snake found in the region of the Pantanal. Amaral (1977) reported that particularly aggressive individuals may pursue people and large animals. If seized, the snake will bite and release a malodorous substance, as previously described by Greene (1988). The diet of the species consists of small vertebrates, mainly mammals, birds, and their eggs. It is an oviparous snake that usually lays 10 to 16 eggs at one time (Amara, 1977).

In addition, a large, fast-moving snake, locally called the *cascavel-do-charco*, and elsewhere the *biru*, with the official name, *Mastigodryas bifossatus* (Raddi, 1820), was observed occasionally on the floodplain after the water had receded during the *vazante* and *seca*. Strussmann and Sazima (1993) observed this species in forested and cleared areas, where it reportedly feeds mainly on frogs but consumes lizards and small mammals, as well.

The indigo snake, *Drymarchon corais* (Boie, 1827), is another resident species in the terrestrial regions surrounding the Pantanal. Strussmann and Sazima (1993) reported finding two specimens in forests in the Pantanal of Poconé. The indigo snake is a rather unselective predator that consumes mainly various kinds of vertebrate. It is a large snake with a range extending northward to the subtropical regions of North America.

A water snake commonly encountered on the floodplain during the high water period is *Helicops leopardinus* (Schlegel, 1837). Strussmann and Sazima (1993), who found it to be one of the most abundant species during their survey, reported that fishes and frogs are its main food.

Among the small reptiles in the Pantanal, *Liophis poecilogyrus* (Wied, 1825) is one of the commonest. These rather small snakes are most often encountered beneath objects at the water's edge or among piles of rotting vegetation. This species subsists on small fishes and frogs, which share its habitat. It is aquatic and generally escapes into the water when threatened. Its range seems to be confined to Mato Grosso and the Amazon region. The species was included in the genus *Liophis* by Dixon (1989), but in the past, it had been referred to as *Leimadophis poecilogyrus*, following the classification system of Peters and Orejas-Miranda (1970).

A snake with somewhat similar habits, *Liophis longiventris* Amaral, 1925, also seems to be common in southern Mato Grosso. It prefers marshy areas with much debris under which to hide and is present on the floodplain and along the small streams draining it. Strussmann and Sazima reported that its diet consists of frogs. According to Dixon (1989), the known range of this species does not extend northward of the southern boundaries of the Pantanal, so its occurrence in the northern part of the wetland extends its known range considerably.

The dark-colored, piscivorous water snake, *Pseudoeryx plicatilis plicatilis* (Linnaeus, 1758), is encountered along the river systems throughout the Pantanal. It was found on the floodplain by Strussmann and Sazima (1993) during the seasons of flooding, as well. It also occurs in the Amazon Valley (Amaral, 1977).

A thin-bodied terrestrial snake, *Psomophis geniculatus* (Boettger, 1885), is a species with a range thought to be limited to the drainage area of the upper Rio Paraguay and adjacent regions of Amazonas and Goiás (Amaral, 1977). It was observed on the Fazenda Ypiranga about 15 km south of Poconé during the *enchente* and apparently is not uncommon in the northern part of the Pantanal. This oviparous species, which reportedly feeds on lizards, frogs, and rodents, is locally called the *jararaca lançada* in Goiás.

In addition to the few wetland species, a considerable number of colubrid snakes are found along the northern border of the Pantanal. *Sibynomorphus mikanii mikanii* (Schlegel, 1887) was abundant on the plateau to the east of the Pantanal Basin as well as in the grassland and *cerrado* to the north of Poconé. Like the crotalids, discussed below, this was a snake that seemed to avoid the wetlands. The congeneric *Sibynomorphus turgidus* (Cope, 1868) is probably endemic to the region of the Upper Paraguay River (Amaral, 1977), but it was not seen during the study.

One of the most widespread of all Brazilian snakes, *Clelia clelia* (Daudin, 1803), can be encountered throughout deforested terrestrial habitats in the region of the Pantanal (Strussmann & Sazima, 1993), where it is represented by the subspecies *C. c. clelia*. Sometimes it appears in the literature under the name *Pseudoboa cloelia* (Daudin, 1803), and in Mato Grosso, it is known by the common name, *muçurana*. It generally remains on the ground, moving at night through fields and forests, where it preys on other snakes, including the poisonous species of *Bothrops*. The females lay from 8 to 14 eggs each time they reproduce (Amaral, 1977). The species is not only found throughout Brazil, its range extends northward into Central America. A second species of the genus was found to be even more abundant in the Pantanal by Strussmann and Sazima (1993). They found *Clelia bicolor* (Peracca, 1904) to be abundant in deforested areas, but it also was observed on the floodplain. Its prey includes frogs, lizards, and snakes.

Another species found throughout all of Brazil, *Waglerophis merremii* (Wagler, 1824), also known as *Xenodon merremii*, is an oviparous snake that can be encountered throughout the entire region of the Pantanal. It is terrestrial, however, and avoids the flooded areas. The species has large teeth on the posterior parts of the jaw bone, which it uses for puncturing the toads on which it preys to deflate them, reducing their size for swallowing. Smith (1963) studied their resistance to the toxins produced in the skin of the toads. The females produce six to eight eggs each time they reproduce (Amaral, 1977). Within its great range in Brazil, this species has accumulated a large number of common names. *Xenodon neuwiedii* (Günther, 1863), known as the *quiriripita* in Mato Grosso do Sul, was not observed in the northern part of the Pantanal and is apparently confined to the South.

Among the species found dead on the road in the pasture and *cerrado* region near Poconé were two moderately long, green snakes, *Philodryas olfersii* (Lichtenstein, 1823) and *Philodryas viridissimus* (Linnaeus, 1758), both of which feed on frogs and probably lizards, as well. *P. viridissimus*, known as the *tucanaboia*, is a diurnal, arboreal species that feeds mainly on lizards, rodents, and frogs (Amaral, 1977). It has an extensive range in South America, being known from Surinam and the Amazon Valley as well as in the watershed of the Rio Paraguay. Other victims of the heavy traffic on the roads through the *cerrado* were *Phimophis guerini* (Duméril, Bibron, & Duméril, 1854) and *Liophis jaegeri* (Günther, 1858). One specimen of *Tantilla melanocephala melanocephala* (Linnaeus, 1758), nearly all white except for the head, was found dead about 20 km south of Varzea Grande on the road to Pôconé. This species is widespread in Brazil, having been reported to live in subterranean retreats in gallery forests of Amazonia (Amaral, 1977).

Lystrophis matogrossensis (Scrocchi & Cruz, 1993), known locally as the *boi-corá*, is a non-poisonous "false coral snake" with an upturned snout that was common at one terrestrial region near the confluence of the Rios Cuiabá and Paraguay, where many cattle ranches are located. One was found alive on the dirt road about 20 km north of the Fazenda São Antônio das Lendas and another was found dead on the same road three months later. This oviparous snake, which is known from Mato Grosso southward to Argentina, reportedly feeds on frogs and insects (Amaral, 1977). Another "false coral snake", *Oxyrhopus rhombifer* Duméril, Bibron, and Duméril, 1854, was found dead on the road near Poconé. It is known as a species of the dry forests, which feeds mainly on rodents (Amaral, 1977).

Another common species in the upland fields of Mato Grosso is *Leimadophis almadensis* (Wagler, 1824), which is thought to make itself valuable to farmers by preying on rodents as well as lizards. However, Strussmann and Sazima (1993) reported only frogs as items in its diet. The species can be recognized by the red coloration of its venter. The females lay 6 to 10 eggs each time they reproduce.

Peters and Orejas-Miranda (1970) reported that the nocturnal *Pseudoboa nigra* (Duméril, Bibron, & Duméril, 1854) is primarily a savanna species. Strussmann and Sazima (1993) reported its presence in forests and deforested regions of the Pantanal and that its diet consists of lizards. It was not personally observed during the study.

Chironius quadricarinatus (Boie, 1827) was reported by Strussmann and Sazima (1993) from forests, where it is a semi-arboreal, diurnal snake that feeds on frogs.

Philodryas nattereri (Steindachner, 1870) occurs from the northeastern states of Brazil through Mato Grosso, where the type specimen was taken. Called the *cobra-cipó*, it is another diurnal, arboreal snake that feeds on lizards, frogs, and birds. The females produce 10 to 14 eggs each time they reproduce (Amaral, 1977). More limited in distribution but with generally the same life style is *Philodryas mattogrossensis* Koslowski, 1898, a snake first described from a specimen taken in the Miranda region of Mato Grosso do Sul. It seems to be quite limited in distribution and was not observed in the northern part of the Pantanal. According to M. Carvalho (pers. comm.).

A thin arboreal snake with a relatively large head, *Dipsas albifrons* (Sauvage, 1884), is also native to the Central Region of Brazil. This arboreal species was not observed during the study. Another thin, arboreal species that is native to Mato Grosso as well as the northeastern states of Brazil, *Taeniophalus occipitalis* (Jan, 1863), also escaped observation, although its presence in the region of the Pantanal is assumed.

A common arboreal snake of woodlands, *Leptophis ahaetulla marginatus* (Cope, 1862), feeds on frogs and lizards. The females produce about 12 eggs each time they reproduce (Amaral, 1977). Strussmann and Sazima (1993) encountered the species in forests and cleared areas in the Pantanal of Poconé.

The *cacaual*, *Leptodeira annulata pulchriceps* Duellman, 1958, is a nocturnal arboreal species, which inhabits the region of the wetland but was not observed during the study.

Several species of *Apostolepis* have been reported from the region of the Pantanal, although the presence of these poorly known snakes was not detected during the study. These include *Apostolepis assimilis* (Reinhardt, 1861), known from the Central Region of Brazil, including the region of the Pantanal; *Apostolepis* cf. *rondoni* Amaral, 1925, known from Chapada dos Guimarães northeast of Cuiabá and otherwise from the Amazon Drainage Area of Mato Grosso; *Apostolepis dimidiata* (Jan, 1862); and *Apostolepis intermedia* Koslowski, 1898.

Strussmann and Sazima (1993) collected specimens identified as *Liophis joberti* (Sauvage, 1884) from forests and deforested pastures in the Pantanal. The classification of this species is problematic, and the authors placed the generic name in quotation marks to stress that the systematic position of this species, which was treated as *incertae sedis* by Dixon (1980), is not presently clear. The authors also reported that the species was illustrated by Amaral (1977) under the name *Liophis genimaculatus* Boettger, 1885. Another taxonomically problematic species found on the floodplain during the rainy season by Strussmann and Sazima (1993) was called *Liophis reginae* (Linnaeus, 1758) in their paper and by Dixon (1989) but had formerly been placed in the genus *Leimadophis* by Peters and Orejas-Miranda (1970). It reportedly feeds mainly of frogs.

In addition to the species encountered, several others of the genus *Liophis* should range throughout southern Mato Grosso (Dixon, 1989). These include *L. meridionalis* (Schenkel, 1901) and *L. typhlus* (Linnaeus, 1758). In addition to these, *L. dilepis* (Cope, 1862) and *L. melanotus* (Shaw, 1802) reportedly occur on the southern borders of the Pantanal in Mato Grosso do Sul (Dixon, 1989).

Atractus reticulatus paraguayensis Werner, 1924, and *Phalotris mertensi* (Hoge, 1955) are other colubrids present in southern Mato Grosso. They were not personally encountered in the Pantanal.

One fossorial colubrid, *Elapomorphus tricolor* Duméril, Bibron, and Duméril, 1854, was reported from the region of the Fazenda Santa Inês between the Rios Paraguaizinho and Bento Gomes near the western border of the Pantanal of Poconé by Strussmann and Sazima (1993).

Taxonomic confusion is also encountered in the case of a secretive species believed to be *Paroxyrhopus undulatus* (Jensen, 1900), which was referred to as *Xenopholis* cf. *undulatus* by Strussmann and Sazima (1993), who collected specimens form cleared forests in the Pantanal and reported it to feed on frogs.

Pseudablabes agassizii (Jan, 1863) is a species found in the upland region to the north of the Pantanal. It has been collected in the city of Cuiabá (M. Carvalho, pers. comm.) and is probably common beyond the borders of the Pantanal.

Among the snakes that seem to be confined to the southern part of the Pantanal region is the *bio-pemi*, *Tomodon dorsatus* Duméril, Bibron, and Duméril, 1854, a species ranging throughout southeast Brazil. It is a rather fast-moving snake encountered in open fields. *Platynion lividum* Amaral, 1923, is a rather plain-looking yellowish brown snake with darkened borders to the scales. It is reportedly found from Mato Grosso to Paraná (Peters & Orejas-Miranda, 1970), but apparently it occurs only in the upland regions of Mato Grosso do Sul to the south of the wetland (M. Carvalho, pers. comm.).

Other species of the southern Pantanal that apparently do not range as far north as southern Mato Grosso include *Philodryas patagoniensis* (Girard, 1857), *Philodryas psammophideus* Günther, 1872, and *Simophis rhinostoma* Schlegel, 1837.

The representation of the family Colubridae in the Pantanal wetland is obviously poor, with only a few species accounting for nearly all of the sightings. In contrast, the fields and forests of the upland regions surrounding the wetland seem rich in species, as the foregoing discussion shows. This fauna is most endangered by thoughtless human activities, including the deliberate killing by many residents and visitors of all snakes encountered.

Elapidae

Elapids are uncommon in southern Mato Grosso. The only species known to occur in the region of Poconé has been recently referred to by various names, including *Micrurus tricolor* Hoge, 1957, by Strussmann and Sazima (1993), *M. frontalis pyrrhocryptus* (Cope, 1862) by Peters and Orejas-Miranda (1970), and *M. pyrrhocryptus* (Cope, 1962) by Hoge and Romano-Hoge (1981). Roze (1983) also considered it to be a subspecies of *M. frontalis* (Duméril, Bibron, & Duméril, 1854). Its taxonomic position remains uncertain. It probably does not occur in the wetland at all, as Strussmann and Sazima (1993) placed it on their list because specimens reportedly from the region were obtained from other collections. This nocturnal species, which seems to feed mainly on snakes, must be generally regarded as rare in southern Mato Grosso.

Viperidae

The natural snake fauna of the Pantanal apparently includes only two poisonous species that venture onto the floodplain. Strussmann and Sazima (1993) reported the presence of *Bothrops moojeni* Hoge, 1966, and *Bothrops neuwiedi* Wegler, 1824, collectively called, on temporarily flooded areas. However, the neighboring *cerrado* is also inhabited by these and at least one additional species of crotalid: the South American rattlesnake or *cascavel*, *Crotalus durissus* (Linnaeus, 1758), represented by the subspecies *C. d. collilineatus* Amaral, 1926. The habits of the South American rattlesnake were studied by Graça Salomão *et al.* (1995), who found that they subsist mainly on rodents. The adequacy this diet for *Bothrops neuwiedi* and *Crotalus durissus* was confirmed in the laboratory, where specimens rescued from the roads readily fed on laboratory mice. One *Bothrops neuwiedi* grew considerably for over a year until the project ended. It was then turned over to a colleague at the Federal University of Mato Grosso, who continued to maintain it successfully on the same diet. Mice died in a few seconds after being bitten. *B. neuwiedi* is represented in southern Mato Grosso by two subspecies: *B. n. matogrossensis* Amaral, 1925, and *B. n. boliviana* Amaral, 1927.

Apparently, the periodic flooding discourages most of these species from entering the Pantanal. These and one elapid (above) seem to be the only poisonous snakes in the region of the wetland. In contrast, venomous snakes seem to be quite common in the *cerrado* and savanna, where they were seen on frequent occasions during the study, especially along the roads at night. None were ever seen by the researchers in the wetland, and none of the residents of the Pantanal ranches ever reported seeing what they recognized to be one of the poisonous species there, although several expressed the belief that various harmless colubrids were poisonous. On the roads through the *cerrado* between Poconé and Cuiabá, however, a considerable number of these viperids were found after they had been killed by cars. Snakes, in general, seek the warmth of the paved roads on cool nights and suffer a high mortality as a result.

Crocodilia

The only crocodilian known from the Pantanal region is *Caiman yacare* (Daudin, 1802), also known as *Caiman crocodylus yacare*. This caiman was reduced to a subspecies of *Caiman crocodylus* (Linnaeus, 1758) by Werner (1933), but this status has not been universally accepted, and *Caiman yacare* has appeared in more recent literature. Müller and Hellmich (1936) speculated whether or not the species should be considered a race of *Caiman crocodylus* based on studies of the skulls. While they found considerable differences, they also noted the great variability of the characteristics and concluded that the overlap in the features is great. They therefore tended to agree with Werner (1933) that the two should be considered conspecific, although clear differences exist. The question seems far from settled, and a practical reason for tentatively regarding this caiman as a distinct species will be discussed further below.

This species is well adapted to the annual cycle. At the onset of the rainy season, the caimans apparently begin their mating activity. Only at this time of year do the males use their voices. When they make their sounds, they arch their backs so that the head and tail are lifted high out of the water. The smallest adults sound almost like small dogs barking, while the largest ones make a very deep resonant sound that can be heard a long distance. As the rainy season continues and the water begins to become deeper, they cease this activity. While spending most of a morning near the Rio Bento Gomes along the Transpantaneira Highway, I heard a considerable number of caimans of various sizes emitting their calls, but I never observed any one of them calling for a second time. Most of them called during the period within an hour after sunrise, and thereafter, this activity diminished. By about 11 o'clock, they had nearly all left the water and were sunning themselves along the shores. No further calls were heard.

The flood and high water periods mean famine for the Pantanal caimans. During this season, the few surviving fishes are able to leave the deeper streams and rivers and spread out over the vast floodplain to breed. They become hard to find, and the caimans normally begin to look emaciated as the flood period progresses. As the fishes of the new generation begin to rapidly increase in size, fishing steadily improves for the caimans. During the high water period, they are rarely observed during the day. They remain concealed in rather deep water among plants, and only at night is their presence detectable through the reddish glow of reflected light from their eyes. It is also during this period that the nests are built and the eggs incubated. The adults seem to remain near the nests to guard them (Cintra, 1985, 1988), as reported for other crocodilians (Magnusson *et al.*, 1989), some of which are known to fast during the weeks that they are guarding their nests. All reproduction does not seem to occur during a particular breeding season, however, and at least the common caiman is capable of producing eggs during almost any season (Magnusson *et al.*, 1989). *Caiman yacare* seems to construct its nests particularly close to the water, and Cintra (1985) reported that the average distance of four nests he observed from the water´s edge was only 2.3 m. This increases the risk of flooding when the water level becomes unusually high, and Campos (1993) reported that, except perhaps for predation, flooding causes the highest mortality among the young. While parental care protects the young from most of the larger predators, the fire ants, *Solenopsis wagneri*, reportedly take a toll on the young in the nests (Cintra, 1985).

The sex of the developing caimans is determined by the temperature rather than by the chromosome complement (Lang, 1989). When the temperature is relatively high, all of the eggs yield males, while females are produced when the temperature is relatively low. Only within an intermediate temperature range are mixed broods of males and females produced. In the Pantanal, Cintra (1985) recorded the conditions in the nests and reported that the maximum temperature variation in any nest during the incubation period was 9°C. The incubation periods for the eggs in the four nests he observed from January through April were 92, 92, 91, and 61 days.

As the water level rapidly begins to fall during the *vazante*, the fishes become more and more crowded together in shrinking shallow water bodies, often with low oxygen concentrations in the water. At this time of year, the caimans gather in these pools and begin to consume huge quantities of fishes from the dying populations. The animals become extremely fat, and the tails show especially large deposits of stored nutrients needed by the animals to survive the next rainy season. The caimans in the Pantanal were observed to feed most intensively during the hours just after sunrise, when the oxygen concentration is lowest (Section 6), forcing the fishes to stay at the surface for respiration. They also feed rather actively just before sundown. During the midday hours, the caimans leave the water and sun themselves on the shore. They usually do this in large groups, and in some locations, great numbers of large caimans can be seen together.

The Pantanal caiman is a species that exhibits a rather complex social behavior and parental care for the young. The hatchlings are more or less sealed within a nest constructed of vegetation, and they are able to emerge only with the help of adults, which hear the grunts of the young and tear open the nest to free them.

The young caimans usually hatch and leave the nest during the early dry season. They often seek the protection of a large caiman, usually the mother, which they remain close to in the water. As the water bodies dry up, the small caimans wander in large groups to seek deeper water bodies, sometimes following a large individual. Cintra (1989) observed the behavior of one of these caiman families in the Pantanal of Poconé for a long period of time and reported that a group of 22 young remained for a period of six months with the mother. The caimans usually flee when approached too closely, but the only time a large caiman ever came toward me was after I had picked up a hatchling that had become separated from the rest of the group. When I put the young caiman down in front of her, the adult went on her way followed by the group of hatchlings. Mortalities occur when such groups of caimans cross the Transpantaneira Highway and are run over by motorists who do not bother to stop or who deliberately try to run them over.

Within the group of young, there is considerable communication by grunts. It was observed several times that one of the hatchlings slightly larger than the others begins to grunt, and the entire group leaves the water at once to rest beneath terrestrial vegetation along the shore. Later, the grunt signals are again given, and the entire group returns simultaneously to the water. During this time, a large individual which the young have chosen as a protector, perhaps the mother, remains in the water. This individual is often concealed beneath the surface, and when the young are frightened, they congregate above this adult. While wading in the water, I accidentally kicked such large caimans at times without provoking a reaction.

Figure 8.12: An adult caiman, *Caiman yacare*, beside the Transpantaneira Highway.

In contrast to the young, the adults rarely make sounds, except for the bull caimans at the time of the freshet, as mentioned above. *Caiman yacare* is an extremely docile crocodilian, and it remains rather small (Fig. 8.12). A piscivorous species, the caiman presents little or no danger to man and apparently little danger to the water birds, which are frequently seen wading close to large adults near shore without evoking any response from the reptiles. Large caimans accidentally kicked or stepped on while submerged did nothing but flee, even when young were present in the water, as mentioned above. On one occasion, while wading to photograph a group of hatchling caimans in shallow water, I kicked a submerged object. At first it was thought to be a log, but as it came slowly to the surface, the head of a large female caiman became visible in the turbid water. The hatchlings quickly congregated at the surface of the water above the head, apparently seeking protection. I photographed the young above the still submerged head of the large caiman before beginning a slow retreat from the shallow water. Although the head of the caiman was only a few centimeters away when it surfaced and only about one to two meters away while being photographed, the animal made no aggressive movements. On another occasion, I was able to remove the leeches attached at the base of the hind legs of a full grown caiman using drops of alcohol applied from the end of a straw. The caiman was approached while it was sunning itself, and by moving slowly, I was able to get close enough to take hold of the back leg. Although the caiman growled and hissed at times, it remained in its place long enough to permit me to remove over a dozen leeches.

Because caimans are mainly piscivorous (Schaller & Crawshaw, 1982), they were deliberately killed off in the Amazon region because they were thought to reduce the number of fish species prized by the fishermen. Fittkau (1973) investigated the fish populations and found that the fishery yield decreased after the caimans were eliminated. He surmised that caimans had played a key role in liberating nutrients necessary to support the productivity of the aquatic community. In the Pantanal, it can be assumed that most of the fishes consumed by the caimans would die anyway due to the deterioration of conditions during the dry season, and the elimination of the caimans would have a detrimental effect on the nutrient supplies for the microorganisms and thereby decrease the total fish production, as it seems to have done in the Amazon Basin.

Because of its presence in large groups near shallow water bodies during the dry season and its docility, the caiman is an ideal object for poaching. Crawshaw (1991) reported the effects of hunting the caimans in the Pantanal. The species is still abundant in many parts of the wetland, but estimates of the numbers killed illegally each year are high. Because its population is rather stable, it is one of the species that could become extinct in a rather short period of time. This problem will be discussed further in Section 13.

Returning to the question of whether the Pantanal caiman is a distinct species or a subspecies of the abundant Amazonian *Caiman crocodylus*, we find that this caiman was generally known as *Caiman yacare* for a great many years, in spite of the publication of Werner (1933) discussed above. Today, although few studies have yet been carried out on its morphology, physiology, genetics, and ecology, specialists have tended to "clump" it with the Amazonian species, calling it *Caiman crocodylus yacare*. This kind of speculative taxonomy seems to follow trends and fads, and until conclusive information is available, I would strongly suggest maintaining this caiman in the literature as a species unless more evidence to the contrary is obtained, through breeding experiments, for example. The reason for this is the improved legal status this caiman would benefit from. Whether it is considered an endangered species or only a subspecies of the more abundant common caiman could in the future determine whether adequate measures are taken to protect it.

It is interesting to note that two other crocodilian species formerly seem to have inhabited the Pantanal. *Caiman latirostris* (Daudin, 1802), inhabits the chaco region to the west of the Pantanal (Gallardo, 1979). Occasionally, it has been reported to occur in the southwestern part of the wetland itself, but it is probably already extinct there (Dubs, 1995). However, the herpetofauna of the chaco, as reported by Gallardo (1979), is quite different from that in the Pantanal, and *Caiman yacare* does not seem to occur there at all. However, during an expedition to Paraguay in the first half of this century, Müller and Helmich (1936) were able to collect two specimens of *Caiman latirostris* along with many *Caiman yacare*. Unfortunately, they reported little about the kinds of habitats frequented by these two species, and it is therefore not certain whether they were ever sympatric within parts of their ranges. At present, it seems that a rather sharp demarcation line exists between the kinds of habitat each prefers, and the distribution of the crocodilian fauna seems to be strictly limited by the availability of a preferred environment.

The smooth-fronted caiman or *jacaré-de-corôa*, *Paleosuchus palpebrosus* (Cuvier, 1807), was reported from the northern part of the Rio Paraguay System at the beginning of this century (Perraca, 1904), but it is now confined to the Amazon Region and northern South America (Wermuth & Mertens, 1961).

8.13.4. Aves

The large numbers of birds is one of the most notable features of the Pantanal. A great many of these species nourish themselves from the rich supply of fishes and amphibians in the wetlands. The majority, however, subsist on other foods, particularly insects. Two of the birds consume mainly snails, while the many parrots are chiefly seed and fruit eaters, and the toucans subsist predominantly on fruit, although their reputed taste for the eggs of other birds makes them the target of frequent attacks by smaller birds.

The species in the Pantanal can be classified according to their ecological relationship to the wetlands. Only a relatively small percentage of the bird species discussed below are among the aquatic avifauna typical of the Pantanal. These species can be observed regularly in the wetland, at least during one of the seasons. Most of the common water birds that are observed in the Pantanal of Poconé are listed in Table 8.3, and their interactions with other groups of species are analyzed in detail in subsequent sections of this book. Other species are common terrestrial birds in the Pantanal or are abundant faunal elements of southern Mato Grosso but are not particularly abundant in the Pantanal. These are discussed with their respective families below. Additional species known from sightings in the northern Pantanal are also included, although they are abundant neither in the wetland itself nor elsewhere in the watershed. Such species may have formerly been abundant but are now rare because of habitat alterations, or they may be species that visit the region from time to time while on their migrations. The complete list of birds observed in the northern Pantanal is relatively short compared to the number known from the Pantanal of Mato Grosso do Sul. At the biological research station of the Universidade Federal de Mato Grosso do Sul near the Passo da Lontra, about 650 bird species have been observed or captured, and new species are still being found during investigations involving routine trapping and banding. The lack of a regular program for collecting and identifying the birds in Mato Grosso may be a partial explanation for a considerably lower number of species being recorded from the northern part of the Pantanal, but the richer variety of habitats in the southern part of the wetland probably also plays a large role in augmenting the avifauna.

The birds of Brazil are better known than the species of most other animal classes, particularly due to the exhaustive work by Sick (1988), who succeeded in providing illustrations for a great many of the rarer species. Many pictures suitable for identifying species in the field were provided by Schauensee and Phelps (1978), but their guide is limited to the species occurring in Venezuela, and a great many of the species observed in Mato Grosso are not included. Dubs (1983, 1992) provided short discussions and illustrations of most of the bird species of the Pantanal region. In addition, information about birds in many individual families is provided in monographic works, such as those on the hummingbirds by Grantsau (1989), the herons by Hancock and Kushlan (1984), and the raptors by Brown and Amadon (1968). Some of the older works provide extensive lists of avifauna with information on general distribution and synonymy but lack the illustrations necessary for reliable identification of the species in the field (Naumburg, 1930; Laubman, 1939, 1940; Pinto, 1939, 1940). A few shorter works deal specifically with the avifauna of specific subregions of the Pantanal, such as the publication on the birds of the Pantanal of Poconé by Cintra and Yamashita (1990) and that on the avifauna of the upland areas of southern Mato Grosso by Silva and Oniki (1988). In addition, some information on the species encountered in the wetland can be obtained from comprehensive works on the birds of neighboring countries, such as the monograph by Steinbacher (1962) on the birds of Paraguay. Various other works, too numerous to mention here, are cited in the text below.

In spite of the considerably greater number of publications on birds than on the other faunal elements in the Pantanal, there is an obvious deficiency of information on the ecology of a great many of the species, and considerably less is known about many of the smaller forest dwellers than about most of the rest of the world's avifauna.

Table 8.3: The typical water birds of the Pantanal, which occur abundantly in or near water during at least one of the seasons, and their main food, as observed in the wetland. The piscivorous species are arranged approximately in order of the size of their prey.

Species	**Seasonal occurrence**	**Feeding habits**
A. Fishes as main food		
Rynchops nigra	Resident	Skimming
Chloroceryle americana	Resident	Diving from above
Chloroceryle amazona	Resident	Diving from above
Ceryle torquata	Resident	Diving from above
Agamia agami	Resident	Spearing from shore
Nycticorax nycticorax	Resident	Spearing from shore*
Egretta thula	Resident	Wading and spearing
Phalacrocorax brasilianus	Resident	Underwater capture
Anhinga anhinga	Resident	Underwater capture
Ardea cocoi	Resident	Spearing from shore
Casmerodius albus	Resident	Wading and spearing
Tigrisoma lineatum	Resident	Spearing from shore
Butorides striatus	Resident	Spearing from shore
Mycteria americana	Migrant	Wading and spearing
Jabiru mycteria	Resident	Wading and spearing
Busarellus nigricollis	Resident	Grabbing in talons
B. Snails as main food		
Harpiprion caerulescens	Resident	Wading, shallow water
Rostrhamus sociabilis	Resident	Grabbing with talons
Aramus guarauna	Resident	Wading and spearing
C. Small aquatic organisms or seeds as main food: omnivores		
Ajaia ajaja	Resident	Spoon bill filtering
Pilherodius pileatus	Resident	Spearing invertebrates
Phimosus infuscatus	Resident	Pobing sediments
Amazonetta brasiliensis	Migrant	Browsing
Neochen jubata	Resident	Presumed omnivore*
Carina moschata	Migrant	Probing the water
Porphyrula flavirostris	Resident	Probing sediments*
Laterallus viridis	Resident	Probing sediments*
Eurypyga helias	Resident	Probing sediments*
Gallinago gallinago	Resident	Probing sediments
Tringa solitaria	Migrant	Probing sediments
Himantopus himantopus	Migrant	Probing sediments
Pseudoleistes guirahuro	Resident	Seeds, small animals
D. Terrestrial and pseudoterrestrial animals as main food		
Bubulcus ibis	Resident	Spearing insects
Theristicus caudatus	Resident	Supralittoral hunter
Jacana jacana	Resident	Spearing insects
Fluvicola pica	Resident	Snapping in flight
Arundinicola leucocephala	Resident	Snapping in flight
Alectrurus tricolor	Resident	Snapping in flight
Tachycineta albiventer	Resident	Snapping in flight

Species	Seasonal occurrence	Feeding habits
E. Mainly herbivorous		
Chauna torquata	Resident	Eats leafy parts
Sarkidiornis melanotos	Resident	General browsing
Dendrocygna autumnalis	Migrant?	Seed eater

* The diet has not been conclusively determined.

As shown in Table 8.3, most of the aquatic birds, like the fishes, belong to two general groups according to their preferred habitats. The first of these encompasses the birds of the shallow water on the floodplains, which move gradually toward the deepest locations as the water level decreases. The second group includes the birds of the gallery forests along the large rivers. In the case of the birds, these groups are rather ill defined, and there is much movement of the birds from one habitat to the other, except in the case of a few of the forest birds, which seem to shun the open countryside. The birds can also be classified according to whether they are residents which breed in the Pantanal, regular migratory visitors, or occasional visitors, which only appear irregularly.

Rheidae

The greater rhea or *ema*, *Rhea americana* (Linnaeus, 1758), which can reach a height of more than 170 cm, is widespread throughout the dry forest region of southern Mato Grosso. However, it generally prefers the pasture lands cleared for the herds of cattle (Fig. 8.13). The nests are constructed on the ground by the male, sometimes for the eggs of more than one female (Sick, 1988). During the dry season, small groups were frequently encountered on formerly flooded areas. Occasionally during the late dry season, flocks of rhea chicks were observed, tended by one or more adults. They travel in small family groups until the young birds have reached adulthood. During the period of flooding, most of the rheas leave the Pantanal for higher ground, but a few persist in more elevated areas of the wetland. Generally, these birds avoid even the shallowest water.

Many observations in and near the Pantanal confirmed that rheas show a decided preference for pasture land cleared for cattle. In the Pantanal, they are common on the floodplain after the water has receded and avoid both the dry woodland in the upland areas and the islands of forest on the desiccated wetland. These birds are not at all disturbed by the barbed wire fences that cross the pastures. In spite of their large size, they are adept at squeezing between two strands of the wire. They do this quite quickly and gracefully, placing one leg and the head through first, then moving the neck and body through and pulling the other leg through last. If they come to a fence on the run, they can slip through in a matter of seconds and continue running. Sometimes, they lose a feather. The agility displayed during this maneuver must be seen to be believed. The rhea can run up to 70 km per hour (Sick, 1988).

The rhea is an omnivore, feeding on small vertebrates, insects and the leaves, fruit, and seeds of plants. Small groups of them often seek food on fields recently burned (Sick, 1988). The range of the species includes the plains of South America, including much of Brazil, Argentina, Paraguay, and Bolivia.

Figure 8.13: The largest bird in the Pantanal is the American rhea, *Rhea americana*, which travels across the floodplain in small family groups only during the *seca*. During the rest of the year, the rheas usually remain on the cattle pastures above the level of the wetland.

Tinamidae

The birds of this exclusively Neotropical family occur throughout the Pantanal, chiefly in the wooded areas slightly above the high water level. They are secretive ground birds that are seldom seen, but they were encountered regularly crossing the Transpantaneira Highway southward from the point it is crossed by the Rio Novo nearly to the Rio Pixaim. A few were observed between the Pantanal and Poconé. The birds were never seen in flight, but they reportedly fly rather short distances (Sick, 1988).

The species in this family are similar in appearance and can apparently best be distinguished by their size. They all seem to be omnivores, feeding primarily on fruits, seeds, and leaves, but they also consume small invertebrates (Sick, 1988).

The largest species of the family, *Tinamus tao* (Temminck, 1815), the gray tinamou or *azulona*, and *Tinamus solitarius* Vieilot, 1819, the solitary tinamou or *macuco*, are frequent inhabitants of river valleys, especially in dry forest regions (Sick, 1988). They are therefore not specifically adapted to the conditions in the Pantanal. However, the presence of both is frequently reported from the wetland, and one of them, probably *Tinamus tao*, was the species observed crossing the Transpantaneira at times during the study. According to Dubs (1992), *Tinamus solitarius* is found mainly or exclusively at the southeastern edge of the wetland, while *Tinamus tao* inhabits the northern outskirts.

The smaller species of the family were well represented in the Pantanal as well as in the upland areas of the watershed. By far the most common of these was the small-billed tinamou or *inhambu-chororó*, *Crypturellus parvirostris* (Wagler, 1827), about 21 cm in height (Sick, 1988), was frequently seen at the water's edge along the Rio Novo as well as in other places throughout the Pantanal. These birds were extremely hard to photograph because as soon as a person or automobile approached, they would deftly disappear into the woods.

Probably the second most abundant species of the family was the brown tinamou or *inhambu-guaçu*, *Crypturellus obsoletus* (Temminck, 1815), which frequented drier localities and was seldom seen near water.

The undulated tinamou or *jaó*, *Crypturellus undulatus* (Temminck, 1825), is a species generally confined to the forested river banks. It was not seen in the areas subject to temporary flooding or adjacent hummocks.

The tiny *Nothura minor* (Spix, 1825), lesser nothura, or *cadorna-buraqueira*, which reaches only about 19.5 cm in height, seems to be abundant in the upland regions of the watershed, but in the Pantanal, it is apparently seen sporadically and only near the southeastern borders of the wetland (Dubs, 1992). In contrast, the little tinamou or *sururina*, *Crypturellus soui* (Brabourne and Chubb, 1914), and the red-winged tinamou or *perdigão*, *Rhynchotus rufescens* (Temminck, 1815), are not at all rare in southern Mato Grosso, but they are not often encountered within the Pantanal wetland.

The presence of three other species in the family Tinamidae has been reported from the region of the Pantanal but could not be confirmed: the Brazilian tinamou, *Crypturellus stringulosus* (Temminck, 1815); the tetaupa tinamou, *Crypturellus tataupa* (Temminck, 1815); and the spotted nothura, *Nothura maculosa* (Spix, 1825). From the distribution data of Sick (1988), only the presence of *Crypturellus stringulosus* is subject to doubt because it has hitherto been reported only from northern Mato Grosso near Alto Xingu, although Dubs (1992) includes it among the species possibly encountered near the northern borders of the Pantanal.

Podicepidae

Grebes are not among the abundant bird species in the Pantanal. They occur only sporadically there, apparently as migratory visitors to the rivers and large lakes. The two species known to be regular inhabitants of the region are the least grebe or *mergulhão-pompom*, *Podiceps dominicus* (Linnaeus, 1766), and the white-tufted grebe or *mergulhão-de-orelha-branca, Podiceps rolland* Quoy and Gaimard, 1824. These species were referred to by Dubs (1992) as *Tachybabtus dominicus* and *Rollandia rolland*, respectively, reflecting some unsettled points of taxonomy. Both species have an extensive range in South America, and *P. dominicus* ranges as far north as the West Indies and North America (Schauensee and Phelps, 1978).

In addition to these species, Dubs (1992) reported that there has been one sighting of the great grebe, *Podiceps major* (Boddaert, 1783) in the Pantanal of Mato Grosso do Sul along the Rio Miranda.

Phalacrocoracidae

The Neotropic cormorant or *biguá*, *Phalacrocorax brasilianus* Gmelin, 1789, often referred to by its synonym, *Phalacrocorax olivaceus* (Gmelin, 1789), is one of the most abundant birds of the Pantanal (Fig. 8.14). During the dry season, large flocks congregate near the few remaining water bodies to fish. They avoid the shallow pools preferred by the egrets and wood storks and seek their prey instead where the water is fairly deep and the underwater visibility somewhat better, in the seasonally lentic water of rivers and streams.

During the high water periods, the cormorants distribute themselves in groups of two to four over a wide area of the Pantanal. At this time, fishes are relatively scarce and must be sought actively. They can been seen diving throughout much of the day or resting on dead trees or fence posts drying their feathers (Plate 7a). The cormorants usually spend periods of about 30 to 90 seconds submerged looking for fishes. They emerge briefly, then dive again. When they catch a large or spiny fish, they emerge to try to swallow it. Sometimes they spend periods of several minutes crushing the bones of the fishes to make them less dangerous to swallow. Occasionally, they are unsuccessful and throw the fish away. Beltzer (1983a) studied the feeding habits in Argentina and identified the fish species consumed, many of which are common in the Pantanal, as well. While fishes constituted 96% of the prey in Argentina, crustaceans were also consumed and accounted for the remaining 4%.

When threatened, the cormorants invariably fly away and do not seek to escape underwater like the snake-neck birds, which fish in a similar way.

Figure 8.14: Among the common piscivorous birds in the Pantanal is the Neotropic cormorant, *Phalacrocorax brasilianus*. Usually, small groups can be observed diving for fishes, but rarely during the *seca*, large numbers congregate in permanent water bodies, as in this river near Porto Cercado.

Anhingidae

The *biguatinga*, anhinga, or snake-neck bird, *Anhinga anhinga* (Linnaeus, 1766), is a darter that can be observed in the Pantanal throughout the year. The individuals station themselves solitarily on tree branches, usually at the edge of rivers and deep permanent ponds, where they are often seen drying their wings after long dives (Plate 7b). Particularly common perching sites are thick branches of dead trees emerging from the water near river banks supporting a gallery forest. When disturbed, these birds usually dive into the water and disappear below the surface rather than fly away. Generally, they seem to remain under water longer than cormorants, with which they share common underwater fishing methods. At times, they extend only their heads and necks above the surface to see if it is safe to emerge.

The snake-neck birds generally remain together in pairs and nest in trees of the gallery forests. Their feather colors are variable, and some of the birds are nearly all black, while others have variable numbers of white feathers arranged among the black ones for forming a light pattern on the head and neck.

Of all of the bird families represented in the Pantanal, the Ardeidae is by far the most prominent, its species being most easily seen in the largest numbers throughout the region. Observations on the fluctuations in the total number of birds in this family present in a wetland along the Rio Paraná in Argentina during a period of nine years revealed some year to year differences. They also demonstrated that the abundance of these birds in the Argentinean habitat investigated was far lower than it is in the Pantanal (Beltzer, 1989).

Ardea cocoi Linnaeus, 1766, the white-necked heron, *maguari*, or *socó-grande*, is the largest species of the family, reaching a height of about 180 cm (Sick, 1988). It is a common species in the Pantanal, occurring during all seasons, generally alone or rarely in pairs. It is not as abundant as the egrets and usually shuns the large flocks of various species that descend upon the isolated pools of water during the dry season. Its usual hunting grounds are fairly shallow water bodies bordered by low trees or shrubs, in which the birds perch (Fig. 8.15). These herons apparently subsist mainly on fishes.

Figure 8.15: The largest heron in the Pantanal is the white-necked heron, *Ardea cocoi*, which generally stands solitary at the water's edge, thereby falling into the category of the "black" water birds, those that do not form flocks, generally remain secretive, and display crepuscular, or in other cases nocturnal habits.

The *garca-branca-grande* or great egret, *Casmerodius albus* (Gmelin, 1789), is a ubiquitous wading species typical of the Pantanal Fig. 8.16). It can be encountered anywhere relatively shallow water is available for wading. During the high water periods, individuals seek fishes solitarily in the relatively shallow water of the flooded fields, while during the dry season, they congregate in large numbers in the remaining water bodies, where fishes have taken refuge as their habitat rapidly shrinks. This species is found associated with other large wading birds in such feeding places during the day and in certain groves of trees chosen by them as rookeries to spend the night (Fig. 8.17). The diet of this species probably consists entirely of fishes and amphibians, with fishes the predominant items. In the Pantanal, amphibians large enough for this egret to feed upon are uncommon, so it can generally be assumed that most egrets live entirely on the fishes they capture with rapid stabs of their beaks into the water. They generally prefer cichlids and other fishes with few strong spines. A particular favorite is *Synbranchus marmoratus*, which seems to be an especially prized food item by many of the other wading birds, as well (Fig. 8.4).

A smaller species with a similar snowy white feather coloration and piscivorous diet is the snowy egret, *Egretta thula* (Molina, 1782), which has a feeding niche slightly different than that of the great egret in that it consumes smaller fishes (Fig. 8.18). The two are easily distinguished by the color of the beak; that of the snowy egret is very dark gray, while that of the great egret is yellow. The snowy egret is distinguished in the same way from the cattle egret, *Bubulcus ibis* (Linnaeus, 1758), which is about the same size and also has a uniform white coat of feathers. The cattle egret has a yellow beak and yellowish feet, while the feet of the snowy egret are dark gray like the beak (Fig. 8.19).

Figure 8.16: The larger of the two most common "white" egrets in the Pantanal is the great egret, *Casmerodius albus*. The "white" water birds are active throughout the day, frequently form large flocks, and roost together in rookeries at night.

The cattle egret is often observed near water in the Pantanal, but its feeding habits are completely different from those of the aquatic herons. As the name implies, the cattle egrets are nearly always associated with cattle, horses, and other ungulates, which they follow closely in order to feed on the insects that are disturbed by the grazers as they move from place to place. Large grasshoppers are particularly prized by these egrets. Surprisingly, a study in Argentina revealed that the prey consumed by the cattle egret were almost exclusively members of the pseudoterrestrial community in the Pantanal (Beltzer et al., 1987). *Cornops aquaticum* was numerically best represented in the egrets examined, followed in abundance by unidentified spiders, *Gryllodes* sp., and *Marellia* spp. The tree frog, *Hyla pulchella*, unidentified hydrophilids and other beetles, as well as *Paulinia acuminata* were also consumed. The cattle egret was introduced from Africa during the 19th century and has enjoyed enormous success in colonizing South America. Apparently, it has maintained its Old World habits in the Pantanal, feeding in company with livestock, mainly in terrestrial parts of the region. It may well be in the process of adapting to a diet of the more abundant insects that live on the emergent aquatic plants in Argentina at the present time. The possibility that it will thereby become a competitor of the native snowy egret, *Egretta thula*, was investigated by Zaccagnini and Beltzer (1982). They found some quantitative differences but few qualitative ones other than the lack of fishes and in the diet of the cattle egret, which consumed more arachnids and grasshoppers than the snowy egret.

Figure 8.17: The herons and wood storks generally spend the night together in large trees. These rookeries, which are occupied for many years, are often so crowded that the birds have difficulty finding a free place to roost.

Another common ardeid of the Pantanal is the capped heron or *garça-real*, *Pilherodius pileatus* Boddaert, 1783, which can be recognized immediately by its blue and yellow crown (Plate 8a). Unlike the other mainly white egrets, it never remains in a flock. Individuals are always seen alone or rarely in pairs. They are strongly territorial, and the same individual can be observed at the same site beside a shallow water body for weeks at a time. When disturbed, the capped heron flys into a nearby tree, generally chosen time and again for refuge if the individual is disturbed repeatedly. These egrets subsist on small aquatic animals, but few direct observations could be made to determine the exact composition of the diet.

A bird of the seashore which also visits the Pantanal is the little blue heron or *garça-azul, Florida caerulea* (Linnaeus, 1758), sometimes referred to as *Egretta caerulea*. This species ranges from Canada and Alaska to Peru, southern Brazil, and Uruguay. Through banding studies, it has been demonstrated that some of these herons complete intercontinental migrations (Schauensee and Phelps, 1978). The diet consists of insects as well as frogs and small fishes.

The *socozinho* or striated heron, *Butorides striatus* (Linnaeus, 1758) is frequently seen on low trees on the floodplains of the Pantanal (Fig. 8.20). Of all the species in its family, it is probably the most widespread, occurring in both North and South America, Africa, Australia, and the islands of the eastern Pacific. The species can be encountered near practically every water body in Brazil (Sick, 1988). On the floodplains along the lower Rio Paraná, Beltzer (1983b) found that 16 species of fish constituted the bulk of the diet by volume, while insects were predominant numerically. In addition, leptodactylid frogs, decapods, and aquatic spiders were included in the diet. Beltzer

(1991) found that the survival rate during the period from hatching until the juveniles can fly is about 75% in Argentina.

Figure 8.18: The smaller of the two common "white" egrets, the snowy egret, *Egretta thula*. It is distinguished from the great egret by its much smaller size and its dark gray beak and legs. The great egret has a yellow beak and dark gray legs, while the terrestrial cattle egret has a yellow beak and somewhat yellowish legs.

The chestnut-billed heron or *garça-da-mata*, *Agamia agami* (Gmelin, 1789), is another fish-eating bird of the Pantanal. This species is less seen than most of the other species of its family due to its habit of concealing itself in gallery forests. The species, which occurs from Mexico to Mato Grosso (Sick, 1988), is at the southern extreme of its range in the Pantanal.

The whistling heron or *maria-faceira*, *Syrigma sibilatrix* (Temminck, 1824), is a species not too frequently seen in the Pantanal. It occurs mainly during the *enchente* in the newly flooded fields, in which individuals wade, generally alone.

The black-crowned night heron, *savacu*, or *soco-dorminhoco*, *Nycticorax nycticorax* (Linnaeus, 1758), is not uncommon in the Pantanal, although it is less frequently seen due to its crepuscular and nocturnal habits. Anatomical and histological studies of the digestive tract revealed a structure typical of piscivorous birds (Mosso *et al.*, 1994).

The rufescent tiger-heron or *socó-boi*, *Tigrisoma lineatum* (Vieillot, 1817), is a common species in the Pantanal throughout the year. A special characteristic of the species is the considerable difference in the coloration of individuals at different ages. The young are mottled, while the older allots have a uniform reddish-brown coloration dorsally on the head and upper body and a darker coloration on the lower body (Plate 8b). Ventrally, they are lighter in color, but the pattern is quite variable. They are abundant along the Transpantaneira Highway where they seek sites with relatively deep water for fishing. The rufescent tiger-heron is not a wading bird like the egret but rather a bird that fishes from the shore. Beltzer (1990a) investigated the diet of this bird in Argentina and found that fishes constituted the bulk of the diet, with *Synbranchus marmoratus* being the most frequently consumed of the fishes. However, one colubrid snake and various insects, including dragonflies, grasshoppers, and hydrophilids in the genus *Tropisternus*, had also been eaten.

Figure 8.19: Since their introduction from Africa, the cattle egret, *Bubulcus ibis*, has spread throughout Brazil. Although their original diet consisted mainly of terrestrial insects, they seem to be adapting to consume insects in the pseudoterrestrial communities on the surface of the shallow water bodies.

Figure 8.20: The striated heron, *Butorides striatus*, a rather secretive bird that is seldom active during the day. Although less observed than the other species in its family, it is quite common in the Pantanal.

In contrast to the abundant rufescent tiger-heron, the fasciated tiger-heron, *Tigrisoma fasciatum* (Such, 1835), is not at all abundant in the Pantanal. Only solitary individuals are observed, generally in trees and shrubs along small rivers and streams. They feed on fishes and large insects (Schauensee and Phelps, 1978).

Another species that is rare in the Pantanal, if present at all, is the zigzag heron, *Zebrilus undulatus* (Gmelin, 1789), which is a small species that was not observed at all during the study.

Two additional species in this family are thought to be infrequent visitors to the Pantanal: the stripe-backed bittern, *Ixobrychus involucris* (Vieillot, 1823), and the least bittern, *Ixobrychus exilis* (Vieillot, 1789). Neither of these species were observed during the study.

The pinnated bittern or *soco-boi-baio*, *Botaurus pinnatus* (Wagler, 1829), is not a common bird in the Pantanal. Individuals are generally solitary and can sometimes be observed resting nearly motionless at the edge of stands of tall plants beside the water.

Cochleariidae

The only Neotropical species in this family is the boat-billed heron, *Cochlearius cochlearius* Linnaeus, 1766, which is found in wetlands throughout Brazil (Sick, 1988). Although it is known from the Pantanal, none were seen during the study. Its habit of remaining in forests near the water, however, makes it less conspicuous than other herons.

Ciconiidae

The stork jabiru, *Jabiru mycteria* (Lichtenstein, 1819), has come to be a symbol of the Pantanal. It is the largest wading bird, and is a resident species that nests in large trees during the dry season, when fishes are most abundant. The successful rearing of as many as three offspring in one nest has been observed (Fig. 8.21).

Figure 8.21: A nest of stork jabirus, *Jabiru mycteria*, in a tall tree near the Rio Pixaim. Each year during the dry season, the same nest was occupied by breeding pairs, such as this one with its fledgling.

A major cause of mortality among this species is the presence of elevated power lines along the northern part of the Transpantaneira Highway. During the dry season, the jabirus seek food in the remaining shallow water bodies, many of which are located beside the highway where fill was scooped out to construct the embankment along which the highway runs. When these birds take off or land during the hours of darkness, they frequently fly into the wires and break their necks. As many as seven cadavers of jabirus have been observed in one day in the vicinity of the Rio Bento Gomes. This mortality could be eliminated by the replacement of the high masts with a cable running in a thick pipe along the highway, either underground or on low posts.

Although the Portuguese word for the species is *jabiru*, it is generally called the *tuiuiu* in Mato Grosso.

The American wood stork or *cabeça seca*, *Mycteria americana* (Linnaeus, 1758), is an abundant wading bird during the dry season, when large groups of individuals crowd together to feed on the fishes trapped in the remaining pools of water. Shortly before noon, they take off and soar on the thermal updrafts until mid-afternoon, in this way escaping the midday heat. The species is sociable, and while present in the Pantanal, they invariable travel around together in small to large groups of individuals. During the rainy season, the species is absent from the Pantanal because the birds have migrated to their breeding grounds.

Another member of the stork family that is reported to inhabit the Pantanal is the maguari stork or *tabuiaia*, *Euxenura maguari* (Gmelin, 1789), but individuals were quite rare during the study.

Threskiornithidae

The roseate spoonbill, *Ajaia ajaja* (Linnaeus, 1758), is one of the common wading birds of the Pantanal. Unlike most of the other wading birds, this species specializes in small prey, which it sifts out with its spoon-shaped bill (Plate 9a). Roseate spoonbills are encountered throughout the Pantanal, but in recent years, they have become relatively scarce in the northernmost part of the wetland. They are still regularly encountered at Campo Jofre, near the southern end of the Transpantaneira Highway not far from the Cuiabá River.

Figure 8.22: The plumbeous ibis, *Harpiprion caerulescens*, is a very abundant species on the floodplain. Typically they appear alone or in groups of two or three near the edge of water bodies.

The ibises are well represented numerically in the Pantanal, although only two species account for most of the sightings. Frequently seen alone or in small groups is the *curicaca cinzenta* or plumbeous ibis, *Harpiprion caerulescens* (Vieillot, 1817), which feeds on the pilid snails by probing the soft sediments in the flooded fields. This species is generally observed alone or in groups of up to four or five (Fig. 8.22).

The bare-faced ibis or *tapicuru*, *Phimosus infuscatus* (Spix, 1825), is a species that was often abundant on the Fazenda Ypiranga, on which sampling sites were located. The birds were observed on many occasions in large flocks of ten or more (Plate 9b). This species feeds on various small organisms which it collects by probing its beak in the shallow water of the flooded fields. The diet includes plant material, such as seeds (Sick, 1988).

A third abundant ibis in the Pantanal is the buff-necked ibis or *curicaca*, *Theristicus caudatus* Boddaert, 1783, which is generally observed in groups of two to four on unflooded areas, but usually near water. According to Sick (1988), this species feeds primarily on terrestrial animals, including reptiles and small mammals as well as large insects and other arthropods. The species is known from Panama and throughout the entire South American continent.

In addition to the species observed, three additional ibises are reported from the Pantanal. The green ibis or *corocoró*, *Mesembrinibis cayennensis* (Gmelin, 1789), might have been overlooked because of its crepuscular habits and preference for gallery forests.

The *caraúna* or *tapicuru*, *Plegadis chihi* (Vieillot, 1817), is a widespread migratory bird in Brazil. Its presence in the Pantanal at times can be assumed, but no individuals were identified during the studies.

The scarlet ibis, *Eudocimus ruber* Linnaeus, 1758, has been reported at times from the Pantanal. Sick (1988) considers the reports to be erroneous, resulting from confusion with the roseate spoonbill. None of these rare birds was observed during the study, and its existence in the Pantanal remains doubtful.

Phoenicopteridae

The only South American flamingo, *Phoenicopterus ruber* (Linnaeus, 1758), was sighted only once over the Baia Piuval during the early rainy season. The species was hitherto unreported from the Pantanal, and the individuals seen were undoubtedly migrating. One was sighted flying northward along the shoreline, and about two minutes later, a second one flew by. None were ever seen on the ground in the Pantanal.

Anhimidae

Only one species in this family is abundant in the Pantanal. This is *Chauna torquata* (Oken, 1816), the *tachã* or southern screamer (Plate 10a). The birds are almost always seen in pairs. They cannot be approached too closely before they take to the air, finding a low tree in which to roost. When disturbed in this way, they begin to scream in a loud, raucous voice, a habit which gives them their English common name. Carboneras (1992a) reported that the green parts of plants, especially succulent aquatic species, are preferred, but a limited amount of animal matter is also consumed.

Another species in this family is reported to occur in the Pantanal (Sick, 1988). *Anhima cornuta* (Linnaeus, 1766), the horned screamer, proved to be rare in the region, an none were observed during the ecological study.

Anatidae

Ducks and geese are not nearly as common as the large wading birds of the Pantanal, and they are very wary and difficult to observe. However, several species are encountered there seasonally. Probably the most abundant species is the black-billed whistling duck or *marreca cabocla*, *Dendrocygna autumnalis* (Linnaeus, 1958), which occurs in small flocks in open areas near water during the dry and early rainy season

(Fig. 8.23). This duck seems to prefer a vegetarian diet, with only a limited amount of animal matter being consumed (Carboneras 1992b). It travels seasonally to seek water, and some of the population may undertake longer migrations. At times, this duck is found in company with the white-faced whistling duck or *ireré*, *Dendrocygna viduata* (Linnaeus, 1766), which also was observed on the flood plain about 15 km south of Poconé, on the Fazenda Ypiranga (Fig. 8.24). *D. viduata*, which often spends the summer in Argentina, was reported by Beltzer and Mosso (1992) to subsist mainly on plant material there, especially *Polygonum* spp. and grasses, although Carboneras (1992b) reported it to be an omnivore with a varied diet.

Figure 8.23: Unlike most other water birds, ducks generally remain far from the roads in the Pantanal, but flocks like this one of black-billed whistling ducks, *Dendrocygna autumnalis*, are often observed on the partially inundated floodplain during the *vazante*.

Also fairly abundant on the flooded areas of the Pantanal during the *vazante* is the Orinoco goose, *Neochen jubata* (Spix, 1825), which is also usually observed in small flocks, sometimes near flocks of whistling ducks. These geese are not known to undertake long migrations (Carboneras, 1992b). More secretive but also fairly common during the same period is *Amazonetta brasiliensis* (Gmelin, 1789), called the *ananai*, *pé-vermelho*, or Brazilian duck. In Argentina, Brazilian ducks were found to feed on plants and beetles, with the seeds of *Polygonum* spp. being predominant (Beltzer and Mosso, 1992).

The Muscovy duck or *pato-do-mato, Carina moschata* (Linnaeus, 1758), is not uncommon in the northern Pantanal. Pairs or groups of four to six are generally observed on the surface of small water bodies or browsing at the water's edge during the dry and early rainy seasons (Fig. 8.25). They move from place to place in the wetland but apparently do not undertake long migrations. Carboneras (1992b) reported them to be onmivores with a particular liking for termites. They also consume fishes and other small vertebrates along with molluscs, arthropods, and various kinds of plant material. Interestingly enough, they are the only ducks observed near the Transpantaneira Highway. Perhaps because of their extreme wariness, the smaller species are almost never observable from that road, although they may be quite abundant near the small side roads leading to ranches.

Figure 8.24: Small flocks of black-billed, *Dendrocygna autumnalis* (left), and white-faced (right) whistling ducks, *Dendrocygna viduata*, are often seen together in the Pantanal, especially during the *vazante*.

Figure 8.25: The Muscovy duck, *Carina moschata*, is seen occasionally in small groups in the Pantanal.

The comb duck, *Sarkidiornis melanotos* (Ihering and Ihering, 1907), was rarely seen in the Pantanal, although it might be expected to be relatively common there. Carboneras (1992b) reported that some of these ducks undertake migrations and that the diet is largely vegetarian.

None of the ducks or geese were ever found on the local rivers or streams, even when these water bodies became lentic during the dry season. Other species that should be expected to occur in the Pantanal were not observed during the study. These include the fulvous whistling duck or *marreca caneleira*, *Dendrocygna bicolor* (Vieillot, 1816), and the masked duck, *Oxyura dominica* (Linnaeus, 1766). It may well be that these ducks have become secretive due to the intensity of hunting in other parts of Brazil. The diet of *Dendrocygna bicolor* is almost entirely vegetarian. *Oxyura dominica* is a non-migratory species with a mainly vegetarian diet (Carboneras, 1992b). In the valley of the Rio Paraná in Argentina, ducks seem to be better represented, with ten species recorded, and more numerous, as well (Beltzer and Mosso, 1992).

Cathartidae

Generally, five true vulture species are reported for the Pantanal. It is doubtful whether they all occur there at the present time. For example, the largest of the vultures, *Vultur gryphus* Linnaeus, 1758, the Andean condor, may well have soared over the Pantanal from time to time in past years, and Sick (1988) reported that it has been known to penetrate into Brazilian territory to the west of Cáceres near the edge of the Pantanal. However, sightings of this condor in or near the Pantanal are rare enough to warrant its exclusion from the native avifauna of the wetland. Usually, the condor is observed in mountainous regions, and a vast, low-lying plain would not be attractive to this bird.

The commonest of the Pantanal cathartids is the black vulture or *urubu-preta-comum*, *Coragyps atratus* (Bechstein, 1793), which is extremely abundant throughout the Pantanal (Fig. 8.26). Flocks of them quickly eliminate the many cadavers of cattle and horses, as well as caimans and other native animals that succumb to the dryness during the *seca* or are run over by cars and trucks on the roads. They perform an efficient refuse disposal service, thereby preventing local water pollution and protecting the aquatic fauna from the severe oxygen depletion that a large dead animal in the shallow water would cause.

A competitor of the black vulture that is common but not nearly as abundant is the turkey vulture, *Cathartes aura* (Spix, 1829), known locally as the *urubu-de-cabeça-vermelha*. This scavenger does not form such large groups on cadavers as the black vulture, apparently because of its relative scarcity (Fig. 8.27).

Two additional vultures should also occur in the northern Pantanal: the king vulture or *urubu-rei*, *Sarcoramphus papa* (Linnaeus, 1758); and the lesser yellow-headed vulture, *Cathartes burrovianus* Pelzeln, 1861. However, neither were observed during the study.

A single member of this family has been reported on a few occasions from the Pantanal. The osprey or *aguia-pesqueira*, *Pandion haliaetus* (Linnaeus, 1758), was reported to be seasonally common in the Pantanal of Poconé by Cintra and Yamashita (1990) and was sighted by Dubs (1992) near Cáceres. This cosmopolitan species is piscivorous. Most probably, the individuals sighted were from North America, spending the northern winter fishing in the Pantanal.

Accipitridae

The commonest species of this raptor family in the Pantanal is the snail kite or *gavião-caramujeiro, Rostrhamus sociabilis* (Vieillot, 1817), which lives mainly on the pilid snails that are extremely abundant in the Pantanal (Fig. 8.28).

Figure 8.26: A group of black vultures, *Coragyps atratus*, feeding of the carcass of a cow. These are by far the most numerous large scavengers in the Pantanal.

Figure 8.27: The widespread turkey vulture, *Cathartes aura*, is a scavenger that is fairly common but much less abundant than the black vulture in the Pantanal.
Pandionidae

Snail kites are territorial, generally picking snails from the shallow water in the Pantanal and flying to one particular tree or fence post, where they bite a small hole in one area of the shell and pull the soft parts out. The favorite roosting places of the snail kite are invariably marked by large piles of empty snail shells, each with a small hole cut in the side. Sometimes, crab shells are also found among this refuse, and occasionally the kites are observed consuming one of the trichodactylid crabs as a change in diet.

Figure 8.28: One of the most abundant raptors in the Pantanal is the snail kite, *Rostrhamus sociabilis*, which collects large pilid snails and feeds on them by biting a small hole in the shell and removing the soft parts. Each individual almost always feeds on the snails while perched on the same tree branch or fence post, under which a large pile of shells accumulates.

Another member of this family that is also abundant in the Pantanal is the black-collared hawk or *gavião-belo*, *Busarellus nigricollis* (Latham, 1790), which preys mainly on fishes by scooping them out of the shallow water with its talons and carrying them to a tree to eat, as personally observed on various occasions. As a change in diet, snails or insects may be eaten. Although they are often seen along the Transpantaneira Highway, they are not nearly as abundant as the snail kite (Fig. 8.29). A population census of this species during the early dry season in southeastern Mato Grosso by Willis and Oniki (1990a) has provided information on its seasonal role in the community.

Other species of this family are generally distributed in southern Mato Grosso and do not particularly seek out the Pantanal as a habitat. A few are just as abundant in the wetland during the dry season as they are elsewhere in the watershed, while most seem to avoid the Pantanal and are only rare visitors.

One of the species that is abundant in the Pantanal is the pearl kite or *gaviãozinho*, *Gampsonyx swainsoni* (Vigons, 1825), which is a species of the *cerrado* but prefers hunting for insects and reptiles along river banks. According to Sick (1988), with a stature of 22 cm, it is the smallest Brazilian species in this family.

Another species that commonly hunts in the Pantanal is the white-tailed hawk or *gavião-de-rabo-branco*, *Buteo albicaudatus* (Vieillot, 1816), which feeds on a variety of reptiles and small mammals. This hawk is also known to feed on toads (Sick, 1988), which most other predators shun because of the poison they secrete.

Figure 8.29: A piscivorous raptor, the black-collared hawk, *Busarellus nigricollis*, in flight over the flooded wetland.

The great black hawk or *gavião-preto*, *Buteogallus urubitinga* (Gmelin, 1788), is another frequent visitor to the Pantanal, where it catches fishes as well as small mammals, reptiles, insects, and dead animals, including the young of large water birds that fall out of the nest. One of these hawks was seen carrying away a snake more than two meters long that it had just killed. The snake entangled in the branches of trees, forcing the hawk to land several times while trying to take to the air with its heavy prey. This species also consumes venomous snakes (Sick, 1988) and has been reported as one of the main predators on young caimans (Cintra, 1985).

Circus buffoni Gmelin, 1788, the long-winged harrier or *gavião-de-mangue*, is a wetland species that hunts mainly frogs and small mammals. It was not frequently seen in the Pantanal.

The crowned solitary eagle or *águia-cinzenta*, *Harpyhaliaetus coronatus* (Vieillot, 1817), is a crepuscular hunter that can often be observed in the Pantanal. Its range is limited, encompassing only northern Argentina, Bolivia, and Brazil from Rio Grande do Sul to southern Mato Grosso (Sick, 1988).

The savanna hawk or *gavião-caboclo*, *Heterospizias meridionalis* (Latham, 1790), was identified in the Pantanal and in the neighboring *cerrado* region about the same number of times. It was not overly abundant in either of these habitats, and generally, individuals were seen in both at the same time of year, usually during the early rainy season.

More frequent in the upland regions near the Pantanal than in the wetland itself is the black hawk-eagle or *gavião-pega-macaco*, *Spizaetus tyrannus* (Friedmann, 1950), which feeds mainly on mammals, including monkeys, accounting for its Portuguese name.

Abundant in the dry forest region but rare in the Pantanal wetlands is the roadside hawk or *gavião-carijó*, *Buteo magnirostris* (Bertoni, 1901), which, according to Sick (1988), feeds on a variety of insects, lizards, snakes, and small birds.

Other small hawk species with generally the same local distribution in southern Mato Grosso as the roadside hawk are the gray hawk, *Buteo nitidus* Latham, 1790, and the short-tailed hawk, *Buteo brachyurus* (Vieillot, 1816).

The white-tailed kite or *peneira*, *Elanus leucurus* (Vieillot, 1818), is a species mainly encountered in upland agricultural areas, where it hunts small mammals and reptiles. It was not seen in the wetland.

Another snail-eating species, the hook-billed kite, *Chondrohierax uncinatus* (Temminck, 1822), is an upland species generally encountered near forested areas. It consumes mainly arthropods and terrestrial snails, which it swallows whole without breaking open the shells (Sick, 1988).

The harpy eagle, *Harpia harpyia* (Linnaeus, 1758), is the largest raptor with a range including the region of the Pantanal. Its diet includes macaws and other large birds, coatis, bush dogs, monkeys, and sloths. Unfortunately, it has become extremely rare, none were seen under natural conditions during the study, and none have recently been reported from the Pantanal.

Other species in this family have ranges including southern Mato Grosso and might be expected at times in the northern Pantanal. Among the species in this group not observed during the study are the ornate hawk-eagle or *gavião-de-penacho, Spizastur ornatus* (Daudin, 1801), and the black and white hawk or *gavião-pato, Spizastur melanoleucus* (Vieillot, 1816), both of which were reported as rare in the Pantanal of Poconé by Cintra and Yamashita (1990). Species with ranges mainly in the drier upland areas to the north and east of the Pantanal include the swallow-tailed kite or *gavião-tesoura, Elanaoides forficatus* (Linnaeus, 1758); the plumbeous kite or *sovi, Ictinia plumbea* (Gmelin, 1788); bay-winged hawk or *gavião-de-asa-de-telha, Parabuteo unicinctus* (Temminck, 1824); the white hawk or *gavião-branco, Leucopternis albicollis* (Latham, 1790); and the crane hawk or *gavião-pernilongo, Geranospiza caerulescens* (Vieillot, 1817). The gray-headed kite or *gavião-de-cabeça-cinza, Leptodon cayanensis* (Latham, 1790); gray-bellied hawk or *tauató, Accipiter poliogaster* (Temminck, 1824); and bicolored hawk or *gavião-bombacho, Accipiter bicolor* (Vieillot, 1817), seem to be species rather common in Mato Grosso do Sul but rare or absent from the northern part of the Pantanal. None of these species were especially sought during the study, and it is strongly suspected that most or all of them could be found in the Pantanal at times during an exhaustive study of the avifauna. The presence of all of these species was reported for Paraguay by Laubmann (1939).

Falconidae

The crested caracara or *caracará-commum*, *Polyborus plancus* (Miller, 1777), is one of the most common raptors found along the Transpantaneira Highway (Plate 10b). These birds are less fearful of people than the other raptors and can be approached fairly closely. This species is generally well adapted to modern conditions, getting much of its food by picking small animals that have been killed by automobiles from the highway. At sunup, these birds are particularly active seeking dead frogs, small mammals, and reptiles that were run over during the night.

This is a resident bird in the Pantanal, and parents with fledgling caracaras are not uncommon along the highway and can be approached within a few meters. The local population of the species is not presently endangered, and because it presently has no market value, either as a producer of some salable product or as a house pet, it will probably not be endangered in the near future. Nevertheless, the foolish idea is repeatedly voiced that the population of these birds should be reduced because caracaras are accused of feeding on smaller animals which must be "protected". If proposals to "thin" the population are given serious consideration, these attractive birds could become seriously endangered. It is not likely that the caracara is much more abundant than other raptors in southern Mato Grosso; it is simply a bird that seeks out highways and human habitations in the wetland, where it finds its food more easily. It therefore appears to be more abundant than the species that inhabit forests and the *cerrado* and give man a wide berth.

The other species in this family are not nearly as frequently observed in the Pantanal, and they apparently occur there only while undertaking occasional excursions in search of food. Most of them are more abundant in the adjacent *cerrado.*

The laughing hawk or *acauã, Herpetotheres cachinnans* (Linnaeus, 1758), is a species that feeds mainly on reptiles and bats (Sick, 1988). It is a temporary visitor to the Pantanal that is not seen often, apparently because its prey is more abundant and easier to capture in the upland forest regions.

Three forest-falcon species are apparently suffering from the general deforestation in the upland regions of Mato Grosso, particularly on the plateau to the east of the Pantanal. The collared forest-falcon or *falcão-relógio, Micrastur semitorquatus* (Vieillot, 1817); the barred forest-falcon or *falcão-carboré, Micrastur ruficollis* (Vieillot, 1817); and the lined forest-falcon, *Micrastur gilvicollis* (Vieillot, 1817), could all be expected in the region of the Pantanal, but only *M. semitorquatus* was observed during the study. It should be noted that according to Dubs (1992), *M. gilvicollis* should be considered to be a northern subspecies of *M. ruficollis* and should not be expected in the region of the Pantanal. However, it should still be maintained in the list of regional avifauna until its taxonomy and range are better defined.

In the region of the Pantanal, four common species of *Falco* occur with varying frequency. They seldom enter the wetlands to hunt during the *cheia* but can be observed during the dry seasons, less frequently than in the upland regions, however. The species included in the avifauna of southern Mato Grosso include the orange-breasted falcon *falcão-de-peito-laranja, Falco deiroleucus* (Temminck, 1825); the bat falcon or *cauré, Falco rufigularis* Daudin, 1800; the American kestrel, sparrow hawk or *quiriquiri, Falco sparverius* (Cory, 1915); and the aplomado falcon or *falcão-de-coleira, Falco femoralis* (Temminck, 1822). Recent investigations to determine the diet of *F. sparverius* in the middle Paraná River Valley of Argentina revealed that much of the diet consists of insects, especially grasshoppers, but hylid frogs and rodents were also consumed (Beltzer, 1990b).

The black caracara or *gavião-de-anta, Daptrius ater* (Vieillot, 1816), is an omnivorous species that feeds on ticks, as well as other arthropods and small vertebrates, including fishes and the young of other bird species. It is an uncommon species in the Pantanal, apparently having declined with the tapir, from which it has often been observed to remove ticks (Sick, 1988).

The red-throated caracara or *canção-grande, Daptrius americanus* (Boddaert, 1783), and yellow-headed caracara or *carrapeteiro, Milvago chimachima* Vieillot, 1816, are both species that could be expected to occur in or near the northern Pantanal but were not observed there during the study.

Cracidae

Two species of cracid are typical birds in the drier parts of the Pantanal and throughout the entire wetland during the dry season. The speckled chachalaca or *aracuã pequeno, Ortalis guttata* (Spix, 1825), is a bird that prefers to live on the ground, although it can fly and often takes refuge in low trees when disturbed, emitting loud cries (Fig. 8.30). It is characteristically terrestrial in its habits and generally avoids water or marshy places. The birds are residents in the Pantanal, and chicks are often seen accompanying the parents. The chaco chachalaca or *arancuã-do-pantanal, Ortalis canicollis pantanalensis* (Cherrie and Reichenberger, 1830), is somewhat larger. It is a species that prefers to remain in trees, especially those of the gallery forests. It is more commonly heard than seen but is probably at least as abundant as the chaco chachalaca.

Another species that can be observed regularly in the Pantanal is the bare-faced currasow or *mutum-de-penacho, Crax fasciolata* Spix, 1825, which may be seen in open areas but prefers the proximity of woodlands near water, such as gallery forests. Unlike the chachalacas, which generally remain in small flocks, these birds generally travel in pairs, the males and females being distinguished by strikingly different plumage. The males are black dorsally, while the females display a checkered pattern. Both are whitish ventrally. They are apparently omnivores, feeding on fruits and seeds or small arthropods according to availability. The pairs migrate locally from one region to another and are seldom encountered more than once at any location.

Figure 8.30: The speckled chachalaca, *Ortalis guttata*, is a terrestrial bird that is very abundant in the woodlands of the Pantanal and rare in the upland parts of the watershed.

The rusty-margined guan or *jacupemba*, *Penelope superciliaris* (Spix, 1825), is a species that somewhat resembles the chaco chachalaca in its habits, flying into low trees when disturbed. It also occurs in the Pantanal but is less abundant than the chachalacas, and only solitary individuals were observed on infrequent occasions. A congeneric species, *Penelope ochrogaster* (Pelzeln, 1870), called the chestnut-bellied guan or *jacupoí*, is known from the plateau region of southern Mato Grosso and probably also occurs in parts of the Pantanal.

A resident bird of the Pantanal which occurs westward as far as Peru is the blue-throated piping guan or *jacopara, Pipile pipile grayi* Pelzeln, 1870, which is not frequently seen. Another subspecies, the red-throated guan or *jacutinga*, *Pipile pipile nattereri* Reichenbach, 1862, is widely distributed in Brazil and may also rarely occur in the region of the Pantanal.

The *mutum-cavalo*, *Mitu mitu* Spix, 1825, has suffered a severe reduction in its former range, apparently due to habitat destruction. It is presently encountered most frequently in the Amazon Basin, including northern Mato Grosso. Its occurrence in the region of the Pantanal could not be confirmed during the study, and it is likely that the species does not occur there anymore.

Phasianidae

The commonest member of this family in the Pantanal is the domestic chicken, *Gallus gallus* (Linnaeus, 1758), which is present on nearly every farm and ranch in the region, on which they generally run freely, ranging through adjacent areas of the Pantanal in search of food.

This family is also represented in the region by quails. They are terrestrial and usually avoid water or marshy places. Because of their secretiveness, they are seldom seen, but they seem to be quite common in the Pantanal. The spot-winged wood-quail, *uru*, or *capoeira*, *Odontophorus capueira* (Spix, 1825), is probably the commonest quail species in southern Mato Grosso. It has been observed along the Transpantaneira Highway near Poconé, at elevations above the Pantanal wetland. It seems to be more abundant at still higher elevations, however. The presence of the marbled wood-quail or *uru-corcovado, Odontophorus gujanensis* (Gmelin, 1789), is locally reported, but its presence could not be confirmed during the study, probably because of its preference for the upland forests.

Numididae

The helmeted Guineafowl or *galinha-d'Angola, Numida meleagris* Linnaeus, 1766, was introduced to Brazil from the Cape Verde Islands off the coast of Africa early during the Portuguese colonial period. It is commonly raised on the farms and ranches in and near the Pantanal, and individuals are sometimes seen more or less feral in the Pantanal near human habitations.

Opisthocomidae

The presence of the hoatzin or *cigana, Opisthocomis hoazin* (P. L. S. Müller, 1776), has been reported from the Pantanal on various occasions, but perhaps some of these reports come from region of the uppermost stretches of the Rio Paraguay north of the wetland (Sick, 1988), where this species presently occurs. During the study, it was not observed.

Aramidae

Among the various species of bird specialized for feeding on the common snails in the Pantanal is the limpkin or *carão*, *Aramus guarauna* Linnaeus, 1766, which is a wading bird that is often seen collecting snails and transporting them intact to one site it chooses for having its meals (Fig. 8.31). The snails are opened with a hammering blow of the pointed beak and eaten only after enough have been collected for a full meal. The limpkin is a resident bird, and young have been observed with a parent during the *enchente*.

Rallidae

The rallids are not as abundant as many of the other shore birds in the Pantanal, and those that are present are secretive and hard to observe. The commonest species is not a littoral one but rather the gray-necked wood-rail or *saracura-três-potes, Aramides cajanea* (Müller, 1776), which runs in woodlands at higher elevations in the Pantanal. They are frequently seen crossing the Transpantaneira Highway in unflooded, wooded areas.

Figure 8.31: One of the typical snail-eating birds of the Pantanal is the limpkin, *Aramus guarauna*. It is a resident bird that breeds in the wetland.

The most abundant of the littoral species, the azure gallinule or *frango-d'água-pequeno, Porphyrula flavirostris* (Gmelin, 1789), was rather often observed along a wooded shoreline near the point the Rio Frio crosses the Transpantaneira Highway. It was not observed at all elsewhere. The congeneric species, the purple gallinule or *frango-d'água-azul, Porphyrula martinica* (Linnaeus, 1766), is a littoral species that is rarely observed in the Pantanal. It is difficult to judge the actual abundance of these species in the wetland because of their secretive habits.

The russet-crowned crake or *sanã-castanha, Laterallus viridis* (P. L. S. Müller, 1776), is a floodplain species that generally conceals itself among marsh plants. It was seen from time to time, but it may be more abundant than the sightings indicate because of it is secretive.

Other inhabitants of the Pantanal which were not specifically identified during the study include the fairly common uniform crake or *saracurinha-lisa, Amaurolimnas concolor* (Gosse, 1847); the locally rare rufous-sided crake or *sanã-parda, Laterallus melanophaius* (Vieillot, 1819); the rare ocellated crake or *sanã-ocellada, Micropygia schomburgkii* (Schomburg, 1848); and the fairly abundant common gallinule or *galinha-d'água, Gallinula chloropus* (Linnaeus, 1758). Cintra and Yamashita (1990) reported the blackish rail or *saracura-preta, Rallus nigricans* Vieillot, 1819, and paint-billed crake or *sanã-de-bico-vermelho, Neocrex erythrops* (Sclater, 1867), to be rare seasonal visitors to the Pantanal of Poconé.

A reported inhabitant of the region of the Pantanal in Mato Grosso do Sul which apparently does not live in the northern part is the ash-throated crake or *sanã-carijo, Porzana albicollis* (Vieillot, 1819).

Heliornithidae

Although the sungrebe or *ipequi, Heliornis fulica* Boddaert, 1791, is a water bird that ranges throughout most of Brazil, it is uncommon in the Pantanal of Poconé. During the period of the study, no individuals were observed, although the species has been reported from southern Mato Grosso, including the region of Poconé during all seasons (Cintra and Yamashita, 1990).

Eurypygidae

One of the most colorful birds seen regularly in the Pantanal, although not in great numbers, is the sunbittern or *povãozinho-do-pará, Eurypyga helias* (Pallas, 1781). The birds usually remain alone. Generally, no more than two are ever seen together. It is one of the most inconspicuous birds that remains near the water's edge because its drab colors blend in with the shades of brown displayed by the dry season vegetation. However, when it takes to flight, it displays a beautiful sun-burst pattern on the open wings (Plate 11a). This is a classic example of the "startle effect" used by various species to elude predators, which are supposed to be so startled by the sudden burst of colors that they stop their attack long enough for the prey to escape. Although never seen in great numbers, the birds were seen repeatedly, usually beside the same water bodies, for long periods of time. It is likely that the same individuals were seen week for week.

Cariamidae

The red-legged roadrunner or *seriema-de-pé-vermelho*, *Cariama cristata* (Linnaeus, 1766), is a fairly common species in the *cerrado* near the Pantanal. It is also seen in the Pantanal on infrequent occasions, especially during the dry season. The birds feed on grasshoppers and other arthropods, rodents, and occasionally snakes. They are sometimes seen alone, more frequently in pairs or small groups of three or four. Generally, this species can only be expected to occur in the wetland after the floodplain has dried out during the *seca*.

Jacanidae

The wattled jacana or *jacana-preta, Jacana jacana* (Linnaeus, 1766), is the smallest of the common wading birds in the Pantanal. It is characterized by enormous feet, which help spread its weight of the surface of the floating plants on which it frequently walks (Fig. 8.32). The remarkably long toes and toe nails distribute the weight over a large area, permitting the birds to patrol through the pseudoterrestrium, even when the plants they walk on are not buoyant enough to support much weight. It feeds mainly on small invertebrates and probably amphibians and occasional fishes, which it captures with its long, probing beak. Jenni (1996) reported that seeds are consumed, as well, whenever they are available.

The jacana is a resident bird of the Pantanal, and the young birds are often seen together with their parents seeking food in the flooded areas of the wetland. The juveniles have a first coat of feathers considerably different from their parents after they shed the down of the chicks. With age, the coloration changes to the typical reddish brown and yellow, which gives the bird its local nickname, *cafezinho*, after a black coffee popular in Brazil.

The jacana is a resident bird in the Pantanal, as small groups of chicks were seen on water plants together with the parents. While the adults can be approached fairly closely, the small family groups are secretive and take cover in dense brush on the shores whenever human beings are within sight.

Figure 8.32: The wattled jacana, *Jacana jacana*, is one of the few birds capable of walking across the mats of floating plants to feed on members of the pseudoterrestrial community. This is possible because its extremely long toes are able to distribute its fairly light weight of a large area of the plant mats. This birds raises its young in the Pantanal.

Charadriidae

The most common lapwing in the Pantanal is *Vanellus chilensis* (Gmelin, 1789), the *quero-quero* or southern lapwing, which can be observed on dried areas of the Pantanal during the South American winter. Chicks are sometimes encountered during the dry season, guarded closely by their parents, which protest loudly and make low swoops over persons or animals getting too close to their offspring. Even when chicks are not present, the lapwings will often defend a particular territory from intrusion.

Two much less common charadriids are present in the region of the Pantanal: the pied lapwing, *batuira-de-esporão* or *mexeriqueira, Hoploxypterus cayanus* (Latham, 1790), and the collared plover or *batuíra-de-coleira, Charadrius collaris* Vieillot, 1818. Neither of these species approach the southern lapwing in abundance and are generally considered to occur only sporadically throughout much of their ranges. In Paraguay, Laubmann (1939) reported *Hoploxypterus cayanus* to occur near streams and rivers, but the species was also regarded in that country as rare. He stated, however, that *Charadrius collaris* was abundant in swamps but rare in mountainous regions.

Scolopacidae

Several species of snipe are known from the northern region of the Pantanal, but they are secretive and difficult to observe, except along the large rivers and lakes. The brown or South American snipe, locally called the *narceja-comum, Gallinago gallinago* (Linnaeus, 1758), is probably the commonest species encountered in the Pantanal of Poconé. Cintra and Yamashita (1990) recorded its presence during all seasons. They also seasonally observed the solitary sandpiper or *maçarico-solitário, Tringa solitaria* Wilson, 1813, a species which breeds in North America and spends the northern winter in South America (Sick, 1988). A less common species with similar breeding and migratory habits is the buff-breasted sandpiper or *maçarico-acanelado, Tryngites subruficollis* (Vieillot, 1819). Probably also a migratory visitor to the Pantanal is the red knot or *ruiva, Calidris canutus* (Linnaeus, 1758), which was reported from wet grasslands near Poconé by Cintra and Yamashita (1990). According to Dubs (1992), rare rainy season migrants to the Pantanal from North America include the spotted sandpiper or *maçarico-pintado, Actitis macularia* (Linnaeus, 1766); Hudsonian godwit or *maçarico-de-bico-virado, Limosa haemastica* (Linnaeus, 1758); lesser yellowlegs or *maçarico-de-perna-amarela, Tringa flavipes* (Gmelin, 1789); greater yellowlegs or *maçarico-tititiu, Tringa melanoleuca* Gmelin, 1789; and possibly the upland sandpiper or *maçarico-do-campo, Bartramia longicauda* (Bechstein, 1812).

Species found to be rare in the northern Pantanal by Cintra and Yamashita (1990) include the white-rumped sandpiper or *maçarico-de-sobre-branco, Erolia fuscicollis* (Vieillot, 1819), pectoral sandpiper or *maçarico-de-colete, Erolia melanotos* (Vieillot, 1819), and least sandpiper or *maçariquinho, Erolia minutilla* (Vieillot, 1819).

The giant snipe or *narcejão, Gallinago undulata* (Boddaert, 1783), is a nocturnal forest species that may be encountered to the northeast of the Pantanal (Dubs, 1992). It is represented by the subspecies *G. undulata gigantea* (Temminck, 1826). Gils and Wiersma (1996) reported that it conceals itself during the day among tall marsh plants, and that little is known about its diet. However, it is suspected to feed on frogs.

Recurvirostridae

The only species of this family reported from the Pantanal is the common stilt or *pernilongo, Himantopus himantopus melanurus* (Vieillot, 1817), which is a cosmopolitan shore bird particularly known to be common along large rivers. It is usually not encountered in the shallow wetlands. During the years of the study, small groups were observed occasionally on sand flats along the Rio Cuiabá. These birds feed mainly on benthic invertebrates.

Laridae

Gulls and terns are typically sea birds, but several species are able to survive near large inland water bodies. Three species have been reported from the Pantanal, although all are obviously migratory visitors to the wetland. Positive identifications of the species could not be made during the study because the individuals were only seen from a distance. According to Sick (1988), the gray-hooded gull or *gaivota-de-cabeza-cinza, Larus cirrocephalus* (Vieillot, 1818), is a marine species known from the coasts of South America and Africa that has been reported with a reasonable amount of certainty from southern Mato Grosso. The large-billed tern or *trinta-reis-grande*, *Phaetusa simplex* (Gmelin, 1789), was most probably the species personally seen flying along the Rio Bento Gomes on several occasions. The presence of the yellow-billed tern of *trinta-reis-anão*, *Sterna superciliaris* (Vieillot, 1819), in the Pantanal was not detected during the study.

Rynchoptidae

The black skimmer, *corta-água*, or *talha-mar, Rynchops nigra* (Saunders, 1895), is an abundant bird in the Pantanal. Its habit of skimming water from the surface during flight makes it unmistakable. Individuals were common over large, open bodies of water, such as the Rio Bento Gomes near the Transpantaneira Highway. This bird skims over the surface of the water with the tip of its lower jaw submerged until it contacts a small fish at the surface. At this moment, the bird ducks its head ventrad to posteriad, closing the beak and snapping up the fish. This complex movement of the head is completed without any change in the flight path. The bird continues flying while swallowing the fish, and if it is still hungry, it makes another run with the lower jaw in the water. The bird spends most of its time in flight. Sick (1988) reports that this species prefers to fish during twilight or even at night. Reports of this species feeding on plankton are false, as affirmed by Sick (1988) and confirmed by the personal studies in the Pantanal, where plankton simply does not exist in the water at the times this bird was observed skimming the water.

Columbidae

A variety of doves are encountered in the Pantanal, but only a few occur frequently. The species are terrestrial and generally do not occur near the water. Many doves can be observed on the Transpantaneira Highway as well as on the roads leading to ranches.

The ruddy ground-dove or *rolinha-roxa, Columbina talpacoti* (Temminck, 1811), is probably the commonest species in the family in southern Mato Grosso, at least during the dry season. It is seen somewhat less in the Pantanal than in and near Cuiabá, but is also occurs near Poconé.

In the Pantanal itself, the urbanized species are more or less displaced by the plain-breasted ground-dove, *rolinha-de-asa-canela*, or *rolinha-caxexa, Columbina minuta* (Linnaeus, 1766), and by the rare blue-eyed ground dove or *rolinha-do-planalto, Columbina cyanopis* (Pelzeln, 1870), which is limited in its range to a relatively small area in Central Brazil (Sick, 1988). Small flocks of four to eight plain-breasted ground doves are frequently seen along the edges of the Transpantaneira Highway and side roads leading to the ranches. Occasionally, the *rolinha-do-planalto* was also seen along the roadsides. The presence of the picui ground dove or *rolinha-picua, Columbina picui* (Temminck, 1813), has also been reported from all parts of the Pantanal (Dubs, 1992).

During their study of the avifauna of the Poconé region, Cintra and Yamashita (1990) found the pale-vented pigeon or *pomba-galega, Columba cayennensis* (Vieillot, 1818); the blue ground-dove or *pararu-azul, Claravis pretiosa* (Ferrari-Perez, 1886); and the long-tailed ground dove or *rola-vaqueira, Uropelia campestris* (Spix, 1825), to be common, non-migratory birds. They have been reported from other parts of the wetland, as well (Dubs, 1992). These species, which sometimes formed small flocks, were considerably more abundant in the upland regions around the borders of the Pantanal than in the wetland itself. However, they tended to congregate on roadways, such as the Transpantaneira, elevated above the level of the floodplain.

The scaled dove or *fogo-apagou, Scardafella squammata* (Lesson, 1831), is sometimes observed near the ranch houses in various parts of the Pantanal. It was usually observed on the ground, displaying feeding habits similar to those of the ground doves.

The Picazuro pidgeon or *pomba-asa-branca, Columba picazuro* (Temminck, 1813), was observed by Cintra and Yamashita (1990) to occur seasonally in the Pantanal of Poconé, from March to December. After the beginning of the flood period, they form flocks and depart from the Pantanal. The eared dove or *avoante, Zenaida auriculata* (Reichenbach, 1847), was also found to occur seasonally from April to September in the Pantanal of Poconé, but it was rare.

The only other columbid species that were nearly as common as the ground doves in the Pantanal were the gray-fronted dove or *juriti-gemedeira, Leptotila rufaxilla* (Richard and Bernard, 1792), and the white-tipped dove or *juriti-pupú, Leptotila verreauxi* (Bonaparte, 1855), which were frequently observed in trees and low shrubs along the Transpantaneira Highway or on the ground feeding on insects and seeds, sometimes together with ground doves.

Other species in the family were seldom, if ever, seen in the Pantanal and were encountered mainly in sparsely to densely forested areas in the upland parts of the watershed. Among the species encountered in southern Mato Grosso but not seen in the wetland were the scaled pigeon or *pomba-pedrês, Columba speciosa*, Gmelin, 1789; the plumbeous pigeon or *pomba-amargosa, Columba plumbea* (Hellmayr, 1908); and the ruddy quail-dove or *jutiri-piranga, Geotrygon montana* (Linnaeus, 1758). None of these species were found to be abundant anywhere in southern Mato Grosso.

Psittacidae

The macaws, parrots, and parakeets are important elements of the Pantanal avifauna. The species most prominent among these is the largest member of the family in the world, the hyancinthine macaw, *Anodorhynchus hyacinthinus* (Latham, 1790), called the *arara-azul-grande* in Portuguese. This species makes a strikingly attractive house pet and commands a price of several thousand dollars on the market in the United States. For this reason, it has been hunted to extinction over a large part of its former range. The Pantanal population is still generally intact, but that could change within a short period of time if active protection is not provided.

The population center of the hyancinthine macaw is not the rainforest but rather the wet-and-dry climatic zone of South America. Generally, these macaws travel in pairs, which often form small groups. They feed, among other things, on the fruit of various palm trees, but there is one report that four of these macaws were observed feeding on aquatic snails, probably *Pomacea* sp. (Roth, 1989). The pairs apparently communicate by loud squawks, which quickly reveal the presence of the small flocks. It was observed on several occasions that imitating the squawks while pairs of these macaws were flying overhead would induce them to circle back and pass low over the source of the sound, perhaps to find out whether a colleague was in need of assistance.

A much smaller macaw species confined to the region of the Pantanal and adjacent areas of Paraguay, Bolivia, and Argentina is the golden-collared macaw or *maracanã-de-colar, Ara auricollis* Cassin, 1853, which is an inhabitant of the hummocks or *capões de mata* and gallery forests of the Pantanal.

The red and green macaw or *arara-vermelha, Ara chloroptera* Gray, 1859, was one of the more common species, observed in the *cerrado* and *cerradão* near the town of Poconé. Dubs (1992) reported it to be common throughout the Pantanal, except at the southwest border. A similar distribution throughout the Pantanal except for the extreme southwest was reported by Dubs (1992) for the red-shouldered macaw or *maracanã-pequena, Ara nobilis* (Linnaeus, 1758), represented in the Pantanal region by *Ara nobilis longipennis* (Neumann, 1931), which is called the long-winged macaw.

Several macaw species more common in other parts of South America are rarely seen in the Pantanal. The blue and yellow macaw or *arara-canindé, Ara ararauna* (Linnaeus, 1758), has been observed in the northern part of the Pantanal but is not at all abundant. Other species reported at times for southern Mato Grosso, including the chestnut-fronted macaw or *maracanã-guaaçu, Ara severa* (Linnaeus, 1758); red-bellied macaw or *maracanã-do-butiti, Ara manilata* (Boddaert, 1783); and blue-winged macaw or *maracanã-verdadeira, Ara maracana* (Vieillot, 1816), were not observed during the study and may be confined to the upland regions to the north of the Pantanal. Reports of their occurrence near the Pantanal have generally been from the chaco of Bolivia and northern Paraguay and the *cerrado* regions of Mato Grosso.

In addition to the macaws, many small parrot and parakeet species are abundant in southern Mato Grosso and frequently enter the Pantanal. These include the canary-winged-parakeet or *periquito-de-asa-amarela, Brotogeris versicolurus* (P. L. S. Müller, 1776), one of the parakeet species commonly observed in large flocks in the trees near the ranch houses; the blue-crowned parakeet or *aratinga-de-testa-azul, Aratinga acuticaudata* (Vieillot, 1817); the white-eyed parakeet or *aratinga-de-bando, Aratinga leucophthalmus* (P. L. S. Müller, 1776), which seems to prefer gallery forests (Latham, 1939) or hummocks (Dubs, 1992); the peach-fronted parakeet or *aratinga-estrela, Aratinga aurea* (Gmelin, 1789), which often seeks fruit near the ranch houses; the *caturrita* or monk parakeet, *Myiopsitta monachus* (Vieillot, 1817), which is common in *cerrado* and dry forests (Latham, 1939) but also frequents the wetland, where, according to Dubs (1992), it consumes the seeds of water plants, including *Pontederia lanceolata* and *Thalia geniculata*; the turquoise-fronted parrot or *papagaio-curau, Amazona aestiva* (Linnaeus, 1758), a widespread parakeet in Central and Southern Brazil, Bolivia, Argentina, and Paraguay; orange-winged parrot or *papagaio-grego, Amazona amazonica* (Linnaeus, 1766), a woodland and forest species; blue-headed parrot or *maitaca-de-cabeça-azul, Pionus menstruus* (Linnaeus, 1766); and the scaly-headed parrot or *maitaca-verde, Pionus maximiliani* (Kuhl, 1820), which is also widespread and abundant in different kinds of habitat in the Pantanal of Poconé (Cintra and Yamashita, 1990) and represented in the region by the subspecies *P. maximiliani siy* (Souancé, 1856).

The black-hooded parakeet or *príncipe-negro, Nandayus nenday* (Vieillot, 1823), was not noted in the Pantanal of Poconé, perhaps because it is generally associated with stands of the palm, *Copernicia alba*, the largest stands of which were located near the confluence of the Rios Cuiabá and Paraguay and at sites beyond the borders of the wetland.

Other species sometimes reported from the Pantanal are uncommon in the region of the wetland and were not identified during the study. These include the white-eyed parakeet or *periquito-de-cabeça-suja, Aratinga weddelli* (Deville, 1851); the reddish-bellied parakeet or *tiriba-de-testa-vermelho, Pyrrhura frontalis* (Laubmann, 1932); crimson-bellied parakeet or *tiriba-de-barriga-vermelha, Pyrrhura rhodogaster* (Sclater, 1870); the green-cheeked parakeet or *tiriba-de-cara-suja, Pyrrhura molinae* (Todd, 1847); the mealy parrot or *papagaio-moleiro, Amazona farinosa* (Boddaert, 1783); and the yellow-faced parrot or *papagaio-galego, Amazona xanthops* (Spix, 1824). The blaze-winged parakeet or *tiriba-fogo, Pyrrhura devillei* (Massena and Souancé, 1854), and the plain parakeet or *periquito-rico, Brotogeris tirica* (Gmelin, 1788) are species inhabiting the southern Pantanal in the state of Mato Grosso do Sul.

Cuculidae

Among the cuculids are two extremely characteristic birds of southern Mato Grosso, which seem to be just as abundant in the Pantanal as in the upland areas, especially during the dry season. One of these is the smooth-billed ani or *anu-preto, Crotophaga ani* Linnaeus, 1758, which is abundant in the Pantanal as well as in sections of the city of Cuiabá. They are usually encountered in groups of two to five (Fig. 8.11), but sometimes larger flocks are encountered. They generally form small flocks for breeding rather than pairs, and each female builds her own nest, sometimes with the help of one or more other members of the flock. They rear their own young and are not parasitic like some of the cuculids. Sick (1988) reported an association between these birds and domestic animals, which attract or reveal the presence of insects while grazing, making it easier for the anis to find them. However, in the Pantanal, the anis did not seem to remain particularly close to the cattle. Two were observed watching an anaconda more than 4 m in length cross a patch of open ground after the pool it had been in became nearly desiccated during the dry season. The birds seemed fascinated by the reptile and moved along with it, remaining near its head, but not too near.

The second of the abundant species in the Pantanal is the guira cuckoo or *anu-branco, Guira guira* (Gmelin, 1788), which is generally a species associated with cleared fields in dry, upland locations (Sick, 1988). However, this species is also frequently observed in the Pantanal, as well as in the city of Cuiabá. Generally, these birds are observed in large flocks of no less than five birds but often of many more. On hot days in the Pantanal, these birds display an interesting behavior, squatting low on the ground and opening large dorsolateral patches of feathers to expose the down beneath, apparently as a way of cooling themselves. Like the smooth-billed ani, they are resident birds that live and reproduce in small communities, even constructing some community nests, the eggs in which are poorly tended or even discarded by one of the females that did not produce them (Sick, 1988).

Although a typical bird of wetlands, the greater ani or *anu-coroca, Crotophaga major* Gmelin, 1788, is much less abundant in the Pantanal than the other two anis. It is usually a species associated with forested areas near water, such as the gallery forests along rivers. It was never observed on the floodplains, and seldom in the gallery forests.

The other members of the cuckoo family were uncommon in the Pantanal. Three of them are parasites, which lay each of their eggs in the nest of a bird of another species and let the unwitting parents raise the fledgling, which pushes its adoptive siblings out of the nest. These are the striped cuckoo or *saci-do-campo*, *Tapera naevia* (Linnaeus, 1766); the pheasant cuckoo or *peixe-frito-verdadeiro, Dromococcyx phasianellus* (Spix, 1824); and the pavonine cuckoo or *peixe-frito-pavonino, Dromococcyx pavoninus* Pelzeln, 1870, which seems to have been encountered mainly in upland regions (Laubmann, 1939). All three of them occur in southern Mato Grosso, including the Pantanal.

In addition to the nest parasites, four additional species are known from southern Mato Grosso. The dark-billed cuckoo or *papa-lagarta, Coccyzus melacoryphus* Vieillot, 1817, was observed on occasion in and near the Pantanal. The pearly-breasted cuckoo or *papa-lagarta-de-euler, Coccyzus euleri* Cabanis, 1873; the squirrel cuckoo or *alma-de-gato, Piaya cayana* (Linnaeus, 1766); and the little cuckoo or *chincoã pequeno, Piaya minuta* (Vieillot, 1817), occur with certainty in the Pantanal but were not positively identified during the study.

8.13.4.35. Tytonidae

The occurrence of the barn owl or *suindara*, *Tyto alba* (Scopoli, 1769), in the Pantanal, as elsewhere in Mato Grosso, depends upon the presence of suitable nesting sites. This species prefers old buildings, where it constructs nests in openings beneath the roof. Such nesting sites are often found in stables and barns. Hollow trees were apparently the natural nesting sites before the arrival of European settlers, but these are presently not in fashion with the barn owls. In the Pantanal, this greatly limits the size of the population since old buildings are generally torn down. The barn owl is presently not abundant in southern Mato Grosso.

8.13.4.36. Strigidae

The most abundant owl in southern Mato Grosso is the burrowing owl or *caruja-do-campo*, *Speotyto cunicularia* (Molina, 1782), which is not a resident of the Pantanal because of its nesting habits. The owls are typical of dry grasslands, where they nest in underground burrows, which may be constructed by the owls or taken over after having been abandoned by mammals (Fig. 8.33). These often consist of a nesting chamber about 1 m deep that must be reached through a winding burrow about 3 m long (Ginn, 1973). The owls are often seen on the roads at night, even in the city of Cuiabá, and they can also be observed close to the Pantanal, where they patrol the outskirts of the wetlands in search of food. The species occurs in western North America and is widespread in South America, as well, ranging from the tropics far into the temperate zone as far as the Straits of Magellan (Laubmann, 1939).

A species sometimes observed in the Pantanal is the great horned owl or *jacurutu*, *Bubo virginianus* (Gmelin, 1788), which, like the barn owl, prefers to nest in old buildings. These owls are generally found on the ranches and do not seem to nest at all in the trees of the wetlands. It is a species with an enormous range in the New World.

A common species in the forests of the Amazon Basin with a range that also includes the entire region of the Pantanal is the black-banded owl or *coruja-preta, Ciccaba huhula* (Daudin, 1800). This owl is common on farms and plantations.

Figure 8.33: The burrowing owl, *Speotyto cunicularia*, is a bird that enters the Pantanal only rarely to feed. In well-drained upland areas, it nests in abandoned burrows. Here, the young have left the nest during the day. When approached, they disappear underground, while the parents remain near the entrance.

The ferrugineous pigmy-owl or *caburé-ferrugem, Glaucidium brasilianum* (Gmelin, 1788). is a species that inhabits gallery forests and woodlands near water, where it is active during the daytime, mainly in the early morning and shortly before sunset (Ginn, 1873). It is a widespread species in the tropics and subtropics of the New World and can be encountered throughout the entire region of the Pantanal (Dubs, 1992).

The striped owl or *coruja-orelhuda, Rhinoptynx clamator* (Vieillot, 1807), occurs throughout most of Brazil, except for the areas covered by Amazonian forests (Sick, 1988). Although this is a common species in the region of the Pantanal, no evidence of it was discovered in the wetland during the study.

One species common in southern Mato Grosso that is rare in or absent from the Pantanal of Poconé (Cintra and Yamashita, 1990) is the tropical screech-owl or *corujinha-de-orelha, Otus choliba* (Vieillot, 1817), which is a native of dry forests and grasslands of central South America (Hekstra, 1973). It ranges from Bolivia, Paraguay and Southern and Central-Western Brazil to Buenos Aires, Santa Fe, Cordoba, San Luis, and Mendoza in Argentina (Olrog, 1963). This owl was not observed during the study.

Other owls with a range including southern Mato Grosso are the tawny-bellied screech-owl or *corujinha-amazônica*, *Otus watsonii* (Cassin, 1848); the Stygean owl or *mocho-diabo*, *Asio stygius* (Wagler, 1832); and the least pigmy owl or *caburezinho*, *Glaucidium minutissimum* (Wied, 1830). All of these species are poorly known. The Stygean owl was recorded near the Pantanal only once near Corumbá in the South, while the other two species are confined to the region northwest of the wetland (Dubs, 1992). The spectacled owl or *mucurututu*, *Pulsatrix perspicillata* (Latham, 1790), is a somewhat better known species that is present near the northern and eastern borders of the Pantanal but a rare visitor during all seasons near Poconé (Cintra and Yamashita, 1990).

The locally rare short-eared owl or *coruja-do-banhado, Asio flammeus* (Pontoppidan, 1763), which is sometimes included among the avifauna of the Pantanal, was not observed during the study and would be expected in the *cerrado* regions rather than the wetlands (Sick, 1988).

Nyctibiidae

Two species of potoo are known from the region of the Pantanal: the great potoo or *mae-de-lua-gigante, Nyctibius grandis* (Gmelin, 1788), and the common potoo or *urutau, Nyctibius griseus* (Gmelin, 1789). Neither of these two species are abundant in the Pantanal itself, but they may enter the forested islands, especially during the dry season. Both species reproduce by depositing a single egg in a depression atop a tree stump (Schauensee and Phelps, 1978). Because of their nocturnal habits, the species were not distinguished during the study.

Caprimulgidae

Several species in this family are abundant in the Pantanal, where they are seen along the edges of the road at night. During nocturnal field trips to the Pantanal, both sides of the Transpantaneira Highway were sometimes lined with small red points of light, which were the reflections of the headlights in the eyes of many pauraques or *curiango, Nyctidromus albicollis* (Gmelin, 1789), which remain motionless at the borders of the road and the herbal vegetation along the sides. They are hardly less common at higher elevations near the wetlands, as well. When approached to closely, the pauraques fly away, often exhibiting a nearly vertical flight.

The little nightjar or *bacurau-pequeno, Caprimulgus parvulus* Gould, 1837, is also seasonally abundant in and near the Pantanal, occurring mainly at the end of the dry and beginning of the rainy season.

The rufous nightjar or *João-corta-pau, Caprimulgus rufus* Boddaert, 1783, is more frequently heard than seen. It occurs abundantly in partially wooded areas along the Rio Bento Gomes at the outskirts of the Pantanal but seldom in the interior of the wetland.

Another species found along the rivers in and near the Pantanal is the band-tailed nighthawk or *bacarau-cauda-barrada, Nyctiprogne leucopyga* Spix, 1825, which is fairly abundant at times along the Rio Bento Gomes near the exposition park of Poconé. These nighthawks, which occur in a variety of habitats, including rainforests, savannas, and gallery forests, spend the days roosting in small groups in thickets (Cherrie, 1916).

A species less common in the Pantanal but abundant elsewhere in southern Mato Grosso is the nacunda nighthawk or *corução, Podager nacunda* (Vieillot, 1817), which is frequently seen catching insects under street lamps in Cuiabá. It is known to breed in open fields, palm groves, and forests (Laubmann, 1940). Similarly, the *bacurau-rabo-branco, Caprimulgus candicans* Pelzeln, 1866, is fairly abundant in and around Cuiabá but is not often seen in the Pantanal. Much less abundant in the region is the *bacurau-tesoura, Hydropsalis brasiliana* (Gmelin, 1788), which is seen infrequently along roads at the outskirts of the Pantanal as well as elsewhere in southern Mato Grosso.

Other species in this family that are found in southern Mato Grosso but were not specifically identified in the Pantanal during the study include the lesser nighthawk or *bacurau-de-asa-fina, Chordeiles acutipennis* (Boddaert, 1783); the least nighthawk or *bacurau-pequeno, Chordeiles pusillus* Gould, 1861; the semicollared nighthawk or *tuju, Lurocalis semitorquatus* (Gmelin, 1789); and the ocellated nighthawk or *bacarau-ocelado, Nyctidromus ocellatus* (Tschudi, 1844). Cintra and Yamashita (1990) reported *Chordeiles pusillus* to occur near Poconé seasonally from May to September. *Nyctidromus ocellatus* reportedly remains deep in the forests, breeding on the ground

(Sick, 1988), and is therefore seldom encountered in the sparsely forested Pantanal. According to Dubs (1992), of the three subspecies, only *Nyctidromus o. ocellatus* occurs in the region of the Pantanal.

Apodidae

The fork-tailed palm-swift or *tesourinha, Reinarda squamata* Cassin, 1853, is an insectivorous bird that is associated with palm trees (Sick, 1988). In the region of the Pantanal, it is generally found in the buriti, *Mauritia vinifera.* In March, these migratory birds leave Brazil and do not return until November.

Other swift species that are encountered in Mato Grosso include the ashy-tailed swift, *Chaetura andrei* Berlepsch and Hartert, 1902; the short-tailed swift, *Chaetura brachyura* Jardine, 1846; *Chaetura cinereiventris* Sclater, 1862, represented in Mato Grosso by *C. c. egregia* Todd, 1916; and the white-collared swift or *taperuçu-de-coleira-branca, Streptoprocne zonaris* (Shaw, 1796). *Chaetura andrei* is reported to be rare in the Poconé district but more common at the northeastern and southwestern borders of the wetland from December to April (Cintra and Yamashita, 1990). *Streptoprocne zonaris* nests along cliffs, explaining why it avoids vast plains without hills, such as the Pantanal wetland (Sick, 1988). It is understandably attracted to the regions northeast of the Pantanal because many good nesting sites are available. All of these swift species are insectivorous. They are strong fliers and can cover long distances in search of food. They seem to have no preferences for wetland habitats, and their occurrence in the Pantanal is sporadic.

Trochilidae

Hummingbirds and their relatives are abundant in the Pantanal as well as in other parts of southern Mato Grosso. However, the species richness is not as great as in other parts of Brazil, apparently because of the heat and dryness. Many of the species known from southern Mato Grosso are rare in the Pantanal, and a few species accounted for most of the sightings in the wetland. Grantsau (1989) listed most of these species as inhabitants of the *cerrado*, the dry forest, open fields, and patches of woodland.

Those species that are encountered most often in the northern part of the wetland include the glittering-bellied emerald or *esmeralda-de-bico-vermenho, Chlorostilbon aureoventris aureoventris* (d'Orbigny and Lafresnaye, 1838), and the glittering-throated emerald or *beija-flor-de-garganta-verde, Amazilia fimbriata* (Gmelin, 1788). In addition, Cintra and Yamashita (1990) reported the rufous-breasted hermit or *beija-flor-besourão, Glaucis hirsuta* (Gmelin, 1788), and the fork-tailed woodnymph or *beija-flor-de-barriga-violeta, Thalurania furcata* (Gmelin, 1788), to be abundant species in the Poconé district from April to September; the planalto hermit or *rabo-branco-acanelado, Phaethornis pretrei* (Lesson and DeLattre, 1839), to be a resident species present throughout most of the Pantanal, and the buff-bellied hermit or *rabo-branco-de-barriga-fulva, Phaethornis subochraceus* Todd, 1915, to be found during all seasons in gallery forests of the Poconé Municipal District.

The versicolored emerald or *beija-flor-de-banda-branca, Amazilia versicolor versicolor* (Vieillot, 1818), is a common species of the dry, forested upland regions. Cintra and Yamashita (1990) reported it to be rare in the region of Poconé, but one hummingbird sighted there during the study was tentatively identified as this species. The sombre hummingbird or *beija-flor-cinza, Aphantochroa cirrhochloris* (Vieillot, 1818), is a species common in upland forests, which is apparently common only in the region of Cáceres. Dubs (1992) reported it from the northwestern border region of the wetland. The blue-tufted starthroat or *bico-reto-azul, Heliomaster furcifer* (Shaw, 1812), is also a species of the *cerrado* rather than wetlands, and it has been reported mainly from the regions to the north and southwest of the wetland (Dubs, 1992). The horned sungem or *chifre-de-ouro, Heliactin cornuta* (Wied, 1821), is a migratory species that seasonally visits the dry forests and grasslands of southern Mato Grosso (Ruschi, 1982) and is known from the uplands northeast of the Pantanal (Dubs, 1992). Another migrant is the black-throated mango or *beija-flor-de-veste-preto, Anthracothorax nigricollis* (Vieillot, 1817), which prefers to feed on nectar and insects from the flowers of trees (Ruschi, 1982) and can be encountered throughout the Pantanal region, although it apparently prefers drier forests.

Also present in southern Mato Grosso but not observed during the study are the cinnamon-throated hermit or *rabo-branco-cinamômeo, Phaethornis nattereri* Berlepsch, 1887, reported from the northern borders of the wetland (Dubs, 1992); swallow-tailed hummingbird or *beija-flor-tesoura, Eupetomena macroura macroura* (Gmelin, 1788), not uncommon throughout the region (Dubs, 1992); dot-eared coquette, *Lophornis gouldii* (Lesson, 1832); gilded sapphire or *beija-flor-dourado, Hylocharis chrysura chrysura* (Shaw, 1812); white-tailed goldenthroat or *beija-flor-de-bico-curvo, Polytmus guainumbi* (Pallas, 1764); and amethyst woodstar or *estrelinha-ametista, Calliphlox amethystina* (Boddaert, 1783), a migratory species (Ruschi, 1982) apparently common in drier locations throughout the region.

All of these nectar-eating hummingbirds congregate at whichever tree or shrub is presently in bloom. During the dry season, there is a sequence of flowering, first of one species then of another, usually while the trees are bare of leaves. In the Pantanal, the species of *Tabebuia* and various legumes bloom at different times during the *seca*. After the start of the rainy season, other species flower, providing a continuous supply of nectar. In spite of the availability of flowers, the hummingbird populations remain sparse in the Pantanal, and those species observed apparently travel great distances in search of food because they are usually observed only once at a tree to which they do not return again, at least for several hours.

In addition to the species known from the region of the Pantanal, the occasional presence of various others has been reported or is suspected. These are known from other parts of Brazil and probably migrate into southern Mato Grosso occasionally. Among these are the scale-throated hermit or *rabo-branco-de-garganta-rajada, Phaethornis eurynome* (Lesson, 1832), which is confined to the southeastern borders of the wetland (Dubs, 1992); the white-bearded hermit or *rabo-branco-cinza, Phaethornis hispidus* (Gould, 1846), which occurs near the northwestern border (Dubs, 1992); reddish hermit or *rabo-branco-rubro, Phaethornis ruber* (Linnaeus, 1758), also confined to the northwestern edges of the wetland (Dubs, 1992); gray-breasted sabrewing of *asa-da-sabre-cinza, Campylopterus largipennis* (Boddaert, 1783), reported by N. Willis (1976) from the region northwest of the Pantanal; white-necked jacobin or *beija-flor-branco, Florisuga melivora* (Linnaeus, 1758); white-vented violetear or *beija-flor-de-canto, Colibri serrirostris* (Vieillot, 1816); ruby-topaz hummingbird or *beija-flor-vermelho, Chrysolampis mosquitus* (Linnaeus, 1758); blue-tailed emerald, *Chlorostilbon mellisugus* (Linnaeus, 1758); violet-capped woodnymph or *beija-flor-de-fronte-violeta, Thalurania glaucopis* (Gmelin, 1788); white-chinned sapphire or *beija-flor-roxo, Hylocharis cyanus* (Vieillot, 1818); white-bellied hummingbird, *Amazilia chionogaster hypoleucus* (Gould, 1846); and long-billed starthroat or *bico-reto-cinzento, Heliomaster longirostris* (Audebert and Vieillot, 1801). None of these species has been reported from southern Mato Grosso by Grantsau (1989), but most of them have been reported from the border areas by Dubs (1992), and complete lists of the avifauna of the wetland are still lacking. Most of these species seem to be elements of the Amazonian fauna that reach the southern limits of their ranges at the northern borders of the Pantanal. Only *Thalurania glaucopis* is apparently a species that reaches the western and northern rather than southern limits of its range at the borders of the Pantanal.

Trogonidae

Trogons are not common throughout most of the Pantanal. They are birds that usually live in forests, and their presence in the wetland is limited mainly to the gallery forests along the larger rivers. The blue-crowned trogon or *surucuá-de-coroa-azul, Trogon curucui* Linnaeus, 1766, is apparently the only species commonly encountered in the wetland itself, albeit in the forested areas at elevations slightly above the floodplain. Dubs (1992) reported seeing individuals on a ranch in the southern part of the Pantanal in February and April and small flocks in September.

The black-tailed trogon or *surucuá-de-cauda-preta, Trogon melanurus* Swainson, 1837, is a species common in rainforests of northern South America that ranges south as far as southern Brazil and Bolivia (Schauensee and Phelps, 1978). The white-tailed trogon or *surucuá-de-barriga-dourada, Trogon viridis* Linnaeus, 1766, has generally the same range, but it is found in somewhat less dense forests as well as clearings, where it feeds of fruit, seeds, insects, and lizards (Schauensee and Phelps, 1978). The violaceous trogon or *surucuá-pequeno, Trogon violaceus* Gmelin, 1788, is a species abundant along streams as well as in woodlands and clearings. The black-throated trogon or *surucuá-de-barriga-amerela, Trogon rufus* Gmelin, 1788, and the collared trogon or *surucuá-de-coleira, Trogon collaris* Vieillot, 1817, occur mainly in forests. Dubs (1992) reported all five species only from the region northwest of the wetland.

A trogon species that inhabits various kinds of forest and open woodlands and seems to be limited to the southeastern borders of the Pantanal is the surrucuá trogon or *surucuá-variado, Trogon surrucura* Vieillot, 1817, which was reported by Sick (1988) to nest in the hollows of mounds constructed by termites on trees.

Alcedinidae

Among the three common species of kingfisher on the floodplain of the Pantanal, the ringed kingfisher or *martim-pescador-grande, Ceryle torquata* (Linnaeus, 1766), is the largest (Plate 11b). This species typically occupies a territory on a high object overlooking a shallow water body and drops swiftly into the water when a small fish moves too close to the surface. The fish is grabbed in the beak and carried back to the high perch, where it is inactivated by a characteristic sideways movement of the kingfisher's head which bashes the head of the fish against a hard object. The fish is then swallowed. The power lines parallel to the Transpantaneira Highway have become favorite roosting places of these kingfishers.

The Amazon kingfisher or *martim-pescador-verde, Chloroceryle amazona* (Latham, 1790), is nearly as common as the ringed kingfisher. Rather than the powerlines, it prefers the wooden bridges, fence posts, and other objects closer to the surface of the water (Fig. 8.34). It catches and inactivates the fishes in the same way as the ringed kingfisher.

The least abundant of the common kingfishers on the Pantanal floodplain is the green kingfisher or *martim-pescador-pequeno, Chloroceryle americana* (Gmelin, 1788), which is widespread in Brazil and displays habits similar to those of the other kingfishers. However, it feeds on somewhat smaller fishes.

Two other species of kingfisher are known from the Pantanal, but they seem to be confined to gallery forests and other wooded areas. Both are at or near the southern limits of their ranges in the Pantanal (Sick, 1988). The green and rufous kingfisher or *martim-pescador-da-mata, Chloroceryle inda* (Linnaeus, 1766), occurs in open woodland and is apparently more common along upland rivers than in the Pantanal. The pigmy kingfisher, *arirambinha, or martim-pescador-anão, Chloroceryle aenea* (Vroeg, 1764), is a bird preferring dense vegetation along rivers (Sick, 1988). It is rather secretive and not often seen, at least in the northern part of the Pantanal.

Figure 8.34: The Amazon kingfisher, *Chloroceryle amazona*, is considerably smaller than the ringed kingfisher. It is somewhat more widespread but is less sociable and is usually not observed together with other members of its species.

Kingfishers in general are birds with some nesting problems. They typically brood their eggs in burrows leading more or less horizontally from holes in steep, eroded riverbanks. This offers them considerable protection from natural enemies, most of which are unable to approach the entrances to the nests, but it makes it difficult for the birds to find suitable nesting places, especially when the river banks are modified by man. Sick (1988) reports that suitable erosional zones of rivers are densely colonized by nesting kingfishers, an indication that such sites are not easy to find. Although three kingfishers are abundant on the Pantanal floodplains, it must be assumed that the nesting places are rather far from their feeding grounds because this temporarily flooded wetland lacks steep erosional river banks. Only the large rivers and upland streams can offer the birds the breeding grounds they require.

Momotidae

The blue-crowned motmot or *udu-de-coroa-azul, Momotus momota* (Linnaeus, 1766), is a species encountered in southern Mato Grosso. Few individuals were encountered in the Pantanal, and these were generally seen near ranches. The diet of these species consists mainly of arthropods but may also include small terrestrial vertebrates (Sick, 1988). The motmots nest in underground burrows constructed by other animals and modified by the motmots.

Apparently, two subspecies may be encountered in the Pantanal, *Momotus momota simplex* Chapman, 1923, and *Momotus momota pilcomajensis* Reichenow, 1919. Dubs (1992) reported that the former is encountered in the northern Pantanal, and the latter, in the southern part of the wetland. Naumburg (1930) reported that specimens of *M. m. simplex* had been taken from Cuiabá and Chapada, apparently referring to Chapada dos Guimarães.

Galbulidae

The jacamars are forest birds not uncommon along the rivers in the Pantanal. They seem to be birds that enjoy the proximity of water, feeding on various insects (Sick, 1988). The brown jacamar or *ariramba-preta, Brachygalba lugubris* (Swainson, 1837), is a widespread species in Brazil. According to Sick (1988), it is a typical species of gallery forests.

The rufous-tailed jacamar or *bico-de-agulha-de-rabo-vermelho, Galbula ruficauda* Cuvier, 1817, is not confined to gallery forests but also ranges into drier forests. However, it remains close enough to water to have earned a local Portuguese name meaning "water hummingbird". It feeds mainly on bees and wasps (Sick, 1988).

The bronzy jacamar or *ariramba-acobreada, Galbula leucogastra* Vieillot, 1817, is a species mainly of the Amazon Basin (Sick, 1988). Reports of its presence in the region of the Pantanal could not be confirmed during the study, although its occurrence there seems likely.

Bucconidae

The species in this family are mainly forest birds, which enter the Pantanal only at locations offering them suitable habitats. The *João-bobo* or *dormião, Nystalus chacuru* (Vieillot, 1816), nests in burrows, generally located in drier locations than the Pantanal (Sick, 1988), although its range includes much of the wetland. Less common congeneric species of southern Mato Grosso include the striolated puffbird or *rapazinho-estriado, Nystalus striolatus* Pelzeln, 1856, and the spot-backed puffbird or *rapazinho-dos-velhos, Nystalus maculatus* (Gmelin, 1788), both species more common in the upland regions of Mato Grosso.

According to Sick (1988), the white-necked puffbird or *capitão-do-mato, Notharchus macrorhynchus* (Gmelin, 1788), occurs in the northern part of Mato Grosso but may not range as far south as the Pantanal. However, Schauensee and Phelps (1978) reported that the species is found as far south as Paraguay and northern Argentina. Unfortunately, the presence of the species in the region of the Pantanal could not be confirmed.

Among the species that do not seem to be regular members of the Pantanal avifauna but may enter some of the gallery forests in the wetland is the spotted puffbird or *rapazinho-carajó, Bucco tamatia* Gmelin, 1788, which is known to feed on insects and fruit (Schauensee and Phelps, 1978). Other rare species in the Pantanal include the white-fronted nunbird or *chora-chuva-de-cara-branca, Monasa morphoeus* (Wagler, 1823); pied puffbird or *macuru-pintado, Notharchus tectus* (Boddaert, 1783); the grey-cheeked nunlet or *freirinha-de coroa-castanha, Nonnula ruficapilla* (Tschudi, 1844); the black-fronted nunbird, *Monasa nigrifrons* (Spix, 1824); and the swallow-wing or *urubuzinho, Chelidoptera tenebrosa* (Pallas, 1782).

Capitonidae

The black-girdled barbet or *capitão-de-bigode-de-cinta, Capito dayi* Cherrie, 1916, is known from northern Mato Grosso (Sick, 1988). Reports of its occurrence in the northern part of the Pantanal could not be confirmed during the study.

Ramphastidae

Toucans are seen regularly in and near the Pantanal. However, of the species known from the region, only one accounts for nearly all of the sightings. The toco toucan or *tucanuçu, Ramphastos toco* P. L. S. Müller, 1776, occurs in abundance throughout southern Mato Grosso, including the Pantanal. It feeds mainly on fruit but may also consume insects and even the eggs and young of smaller birds (Andrade, 1992). Flying toucans are often observed being harassed by small birds, which repeatedly attack the cumbersome flyers, forcing them to flee or to land. These attacks suggest that the toucans are recognized as a danger by the smaller bird species, the nests of which they rob.

A variety of other toucan and toucanette species are also known from Brazil. During the study, none of these were observed. Although they are often cited for southern Mato Grosso, it is likely that they are presently rare there. Among these species are the chestnut-eared araçari or *araçuri-castanho, Pteroglossus castanotis* Gould, 1834; yellow-ridged toucan, *Ramphastos culminatus* Gould, 1833; Cuvier's toucan, *Ramphastos tucanus cuvieri* (Wagler, 1827); the green aracari, *Pteroglossus inscriptus* Swainson, 1822; red-necked aracari or *araçari-de-nuca-vermehla, Pteroglossus bitorquatus* Vigors, 1826; spot-billed toucanet or *saripoca-de-bico-riscado, Selenidera maculirostris* (Lichtenstein, 1823); and red-breasted toucan or *tucano-de-bico-verde, Ramphastos dicolorus* Linnaeus, 1766.

Picidae

Woodpeckers are common in the Pantanal as well as throughout southern Mato Grosso. The members of this family are most threatened by the practice of burning the dry woodland as well as by the conversion of vast areas on the plateau of southeastern Mato Grosso into fields for growing soybeans, sugar cane, and other crops. These birds obviously depend mainly on insects in dead wood for their nourishment, and when the woodland is cleared, they have no chance to survive.

Probably the most abundant species in the Pantanal and its environs is the *pica-pau-do-campo, Colaptes campestris* (Vieillot, 1818), which, according to Sick (1988), feeds to a large extent on ants and termites.

Nearly as abundant as C. campestris is the white woodpecker, *bilro*, or *pica-pau-branco, Leuconerpes candidus* (Otto, 1796), a species often observed on trees near the ranch houses.

The lineated woodpecker or *pica-pau-branco-de-banda-branca, Dryocopus lineatus* (Linnaeus, 1766), and the yellow-tufted woodpecker or *pica-pau-de-barriga-vermelha, Melanerpes cruentatus* (Boddaert, 1783), are species observed from time to time on the hummocks above the high water level in the Pantanal. They were observed only during the dry season.

Species observed occasionally in gallery forests and hummocks on the floodplain include the chestnut woodpecker or *pica-pau-chocolate, Celeus elegans* (P. L. S. Müller, 1776); ringed woodpecker or *pica-pau-de-coleira, Celeus torquatus* (Boddaert, 1783), little woodpecker or *pica-pau-pequeno, Veniliornis passerinus* (Linnaeus, 1766); crimson-crested woodpecker or *pica-pau-de-garganta-preta, Phloeoceastes melanoleucos* (Gmelin, 1788); arrowhead piculet, *Picumnus minutissimus* (Pallas, 1782); pale-crested woodpecker or *pica-pau-louro, Celeus lugubris* (Malherbe, 1851); golden-green woodpecker or *pica-pau-da-copa, Piculus chrysochloros* (Vieillot, 1818); green-barred woodpecker or *pica-pau-carijó, Chrysoptilus melanochloros* (Gmelin, 1788); and cream-backed woodpecker or *pica-pau-de-barriga-preta, Phloeoceastes leucopogon* (Valenciennes, 1826).

Several woodpecker species have been reported from the region of the Pantanal, although their known ranges usually do not encompass the wetland. These include the white-barred piculet or *pica-pau-anão-barrado, Picumnus cirratus* Temminck, 1825; gold fronted piculet or *pica-pau-anão-amarelo, Picumnus aurifrons* Pelzeln, 1870; rusty-necked piculet or *pica-pau-anão-fusco, Picumnus fuscus* Pelzeln, 1870; red-necked woodpecker or *pica-pau-de-penacho, Phloeoceastes rubricollis* (Boddaert, 1783); blond-crested woodpecker or *pica-pau-velho, Celeus flavescens* (Gmelin, 1788); yellow-fronted woodpecker or *Benedito, Melanerpes flavifrons* (Vieillot, 1818); white-fronted woodpecker or *pica-pau-de-testa-branca, Trichopicus cactorum* (d'Orbigny, 1847); checkered woodpecker or *picapau-chorão, Dendrocopos mixtus* (Boddaert, 1783); robust woodpecker or *picapau-rei, Phloeoceastes robustus* (Lichtenstein, 1819), found to the southeast; and red-stained woodpecker or *picapau-de-asa-vermelha, Veniliornis affinis* (Swainson, 1821), confined to the northwestern margins of the Pantanal (Dubs, 1992). None of these locally rare species were identified during the study.

Dendrocolaptidae

Woodcreepers are abundant birds on the hummocks and gallery forests of the Pantanal. Most of these species are easily distinguished from woodpeckers by their curved beaks. They are insectivorous, picking insects from their places of concealment in the bark of the trees. Many of the species prefer dense forested areas, such as the gallery forests along large rivers. Hence, it is difficult to judge their abundance in relation to other species because the gallery forests occupy much less area in the Pantanal than the inundated habitats. Judging from the distribution data provided by Sick (1988), many of the woodcreepers are near the geographic limits of their ranges in the Pantanal.

One of the birds commonly seen on the hummocks in the Pantanal is the red-billed scythebill or *arapaçu-de-bico-torto, Campylorhamphus trochilirostris* (Lichtenstein, 1820), which is unmistakable because of its long, curved beak. The species was observed most often near ranch houses and in a wooded area near the Baia Piuval. Also widespread but not abundant in the Pantanal as well as other wooded regions of southern Mato Grosso is the plain brown woodcreeper or *arapaçu-pardo, Dendrocincla fuliginosa* (Vieillot, 1818), which sometimes travels in small flocks.

The buff-throated woodcreeper or *arapaçu-de-garganta-camurça, Xiphorhynchus guttatus* Lichtenstein, 1820; wedge-billed woodcreeper or *arapaçu-de-bico-de-cunha, Glyphorynchus spirulus* (Vieillot, 1819); and olivaceous woodcreeper or *arapaçu-de-cabeça-cinza*, *Sittasomus griseicapillus* (Vieillot, 1818), visit all kinds of habitats, from rainforests to arid regions (Schauensee and Phelps, 1978). All three species are present in the northern part of the Pantanal (Dubs, 1992) but are not often encountered.

Other woodcreepers reportedly occur in the forested regions in and near the Pantanal, but their presence was not confirmed during the study. The black-banded woodcreeper or *arapaçu-marrom*, *Dendrocolaptes picumnus* Lichtenstein, 1820, is a non-migratory species widely distributed in Central and South America, which inhabits dense forests and avoids the open wetlands of the Pantanal, being known only from its northwestern and southeastern outskirts (Dubs, 1992). Those poorly known species preferring forested areas of the Pantanal and which probably occur most frequently in the gallery forests include the striped woodcreeper or *arapaçu-riscado*, *Xiphorhynchus obsoletus* (Lichtenstein, 1822); the elegant woodcreeper or *arapaçu-elegante*; *Xiphorhynchus elegans* (Pelzeln, 1868); and Hoffmann's woodcreeper or *arapaçu-marrom, Dendrocolaptes hoffmannsi* Hellmayr, 1909, all of which are reported only from locations near the northwestern borders of the Pantanal and are otherwise distributed mainly in Amazonia (N. Willis, 1976; Dubs, 1992). The lineated woodcreeper or *arapaçu-de-listras-brancas*, *Lepidocolaptes albolineatus* (Lafresnaye, 1846); *Dendrocolaptes certhia* (Boddaert, 1783), represented by *D. c. concolor* Pelzeln, 1868 in Mato Grosso; *Xiphocolaptas major* (Vieillot, 1818); red-billed woodcreeper or *arapaçu-de-bico-vermelho*, *Hylexetastes perrotii* (Lafresnaye, 1844); and white-chinned woodcreeper or *arapaçu-da-taoca*, *Dendrocincla merula* (Lichtenstein, 1818) are poorly known species, apparently with ranges limited to southern Amazonia, eastern Bolivia, and the forests northwest of the Pantanal. The straight-billed woodcreeper of *arapaçu-de-bico-reto*, *Xiphorhynchus picus* (Gmelin, 1788), is an insectivorous species that seems to prefer the *cerrado* (Dubs, 1992).

A few of the woodcreepers prefer the *cerrado* to the gallery forest. One of these is the planalto woodcreeper or *arapaçu-grande*, *Dendrocolaptes platyrostris* Spix, 1824, which seems to be commoner in upland areas than in the Pantanal. Another is the *arapaçu-do-cerrado, Dendrocolaptes angustirostris* (Vieillot, 1818), which is a typical species of dry woodlands. It was not observed in the Pantanal, although it probably occurs there, at least during the dry season.

Furnariidae

The horneros and spinetails are distributed throughout Brazil and are not at all confined to the region of the Pantanal. The most common of these near human habitations is the rufous hornero, ovenbird, or *joão-de-barro, Furnarius rufus* (Gmelin, 1788), which constructs its nest out of mud. The nest bakes to a hard, dome-shaped shelter in the sun. There is one small opening on the side. This characteristic type of nest is observed throughout the Pantanal. The masts of power lines are among the favorite sites chosen for nest construction, but trees are still utilized, as well.

A typical wetland species sometimes seen among the emergent vegetation along the small streams draining the Pantanal is the yellow-throated spinetail or *curutié, Certiaxis cinnamomea* (Gmelin, 1788), which builds nearly spherical nests of coarse sticks near the edge of the water. The nests are seen more often than the birds, which generally remain concealed in the vegetation.

The yellow-chinned spinetail or *garrincha-do-buriti, Phacellodromus ruber* (Vieillot, 1817), is another species occasionally observed in the floodplain of the Pantanal. It received its Portuguese name because it is said to use a leaf of the buriti palm as a base for its nest (Andrade, 1992).

The chotoy spinetail or *bichoita, Schoeniophylax phryganophila* Vieillot, 1817, is a widespread species that is well represented in the Pantanal as well as elsewhere in southern Mato Grosso.

Throughout the region of the Pantanal, the ochre-cheeked spinetail or *viu-vi, Poecilurus scutatus* (Shaw, 1792), can be encountered, mainly among the *cerrado* vegetation. It occurs in the tropics and subtropics to the south of the Amazonian forests as far south as northern Argentina as far west as eastern Bolivia (Olrog, 1963). It is represented by its subspecies, *Poecilurus scutatus whitii* (Sclater, 1881), near Corumbá in the southern Pantanal and adjacent eastern Bolivia (Olreg, 1963; Dubs, 1992) and by *P. s. scutatus* to the northeast (Dubs, 1992).

Other species in this family are not abundant in the Pantanal and can be considered only occasional visitors. However, this group includes a considerable number of species, and one or more of the less common ones are observed during almost every visit to the Pantanal, except during the high water period. Among the horneros in the region of the Pantanal are the pale-breasted spinetail or *ui-pi, Synallaxis albescens* Temminck, 1823; pale-crowned spinetail or *becuá, Synallaxis gujanensis* (Gmelin, 1789); plain-fronted or rufous-fronted hornbird, known locally as *joão-graveto, Phacellodromus rufifrons* (Wied, 1821); ruddy spinetail or *joão-castanho, Synallaxis rutilans* Temminck, 1823; sooty-fronted spinetail or *tifli, Synallaxis frontalis* Pelzeln, 1859; cinereous-breasted spinetail or *joão-grilo, Synallaxis hypospodia* Sclater, 1874; buff-throated foliage-gleaner or *barranqueiro-camurça, Automolus ochrolaemus* (Tschudi, 1844); chestnut-winged foliage-gleaner or *vira-folha-de-asa-castanha, Philydor erythropterus* (Sclater, 1856); the russet-mantled foliage gleaner or *limpa-folha-ferrugem, Philydor dimidiatus* (Pelzeln, 1859); the rufous-crested chacalote or *casaca-do-couro, Pseudoseisura cristata* (Spix, 1824); the rusty-backed spinetail or *joão-do-rio, Cranioleuca vulpina* (Pelzeln, 1856); and the chestnut-capped foliage-gleaner or *barranqueiro-de-bico-reto, Hylocryptus rectirostris* (Wied, 1831).

Among the species that have been reported from the region of the Pantanal but generally prefer drier locations and are not abundant in southern Mato Grosso are the pale-legged hornero or *amassa-barro, Furnarius leucopus* Swainson, 1838; streaked xenops or *bico-virado-carijó, Xenops rutilans* Temminck, 1821; plain xenops or *bico-virado-miúdo, Xenops minutus* (Sparrmann, 1788); campo miner or *andarilho, Geobates poecilopterus* (Wied, 1838); rufous-capped spinetail or *pichororé, Synallaxis ruficapilla* Vieillot, 1819; rufous-rumped foliage-gleaner or *limpa-folia-de-sobre-ruiva, Philydor erythrocerus* (Pelzeln, 1859); ochre-breasted foliage-gleaner or *limpa-folia-de-coroa-cinza, Philydor lichtensteini* Cabanis and Heine, 1859; and rufous-breasted leafscraper or *vira-folha-vermelho, Sclerurus scansor* (Ménétriès, 1835).

The sharp-tailed streamcreeper or *joão-de-riacho, Lochmias nematura* (Lichtenstein, 1823), is a moisture-loving species occasionally observed in woodland along upland streams to the north of the Pantanal. It reportedly nests in holes in the banks of streams and rivers bordered by gallery forests (Dubs, 1992). A species preferring terrestrial habitats in hilly forested regions is the buff-fronted foliage-gleaner or *limpa-folia-de-testa-canela, Philydor rufus* (Vieillot, 1818). Neither or these species were observed in the Pantanal during the study.

Formicariidae

This family is second only to the Tyrannidae in the number of species known from South America (Sick, 1988). In the Pantanal, these birds are seldom seen, although the large ranges of many of the species certainly guarantee them a place in the avifauna. None of them were at all abundant on the floodplains, and those that are present in the wetland usually remain in the wooded areas, where they are better able to secure their food, which consists of insects, arachnids, diplopods, and small vertebrates, with the preferences depending upon the species.

The Pantanal and southern Mato Grosso have one regionally endemic species, the Mato Grosso antbird or *chororó-do-Pantanal, Cercomacra melanaria* (Ménétriés, 1935). It is a species that lives on or near the ground in forests and does not migrate. The type specimen was collected at or near Cuiabá, and the range is limited to the Pantanal and northern Bolivia (Dubs, 1992).

The formicariid most frequently observed along the Transpantaneira Highway was the barred antshrike or *choca-barrada, Thamnophilus doliatus* (Linnaeus, 1764). Although other antbirds may well have been more abundant, this one was most often noticed because of the striking coloration of the male. According to Dubs (1992), it is represented in the Pantanal by *T. doliatus radiatus* Vieillot, 1816.

The great antshrike or *choró-boi, Taraba major* (Vieillot, 1816), was also widespread in the Pantanal, but it is a non-migratory species that generally remains concealed in dense underbrush, making it difficult to observe.

The rusty-backed antwren or *formigueiro-ruivo, Formicivora rufa* (Wied, 1831), is found throughout the region of the Pantanal. It can be encountered in many kinds of habitat.

Species found to be rare in the Pantanal of Poconé by Cintra and Yamashita (1990) include the stripe-backed antbird or *piu-piu, Myrmorcheilus strigilatus* (Wied, 1831), represented from Minas Gerais to the chaco of Argentina, Paraguay, and Bolivia by *Myrmorcheilus strigilatus suspicax* Wetmore, 1922, and the large-billed antwren or *chrozinha-de-bico-comprido, Herpsilochmus longirostris* Pelzeln, 1869, which has a range limited to the Brazilian states of Piauí, Goiás, São Paulo, Mato Grosso do Sul, and Mato Grosso (Dubs, 1992).

Among the antshrikes often reported from around the borders of the Pantanal but not observed during the study are two subspecies of the slaty antshrike or *choca-bate-cabo, Thamnophilus punctatus* (Shaw, 1809); *Thamnophilus punctatus stricturus* Pelzeln, 1868, and *Thamnophilus punctatus pelzelni* Hellmayr, 1924. Although Dubs (1992) listed only the *T. p. stricturus* among the subspecies present in the region of the

Pantanal, Naumburg (1930) reported T. p. pelzelni from "Chapada", apparently Chapada dos Guimarães, which is in the Pantanal Watershed.

The rufous-capped antthrush or *pinto-da-mata-coroado, Formicarius colma* (Boddaert, 1783), runs on the ground in dense forests, somewhat resembling a rail. It is known from the border regions to the northwest and southeast of the Pantanal (Dubs, 1992). A species with similar habits, the black-faced antthrush or *pinto-da-mata-de-cara-preta, Formicarius analis* (d'Orbigny and Lafresnaye, 1837), lives in small groups on the floor of tall forests (Schauensee and Phelps, 1978), mainly to the northwest of the wetland.

Other less abundant species found around the margins of the Pantanal include the fasciated antshrike or *choca-zebrada, Cymbelaimus lineatus* (Leach, 1814), which Silva and Oniki (1988) observed rarely in the Serra das Araras and N. Willis (1976) recorded from other woodland savanna, wooded, and cultivated regions to the northwest of the Pantanal. The plain antvireo or *choquinha-lisa, Dysithamnus mentalis* (Temminck, 1823), is reported from the wetland and bordering uplands to the north, where, according to Dubs (1992), it is represented by the subspecies *D. mentalis affinis* (d'Orbigny & Lafresnaye, 1837). The region to the north is also the local range of the gray antwren or *chocinha-de-garganta-cinza, Myrmotherula menetriesii* d'Orbigny, 1837, which associates with other species of the family while hunting for insects in the treetops (Dubs, 1992).

Species best known from the Amazon Region and northern South America may occasionally be observed in southern Mato Grosso but were not observed during the study. These include the warbling antbird or *cantador-comum, Hypocnemis cantator* (Boddaert, 1783), which prefers rainforests and moist hillside woodlands, and the gray antbird or *chororó-pocuá, Cercomacra cinerascens* (Sclater, 1857), from the middle levels of the same kinds of forest (Schauensee and Phelps, 1978). The band-tailed antbird or *solta-asa-do-sul, Hypocnemoides maculicauda* (Pelzeln, 1869), is a species that generally remains on the floor of gallery forests (Naumburg, 1930) and may enter the northern part of the wetland, at times. The scale-backed antbird or *rendadinho, Hylophylax poecilonota* (Cabanis, 1847), is another forest floor dweller noted for stalking army ants in tall rainforests (Schauensee and Phelps, 1978). Other formicariids that lives on or near the floor of forests include the black-spotted bare-eye or *mãe-de-taoca-pintada, Phlegopsis nigromaculata* (d'Orbigny and Lafresnaye, 1837), which was reported by N. Willis (1976) near the northern edges of the Pantanal, and the Amazonian antpitta or *torom-torom, Grallaria berlepschi* Hellmayr, 1903, represented in the Pantanal by *G. berlepschi berlepschi*. One mainly Amazonian species that is rather common in the Serra das Araras (Silva and Oniki, 1988) is the black-throated antbird or *formigeiro-de-peito-preto,, Myrmeciza atrothorax* (Lesson, 1830), most abundant in the ecotone between dense forests and savanna (Schauensee and Phelps, 1978). The silvered antbird or *formigeiro-de-igarapé, Sclateria naevia* (Gmelin, 1788), which lives in swampy forests and open woods as well as in dense forests (Sick, 1988), and the chestnut-tailed antbird or *formigueiro-de-cauda-castanha, Myrmeciza hemimelaena* Sclater, 1857, which generally remains in dense undergrowth, seem confined locally to the region northwest of the Pantanal. The black-faced antbird or *formigueiro-de-cara-preta, Myromoborus myotherinus* (Spix, 1825), which is found mainly in dense rainforests between 100 m and 800 m in elevation in the Amazon region (Schauensee and Phelps, 1978), was reported to be rare in the Serra das Araras (Silva and Oniki, 1988). A bird also reported to be rare in these uplands (Silva and Oniki, 1988), the blackish antbird or *chororó-negro, Cercomacra nigrescens* (Cabanis and Heine, 1859), is a denizen of gallery forests. The dot-winged antwren or *choquinha-de-bando, Microrhopias quixensis* (E. Cornalia, 1849), is apparently even rarer locally. Only a single sighting in the Serra das Araras was listed by Willis and Oniki (1990b).

A number of species observed in the Pantanal seldom or not at all but present in regions adjacent to the wetland are inhabitants by preference of the *cerrado* and dry forest regions. Some of the species known locally only to the northwest of the wetland (N. Willis, 1976; Dubs, 1992) include the lined antshrike or *choca-listrada,*

Thamnophilus palliatus Lichtenstein, 1823; white-shouldered antshrike or *choca-lisa, Thamnophilus aethiops* Sclater, 1858; Amazonian antshrike or *choca-canela, Thamnophilus amazonicus* Sclater, 1858; rufous-winged antwren or *chorozinho-de-asa-ruiva, Herpsilochmus rufimarginatus* (Temminck, 1822); striated antbird or *trovoada-listrada, Drymophila devillei* (Menegaux and Hellmayr, 1906); white-browed antbird or *formigueiro-de-sobrancelha, Myromoborus leucophrys* (Temminck, 1827); white-flanked antwren or *choquinha-de-flanco-branco, Myrmotherula axillaris* (Vieillot, 1817); ornate antwren or *choquinha ornada, Myrmotherula ornata* (Sclater, 1853); plain-throated antwren or *choquinha-de-garganta-clara, Myrmotherula hauxwelli* (Sclater, 1857); stipple-throated antwren or *choquinha-de-garganta-carijó, Myrmotherula haematonota* (Sclater, 1857); pigmy antwren or *choquinha-miúda, Myrmotherula brachyura* (Linnaeus, 1766), which is found in a variety of forest and savanna habitats and along rivers; white-breasted antbird or *mãe-de-taoca-papuda, Rhegmatorhina hoffmannsi* (Hellmayr, 1907); the variegated antpitta or *tovacuçu-malhado, Grallaria varia* (Boddaert, 1783); and cinereous antshrike or *uirapuru-de-bando, Thamnomanes caesius* (Temminck, 1820), which is usually observed in dense and open woodlands, often near streams.

Dubs (1992) reported that the white-fringed antwren or *formigueiro-pardo, Formicivora grisea* (Gmelin, 1788), lives in thickets near the ground in arid places (Sick, 1988) only along the northern margins of the Pantanal. In contrast, the black-bellied antwren or *formigeiro-de-barriga-preta, Formicivora melanogaster* Pelzeln, 1869, seems to be a species typical for the Pantanal of Mato Grosso do Sul. It seldom reaches the northern part of the wetland. The white-backed fire-eye or *olho-de-fogo-selado, Pyriglena leuconota* (Spix, 1824) has a range on the periphery of the Pantanal that seems to have become disjunct because of the wetland, perhaps because it is a non-migratory inhabitant of dense forests. It is known only to the north and southwest of the Pantanal (Dubs, 1992).

The rufous-winged antshrike or *choca-da-asa-ruiva, Thamnophilus torquatus* Swainson, 1825, was commonly observed by Silva and Oniki (1988) in the Serra das Araras, and it is also known from other upland locations near the Pantanal (Dubs, 1992).

The central part of the range of the black-capped antshrike or *choca-de-olho-vermelho, Thamnophilus schistaceus* (d'Orbigny, 1837), is Bolivia, from which it ranges into the dry forest regions of southeastern Peru, northwestern Argentina, and western Brazil (Olrog, 1963). It has been reported only from the upland areas to the northwest of the wetland (N. Willis, 1976).

According to the reports of Laubmann (1940) on the avifauna of northern Paraguay, the variable antshrike or *choca-da-mata, Thamnophilus caerulescens* (Vieillot, 1816), is represented in the region of the Pantanal near the border of Paraguay and Bolivia by two subspecies: *T. c. paraguayensis* Hellmayr, 1901, and *T. c. dinelli* Berlepsch, 1906. Dubs (1992) also reported one of the subspecies, the non-migratory *T. c. paraguayensis*, from various kinds of habitat to the north of the Pantanal.

The short-tailed antthrush or *tovaca-campainha, Chamaeza campanisona* (Lichtenstein, 1823), runs on the ground and in low bushes in moist, hillside forests (Schauensee and Phelps, 1978), and is thus rarely seen in the Pantanal. It was never observed on the floodplain. According to Brown (1986), it is known only from the southeastern borders of the Pantanal.

Other species still known only from the southern subareas of the Pantanal in Mato Grosso do Sul include the spot-backed antshrike or *chocão-carijó, Hypoedaleus guttatus* (Vieillot, 1816); spot-breasted antvireo or *choquinha-de-peito-pintado, Dysithamnis strictothorax* (Temminck, 1823); black-capped antwren or *chorozinho-de-boné, Herpsilochmus pileatus* (Lichtenstein, 1823); the streak-capped antwren or *zídedê-do-sul, Terenura maculata* (Wied, 1831); and the white-shouldered fire-eye or *olho-de-fogo-do-sul, Pyriglena leucoptera* (Vieillot, 1818).

Conopophagidae

The rufous gnateater or *chupa-dente-marrom, Conopophaga lineatus* (Wied, 1831), is known from forrested regions of the northern Pantanal. It is locally uncommon and was not observed in the wetland during the study.

Rhynocriptidae

The only species in this family that is known to occur in the Pantanal is the *tapaculo-de-colarinho, Melanopareia torquata* (Wied, 1831), a species typical of the hot dry forest areas of Central Brazil. The species occurs in the drier parts of the wetland and on its outskirts, most abundantly during the dry season. It feeds on various invertebrates, especially insects, which it catches on the ground or among low plants. Silva and Oniki (1988) reported it to be abundant in the upland Serra das Araras region.

Cotingidae

The birds of this family are primarily fruit-eaters. They are most frequently observed near ranch houses in the Pantanal where fruit trees have been planted. They may also be observed in stands of palm trees.

The screaming piha, *cricrió-seringueiro,* or *tropeiro, Lipaugus vociferans* (Wied, 1820), is a species of tall forests that occurs sparsely in the gallery forests near large rivers in southern Mato Grosso.

The white-winged becard, *Pachyrhamphus polychopterus* (Vieillot, 1818), is a widespread species found throughout Brazil, including the region of the Pantanal.

The green-backed becard or *canaleirinho-verde, Pachyrhamphus viridis* (Vieillot, 1816), occurs in various localities throughout much of Brazil. It was infrequently seen near the borders of the Pantanal, but is not a common species in the wetlands.

The black-crowned tityra or *anambé-branco-de-bochecha-parda, Tityra inquisitor* (Lichtenstein, 1823), is a migratory species that can be observed occasionally in the Pantanal. A common congeneric species, the black-tailed tityra or *anambé-branco-de-rabo-preto, Tityra cayana* (Linnaeus, 1766), occurs throughout much of Mato Grosso, including the region of the Pantanal. A third species, the masked tityra or *araponginha-de-rabo-cintado, Tityra semifasciata* (Spix, 1825), is a species of open, upland forests that was not observed in the Pantanal during the study.

Other species that are possibly occasional members of the Pantanal avifauna are the pompadour cotinga, *Xipholena punicea* (Pallas, 1764); black-capped becard or *caneleiro bordado, Pachyrhamphus marginatus* (Lichtenstein, 1823); pink-throated becard or *caneleiro-pequeno, Platypsaris minor* (Lesson, 1830); bare-necked fruitcrow or *anamvé-pombo, Gymnoderus foetidus* (Linnaeus, 1758); and *Platypsaris rufus* (Vieillot, 1816).

The Amazonian umbrellabird or *anambé-preto, Cephalopterus ornatus* Geoffroy Saint-Hilaire, 1809, is an endangered species inhabiting forests and the ecotone between forests and open countryside. It is known from the northern part of the Pantanal (Dubs, 1992).

The naked-throated bellbird or *araponga, Procnias nudicollis* (Vieillot, 1817), is known only from the southern subregions of the Pantanal in Mato Grosso do Sul.

Pipridae

Several manakins are among the sparrow-sized bird species encountered from time to time in the Pantanal. They are onmivores, feeding on insects and fruit. The white-barred manakin or *caleleirinho-cantor, Piprites chloris* (Temminck, 1822), is abundant in the Pantanal, frequenting places near human habitations where foods, such as chicken feed and fruit from orchards, are more abundant, and there is some protection from predators.

The pale-bellied tyrant-manakin or *fruxu-de-coroa-sulfúrea, Neopelma pallescens* (Lafresnaye, 1853), is a common species in the woodlands of Mato Grosso and Mato Grosso do Sul. It is not uncommon in the region of the Pantanal, but it is not abundant in the wetland itself.

The white-bearded manakin or *rendeira-branca, Manacus manacus* (Linnaeus, 1766), is a widespread bird in Brazil, which frequents wooded areas (Andrade, 1992). It is seen from time to time in the Pantanal but is not at all abundant.

Another species known to occur in Mato Grosso (Andrade, 1992) is the helmeted manakin or *tangará-de-cabeça-encarnada, Antilophia galeata* (Lichtenstein, 1823), which is a bird that feeds on insects and small fruits of trees in dry forests. It is commonly encountered along water courses and in wetlands, but it also occurs in the dry forests. It is present in the Pantanal but not often seen.

The swallow-tailed manakin, *tangará*, or *dançador, Chiroxiphia caudata* (Shaw and Nodder, 1793), is a species common in southern Brazil (Andrade, 1992). It probably occurs in the region of the Pantanal but was not seen during the study.

The dry forest areas to the north of the wetland are preferred by the fiery-capped manakin or *dançarinho-perereca, Machaeropterus pyrocephalus* (Sclater, 1852). It seems to be rare in the Pantanal, because it was not seen during the study.

Several species of the Amazonian region may occasionally range as far south as the Pantanal catchment area. These include the dwarf tyrant-manakin or *supi, Tyranneutes stolzmanni* (Hellmayr, 1906); thrush-like manakin or *dançador-marrom, Schiffornis turdinus* (Wied, 1831); red-headed manakin or *dançador-de-cabeça-encarnada, Pipra rubrocapilla* Temminck, 1821; snow-capped manakin or *dançador-de-coroa-de-neve, Pipra nattereri* Sclater, 1865; band-tailed manakin or *dançador-de-cauda-fina, Pipra fasciicauda* Hellmayr, 1906; flame-crowned-manakin or *dançador-coroa-de-fogo, Heterocerus lineatus* (Strickland, 1850); and sulfur-bellied tyrant-manakin or *furchu-da-barriga-amarela, Neopelma sulphureiventer* (Hellmayr, 1903).

The greenish manakin or *flautim-verde, Schiffornis virescens* (Wied, 1831), may be another rare visitor to the Pantanal. It is known mainly from the forests of southeastern Brazil, but it has also been found in Paraguay (Laubmann, 1940).

Tyrannidae

The tyrant family is the bird family richest in species in South America. An impressive list of tyrants that are at least temporary visitors to the Pantanal can be presented. However, few of them can be considered typical species of the wetland, even for part of the year. One of these exceptional wetland species is the pied water-tyrant or *lavadeira-de-cara-branca, Fluvicola pica* (Boddaert, 1783), which is encountered throughout the Pantanal, generally on trees overhanging the shallow water of streams and small rivers or on large emergent water plants. This species, like the other members of this family, seems to subsist almost exclusively on insects.

The white-headed marsh tyrant, *viuvinha*, or *Maria-velhinha, Arundinicola leucocephala* (Linnaeus, 1764), is another of the few wetland species that occurs over a wide range from Trinidad and northern South America as far south as Paraguay and northern Argentina (Laubmann, 1940). It was often observed on emergent marsh plants along the Transpantaneira Highway.

A third species that sometimes frequents marshy regions as well as forests is the cock-tailed tyrant or *galito, Alectrurus tricolor* (Vieillot, 1816). The species is not widely distributed, having a range including only the central region of Brazil, eastern Bolivia, northern Argentina, and probably Paraguay (Dubs, 1992).

Another wetland species, the bearded tachuri, *Polystictus pectoralis* (Vieillot, 1817), has an extensive range in tropical and subtropical South America. However, it seems to be uncommon in the northern part of the Pantanal and was not seen there during the study.

The yellow-browed tyrant or *suiriri-pequeno, Satrapa icterophrys* (Vieillot, 1818), is a water-loving species that is nevertheless not often seen in the northern part of the Pantanal. Cintra and Yamashita (1990) reported it to be rare in the District of Poconé, where it was only encountered from July to January.

In addition to the marshland species, a large number of terrestrial tyrants are known from southern Mato Grosso. The fuscous flycatcher or *guaracavuçu-quieto, Cnemotriccus fuscatus* (Wied, 1831), is a resident species (Short, 1975), which can be encountered along the edges of forests in most peripheral areas of the Pantanal. Cinta and Yamashita (1990) reported it to be a common bird in the region of Poconé.

The sharp-tailed tyrant, *Cilicivora caudacuta* (Vieillot, 1818), is a bird that seem to be confined to the Central Western part of Brazil and adjacent parts of Bolivia and Paraguay. It is known throughout the Pantanal catchment area and only as far east as the western parts of São Paulo and Paraná. Little has been reported about its habits.

Flocks of the fork-tailed flycatcher, *Muscivora tyrannus* (Linnaeus, 1766), are temporary visitors to the Pantanal. They breed throughout southern South America during the summer and spend the winters in northern South America. Some also migrate to North America or the West Indies. A large flock was present in the Pantanal near Poconé for several weeks during the rainy season of 1991-92. During subsequent years of the study, the species was rarely seen.

Some of the other tyrants are widely distributed in tropical South America and are generally rarer in the Pantanal than elsewhere within their ranges. The white-rumped monjita or *noivinha-das-almas, Xolmis velata* (Lichtenstein, 1823), occurs from São Paulo and Minas Gerais westward into Paraguay (Laubmann, 1940). It is usually an inhabitant of the *cerrado*. Another species of the woodland savanna and open fields is the strange-tailed tyrant or *bandeira-do-campo, Yetapa risora* (Vieillot, 1824), which is locally confined to the regions to the northwest and southeast of the Pantanal (Dubs, 1992). The streamer-tailed tyrant or *tesoura-do-brejo*, *Gubernetes yetapa* (Vieillot, 1818), was reported to prefer swampy forests by Dubs (1992), but it is nevertheless encountered only at the northwestern and southeastern borders of the wetland. The pale-bellied mourner or *Maria-cantora*, *Rhytipterna immunda* (Sclater and Salvin, 1873), is a rare species of the *cerrado* and *cerradão* that consumes lizards, as well as spiders, insects, and fruit (Sick, 1988). It is a species of the drier parts of the Amazonian region that apparently reaches the southern limit of its range to the northwest of the Pantanal. The stripe-necked tody-tyrant or *Maria-de-papo-branco, Idioptilon striaticolle* (Lafresnaye, 1853), is another non-migratory species of the semideciduous forest regions of Amazonia that reaches the southern limit of its range in the northern Pantanal.

The tropical kingbird or *suiriri-tropical, Tyrannus melancholicus* Vieillot, 1819, which was sometimes seen near the ranch houses in the Pantanal of Poconé, ranges from southwestern North America through the Caribbean Islands to Peru and Argentina. It generally catches insects in flight, and sometimes harasses larger birds. During the dry season, which is the southern autumn and winter, it migrates to northern South America (Dubs, 1992). The streaked flycatcher or *bem-te-vi-rajado, Myriodynastes maculatus* (P. L. S. Müller, 1776), which is found throughout the region of the Pantanal, also catches insects in flight. It is a migratory species ranging from Mexico to central Argentina and Chile. The population in South America migrates to the northern part of the continent during the northern summer and to the southern part during the southern summer (Dubs, 1992). It thus may be encountered anywhere in the Pantanal during its travels.

Probably the most common of the tyrants in the southern part of Mato Grosso is the great kiskadee or *bem-te-vi, Pitangus sulphuratus* (Linnaeus, 1766, which occurs in and near cities and towns as well as in the open countryside. In the Pantanal, it can be observed during the entire year, especially near the ranch houses, although it is not nearly as abundant there as it is in the city of Cuiabá and smaller towns nearby. The species has a large range extending from the southern United States to northern

Argentina. In Mato Grosso, only the subspecies *P. sulphuratus maximiliani* (Cabanis and Heine, 1859) is present (Dubs, 1992).

The lesser kiskadee or *bem-te-vizinho-do-brejo, Pitangus lictor* (Lichtenstein, 1823), is a water-loving bird that is not nearly as common in southern Mato Grosso as the great kiskadee, since it is near the southern limit of its range there (Schauensee and Phelps, 1978). It was seldom seen in the Pantanal.

A less strikingly colored yellow and brown tyrannid is the cattle tyrant or *suiriri-cavaleiro, Machetornis rixosus* (Vieillot, 1819), which is widespread and abundant in the Pantanal, where it occupies nests constructed by other bird species. In the wetland, it is more common than the great kiskadee, apparently because of the great abundance of cattle, which this bird habitually accompanies.

The sulphury flycather or *suiriri-de-garganta-rajada, Tyrannopsis sulphurea* (Spix, 1825), is apparently not a member of the wetland avifauna. The species is reportedly abundant in the Amazonian region and northern South America (Schauensee and Phelps, 1978) and reaches the southern limits of its range to the northwest of the Pantanal.

The boat-billed flycatcher or *bem-te-vi-de-bico-chato, Megarhynchus pitangua* (Linnaeus, 1766), is a widespread bird in South America that was not notably abundant in the region of the Pantanal. It is otherwise found as far north as Mexico and as far south as northern Argentina.

One of the species reported from the entire region of the Pantanal by Dubs (1992), the rusty-fronted toady-flycatcher or *ferreirinho-de-cara-parda, Todirostrum latirostre* (Pelzeln, 1868), is one of the tyrants confined in its range to tropical South America. Although it can be encountered almost anywhere, it is not a common species in the wetland.

The white-throated kingbird or *suiriri-de-papo-branco, Tyrannus albogularis* Burmeister, 1856, has a range in South America apparently limited to the open woodland regions and savannas of the equatorial zone with some extension into the tropical wet-and-dry region. It was not observed in the Pantanal during the study and may reach the southern limit of its range at the northern border of the Pantanal (Dubs, 1992).

The vermilion flycatcher, *verão,* or *príncipe, Pyrocephalus rubinus* (Boddaert, 1783), is a winter visitor to Brazil (Andrade, 1992), which was observed occasionally on or near the ranch houses in the Pantanal.

Another migratory species, the crowned-slaty flycatcher or *bem-te-vi-cinza, Empidonomus aurantioatrocristatus* (Lafresnaye and d'Orbigny, 1837), generally flies inland and northward as far as Peru from its summer home in northern Argentina. The species is apparently only a transient visitor to the Pantanal.

The range of the piratic flycatcher or *bem-te-vi-pirata, Legatus leucophaius* (Vieillot, 1818), during the southern summer includes northern Argentina, southern Brazil, Paraguay, and Bolivia. In the southern autumn, these birds migrate to Venezuela, Colombia, and Nicaragua. They may be encountered seasonally anywhere in the Pantanal. The brown-crested flycatcher or *Maria-de-asa-ferrugem, Myiarchus tyrannulus* (P. L. S. Müller, 1776), has generally the same migration habits but spends its winters somewhat farther to the east, visiting the Guianas rather than Nicaragua (Olrog, 1963).

The dusky-capped flycatcher or *Maria-triste, Myiarchus tuberculifer* (d'Orbigny and Lafresnaye, 1837), spends the rainy season in Peru, Bolivia, and western Brazil, then migrates at the beginning of the dry season to Colombia, Venezuela, and the Guianas (Olrog, 1963). It does not seem to enter the wetland but can be found to the northwest (Dubs, 1992).

Swainson's flycatcher or the *Maria-irré, Myiarchus swainsoni* Cabanis and Heine, 1859, and Euler's flycatcher or *enferrujado, Empidonax euleri* (Cabanis, 1868), apparently spend their summers mainly in the South Temperate Zone of northern Argentina and Paraguay and range eastward to Uruguay and southeastern Brazil. In the autumn, the former migrates to the Guianas and Venezuela, and the latter to Amazonas (Olrog, 1963). Therefore, they are only brief transient visitors to the region of the

Pantanal. The migrations of the fork-tailed flycatcher or *tesourinha-do-campo, Tyrannus savana* Vieillot, 1808, are even longer, sometimes covering the distance from breeding grounds in southern Argentina to northern South America. The species may be encountered almost anywhere in the wetland, except for the northwestern and southeastern borders (Dubs, 1992).

The white-crested tyrannulet or *alegrinho-de-leste, Serpophaga subcristata* (Vieillot, 1817) and *Serpophaga munda* Berlepsch, 1893, spend their summers in the cooler parts of the South Temperate Zone and migrate northward to the region of the Pantanal, including Paraguay and Bolivia, during the winters.

The tawny-crowned pygmy-tyrant, *Euscarthmus meloryphus* (Vieillot, 1818), is widely distributed from Argentina, Uruguay, Paraguay, and eastern Bolivia northward to Colombia and Venezuela. The species can be encountered in all parts of Brazil. Olrog (1963) speculated that it is a migratory species, but details of its travels have not been reported. *Pseudocolopteryx sclateri* (Oustalet, 1892) is another probable migrant that occurs between northern Argentina and the Guianas. It is known to occur in Mato Grosso (Olrog, 1963).

The variegated flycatcher or *bem-te-vi-peitica, Empidonomus varius* (Vieillot, 1818), is a species that prefers the *cerrado* of southern Mato Grosso to the wetland. Schauensee and Phelps (1978) seem to suggest that of the two subspecies, *E. v. varius* (Vieillot, 1818) is a migrant that travels between northern South America and Argentina, Uruguay, and southern Brazil, while *E. v. rufinus* (Vieillot, 1822) generally remains in one general region and may not undertake seasonal migrations at all.

A common species throughout southern Mato Grosso, the yellow-breasted flycatcher or *bico-chato-amarelo, Tolmomyias flaviventris* (Wied, 1831), has been observed to the northwest of the Pantanal occasionally. It is found in many kinds of habitat throughout tropical South America (Schauensee and Phelps, 1978) but reaches the southern limit of its range in Mato Grosso (Dubs, 1992). Another species found in a wide variety of habitats in and near the Pantanal is the short-crested flycatcher or *Maria-caveleira, Myiarchus ferox* (Gmelin, 1789), which occurs mainly in the forests and woodland savanna of lowland South America. The yellow-olive flycatcher or *bico-chato-de-orelha-preta, Tolmomyias sulphurescens* (Spix, 1825), is a non-migratory species that occurs throughout the entire region of the Pantanal and otherwise ranges from southern North America to northern Argentina.

Suiriri suiriri (Vieillot, 1818) is believed to be a species that undertakes short migrations between northern Argentina and the region of the Pantanal (Olrog, 1963). It occurs mainly in open fields and woodland savanna.

Inezia inornata (Salvadori, 1897) is found mainly in the *cerrado* region of Bolivia and Paraguay. A few of these birds range into the region of the Pantanal, and a single specimen was found as far south as Salta, Argentina (Olrog, 1963).

The crested black-tyrant or *Maria-preta-de-penacho, Knipolegus lophotes* Boie, 1828, is observed occasionally along the roads near the Pantanal, especially in areas of cattle ranching in the district of Cáceres. In addition, Dubs (1992) reported it from the northeastern margins of the Pantanal. It is not usually seen in the wetland.

The sepia-capped flycatcher, *Leptopogon amaurocephalus* Tschudi, 1846, and the pearly-vented tody-tyrant or *Maria-olho-de-ouro, Idioptilon margaritaceiventer* (d'Orbigny and Lafresnaye, 1837), are residents of northern Argentina, the *cerrado* regions of Paraguay and eastern Bolivia, and South and Central Brazil (Olrog, 1963). These species are uncommon in the wetland but was reported by Dubs (1992) from forests, fields, and *cerrado* throughout the entire region of the Pantanal.

The bran-colored flycatcher or *Filipe-de-peito-riscado, Myiophobus fasciatus* (P. L. S. Müller, 1776), is a small, wide ranging bird that hunts insects in open fields, scrub, and grasslands. It is occasionally seen in southern Mato Grosso, but it seems to enter the Pantanal rarely and only during the dry season. It prefers the *cerrado*.

One species found mainly in moist woodlands along streams and on hillsides in the Pantanal watershed is the ochre-bellied flycatcher, *Pipromorpha oleaginea* (Sclater, 1883), referred to as *Syndactyla rufosuperciliata oleaginea* (Sclater, 1883) by Olrog (1963). It has an extensive range in tropical South America. Another forest dweller is the ruddy-tailed flycatcher, *Terenotriccus erythrurus* (Cabanis, 1847), which is also encountered mainly in the upland parts of the watershed. Wolters (1975-1982) reported the range of this species to extend southward only as far as the northern parts of Mato Grosso and Bolivia, indicating that its occurrence in the northern part of the Pantanal catchment area is unusual.

A few resident birds in this family inhabit dense forests and woodland and are therefore seldom seen. One of these, the rufous-tailed attila or *tinguaçu-castanho, Pseudattila phoenicurus* (Pelzeln, 1816), is found in the canopies of forests and in woodlands throughout the Amazon and Pantanal regions as far north as Venezuela and south to Paraguay and northeastern Argentina (Schauensee and Phelps, 1978). The grayish mourner or *Maria-cinza, Rhytipterna simplex* (Lichtenstein, 1823), has generally the same habitat and range and is also not often seen, although it may be more abundant than it seems to be. The upper forest canopy is also the hunting ground of the sirystes or *Maria-assobiadeira, Sirystes sibilator* (Vieillot, 1818), which may be encountered throughout the entire Pantanal region. Another forest species is the rufous attila or *bate-para, Attila bolivianus* Lafresnaye, 1848, which lives in the understories of semi-deciduous forests and woodlands from Colombia to western Brazil. A woodland bird observed rarely in the District of Poconé from July to January (Cintra and Yamashita, 1990) is the rufous casiornis or *Maria-ferrugem, Casiornis rufa* (Vieillot, 1816), which is apparently an uncommon guest in the Pantanal.

A distinctive-looking bird, the long-tailed tyrant or *Maria viuvinha, Colonia colonus* (Vieillot, 1818), is another example of a species in the Pantanal that resides in dense forests and woodlands and is seldom seen. Dubs (1992) reported that it occurs throughout the entire region of the Pantanal. The sulphur-rumped flycatcher or *assanhadinho-de-peito-dourado, Myiobius barbatus* (Gmelin, 1789), prefers the shady undergrowth of tall forests and is therefore also a species that is seldom seen. It is known from the northern part of the Pantanal and otherwise ranges from Central America to southern Brazil. The planalto tyrannulet or *poaieiro-triste, Phyllomyias fasciatus* (Thunberg, 1822), and the greenish tyrannulet or *poaieiro-verde, Xanthomyias virescens* (Temminck, 1824), are species that live in the upper canopies of forests (Sick, 1988) and seem to occur only to the Northeast of the Pantanal (Dubs, 1992).

Various species seem to show a preference for the drier pasture and *cerrado* areas but can still be observed in the wetland. These include the southern beardless tyrannulet or *risadinha, Camptostoma obsoletum* (Temminck, 1824); and gray monjita or *primaveira, Xolmis cinerea* (Vieillot, 1816).

The tyrants that show preference for open woodlands and dense shrub growth seem to be most common in the subtropics to the north and south of the Amazonian region and may avoid the equatorial rainforest zone. One of these, the common tody-flycatcher or *ferreirinho-relógio, Todirostrum cinereum* (Linnaeus, 1766), occurs as far north as southern Mexico and southward to Bolivia and southeastern Brazil (Schauensee and Phelps, 1978). It is a non-migratory species that can be encountered almost anywhere in the region of the Pantanal (Dubs, 1992). The tropical pewee or *piuí-cinza, Contopus cinereus* (Spix, 1825), is a forest species ranging from southern North America to northern Argentina and is encountered to the northeast and southwest of the wetland (Dubs, 1992). The social flycatcher, *Myiozetetes similis* (Spix, 1825), is also reported to inhabit open woodland, especially near water, from Mexico to Argentina. However, it was not observed in the Pantanal during the study, although it most certainly occurs there. The scrub flycatcher, *Sublegatus modestus* (Wied, 1831), and the mouse-colored tyrannulet or *bagageiro, Phaeomyias murina* (Spix, 1825), occur mainly in dry shrub growth and sparce woodlands. It was suggested that some scrub flycatchers seasonally migrate to northern South America (Schauensee and Phelps, 1978), probably from regions south of the equatorial rainforests.

Among the avifauna were several species of elaenias, which were not distinguished with certainty in the field. Those species known or thought to occur in the Pantanal and its surroundings include the yellow-bellied elaenia or *Maria-e-dia, Elaenia flavogaster* (Thunberg, 1822); small-billed elaenia or *guaracava-verde, Elaenia parvirostris* Pelzeln, 1868; plain-crested elaenia or *guaracava-de-topete, Elaenia cristata* Pelzeln,

1868; lesser elaenia or *chibum, Elaenia chiriquensis* Lawrence, 1867; large elaenia or *guaracava-grande, Elaenia spectabilis* Pelzeln, 1868; white-crested elaenia or *guaracave-de-crista-branca, Elaenia albiceps* (d'Orbigny and Lafresnaye, 1837); highland elaenia or *guaracava-de-óculos, Elaenia obscura* (d'Orbigny and Lafresnaye, 1837); gray elaenia or *Maria-da-copa, Myiopagis caniceps* (Swainson, 1836); forest elaenia or *Maria-pechim, Myiopagis gaimardii* (d'Orbigny, 1840); and greenish elaenia or *Maria-verde, Myiopagis viridicata* (Vieillot, 1817). Most of these species, except for *Elaenia flavogaster* and perhaps *Elaenia chiriquensis*, are uncommon to rare in the northern parts of the Pantanal (Cintra and Yamashita, 1990; Dubs, 1992).

Some of the species seem to range only as far northward as the Pantanal and are confined mainly to the southern part of the wetland. These include the southern beardless tyrannulet, *Pogonotriccus eximius* (Temminck, 1822); *Phylloscartes ventralis* (Temminck, 1824); and *Myiornis auricularis* (Vieillot, 1818), which are known from northern Argentina, eastern Paraguay, and southeastern and eastern Brazil (Olrog, 1963).

A few rainforest species found mainly in the Amazon region are also occasional visitors to southern Mato Grosso. The white-throated spadebill or *patinho-de-garganta-branca, Platyrinchus mystaceus* Vieillot, 1818, inhabits forests where it hides in dense underbrush. Its ranges from Central America to northern Argentina. The rusty-margined flycatcher or *bem-te-vi-assobiador, Myiozetetes cayanensis* (Linnaeus, 1766), is a species usually observed catching insects in the air beside ponds and streams in open woodland (Schauensee and Phelps, 1978). It was not observed in the northern part of the Pantanal, but it is reported to occur there (Dubs, 1992). The short-tailed pygmy-tyrant or *Maria-caçula, Myiornis ecaudatus* (d'Orbigny and Lafresnaye, 1837); and the white-eyed tody-tyrant or *Maria-de-olho-branco, Idioptilon zosterops* (Pelzeln, 1868), are both common in rainforests from Trinidad and the Guianas to Ecuador and southeastern Colombia, respectively, and south through the Amazon region to Bolivia with the forests to the north of the Pantanal at the southern edge of their ranges. Schauensee and Phelps (1978) reported the usual habitat of the rufous-tailed flatbill or *Maria-de-cauda-ferrugem, Ramphotrigon ruficauda* (Spix, 1825), to be dense rainforest. Its occurrence in southern Mato Grosso thus seems to be limited to gallery forests northwest of the Pantanal (Dubs, 1992).

A resident subtropical species that prefers to hunt insects along mountainsides (Sick, 1988), the cliff flycatcher or *gibão-de-couro, Hirundinea ferruginea* (Gmelin, 1788), is found at times throughout the tropics. It is not normally encountered in the wetlands but may occur along the cliffs at the border between the plateau and the lowland plain in which the Pantanal is located. In Mato Grosso, it is represented by the subspecies *H. ferruginea bellicosa* (Vieillot, 1819).

Several tyrants are locally known only from the region of the Pantanal of Mato Grosso do Sul and are not usually seen in Mato Grosso. These include the white monjita or *noivinha-branca, Xolmis irupero* (Vieillot, 1823); cinereous tyrant or *Maria-preta-acinzentada, Entotriccus striaticeps* (d'Orbigny and Lafresnaye, 1837); blue-billed black-tyrant or *Maria-preta-pequena, Knipolegus cyanirostris* (Vieillot, 1818); *Corythopis delalandi* (Lesson, 1830), rufous-sided pygmy tyrant, *Euscarthmus rufomarginatus* (Pelzeln, 1868), and three-striped flycatcher or *bem-te-vi-de-três-riscas, Conopias trivirgata* (Wied, 1831), an upper forest canopy species which is found as far south as northeastern Argentina and eastern Paraguay.

One wetland species from regions well to the south of the Pantanal, the spectacled tyrant or *Maria-preta-de-óculos, Hymenops perspicillata* (Gmelin, 1789), has been reported only once from the region to the northwest of the Pantanal (Dubs, 1992).

Hirundinidae

The white-winged swallow or *andorinha-do-rio*, *Tachycineta albiventer* (Boddaert, 1783), is an abundant swallow throughout the Pantanal. It is frequently seen flying low over the water in search of flying insects or resting on fence posts or bridges along the Transpantaneira Highway.

Other swallows observed in the Pantanal are generally uncommon, and most of them seem to be more at home in the upland areas of the watershed. These include the brown-chested martin, *Phaeoprogne tapera* (Linnaeus, 1766); gray-chested martin, *Progne chalybea* (Gmelin, 1789); blue and white swallow, *Natiochelidon cyanoleuca* (Vieillot, 1817); tawny-headed swallow, *Alopochelidon fucata* (Temminck, 1822); rough-winged swallow or *andorinha-serrador*, *Stelgidopteryx ruficollis* (Vieillot, 1817); and *Tachycineta leucorrhoa* (Vieillot, 1817).

Corvidae

This family is not well-represented in the Pantanal. Generally, the tropical corvids are brightly colored or strikingly feathered forest birds. Among the species that may be included among the Pantanal fauna are the curl-crested jay or *gralha-do-campo*, *Cyanocorax cristatellus* (Temminck, 1823); the plush-crested jay or *gralha-picaça*, *Cyanocorax chrysops* (Vieillot, 1818); and the purplish jay or *gralha-cinza, Cyanocorax cyanomelas* (Vieillot, 1818).

Troglodytidae

The wrens are not abundant in the Pantanal wetlands, although several species temporarily visit the region to feed on insects that appear seasonally. Most prefer wooded areas, especially those near houses. The house wren or *corruíra-de-casa, Troglodytes aedon* Vieillot, 1808, is sometimes observed near the ranch houses. The fawn-breasted wren or *garrincha-do oeste, Thryothorus guarayanus* (Lafresnaye and d'Orbigny, 1837), has the same habits but seems to be somewhat less abundant. The more secretive species that also occur in or near the Pantanal, especially in the forests to the northwest, include the moustached wren or *garrincha-de-bigode, Thryothorus genibarbis* Swainson, 1837; buff-breasted wren or *garrincha-trovão, Thryothorus leucotis* Lafresnaye, 1845; nightingale wren or *flautista-da-mata, Microcirculus marginatus* (Sclater, 1855); and thrush-like wren or *nicolau*, *Campylorhynchus turdinus* (Wied, 1821).

The tooth-billed wren or *cambaxirra-da-copa, Odontocheilus cinereus* Pelzeln, 1868, is a species endemic to the State of Mato Grosso. Although the main part of its range seems to be in the Amazon Basin to the north, N. Willis (1976) reported it near the northwestern border of the Pantanal.

Mimidae

One of the species in this family, the black-capped mockingthrush or *japacanim*, *Donacobius atricapillus* (Linnaeus, 1766), is a wetland species commonly observed in the Pantanal. It can be encountered in gallery forests, in the open countryside of the floodplain, and rarely in the upland *cerrado*.

Two other species in this family show no preference for the wetland habitat, and their presence in the Pantanal is infrequent. One of these is the chalk-browed mockingbird or *sabiá-do-campo*, *Mimus saturninus* (Lichtenstein, 1823), and the other, *Mimus triurus* (Vieillot, 1818).

Turdidae

Thrushes are uncommon among the avifauna of the Pantanal, although several species occur in southern Mato Grosso and may enter the Pantanal occasionally. The one most often seen is the rufous-bellied thrush or *sabiá-laranjeira*, *Turdus rufiventris* Vieillot, 1818, a species that consumes oranges and other cultivated fruits. It is generally encountered near ranch houses around which fruit trees have been planted.

A species of forests and moist woodlands, the cocoa thrush or *sabiá-vermelho, Turdus fumigatus* Lichtenstein, 1823, was observed rarely near gallery forests and in shaded places at the edge of woodlands near the northern borders of the wetland.

The pale-breasted thrush, *sabiá-branco,* or *sabiá-de-cabeça-cinza, Turdus leucomelas* Vieillot, 1818, is a less common species that was noted only rarely during the study. Species which might also be expected to occur in the Pantanal but were not seen during the study include the creamy-bellied thrush or *sabiá-poca, Turdus amaurochalinus* Cabanis, 1851, and the white-necked thrush, *Turdus albicollis* Vieillot, 1818, both of which should be expected to occur in southern Mato Grosso.

Turdus nigriceps Cabanis, 1874, is a thrush found from northern Argentina to Peru and Ecuador, which was not noted in the region during the study.

Sylviidae

The gnatwrens are rather small insectivorous birds that are more numerous in the Old than the New World (Schauensee and Phelps, 1978). In the Pantanal, this family is represented by two of the three subspecies in the region of the Pantanal. *Polioptila dumicola dumicola* (Vieillot, 1817) is encountered in Mato Grosso do Sul and has been observed in the Pantanal of Corumbá and Aquidauana, while *Polioptila dumicola berlepschi* Hellmayr, 1901, is the subspecies present in the northern part of the wetland (Naumburg, 1930).

The occurrence of the long-billed gnatwren or *balança-rabo-de-bico-longo, Ramphocaenus melanurus* Vieillot, 1819, in the Pantanal wetland itself is uncertain. It is known from Mexico to the Amazon region in Mato Grosso and eastern Brazil, and Dubs (1992) reported it from the northern border areas of the wetland.

Moticillidae

Only one member of this family occurs in the region of the northern Pantanal: the yellowish pipit or *caminheiro-zumbidor, Anthus lutescens* Pucheron, 1855, which hunts insects on damp grassland and cow pastures in and near the Pantanal. It is not abundant, probably because it is subject to predation by the many raptors in the wetland. The yellowish pipit is a migratory species that frequents the floodplain when it is dry or nearly so, seeking out the edges of the remaining pools of water to hunt food left behind by the receding water.

Vireonidae

This family is represented in the Pantanal by several omnivorous species that do not seem to show any preference for the wetland. The rufous-browed peppershrike or *pitiguari, Cyclarhis gujanensis* (Gmelin, 1789), is present throughout the region but not abundant.

The red-eyed vireo or *juruviara-oliva, Vireo olivaceus* (Linnaeus, 1766), a migratory species with an extremely extensive range from the North Temperate Zone in eastern

North America to Uruguay and Argentina, is an occasional visitor throughout southern Mato Grosso.

Two species typically encountered in dense forest and in small clearings, the ashy-headed greenlet or *vite-vite-de-cabeça-cinza, Hylophilus pectoralis* Sclater, 1866, and the buff-cheeked greenlet, *Hylophilus muscicapinus* Sclater and Salvin, 1873, are not typically present in the Pantanal. *H. pectoralis* has long been known to occur in southern Mato Grosso (Schauensee and Phelps, 1978), *H. muscicapinus* is not often observed south of the Amazon region, but Silva and Oniki (1988) reported it to be a rare species in the Serra das Araras, where is has been seen in pairs seeking insects in the understories of gallery forests.

Hylophilus poicilotis Temminck, 1822, seems to be a species confined mainly to Argentina, Paraguay, southeastern Brazil, and perhaps the southern part of the Pantanal (Olrog, 1963), but Dubs (1992) recently reported it near both the northern and the southwestern borders of the wetland.

Icteridae

This family is well represented in the region of the Pantanal, and several species are specialized for living in the wetland habitats. The yellow-rumped marshbird or *chopim-do-brejo, Pseudoleistes guirahuro* (Vieillot, 1819), is frequently seen among the emergent aquatic plants, especially *Thalia geniculata* and *Aeschynomene fluminensis.*

The shiny cowbird, *Molothrus bonariensis* (Gmelin, 1789), is frequently seen in groups to two to five at places along the Transpantaneira Highway where local residents maintain feeding stations or tourists stops. It is often seen in association with the yellow-billed cardinal, which may greatly outnumber the cowbirds. The species is widespread in central South America, ranging from Amazonas to Argentina (Olrog, 1963).

A bird commonly seen on palms and fruit trees in the Pantanal, often near ranch houses, is the yellow-rumped cacique or *japuíra, Cacicus cela* (Linnaeus, 1758). The birds are almost always seen in pairs. They are nest weavers. Somewhat less abundant but with similar habits is the red-rumped cacique, *Cacicus haemorrhous* (Linnaeus, 1766), a somewhat larger nest weaver.

The solitary black cacique, *Cacicus solitarius* Vieillot, 1816, is a widespread species in South America, occurring from northern Argentina and Uruguay as far northward as Amazonas, Colombia, and Ecuador. It is an infrequent visitor to the Pantanal.

Icterus cayanensis pyrrhopterus (Vieillot, 1819) is a subspecies that is generally confined to the drainage area of the Rio Paraná System, from Bolivia southward to Paraguay, southern Brazil, northern Argentina, and Uruguay. The rarer subspecies, *Icterus cayanensis periporphyrus* (Bonaparte, 1850), seems to be known only from the region between southern Bolivia and northern Argentina. The general range is shared with the garibaldi or chestnut-capped blackbird, *Agelaius ruficapillus* (Vieillot, 1819).

The orange-backed oriole or *Joao-Pinto, Icterus icterus croconotus* (Wagler, 1829), can be found occasionally in southern Mato Grosso and at the borders of the Pantanal, but it was not found to be common. A similarly uncommon species, the crested oropendola or *japu-preto, Psarocolius decumanus* (Pallas, 1769), is also encountered occasionally in southern Mato Grosso. Dubs (1992) reported its presence throughout the entire region of the Pantanal.

A bird most commonly encountered in the Amazon region, the velvet-fronted grackle or *ireúna-veluda, Lampropsar tanagrinus* (Spix, 1824), seems to be only a rare visitor to southern Mato Grosso. It generally forages for insects in forests (Schauensee and Phelps, 1978). Similarly, the olive oropendola or *japugaçu, Gymnostinops yuracares* (Lafresnaye and d'Orbigny, 1838), is a species of the Amazon region that may rarely be encountered in the region of the Pantanal.

Molothrus badius (Vieillot, 1819) is a resident bird of northern Argentina, Uruguay, southeastern Brazil, and Paraguay. It may be confined to the southern part of the Pantanal.

The giant cowbird, *Scaphidura oryzivora* (Gmelin, 1788), *Amblyrhamphus holosericeus* (Scopoli, 1786), as well as *Gnorimopsar chopi* (Vieillot, 1819) and *Agelaius cyanopus* (Vieillot, 1819), seem to be species limited mainly to eastern Bolivia, Paraguay, and southern Brazil. They also range into northern Argentina (Olrog, 1963). They seem to be species of the drier forested areas and were not observed in the wetland during the study. However, an investigation by Beltzer and Paporello (1983) showed that *Agelaius cyanopus cyanopus* feeds mainly on seeds of grasses and *Polygonum* spp. together with aquatic insects, crustaceans, molluscs, and a few fishes. The birds proved to be selective according to the size of the prey, which consisted of species of *Belostoma*, *Tropisternus, Eulalia,* and *Trichodactylus*. Hence, judging from the species consumed, the birds examined must have spent considerable time seeking food in wetlands.

Leistes superciliaris (Bonaparte, 1850) is a species that spends its summers in central Argentina. During the winter, it migrates into the region of the Pantanal as far north as Bolivia. It may be seen at times in the Pantanal during the dry season, but it is not abundant.

The golden-winged cacique or *japiim-soldado, Cacicus chrysopterus* (Vigors, 1825), apparently does not range as far northward as Poconé. Dubs (1992) reported it only from the southern Pantanal and adjacent southeastern uplands.

Parulidae

Small wood warblers are present in the Pantanal but not abundant. Several species seem to prefer the proximity of human habitations because of the relative safety from predators and probably due to an abundance of insects, on which they feed. One of the most common species is the golden-crowned warbler, *Basileuterus culicivorus* (Lichtenstein, 1830), a species found throughout tropical South America east of the Andes.

Small flocks of masked yellowthroats or *pia-cobras*, *Geothlypis aequinoctialis* (Gmelin, 1789), were seen in the tall grasses along the Transpantaneira Highway between Poconé and the Pantanal together with various other small bird species at different times during the study. In the wetland itself, the species was uncommon.

The river warbler, *Phaeothlypis rivularis* (Wied, 1821), formerly called *Basileuterus rivularis*, is a species that lives in thick woods along streams. It was present in the Pantanal but not often observed.

The typical habitat of the flavescent warbler, *Basileuterus flaveolus* (Baird, 1865), is deciduous forest (Schauensee and Phelps, 1978). In southern Mato Grosso, it occurs in the upland parts of the watershed but seldom in the wetland.

A species also seen in grasses and low shrubs from time to time was the tropical parula, *Parula pitiayumi* (Vieillot, 1817), which is found in all kinds of habitats from North America to Argentina.

The occasional presence of the rose-breasted chat or *policial-do-sul, Granatelius pelzelni* Sclater, 1865, in the region of the Pantanal is suspected, but no individuals were observed during the study. It was reported by N. Willis (1976) from the region to the northwest of the wetland.

Other parulids of southern Mato Grosso that are uncommon in the catchment area and probably absent from the wetland include the white-bellied warbler or *pula-pula-pichito, Basileuterus hypoleucus* (Bonaparte, 1850); white-striped warbler or *pula-pula-branco, Basileuterus leucophrys* Pelzeln, 1868; and white-browed warbler or *pula-pula assobiador, Basileuterus leucoblepharus* (Vieillot, 1817). *B. hypoleucus* and *B. leucophrys* are known to occur to the north and east of the wetland, while *B. leucoblepharus* is known only from the southeast (Silva & Oniki, 1988; Dubs, 1992).

Coerebiidae

The honeycreepers are presently placed in the swallow-tanager family, Tersinidae, by some authors (Schauensee and Phelps, 1978) and in the tanager family, Thraupidae, by others (Dubs, 1992). Of the species known from southern Mato Grosso, the males of all but one are a bright blue or purple color. These easily observed species were conspicuous for their absence in the Pantanal, and most were also not common in the dry upland forests in the watershed.

Perhaps the most abundant upland species in southern Mato Grosso are the blue dacnis or *saí-azul, Dacnis cayana* (Linnaeus, 1766); the red-legged honeycreeper or *saí-beija-flor, Cyanerpes cyaneus* (Linnaeus, 1766); and the green honeycreeper or *saí-verde, Chlorophanes spiza* (Linnaeus, 1758), species reported to be rather common in orchards and the canopies of gallery forests in the Serra das Araras by Silva and Oniki (1988). The same authors also reported the presence of the purple honeycreeper or *saí-roxa, Cyanerpes caeruleus* (Linnaeus, 1758), which was rather rare. Another rare species known to occur northwest of the Pantanal is the black-faced dacnis or *saíra-de-cara-preta*, *Dacnis lineata* (Gmelin, 1789), which appeared in the list of avifauna provided by Brown (1986).

The only yellow and black species, the bananaquit or *cambacica, Coereba flaveola* (Linnaeus, 1758), is a necter-eating bird found throughout nearly all of Brazil. In the Pantanal, the species is not at all abundant.

Another honeycreeper with a range encompassing southern Mato Grosso is the chestnut-vented conebill, *Conirostrum speciosum* (Temminck, 1824), which inhabits gallery forests and dry woodland.

Tersinidae

The swallow-tanager or *saí-andorinha, Tersina viridis* (Illiger, 1811), is a bird that is often seen at the edge of woods or in gallery forests. It is widespread in the area of the Pantanal and can be included with the group of species found along the edges of the large rivers. The species is omnivorous, with fruit and insects included in the diet (Sick, 1988).

Thraupidae

Tanagers and their relatives are not uncommon in the Pantanal, but none are abundant and most are seldom seen. The birds in this family are mainly omnivorous, but some are herbivorous, feeding on small fruits.

The silver-beaked tanager, *rico-de-prata*, or *pipira-vermelha*, *Ramphocelus carbo* (Pallas, 1764), is a common species found throughout the Pantanal. Dubs (1992) reported that these birds feed on nectar, and observed them splitting the corolla tubes of the flowers of *Tabebuia* spp. to get it.

Also abundant throughout southern Mato Grosso is the sayaca tanager or *sanhaço-cinza, Thraupis sayaca* (Linnaeus, 1766), which is locally represented by the subspecies *T. sayaca sayaca*. According to Dubs (1992), it feeds on fruit and insects. Similarly common is the palm tanager or *sanhaço-do-coqueiro, Thraupis palmarum* (Wied, 1821), represented in the Pantanal by *T. palmarum palmarum*. A second subspecies, *Thraupis palmarum melanoptera* (Sclater, 1856) was reported by Naumburg (1930) to inhabit upland areas to the north of the wetland. The name palm tanager was earned by the birds for usually constructing their nests in palm trees.

The gray-headed tanager or *pipira-de-taoca, Eucometis penicillata* (Spix, 1825), is one of the common species in the Pantanal of Poconé. Cintra and Yamashita (1990) usually found it in patches of dense forest.

A wide ranging species found in many kinds of habitat from Central America to northern Argentina is the white-lined tanager or *pipira-preta, Tachyphonus rufus* (Boddaert, 1783), which hunts its insect prey in flocks together with other species. Another bird with an even larger range from the southern United States to northern Argentina is the hepatic tanager or *sanhaço-de-fogo, Piranga flava* (Vieillot, 1822), represented in the region by the subspecies *P. flava saira* (Spix, 1825). While Dubs (1992) reported that the species was not at all abundant near the Pantanal, Naumburg (1930) mentioned that a number of specimens had been collected at Cuiabá, "Chapada", and other parts of southern Mato Grosso in the late 19th and early 20th centuries. The fawn-breasted tanager, *Pipraeidea melanonota* (Vieillot, 1819), also has a wide range throughout tropical South America, but it seems to avoid the wetland. Little is know about the white-banded tanager or *tiê-do-cerrado, Neothraupis fasciata* (Lichtenstein, 1823), which seems to have the center of its range in Mato Grosso. Wolters (1979-1982) reported that this bird ranges into eastern Brazil and westward to eastern Bolivia and northeastern Paraguay.

The blue-hooded euphonia or *gatoramo-rei, Euphonia musica* (Gmelin, 1789); the hooded tanager or *saíra-de-chapeau-preto, Nemosia pileata* (Boddaert, 1783); and the guira tanager or *saíra-de-papa-preto, Hemithraupis guira* (Linnaeus, 1766), are species found throughout the region of the Pantanal, but apparently not in the central part of the wetland itself.

The violaceous euphonia or *gaturamo-verdadeiro, Euphonia violacea* (Linnaeus, 1758), seems to be a locally uncommon species that was recently sighted in the Serra das Araras (Silva and Oniki, 1988) and near Poconé (Cintra and Yamashita, 1990). It frequents all kinds of habitat.

The orange-headed tanager, *Thlypopsis sordida* (d'Orbigny and Lafresnaye, 1837), seems to be a species preferring the dry forests of Central South America, occurring from northern Argentina northward throughout the entire region of the Pantanal to Paraguay and Bolivia.

Some of the species in southern Mato Grosso seem to avoid the wetland and occur exclusively in the upland border regions. The white-rumped tanager or *bandoleta, Cypsnaga hirundinacea* (Lesson, 1831), is a common insectivorous bird in the upland regions of southern Mato Grosso. The black-faced tanager or *tiê-cinza, Schistochlamys melanopis* (Latham, 1790); turquoise tanager or *saíra-de-bando, Tangara mexicana* (Linnaeus, 1766); paradise tanager or *saíra-do-paradiso, Tangara chilensis* (Vigors, 1832); blue-necked tanager or *saíra-de-cabeça-azul, Tangara cyanicollis* (d'Orbigny and Lafresnaye, 1837); and yellow-backed tanager or *saíra-galega, Hemithraupis flavicollis* (Vieillot, 1818), were all reported to be relatively rare birds in the Serra das Araras by Silva and Oniki (1988). Both *C. hirundinacea* and *S. melanopis* are denizens of open fields and *cerrado*, while *H. flavicollis* is a non-migratory omnivore that generally forms flocks in gallery forests (Dubs, 1992). A widespread species found in almost all of the border regions around the Pantanal in Brazil is the black-goggled tanager or *tiê-do-topete, Trichothraupis melanops* (Vieillot, 1818), which occurs in various kinds of forest and woodland. The white-winged shrike-tanager or *pipira-barulhenta, Lanio versicolor* (d'Orbigny and Lafresnaye, 1837), is a bird that generally remains in the upper canopies of forests, often together with other species. N. Willis (1976) reported this species from regions to the northwest of the Pantanal along with the bay-headed tanager or *saíra-de-cabeça-castanha, Tangara gyrola* (Linnaeus, 1758), a species that remains near the ground at the edge of forests (Sick, 1988), and the masked tanager or *saíra mascarada, Tangara nigrocincta* (Bonaparte, 1838).

Also common in the upper canopies of gallery forests to the north of the Pantanal (N. Willis, 1976; Silva and Oniki, 1988) are the rufous-bellied euphonia or *gaturamo-serrador, Euphonia rufiventris* (Vieillot, 1819), and the burnished-buff tanager or *saíra-cabocla, Tangara cayana* (Linnaeus, 1766), represented to the north of the wetland by *T. cayana margaritae*(Allen, 1891) and possibly also by *T. cayana chloroptera* (Vieillot, 1819). Naumburg (1930) believed that *T. c. margaritae* was endemic to the State of Mato Grosso, which included Mato Grosso do Sul at the time.

The white-shouldered tanager or *tiê-de-encontro-branco, Tachyphonus luctuosus* (d'Orbigny and Lafresnaye, 1837), is another forest species encountered northwest of the Pantanal (Dubs, 1992), while the regionally widespread purple-throated euphonia or *gaturamo-fifi, Euphonia chlorotica* (Linnaeus, 1766) is noted for remaining in small flocks and nesting among epiphytic bromelias in dry forests and *cerrado* (Oniki and Willis, 1983). Another species that inhabits the same region and also constructs its nests in epiphytic bromelias is the white-vented euphonia or *gaturamo-de-barriga-branca, Euphonia minuta* Cabanis, 1849. Also found in dry forests (Sick, 1988) and reported to the north of the Pantanal (Dubs, 1992) is the thick-billed euphonia or *gaturamo-bicudo, Euphonia laniirostris* (d'Orbigny and Lafresnaye, 1837). It is locally represented by the subspecies *E. laniirostris laniirostris*.

A bird that prefers to remain along the edges of ponds and rivers is the flame-crested tanager or *tiê-galo, Tachyphonus cristatus* (Linnaeus, 1766), which, however, does not seem to occur in the central parts of the Pantanal (Dubs, 1992). It is apparently represented in southern Mato Grosso by two subspecies: *Tachyphonus cristatus nattereri* Pelzeln, 1870, which is perhaps an independent species (Wolters, 1975-1982) and *T. cristatus madeirae* Hellmayr, 1910.

The magpie tanager or *tiêtinga, Cissopis leveriana* (Gmelin, 1788), was observed by Dubs (1992) to be common in the Serra das Araras to the northwest of the Pantanal. It is an omnivorous, non-migratory species that generally forages in flocks.

The cone-billed tanager or *tiê-bicudo, Conothraupis mesoleuca* Berlioz, 1939, another species from the dry, open woodland, is known only from the locality at which the type was collected, in the city of Cuiabá.

A few thraupid species are known only from Mato Grosso do Sul and have not yet been reported from Mato Grosso. These include the ruby-crowned tanager or *gurundi, Tachyphonus coronatus* (Vieillot, 1822); red-crowned ant-tanager or *tiê-da-mata, Habia rubica* (Vieillot, 1817); and green-headed tanager or saíra-de-sete-cores, *Tangara seledon* (P. L. S. Müller, 1776).

Fringillidae

One of the most abundant small birds observed near buildings in the Pantanal is the yellow-billed cardinal, *cavalaria* or *cardeal-do-pantanal, Paroaria capitata* (d'Orbigny and Lafresnaye, 1837), which forms enormous flocks every year near the Rio Pixaim. Normally at least 50 line up on the wooden fence railings alongside the restaurant during the dry season, and the flock can grow to nearly 100 at times. The birds have remarkably little fear of people, and a flock will gather just to listen to someone whistle. The birds also accept bread crumbs and other table scraps, which they will practically take from the hand (Plate 12a). During the rainy season, these birds disperse to various parts of the Pantanal.

Another red-headed cardinal frequently seen in the Pantanal is the red-crested cardinal or *cardeal-do-sul, Paroaria coronata* (J. F. Miller, 1776), easily distinguished by its prominent crest. This species is much less abundant than the yellow-billed cardinal, and more than one or two together were seldom seen. It is much more timid and cannot be approached closely. Reports of the occurrence of the red-capped cardinal or *cardeal-da-Amazônia, Paroaria gularis* (Linnaeus, 1766), in the Pantanal could not be confirmed. Although Hilty and Brown (1986) reported the species to prefer wetlands, Dubs (1992) listed its local range to include only the northwestern border areas. Wolters (1975-1982) assigned the species to the genus *Coccopsis*.

The wedge-tailed grass finch or *canário-do-campo, Emberizoides herbicola* (Vieillot, 1817), is abundant in the Pantanal, where it is frequently seen eating grass seeds along the sides of the road or at a feeding station set up at an open air refreshment stand called Figeiro's Bar beside the Transpantaneira Highway not far south of the Rio Novo. This widespread finch has obviously benefited from the introduction of many different grass species, the seeds of which this and several other fringillid species thrive upon. It was also reported to be abundant in fields near the Serra das Araras, located in the upland region to the north of the wetland (Silva and Oniki, 1988).

Also numerous near the Serra das Araras (Silva and Oniki, 1988) but less common near Poconé is the yellow-bellied seedeater or *baiano, Sporophila nigricollis* (Vieillot, 1823).

Oryzoborus maximiliani (Cabanis, 1851), the greater large-billed seed-finch or *bicudinho*, which Dubs (1992) regards as a subspecies of *Oryzoborus crassirostris* (Gmelin, 1789), is a bird of wooded wetlands that is known from the northern part of the Pantanal. This species, or subspecies, is otherwise known from regions of eastern and central Brazil (Sick, 1988).

An abundant seedeater in the Pantanal is the capped seedeater or *caboclinho, Sporophila bouvreuil* (P. L. S. Müller, 1776), which is not generally seen near human habitations but remains in the shrubs and tall grasses beside the roads. Three subspecies are known in Brazil (Sick, 1988), but the one occurring in the Pantanal is not known with certainty. Cintra and Yamashita (1990) observed it in the region of Poconé only from February through May.

A non-migratory bird that was sometimes observed near the ranch houses in the Pantanal is the grayish saltator or *gonga, Saltator coerulescens* Vieillot, 1817. According to Dubs (1992), it is represented in eastern Bolivia, Mato Grosso, and northwestern and central Argentina by *S. coerulescens coerulescens*.

Cintra and Yamashita (1990) found the stripe-tailed yellow-finch or *canário-rasteiro, Sicalis citrina* Pelzeln, 1870, to be common in moist grasslands near Poconé, where it is represented by the subspecies *S. citrina citrina*. Otherwise, it inhabits open fields in the upland regions north of the Pantanal, often near human habitations.

Small flocks of the saffron finch or *canário-da-terra, Sicalis flaveola* (Linnaeus, 1766), are found in various kinds of habitat throughout the Pantanal and its surroundings, represented by the subspecies *S. flaveola pelzelni* (Dubs, 1992). Other species that form small flocks in the grassland and *cerrado*, including the yellow-finch or *canário-tipio, Sicalis luteola* (Sparrman, 1789), and the coal-crested finch or *mineirinho, Charitospiza eucosma* Oberholser, 1905, were reported to be semi-migratory respectively by Short (1975) and Sick (1988). They both inhabit the northeastern border regions of the wetland, and *S. luteola* sometimes visits the southern part of the Pantanal, as well (Dubs, 1992).

The black-backed grosbeak or *rei-do-bosque, Pheucticus aureoventris* (d'Orbigny and Lafresnaye, 1837), was reported by various authors to inhabit widely varying kinds of habitat (Short, 1975; Sick, 1988; Cintra and Yamashita, 1990). It occurs regularly in the region of Poconé. Dubs (1992) reported it to be regionally confined to the northern borders of the wetland but to otherwise occur from Colombia to northwestern Argentina.

The lesser seed-finch or *curió, Oryzoborus angolensis* (Linnaeus, 1766), is also a wetland species that has been reported to be rare in some of the upland regions investigated, such as the Serra das Araras (Silva and Oniki, 1988).

Sick (1988) also reported the sooty grassquit or *cigarrinha-do-coqueiro, Tiaris fuliginosa* Wied, 1830, to be a typical marshland species. Dubs (1992) reported its occurrence throughout the Pantanal to in adjacent areas to the northeast.

The blue-black grassquit or *tiziu, Volatinia jacarina* (Linnaeus, 1766), the lined seedeater or *bigodinho, Sporophila lineola* (Linnaeus, 1758), the white-bellied seedeater or *chorão, Sporophila leucoptera* (Vieillot, 1817); and the rusty-collared seedeater or *coleiro-do-brejo, Sporophila collaris* Boddaert, 1783, have been reported from the region of Poconé from January through May (Cintra and Yamashita, 1990). Dubs (1992) reported them to feed on seeds and insects and to occur throughout the entire region of the Pantanal. Two subspecies of *Sporophila collaris* inhabit the wetland: *S. collaris melanocephala* (Vieillot, 1817) and *S. collaris ochrascens* (Hellmayr, 1904). According to Dubs (1992), *Sporophila c. ochrascens,* which Naumburg (1930) observed near Cuiabá, inhabits the northern part of the Pantanal, while *S. c. melanocephala* is encountered farther south.

Relatively few fringillids of the Pantanal region prefer forests to open grassland and *cerrado*. One of these, the pectoral sparrow or *tico-tico-da-mata, Arremon taciturnus* (Herrmann, 1783), is a species commonly inhabiting rather dense woodland, particularly gallery forests (Silva and Oniki, 1988). It is therefore not seen as often as many of the other species of its family.

The buff-throated saltator or *tempera-viola*, *Saltator maximus* (P. L. S. Müller, 1776), is a species that lives at the edges of forests at the borders of the Pantanal, but apparently, it is uncommon in southern Mato Grosso (Silva and Oniki, 1988).

A third woodland species, the saffron-billed sparrow or *tico-tico-de-bico-amerelo, Arremon flavirostris* Swainson, 1838, is a non-migratory species living near the forest floor (Short, 1975), which, according to Dubs (1992), has not been recorded from the central part of the Pantanal but is present throughout the peripheral regions.

The indigo grosbeak or *azulinho-do-sul, Cyanoloxia glaucocaerulea* (d'Orbigny and Lafresnaye, 1837), is a species that lives at the edges of forests near water. However, according to Dubs (1992), it does not frequent the Pantanal wetland but rather is confined to the regions to the northwest and southeast.

Many fringillids are non-migratory species of open grassland and *cerrado*, which sometimes enter the Pantanal but are usually more common at its outskirts. These include the grassland sparrow or *tico-tico-do-campo*, *Myospiza humeralis* (Bosc, 1792), called *Ammodramus humeralis* by Dubs (1992), who reported that it is represented only by the subspecies *M. humeralis xanthornus* (Darwin, 1839) in Mato Grosso. Another upland species is the rufous-collared sparrow or *tico-tico-verdadeiro, Zonotrichia capensis* (P. L. S. Müller, 1776), which is a common species near human habitations and can sometimes be seen in the wetland. According to Dubs (1992), the species is represented by *Z. capensis matutina* (Lichtenstein, 1823) in the northern part of the Pantanal, including Poconé (Naumburg, 1930), and by *Z. capensis subtorquata* Swainson, 1837, in the southern districts. The cinereous warbling finch or *capacetinho-do-sul, Poospiza melanoleuca* (d'Orbigny and Lafresnaye, 1837), is another grassland species that seems to avoid the wetland. It is found along the northern margins of the Pantanal, and Dubs (1992) reported that it also inhabits regions to the east and is known from Corumbá in the South. In the Pantanal of Mato Grosso, it is represented by the subspecies *Poospiza melanoleuca cinerea* Bonaparte, 1850.

The red-crested finch or *tico-tico-rei-vermelho, Coryphospingus cucullatus* (P. L. S. Müller, 1776); the blue finch or *campainha-azul, Porphyrospiza caerulescens* (Wied, 1830); and the black-throated saltator, *Saltator atricollis* Vieillot, 1817, are species of the *cerrado*, which seem to be rather abundant in the upland areas of the wetland itself, as well.

The orange-fronted yellow-finch or *canário-do-campo, Sicalis columbiana* Cabanis, 1851, is a species Sick (1988) reported from *cerrado* and grassland. It has been reported from the northern part of the Pantanal near Poconé (Cintra and Yamashita, 1990) and north of Cáceres (N. Willis, 1976).

Dubs (1992) reported the presence of the green-winged saltator or *trinca-ferro-de-asa-verde*, *Saltator similis* (d'Orbigny and Lafresnaye, 1837), from all parts of the Pantanal, but Cintra and Yamashita (1990) listed it as rare in the District of Poconé. It was not observed during the study.

Other members of the genus occur in regions of southern Mato Grosso and are occasionally seen in or near the Pantanal. They include the dark-throated seedeater or *caboclinho-paraguai*, *Sporophila ruficollis* Cabanis, 1851; the plumbeous seedeater or *patativa-verdadeira*, *Sporophila plumbea* (Wied, 1830); the double-collared seedeater, *coleirinho*, or *papa-capim-capuchinho*, *Sporophila caerulescens* (Vieillot, 1817); the ruddy-breasted seedeater or *caboclinho-de-berriga-vermelha, Sporophila minuta hypoxantha* Cabanis and Heine, 1851; the chestnut seedeater or *caboclinho-de-chapéu-cinzento*, *Sporophila cinnamomea* (Lafresnaye, 1839); the black-and-tawny seedeater or *caboclinho-do-sertão, Sporophila nigrorufa* (d'Orbigny and Lafresnaye, 1837); the black-masked finch or *tico-tico-mascarado, Coryphaspiza melanotis* (Temminck, 1822); the slate-colored grosbeak or *bico-encarnado, Pitylus grossus* (Linnaeus, 1766);

and the blue-black grosbeak or *azulão-da-mata, Cyanocompsa cyanoides* (Lafresnaye, 1847).

Among the migratory species that are sometimes encountered in southern Mato Grosso is the hooded siskin or *pintassilgo-de-cabeça-preta, Spinus magellanicus* (Vieillot, 1805), which is apparently known only from the plateau region near Rondonopolis, to the east of the Pantanal, where it was captured in July and August about a half century ago (Pinto, 1944).

Several fringillids are known from the Pantanal in Mato Grosso do Sul but have not yet been reported from Mato Grosso (Olrog, 1963; Dubs, 1992). These include the golden-billed saltator or *trinca-ferro-patetão, Saltator aurantiirostris* Vieillot, 1817; thick-billed saltator or *trinca-ferro-bicudo, Saltator maxillosus* Cabanis, 1851; and *Cyanocompsa cyanea* (Linnaeus, 1758), represented by *Cyanocompsa cyanea sterea* Oberholser, 1901.

Ploceidae

The only representative of this family that has reached the Pantanal is the house sparrow, *Passer domesticus* (Linnaeus, 1758). Introduced to Argentina in 1872 (Sick, 1988), the house sparrow has accomplished a nearly complete conquest of the South American continent. It is an anthropophilic species, thriving best in cities, but also frequenting human habitations in rural areas. The house sparrow is not at all abundant in the Pantanal because human habitations are scarce, but it occurs in the towns at the edge of the Pantanal and from there reaches some of the ranch houses. On ranches in the Pantanal far from towns, as well as in the uninhabited areas, this species never seems to occur.

8.13.5. Mammalia

The large mammals of the Pantanal are relatively well known. There is still much to learn about the small mammal fauna, however. The number of mammalian species is not great, so the most characteristic species can be discussed individually.

Marsupiala

The marsupials in the Pantanal are represented only by members of the family Didelphidae. The only species observed during the study was the white-eared opossum or *gambá, Didelphis albiventris* Lund, 1839, which was seen near the Pantanal but not in the wetland itself. According to Magalhães (1992), the common opossum, *Didelphis marsupialis* Linnaeus, 1758, is a second species of the genus known from the Pantanal. Opossums are not at all abundant in Mato Grosso, perhaps because of predation by various carnivores and capricious killing of the animals by man. Opossums are considered pests by the rural population because they eat the eggs and chicks of domestic fowl, which are allowed to roam freely. The opossums are omnivores, but they subsist mainly on animal matter. The survival of this species seems to depend on its rapid rate of reproduction which offsets the great mortality it suffers.

At least six additional species of Didelphidae are among the fauna reported for parts of southern Mato Grosso and the Pantanal by Emmons (1990): the western woolly opossum, *Caluromys lanatus* (Retzius, 1783); common gray four-eyed opossum, *Philander opossum* (Linnaeus, 1758); brown four-eyed opossum, *Metachirus nudicaudatus* (Geoffroy, 1817); murine mouse opossum, *Marmosa murina* (Linnaeus, 1758), agile gracile mouse opossum, *Gracilinanus agilis* (Villers, 1789); and gray short-tailed opossum, *Monodelphis domestica* (Hufnagel, 1767).

According to Emmons (1990), only the common gray four-eyed opossum shows a preference for forests bordering streams, where it seeks frogs by listening to their calls. The other species would not be expected to show any preference for wetlands. All are omnivorous, with the western woolly opossum subsisting primarily on fruit or nectar supplemented by insects, and the other species seeking primarily insects or small vertebrates. The brown four-eyed opossum shows a special preference for termites. Locally known as *cuicas* or *catitas*, the mouse opossums are secretive, nocturnal marsupials, the habits of which are poorly known. None were observed during the study, and their presence is probably confined to the few forested areas in the Pantanal, mainly along the rivers. Magalhães (1992) reported the presence to two unidentified mouse opossums: *Peramys* sp. and *Marmosa* sp., which may actually have been two of the species mentioned above.

A rare didelphid, the water opossum, *Chironectes minimus* (Zimmermann, 1780), may be present in the Pantanal of Mato Grosso do Sul. It is an aquatic species with webbed feet, which has never been reported from the northern part of the Pantanal (Marshall, 1978).

Edentata

Myrmecophagidae

The giant anteater or *tamanduá-bandeira*, *Myrmecophaga tridactyla* Linnaeus, 1758, is encountered in the Pantanal on rare occasions. It is actually a species that should be rather abundant, considering the large number of available termite nests, which characterize the landscape form known as the *campo morundu*, or field of small hills. When trees die naturally, are cut, or die after one of the frequent fires, many set deliberately to clear fields, the stumps are invariably covered by a layer of mud by the termite species, including *Armites euamignathus* and *Rotunditermes bragantinus* (Section 8.12.4), and are slowly converted to mud hills as the wood is systematically consumed. The giant anteater is adept at breaking open these nests and consuming the termites with its long tongue. The reason for the rarity of the giant anteater is probably human activity, including deliberate and accidental arson, which exterminates or drives the anteaters away from the areas rich in termite colonies. Deliberately killing the animals, which have little chance to conceal themselves, may also be a factor in the rarity of the giant anteater. An active protection program for this species should be undertaken. Preserving the species in captivity is difficult because of its diet.

Reported by (Magalhães, 1992) but not observed in the Pantanal during the study was the smaller tamandua or *tamanduá-mirim*, *Tamandua tetradactyla* (Linnaeus, 1758), an arboreal species that would not be expected to occur on the floodplain, although it might be found occasionally in gallery forests.

Dasypodidae

More abundant than the anteaters throughout Mato Grosso, including the Pantanal, are three armadillo species: the nine-banded armadillo or *tatu-galinha*, *Dasypus novemcinctus* Linnaeus, 1758; the six-banded or yellow armadillo, *Euphractus sexcinctus* (Linnaeus, 1758), called the *tatu-peba* in Portuguese and represented in southern Mato Grosso by the subspecies *E. s. flavimanus* (Desmarest, 1804); and the three-banded armadillo, *Tolypeutes matacus* (Desmarest, 1804). Of these, the nine-banded armadillo is much more common in the wetland regions, where Schaller (1983) found it to be almost the only species present, as well as in forests, where it accounted for 88% of his sightings. In contrast, he found the six and three-banded armadillos to be almost equally abundant in the dry, upland areas to the west of the Pantanal, each outnumbering the nine-banded species by about two to one. In some of the upland areas of Mato Grosso, the six-banded armadillo was found to account for ⅔ of the armadillo "biomass" (Schaller, 1983). Because it is larger than the other species, however, the six-banded armadillo does not have to be numerically more abundant to account for the bulk of the "biomass". All of these species seem considerably more numerous on the "*cordilleiras*" and hummocks as well as on plains at higher elevations than on the floodplain because they are burrowing animals and cannot inhabit tunnels that are periodically underwater. In the Pantanal, they must seek out the hummocks, while outside of the Pantanal, they generally burrow into hillsides, eroded cliffs, and the sides of dry stream beds. The success of each species in adapting to the wetland conditions is reflected in their relative abundances in each kind of habitat. Because of their burrowing habits, they are somewhat better protected from the fires frequently set to eliminate woody vegetation in the *cerrado*.

The armadillos are generally omnivorous, but animal matter makes up the great bulk of the diet of *Dasypus novemcinctus*: 90% of the stomach contents of specimens from the southern United States (Fitch *et al.*, 1952). The diet was found to include insects and other small arthropods, eggs, and carrion. The six-banded armadillo also preys on small mammals and has been known to kill rats (Redford and Wetzel, 1985), but Schaller (1983) reported that 90% of the stomach contents of ten specimens examined was of plant origin. The three-banded armadillo seems to have the most specialized diet, which consists mainly of termites and ants (Schaller, 1983).

Nine-banded armadillos are crepuscular (McBee and Baker, 1982), leaving their burrows at sunset to seek food, while the six-banded species is mainly diurnal (Redford and Wetzel, 1985).

In addition to the three common species, *the tatu-de-rabo-mole*, *Cabassous unicinctus* (Linnaeus, 1758) is also reported for the region of the Pantanal, and the giant armadillo or *tatu-canastra*, *Priodontes maximus* (Kerr, 1792), a species close to extinction, is presently known in the Pantanal drainage area from only one forest in the Bodoquena Mountains (Magalhães, 1992).

Chiroptera

Emballonuridae

The bat fauna of Mato Grosso is rich, but most species avoid the Pantanal because of the lack of hiding places to spend the days. An exception is the long-nosed bat, *Rhynchonycteris naso* (Wied, 1820), which was frequently seen along the Transpantaneira Highway in the evening. A large colony lived under the wooden bridge over which the highway crossed the Rio Bento Gomes (Plate 12b). These bats were observed regularly throughout the year at this site, except during *cheia*, when the water level in the river was highest and the places under the bridge usually occupied by the bats were under water.

This species loves warmth, and the bats often left their hiding place under the bridge during the day to hang on the outside of the wooden beams and sun themselves. On cool mornings during the dry season, they could be taken in the hand. When released, they were reluctant to fly away, waiting instead several minutes to warm themselves to the temperature of the hand.

Noctilionidae

The greater bulldog bat of *morcego-pescador*, *Noctilio leporinus* (Linnaeus, 1758), occurs in the Pantanal (Magalhães, 1992). The bats capture small fishes that remain at the surface of the water during the night. According to Hood and Jones (1984), these bats may not range all the way to the northern limits of the Pantanal, but they are represented in the central part of the wetland by *N. l. rufescens* Olfers, 1818. They roost in hollow trees, where colonies of several hundred individuals gather.

A smaller fishing species, the lesser bulldog bat, *Noctilio albiventris* Desmarest, 1818, is also a likely inhabitant of the Pantanal. However, it was not observed during the study. According to Davis (1976), it spends its days in hollow trees and buildings in tropical forest regions and forages at night by flying over rivers, streams, and marshes. Unlike the greater bulldog bat, it apparently feeds mainly on insects rather than fishes (Davis *et al.*, 1964; Fleming *et al.*, 1972). Dytiscid beetles were found to be prominent in the diet of this bat (Whitaker and Findley, 1980). However, most of the information about the species was obtained by studies of Central American populations (Hood and Pitocchelli, 1983), and there is little information on the behavior of this bat in South America.

Phyllostomidae

The family Phyllostomidae is found exclusively in the New World, where it is represented by a considerable number of species. The ranges of many of them include the Pantanal or adjacent habitats in the watershed. During the study, many more bats were observed in the upland parts of the watershed than in the Pantanal. The city of Cuiabá was particularly rich in insectivorous bat species, which could be seen every night catching insects around the street lamps. The species discussed are those known to live in the region, but their temporary or permanent presence in the wetland itself is suspected but not confirmed. The ecology of many of these species is poorly known.

Geoffroy's rayed bat, *Vampyrops lineatus* (E. Geoffroy St.-Hilaire, 1810), is found throughout the region of the Pantanal (Willig and Hollander, 1987). It roosts among palm leaves and feeds not only on fruit and nectar, but also on insects, especially sphingid moths (Gardner, 1977).

Also known from the southern part of Mato Grosso were the fructivorous *morcego-de-cara-branca*, *Artibeus lituratus* (Clerck, 1759), and the nectar-loving Pallas's long-tongued bat or *morcego-beija-flor*, *Glossophaga soricina* (Pallas, 1766), neither of which were numerous in the Pantanal wetlands. Both of these species have extensive ranges in tropical South America and may be encountered anywhere in the general region of the Pantanal (Emmons, 1990).

The range of the vampire bat, *Desmodius rotundus rotundus* (E. Geoffroy St.-Hilaire, 1810), appears to include the entire Pantanal (Greenhall *et al.*, 1983). However, in spite of the availability of numerous cattle and horses, on which this bat feeds, it does not seem to occur at all in the wetland. The reason for this may be the high temperatures and lack of caves, in which the bats could escape the heat of the day. Studies by Wimsatt (1962) revealed that the upper tolerable temperature limit for the bats is 27°C to 30°C, and temperatures above 37°C or 38°C are lethal. Therefore, the tolerable temperatures are exceeded daily in the Pantanal throughout most of the year, and the lethal temperature is often reached during October and November. This may well be the reason for the absence of these bats.

The woolly false vampire bat, *Chrotopterus auritus* (Peters, 1856), lives in caves and mines during the day and feeds on insects and small vertebrates at night (Medellín, 1989). It is more likely to be encountered in the upland parts of the watershed than in the Pantanal.

Tomes's long-eared bat, *Lonchorhina aurita* Tomes, 1863, is mainly a rainforest species that also inhabits the border areas of the dry forests. The diet consists mainly of insects, although there are reports that fruit may also be consumed. It roosts in large groups of as many as 500 individuals (Lassier and Wilson, 1989).

The common yellow-shouldered bat, *Sturnira lilium* (E. Geoffroy St.-Hilaire, 1810), inhabits both humid and semi-arid forests, preferring the humid ones as well as open areas (Handley, 1976). It consumes insects, pollen, and fruit, including dates, bananas, and wild figs (Gannon *et al.*, 1989). Its days are spent in both natural and man-made shelters.

The autecology of *Tonatia silvicola* (d'Orbigny, 1836) is still poorly known. During the day, it is reported to roost in hollow termite nests (Medellín and Avrita, 1989). Its diet is thought to consist of fruit and insects (Gardner, 1977), and it was found to be attracted to the sounds made by insects but not to frog calls (Medellín and Arita, 1989).

The little big-eared bat, *Micronycteris megalotis* (Gray, 1842), is widely distributed in many habitats, both wet and dry, including wetlands (Alonso-Mejía and Medellín, 1991). It consumes both insects and fruit, and its feeding habits have been classified as "gleaning carnivorous", making it a member of the "foliage gleaning guild" (LaVal and LaVal, 1980; Belwood and Morris, 1987). This means that it gleans its food from the ground or from foliage rather than catching it in the air.

For two of the phyllostomid species, the Pantanal represents the geographic limit of their ranges. There is little ecological information available about either. *Pygoderma bilabiatum* (Wagner, 1843) ranges only as far westward as the southern edge of the Pantanal (Webster and Owen, 1984), while Heller's broad-nosed bat, *Platyrrhinus helleri* (Peters, 1966), has been reported in southern Mato Grosso only from the Serra do Roncador by Pine *et al.* (1970). The presence of these species in the wetland is doubtful, although this order of animals has not been studied enough to sufficiently determine the bat fauna of the region. Heller's broad-nosed bat is a small canopy frugivore (Bonaccorso, 1978) that reaches a total length of only 50 to 66 mm (Ferrell and Wilson, 1991). It roosts in the widest variety of natural and artificial shelters, including caves, tunnels, culverts, under bridges, in hollow trees and buildings, and beneath palm leaves.

In addition, the Pantanal seems to be the southern limit of the hairy-legged vampire bat, *Diphylla ecaudata* Spix, 1823. Studies have shown that this vampire consumes only the blood of birds, including chickens (Greenhall *et al.*, 1984).

Vespertilionidae

The bats in this family were not at all common in the wetland. In contrast, they could be seen capturing insects almost every night around the street lamps in the city of Cuiabá. According to Emmons (1990), the species encountered in southern Mato Grosso belong to the genera *Lasiurus*, *Eptesicus*, and *Myotis*.

Lasiurus cinereus (de Beauvois, 1796) is represented in southern Mato Grosso by the subspecies *L. c. villosissimus* (Geoffroy St. Hilaire, 1806). Information on other subspecies suggests that the bats remain solitary among tree foliage or in woodpecker holes. The diet is poorly known, but it has been reported that moths are important food items (Shump, 1982).

Lasiurus borealis (P. L. S. Müller, 1776) is a species found throughout the Americas, except in the cold regions of Canada and southern South America. It is represented by the subspecies *L. b. blossevillii* (Lesson and Garnot, 1826) in Mato Grosso. The North American subspecies are known to roost in trees and shrubs (Shump and Shump, 1982).

Myotis nigricans (Schinz, 1821) is another vespertilionid bat that occurs in Mato Grosso. Although it subsists mainly on insects, including moths, vegetable matter was found in the stomachs of some specimens from Trinidad (Goodwin and Greenhall, 1961). Because the species is known to carry rabies and roost in buildings, sometimes together with molossid bats, it has sometimes been subject to eradication programs in Central America (Wilson and LaVal, 1974).

Molossidae

The free-tailed or mastiff bats have become associated with man, inhabiting buildings during the day. In the Pantanal, they lack suitable daytime shelters, but in the city, they were able to find hiding places in the narrow spaces behind the concrete plates of recently constructed buildings. The dense crowds of *Molossus molossus* (Pallas, 1767) inhabiting the spaces between the concrete plates in the university buildings in Cuiabá caused considerable consternation. Their presence caused problems, but the elimination of their shelters would endanger their local population. The species has been reported to roost together with *Myotis nigricans* by Wilson and LaVal (1974).

Peale's free-tailed bat, *Nyctinomops aurispinosus* (Peale, 1848), is a species reported to prefer lowlands. Its diet seems to consist mainly of soft-bodied insects (Jones and Arroyo-Cabrales, 1990). The congeneric big free-tailed bat, *Nyctinomops macrotis* (Gray, 1839), seems to feed almost exclusively on large moths, but grasshoppers, crickets, flying ants, stink bugs, and various homopterans are also consumed (Easterla and Whitaker, 1972; Easterla, 1973). However, Ochoa *et al.* (1988) found that the stomach of an individual captured in Venezuela contained mainly species of Coleoptera and Hemiptera, as well as parts of unidentified insects. The preferred habitat seems to be rocky country, but this was determined by extensive studies in North America. Little is known about its habits in South America. It is a seasonal migrant throughout much of its range (Jones and Arroyo-Cabrales, 1990). A specimen on which an early description of this bat was based was captured in Cuiabá prior to 1843 (Milner *et al.*, 1990).

The northern edge of the Pantanal seems to be the southern limit of the range of the South American flat-headed bat, *Neoplatymops matogrossensis* (Vieira, 1942), but its actual range is poorly known (Willig and Jones, 1985). This insectivorous bat roosts in rocky outcroppings, which are lacking in the Pantanal but plentiful to the northeast of Cuiabá. This may help explain the distribution pattern.

Mormoopidae

The northern edge of the Pantanal is the southern range limit of Parnell's mustached bat, *Pteronotus parnellii* (Gray, 1843), an insectivorous species that apparently feeds mainly on species belonging to the Lepidoptera and Coleoptera. Because it roosts in large caves (Herd, 1983), this bat apparently fails to find adequate living conditions in the wetland.

Figure 8.35: One of the monkeys that lives around the edges of the Pantanal is the tufted-ear marmoset, *Callithrix jacchus*. This one frequented a small private zoo near Cuiabá, where it begged food from the visitors, although it was not in the collection but rather ran free through the trees.

Primata

Callitrichidae

The Pantanal is inhabited by several species of marmoset. The smallest of the common species is the tufted-ear marmoset, *Callithrix jacchus jacchus* (Linnaeus, 1758). It is most frequently seen alone or in small groups of two to three, and it does not seem to form the large clans typical for the cebid monkeys (Fig. 8.35).

Cebidae

One of the common cebid species in the Pantanal is the black howler monkey, locally called the *guariba*, *Alouatta caraya* (Humboldt, 1811), which can be heard over distances of several kilometers. A large tribe of these monkeys ranged through the forested areas along the northern margin of the Pantanal wetland near Poconé. They accustomed themselves quickly to the presence their human observers and could be photographed from close range on many occasions (Fig. 8.36). Black howler monkeys subsist largely on plant material, including buds, leaves, and flowers (C. Smith, 1977). For unknown reasons, adult males in the groups seem to significantly outnumber adult females, an observation reported in northern Argentina (Pope, 1966) as well as in the western part of the Pantanal (Schaller, 1983).

Figure 8.36: Black howler monkeys, *Alouatta caraya*, live in large groups that travel through the Pantanal feeding in the trees on the hummocks. The strange sound that they make can be heard for many kilometers.

The common squirrel monkey, *Saimiri scuireus* (Linnaeus, 1758), was observed almost as often as the howlers in the Pantanal. It was generally encountered in groups of no more than five. Unlike the howler monkeys, they did not remain in any one place for long, seldom appearing at the same location more than once.

The black-capped or brown capuchin monkey, locally called the *macaco-prego*, *Cebus apella* (Linnaeus, 1758), is another species that is fairly common in the northern part of the Pantanal. It has a wide range in tropical South America (Emmons, 1990). Schaller (1983) reported it to be omnivorous, with considerably more animal material in its diet than that of the howler monkeys.

The dusky titi monkey, *Callicebus moloch* (Hoffmannsegg, 1807), represented in southern Mato Grosso by the subspecies *C. m. donacophilus* (d'Orbigny, 1836), is typically a species of tropical forests along river banks (Jones and Anderson, 1978). In addition to fruit, the main component of its diet, this monkey consumes insects, eggs, and small birds (Napier and Napier, 1967).

The black spider monkey or *macaco aranha*, *Ateles paniscus* (Linnaeus, 1758), has been reported from the region of the Pantanal, which is just beyond the border of its known range. It is unlikely that this species occurs in the wetland, but it is probably present in the upland forests to the north of Cáceres as well as in the Amazon drainage area to the north.

Rodenta

Hydrochaeridae

Among the mammalian fauna of the Pantanal is the largest known rodent in the world, the capybara or *capyvara*, *Hydrochoerus hydrochaeris* (Linnaeus, 1766), which is one of the most abundant of the mammals in the Pantanal (Fig. 8.37). It is aquatic and specialized to the wetland conditions that prevail in the Pantanal (Ojasti, 1983). The capybaras typically move in large family groups, with young individuals usually accompanying the larger, older ones. Each group is led by a dominant male, and a hierarchy of dominance was assumed to exist by Azcárate (1980), who studied the social behavior of the species. The capybara is herbivorous, subsisting mainly on water plants but also grazing along the shores of the water bodies. Among its favorite foods are the aquatic grasses that are abundant in the northern Pantanal, including *Leersia hexandra*, *Hymenachne amplexicaulis*, and *Panicum laxum*. Sedges are also consumed, but, according to Gonzáles-Jiménes (1978), species of *Eichhornia* are seldom eaten. However, see Figure 8.37! There have been reports of capybaras feeding on fishes, but the importance of fish in the diet is presently thought to be minimal (Mores and Ojasti, 1986).

Studies by Schaller and Crawshaw (1981) revealed that the capybaras reproduce throughout the year. Females usually bear from one to eight young at a time, and Ojasti (1973) calculated the average number of young in each litter to be 4.8. However, breeding is most intensive at the beginning of the rainy season. In contrast to the predatory mammals, herbivores such as the capybara encounter optimal conditions during the high water period in the Pantanal, when food is plentiful and predators relatively easy to avoid.

Although relatively little is known about parasites and diseases of most native South American mammals, the capybara seems to be an exception. A disease called surra or *mal de cadeiras* caused by *Trypanosoma evansi* (Steel, 1885), a protozoan transmitted to many animals by tabanid flies (Nieschulz, 1928), seems to be a frequent killer of the capybara (Magalhães, 1992). In addition, Costa and Catto (1994) identified 12 different parasitic worms in a single population from the Pantanal of Nhecolândia.

Capybaras spend much time in the water, but like the caimans, they often dry off on the shore during the day. Their natural enemies apparently do not include the caiman, and sometimes the two animals were seen in close proximity without one disturbing the other. The young are certainly suitable food for a variety of predators, almost certainly including the anaconda, the jaguarondi, and the lobinho. The older individuals are quite large, but would be no match for the jaguar, for which the capybara is an important prey (Schaller and Vasconcelos, 1978a). The capybaras are killed by poachers, apparently for no other reason than to get the feet to sell as souvenirs. Many proposals of international developers to raise these rodents commercially for their meat have been received with little local interest because the people inhabiting the Pantanal have no desire to eat the meat of this animal. Elsewhere in Brazil, studies have shown that the reduction in the amount of their available food resources have caused changes in the natural social behavior of the capybara (MacDonald, 1981). This indicates that in the Pantanal, the existence of some predators that feed on the capybaras help to maintain a healthy stock.

Figure 8.37: The capybara, *Hydrochoerus hydrochaeris*, the largest rodent in the world, is still abundant in the Pantanal. The photo shows that they consume not only their favorite foods, grasses, but also the leaves of *Eichhornia azurea*.

Agoutidae

The paca, *Agouti paca* Lacépède, 1799, is a fairly abundant species in the Pantanal. It occurs in forested regions above the high water level of the rainy season. The gallery forests along the larger rivers are apparently the preferred habitat of the species, although pacas may be vulnerable to poachers in such locations.

Caviidae

One of the most abundant small mammals in southern Mato Grosso is the yellow-toothed cavy, *Galea spixii* Wagler, 1847, which is typical of the dry forests and grasslands in Western Central Brazil and Paraguay but which also inhabits the outskirts of the Pantanal. It is frequently seen during early morning and evening along the edges of the roads.

Dasyproctidae

Azara's agouti or *cutía*, *Dasyprocta azarae* Lichtenstein, 1823, is frequently seen along the roads in and near the Pantanal. Unlike the capybara, it does not usually enter the water and is not a true wetland species. Nevertheless, the forested areas of the Pantanal offer it a suitable habitat.

In addition to this species, Cunha Vieira (1951) reported *Dasyprocta aurea* Cope, 1889, from southern Mato Grosso, and Por (1995) also recorded *Dasyprocta variegata* Tschudi, 1845, from the region.

Erethizontidae

Porcupines inhabit the wooded regions in and near the Pantanal. However, they are presently rare. One Brazilian porcupine or *ouriço*, *Coendou prehensilis* Lacépède, 1799, was found dead on the road about 10 km north of Poconé. According to Cunha Vieira (1951), the local subspecies is *C. prehensilis centralis* O. Thomas, 1903. These porcupines do not seem to be abundant in the wetland but are rather typical species of the *cerrado*. They do not seem to be at all abundant in the region.

Muridae

There is one typical murid in the Pantanal, the Neotropical water rat, *Nectomys squamipes* (Brants, 1827), which has some webbing between the toes to help it swim (Ernest, 1986). This rat is secretive, nocturnal, and therefore seldom seen. Its habits help it escape the many predators that seek prey of its size. These rats are common in grassy areas, building nests sometimes with several underwater entrances in tree stumps and logs along the edge of water bodies (Emmons, 1990).

The type specimen for the subspecies *N. s. mattensis* (Thomas, 1903) was taken about 30 km northeast of Cuiabá at the town of St. Anna de Chapada. At the present time, this form is generally not considered to merit a subspecific status, and the name is treated by most authors as a synonym of *N. squamipes*. Through studies in Venezuela, Handley (1976) found that the Neotropical water rat prefers moist forests, including temporarily flooded gallery forests, and it is generally not encountered in dry forest regions or in the habitats called *campo úmido* and *campo seco*, meaning wet and dry fields, respectively (Section 2).

Lagomorpha

The only lagomorph known from the Pantanal is the Brazilian rabbit or *tapiti*, *Silvilagus brasiliensis* (Linnaeus, 1758), a crepuscular leporid often observed shortly after sunset along forested stretches of road at elevations above the floodplain. In the Pantanal itself, it is rare, probably because it cannot burrow deeply enough in the wetland to escape its enemies without reaching the upper layer of ground water or without being flooded out of its hiding place by heavy thunderstorms. Its relative scarcity attests to the continued existence of sufficient numbers of predators to exert a natural control on the herbivore populations.

Perissodactyla

The only native member of the Perissodactyla in the Pantanal is the largest native ungulate in the region by weight, the Brazilian tapir or *anta*, *Tapirus terrestris*

(Linnaeus, 1758), which has become rare in the wetland as well as elsewhere in Mato Grosso (Fig. 8.38). This species is a rather docile herbivore, which has generally been displaced from its original habitats by the domestic cattle. The tapirs have generally been confined to the edges of the wetlands, and the only surviving individuals in the region are reputed to live in remote parts of the plateau that have not yet been too greatly modified for agriculture. The species today has to be considered locally endangered.

Figure 8.38: The tapir, *Tapirus terrestris*, has been largely exterminated in the region of the Pantanal, but a few individuals, such as this unusally light colored one, are maintained in small zoos in southern Mato Grosso.

Artiodactyla

Tayassuidae

The white-lipped peccary or *queixada*, *Tayassu pecari* (Link, 1795), is present throughout the Pantanal, but it is not abundant. Some of these animals are kept under semi-domestic conditions on ranches, where they are raised in small herds. Mayer and Wetzel (1987) reported that the wild herds of this species may range from as few as 5 to over 200 individuals. Generally, however, the peccaries are somewhat more common in the upland regions of the watershed, although their numbers are limited, apparently due to hunting, which wiped out this species in many parts of Brazil during earlier years.

Por (1995) reported the presence of a smaller species, the collared peccary or *cateto*, *Tayassu tajacu* (Linnaeus, 1758), which is a good swimmer and inhabits forests along river banks. It frequents the edges of other forests, as well. It was not observed during the study, but it was found to be rather common in the upland regions to the northeast of the Pantanal (Cunha Vieira, 1951).

Suidae

Since its introduction as a farm animal, the domestic pig, *Sus scrofa* Linnaeus, 1758, has developed a feral population and is not uncommon in the Pantanal. Native species in the pig family are not present in the Neotropical Region.

Cervidae

The marsh deer, often called the Pantanal deer or *cervo-do-Pantanal*, *Blastocerus dichotomus* (Illiger, 1815), is the largest South American cervid (Magalhães, 1992). It was observed on many occasions near the Transpantaneira Highway during the study (Plate 13a). Individuals are generally seen alone or in pairs, but, according to Pinder and Grosse (1991), groups of as many as five may form seasonally. They frequently feed on water plants, including *Eichhornia azurea*, and often enter the water to feed. Moving across soft mud is facilitated by their modified feet, consisting of two toes connected by a strong membrane to increase the area of contact with the sediment. They are less shy than the smaller deer species, and can sometimes be approached to within about 20 m, or even closer in a vehicle, before they flee. This lack of caution was first noted by Miller (1930).

The species formerly had a large range in Brazil, but because of habitat destruction, only isolated populations of the species still survive. One of these occurs in a protected area along the Araguaia River System, which encompasses the Bananal wetland, and another is present in the Pantanal (Thornback and Jenkins, 1982). Schaller and Vasconcelos (1978b) estimated the population in the northern Pantanal to be 4000 and decreasing during the 1970s. They reported the entire Pantanal population to consist of about 7000 individuals. Besides habitat destruction, poaching, and competition from domestic animals, the marsh deer suffers from introduced diseases, including the foot and mouth disease.

The red-brocket deer or *veado mateiro*, *Mazama americana* (Erxleben, 1777), is small and much more difficult to approach than the Pantanal deer. The species is abundant in the Pantanal, more so than in the adjacent *cerrado* areas. Generally, they are seen in pairs or, more rarely, in small herds. They remain in wooded areas, crossing open grassland on the floodplain rapidly and cautiously.

The crepuscular *veado campeiro*, *Ozotocerus bezoarticus* (Linnaeus, 1758), also occurs in the Pantanal, although it has become rare all over Brazil due to excessive hunting. Por (1995) reported its presence in the southern part of the Pantanal, and Tomás (1995) observed a considerable number of this species in the Nhecolândia subregion, where the antlers are shed seasonally. He found that the velvet antlers appear overwhelmingly during the months of June through August. This species seems considerably scarcer in the northern part of the Pantanal, although Cunha Vieira (1951) found this species in the upland regions of Mato Grosso to the northeast of the Pantanal in the first half of this century.

A deer of similar size, the gray-brocket or brown-brocket deer, called the *veado catingeiro* in Portuguese, *Mazama gouazoubira* (Fischer, 1814), also occurs in the Pantanal, especially in gallery forests. It is not frequently observed. It should be noted that in the literature, a few authors spell the specific name *gouazoupira*. Apparently, Fischer misspelled the name he intended to use for the species, substituting a "p" for a "b", and since then, there have been two interpretations about which spelling is correct.

Bovidae

Domestic cattle, *Bos taurus* Linnaeus, 1758, were introduced to the Pantanal during the 18th century from the neighboring lands of the Guarani, the present-day Paraguay, which had been organized for the defense of the native South American people against the slave traders by the Jesuits. Since the arrival of the first cattle, which have developed into a typical Pantanal race, several other varieties have been introduced. Among the most abundant in the northern part of the Pantanal at the present time are Brahma cattle. They are presently the dominant terrestrial herbivores, and cattle ranching has had a profound effect on the wetland, as discussed in Section 12.2. In the terrestrial food web, the cattle serve as occasional prey for the jaguar, prompting some ranchers to have these predators killed. While the jaguar is strictly protected by law, enforcement is difficult (Section 13).

A second domestic species has recently been introduced to the Pantanal: the water buffalo, *Bubalus bubalus* (Linnaeus, 1758). These large amphibious mammals are much better suited to life in the wetland than domestic cattle. Nevertheless, experience in Amazonas has shown that their milk and flesh may not be accepted as well as beef products, although they are superior in many ways.

Carnivora

Procyonidae

The racoon family is represented by three species in the region of the Pantanal. The most abundant of these is the South American or ring-tailed coati or *quati, Nasua nasua* (Linnaeus, 1766), which crosses the Pantanal in large family groups and can be encountered almost anywhere. Occasionally, coatis are encountered alone (Fig. 8.39), but more frequently, they remain with their groups. They are omnivores with a preference for food from animal sources, such as eggs, honey, and various invertebrates. They travel both on the ground and in trees, in which they usually rest. They are active in the daytime, but usually sleep during the hot midday hours. In the late afternoon and evening, they are generally under way.

Figure 8.39: Coatis, *Nasua nasua*, usually run in small groups but are sometimes found alone in the Pantanal, such as this female that took refuge in a tree.

A second member of this family is the crab-eating racoon or *guaxinim*, *Procyon cancrivorus* (Cuvier, 1798), which, like its North American congener, frequently seeks food in the water. Cabrera and Yepes (1960) reported its diet to consist of insects, crabs, small vertebrates, and fruit. It usually travels alone and is not frequently seen. Dead individuals were occasionally observed on the highway near Poconé.

The third procyonid that occurs in the region of the Pantanal is the kinkajou, *Potos flavus* (Schreber, 1774), a strictly nocturnal species that remains concealed during the day. The type specimen of the regional subspecies, *P. flavus chapadensis* J. A. Allen, 1904, was probably captured near Chapada dos Guimarães in the watershed of the Cuiabá River (Ford and Hoffmann, 1988).

Felidae

There are several native cat species in southern Mato Grosso, including the largest predatory felid in South America, the jaguar or *onça-pintada*, *Panthera onca* (Linnaeus, 1758). According to Seymour (1989), the subspecies found in the Pantanal region is *P. onca paraguensis* Hollister, 1914. The species originally occurred throughout Brazil, and it is known from regions of all vegetation types in the country. The jaguars are endangered from hunting by man because they have come to feed frequently on cattle. However, the bulk of the diet usually consists of smaller mammals and birds. The movements of the individual or small family groups in the Pantanal can usually be followed by the reports of losses from various ranches and from the footprint sightings. The animals seldom stay for more than a few days in any one area. Using radio telemetry, Schaller and Crawshaw (1980) showed that females moved around within a territory from 25 to 38 km^2, while males wandered within an area of 90 km^2. Schaller (1983) also estimated the population density of jaguars in Brazil to be 1 per 12.5 to 25 km^2.

The jaguar is arboreal, remaining in trees for much of the time and coming to the ground mainly to travel or hunt. It is not shy of water, and it may cross flooded wetlands if the water is not too deep. It is even found on the floating islands of vegetation that form on water bodies (Seymour, 1989). Schaller (1983) reported that caimans, anacondas, stork jabirus, and even duck eggs are consumed by the large cats of the Pantanal, although mammals seem to be of prime importance in the diet.

Another large predator that inhabits southern Mato Grosso is the puma, cougar, mountain lion, or *onça-parda*, *Felis concolor* Linnaeus, 1771, sometimes called *Puma concolor*, represented in Mato Grosso by the subspecies *F. c. acrocodia* Goldman, 1943. This species was not observed in the Pantanal during the study, and its occurrence there is doubtful. It is a species that prefers dry habitats and seems to shun wetlands. It has also suffered reductions in the size of its population due to hunting. Currier (1983) provided information on the diet of the species, indicating that mammals, including domestic species, are the main items in the diet, but birds, fishes, and even insects are also consumed.

Among the smaller cats of the Pantanal are *Herpailurus yagouaroundi* (Desmarest, 1820), the jaguarondi or *gato-mourisco*, which is occasionally seen along the Transpantaneira Highway, especially in the evening. These cats are general predators, and they are not uncommon in southern Mato Grosso. They seem to get along well in the wetlands.

A slightly smaller predator of the Pantanal is the ocelot or *jaguatirica*, *Leopardis pardalis* (Linnaeus, 1758). This spotted cat is normally encountered in forested regions, and it is generally more common in the equatorial regions of South America than in southern Mato Grosso. The limited extent of the forested areas in the Pantanal precludes the development of a large population of this species in the wetland. Ocelots were not observed during the study.

Canidae

There are two rather common dog species in the Pantanal. By far the most common is the crab-eating fox or *cachorro-do-mato*, *Dusicyon thous* Hamilton Smith, 1839, also known as *Cerdocyon thous* (Linnaeus, 1766), according to Ewer (1973). From a distance, it resembles a fox more than a dog, and it has a problematic phylogenetic relationship within the family Canidae (Plate 13b). According to Berta (1982), two subspecies may be encountered in the region of the Pantanal: *D. t. azarae* (Wied, 1824) at the northern edge of the wetland, and *D. t. entrerianus* (Burmeister, 1861) in the central part of the wetland. The great abundance of the species in southern Mato Grosso is attested to by the large number that were observed dead on the highway. In the Pantanal, they were seen traveling during the day, alone or in pairs. Bush dogs are generally observed in woodland, especially in ecotones at the edge of the woods, where it seeks small mammals, frogs, lizards, crabs, eggs, fruit, insects, and carrion (Brady, 1979; Berta, 1982).

The other common canid in the Pantanal is the maned wolf or *lobo-guará*, *Chrysocyon brachyurus* (Illiger, 1804), which is perhaps slightly less tolerant of the moisture than the bush dog. The *lobo-guará* is known as a species of grassland and scrub forest, and it apparently enters the Pantanal during the dry seasons and retreats back to the surrounding high ground as the flood waters advance. This species, although it can still be seen fairly often along the roads just after sundown in the Pantanal, is internationally considered "vulnerable" and is on the endangered species list in Brazil. It is the largest South American canid, reaching a weight of 23 kg and a length of 147 cm (Dietz, 1985). It is an omnivore that feeds mainly on fruit and small mammals.

Another endangered species with a fairly large population in the region of the northern Pantanal is the bush dog, *lobinho*, or *cachorro vinagre*, *Speothos venaticus* (Lund, 1842), which travels in small packs on the floodplains. Individuals were seen from time to time after sunset along the Transpantaneira Highway.

The short-eared dog or *cachorro da mata*, *Atelocynus microtis* (Sclater, 1882), ranges as far south as the northern Pantanal. It is present but uncommon in the Pantanal of Poconé. The species is endangered in Brazil (Berta, 1986).

Mustelidae

The mustelids are well represented in the State of Mato Grosso, but most of the species are terrestrial and seldom enter the Pantanal. Two aquatic species, however, are not uncommon in the region. The largest is *Pteronura brasiliensis* (Gmelin, 1788), the giant otter or *ariranha*. The world population of this species is small enough to cause concern (Bisbal, 1993), but they are commonly observed along the rivers of the northern Pantanal (Schweizer, 1992).

One large family group inhabited the Rio Bento Gomes between Poconé and the place where this river dispersed beneath dense vegetation before entering the Baia Piuval. Members of the group were seen each year, and on one occasion during the dry season, a large part of the group, four adults and five young, was observed on the shore (Plate 14a).

The largest member of the group approached the boat, feigning to attack, while the rest of the otters fled under water. After a short time, several of them began giving their high-pitched call from a good distance away around the river bend, and the one that was watching the boat from a distance of about five to ten meters disappeared underwater and followed them. This kind of behavior was repeated on another occasion, but this time the otters all left the water and disappeared into the dense gallery forest along the shore. A few of them watched the boat for a while and then disappeared into the undergrowth.

The giant otter is a piscivorous mammal that could thrive in the small rivers and permanent lakes of the Pantanal thanks to the large supply of fishes. Its present scarcity can certainly be attributed to the actions of man, including hunting and destruction of the gallery forests in which these animals often seek shelter.

The second aquatic mustelid in the Pantanal is *Lutra platensis* Waterhouse, 1838, the southern river otter or *lontra*. Only individual members of this species were seen in the Rio Bento Gomes near the Transpantaneira Highway. Pairs of family groups were never observed. The species seems somewhat rarer than the giant otter, but it is much smaller and perhaps usually goes unobserved. One individual was seen regularly for about a month on the surface of the water near the highway bridge over the river. Apparently, the otters of this species wander alone over a large territory and do not stay in one place for long.

In addition to the otters, the family Mustilidae is represented by several terrestrial species that are generally more abundant in the dry forest regions than in the Pantanal. The grison or *furão*, *Galictis vittata* (Schreber, 1776), was observed on several occasions on or near the highways in the upland regions of the Pantanal watershed. Apparently, pairs of grisons have the habit of running close side by side, keeping more of less, „in step", with the backs arching at the same time. This interesting behavior was observed twice.

An arboreal species of the family, the tayra or *irara*, *Eira barbara* (Linnaeus, 1758), inhabits forested regions, where it is crepuscular or nocturnal (Magalhães, 1992). There is some confusion about its correct generic name, and Ewer (1973) proposed retaining the incorrect name "*Tayra*", which was originally proposed. The diet consists of much fruit, as well as lizards, agoutis, and small monkeys (Kaufmann and Kaufmann, 1965). In captivity, males reached a weight of 5.3 kg, and females, 3.9 kg (Poglayen-Neuwall, 1978). Tayras were not observed during the study, although their presence in the Pantanal is assumed on the basis of earlier reports from southern Mato Grosso (Cunha Vieira, 1951).

9. Biotic communities

From among the species included in the long lists of flora and fauna encountered in and near the Pantanal, only a minority are actually characteristic species of the wetland. The others were introduced through human activities to special man-made habitats, such as roadsides, cattle pastures, and the grounds of ranch houses, or they are temporary visitors, such as the migratory terrestrial bird species and fishes making long spawning migrations in the rivers, as well as animals from adjacent terrestrial habitats that stray into the wetland, especially during the dry season. In the investigation of the permanent ecological relationships within the Pantanal ecosystem, these transient species and temporary visitors can be omitted. However, it should be noted that these non-resident species may well be surviving and prospering because the Pantanal is there. Even their short-term, temporary interactions with the resident species can be vital to their well being, where they are able to grow in a seral stage maintained by continuous human interference with the climax, take meals en route during long migratory swims or flights, or benefit from occasional incursions into a productive wetland habitat bordering on their own drier one. The zone between the wetlands and neighboring *cerrado* represents a classical ecotone, and this kind of border zone is noted to be particularly rich in biota (Odum, 1971).

In this section, however, we consciously omit the locally rarer plants, terrestrial insects, fishes that pass through the Pantanal on migrations covering hundreds of kilometers in the river systems, reptiles primarily of the *cerrado*, and birds that fly into the region from more preferred habitats elsewhere, whether it be frequently or rarely. We are left with locally abundant species typically associated with the wetland, which can be grouped into well defined biotic communities inhabiting the water bodies and moist adjacent parts of the floodplain, as well as the rivers and streams draining the region.

The biotic communities in the Pantanal are quite distinct from those in most other water bodies and wetlands of the world. There are large contiguous areas covered by water during the wet season that become separated and remain as rivers and shallow lakes when the water level falls during the dry season. There are thus few stable communities that remain relatively unchanged the year round. Apparently, these exceptionally stable ecological units include the gallery forests and the forested islands, called hummocks or *capões de mata* in Portuguese, that perpetually remain above the high water level because their terrain elevation is two to three meters higher than that of the rest of the plain.

The only aquatic communities that can depend upon a degree of habitat stability throughout the year are those inhabiting the Paraguay and Cuiabá Rivers, which are large enough to maintain a steady flow of water throughout the year. These communities, however, must contend with considerable changes in water level and flow rate, and many of the larger species are migratory.

The communities in the other permanent water courses of the Pantanal are subject to fundamental changes during the annual cycle. Even the larger tributaries of the Rio Paraguay crossing the Pantanal become standing water bodies for long periods during the rainless months of the year. This eliminates those elements that require a steady water flow, such as stoneflies, most mayflies, various dragonflies, and dipterans of certain families, including Blephariceridae, Simuliidae, and the majority of the Ceratopogonidae (Hynes, 1970).

During the rainy season, and especially during the *cheia* or high water season, these water courses become deep, fast-flowing streams and rivers. Their flow continues long enough to disrupt the lentic water communities that begin to develop during the dry season. Thus, the smaller rivers present no stable aquatic environment for the establishment of permanent aquatic communities. Under these circumstances, there is a well defined seasonal succession of biotic communities, as discussed in Section 10. Therefore, many of the species aggregations form on a temporary basis, and some of them exist for only a short time during the annual cycle. This should be kept in mind so that the dynamic nature of the system can be understood.

Great areas permanently or temporarily covered by water support aquatic communities that are relatively stable in their activity patterns within the seasonal cycle and are most certainly in more or less the same condition as they were before the arrival of man. The great majority of the aquatic and semi-aquatic plant species are native to South America, and it seems probable that the cosmopolitan species of microorganism had found their way to the region without the help of man. The terrestrial communities, on the other hand, have been greatly modified in many parts of the wetland in attempts to improve the pastures for the cattle. Furthermore, there have been many unintentional introductions of species associated with the increasing tourist traffic, through which the seeds of plants from many other parts of Brazil have found their way into the Pantanal. As a limnological study, the greatest attention will be given to the natural aquatic communities, but mention must also be made of the modified terrestrial communities, some of which occupy the same physical areas as the aquatic ones during the dry seasons.

9.1. Natural communities

Since rice cultivation has still not made inroads into the northern part of the Pantanal and aquaculture in the wetland is still in the experimental phase, all of the aquatic communities described below can be considered to be natural ones. At most, they may be influenced marginally by slight local eutrophication resulting from cattle ranching on the adjacent high ground and on the dry floodplain during the *seca* as well as from the input of sewage into the large rivers and the leaching of minerals from soils greatly enriched by dry season fires. Even the communities in artificial ponds, most of which were left after open pit gold mines were abandoned, show no differences from those in natural water bodies, although there is an obvious succession beginning with the pioneer community that settles in the pools of water that form almost immediately after the open mines are abandoned.

The failure of the aquatic communities present in the Pantanal to conform to the definitions and descriptions of those considered most important in the temperate zones is obvious even after only a cursory inspection (Heckman, 1991). It has long been known that the presently utilized limnological nomenclature, in general, is not satisfactory for use in the tropics (Zafar, 1959). Hence, new names were proposed for the most important biotic communities in the Pantanal that had for many years remained undefined although not unknown (Heckman, 1994b). Individual descriptions of the most prominent aquatic communities are therefore provided together with information on their seasonal development below.

9.1.1. Phytoplankton

Generally, phytoplankton is one of the most stable communities in water bodies of the temperate zone, where it has become adapted to a regular seasonal cycle of activity. In the Pantanal, as well as in much of the tropical wet-and-dry zone throughout the world, this community is only of minor quantitative importance to the ecosystem, being virtually absent from the water bodies for much of the year (Heckman, 1994b). The reason for the impoverishment of the phytoplankton in communities that have stabilized to climax states is the extreme shortage of mineral nutrients in the water during the seasons of heavy rains. These substances are almost completely incorporation into aquatic and littoral tracheophytes, which are adapted to a seasonal period of growth coinciding with the season of flooding. This incorporation of nutrients is accelerated by the great daily supply of solar energy and the adaptation of the plants to utilize the high light intensities it provides for growth at the relatively high ambient temperatures. This growth, discussed in more detail below in the discussion of katharobic communities, withdraws the free nutrients from the water so quickly after the rainy season has begun that the algae have scarcely any time to develop.

In the Pantanal, typically phytoplanktonic algae appear and flourish for a rather brief period after the macrophytes have begun to die off during the dry season. The rapid decomposition of the tracheophytes releases enough nutrients shortly before the water bodies dry up to support a luxuriant bloom of these algae in the shallow pools of water, in spite of water temperatures that sometimes exceed 40°C (Heckman *et al.*, 1993). The species encountered differ from pool to pool, but the most consistently abundant ones belonged to the Chlorococcales and Euglenophyceae, sometimes accompanied by such Cyanobacteria species as *Microcystis aeruginosa* and *Oscillatoria agardhii.* In some of these water bodies, mixed communities of many Chlorococcales species develop, while in others, nearly monospecific blooms of *Microcystis aeruginosa* appear, sometimes together with a surface tension layer bloom of *Euglena sanguinea* or *E. gracilis* (Section 7.2.5.).

In the permanent water bodies along or adjacent to the main rivers, the phytoplankton can develop in a manner reminiscent of that in the temperate zones. However, on the floodplain of the Pantanal itself, the development is quite different from that in the deeper water bodies. Instead of the cells being suspended in the water, they form a green, mud-like sludge on the sediment of pools of water no more than a few centimeters deep, which remain in the temporary streams and drainage channels through which the water passes as the flood waters flow away toward the rivers during the *vazante.* The species present are mainly cosmopolitan and many of them are generally found in phytoplankton (Table 9.1). Intensive growth of phytoplankton organisms in such shallow, warm habitats is scarcely known in the temperate zones. That the algae are active is attested to by the extremely high oxygen concentrations that prevail during the daytime (Heckman *et al.*, 1993). Obviously, the methods frequently used to study planktonic algae in the temperate zones, such as net sampling (Rott, 1981), chlorophyll determination (Nusch & Palme, 1975), and protein analysis (Rausch, 1981), would not prove reliable in this situation because of the shallowness of the flood plain waters and the concentration of the algae on the sediment or on the surface tension layer of the water.

Table 9.1. The dry season assemblage of species that develop in enormous numbers in drying pools of water in the small drainage channels and temporary streams on the floodplain, where the water temperature sometimes exceeds 40°C (Heckman *et al.*, 1993). The assignment of the euglenophyte species to communities is somewhat arbitrary because most of them appear in several different ones at various times of year.

Species	Usual community during other seasons
Cyanobacteria	
Microcystis aeruginosa	Phytoplankton
Agmenellum quadruplicatum	Phytoplankton
Oscillatoria agardhi	Phytoplankton
Oscillatoria lacustris	Benthos
Oscillatoria princeps	Benthos
Oscillatoria limosa	Benthos
Oscillatoria tenuis	Katharobic
Spirulina jenneri	Benthos
Pseudanabaena cf. *constricta*	Katharobic

Species	Usual community during other seasons
Chlorophyta	
Actinastrum hantzschii	Phytoplankton
Coelastrum cambricum	Phytoplankton
Coelastrum intermedium	Phytoplankton
Coelastrum asteroideum	Phytoplankton
Coelastrum microporum	Phytoplankton
Coelastrum pseudomicroporum	Phytoplankton
Crucigenia fenestrata	Phytoplankton
Crucigeniella rectangularis	Phytoplankton
Dicloster arcuatus	Phytoplankton
Eutetramorus fottii	Phytoplankton
Eutetramorus planctonicus	Phytoplankton
Kirchneriella lunaris	Phytoplankton
Pediastrum duplex	Phytoplankton
Pediastrum tetras	Phytoplankton
Scenedesmus armatus	Phytoplankton
Scenedesmus acuminatus	Phytoplankton
Scenedesmus bicaudatus	Phytoplankton
Scenedesmus brasiliensis	Phytoplankton
Scenedesmus carinatus	Phytoplankton
Scenedesmus denticulatus	Phytoplankton
Scenedesmus granulatus	Phytoplankton
Scenedesmus intermedius	Phytoplankton
Selenastrum rinoi	Phytoplankton
Sphaerocystis schroeteri	Phytoplankton
Sphaerocystis planctonica	Phytoplankton
Euglenophyta	
Euglena acus	Katharobic
Euglena deses	Benthos
Euglena fusca	Katharobic
Euglena gracilis	Katharobic
Euglena oxyuris	Benthos
Euglena sanguinea	Epineuston
Euglena spirogyra	Katharobic
Euglena tripteris	Katharobic
Lepocinclis ovum	Katharobic
Phacus brachykentron	Katharobic
Phacus longicauda	Katharobic
Phacus parvulus	Katharobic
Phacus plataleus	Katharobic
Phacus pleuronectes	Katharobic
Phacus suecicus	Katharobic
Phacus tortus	Katharobic
Strombomonas australica	Detrital
Strombomonas chodati	Detrital
Strombomonas encifera	Detrital
Strombomonas gibberosa	Detrital
Strombomonas scabra	Detrital
Trachelomonas abrupta	Detrital
Trachelomonas acanthophora	Detrital
Trachelomonas armata	Katharobic
Trachelomonas bacillifera	Detrital
Trachelomonas cervicula	Katharobic

Species	Usual community during other seasons
Trachelomonas crispa	Detrital
Trachelomonas dubia	Detrital
Trachelomonas elliptica	Detrital
Trachelomonas hispida	Katharobic
Trachelomonas lemmermannii	Detrital
Trachelomonas megalacantha	Detrital
Trachelomonas molesta	Detrital
Trachelomonas robusta	Detrital
Trachelomonas superba	Detrital
Trachelomonas varians	Katharobic
Trachelomonas volvocina	Katharobic
Trachelomonas volvocinopsis	Katharobic

Elsewhere in Brazil, a dense phytoplankton, sometimes consisting of only a few species, is often associated with pioneer communities in impoundments shortly after dams are constructed (Giani & Leonardo, 1988; Altafin et al., 1995). This kind of early seral stage is not encountered anywhere in the northern Pantanal because no artificial lakes or impoundments exist. In the watershed upstream from the Pantanal, however, several small impoundments existed for a short time before they were drained again, and ponds and lakes have formed in former gold mines. These support seral communities with quite a different algal flora, but in southern Mato Grosso, they generally go through a rapid process of eutrophication and transition to terrestrial habitats again after only a few years.

In summary, the role of phytoplankton can be classified as very minor in the nutrient and energy cycles of the Pantanal. It occurrence is seasonally limited to a relatively short period during the dry season when the decomposition of the aquatic macrophytes and sometimes dead fishes and invertebrates, as well, liberates large quantities of nutrients. Its presence during the freshet, when much reproduction of the aquatic fauna occurs, is nevertheless important for the functioning of the ecosystem because it helps nourish the zooplankton, which serves as the main food of many of the hatchling fishes during its own brief period of abundance.

9.1.2. Zooplankton

Like phytoplankton, zooplankton has a limited spatial and temporal distribution in the Pantanal system. It is eliminated during much of the year by a combination of factors, including a great scarcity of nutrients, particularly algae and bacteria, during the high water periods and an enormous predator pressure from the large schools of small fishes that prowl the waters of the Pantanal. The occurrence of zooplankton in notable quantities is thus limited to a short seasonal period during the dry season, when many bacteria and algae are present in the water column and when the predatory activity of the fishes is reduced by considerable mortalities among the stocks, which lose much of their habitat due to desiccation and become subject to increased predation by birds and larger fishes when they take refuge in the permanent water bodies.

The seasonal occurrence of zooplankton is, however, somewhat different than that of phytoplankton. The micro-invertebrates that occur together with the algae in the drying pools are considerably limited in diversity, apparently due to the high water temperatures. This community, somewhat reminiscent of zooplankton, includes several rotifer species which are sometimes encountered in plankton (Kretzschmar *et al.*, 1993), and cyclopoid copepods, which are not planktonic in their habits. The dominant rotifer in this aggregation is *Asplanchna sieboldi*, which is noted as a warm stenothermal element (Table 9.2).

The zooplankton encountered in the permanent water bodies along the main rivers of the Pantanal is quite different in structure (Table 9.3) than the species aggregations in the drying pools (Table 9.2). It includes considerably more crustacean species as well as rotifers that do not appear in the shallow pools on the drying floodplain. This community includes species that are common in South American impoundments. Like the phytoplankton, both of these zooplankton aggregations are subject to severe top-down and bottom-up pressure during much of the year, which virtually eliminates them from the ecosystem. The peak season for zooplankton is the late dry season, when many aquatic macrophytes and animals have died and are decomposing in the water, providing nutrients for both bacteria and small algae. After the freshet, massive fish reproduction begins, subjecting the zooplankton to an ever increasing predator pressure from the large numbers of fry. In addition, the nutrient supply dwindles as the water becomes deeper and the rapidly growing macrophytes come to monopolize the available nutrients once again. During the *cheia*, the standing water on the floodplain is extremely clear and free from both plankton and suspended matter. In the streams and rivers, strong currents carry large amounts of material downstream, making them unsuitable for a plankton development, as well. Only in the peripheral lakes and backwaters of the rivers is the existence of any zooplankton possible during this period, but water samples taken at such locations generally failed to contain any of the planktonic species.

Table 9.2. The predominant microfauna species in the shallow pools of water left in the small drainage channels and temporary streams crossing the floodplain during the *seca*, in which the water temperature sometimes exceeds 40°C. They are mainly rotifers that are abundant elsewhere in zooplankton (Kretzschmar *et al.*, 1993).

Species	Seasons observed
Rotifera	
Brachionus bidentata inermis	*Seca, enchente*
Brachionus patulus	*Seca, enchente*
Brachionus urceolaris	*Seca, enchente*
Keratella americana	*Seca, enchente*
Keratella cochlearis	*Seca, enchente*
Euchlanis dilatata	*Seca*
Asplanchna sieboldi	*Seca, enchente*
Lepadella patella	*Seca, enchente*
Polyarthra vulgaris	*Vazante, seca, enchente*
Trichocerca braziliensis	*Seca, enchente*
Lecane ungulata	*Seca*
Lecane bulla	*Seca*
Lecane lunaris lunaris	*Seca*
Lecane lunaris crenata	*Seca, enchente*
Acari	
Eylais sp.	*Seca*
Arrenurus sp.	*Seca, enchente*

In spite of the relatively minor importance of zooplankton in the dynamics of the ecosystem, it has traditionally been an important element for study in the temperate zones, and the methods to study it have been uncritically transferred to the tropics. Investigations of this community had already begun by the beginning of the century. Daday (1904) investigated the protozoa and metazoans in the water bodies of Paraguay, some distance downstream from the Pantanal, where the water is apparently richer in such organisms. Nevertheless, an extensive literature on the organisms has not been produced, apparently because of the meager yields of the sampling. In addition to the few truly planktonic species collected, a good many organisms that are apparently micronekton are generally in the collections (see below). In the vast Amazon region, many planktonic rotifers and microcrustaceans occur, but relatively few are among the predominant species that exert any quantitative impact on the biotic communities (Robertson and Hardy, 1984). Significantly, it was also found that the zooplankton was most abundant during the dry season, as it is in the Pantanal.

Table 9.3. The species in the zooplankton of the backwaters and lakes along the main rivers and in smaller permanent water bodies. The species of this community become abundant for only a few weeks at the end of the *seca* and beginning of the *enchente* (Heckman, 1994b).

Species	Seasons observed
Rotifera	
Brachionus ahlstromi	*Enchente*
Brachionus quadridentatus	*Enchente*
Keratella americana	*Seca, enchente*
Keratella cochlearis	*Seca, enchente*
Lecane lunaris lunaris	*Seca*
Polyarthra vulgaris	*Vazante, seca, enchente*
Cladocera	
Diaphanosoma birgei	*Seca, enchente*
Sida crystallina	*Seca, enchente*
Ceriodaphnia cornuta	*Seca, enchente*
Bosmina longirostris	*Seca, enchente*
Copepoda	
Notodiaptomus coniferoides	*Seca, enchente*
Thermocyclops minutus	*Seca*
Mesocyclops meridianus	*Seca*

A more detailed study along the floodplains of the Rio Paraná between the Brazilian states of Paraná and Mato Grosso do Sul also demonstrated that there is a great decline of planktonic species during the *cheia* (Lansac Tôha *et al.*, 1993). The disappearance of cladocerans and adult copepods is accompanied by an increase in the abundance of testaceans and the appearance of the nauplii of cyclopoid copepods. The authors, citing the findings of Hardy *et al.* (1984), speculated that the nauplii had hatched from dormant eggs because adults were absent. This would indicate that the active phase of the annual life cycle begins during the *cheia*, explaining the occurrence of the main reproductive period of the adults and probable production of dormant eggs at the end of the *seca* or beginning of the *enchente.*

The reduced importance of zooplankton to the ecosystem in the tropics has also been observed in other tropical regions. For example, Fernando (1980) noted that tropical limnetic zooplankton was less diverse on Sri Lanka than in comparable water bodies of the temperate zones and that cladocerans were particularly poorly represented. It is noteworthy with respect to the reliability of this information that the freshwater fauna of Sri Lanka is better known systematically than that in any other tropical country (Fernando, 1984).

It must be concluded that zooplankton plays a small role in the material and energy cycles of the Pantanal and does not even exist as a detectable community for much of the year. However, its role is important for the dynamics of the system because the peak season of abundance of these species comes at the time of the freshet, when enormous numbers of newly hatched fishes of many species depend on them as a first food after completing the yolk-sac stage.

9.1.3. Katharobic communities

Throughout much of the flood period, a dense underwater growth of macrophytes is observed in the temporarily flooded fields of the Pantanal. This vegetation includes a number of aquatic macrophyte species that develop entirely underwater except for the flowers, which appear above the water surface and submerge again after they are fertilized (Fig. 9.1).

The substrate for this commumity include such species as *Cabomba furcata, Myriophyllum mattogrossense*, *Ludwigia inclinata*, *Utricularia foliosa*, *Utricularia poconensis*, *Utricularia gibba*, *Eleocharis minima*, and *Diodia kuntzei*. The submerged forms of these water plants develop into dense beds among which microorganisms can swim with some protection against the hungry fishes that patrol rapidly through the flood waters. Some of the species, such as the ciliates, *Mesodinium pulex*, *Coleps hirtus*, and *Coleps octospinus*, are somewhat similar to planktonic organisms in their movements, but during the high water period, they are not encountered in open water, but only within or near the macrophyte beds (Hardoim and Heckman, 1996). The *Coleps* species remain especially close to the plants and seldom venture at all into the open water.

Figure 9.1: The katharobic community develops among the masses of submerged plants that develop rapidly on the Pantanal floodplain as soon as they become inundated after the beginning of the rainy season. Here, the submerged leaves of *Ludwigia inclinata* form a reddish mass in which fishes and large invertebrates conceal themselves.

Also among the water plants are some sessile species, organisms that creep on the surfaces of the plants, and species of algae that usually remain enmeshed in the submerged macrophtye masses. These are the dominant elements of the microbiota in the Pantanal for much of the year, including the later half of the *enchente*, the *cheia*, and the early part of the *vazante*. The most abundant algae of these aggregations are solitary desmids, which generally anchor themselves in the plant masses by means of their spines or gelatinous envelopes, filamentous algae including desmids and species of other taxa, testaceans, ciliates with cirri modified for creeping along surfaces, sessile ciliates, rotifers which generally move along surfaces to feed, sessile rotifers, chydorid cladocerans, and pulmonate snails. A list of the species in this community is provided in Table 9.4. Surprisingly, sessile ciliates and ostracods proved to be rare among this species aggregation, although a few species belonging to both of these groups were encountered seasonally.

Table 9.4. The most prominent species in the katharobic community, which are frequently encountered during the period of flooding but seldom become numerous. Many species that are more abundant in other communities but may sometimes also be present among the submerged plants are omitted, although some of the algae belong to this community as well as to the interrhizon. Some of the data are taken from De-Lamonica-Freire and Heckman (1996), Hardoim and Heckman (1996), and Heckman (1998).

Species	Position in the food web
Bacillariophyta	
Melosira varians	Autotroph
Diatoma mesodon	Autotroph
Fragilaria ulna	Autotroph
Gomphonema augur	Autotroph
Gomphonema clavatum	Autotroph
Rhoicosphenia abbreviata	Autotroph
Chlorophyta	
Spirogyra sp.	Autotroph
Mesotaenium mirificum	Autotroph
Spirotaenia condensata	Autotroph
Gonatozygon monotaenium	Autotroph
Gonatozygon pilosum	Autotroph
Gonatozygon aculeatum	Autotroph
Closterium kuetzingii	Autotroph
Closterium lunula	Autotroph
Closterium parvulum	Autotroph
Closterium ralfsii	Autotroph
Closterium setaceum	Autotroph
Closterium tumidum	Autotroph
Closterium turgidum	Autotroph
Closterium venus	Autotroph
Cosmarium candianum	Autotroph
Cosmarium denticulatum	Autotroph
Cosmarium margaritatum	Autotroph
Cosmarium moerlianum	Autotroph
Cosmarium moniliforme	Autotroph
Cosmarium ornatum	Autotroph
Cosmarium pseudopyramidatum	Autotroph
Cosmarium vitiosum	Autotroph
Desmidium aptogonum	Autotroph
Desmidium laticeps	Autotroph
Desmidium quadratum	Autotroph
Euastrum ansatum	Autotroph
Euastrum elegans	Autotroph

Species	Position in the food web
Euastrum evolutum	Autotroph
Euastrum fissum	Autotroph
Euastrum subintegrum	Autotroph
Micrasterias abrupta	Autotroph
Micrasterias borgei	Autotroph
Micrasterias furcata	Autotroph
Micrasterias laticeps	Autotroph
Micrasterias mahabuleshwarensis	Autotroph
Micrasterias radians	Autotroph
Micrasterias radiosa	Autotroph
Onychonema laeve	Autotroph
Pleurotaenium minutum	Autotroph
Pleurotaenium nodulosum	Autotroph
Staurastrum erinacoideum	Autotroph
Staurastrum leptocladum	Autotroph
Staurastrum paradoxum	Autotroph
Staurastrum quadrangulare	Autotroph
Staurastrum rotula	Autotroph
Staurastrum sebaldi	Autotroph
Staurastrum setigerum	Autotroph
Staurastrum striolatum	Autotroph
Staurastrum tohopekaligense	Autotroph
Staurastrum trachytithophorum	Autotroph
Staurastrum trifidum	Autotroph
Staurodesmus convergens	Autotroph
Staurodesmus dickiei	Autotroph
Staurodesmus mamillatus	Autotroph
Staurodesmus maximus	Autotroph
Staurodesmus subulatus	Autotroph
Staurodesmus validus	Autotroph
Xanthidium antilopaeum	Autotroph
Xanthidium convergens	Autotroph
Xanthidium mamillosum	Autotroph
Xanthidium mucronulatum	Autotroph
Xanthidium trilobum	Autotroph
Euglenophyta	
Euglena tripteris	Autotroph, heterotroph
Phacus tortus	Autotroph, heterotroph
Sarcodina	
Arcella conica	Consumes microflora
Arcella megastoma	Consumes microflora
Arcella vulgaris	Consumes microflora
Centropyxis discoides	Consumes microflora
Centropyxis platystoma	Consumes microflora
Centropyxis spinosa	Consumes microflora
Cucurbitella mespiliformis	Consumes microflora
Difflugia corona	Consumes microflora
Difflugia oblonga	Consumes microflora
Difflugia tuberculata	Consumes microflora
Awerintzewia cyclostoma	Consumes microflora
Euglypha acanthophora	Consumes microflora
Actinosphaerium eichhorni	Consumes microfauna and algae

Species	Position in the food web
Ciliophora	
Coleps hirtus	Consumes microflora
Coleps octospinus	Consumes microflora
Coleps nolandi	Consumes microflora
Mesodinium pulex	Consumes microflora
Frontonia leucas	Consumes algae
Frontonia acuminata	Consumes microorganisms
Urocentrum turbo	Consumes bacteria
Halteria grandinella	Consumes microflora
Strobilidium gyrans	Consumes bacteria
Paruroleptus piscis	Consumes bacteria
Euplotes patella	Consumes microflora
Aspidisca lynceus	Consumes bacteria
Aspidisca costata	Consumes bacteria
Rotifera	
Lepadella triptera	Consumes microflora
Colurella uncinata	Consumes microflora
Colurella obtusa	Consumes microflora
Lecane lunaris	Consumes microflora
Annelida	
Stylaria fossularis	Consumes algae and microfauna
Conchostraca	
Cyclestheria hislopi	Consumes microorganisms
Cladocera	
Simocephalus exspinosus	Consumes microflora
Moina micrura	Consumes microflora
Chydorus sp.	Consumes microflora
Decapoda	
Macrobrachium amazonicum	Consumes microorganisms
Insecta	
Argia sp.	Preys on arthropods
Buenoa salutis	Preys on arthropods
Buenoa amnigenus	Preys on arthropods
Martarega gonostyla	Preys on arthropods
Pelocoris impicticollis	Preys on arthropods

9.1.4. Benthos

The smaller benthic organisms in the Pantanal are generally forced by predator pressure to live beneath the surface of the sediment in endobenthic habitats. Larger species with various forms of protection, such as the pilid snails and trichodactylid crabs, are able to live as epibenthic species, at least part of the time. These free-moving species are discussed with the tacheion (Section 9.1.7). Epibenthic algae are also rare in the Pantanal because the sediment is fine and provides no secure surface for development. Furthermore, the bottom is often strongly shaded by macrophytes, limiting photosynthesis by benthic algae. That the formation of a benthic algal community is not prevented by an absence of suitable species was demonstrated at a site in the Rio Bento Gomes near the Poconé exposition park, where small stones were dumped along the shore to provide a secure bottom for launching boats. On the surface of these stones, a rich community of epibenthic algae developed. It included mainly diatoms (Section 7.2.3.). Similar epibenthic communities form immediately after the flooding of seasonally dry areas during the early *enchente*. These disappear after the hard-baked inundated sediment has become soft again and the rapid growth of the submerged macrophytes has robbed the epibenthic community of both light and nutrients.

At sites where human activity has either largely eliminated the natural aquatic communities, such as at sites from which fill was obtained, or produced an artificial water body where none had existed before, such as at abandoned open pit gold mines, a distinct succession is observed, during which a short-lived benthic community of pioneer species develops. Among the first organisms to arrive are benthic chironomids, which appear in large numbers, although they are otherwise not abundant in the Pantanal. Other aquatic insects can be encountered in these water bodies, as well. Adult corixids and water beetles together with larval dragonflies are among the first arrivals. Various benthic algae also begin to develop, generally large numbers of few species. Large numbers of amphibian larvae usually appear early in the succession, as well.

After the community has become better established, it usually includes various an increasing number of microorganisms that feed on algae as well as the pioneer insect and amphibian larvae. On the floodplain, where the water bodies usually dry out completely during the dry season, this benthic community remains much less diverse in species than the upland streams flowing toward the Pantanal. Generally, the invertebrates specialized to life in ephemeral water bodies have the ability to burrow into the sediment before it dries up and remain dormant until the next period of flooding (Section 8; Kretzschmar and Heckman, 1995; Heckman, 1998).

Benthic algae were obtained on glass cover slips inserted in the sediments of streams to the north of the Pantanal, most frequently in the Corrego Guanandi near a large alcohol distillery that produced fuel for automobiles before it was closed down during the final year of the project. Such artificial substrates have long proven satisfactory for investigating aufwuchs organisms (Grzenda and Brehmer, 1960; Sládecek and Sládeckova, 1964; Kevern *et al.*, 1966; Neal *et al.*, 1967; Dumont, 1969; Riedel-Lorje, 1981)). However, inserted in the sediment, they also served to attract many vagile benthic organisms, including pennate diatoms, motile ciliates, and rotifers. These organisms were somewhat less common on the sediments of the small permanent water courses draining the Pantanal wetland during the high water period and much less abundant on the parts of the floodplain that are inundated for less than half of the year. Algae were especially poorly represented, but a few species developed massive populations during the early dry season and other appeared only a matter of hours after the areas were flooded again by rainwater. That many of these species are euglenophytes may reflect the ability of some species in this group to seek an aestivation site in the sediment when conditions become adverse. Complex migration patterns into and out of sediments by species of *Euglena* in estuaries were related to the tidal water movements in Europe (Heckman, 1985). An examination of many glass slides left for various periods of time in streams and rivers of the Pantanal and its watershed revealed a special feature related to the geological peculiarities of the region: the presence of many iron bacteria (Section 7.1.2).

Several estimates of the zoobenthos standing stock in the Pantanal of Mato Grosso do Sul have been provided by Fukuhara and Mitamura (1985) and by Fukuhara and Henry (1987) for a Pantanal lake in Mato Grosso. Because of the great seasonal variability and the existence of many special microhabitats, the value of any numbers produced by such studies is minimal. These values can be expected to vary enormously not only from place to place but also according to the season. Por (1995) judged these results to be "inconclusive". His comments suggest once again that methods developed in the temperate zones are fully inadequate for investigating tropical systems, and many of them are questionable in the temperate zones, as well.

9.1.5. Interrhizon

During the flood periods, most activity by small invertebrates and some small fishes takes place within the shelter of the root or rhizoid systems of the floating macrophytes (Plate 3a). This habitat offers distinct advantages to its biotic community, which was named interrhizon and described in detail by Heckman (1994b). It was recognized as one of the most important biotic communities in tropical water bodies. The importance of the roots of floating plants for the populations of microorganisms had already been noted by many authors (Dioni, 1967; Koste, 1975; Heckman, 1979). There are considerable differences in the variety and abundance of species that inhabit the rhizoid or root systems of the different water plant species, with the richest interrhizon community of the Pantanal inhabiting the rhizoid masses of *Salvinia auriculata*. These rhizoids are particularly dense and soft, permitting the small organisms to conceal themselves well from the many predators in the open water. From the density of the floating plant masses and the large areas of the water's surface that they cover (Fig. 9.2), it is obvious why the interrhizon community has such a great importance in tropical water bodies. In the Pantanal, the density of the root and rhizoid masses in the water becomes continually greater as the water recedes during the *vazante*, and during particularly dry years, much of the floating vegetation dies off. However, recovery is rapid during the *enchente,* and a suitable habitat for the interrhizon community is present throughout the entire year.

Pistia stratiotes is also a floating macrophyte that shelters a variety of species in its interrhizon community (Dunn, 1934). Among the organisms noted among the roots of this plant in Argentina was *Belostoma micantulum*, a species also encountered in the Pantanal (Poi-de-Neiff, 1983). In addition, Fairchild (1940) noted that the larvae of the widespread tabanid, *Lepiselaga crassipes*, are typically associated with *P. stratiotes*.

Figure 9.2: The interrhizon community develops among the roots and rhizoids of the floating plants, which cover vast areas of the Pantanal. The submerged parts of the different plants species harbor different animals according to the density of the roots. Here, these plants completely cover the Rio Bento Gomes near one of the bridges of the Transpantaneira Highway during the dry season. The predominant species, *Salvinia auriculata*, has particularly dense rhizoids among which a variety of invertebrate species seek hiding places from predators.

The habitat of the interrhizon provides many of the species inhabiting it with food and oxygen as well as concealment. Food is present in the form of detritus produced by the plants or filtered from the water by their feathery underwater structures. In addition, the roots of some of the plants transport atmospheric oxygen through the aerenchyme tissue. Enough oxygen passes through this tissue to supply the plant cells during periods of oxygen deficiency in the water and to diffuse into the surrounding water, as well. The importance of these processes for fishes and other aquatic animals that must survive nocturnal oxygen shortages at various times of year was demonstrated by Jedicke *et al.* (1989).

Some of the species most frequently found among the interrhizon in the Pantanal are listed in Table 9.5. Few of these species were ever observed in the open water or on the sediment, but they were often collected by removing the floating plants from the water and beating the submerged parts vigorously on the bottom of a shallow pan. The animals so removed attempted to return to the nearest plant mass as quickly as possible, showing a strong affinity for their habitat. Survival of these species in the open water during the *cheia* and *vazante* for more than a few seconds would have been doubtful due to the great abundance of hungry fishes. Thus, this behavior pattern is an essential one for survival, adapting the animals to life in the interrhizon.

Table 9.5. Typical species encountered in the interrhizon community in the Pantanal of Poconé and their observed or presumed roles in the food web. Some of the species also occur in the katharobic community among the submerged macrophytes. Sources of the data include De-Lamonica-Freire and Heckman (1996) and Heckman (1998).

Species	Food source	Consumed by
Anabaena azollae	Autotroph	Unknown
Gomphonema augur	Autotroph	Microfauna
Surirella biseriata	Autotroph	Microfauna
Spirotaenia condensata	Autotroph	Microfauna
Gonatozygon pilosum	Autotroph	Microfauna
Gonatozygon monotaenium	Autotroph	Microfauna
Closterium abruptum	Autotroph	Microfauna
Closterium dianae	Autotroph	Microfauna
Closterium lunula	Autotroph	Microfauna
Closterium venus	Autotroph	Microfauna
Bambusina borreri	Autotroph	Microfauna
Cosmarium bireme	Autotroph	Microfauna
Cosmarium candianum	Autotroph	Microfauna
Cosmarium commissurale	Autotroph	Microfauna
Cosmarium denticulatum	Autotroph	Microfauna
Cosmarium ginzbergeri	Autotroph	Microfauna
Cosmarium hexagonum	Autotroph	Microfauna
Cosmarium margaritatum	Autotroph	Microfauna
Cosmarium moniliforme	Autotroph	Microfauna
Cosmarium ornatum	Autotroph	Microfauna
Cosmarium orthostichum	Autotroph	Microfauna
Cosmarium pseudoconnatum	Autotroph	Microfauna
Cosmarium pseudopyramidatum	Autotroph	Microfauna
Cosmarium punctulatum	Autotroph	Microfauna
Cosmarium refringens	Autotroph	Microfauna
Cosmarium subspeciosum	Autotroph	Microfauna
Cosmarium supraspeciosum	Autotroph	Microfauna
Cosmarium tinctum	Autotroph	Microfauna
Cosmarium triangulare	Autotroph	Microfauna
Cosmarium trilobatum	Autotroph	Microfauna
Desmidium aequale	Autotroph	Unknown

Species	Food source	Consumed by
Desmidium aptogonum	Autotroph	Unknown
Desmidium baileyi	Autotroph	Unknown
Desmidium graciliceps	Autotroph	Unknown
Desmidium laticeps	Autotroph	Unknown
Docidium baculum	Autotroph	Unknown
Euastrum elegans	Autotroph	Microfauna
Euastrum evolutum	Autotroph	Microfauna
Euastrum fissum	Autotroph	Microfauna
Euastrum gemmatum	Autotroph	Unknown
Euastrum pectinatum	Autotroph	Unknown
Euastrum subintegrum	Autotroph	Microfauna
Groenbladia undulata	Autotroph	Unknown
Hyalotheca dissiliens	Autotroph	Unknown
Micrasterias abrupta	Autotroph	Unknown
Micrasterias arcuata	Autotroph	Unknown
Micrasterias borgei	Autotroph	Unknown
Micrasterias furcata	Autotroph	Unknown
Micrasterias laticeps	Autotroph	Unknown
Micrasterias papillifera	Autotroph	Unknown
Micrasterias pinnatifida	Autotroph	Unknown
Micrasterias radians	Autotroph	Unknown
Micrasterias radiosa	Autotroph	Unknown
Micrasterias ringens	Autotroph	Unknown
Micrasterias thomasiana	Autotroph	Unknown
Micrasterias torreyi	Autotroph	Unknown
Micrasterias truncata	Autotroph	Unknown
Onychonema laeve	Autotroph	Unknown
Pleurotaenium minutum	Autotroph	Microfauna
Pleurotaenium nodosum	Autotroph	Microfauna
Pleurotaenium nodulosum	Autotroph	Microfauna
Pleurotaenium ovatum	Autotroph	Microfauna
Spondylosium ellipticum	Autotroph	Microfauna
Staurastrum erinacoideum	Autotroph	Unknown
Staurastrum gladiosum	Autotroph	Unknown
Staurastrum leptocladum	Autotroph	Unknown
Staurastrum muticum	Autotroph	Unknown
Staurastrum nudibrachiatum	Autotroph	Unknown
Staurastrum quadrangulare	Autotroph	Unknown
Staurastrum rotula	Autotroph	Unknown
Staurastrum sebaldi	Autotroph	Unknown
Staurastrum setigerum	Autotroph	Unknown
Staurastrum striolatum	Autotroph	Unknown
Staurastrum tohopekaligense	Autotroph	Unknown
Staurastrum trifidum	Autotroph	Unknown
Staurodesmus dickiei	Autotroph	Unknown
Staurodesmus mamillatus	Autotroph	Unknown
Staurodesmus subulatus	Autotroph	Unknown
Staurodesmus validus	Autotroph	Unknown
Teilingia granulata	Autotroph	Unknown
Xanthidium antilopaeum	Autotroph	Unknown
Xanthidium mamillosum	Autotroph	Unknown
Xanthidium mucronulatum	Autotroph	Unknown
Xanthidium trilobum	Autotroph	Unknown
Penium spirostriolatum	Autotroph	Unknown
Euglena acus	Autotroph, heterotroph	Microfauna

Species	Food source	Consumed by
Euglena fusca	Autotroph	Microfauna
Euglena spirogyra	Autotroph	Microfauna
Amoeba proteus	Microorganisms	Microfauna
Difflugia corona	Microorganisms	Fishes
Spirostomum minus	Bacteria	Microfauna
Vorticella campanula	Bacteria	Microfauna
Catenula lemnae	Microfauna	Unknown
Microstomum lineare	Microfauna	Unknown
Gyratrix hermaphrotidus	Microfauna	Unknown
Trichotria tetractis	Microscopic particles	Small fishes
Mytilina mucronata	Microscopic particles	Small fishes
Macrochaetus sericus	Microscopic particles	Small fishes
Lepadella patella	Microscopic particles	Small fishes
Lepadella triptera	Microscopic particles	Small fishes
Colurella obtusa	Aufwuchs	Invertebrates
Colurella uncinata	Aufwuchs	Invertebrates
Scaridium longicaudum	Microscopic particles	Invertebrates
Floscularia melicerta	Microscopic particles	Unknown
Rotaria rotatoria	Microscopic particles	Small fishes
Chaetonotus murrayi	Microscopic particles	Unknown
Chaetonotus larus	Microscopic particles	Unknown
Lepidodermella squamata	Microscopic particles	Unknown
Polymerurus callosus	Microscopic particles	Unknown
Dero (Aulophorus) borellii	Microscopic particles	Unknown
Macrobrachium amazonicum	Filtered particles	Large fishes
Helveciagrion simulacrum	Arthropods, Annelids	Fishes
Homeoura chelifera	Arthropods, Annelids	Fishes
Helveciagrion chirihuanum	Arthropods, Annelids	Fishes
Erythrodiplax umbrata	Arthropods, Annelids	Fishes
Tenagobia incerta	Tiny animals	Unknown
Pelocoris politus	Small animals	Hemipterans
Belostoma micantulum	Insects, Vertebrates	Hemipterans
Suphisellus rufipes	Arthropods	Fishes
Suphisellus grammicus	Arthropods	Fishes
Suphisellus rufulus	Arthropods	Fishes
Hydrocanthus debilis	Arthropods	Fishes
Hydrocanthus laevigatus	Arthropods	Fishes
Hydrocanthus socius	Arthropods	Fishes
Suphis minutus	Arthropods	Fishes
Anodocheilus silvestrii	Arthropods	Fishes
Anodocheilus sarae	Arthropods	Fishes
Berosus truncatipennis	Detritus, algae	Fishes
Berosus fratellus	Detritus, algae	Fishes
Berosus pleuropunctatus	Detritus, algae	Fishes
Derallus rudis	Detritus, algae	Fishes
Lepiselaga crassipes	Detritus	Fishes
Serrasalmus spilopleura	Fishes	Fishes, Birds
Rineloricaria parva	Microorganisms	Caiman, Birds

Unexpectedly, the number of aufwuchs species in the interrhizon community proved to be low, and their role was minor. Many more sessile species were found attached to the rocks in streams than on the submerged parts of the macrophytes, as discussed above in the section on the benthic community. During the period of flooding, desmids are the commonest algal species encountered, and these enmesh themselves among the tracheophytes, their spines providing some anchorage. The filamentous species are typically entwined among the roots, stems, and rhizoids of the macrophytes. Glass slides exposed in the open water without contact with the sediment during this season were poorly settled by sessile species larger than the iron bacteria, *Siderocapsa* spp.

Only a few of the fishes have adapted to feeding directly within the interrhizon as adults, although the hatchlings of various species can often be found within the root masses, benefiting from the presence of edible microorganisms and from the same protection against predation by larger fishes afforded the invertebrates. Even young piranhas, *Serrasalmus spilopleura*, were found by Sazima and Zamprogno (1984) to seek out the interrhizon beneath floating beds of *Eichhornia crassipes* for shelter and foraging. One of the reasons this community is shunned by medium-sized fishes is the abundance of predators specializing in catching relatively large prey that are camouflaged for concealment among vegetation on the surface of the water. A variety of belostomatids, including *Lethocerus grandis* and *L. maximus*, which can exceed 10 cm in length, remain motionless among the plants at the surface to await fishes that approach the plant masses too closely. Species of *Ranatra* and naucorids have the same habits but capture prey in smaller size classes.

The interrhizon is the nursery for the juveniles and larvae of many species capable of surviving in the open water or in the terrestrial community as adults. In addition to the young fishes, mentioned above, larval prawns, *Macrobrachium amazonicum*, seek out this community for shelter (Collart and Moreira, 1993), and the aquatic larvae of the lampyrid beetle, *Photinus* sp., are also abundant together with many insects that remain in the community both as larvae and as adults (Section 8.12).

9.1.6. Pseudoterrestrium

A special kind of aquatic community in the Pantanal inhabits the emergent parts of aquatic macrophytes (Plate 6b & Fig. 9.3). The species present belong mainly to higher taxa comprised mainly of terrestrial species. However, the members of this community must be considered truly aquatic, like the plants on which they live. They are not found

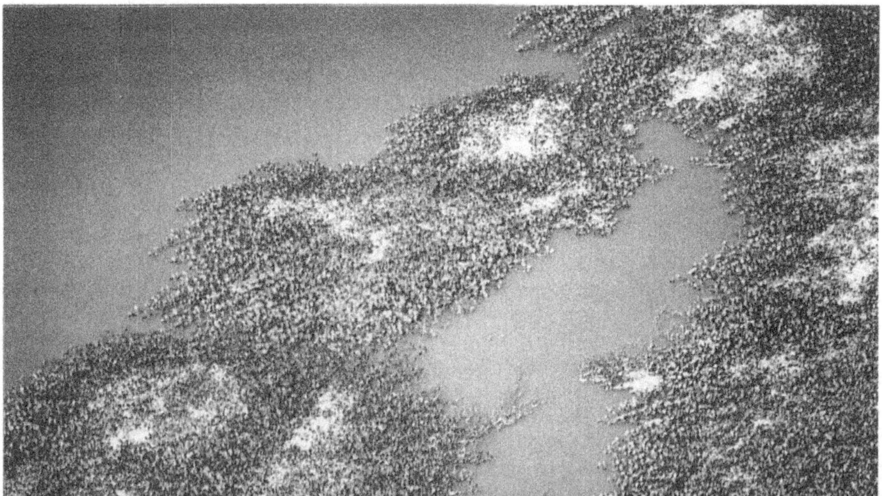

Figure 9.3: The pseudoterrestrial community develops among the emergent parts of the aquatic plants that cover vast shallow water areas of the Pantanal while it is flooded and often form floating islands on deeper water bodies.

on shore and generally do not flee onto the shore if disturbed. Nearly all of them enter the water and hide among the rhizoids of the plants if pursued, and many of the insects can sometimes be seen swimming from plant to plant underwater. They therefore must be considered highly specialized inhabitants of the floating islands of plants and other emergent, aquatic vegetation. The most prominent species of this community are shown in Table 9.6.

While the great majority of the species in the interrhizon among the submerged parts of the plants belong to food webs which depend on algae or saprobic bacteria and fungi as their prime source of energy, the macrophytes themselves are the primary producers that supply their energy directly to the members of the pseudoterrestrium. Grasshoppers, aphids, weevils, and chrysomelids are among the herbivores that live directly from the upper parts of the emergent plants and pass their energy along to the spiders, coccinelid beetles, tree frogs, and jacanas that feed on them. Some interesting interactions with the underwater species occur, however. The two common water snails of the genus *Pomacea* leave the water at night to deposit their egg masses on the emergent parts of aquatic plants, where the eggs remain until they hatch (Fig. 9.4).

Table 9.6. The most prominent members of the pseudoterrestrium community in the Pantanal and their positions in the food web. In addition to the permanent members of the community, short-lived adults of many aquatic insects, such as mayflies, caddisflies, and chaoborids, rest in the pseudoterrestrium during their reproductive periods.

Species	Position in the food web
Araneae	
Actinosoma pentacanthum	Consumer of arthropods
Lycosidae sp. indet.	Consumer of arthropods
Tetragnatha nitens	Consumer of arthropods
Insecta	
Deuterosminthurus richardsi	Consumes pollen and detritus
Helveciagrion simulacrum	Consumes small arthropods
Helveciagrion chirihuanum	Consumes small arthropods
Ischnura capreolus	Consumes small arthropods
Homeoura chelifera	Consumes small arthropods
Diastatops intensa	Consumes arthropods
Cornops aquaticum	Herbivore, consumes macrophytes
Marellia remipes	Herbivore, consumes macrophytes
Paulinia acuminata	Herbivore, consumes macrophytes
Argizala sp.	Herbivore, detritivore
Cicadellidae, spp. indet.	Herbivores, juices of macrophytes
Aphididae, spp. indet.	Herbivores, juices of macrophytes
Coleomegilla maculata	Consume aphids
? *Adelia* sp.	Consumes aphids
Staphylinidae, spp. indet.	Consume arthropods, detritus
Neochetina eichhorniae	Herbivore, consumes *Eichhornia* spp.
Ludovix cf. *longirostris*	Consumes insect eggs
Amphibia	
Hyla raniceps	Consumer of arthropods
Aves	
Jacana jacana	Consumer of arthropods, amphibians

The pseudoterrestrial community is simpler to investigate than the submerged communities, not only because its trophic relationships are less complex and do not include many microorganisms but also because it is easier to observe from a boat or even from the shore. Most of the animals are large enough to be seen with binoculars, and those that are not, can be observed under a dissecting microscope on floating plants freshly collected in the field. Preliminary experiments in the field showed that quantitative studies can be conducted using insects marked with harmless colors. Statistical evaluations of the results are possible using the same methods as those used for marking fishes (Pauly, 1984; Burnham *et al.*, 1987).

The pseudoterrestrium is a well defined community without much contact with other species in adjacent habitats. The animals are extremely rare on terrestrial plants, even those in the littoral communities beside the permanent water bodies, and the only interrelationships with other food webs occur when a passing bird or a large fish or snake consumes one of the larger insects or tree frogs in this community. As already discussed in Section 8, one of the common terrestrial egrets in the Pantanal, *Bubulcus ibis*, seems to have become adapted to feeding on pseudoterrestrial animals in Argentina since its introduction from Africa.

Figure 9.4: The egg mass of *Pomacea lineata* is attached to the emergent parts of water plants well above the water level by the snail during the night.

9.1.7. Tacheion

The tacheion includes freely moving species, mainly fishes and prawns, as well as those organisms that creep along the bottom or over plants. The term encompasses nektonic species as well as large, free-moving epibenthic elements. Nearly all of the fish species belong in this class of organisms, and, strictly speaking, so do the caimans and birds that dive and catch fishes underwater. Some of the small tacheion species, particularly cyclopid copepods, are often confused with plankton, and it is frequently not easy to decide whether a particular species should be considered planktonic or nektonic. Generally, the calanoids can be considered planktonic, with a few exceptions, while most cyclopoids and all harpacticoids should be classified as micronekton according to their habits.

The most important elements of the macrotacheion in the Pantanal can be considered nektonic. The nekton community shows a pronounced seasonal succession, with small micronekton elements becoming predominant during the dry season, then giving way to rapidly growing juvenile fishes during the *enchente.*

The great majority of the ciliate, rotifer, gastrotrich, and microcrustacean species observed during the study are members of the microtacheion. They do not remain suspended in the water column but rather move among water plants, on surfaces, or along the bottom, entering the water column only to seek food. Within this group are animals that occupy a variety of ecological niches, which should be defined and described in detail. Classification systems are still far from perfect, and it is evident that considerable conceptual overlap exists between tacheion and both the katharobic and epibenthic communities. In fact, tacheion is a concept based on the habits of the species, while the definitions of the communities are derived from the interrelationships of organisms living in close proximity. The separate discussion of tacheion is necessary because it includes some large organisms that do not clearly fit into any well-defined community.

Like the planktonic organisms, the elements of the micronekton enjoy a peak period of development during the late dry and early rainy season after mass mortalities have eliminated large numbers of the fishes and before the next generation of young fishes has appeared. Among the most abundant species are the small herbivores and predators that conceal themselves among beds of detritus, submerged water plants, or in the substrate and swim into the open water to seize food objects. Prominent among these are many of the copepods and a few of the rotifers. They are much more abundant in the permanent water bodies than in the shallow waters of the flooded plain and are apparently absent or nearly so during the high water period, when there are strong currents in the streams and rivers. Among the species with these habits encountered in the Pantanal are *Macrocyclops albidus*, *Eucyclops neumani*, *Paracyclops fimbriatus*, *Mesocyclops longisetus*, and species of *Microcyclops*. Extensive species lists for South America were provided by Reid (1985), and the geographical distribution of the species outlined in this work suggests that there is a rich copepod fauna in the Pantanal, although many of the species are probably confined to the southern part of the wetland. One abundant cladoceran, *Chydorus* sp., and one conchostracan, *Cyclestheria hislopi*, also display these habits. Several other species of the family Chydoridae are also present among the facultative micronekton.

Among the prominent micronektonic rotifer species are *Testudinella patina*, *Lecane lunaris*, and *Lecane bulla*. These species are found most frequently in the open water among submerged plants, which apparently provide them with some concealment from larger predators. Throughout most of the year, rotifers are numerically scarce in the water samples, and the number of species observed was very low. This is understandable in view of the large schools of small fishes that were constantly patrolling the shallow waters in search of any edible items.

Like the planktonic species, most of the micronektonic species seem to be omnivores, consuming combinations of algae and small animals or bacteria and algae, depending upon the size of the food particles they prefer. An exception to this seems to be the water mites, the most common of which are the *Arrenurus* species, which are parasitic on insects as larvae and predators as adults.

The larger nektonic species include occupants of a variety of feeding niches. The prawn, *Macrobrachium amazonicum*, filters small particulate matter, while most of the small fishes consume items somewhat larger in size. Among the relatively small fish species, the majority seem to be omnivores, consuming a variety of aquatic prey as well as plant and animal matter that falls into the water. The larger characoids seem to consume more animal than plant material, and the piranhas are apparently almost exclusively carnivorous. The habit of biting chunks out of the flesh of other fishes or feeding on their fins and scales is a highly specialized feeding habit displayed by a few species, such as *Serrasalmus marginatus*, as mentioned earlier (Section 8.13).

In addition to the nektonic species, the tacheion encompasses vagile epibenthic elements, some of which burrow at times into the sediment. Among the most abundant of these burrowing species are the trichodactylid crabs, of which *Trichodactylus parvus* seems to be the most abundant in the wetlands near Poconé. The burrowing fish, *Synbranchus marmoratus*, is another such benthic species. The abundant, large herbivorous snails, *Pomacea linearis*, *P. scalaris*, and *Marisa planogyra*, seem to burrow only during the dry season to escape desiccation. During the rest of the year, they creep over the bottom in shallow water or move across the underside of the floating vegetation or on the submerged parts of the emergent rooted plants. At night, the *Pomacea* species leave the water to deposit their egg masses on the emergent parts of the plants (Fig. 9.4), while *Marisa planogyra* spawns on the submerged parts of the same plant species.

9.1.8. Natural terrestrial communities of the wetland

There are a considerable number of terrestrial communities that should be investigated intensively on an individual basis. Cursory descriptions of the most prominent of these can already be provided, however (Fig. 9.5), and many of the species constituting their biota have been mentioned in the preceding sections.

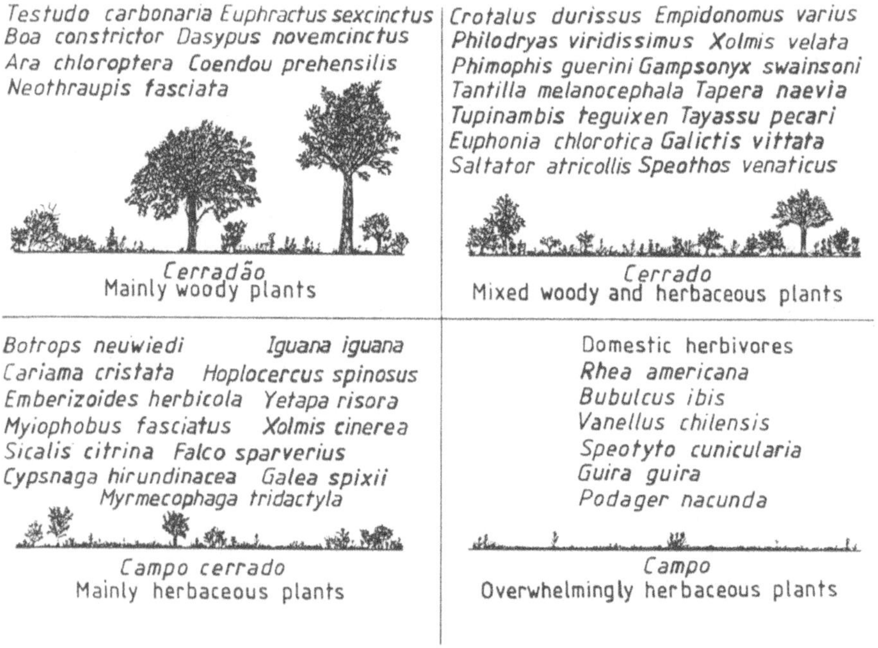

Figure 9.5: Four kinds of terrestrial habitat referred to generally as *cerrado* in the region of the Pantanal. However, the *cerrado* sensu lato has been subdivided into the seral stages illustrated here, as defined by Coutinho (1990). Some of the characteristic vertebrate species that are most abundant in each kind of habitat depicted here are listed.

As mentioned earlier, the terrestrial community that probably covers the greatest area in the Pantanal consists of short-lived annual herbs and modified aquatic plants that occupy the floodplain at the time it is dry during the rainless seasons. This natural community, however, has been considerably modified in many places by the introduction of plants, mainly grasses, from other continents in the hopes of improving the pasture. However, the majority of the species in this community, which forms a rapidly moving, dynamic ecotone (Heckman, 1997b), are specialized macrophytes and a few accompanying littoral animals, which are native to the region of the Pantanal. Generally, the higher the elevation of the site, the greater is the number of introduced species. In contrast, the somewhat lower areas that remain flooded for most of the year support few, if any, introduced species among the flora.

The main terrestrial communities in the upland parts of the catchment areas have been discussed in Section 2.1.6, while those predominant at higher elevations in the wetland itself were mentioned in Section 3.1.2. The problem in recognizing which communities are natural and which artificial is created by the deliberate burning of large areas during the dry season. Fires have been used to clear the land since colonial times and even perhaps in the Pre-Columbian period. Hence, the term "fire climax" has been used for the *cerrado*, as mentioned earlier.

Figure 9.5 provides a diagrammatic depiction of some species most characteristic of the upland *cerradão*, *cerrado*, *campo cerrado*, and *campo* communities. Typical vegetation is encountered in each of the various upland habitats. From Figure 9.6, showing the richness of the flora along water courses contrasted with that of the adjacent savanna woodland, the impact of the long dry periods is evident.

Figure 9.6: In the catchment area of the Pantanal, streams are invariably lined by a vegetation particularly rich in species, showing the effect of the dry months on the species diversity of areas watered only by rainfall.

9.1.9. The land-water ecotone

The zone between the aquatic and terrestrial habitats has long been known among scientists to be of immense importance for the ecological processes keeping the biota of both rich and diverse (Heckman, 1982b). The periodic flooding of extremely large areas along the shores of the water courses flowing through the Pantanal serves to make this wetland an extreme case, illustrating the great role seasonal flooding plays in the distribution of nutrient elements and promoting reproduction and growth. This was recognized over half a century ago by Krieg (1936). The impact of the surrounding terrestrial environment on the biota in the rivers was described later in the Amazonian region by Sioli (1957), and faunistic studies by Adis and Righi (1990) in the same region have revealed some of the special adaptations that enable animals to take advantage of the rich but seasonally varying nutrient supplies. The special richness of the biota in the Rio Negro System in spite of its extremely nutrient-poor water, for example, is in no small way explained by the seasonal flooding of the forests along its margins, called the *igapó* (Goulding *et al.*, 1988). Although the *igapó* vegetation displays a relatively low diversity, it supports a rich invertebrate fauna (Adis, 1984). When compared to the *várzea*, however, the *igapó* clearly shows an effect of the low total nutrient supply, although it is considered particularly rich in endemic species and a center of évolutionary activity (Erwin and Adis, 1982). The *várzea* is the corresponding floodplain along the white water rivers of Amazonia, which are much richer is mineral nutrients (Sioli, 1984).

Although the general importance of the littoral ecotone is confirmed by the studies in the Amazon Basin, it would be misleading to try to apply any findings from one region to the other. The Pantanal is influenced by neither white nor black water rivers. Its water sources can be classified as clear water (Sioli, 1984), and no communities were observed in the northern part of the wetland resembling either an *igapó* or a *várzea*. In the Pantanal of Poconé and its surroundings, a great many biotic communities have developed in the littoral zones along the water courses and on the floodplain. These include gallery forests on elevated river banks, hummocks, the wet fields or *campos úmidos*, and vast areas of the floodplains themselves, which are subject to a pronounced seasonal succession (Heckman, 1994a, 1998; Prado *et al.*, 1994; De-Lamonica-Freire and Heckman, 1996; Hardoim and Heckman, 1996). The floodplain community, which covers vast areas of the Pantanal, bears a greater resemblance to the cultivated wetlands of Southeast Asia, which lie in the same climatic zone (Fernando, 1977; Heckman, 1974, 1979), than they do to the wooded wetlands along the Amazon. The littoral communities are far more varied and diverse in southern Mato Grosso than the purely aquatic and terrestrial ones, and their study must be the main focus of any research program in the Pantanal. The features of the littoral communities, strongly influenced by the seasonal succession, will be given special attention in Section 10.

9.2. Artificially modified species aggregations

The communities substantially modified by human activities, especially by the deliberate introduction of species from other regions of the world, are almost exclusively terrestrial. It may not be correct to refer to all of the species aggregations dominated by recently introduced plants and animals as communities, although the cattle pastures have already become somewhat stable in their features and may be considered communities, albeit ones that are certainly still in a state of long-term flux.

Most pastures have been prepared by killing the trees in wooded areas and burning the rest of the vegetation during the dry season. The cleared areas are then seeded with various grasses, including species that were introduced from Africa. Apparently the seeds of the grasses desired were mixed with those of other species, and a number of introduced plants have appeared on the pastures that the cattle do not like to eat. These come to dominate the pastures as the cattle selectively graze away the desirable species.

9.2.1. Pastures

Ranchers and speculators in real estate often try to enhance the value of their land for cattle production by converting it to pasture. In and along the borders of the Pantanal, various areas above the level of the flooded plain serve as ideal pasture land. Some of these are on the so-called *cordilheiras* and on former *capões de mata*, while others are on areas of the floodplain that are only under a thin layer of water for a few weeks when the highest water level is reached during the *cheia*. The uppermost of these areas were formerly covered by *cerrado*, while the others were occupied by trees more tolerant of temporarily waterlogged soils. To clear the pastures for herbal vegetation, the trees have generally been cut or girdled and killed, and the shrubs burned during the dry season. The cleared land was soon after covered by grasses and a few dicotyledonous herbs, some species of which had been imported from other continents, mainly Africa, to improve the nutrition of the cattle (Section 7.7 & 7.8).

Because grazing is extremely intensive on the few dry areas during the high water periods, the selection mentioned above has taken place, and the species most liked by the cattle have been eliminated. This has left space for the development of dense stands of grass and herb species that the cattle cannot or do not like to eat. This has caused the ranchers much consternation, and they often seek help to find ways to eliminate the "weeds" or "*invasoras.*" Some of the most common introduced plant species that have come to dominate the pastures are shown together with selected species from other continents that have had more favorable impacts in Figure 9.7. A selective process favoring unpalatable plants for cattle and horses is normally a sign of overgrazing, but in the Pantanal, this process can occur even when only a moderate number of grazing ungulates are present. The reason is the annual flood, which forces the animals to congregate in rather small areas for several weeks each year, thereby putting a particular stress on the terrestrial vegetation at the time the best conditions for plant growth are encountered.

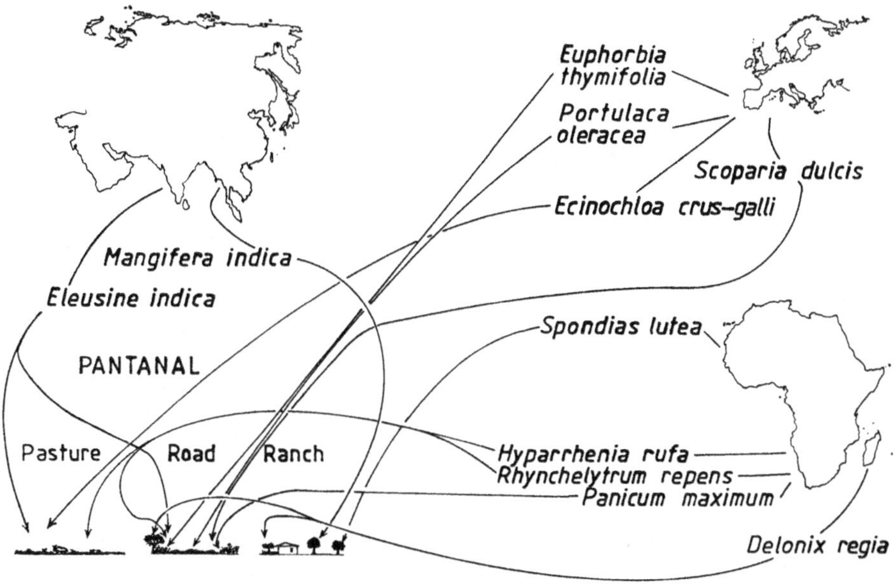

Figure 9.7: A few of the plants introduced from other continents by man to the region of the Pantanal, their probable origins, and the locations in the wetland at which they are most often encountered.

In addition to the plant species, some of the pastures are characterized by large mounds of termite species, including *Armites euamignathus*, *Cornitermes snyderi*, *Cornitermes bequaerti*, *Cornitermes cumulans*, and *Armitermes cerradoensis*, as well as ants, such as *Paratrechina fulva* and *Camponotus* sp. The reason for the abundance of these mounds is apparently the presence of many tree stumps left when the dry forests were cleared. These mounds have given the fields the local name, "*campo dos murundus*", or "fields of mounds", which are generally the equivalent to the *murundu* savanna but without the trees and shrubs.

Locally, two kinds of open fields are recognized by the botanists who study the distribution of the flora: the *campos úmidos* or wet fields and the *campos secos* or dry fields. It is presently difficult to determine whether any of these fields are natural habitats for vegetation because they may develop after areas artificially cleared of trees have progressed in the seral succession to this condition. It seems most probable that at least the *campos secos* develop under natural conditions only in the drier chaco areas to the west of the Pantanal and are stages in an artificially initiated seral succession that began with land burned over to produce pastures.

Campo úmido means "moist field", a term used to describe the kind of habitat encountered where ground water seeps onto the surface, producing a pepetually wet soil. *Campos úmidos* were encountered at various locations in the catchment area of the Pantanal and, more rarely, in the wetland itself. Their typical vegetation is quite distinct from the flora elsewhere and includes species such as *Xyris* cf. *savanensis*, *Xyris laxifolia*, *Lipocarpha humboldtiana*, and *Lipocarpha sphacelata*. In the Pantanal, the areas moist during the dry season and inundated during the *cheia*, such as a rather large *campo úmido* beside the Baia Porta de Fora near Barão de Melgaço, often support stranded aquatic plants, which are able to survive on moist soil until the onset of the next rainy season.

9.2.2. Ranch houses

Many of the ranch houses in the Pantanal were constructed on former natural hummocks, the *capões de mata*, from which the trees were cleared. On these sites, the elevation is sufficient to keep the buildings dry at the time the highest water levels on the floodplain are reached (Plate 14b).

In the immediate vicinity of the ranch houses, a flora greatly enriched by introduced ornamental plants and fruit trees (Plate 14b) together with a fauna of domestic animals have developed. Generally, the species typical in the ranch aggregations are found only near human habitations. Examples of the commonest plants present are trees bearing mangos, lemons and other citrus fruits, custard apples, papayas, and guavas. Only the guava trees were ever observed growing far from ranch houses along the Transpantaneira.

Common ornamental plants include bougainvillea vines, *Bougainvillea glabra*, and cacti of the genus *Cereus*, but almost any kind of tropical plant might be encountered in the gardens near the houses.

The most abundant animals associated with every ranch house include cattle, horses, and dogs. Also present are chickens, turkeys, helmeted Guineafowl, cats, donkeys, goats, and pigs. Among the animals adapted to human dwellings that are not purposely cultivated are the cockroach, *Periplaneta americana*; psychodid flies, *Pericoma* sp.; house flies, *Musca domestica* Linnaeus, 1758; the house geckos, *Phyllopezus pollicaris*; and house sparrows, *Passer domesticus*. In addition to the cattle, only the pigs and, to some extent, the helmeted Guineafowl have succeeded in developing feral populations capable of surviving independently from human beings. Apparently, the others are too poorly equipped to protect themselves against predators and parasites to live outside the zone of human protection.

9.2.3. The Transpantaneira Highway

A group promoting tourism in the Pantanal of Mato Grosso do Sul advised visitors to the Pantanal to carry seeds of plants native to other parts of Brazil and scatter them along the roads in the wetland (A. Pott, pers. comm.). Although no such advice has been given to the visitors arriving in Poconé with plans to travel along the Transpantaneira Highway and observe the great wealth of natural wildlife in the wetland, the transmission of seeds from all parts of Brazil is an obvious effect of the motor vehicle traffic. The presence of the Transpantaneira has provoked the development of a long band of extremely varied fauna and flora that extends from Poconé to Porto Jofre on the Rio Cuiabá. Along this band, not only can the natural biota of the Pantanal be observed but also an extremely rich flora, which can include almost any Brazilian plant species.

When the Transpantaneira was constructed, large amounts of fill were scooped from both sides of the embankment along which the highway runs. To minimize the chance that the road would be inundated during the *cheia* (Fig. 9.8), the embankment was made very high, and many wooden bridges were constructed to permit the flood waters to freely flow underneath it. This produced considerably deepened trenches on both sides of the road, in which water remains much longer during the dry season than in the surrounding parts of the floodplain. As a result, fishes, birds, and aquatic reptiles congregate in these trenches during the dry season, producing an incomparable pageant of wetland life for the tourists who traverse this unfinished highway. Most of these animals do not flee from vehicles. Generally, however, the animals fly away or seek cover beneath the water or in the forests when the passengers leave the vehicles. Thus, it is possible to come quite close to the animals to photograph them from the windows of the cars and busses as long as the tourists remain inside.

Figure 9.8: It is rare for the water level in the Pantanal to become so high that it flows across the Transpantaneira, as it did during January, 1995. Elements of the flora along the road have been introduced from other parts of the country, and many of the these species are not encountered elsewhere in the Pantanal.

Particularly abundant in the ditches along the Transpantaneira are the caimans, cormorants, egrets, and, especially along the southern stretches of the highway, the capybaras. However, many of the less common species can also be observed at times. Many fishermen attempt to catch the fishes that congregate in these ditches during the dry season, and I never noted that any of them ever left the bridges empty-handed.

All of the aquatic species belong to intact natural communities along the Transpantaneira. There are simply more of them beside the road during the dry season because water remains long after it has dried up on the adjacent floodplain. The terrestrial vegetation, however, is unique for the region. The embankment supports a sharp ecotone transitioning from the aquatic community in the ditches beside the road to the dry dirt and stone surface of the road itself. The most prominent introduced species along the edges of the road include such fruits as guava, passion fruits, and pineapple, which almost certainly grew from seeds and vegetative propagules in the wastes thrown from the windows of the autos.

Among the species observed along the Transpantaneira that are apparently otherwise absent from the Pantanal are pioneer species found at the edge of the road, such as *Tridax procumbens*, *Bidens gardneri*, *Ageratum conyzoides*, *Eupatorium candolleanum*, *Ipomoea quamoclit*, *Euphorbia thymifolia*, *Aeschynomene sensitiva*, *Chamaecrista nictitans patellaria*, *Indigofera sabulicola*, *Cuphea melvilla*, *Cuphea speciosa*, *Polygonum persicaria*, *Portulaca oleracea*, *Scoparia dulcis*, *Solanum lycocarpum*, *Eleusine indica*, and *Rhynchelytrum repens*. In addition, several plants that bear edible fruit often transported by travelers have apparently grown from discarded seeds along the roads. Among these are *Ananas sativus*, *Passiflora cincinnata*, *Passiflora giberti*, and *Psidium guajava*.

Although obvious aggregations of plants unusual in the region are encountered along the Transpantaneira, there are no equivalent animal populations. While most of the native animal species can be observed crossing the Transpantaneira at times, no species introduced from other parts of Brazil are evident, other than the domestic animals raised on the ranches. The most common species that have adopted the habits of seeking food along the edges of the road are the lizard, *Ameiva ameiva*, and several ant species, which take advantage of the road's elevation to protect their underground nests from flooding. During the early rainy season, the surface of the road also attracts enormous numbers of butterflies, which gather to seek the salt excreted by the numerous cattle that graze along the edges of the road.

9.2.4. Gold mines

The gold mines near the northern edge of the Pantanal in the districts of Poconé and Cangas are pits dug down to the layers of gold-bearing sand. Most of them are constructed illegally and have been periodically closed down by the police. However, there are not enough police to enforce the law, and if interfered with in one location, the miners can simply move to another and start digging a new pit. Children in Poconé are even able to collect bags of sand along the edges of the Transpantaneira Highway a few kilometers from Poconé and extract a considerable amount of gold sand to earn pocket money.

However, little of the gold mined comes from sediments at the surface. The gold prospectors scraping the material from the bottom of the rivers and streams to remove the gold, which are frequently seen in other parts of Brazil, are almost completely absent from the northern sections of the Pantanal. Instead, the miners dig away the upper layers of the soil, generally using giant earth-moving machines, and expose gold-rich subsurface layers of sand (Fig. 9.9). These are then treated with metallic mercury to remove the gold dust and discarded in large piles on the surface. The discarded sediments may be considerably contaminated with mercury, although the contamination is not evenly distributed in the piles (Tümpling *et al.*, 1995). The sediments collected by the children along the highway were apparently dumped there by the mining concerns because they did not bear enough gold to be worth processing.

After the gold-rich sand layers have been removed, the mines are simply abandoned. Thus, many deep pits remain in the gold mining districts. These fill up with water during the next rainy season and become standing water bodies. Particularly large concentrations of mercury are a common feature of the sediments of these water bodies (Esteves, 1988).

In the temporarily flooded active mines and in all abandoned ones, a group of pioneer species are able to develop. Those that settle most rapidly include mosquitos and chironomids and the dragonflies that feed on their larvae. In addition, many of the other aquatic insects appear in the pools of water that form in the mines after the freshet, just as they do in the more permanent pools of water that collect in holes in the dirt roads. Many of the insects wind up in these pools of water after the massive flights of many species that usually precede the freshet. The gold mines were not visited often enough to permit exhaustive studies during the project, but some of the more common insects to be encountered there are listed in Table 9.7. Breeding choruses of frogs assemble soon after the depressions are flooded, as well.

Figure 9.9: The Dois Paderes gold mine in Poconé, one of the largest gold mines at the northern edge of the Pantanal. The environmental effects of large building machines, which have given man the potential to indiscriminately destroy large areas of special habitats in minimum time without having to pay for human labor, have never been adequately assessed.

After a period of time, the abandoned mines are settled by most of the typical species of the Pantanal. The fish fauna apparently increases in diversity considerably slower than the groups of other taxa, but several species of fish, notably the swamp eel, *Synbranchus marmoratus*, and several cichlids apparently move overland into the new water bodies rather rapidly.

The natural recuperation of the pits dug to reach the gold-rich sand layers is slow. These pits are deep, and the inorganic nutrients required for plant growth are not present in a supply sufficient for a rapid eutrophication and succession to a terrestrial kind of habitat. Furthermore, since the resurgence of gold mining in the 1980s, the pits have appeared far faster than they can be eliminated by natural processes. Although restoration is the legal duty of the miners, none of them can be expected to fill in the pits again unless compelled to do so by the local authorities, and mechanisms for enforcement of the laws are practically non-existent. Furthermore, since the mines themselves are illegal, it would hardly be possible to even identify the person responsible for financing the restoration.

Since these artificial water bodies are increasingly colonized by the locally common aquatic species, including many fishes, the mobilization and biomagnification of the mercury deposited in the sediments presents a continual danger to the birds and mammals, including man, which feed on the fishes from these water bodies. This will be discussed in more detail in Section 13.2.1.

The scars in the landscape appear not only in the form of deep pits (Fig. 9.10) but also as moving rivers of soft mud left on the surface after the processing for gold (Plate 15a).

Figure 9.10: Gold mining as practiced in the district of Poconé produces vast, deep pits scattered over wide areas at the northern edge of the Pantanal.

Table 9.7. Some of the aquatic insects to reach new water bodies formed when the abandoned gold mine excavations are abandoned and become flooded.

Species	Other habitats
Ephemeroptera	
Campsurus striatus	Slow-flowing streams
Baetodes sp.	Slow-flowing streams
Cloeodes hydation	Flood plain
Odonata	
Erythrodiplax castanea	Lentic water bodies
Erythrodiplax umbrata	Lentic water bodies
Heteroptera	
Tenagobia incerta	All kinds of water bodies
Gelastocoris flavus	Margins of water bodies
Buenoa salutis	Inundated flood plain
Lethocerus spp.	Shallow lentic water
Belostoma spp.	Shallow lentic water
Microvelia mimula	Inundated flood plain
Limnogonus aduncus	Lentic and slow-flowing water
Limnogonus ignotus	Lentic and lotic water bodies
Limnogonus cf. *franciscanus*	Lentic and slow-flowing water
Coleoptera	
Berosus spp.	Lentic and slow-flowing water
Hydrophilus ensifer	Unknown

One such slowly moving field of mud from the Dois Paderes Mine in Poconé approached the Transpantaneira during the late 1980s and had to be carried away in trucks to prevent it from inundating the highway, as well as for cosmetic purposes. This kind of sediment is too soft to walk on safely, and when its surface caves in, a person can sink rapidly. However, pools of water that form on its surface may be inhabited by algae, aquatic insects, and frog larvae among the pioneer community, as discussed above.

From an ecological point of view, the construction of artificial lakes around the periphery of the Pantanal is by no means detrimental to man and wildlife. The lakes that form in particularly deep mines will exist for a very long time, and some of them may come to be used as reservoirs and recreation centers for the local human population, They will also be suitable for colonization by fishes and other aquatic animals that require more permanent water supplies than those available on the Pantanal floodplain. Some of the smaller ponds that form could be used for aquaculture. A major constraint which will limit all of these uses of the old mines by man and wildlife, however, is the moderate to heavy contamination of the sediments with mercury. This promises to be a major health hazard that could become acute through changes in water chemistry or disturbance of the sediment for a long time to come.

10. Seasonal succession

Being located in the tropical wet-and-dry climatic zone, the Pantanal is subject to great seasonal differences in the amount of rainfall (Fig. 4.2). This produces a marked seasonal succession of species, with the life cycles of many plants and animals being determined primarily by the season. The definition of the seasonal succession and its distinction from the seral succession have been provided in publications on theoretical ecology (Odum, 1971; Anderson, 1986), and Gopal (1986) discussed successions in shallow and temporary water bodies. In many locations, a terrestrial community of drought and flood resistant species develops during the dry season and is then largely replaced by aquatic and wetland biota that require standing water or waterlogged soils to survive. This community is, in turn, drowned by the annual flooding but reappears, mainly through the germination of water-resistant seeds, when the flood waters recede. Slight irregularities in the terrain elevation therefore produce enormous differences in the conditions encountered by the biota, as shown by studies of the floodplain macrophytes (Prado *et al.*, 1994). Which plant species become prodominant during a given season depends mainly on how long the area remains flooded, how deep the water becomes during the *cheia*, and how far from water courses and hummocks it is located. Characteristic of the floodplain is the dominance of single species over vast areas. Among the most prominent of these are *Pontederia lanceolata* at sites flooded by a relatively shallow layer of water, *Eichhornia azurea* where the water becomes deeper, and *Salvinia auriculata* in still deeper water. The studies of Prado *et al.* (1994) suggested that the border between the *P. lanceolata* and *E. azurea* populations is approximately the 1 m isobath when the water level has reached its maximum. Since this varies somewhat from year to year, the extent of the areas covered by these species also varies somewhat. The uppermost extent of the *P. lanceolata* beds is roughly at the level at which the ground remains flooded for at least half of the year. The *E. azurea* can persist in water that becomes somewhat deeper than 2 m. In deeper waters, floating plants become dominant, with *Eichhornia crassipes* replacing *E. azurea*. However, *E. crassipes* is not nearly as abundant in the Pantanal as *E. azurea*. The open water areas are much more frequently covered by the water fern, *Salvinia auriculata*, together with much smaller numbers of *Pistia stratiotes* and *Azolla mexicana*. During the dry season, the vegetative stages of *P. lanceolata* disappear nearly completely from the floodplain of the Pantanal. The other species persist, but their populations are greatly reduced as a result of desiccation. During all four seasons, Prado *et al.* (1994) demonstrated that there was a significant negative relationship between the water depth and the number of macrophyte species present per unit area.

The radical changes in the physical and chemical conditions during the course of each year also place the biotic community under stress and eliminate many species that require more stable conditions to survive. Hence, the Pantanal is not characterized by the great species diversity that would normally be expected in wetlands (Ewel, 1991) and as parts of the Amazon region are, but rather by large numbers of relatively few species, which have been able to adapt to the prevailing conditions (Guarim Neto, 1992; Brown, 1986). For most of the terrestrial plants, this means an ability to survive long periods in dry soil and then to cope with a submergence of the roots beneath flood waters or in waterlogged soils. Those species studied have been shown to have various structural adaptations to cope with the temporary flooding (Worbes, 1985). Most of the tree species in the Pantanal can only survive on the areas at the highest elevations, which are flooded at most for brief periods, and along river banks, where they do not have to contend with long periods of dryness. Many of the regional plant species, however, are herbaceous and have the capability to grow and produce many seeds within a short period of time (Heckman, 1997b). This assures the survival of the species, even when particularly severe conditions completely eliminate the vegetative forms of the plant from a region (Prado *et al.*, 1994). Conversely, the species well adapted to the wet-and-dry climatic zone generally maintain their seasonal growth patterns even in water bodies and wetlands in which the rainfall patterns are less extreme. For example, *Oxycaryum cubense* shows a pronounced seasonal pattern of growth and seed production in eastern Brazil, although its establishment on floating islands of vegetation gives it access to an ample water supply the year round (Moschini-Carlos *et al.*, 1995).

The seasonal succession of the macrophyte species is depicted in Figure 10.1. Similar successions have been documented in other parts of the world subject to the tropical wet-and-dry climate, including Southeast Asia (Heckman, 1974, 1979) and Northern Australia (Finlayson, 1993). The aquatic macrophytes are of special importance in the Pantanal because their growth and death each year are determinants of the nutrient cycles for the microorganisms, and because of their role as temporary habitats for the many animal species with life cycles timed to the annual climatic cycle. This section will outline the effects of the seasonal succession on the biotic communities of the Pantanal as a whole and supplement the much more detailed information provided in the sections on the flora and fauna.

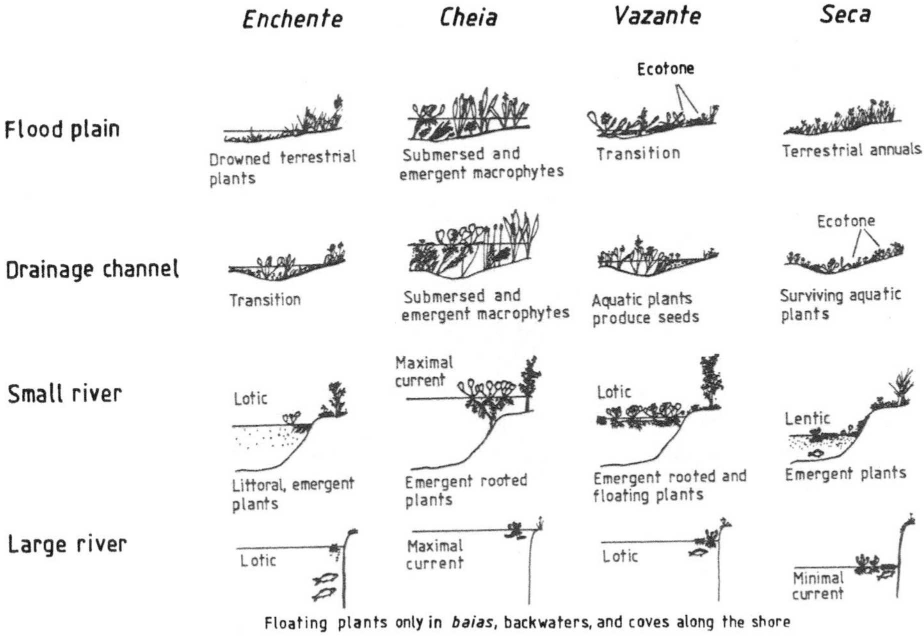

Figure 10.1: A diagrammatic depiction of the seasonal succession of macrophytes in four kinds of habitat in the northern Pantanal.

10.1. *Enchente* transition period

10.1.1. Characteristics of the season

If we commence our observation of the annual cycle with the onset of the rainy season, we find the conditions in the Pantanal reminiscent to those in a semi-arid habitat. The wind patterns generated by the interaction between the cold land mass in the south temperate zone and the warmer oceans usually bring two to three rainless months, during which nearly all standing water bodies dry up, and all streams and rivers, except for the large Rios Paraguay and Cuiabá, become isolated lentic water bodies. In the northern Pantanal, only a few weeks go by between the onset of the rains and the first flooding caused by the overflow of the rivers and streams. The situation in the central and southern Pantanal is quite different since the flooding comes with a considerable delay (Section 5.1).

If the freshet comes early and is not followed soon by more rains, the land may dry up again. In any case, the terrestrial plants lose their brown color, and the landscape becomes a lush green. With the flooding begins the remarkably rapid growth of the aquatic plant species. The productivity recorded by methods developed elsewhere seems amazing, but it continues only as long as the nutrients are available. However, during the *enchente*, the inorganic nutrient content rises with the water level, as the flooded terrestrial vegetation and detritus remaining from the previous year rapidly decompose at the high prevailing water temperatures.

The seasonal succession during the *enchente* is rapid and extremely complex (Fig. 10.2). This must be broken down into events, some of which begin and end within a matter of days and vary considerably locally. These are compiled in Table 10.1.

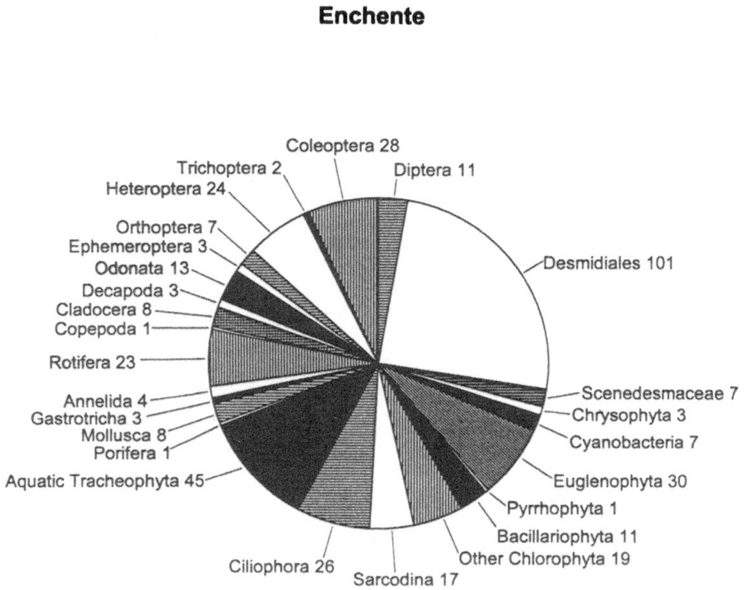

Figure 10.2: The numbers of species belonging to various major taxa that are active during the *enchente*.

It must be noted that algae and other microorganisms play a significant role for only a short period of time at the beginning of the *enchente* before the macrophytes and metazoans have come to monopolize the available nutrient elements in the system. Nevertheless, this brief period of development produces a food net that is vital to the survival of many juvenile fish and aquatic invertebrate species, which are produced in enormous numbers at the beginning of the rainy season.

The algae and bacteria provide the nutritional basis for a brief period of growth by small invertebrates, mainly rotifers, cladocerans, and copepods, which belong to various species, including those generally considered planktonic in other parts of the world (Section 8). This is the annual, short-term production of phyto and zooplankton communities, which disappear again as the water level nears its high point and the concentrations of available nutrients for the algae fall precipitously (Heckman, 1994a, b).

Table 10.1: Events of the *enchente*, characterized by great day to day variability dependent upon the weather. The exact dates of the beginning and end of the phases described below vary considerably from year to year. Two frequently visited sampling sites are considered. The first is a low-lying area on the floodplain, the lowest area of which does not completely dry up during the *seca*, and the second is a shallow section of a river in which the water is lotic during the period of flooding and lentic during the *seca*.

A. On the floodplain at Birici, Fazenda Ypiranga

	Feshet	First flooding	Rising water	Start of *cheia*
Rainfall	Heavy thunderstorms (infrequent)	Thunderstorms (sporadic)	Thunderstorms (frequent)	Thunderstroms (frequent)
Water level (cm)	rainwater pools 0 - <20	rising 0 - c. 50	rising c. 50 - 100	almost stable >100
Flow	none	slight	none	slight
Turbidity	slight to moderate	none	none	none
Color (Hazen)	1 - 10	>100	decreasing	c. 0
Conductivity ($\mu S \cdot cm^{-1}$)	100 - 200	60 - 100	30 - 60	30 - 40
O_2 saturation	wildly fluctuating from 30% - >200%	10% - 50%	10% - 80%	15% - 90%
Inorganic nutrients	high concentrations	highly variable	moderate supply	poor supply
Dominant algal taxa	Chlorococcales Euglenophyta	Transition	Desmidiales Bacillariophyta	Desmidiales
Macrophytes	terrestrial (dying)	aquatic (start of vegetative growth)	aquatic (some flowering)	aquatic (flowering)
Aquatic fauna	inactive	intense reproduction	many juveniles actively growing	

B. In the Rio Bento Gomes at the Transpantaneira Highway

	Feshet	First flooding	Rising water	Start of *cheia*
Water level (cm)	Rainwater pools 0 - 200	rising <300	rising <350	high, stable <400
Flow	none	moderate	increasing	variable, fast
Turbidity	moderate	slight	slight	very slight
Color (Hazen)	40 - 100	>150	decreasing	0 - 30
Conductivity ($\mu S \cdot cm^{-1}$)	80 - 200	<80	<60	30 - 50
O_2 saturation	variable from c. 0% to >300%	10% - 50%	20% - 80%	40% - 70%
Inorganic nutrients	moderate to rich supply	moderate supply	decreasing supply	low concentrations
Dominant algal taxa	Chlorococcales Euglenophytes	transition	Desmidiales Bacillariophyta	Desmidiales
Macrophytes	floating aquatic (moribund)	much drift	intensive growth	Dense emergent and submersed flora
Aquatic fauna	inactive	intense reproduction	rapid growth	much migration

The next stage of the succession is dominated by a few species of aquatic macrophytes, which serve as the primary producers and supply substrate for colonization by various biotic communities, and which monopolize virtually all of the available nutrients in the system. Among the algae, this period is characterized by a rapid disappearance of the Chlorococcales, Euglenophyta, and other groups that were dominant numerically during the dry season, and their replacement by relatively few desmids belonging to a great number of different species (De-Lamonica-Freire and Heckman, 1996). The general representation of various groups during the *enchente* is depicted in Figure 10.2. Because it is a transition period, typical elements of both the dry and wet periods influence the numbers of species reported in the figure, although these species may not interact with one another.

10.1.2. Seasonal peak reproduction

For the animal species, this period of the succession is characterized by the appearance of the survivors of the preceeding dry season, which leave their places of refuge, either in the dry sediment or in the larger rivers and streams that did not dry up, and almost immediately begin a period of intense reproduction. This seasonal activity is observed among nearly all of the large aquatic invertebrates, most of the fishes, and the amphibians, as well. Freshly flooded fields attract large numbers of spawning dragonflies, are the sites of the evening choruses of various frog and toad species, and are adorned with the pink egg clutches of snails in the genus *Pomacea* attached to the emergent stems of the drowning terrestrial plants. The trichodactylid crabs carry young under their abdomens, and the fishes display their courtship colors during this time of year. The first part of the *enchente* is typically the time of breeding, while the latter part of the season is characterized by large numbers of young animals in the water. These consume the rich plankton supply, atypically concentrated in shallow layers of water, until it is no longer available.

It is moot to discuss whether the disappearance of the plankton is caused by "top down" or "bottom up" factors; the fact is that the detritus that supports the bacteria disappears and is initially replaced by nothing from the rapidly growing aquatic macrophytes. Most algae also lack enough nutrients to develop, as the amounts of inorganic substances drop to the detection limits of the standard analytical methods or below, and without suspended bacteria or algae, the planktonic rotifers and cladocerans face starvation. At the same time, huge numbers of tiny predatory fishes greedily search every part of the open water for food, devouring everything that lacks strong chemical repellents, such as the heteropterans. Planktonic species have no more chance to survive, and it is suspected that they produce spores or resting eggs during this period which are not activated again until the next dry season. The invertebrate community relocates during this season to the interstices among the roots and rhizoids of the floating plants, where they can share some of the nutrients stored in the macrophytes and into which most of the hungry fishes cannot penetrate. At the end of the *enchente*, conditions typical of the next season, the *cheia*, have already developed.

One phenomenon that occurs a few hours before the freshet or before other particularly heavy thunderstorms at the beginning of this season is the massive mating flights of insects, especially many aquatic species. These generally involve adults of many different insect orders, with terrestrial lepadopterans being among the most abundant in terms of both individuals and species, but with extremely large numbers of aquatic coleopterans and heteropterans of numerous species, as well. In the pools of rainwater left in many parts of the Pantanal, mixed choruses of many frog species are encountered, as well. These phenomena had already been mentioned in the sections on the individual species involved, but it is worthy of special mention as a seasonal characteristic (Lutz, 1946).

10.2. *Cheia* stability

The *cheia* is the period of high water during which the aquatic macrophytes and relatively large animals monopolize the entire nutrient element supply. The water is crystal clear and contains amounts of dissolved inorganic substances similar to or less than those in rainwater. Inorganic plant nutrients containing nitrogen and phosphorus are no longer present in detectable concentrations. Water movements are strong, and the streams and rivers across the Pantanal become typical lotic water courses. The structure of the biotic community is therefore radically different during this period than it is during the rainless seasons. The representation by species belonging to various major taxa is depicted in Figure 10.3.

Because of the paucity of free inorganic plant nutrients, the desmids come to dominate the microflora. These species are known for their occurrence in waters containing low concentrations of nutrients, such as the moors of Europe (Rose, 1953). The number of individual algae per unit volume of water reaches a minimum during the *cheia*, but the species diversity is obviously higher than at any other time of year, especially due to the large number of desmid species present (Section 7.2). In the water samples, nearly every individual algal cell belongs to a different species, and on single glass slides prepared during this season, many different desmids could be photographed. It is notable that the role of the desmids in the food web remains enigmatic. Brook (1981) noted that little information is available about organisms capable of feeding on desmids. Few animals consume them, and those that do are often incapable of digesting them.

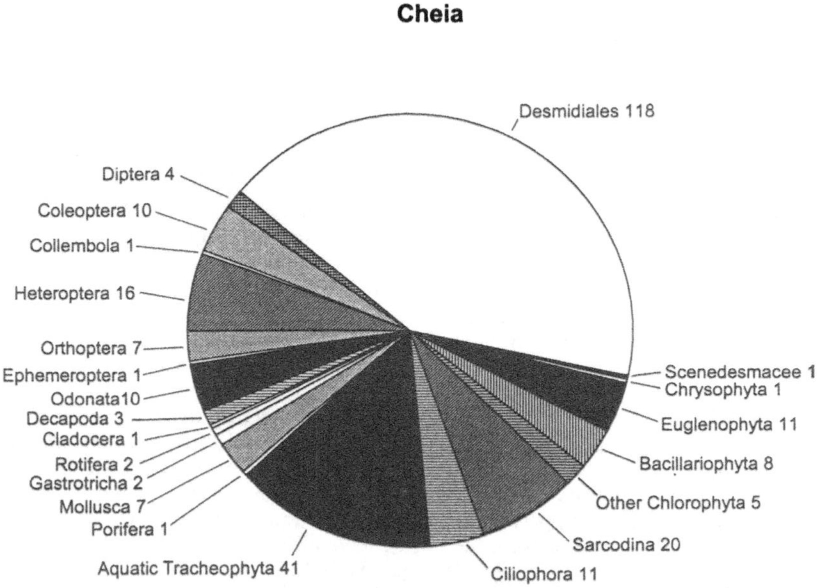

Figure 10.3: The numbers of species belonging to various major taxa that are active during the *cheia*.

In addition to the desmids, a few species belonging to other groups could also be found, but these were generally adaptable species capable of surviving throughout the year, and they were more abundant during other seasons. An exception to this was *Chaetophora elegans*, which formed large gelatinous clumps on the stems of woody plants and other objects in the fast flowing water of the streams. A more detailed discussion of the seasonal occurrences of individual species is provided in Section 11. It must be remembered, however, that special local conditions can promote the development of algae in seasons during which they are normally not active. For example, wastewater outflows from towns or cattle ranches can enrich natural water bodies near the outflow points far more than they ever would be under natural conditions. However, it was observed that the excess nutrients are rapidly absorbed by the macrophytes, and their influence does not extend far from the point sources of the pollution. Evidence of such local enrichment included large mats of species of the Zygnemataceae, mainly *Spirogyra* spp., near the wastewater outlets from the exposition park of Poconé and a rich community of microorganisms at the gate to the Fazenda Ypiranga.

Like the algae, the protozoa show a reduction in abundance during the *cheia*, apparently because of the lack of edible algae and bacteria in the water. The small metazoan species that feed on particles are also relatively rare because of the general shortage of edible particles and the ever increasing predation by the growing juvenile fishes, which evidently suffer a period of hunger during the *cheia*. These form small schools and move rapidly through the water seeking any edible material. Dust and other objects that falls into the water at this time is hurredly investigated by large numbers of fishes, obviously very hungry.

For the larger organisms, including the caimans and large piscivorous birds, the *cheia* is also a time of hunger. The stored fat in the tails of the caimans, which is obvious during the *seca*, becomes exhausted during this period, and the animals appear emaciated. This occurs not only because is food scarce, but also because the caimans build nests and guard their eggs during the *cheia*. They are thus unable to devote their full attention to finding food, and indeed, while studies of these caimans are still only in the preliminary phases, it is not unlikely that the females fast during the nesting period in order to devote more time to guarding the nests and taking care of the young (Cintra, 1985, 1988; Campos, 1993). In any case, the *cheia* is the ideal breeding season for these reptiles because their fasting period while guarding the nests coincides with the shortage of fishes large enough to eat.

Since the *cheia* corresponds to the South American summer, many of the aquatic bird species fly to more food-rich locations in the southern part of the Pantanal or farther away to feed or nest. In contrast, various terrestrial species may migrate to the Pantanal during those months (Cintra and Yamashita, 1990). Some of the species congregate in large colonies for breeding, including the American wood stork, *Mycteria americana*, as noted by Naumburg (1930). Others, such as the cattle egret, *Bubulcus ibis*, and the white-necked heron, *Ardea cocoi*, are semi-migratory and simply become less abundant because they fly to temporary water bodies in the more elevated parts of the catchment area. The herons with nighttime rookeries leave them each morning to reach distant water bodies that disappear after the rainfall has ceased and the water level has begun to decrease. They must travel far and search hard for their meals, so the *cheia* is not a good season for nesting by the resident piscivorous birds of the region.

While the food supply during the *cheia* is not optimal for many of the predatory species, most wetland herbivores are well supplied with fresh water plants, and animals of all kinds are optimally protected from natural enemies in the Pantanal, including man. Concealment is available under the deep water and among the dense beds of water plants. For herbivorous animals, such as the capybara, the *cheia* is a good season in all respects, and young may be brought into the world in relative safety. In contrast to the predatory species that must hunt small mammals, reptiles, amphibians, fishes, and invertebrates, the top predators find conditions optimal, and while aquatic species, such as the anaconda, can appoach their prey under water without being seen, terrestrial ones, such as the jaguar, find particularly lush vegetation on the hummocks to conceal them while waiting to pounce on mammals that take shelter there.

How well the aquatic animals can conceal themselves during the *cheia* was often demonstrated by searches for the caimans at the locations of their greatest abundance during the dry season. At sites where not a single one could be seen during the day, in spite of a careful search, the lights of the project bus were able to reveal dozens of eye reflections after sunset, including those from some large individuals. The *cheia* is therefore not a season optimal for visits to the Pantanal by tourists seeking to observe wildlife.

10.3. *Vazante* transition period

10.3.1. Seasonal changes

The transition to the *vazante* brings some relief to the starving community of intermediate predators. As the water level decreases, the fishes begin to become concentrated in smaller areas, and they have grown considerably since the early rainy season. The caimans and water birds are able to catch increasing numbers of larger fishes. As the season continues, vast areas fall dry, and the aquatic vegetation begins to die off. This brings about an increase in saprophytic activity. Bacteria begin to multiply, providing nutrients for protozoa and small invertebrates and releasing inorganic nutrients for algae. The desmids begin to decline in importance, and algae of many other groups appear in steadily increasing numbers. The fishes are also able to find food easier, and they begin to grow faster, which gives them some advantage for surviving the subsequent dry period. The water become more turbid, and the concentrations of inorganic nutrient elements begins to show a slight increase, although the concentrations fluctuate much with the weather and other variable local conditions.

Most of the species in the aquatic macrophyte community begin a period of intensive blooming, which usually continues until the areas have completely dried up. However, flowers may continue to appear on the dry areas if the species is particularly drought resistant, as in the case of *Hydrolea spinosa*. When the transition from the conditions of the *vazante* to those of the *seca* actually occur depends to a great extent on the terrain elevation (Prado *et al.*, 1994). Some areas become dry long before those just a few decimeters lower. As already mentioned, during the transition period, an interesting community of short-lived wetland species blooms intensively and go to seed before the area completely dries up or becomes flooded again (Heckman, 1997b).

For the smaller aquatic species, the coming of the dry season means entering the sediment for a period of aestivation, migrating to deeper water bodies, or perishing. Many fail to prepare for the desiccation of the habitat in time and die off. Fishes trapped in small drying pools are easily caught by water birds (Fig. 10.4), but in spite of intensive feeding by large flocks of herons, egrets, wood storks, and stork jabirus, many of the fishes do not die until the water bodies completely dry up (Plate 15b). The swamp eel, *Synbranchus marmoratus*, is one of the exceptional fish species that normally survives the period of desiccation encased in the dry mud. Individuals of this species were often found in burrows under pieces of wood during the dry period. In the case of the *Bufo paracnemis* tadpoles, the drying of the pools in which they are developing induces their rapid transformation to adults. As the water rapidly evaporates, a race against time for the tadpoles begins. Some of them lose the race and die before completing their metamorphosis, forming masses of dead toad larvae in small depressions in the dry sediment. Siblings in the same groups succeed in completing their transformations and can be seen hopping away from the muddy patches where the pools had formerly been (Fig. 10.5).

The transition zone between the receding aquatic and advancing terrestrial communities is occupied by the ecotone which moves across the floodplain (Heckman, 1997b). This movement, during which the zonation of the species remains more or less intact, results from the presence of several small herbal plants with reproductive stages so short-lived that the plants are able to sprout, bloom, and go to seed in a matter of a few weeks, or sometimes within a few days. They sprout as the receding water exposes the sediment and die off as the sediment becomes dry or heavily overgrown with other herbal species that persist until the flooding of the subsequent rainy season.

Figure 10.4: In spite of intensive predation by caimans and water birds, many fishes survive long enough to die of desiccation when the last pools of water dry up.

Figure 10.5: For the fast-growing tadpoles of *Bufo parocnemis*, rapid metamorphosis is vital for survival. As the pools of water dry up during the *seca*, many tiny toadlets are seen hopping toward cover, while many tadpoles perish as the water disappears before their metamorphosis is complete.

While the *vazante* is a period of decline for the active stages of the aquatic species, it is the season during which the short-lived terrestrial community on the floodplain begins its development. The receding water first exposes the most elevated areas, generally those located near hummocks and high river banks. These areas remain dry long enough for large terrestrial grasses to complete their life cycles. Some of these areas are therefore optimal for cattle grazing. However, where many cattle selectively eliminate the plant species they like best, various grasses that cattle find unpalatable, such as *Andropogon bicornis*, become the dominant species. Those areas first exposed late in the *vazante*, however, are generally free from these tall grasses because they do not have sufficient time to grow, bloom, and produce seeds before the floods of the following rainy season. Thus, a vegetation of small herbs develops with a distinctive structure typical of the dynamic ecotone mentioned above. The numbers of species representing various major taxa that are active in the biotic communities in the water and on the wetlands during the *vazante* are depicted in Figure 10.6

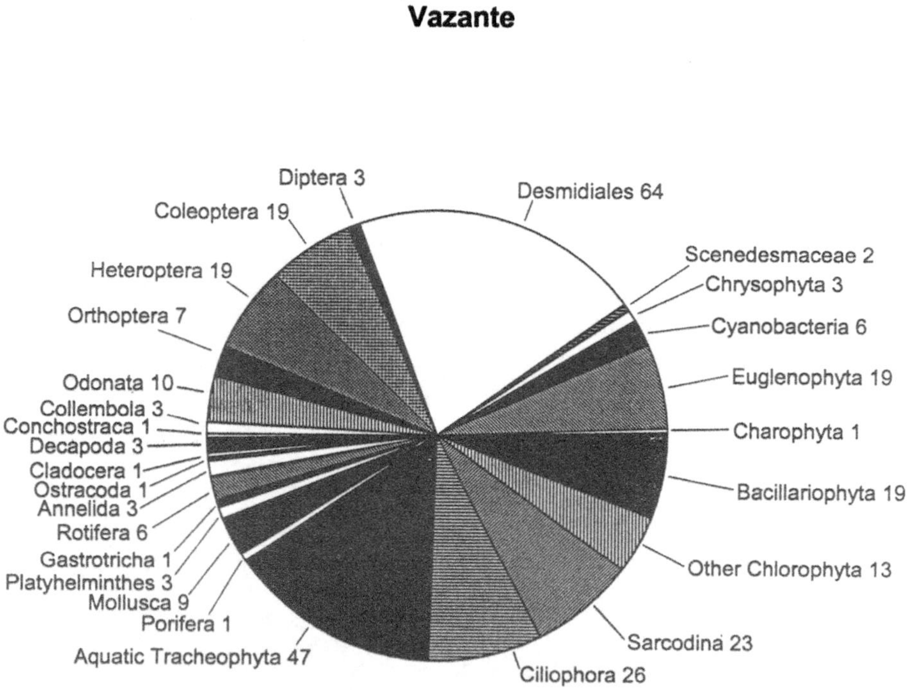

Figure 10.6: The numbers of species belonging to various major taxa that are active during the *vazante*.

10.3.2. Development of the saprobic communities

During the *cheia*, the activity of the aquatic microorganisms in the Pantanal reaches a minimum. The inorganic nutrients remain incorporated in the tissues of the macrophytes, which make organic substances available almost exclusively to those herbivores capable of consuming them and the predators that consume the herbivores.

In other words, the living macrophytes directly support the food web. As the *vazante* advances and the water level recedes, however, the macrophytes begin to die off. At first, an increasing number of leaves begin to die. Then, some of the plant species produce fruit, and the vegetative parts die off completely. *Myriophyllum mattogrossense* is one of the first submerged macrophytes to disappear from the community after it has produced it seeds (Prado *et al.*, 1994). Other species follow, while the vegetative stages of still others remain alive until the area has completely dried out during the *seca*, or, if conditions permit, until they can resume growth during the subsequent rainy season.

The increasing amounts of plant detritus liberate more and more nutrient material for microorganisms. As the number of bacteria begins to increase, protozoans and small metazoans that feed on them begin to develop. The release of inorganic nutrients begins to become detectable (Heckman, 1994a), and euglenoid and chlorococcal chlorophyte species begin to appear in considerable numbers. Thus, the *vazante* is a period of development for a saprobic community, which becomes increasingly important until the water bodies completely dry up during the *seca*.

10.4. *Seca* instability

The conditions during the dry season in the Pantanal are much more dynamic than they are during the rest of the year. Extreme changes from wet conditions to dry are only one feature of the season. During the *seca*, both the highest and the lowest annual air and water temperatures are recorded (Section 4). The water courses become standing water bodies, and there are enormous fluctuations in the concentrations of oxygen and inorganic nutrient substances in the water bodies, mainly due to the alternation between the actions of the saprobic community breaking down the dead plant and animal material and the growth of algae, which often appear in enormous numbers in the isolated pools of water on the floodplain. The contribution of various major taxa to the structure of the biotic communities during this period is depicted in Figure 10.7.

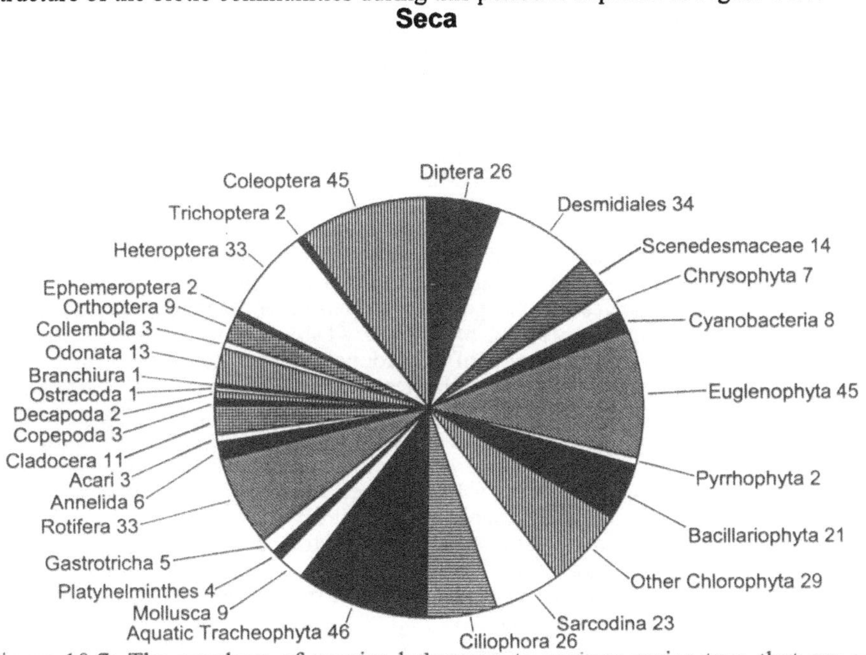

Figure 10.7: The numbers of species belonging to various major taxa that are active during the *seca*.

What is remarkable about these algal aggregations is their ability to thrive at temperatures that are well above 35°C nearly every day for several weeks and sometimes exceed 40°C (Heckman *et al.*, 1993). The vitality of the algae is demonstrated by the daily rise in the oxygen saturation, sometimes to several hundred percent, as reported in Section 6.2.3. Studies of the growth of these same species in the temperate zones seldom employ cultures maintained above 30°C, and special reviews of thermophilic microorganisms have scarcely considered the activity of algae above 40°C. For example, Brock (1978) reported the occurrence of temperatures as high as those recorded in the shallow waters of the Pantanal only in salt marshes. Nevertheless, Sorokin (1967), Hegewald (1984), and Partzsch (1987) demonstrated in the laboratory that strains of common chlorococcal green algae are physiologically adapted to the temperatures prevailing in the Pantanal during the dry season. Moreover, there are significant changes in the anabolic and catabolic processes of plants induced by temperature, manifested, among other things, by a decreased proportion of polyunsaturated fatty acids produced as the temperature increases, documented for photosynthetic organisms from Cyanobacteria (Murata and Nishida, 1987) to Tracheophyta (Stymne and Stobart, 1987).

Generally, the reproduction of algae increases considerably with the temperature until the tolerance threshold of the population is exceeded (Payer *et al.*, 1980). Thus, not only the great availability of nutrients but also the temperature is a determining factor for the development of the algal blooms that are characteristic of the late dry season (Section 11.1.4).

Similarly, bacterial activity is influenced by temperature. Mara (1976) designated thermophils as those species having an optimal growth temperature above 45°C. Those for which the optimum falls between 20°C and 45°C were designated mesophils, while those preferring temperatures below 20°C fall within the psychrophil category. Hence, all three groups could occur during the dry season in the Pantanal, but the thermophils would not be able to develop at their optimal temperature. In any case, the activity of bacteria in general increases with temperature, although warming the water progressively above 20°C produces a steady increase in the number of species eliminated from the community as their maximal tolerable temperatures are exceeded. At the same time, dominance is assumed first by the mesophils and then by the thermophils, which have capabilities of breaking down detritus with increasing speed. The expected result is an extremely rapid decomposition of organic detritus in the water bodies accompanied by high rates of bacterial reproduction. This was actually observed on numerous occasions.

Another aspect of the elevation of the water temperature on the floodplain to above 40°C during the *seca* is the activity of the testaceans and ciliates ciliates (Section 8.3). In the warm water aggregations, the protozoans were scarce or absent (Hardoim and Heckman, 1996). Similar absences of the protozoan species were observed in African water bodies and also attributed to temperature by Dargesco (1980). The high water temparature best accounts for the scarcity of the ciliates, since Bick and Kunze (1971) established temperature ranges for the species observed in the Pantanal much below the prevailing water temperatures there during the *seca* (Table 10.2). In sewage treatment plants, ciliates greatly enhance the final breakdown process of the organic wastes through their grazing away of the excess bacteria (Mara, 1976). Their absence from shallow tropical water bodies could therefore be expected to have profound effects on the breakdown processes, although their functions in the food web may be partially taken over by rotifers. Unfortunately, much information on the ecology of the protozoans under tropical conditions is lacking because this group has been largely ignored by limnologists in the torrid zone (Serruya & Pollingher, 1983).

Unlike the ciliates, several rotifer species seem to have adapted well to water temperatures as high as those in the drying pools of water on the floodplain (Kretzschmar *et al.*, 1993). These species may fulfill the normal role of the ciliates to some extent by grazing on the bacteria.

Table 10.2: Temperature and pH tolerance limits of ciliate species reported by Bick and Kunze (1971) compared with the the values determined for the water in which they occurred in the Pantanal. All of the species in the table were observed in water with oxygen and ammonium concentrations within the range reported for the temperate zone.

	Maximum Temperature (°C)		pH	
	Temperate zone Reported	Pantanal Observed	Reported range	Species occurred at pH outside of this range in the Pantanal?
Lacrymaria olor	25	33.3	6.4 - 9	No
Prorodon teres	25.5	27.8	6.4 - 7.2	No
Coleps hirtus	30	33.6	4.7 - 9.5	No
Didinium nasutum	25	32.5	6 - 8	Yes - 8.9
Mesodinium pulex	21	34.8	Not reported	--
Litonotus lamella	35	36.9	6.0 - 9.4	No
Litonotus fasciola	24.8	32.8	7.1 - 8	No
Trachelius ovum	16	32.5	Not reported	--
Loxodes rostrum	21	33.1	6.4 - 7.2	No
Chilodonella uncinata	50	--	4 - 9.5	No
Paramecium caudatum	40	41.4	4 - 9.4	No
Paramecium bursaria	26	29.3	6 - 9.2	No
Frontonia leucas	32	33.1	7.2 - 9	Yes - 6.6
Frontonia acuminata	31	33.1	6.2 - 9.3	No
Urocentrum turbo	28.5	33.1	5.8 - 9.2	No
Cyclidium glaucoma	51	30.8	7.4 - 9	Yes - 6.4
Spirostomum teres	25	33.3	6.7 - 8.1	No
Stentor roeseli	25	33.1	5.9 - 9.4	No
Stentor igneus	19.9	33.1	7.4 - 9.2	Yes - 6.5
Halteria grandinella	35	36.9	5.2 - 9.5	No
Strobilidium gyrans	22	30.7	6.4 - 7.2	No
Stylonychia mytilus	25	31.0	4 - 8.4	No
Euplotes patella	25	29.3	5.7 - 9.3	No
Aspidisca lynceus	25	33.1	6.5 - 9.8	No
Aspidisca costata	30	34.0	5.4 - 9.4	No
Vorticella campanula	25	29.3	6.9 - 9.5	Yes - 6.5

Another aspect of the dry season activity in the Pantanal is the gradual replacement of aquatic communities by terrestrial ones. This is best observed among the macrophyte species, which are subject to a distinct seasonal succession, greatly influenced by the relative elevation of the individual sites on the floodplain. These changes were monitored during the study by a series of photographs and video images, which show the seasonal changes in the zonation, as described by Heckman (1989). They were also investigated by surveys of the plant species on a 50 m x 50 m experimental area during each of the four seasons (Prado *et al.*, 1994).

The results of these studies revealed that there is a correlation between the relative water depth, which depends on the terrain elevation, and the number of macrophyte species present. Generally, deeper water is dominated by few plants, while shallow water only a few centimeters deep is able to support a greater variety of aquatic and wetland species. This correlation was found throughout all seasons of the year. In the deepest locations, *Eichhornia azurea* was the predominant species throughout the year, while on the shallower parts of the square, a pronounced seasonal succession was observed, which included the moving ecotone mentioned above (Heckman, 1997b).

In the permanent water courses, the water, which becomes lentic during the dry season, displayed characteristically unstable conditions reflected by many variables. The most vital of these for the large invertebrates and aquatic vertebrates is the oxygen concentration. At times, blooms of algae are responsible for elevating the oxygen saturation to over 400% during the daytime, while at other times, bacterial activity reduces the oxygen concentration to nearly 0% for days at a time. Nevertheless, the fishes and invertebrates survive. The pilid snails are able to utilize atmospheric oxygen by means of lung-like organs, while the fishes are observed at such times gulping mixtures of air and water at the surface. Unfortunately, details of all of the many physiological adaptations for respiration in oxygen-poor water remain unknown, but selected studies have revealed details about such adaptations (Saint-Paul, 1984; Braum, 1983; Braum and Bock, 1985). The stategies and morphological modifications of the native fishes in the Pantanal are largely unknown. Because long periods of oxygen deficiency in the water occur most frequently after, rather than before, mass mortalities of fishes are observed and because virtually no natural mortalities of fishes were noted during many periods of long-term dissolved oxygen shortages, it can be assumed that all common fishes in the shallow waters of the wetland are adapted for utilizing atmospheric oxygen.

During the late dry season, a high mortality rate among the fishes results from predation by large mixed flocks of water birds and caimans, which are able to easily capture the fishes gasping air at the surface. Complete mortality of those species not able to aestivate in the mud occurs when the isolated pools of water on the floodplain completely dry out. The somewhat rare mass mortalities in the permanent water bodies occur for other reasons, still not well understood (Section 10.4.2).

10.4.1. Seasonal characteristics

The exact dates on which the transitions to the *seca*, often called the *estagiam*, occur depend not only upon the relative dryness of the particular year but also on the elevation of the sites above sea level. An area only a few centimeters higher than an adjacent one can dry up more than a week later. Hence, only an approximate time period can be given to delimit this season, and this coincides more or less with the southern winter from the shortest day in June through the equinox in September.

The species of the aquatic phase die off after blooming and producing seeds or some form of vegetative propagules, or they produce vegetative plants with characteristics suitable for survival in terrestrial habitats and usually differing morphologically to a great degree from the aquatic forms that developed in the water. The species undergoing the most extreme changes include *Ludwigia inclinata*, *Nymphoides grayana*, and *Eleocharis minima*. Some plants survive maintaining their aquatic form. *Eichhornia azurea* loses most of its band-shaped underwater leaves and some of its emergent foliage but can survive on the exposed sediment for several weeks. Its characteristic rows of parallel-sided, band-like leaves develop anew under the flood waters early in the *enchente* and persist until the *vazante* or *seca*, when the water again recedes from their habitat. If the dryness persists long enough for the sediment to dry out completely, however, the plants die.

The decline of the aquatic habitat is accompanied by a corresponding increase in the terrestrial area of the Pantanal. After a transition phase dominated by a wetland community, plant species adapted to semi-arid conditions occupy the exposed areas. This community is better developed and more stable on the higher areas of the floodplain from which the water recedes first and is dry for the longest period. This community has come to be dominated by introduced grass species, such as *Andropogon bicornis*. It is surmised that prior to the introduction of these species, mainly from Africa, the floodplain was dominated by species of other families, including many in the family Cyperaceae. Many of the introduced species are not liked by the cattle ranchers because they are unsuitable for cattle forage. As already mentioned, heavy grazing has reduced or eliminated the numbers of the species that the cattle prefer and left the floodplain dominated by the plants that they do not like to eat. Among the methods employed in their attempts to eliminate the introduced weeds, the ranchers and land speculators employ fires most frequently. Major brush fires affecting large areas have become characteristic events during the dry season (Section 12).

10.4.2. Mass mortalities of aquatic biota

In August 1992, thousands of dead fishes appeared on the surface of the water where the Rio Bento Gomes crosses beneath the Transpantaneira Highway. Such deaths of large numbers of fishes are often been attributed by scientists from the temperate zones to low oxygen concentrations. However, during other years, the water at the same location was nearly free of dissolved oxygen for weeks at a time during the dry season. Especially at night and during the early morning hours, hundreds of fishes could be seen gasping air at the surface of the water. However, during these years, no mass mortalities occurred. The fishes were easy prey for the many caimans and water birds that congregated at these few permanent water bodies, and the few dead fishes observed had obviously been killed by birds that had captured them but had been unable to crush their spines enough to swallow them. Species such as *Platydoras costatus*, for example, have thick, barbed spines, which can be locked in an extended position. Cormorants have to crush these and other catfishes in their beaks for several minutes before the bones and spines are fractured sufficiently for the fishes to be safely swallowed. Many times, after the birds have worked on crushing the spines for about five minutes without success, the fish is discarded. The few dead fishes in the water during these years showed the unmistakable injuries inflicted by the beaks of the water birds. It is safe to say that nearly all of the fishes present in these small rivers are capable of utilizing atmospheric oxygen in some way sufficient to permit their survival in water nearly free of oxygen.

The most likely cause of the occasional mass mortalities must be sought elsewhere. It is considered most probable that a botulism-like toxin or similar poisonous substance produced by bacteria killed the fishes. Each year during the dry season, the fishes that are killed but not swallowed by water birds decompose in the water under nearly anaerobic conditions. These fish cadavers decompose in stagnant, nearly oxygen-free water, promoting the development of massive populations of anaerobic bacteria and other microorganisms. Such microbes are known to produce a variety of toxins, which work specifically on certain groups of animals. Botulism is a deadly form of intoxication caused in human beings by one of the most toxic substances known, which is produced by anaerobic bacteria, including *Clostridium botulinum* (Van Ermengem) and related species. Similar substances are deadly for birds in extremely small amounts but relatively harmless to man. Still others are highly toxic to other forms of life. Obviously, the conditions in the Rio Bento Gomes during the dry season are suitable for the production of similarly toxic substances, and when a strain of suitable bacteria becomes common in the cadavers of many fishes killed and left uneaten by the birds, toxic substances can be produced that kill fishes, even when present in extremely small amounts. This seems the most likely explanation for the mass death of the fishes. Unfortunately, inputs of domestic sewage and similar substances into the rivers can increase both the amounts of suitable substrates and the number of different microorganisms, thereby increasing the frequency of mass fish mortalities in the rivers. This phenomenon urgently requires microbiological investigation and should not be explained away as being caused by oxygen deficiencies or related phenomena without exhaustive studies to rule out other factors as likely causes of the particular cases under study.

Closer to human settlements, anthropogenic pollution might be suspected as a possible cause of fish mortalities, but in the Pantanal itself, there are still no sources of toxic substances large enough to be responsible. If pesticides are introduced, the situation will certainly change (Heckman, 1993).

Occasional mass mortalities among the invertebrates are also observed. During the dry season of 1994, particularly low water levels in the Rio Pixaim resulted in mass mortalities of the *Anodontites* species in the sediment near the shore. These bivalves remained buried in the firm, fine sediment just below the annual low water level during the other years of the study. However, when the water level sank below its usual minimum level, large numbers of the bivalves died, apparently from overheating when exposed to direct sunlight for several weeks.

Mass mortalities of water plants changing the community structure during subsequent years is another phenomenon frequently observed in the Pantanal. This results from the pronounced year to year differences in the weather, which cause a particular site to remain under water for one entire year and completely dry out during the next. During unusually dry years, low-lying sites dominated by species such as *Eichhornia azurea* or *Salvinia auriculata*, completely dry up for periods long enough to cause the deaths of the entire aquatic plant populations. During the subsequent rainy season, the formerly dominant plant species do not rapidly return, and other aquatic plants, such as *Pontederia lanceolata* or *Ludwigia sedoides*, come to dominate the habitats. After a number of relatively wet years, the *Eichhornia azurea* and *Salvinia auriculata* may regenerate and become dominant again, but if several relatively dry years occur in sequence, still other species may infiltrate the aquatic community.

Similarly, consecutive wet years may favor the spread of such species as *Thalia geniculata* and *Cyperus giganteus*, which require more time in relatively deep water to establish themselves. During dry years, extensive stands of these species may be eliminated over rather large areas.

Like the aquatic flora, the terrestrial plants are subject to mass mortality each year during the course of the annual cycle. Hence, nearly all of the species on the desiccated floodplain are annuals with life cycles lasting approximately as long as the terrestrial conditions at any particular site. Some of the species are drowned by the flood waters, while others have already gone to seed and died in the courses of their natural life cycles before the flood waters arrive.

Although a study spanning a considerable period of time would be necessary to fully clarify the processes functioning, the occurrence of pronounced year to year changes in the structure of the biotic communities could be unequivocally documented during the three complete annual cycles observed.

A general phenomenon apparent from the records of mass mortalities and subsequent recovery of the flora and fauna is the characteristically high reproductive capacity of most of the plant and invertebrate species, enabling them to survive in such an unstable habitat. As a result, the Pantanal can be characterized as a network of habitats suitable mainly for the r-strategists, that is, those species that survive by producing large numbers of offspring to compensate for high mortality. This applies to the aquatic invertebrates, fishes, and some of the mammals and reptiles, but not to the birds. The great mobility of birds and large mammals permits them to avoid the region while conditions are unfavorable to them and take advantage of large seasonal food supplies wherever available.

The problem of defining the strategy used to overcome especially high rates of mortality must be given some special attention. The hydrological conditions in the Pantanal, especially the strong water currents in the streams and rivers during the *cheia* and the relative shallowness of the lentic water on the floodplain, preclude such extremely high reproductive rates as those that occur in the sea. What is essentially important for survival in the Pantanal is a high rate of recruitment, that is, a high rate of production of young at a stage of development advanced enought to allow them a high probability of survival when they begin activity independent of the parents. They must be able to grow fast enough to have reached maturity by the time conditions conducive to reproduction again occur. This means that many young must be produced and protected in a way that a maximum number survive until the annual season of mass mortality so that at least a small parental stock can survive until favorable conditions return.

Among the fishes, the successful species include cichlids which guard their broods. They not only produce many young but also protect them so that they attain a maximum population growth during the favorable rainy seasons. The crabs adopt a similar strategy of parental care of the moderately large numbers of juveniles in the brood pouch beneath the abdomen. In these ways, the larval periods are passed under parental protection, and the young that emerge are already fully-formed juveniles. Thus, the rapid rates of reproduction are frequently augmented by some behavior patterns characteristic of k-strategists but which produce juvenile populations of maximum size. In this way, the probability of at least a few individuals surviving the adverse conditions during the dry season is considerably increased. The meaning of the enormous population fluctuations for the dynamics of the entire wetland system is apparent from the discussions in Sections 7 and 8, as well in this section.

11. Ecosystem dynamics

As in the case of the communities studied, the dynamic systems must be investigated according to the activities of the most abundant species. Those that are predominant in the ecosystem will therefore be given the most consideration, while the rarer species must be somewhat neglected. This, of course, tends to undermine the estimation of the wetland as a habitat for rare and unusual species, which is one of its great assets. Additional reference will therefore be made to the relationships of the rarer species to the main biological energy fluxes within the system.

A fundamental problem arises in the interpretation of the data, specifically, from the lack of a precise definition of gross and net primary production. Although these terms appear in myriad publications, numerous authors have proposed many different definitions for them. To eliminate problems in understanding what is being discussed, terms such as gross and net primary production alone will be avoided or specifically defined each time they are used.

It should not be forgotten that the ecosystem in the Pantanal is extremely complex, and even limiting the investigation to a small part of the wetland does not negate the fact that the processes are far more complex than similar processes in the temperate zones, if, for no other reason, because they proceed so rapidly at such high temperatures and with such an overabundance of solar energy. Furthermore, all processes not only occur within a limited space in the complex biomosaic of the Pantanal but also for relatively short periods of time, after which other organisms assume dominance and other processes take place.

11.1. Primary production

Preliminary evaluations of primary production in the Pantanal of Poconé revealed that the processes involved are extraordinarily complex, far more so than in wetlands of the temperate zones. Several features of the production of organic material are obvious at the outset. Together, they indicate that intriguing processes are taking place which make the Pantanal, and apparently other wetlands in the tropical wet-and-dry climatic zone, as well, fundamentally different from wetlands and water bodies of both temperate and equatorial regions.

Until recently, little information had been available on the standing stock and production in the Pantanal. Some of the physical and chemical data and a general survey of the flora of shallow waters along the Transpantaneira Highway were provided by Da Silva and Pinto Silva (1989). The biomass of three macrophyte species in the Pantanal of Barão de Melgaço has also been reported (Da Silva and Esteves, 1993). The general methods applied in traditional "biomass" studies, however, are poorly suited for determining biotic processes quantitatively in the tropical wet-and-dry climatic zone. This becomes obvious when the same site in the wetland is observed for a period of only a few weeks. When areas are initially flooded and immediately after the floodwaters recede, a nutrient turnover occurs, provoking rapid plant growth. On the other hand, during the dry season, the growth of terrestrial plants is greatly limited over most of the floodplain by a lack of water, while a lack of inorganic nutrients causes the growth to stagnate during the *cheia* (Heckman, 1994a). Moreover, the exact dates on which the conditions conducive to growth are encountered differ from place to place and from year to year.

The growth of the aquatic plants, both algae and macrophytes, occurs amazingly rapidly during the early part of the rainy season, the *enchente*. The rate of photosynthesis is obviously far greater at this time than the rates encountered by limnologists investigating wetlands in the temperate zones. This is possible because of the adaptation of the macrophytes to the strong illumination. Plants in Brazilian wetlands have two distinct kinds of photosynthetic metabolism, one best adapted to growth in shade and utilization of low levels of illumination and the other well adapted to intense insolation. Examples are grasses referred to as C_3 and C_4. C_4 grasses, which are efficient at exploiting the intense sunlight available in the tropics, are known to be able to produce enormous amounts of organic mass in relatively short periods of time. Piedade (1993) reported that one C_4 grass, *Echinochloa polystachya* (H. B. K.) Hitchcock, was able to produce 99 $t \cdot ha^{-1}$ (= 9,900 $g \cdot m^{-2}$) annually, and 2.7% of the organic substance produced is protein. In the Pantanal of Barão de Melgaço, a production of *Eichhornia azurea* of 600 g of dry weight per m^2 was reported during the *enchente*, in the months of November and December, while elsewhere in the Pantanal, 9.7 tons of *Pontederia lanceolata* per hectare (= 970 $g \cdot m^{-2}$) were reportedly produced annually (Camargo and Esteves, 1995). The latter figures are assumed to be dry weight, but this was not specified by the authors. These figures are clearly greater than those recorded in the temperate zones, where the net production does not exceed 500 $g \cdot m^{-2}$ of submerged plants or 750 $g \cdot m^{-2}$ of emergent species for the entire year, recorded as ash-free dry weight (Bican *et al.*, 1986). Considering the fact that many water bodies in the temperate zones are far richer in inorganic nutrients than the water in the Pantanal, it is evident that primary production in the tropics can be much more rapid over the short term. As already mentioned, however, the significance of the comparisons is severely reduced by the methodological differences of the various authors and deficiencies in precisely reporting the method used. It is also evident that where "net production" of the plant material is spoken of, this is actually gross production when considering the ecosystem as a whole.

From the information already available, an extremely high gross primary productivity, defined according to the total accumulation of organic materials in the ecosystem, can be assumed for the Pantanal at this time. Nevertheless, the average net annual primary production on the floodplain, defined as the net increase in organic material after the passage of one year, averages zero. Everything produced is broken down again during the course of the year. There is no accumulation of organic material anywhere in the aquatic system. It must be stressed that these findings do not apply to gallery forests, the wooded islands, called hummocks or *capões de mata*, and other biotic communities above the high water level of the *cheia*. The pronounced seasonal succession described in the previous section is characteristic of the water bodies and adjacent wetlands. During this succession, primary production by algae is intense for short periods of time during the *seca* and early *enchente*. At other times, algae, especially planktonic species, play almost no role at all in the community production. Submerged plants are productive throughout the period of flooding, but they gradually lose their importance as oxygen producers in the water column as the flood period advances. They are increasingly displaced by emergent plants, especially floating ones. Primary production is strongly limited by the nutrient supply during most of the year, but inorganic nutrients are present in excess at the time of the initial flooding during the *enchente*. The anchored water plants and the animals that live among them have stored virtually all of the available nutrients by the time the water level has reached its maximum, and they therefore act as effective conservative agents for preventing the loss of nutrients downstream in the river system. Each of these aspects of the seasonal cycle requires some explanation in detail. Terrestrial plants of the floodplain grow according to an annual schedule similar to those of the aquatic species, but during the rainless rather than the rainy seasons.

The great seasonal variability in the gross primary production throughout the entire river system was confirmed by the review of Neiff (1990), who reported his productivity values as g $C \cdot m^{-2}$ per day. His review treated many groups of primary producers together, including phytoplankton, periphyton, and macrophytes. In the lower sections of the Rio Paraguay at Puerto Bermejo, Argentina, the lowest productivity value of 0.004 g $C \cdot m^{-2}$ per day was recorded (Bonetto, unpublished thesis, cited by Neiff, 1990), while the highest value of 0.750 g $C \cdot m^{-2}$ per day was determined in the river by Bonetto *et al.* (1981). A somewhat similar range of values was recorded in the lower Rio Paraná upstream from its confluence with the Rio Paraguay, with a minimum of 0.000 being recorded near Corrientes, Argentina (Bonetto *et al.*, 1979), and a maximum of 0.850, recorded at a nearby site in the same city (Bonetto, 1983). With differences of two to more than three orders of magnitude, it is apparent that a much more extensive data set is necessary before temporal and spatial patterns can be established. It is noteworthy that the maximum values determined were significantly below the maxima recorded in the Amazon System, where 1.04 g C·m^{-2} per day was reported in the black water Lake Cristalino by Rai and Hill (1984), while 2.41 g C·m^{-2} per day was detected in the Rio Tapajos (Welcomme, 1985).

Considering the phytoplankton alone, Perotti de Jorda (1984) determined its primary production in the middle section of the Rio Paraná in Argentina to be 34.4 g C·m^{-2} per year in 1977 and 39.9 g C·m^{-2} for the year 1978. The seasonal differences in the daily rate showed extreme variability, with the highest values occurring in the southern spring, the time of the early *enchente* in the Pantanal. During the two years of the study, 0.8 g C·m^{-2} was the absolute maximum. The lowest values were encountered in late autumn, when an absolute minimum of 0.001 g C·m^{-2} was recorded. The daily values were reported to be gross productivity, which could also be defined as a short-term net productivity of the plankton community, depending upon definition. At any rate, the integration of the daily values to determine annual productivity is risky because it must be remembered that the short-lived algal species assimilate and release nutrient elements much more rapidly than do vascular plants. Hence, the same molecules could be incorporated in the algae and released again during the course of a single year, giving a false impression about what can be "produced" annually. Furthermore, the nutrient supply in the middle section of the Rio Paraná is quite richer than it is in the Pantanal (Perotti de Jorda, 1984), making specific comparisons of the primary production invalid.

The maximum primary production of a single floating species, *Pistia stratiotes*, was estimated to be 0.58 g C·m^{-2} per day in an impoundment near the Rio de la Plata in Uruguay during the spring (Mazzeo *et al.*, 1993). It can be surmised that the much greater density of this species in the artificial water body, the Cisne Reservoir, than in the Pantanal is a result of the large amounts of inorganic nutrients that became available as the impoundment filled with water after the dam was completed. Mazzeo *et al.* (1993) estimated that the plant covered 33% of the water surface. In contrast to the situation in the Pantanal, where the maximum numbers of this plant and other floating macrophytes are present during the *vazante*, corresponding to the southern autumn, the maximum standing crop of this species in Uruguay, 708.8 g dry weight·m^{-2}, was encountered in September, during the transition from winter to spring conditions.

Table 11.1 shows the main species participating in the primary production within some of the most prominent aquatic and semi-aquatic biotic communities in the wetland. The table also indicates some of the physiological adaptations, such as the ability to live heterotrophically, the medium of gas exchange, and the possible sources of the nutrients. However, a table cannot properly reflect the complexity of the different physiological adaptations giving each species claim to a niche within the diverse network of biotic communities that develops in different localities within the wetland at various times of year.

A few of the more obvious special adaptations that give certain species advantages for survival under unusual conditions deserve mention. Some of these blur the distinctions between submerged and emergent plants. For example, *Euglena sanguinea* and *E. gracilis* form thick mats on the surface tension layer of the water during the day, giving shallow pools of water a blood red or bright green coloration, respectively. A careful look at the undisturbed surfaces of such pools reveals that these organisms actually move through the surface tension layer and rest on the upper side, that is, directly in contact with the atmosphere. At even the slightest disturbance, they disperse into the underlying water. In this way, they can obtain carbon dioxide directly from the atmosphere as well as from the water and release oxygen directly into the atmosphere, as well. At the time these species are commonly observed, massive numbers of submerged algae often produce conditions of supersaturation in the water. At such times, it might well be a great advantage to be able to release the oxygen produced by photosynthesis directly into the atmosphere. Studies of the implications of this phenomenon have never been undertaken, and although the euglenas involved must have special adaptations to extremely bright solar radiation, their photosynthetic metabolism has not been studied under tropical conditions.

Table 11.1: Growth patterns of the most abundant primary producers in the water bodies of the Pantanal of Poconé, a synopsis of the studies by Heckman (1994a), Prado *et al.* (1994), and De-Lamonica-Freire and Heckman (1996). E = *enchente*; C = *cheia*; V = *vazante*; S = *seca*; R = nutrient-rich water; M = water with moderate nutrient content; P = nutrient-poor water; PP = nutrient poor water with supplemental nutrients through trapping small animals or nitrogen fixation by symbiotic microorganisms; W = ambient water; A = atmosphere. The standing stock of a species is rated for during its season of greatest abundance as follows: +++ = a major quantitative role in primary production by algae that produce or obviously participate in blooms or form abundant colonies and macrophytes that cover large areas or are otherwise locally dominant; ++ = never present in massive numbers but abundant in some communities; + = never abundant and not dominant in any community.

Species	Main season of vegetative growth	Nutrient source	Gas exchange	Stock
Cyanobacteria				
Agmenellum quadruplicatum	S-E	R	W	+
Microcystis aeruginosa	S	R	W	+++
Microcystis flos-aquae	S	R	W	++
Crinalium magnum	E	R	W	+
Hormoscilla sp.	V	M	W	+
Oscillatoria agardhii	E	R	W	++
Oscillatoria lacustris	S	R	W	+
Oscillatoria limosa	S	R	W	++
Oscillatoria princeps	S	R	W	++
Oscillatoria tenuis	S	R	W	++
Phormidium sp.	E	R	A	+
Spirulina jenneri	S	R	W	+
Spirulina subsalsa	V	M	W	+
Anabaena sphaerica	V	M	W	+
Anabaena spiroides	S-E	R	W	++
Anabaena azollae	V-E	M	W	+
Pseudanabaena cf. *constricta*	E	R	W	+
Calothrix gypsophila	V	M	W	+
Chrysophyta				
Chrysococcus rufescens	S	R	W	+
Dinobryon behningii	S	M	W	+
Cryptomonas erosa	S-E	R	W	++
Cryptomonas reflexa	S	R	W	+
Synura sp.	all	M	W	++
Mallomonas sp.	S	M	W	+
Bacillariophyta				
Aulacoseira herzogii	V-E	M	W	+
Aulacoseira italica	V-S	M	W	++
Aulacoseira granulata	V-S	M	W	+
Melosira varians	E-S	M	W	++
Diatoma mesodon	E-S	M	W	+
Fragilaria capucina	S	M	W	++
Fragilaria ulna	all	P-M	W	++
Eunotia bilunaris	S	M	W	+
Eunotia camelus	V-S	M	W	+
Eunotia flexuosa	V-S	M	W	+
Eunotia serra	V	M	W	+
Cymbela silesiaca	all	M	W	+

Species	Main season of vegetative growth	Nutrient source	Gas exchange	Stock
Gomphonema angustatum	S	M	W	+
Gomphonema augur	all	M	W	++
Gomphonema clavatum	V-S	M	W	+
Navicula menisculus	V	M	W	+
Navicula cuspidata	V-S	M	W	+
Navicula radiosa	V	M	W	+
Pinnularia dactylus	V	M	W	+
Pinnularia major	all	M	W	+
Pinnularia viridis	all	M	W	+
Rhoicosphenia abbreviata	V-S	M	W	+
Stauroneis phoenicenteron	E & V	M	W	+
Hantzschia amphioxys	E-C	P-M	W	+
Surirella biseriata	C	P	W	+
Surirella robusta	all	M	W	++
Surirella tenera	S	M	W	+
Pyrrhophyta				
Gymnodinium cneoides	S	R	W	+
Sphaerodinium cinctum	S-E	R	W	++
Chlorophyta				
Eudorina elegans	S-E	M-R	W	++
Gonium pectorale	V-E	M-R	W	+
Pandorina morum	V-E	M-R	W	+++
Spondylomorum quarternarium	S	R	W	+
Volvox aureus	E	M	W	++
Sphaerocystis schroeteri	S-E	R	W	++
Sphaerocystis planctonica	S	R	W	+
Actinastrum hantzschii	S-E	R	W	+
Crucigenia fenestrata	S	R	W	+
Crucigeniella rectangularis	S-E	R	W	+
Dicloster acuatus	S	R	W	+
Scenedesmus acuminatus	V-E	R	W	+++
Scenedesmus armatus	S-E	R	W	+++
Scenedesmus bicaudatus	S	R	W	++
Scenedesmus brasiliensis	S	R	W	+
Scenedesmus carinatus	S	R	W	+
Scenedesmus denticulatus	S	R	W	+
Scenedesmus granulatus	S	R	W	+
Scenedesmus intermedius	S-E	R	W	+++
Scenedesmus opoliensis	S-E	R	W	+
Scenedesmus quadricauda	all	M-R	W	+++
Ankistrodesmus bernardii	S	R	W	+
Ankistrodesmus falcatus	S	R	W	++
Ankistrodesmus spiralis	S	R	W	+
Closteriopsis acicularis	V	M	W	+
Kirchneriella lunaris	S-E	R	W	++
Oonephris obesa	S	R	W	+
Selenastrum rinoi	S-E	R	W	+++
Tetraedron trigonum	E	R	W	++
Coelastrum asteroideum	S	R	W	+
Coelastrum cambricum	S-E	R	W	++
Coelastrum indicum	S	R	W	+
Coelastrum microporum	S	R	W	+

Species	Main season of vegetative growth	Nutrient source	Gas exchange	Stock
Coelastrum pseudomicroporum	S	R	W	+
Dictyosphaerium pulchellum	S	R	W	+
Eutetramorus fottii	S	R	W	+
Eutetramorus planctonicus	S	R	W	+
Acanthosphaera zackariasii	S	R	W	++
Golenkinia radiata	S-E	R	W	+++
Pediastrum duplex	V-E	R	W	+++
Pediastrum tetras	S-E	R	W	+++
Zygnema sp.	V-E	M	W	+++
Mougeotia scalaris	V	M	W	++
Spirogyra sp. 1	E-V	P-M	W	++
Spirogyra sp. 2	all	M	W	+++
Spirogyra sp. 3	all	M	W	++
Spirogyra sp. 4	E	M	W	+
Sirogonium sp.	C	P	W	+
Mesotaenium mirificum	C	P	W	+
Netrium digitus	E-V	P	W	+
Spirotaenia condensata	C	P	W	+
Gonatozygon monotaenium	C	P	W	++
Gonatozygon pilosum	C	P	W	++
Gonatozygon aculeatum	C	P-M	W	++
Closterium abruptum	C	P	W	+
Closterium angustatum	E-C	P	W	++
Closterium archerianum	E	M	W	+
Closterium closterioides	E	M	W	+
Closterium cynthia	E	M	W	+
Closterium dianae	C	P	W	+
Closterium ehrenbergii	all	M-R	W	++
Closterium kuetzingii	all	P-M	W	+++
Closterium lagoense	E	M	W	+
Closterium limneticum	C	P	W	++
Closterium lineatum	all	M	W	++
Closterium lunula	E-V	P-M	W	++
Closterium malmei	E-C	P	W	+
Closterium moniliferum	S-E	M-R	W	++
Closterium nasutum	E-C	P	W	+
Closterium nematodes	C	P	W	+
Closterium parvulum	C	P	W	++
Closterium porrectum	E	M	W	+
Closterium praelongum	E-C	P-M	W	++
Closterium pritchardianum	E-C	P-M	W	++
Closterium ralfsii	E-V	P-M	W	+
Closterium setaceum	all	P-M	W	++
Closterium tortum	E-C	P-M	W	+
Closterium tumidum	C	P	W	+
Closterium turgidum	all	P-M	W	++
Closterium venus	C	P	W	+
Actinotaenium turgidum	S	M	W	+
Bambusina borreri	E-C	P-M	W	++
Cosmarium bireme	E-C	P-M	W	+
Cosmarium candianum	all	P-M	W	++
Cosmarium commissurale	E-C	P-M	W	++
Cosmarium conspersum	V	M	W	+
Cosmarium contractum	E	M	W	+

Species	Main season of vegetative growth	Nutrient source	Gas exchange	Stock
Cosmarium decedens	C	P	W	+
Cosmarium denticulatum	S-C	P-M	W	+++
Cosmarium depressum	E-C	P-M	W	++
Cosmarium exiguum	E-C	P-M	W	++
Cosmarium galeritum	E & V	M	W	++
Cosmarium galiciense	C	P	W	+
Cosmarium ginzbergeri	all	P-M	W	++
Cosmarium hexagonum	C	P	W	+
Cosmarium impressulum	E	M	W	+
Cosmarium margaritatum	all	P-M	W	++
Cosmarium moerlianum	E-V	P-M	W	++
Cosmarium moniliforme	all	P-M	W	++
Cosmarium norimbergense	E	M	W	+
Cosmarium obtusatum	S-E	M	W	++
Cosmarium ornatum	C	P	W	+
Cosmarium orthostichum	C	P	W	+
Cosmarium pachydermum	V-E	M	W	++
Cosmarium paraguayense	S	M	W	+
Cosmarium pseudoconnatum	E-V	P-M	W	++
Cosmarium pseudoprotuberans	E-C	P-M	W	++
Cosmarium pseudopyramidatum	E-V	P-M	W	++
Cosmarium punctulatum	C-V	P-M	W	+
Cosmarium pyramidatum	S-C	P-M	W	++
Cosmarium quadrum	V	M	W	+
Cosmarium rectangulare	E	M	W	+
Cosmarium refringens	C	P	W	+
Cosmarium speciosum	S-E	M	W	++
Cosmarium subspeciosum	C	P	W	+
Cosmarium subtumidum	S	M	W	+
Cosmarium supraspeciosum	C	P	W	+
Cosmarium tinctum	C	P	W	+
Cosmarium triangulare	C	P	W	+
Cosmarium trilobatum	C	P	W	+
Cosmarium trilobulatum	E	M	W	+
Cosmarium urnigerum	E	M	W	+
Cosmarium vitiosum	E-V	P-M	W	+
Desmidium aequale	C	P	W	+
Desmidium aptogonum	E-C	P-M	W	++
Desmidium baileyi	E-V	P-M	W	++
Desmidium graciliceps	C	P	W	+
Desmidium laticeps	C	P	W	+
Desmidium quadratum	C-V	P-M	W	++
Docidium baculum	E-V	P-M	W	++
Euastrum abruptum	E	M	W	+
Euastrum ansatum	all	P-M	W	++
Euastrum attenuatum	E	M	W	+
Euastrum didelta	E & V	M	W	++
Euastrum elegans	E-C	M	W	+
Euastrum evolutum	E-C	M	W	+
Euastrum fissum	C	P	W	+
Euastrum gemmatum	E-C	P-M	W	++
Euastrum pectinatum	C	P	W	+
Euastrum platycerum	V	M	W	+
Euastrum rectangulare	E	M	W	+

Species	Main season of vegetative growth	Nutrient source	Gas exchange	Stock
Euastrum subintegrum	E-C	P-M	W	++
Euastrum verrucosum	E & V	M	W	++
Groenbladia undulata	C	P	W	+
Hyalotheca dissiliens	C	P	W	+
Micrasterias abrupta	E-V	P-M	W	++
Micrasterias arcuata	all	P-M	W	++
Micrasterias borgei	E-V	P-M	W	++
Micrasterias furcata	all	P-M	W	++
Micrasterias incisa	E	M	W	+
Micrasterias laticeps	E-V	P-M	W	++
Micrasterias mahabuleshwarensis	E-V	P-M	W	++
Micrasterias papillifera	C	P	W	+
Micrasterias pinnatifida	V	M	W	++
Micrasterias radians	E-V	P-M	W	++
Micrasterias radiosa	S-C	P-M	W	++
Micrasterias ringens	C	P	W	+
Micrasterias thomasiana	C	P	W	+
Micrasterias torreyi	E-V	P-M	W	++
Micrasterias tropica	S-C	P-M	W	++
Micrasterias truncata	all	P-M	W	++
Onychonema filiformis	S	M	W	+
Onychonema laeve	E-V	P-M	W	++
Pleurotaenium baculoides	E & V	M	W	+
Pleurotaenium coroniferum	V-S	M	W	+
Pleurotaenium cylindricum	V	M	W	+
Pleurotaenium ehrenbergii	C & S	P-M	W	++
Pleurotaenium eugeneum	V-S	M	W	++
Pleurotaenium maskellii	V-E	M	W	++
Pleurotaenium minutum	E-V	P-M	W	++
Pleurotaenium nodosum	C-V	P-M	W	++
Pleurotaenium nodulosum	C	P	W	++
Pleurotaenium ovatum	C	P	W	+
Pleurotaenium subalternans	V	M	W	+
Pleurotaenium trabecula	all	P-M	W	++
Spondylosium ellipticum	C	P	W	+
Spondylosium planum	E	M	W	+
Spondylosium pulchrum	V	M	W	+
Staurastrum arctiscon	E	M	W	+
Staurastrum asteroideum	E	M	W	+
Staurastrum claviferum	E	M	W	+
Staurastrum disputatum	S-E	M	W	++
Staurastrum erinacoideum	C	P	W	+
Staurastrum gladiosum	C	P	W	+
Staurastrum leptocladum	C	P	W	++
Staurastrum leptacanthum	V	M	W	+
Staurastrum margaritaceum	E	M	W	++
Staurastrum minnesotense	E	M	W	+
Staurastrum muticum	C	P	W	+
Staurastrum novae-caesareae	V-E	M	W	++
Staurastrum nudibrachiatum	C-V	P-M	W	++
Staurastrum orbiculare	E	M	W	+
Staurastrum paradoxum	E-V	P-M	W	++
Staurastrum quadrangulare	C	P	W	+
Staurastrum rotula	C	P	W	+

Species	Main season of vegetative growth	Nutrient source	Gas exchange	Stock
Staurastrum sebaldi	C	P	W	+
Staurastrum setigerum	C	P	W	+
Staurastrum striolatum	C	P	W	+
Staurastrum tohopekaligense	C	P	W	++
Staurastrum trachytithophorum	C	P	W	+
Staurastrum wolleanum	E	M	W	+
Staurastrum trifidum	C	P	W	+
Staurodesmus convergens	all	P-M	W	++
Staurodesmus crispus	E & V	M	W	++
Staurodesmus cuspidatus	E	M	W	+
Staurodesmus dejectus	E	M	W	+
Staurodesmus dickiei	E-V	P-M	W	++
Staurodesmus lobatus	V-E	M	W	++
Staurodesmus mamillatus	E-C	P-M	W	++
Staurodesmus maximus	C	P	W	+
Staurodesmus subulatus	E-V	P-M	W	++
Staurodesmus validus	E-V	P-M	W	++
Teilingia granulata	E & V	M	W	++
Xanthidium antilopaeum	C-V	P	W	++
Xanthidium convergens	E-V	P-M	W	++
Xanthidium mamillosum	C	P	W	+
Xanthidium mucronulatum	E-V	P-M	W	++
Xanthidium trilobum	C	P	W	++
Penium spirostriolatum	C	P	W	++
Penium subspirostriolatiforme	E	M	W	+
Chaetophora elegans	C	P	W	++
Draparnaldia sp. 1	V	M	W	+
Draparnaldia sp. 2	V	M	W	+
Bulbochaete sp.	E & V	M	W	++
Oedogonium sp. 1	S-E	M-R	W	++
Oedogonium sp. 2	V	M	W	+
Euglenophyta				
Euglena acus	all	P-R	W	+++
Euglena deses	S	R	W	+
Euglena ehrenbergii	V	M	W	++
Euglena fusca	S	R	W	++
Euglena gracilis	V-S	M-R	W & A	+++
Euglena intermedia	V	M	W	+
Euglena limnophila	S	R	W	++
Euglena oxyuris	S	R	W	++
Euglena tripteris	V-E	M-R	W	++
Euglena sanguinea	S-E	R	A & W	+++
Euglena spirogyra	all	P-R	W	++
Lepocinclis fusiformis	E	M	W	+
Lepocinclis ovum	S	R	W	++
Phacus brachykentron	S	R	W	+
Phacus curvicauda	S-E	M-R	W	++
Phacus lismorensis	S-E	M-R	W	++
Phacus longicauda	all	P-R	W	++
Phacus mariae	S-E	M-R	W	++
Phacus orbicularis	S-E	M-R	W	++
Phacus parvulus	S	R	W	+
Phacus platalea	S	R	W	+

Species	Main season of vegetative growth	Nutrient source	Gas exchange	Stock
Phacus pleuronectes	all	P-R	W	++
Phacus pyrum	E	M-R	W	+
Phacus suecicus	V-E	M-R	W	++
Phacus tortus	all	P-R	W	++
Trachelomonas abrupta	S	R	W	++
Trachelomonas acanthophora	S-E	R	W	++
Trachelomonas armata	all	P-R	W	++
Trachelomonas bacillifera	E	R	W	++
Trachelomonas crispa	S	R	W	++
Trachelomonas cervicula	S	R	W	+
Trachelomonas dubia	S	R	W	++
Trachelomonas elliptica	S	R	W	+
Trachelomonas hispida	all	P-R	W	+++
Trachelomonas lacustris	S-E	M-R	W	++
Trachelomonas lemmermannii	S	R	W	++
Trachelomonas magdaleniana	E	R	W	++
Trachelomonas megalacantha	V-E	M-R	W	++
Trachelomonas molesta	S	R	W	+++
Trachelomonas planctonica	E	R	W	++
Trachelomonas pulcherrima	E	R	W	++
Trachelomonas robusta	S	R	W	++
Trachelomonas superba	S-E	R	W	++
Trachelomonas sydneyensis	S-E	R	W	++
Trachelomonas varians	S	R	W	+
Trachelomonas volvocina	all	P-R	W	+++
Trachelomonas volvocinopsis	all	P-R	W	++
Strombomonas australica	E	R	W	+
Strombomonas chodati	S	R	W	+
Strombomonas encifera	S	R	W	++
Strombomonas gibberosa	S	R	W	++
Strombomonas scabra	S	R	W	++
Charophyta				
Nitella acuminata	V	M	W	+
Bryophyta				
Ricciocarpus natans	V-S	R	A	+
Tracheophyta				
Azolla mexicana	all	PP-M	A	+
Azolla filiculoides	all	PP-M	A	+
Salvinia auriculata	all	P-M	A	+++
Marsilea polycarpa	E-V	M-R	A	+
Nymphaea amazonum	all	P-M	A	++
Victoria cruziana	E-V	P-M	A	+
Polygonum punctatum	all	P-M	A	+++
Polygonum hydropiperoides	all	P-M	A	+++
Alternanthera philoxeroides	E-V	P-M	A	++
Ludwigia sedoides	E-V	P-M	A	++
Ludwigia inclinata	all	P-M	W-A	+++
Hydrolea spinosa	all	P-M	A	+++
Aeschynomene fluminensis	all	PP-M	A	+
Caperonia castaneifolia	all	P-M	A	++
Myriophyllum mattogrossense	E-V	P-M	W	+++

Species	Main season of vegetative growth	Nutrient source	Gas exchange	Stock
Cabomba furcata	E-V	P-M	W	+++
Justicia laevilinguis	E-V	P-M	A	++
Nymphoides grayana	all	P-M	A	+++
Diodia kuntzei	V-E	P-M	A	+++
Ipomoea carnea	all	P-M	A	+++
Hyptis lorentziana	E-V	P-M	A	++
Bacopa myriophylloides	V-E	P-M	A	+
Utricularia foliosa	E-V	PP-M	W	+++
Utricularia poconensis	E-V	PP-M	W	+++
Utricularia gibba	E-V	PP-M	W	+
Echinodorus paniculatus	all	P-M	A	+++
Echinodorus macrophyllus	all	P-M	A	+++
Echinodorus glaucus	all	P-M	A	+
Echinodorus tenellus	V-E	P-M	A	++
Sagittaria rhombifolia	E-V	P-M	A	+
Sagittaria guyanensis	E-V	P-M	A	+
Hydrocleys nymphoides	E-V	P-M	A	+
Eichhornia azurea	all	P-M	W-A	+++
Eichhornia crassipes	all	P-M	A	++
Pontederia lanceolata	all	P-M	A	+++
Pontederia cordata	E-V	P-M	A	+
Reussia subovata	E-V	P-M	A	++
Reussia rotundifolia	E-V	P-M	A	++
Eleocharis elegans	E-V	P-M	A	+++
Eleocharis acutangula	E-V	P-M	A	+++
Eleocharis minima	all	P-M	A	+++
Oxycaryum cubense	E-V	P-M	A	+++
Lipocarpha humboldtiana	V-E	P-M	A	+
Cyperus giganteus	all	P-M	A	++
Cyperus haspan	all	P-M	A	++
Cyperus brevifolius	V-E	P-M	A	++
Cyperus ferax	V-E	P-M	A	+
Cyperus luzulae	V-E	P-M	A	+
Fimbristylis dichotoma	V-E	P-M	A	+
Hymenachne amplexicaulis	E-V	P-M	A	++
Leersia hexandra	all	P-M	A	+++
Panicum laxum	all	P-M	A	+++
Panicum repens	all	P-M	A	++
Paspalidium paludivagum	E-V	P-M	A	+
Paspalum repens	all	P-M	A	++
Reimarochloa brasiliensis	E-V	P-M	A	++
Setaria parviflora	V-E	P-M	A	+++

While most emergent aquatic macrophytes exchange gases with the atmosphere through the stomata of their leaves, it is now known that a considerable amount of the gaseous oxygen is conveyed through the aerenchyme tissue in the roots to the underlying water or sediment (Jedicke *et al.*, 1989). Apparently, the amount of oxygen reaching the water in this manner depends mainly upon the oxygen saturation. Thus, at times that the oxygen concentration in the water is minimal, this process can be an important mechanism for supplying aquatic animals with their respiratory needs.

There are important implications of these and other adaptations for the methodologies used to estimate primary production in the water bodies of the Pantanal. Generally, such estimates are based on the ^{14}C method or on the amount of oxygen released into the water. In the Pantanal, the distinction between photosynthesis underwater and that by the emergent plants above the water level is not so clear. The ^{14}C method can be used to calculate the amount of carbon fixed by plants underwater, presuming that all gas exchange takes place between the plants and the water. In the case of species that can obtain carbon dioxide directly from the atmosphere, such as the euglenas that remain on the surface tension layer, the primary production would be underestimated. In addition, these species are able to feed heterotrophically on simple organic compounds, providing them with additional carbon sources that are unaccounted for in the ^{14}C calculations.

Even more problems arise when calculating primary production from the amount of oxygen produced. This method naturally neglects all photosynthesis involving gas exchange directly with the atmosphere. Thus, it might be concluded that during the early *cheia*, scarcely any photosynthesis is taking place because little oxygen is being released into the water. However, a considerable amount of primary production is occurring in the leaves of the emergent macrophytes utilizing inorganic nutrients obtained through the roots from the sediment. Some of this production might be falsely attributed to the community of submerged plants due to a release of O_2 from the root aerenchyme into the water, but only if the oxygen saturation in the water were well below 100%.

The most reliable way to determine production and productivity are actual determinations of the various standard factors used as measures of "biomass". Such studies, however, invariably encounter the problems of obtaining a statistically significant sample. Local conditions can greatly influence the rate of plant growth, so the results of production and productivity studies are necessarily applicable only to the place and time of the determinations. For example, the net annual primary productivity values summarized for *Eichhornia crassipes* by Neiff (1990) varied from 2.71 $t \cdot ha^{-1}$ in India (Sahai and Sinha, 1970) to 13.8 $t \cdot ha^{-1}$ in Argentina (Lallana, 1980). In the Pantanal, *E. crassipes* is much less abundant than *Eichhornia azurea*, the productivity of which varies greatly from place to place and even more greatly at any one site from year to year, as determined by personal observation.

11.1.1. Seasonal changes in gross primary production

By correlating the results of the chemical analyses, including those for the concentrations of inorganic plant nutrients and oxygen, with the observed rates of plant growth and development, it was possible to detect some of the complex seasonal fluctuations in gross primary production. Although the many observed variations during the course of the annual cycle preclude the establishment of universal "rules" that apply to the entire Pantanal, a number of observations were made often enough to demonstrate that in most cases, they can be considered characteristic traits of the plant communities in the wetland.

First, light is seldom, if ever, and temperature is never a limiting factor that determines the gross primary productivity. In contrast to the water bodies of the temperate zones, in which wintertime illumination and temperatures may severely reduce the rate of photosynthesis, far more light is available to the algae and aquatic macrophytes in the Pantanal at suitable water temperatures than these plants can make use of. The absolute limiting factor during the dry season is generally water, while during the rainy season, the availability of inorganic nutrients determines how much solar energy the plants are able to fix and how fast they can grow. Therefore, during the high water period, the rate of release of inorganic nutrients and simple organic substances through the action of animals that feed on the plants or detritus and by the microorganisms that decompose organic material determines to a great extent the rate of primary production. The productivity may be augmented, however, by the fixation of additional nitrogen by lightning discharges and by microorganisms as well as by leaching of nutrients from adjacent terrestrial communities into the water bodies.

Great pulses of extremely rapid primary production occur during the seasons of transition, during which inorganic nutrients are released in large amounts after mass mortalities of plants and animals have occurred. Generally, such events occur during the period of initial flooding, the *enchente*, when the terrestrial community is inundated and its vegetation dies off, and at the time the temporary water bodies on the floodplain are drying up and filled with moribund aquatic plants and animals during the *vazante* and *seca*.

This simple general pattern repeats itself each year. The specific details of these processes, however, appear extremely complicated and may vary considerably at any particular site from year to year. Most of the complexities result from the fact that the total primary production is the sum of photosynthetic activities of a great variety of biotic communities. The algae, for example, show a seasonal productivity cycle quite different from those characteristic for the vascular plants. Generally, the macrophytes are much better able to store up nutrients in their tissues than the algae can in their cells. Thus, the vascular plants have an advantage in the competition for the inorganic nutrients as long as their other needs are being met.

The ability of aquatic macrophytes to store available nutrients is shared by few algae. Exceptions, however, include the charophytes, which were not abundant in the parts of the wetland studied, as well as *Chaetophora elegans*, which forms gelatinous clumps attached to objects in the water, and to a lesser extent, by some of the filamentous green algae. The other algal species common during the high water period belong to those species that characteristically develop in nutrient-poor water. Although relatively little specific information has been published on the amounts of the various nutrient elements contained in the tissues of tropical plants, a few data on *Eichhornia azurea* in an oxbow lake in the Rio Mogi-Guaçu, a tributary of the Rio Paraná, were provided by Camargo and Esteves (1996). They reported that in September, 0.787% of the dry weight is accounted for by nitrogen, while 0.086% of the weight is phosphorus. By January, these figures had increased to 1.422% and 0.191%, respectively. During these months, the high water level was reached in the region investigated, showing the ability of these plants to store up the available nutrients. The total amounts of essential elements withdrawn from the general pool in the water body are determined by the density of the macrophyte beds. In the region studied by these authors, the total amount of nitrogen accumulated by *E. azurea* from September to January increased from 0.42 $g{\cdot}m^{-2}$ to 1.45 $g{\cdot}m^{-2}$, while the corresponding amount of phosphorus in the plants had risen from 0.05 $g{\cdot}m^{-2}$ to 0.19 $g{\cdot}m^{-2}$.

The competition between algae and vascular plants for the available nutrients explains the seasonal succession, as discussed in the last section. Generally, the observation that the available inorganic nutrients are recycled between the aquatic and terrestrial communities twice annually does not reflect the actual complexity of the situation. The mineral nutrients can be recycled from several to many times among the members of the aquatic and terrestrial communities each year. For example, summarizing the processes that will be discussed in more detail in Section 11.1.4, the photosynthesis during the dry season produces plant material that is consumed by grazing animals, most commonly cattle on the ranches near Poconé. Considerable research has been done to determine the food value of the plant material for the cattle (Barros Neto, 1979), with Allem and Valls (1987) reviewing the relative values mainly of the grasses and Pott and Pott (1989) reporting the nutrients contained in plants of other families. The excreta from the cattle may be recycled and again be incorporated into plants several times during the same dry season. With the arrival of the flood waters, the plants die, and the release of their inorganic nutrients are effected by microorganisms. After some of the material is incorporated into the rapidly reproducing algae of the late dry and early rainy season, it may again be released into the water through the action of the small animals that feed on the algae. Finally, it is incorporated into the aquatic macrophytes, from which recycling through the action of aquatic grazers is possible. Thereafter, as the water level falls and the macrophytes begin to die off, the mineral nutrients are again incorporated into algae, which pass it on to small grazers, resulting in its rapid recycling until the water on the floodplains dries up, making the nutrients again available to terrestrial macrophytes.

Obviously, any particular molecule of phosphate or a nitrogen compound may become part of numerous plants and animals during the course of a single year, or it may remain in a single macrophyte for half of the year. The rate of gross primary production depends upon how much of the finite supply of inorganic compounds becomes available, or how often the materials are recycled. Therefore, the rate of grazing and decomposition is the primary determinant of how much primary production occurs. A corollary to this is that an increased grazing rate by cattle and other herbivores increases the gross primary productivity. Thus, the production is regulated by the consumption, as long as the mineral nutrients are not removed from the system.

The implications of this observation are interesting. Normally, the amount of food available for consumption is thought to depend upon the productivity of the plant communities. At least in tropical systems, however, the production increases with the consumption. In other words, consuming more food increases the amount of food produced because it promotes recycling of the scarce nutrient elements. This was demonstrated forcefully by the study of Fittkau (1973), which showed that the slaughter of the piscivorous caimans in the Amazon region decreased fish production rather than increasing it, as expected.

11.1.2. Net primary production by the aquatic communities

Like gross primary production, the concept of net primary production failed to find a consensus for a single definition. It has been alternatively defined as the amount of material produced after the metabolic needs of the plants have been met, which also serves as one of the definitions of gross primary production, and as the material remaining after the plant has met its own needs and those of the species in the community that feed on it. In addition, it sometimes refers to the total material produced by a biotic community during a particular year which remains as "living" plant material, such as wood and leaves, or as detritus. In the last case, the P:R ratio is used as a measure of net productivity and indicates how much long-term accumulation of organic material occurs. A continual accumulation of such substances is a characteristic of moors and developing forests (Odum, 1971).

When defined according to the needs of the plant, net productivity is generally given a positive value. However, when based on the activity of the entire community, an apparent negative value is encountered whenever the P:R ratio is less than one. That is, more organic substances are broken down than produced. A P:R ratio of one indicates that all organic substances produced are broken down again. On the floodplain of the Pantanal, this is the case during the annual cycle. There are neither depositions of organic material nor major inputs of allochthonous organic substances subject to local breakdown processes.

The P:R ratio may not be nearly 1:1 in terrestrial woodlands, including those on the floodplain hummocks and the gallery forests along the rivers. It may also not be the case in the large rivers themselves, which receive sewage from human settlements along their shores. The largest of these inputs come from the twin cities of Cuiabá and Varzea Grande on the Rio Cuiabá. In the case of the woodlands, growing trees and increasing populations of large animals may accumulate organic substances in quantities greater than the amounts decomposing in the soil or water. Thus P becomes greater than R. In rivers subject to anthropogenic pollution, such as the Rio Cuiabá, greater amounts of organic material may be broken down than are produced by the biotic communities. Thus, R becomes greater than P.

The great caution required in interpreting primary productivity figures will be mentioned in the subsequent sections. Even a cursory consideration of the circumstances in the Pantanal should make it obvious that the plant production rates would not be sustainable if the "standing crop" were to be harvested. The removal of any of the nutrients would proportionally reduce the amount of subsequent production. Should international developers see some of the productivity figures reported and decide that the region could become a great food producer, the result could be catastrophic.

Floodplains along the middle and lower stretches of major river systems, such as the Nile and Mekong Deltas, have been highly productive because the annual flood waters have deposited minerals from vast upstream catchment areas each year. The water flooding the Pantanal has the mineral content of rainwater, and the wetland therefore has no potential for sustainable agricultural production without a considerable input of fertilizers.

11.1.3. Mineral cycles

From the analyses of the water and the observations of the floral and faunal development, it is obvious that the mineral nutrients in the great temporarily flooded expanses of the Pantanal are recycled at least twice each year. During the *enchente*, the terrestrial plants are drowned and break down, releasing minerals and humic matter. These substances may first be incorporated by a short-lived algal community and then by the animals that obtain them through the food web. When released again by catabolic processes, they are absorbed by the rapidly growing aquatic macrophytes. These plants generally store the materials during the period of flooding and give them up again when they are eaten by animals or their detritus is broken down by microorganisms during the *vazante* and *seca*. They are then incorporated again into the terrestrial or wetland vegetation that develops on the floodplain after the waters have receded. This again breaks down when the flood waters return during the next *enchente*, as the cycle begins to repeat itself.

The general description of the processes provided here partially explains the chemical data presented in Section 6. The sharp increase in the concentrations of dissolved phosphorus and nitrogen compounds as the water rises during the early *enchente*, the great reduction in these concentrations during the late *enchente*, their minimum values during the *cheia*, and their increase again during the *vazante* and *seca* are general phenomena observed in the water bodies of the floodplain. Unfortunately, the quantitative chemical data for the dry season cannot be determined by the same methods as those used during the other seasons because after the water has evaporated, the free inorganic materials are partitioned, some as salts in the soil and others within the tissues of the terrestrial plants or in the desiccated parts of aquatic plants and animals that do not decompose completely until the subsequent rainy season. It is surmised that greater amounts of these substances remain in the soil than in the terrestrial plant material, the growth of which is limited by shortages of water during the long rainless period.

From the chemical data in Section 6, it is clear that the processes occurring during all seasons, except the stable *cheia*, are subject to considerable short-term local variations that occur at various times at different locations, usually dependent upon the local weather. These events also vary considerably from year to year. The apparent scattering of the data points in Figures 6.1 through 6.8 and 6.9 through 6.41 during the *enchente* and *vazante* are by no means random fluctuations but are rather related to observable events, which can be correlated descriptively with the chemical data. For example, extremely high values for ammonium and phosphate and low oxygen concentrations recorded in the Rio Bento Gomes where it crosses beneath the Transpantaneira Highway during the year 1992 resulted from a mass mortality of thousands of fishes in that stretch of the river during August.

Nitrogen cycle

In the Pantanal, the nitrogen cycle generally follows the same pattern as that in the temperate zones, except for the much greater role of lightning in fixing atmospheric nitrogen, as inferred from studies in other parts of the tropics (Fig. 11.1). Jones and Bromfield (1970) estimated the amount of mineral nitrogen supplied annually to the land in tropical Africa to be 4.5 kg per ha. This was mainly in the form of ammonium and nitrate. While some of these substances may have originated with the dust in the air, it is likely that most was produced by electrical discharges in and near cumulonimbus clouds.

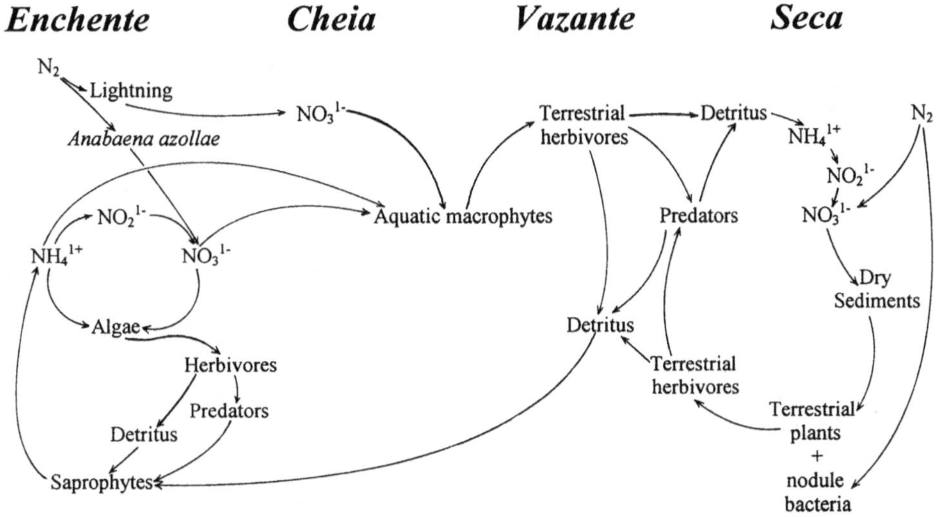

Figure 11.1: A simplified depiction of the nitrogen cycle in the Pantanal during the course of a year.

In addition to nitrogen fixation by physical processes, much inorganic nitrogen is fixed by microorganisms. In the aquatic system, nitrogen-fixing cyanobacteria are not abundant. The species with this capability most frequently observed was the *Anabaena azollae* associated with *Azolla* spp. (Holst and Yopp, 1979), the mosquito fern species that were also not abundant in the vicinity of the Pantanal. Hence, there is apparently little nitrogen fixation by cyanobacteria in the Pantanal.

However, the nitrogen fixation by bacteria is assumed to be considerably more important in the wetland. This is surmised from the great abundance of legumes among the macrophyte vegetation. Of these species, few are mainly aquatic. Only *Aeschynomeme fluminensis*, *Neptunia prostrata*, *Sesbania exasperata*, and *Mimosa pellita* are encountered mainly in aquatic habitats. The other species are terrestrial with varying degrees of preferences for soil humidity. A study of *Aeschynomeme fluminensis,* also referred to as *A. flumensis,* from the Pantanal revealed that this species not only forms root nodules, it can also form nodules for the nitrogen-fixing bacteria on the stems during periods of inundation (Loureiro *et al.*, 1995).

Some soil bacteria have the capability of fixing nitrogen, while others are responsible for mineral nitrogen loss through denitrification processes. Other important contributions of the bacteria to the nitrogen cycle include the breakdown of a variety of organic substances to release ammonium or nitrate and the conversion of ammonium to nitrite and nitrite to nitrate, as known from studies conducted in the temperate zones.

An interesting sidelight of the nitrogen cycle is the presence of plants capable of obtaining organic nitrogen compounds from animals, which they capture and digest. Among the submerged carnivorous plants are the species of *Utricularia,* which are quite common in many parts of the Pantanal. In wetlands in the upland parts of the watershed, the *Drosera* species take advantage of this nitrogen source.

Figure 11.1 is considerably simplified, and many of the cycles shown only once can be completed several times during a single season. For example, during the *enchente*, the incorporation of free nitrogen compounds dissolved in the water into algae and their subsequent transfer to microscopic herbivores and predators may occur frequently. Each time, the herbivores and predators release nitrogen-rich detritus, which is quickly converted by bacteria into free ammonium. Apparently little of the ammonium is converted to nitrite and then to nitrate because it is reabsorbed directly by the algae or macrophytes so quickly that the conversion does not occur. The low concentrations of nitrate recorded in the surface waters during all season attest to this.

The termination of this short-term cycling of the nitrogen compounds is effected by the aquatic macrophytes, which tend to monopolize the available supply of these substances by storing them in their tissues. This cannot occur during the *enchente* because each time the water level rises slightly, additional large areas of the floodplain become inundated, and the rapid decomposition of the dry detritus and drowned terrestrial annuals, as well as leaching from the terrestrial soils, brings large additional quantities of nitrogen compounds into the water. This explains the rapidly fluctuating ammonium concentrations detected during the *enchente* and shown in Figures 6.31 through 6.34. The opposite effects of the decomposition processes on the one hand and rapid photosynthesis by the algae and submerged macrophytes on the other also explains why the oxygen concentrations fluctuate so greatly (Figs. 6.5 through 6.8).

Direct observations of the water level in the field and the weekly changes in the ammonium concentrations suggest a short-term pattern for the nitrogen cycle in which slight rises in the water level are followed by great increases in the ammonium concentration. These events are followed by a few days of relative stability in the water level, during which a remarkably rapid plant growth is accompanied by sharply declining concentrations of ammonium and nitrite. After a short period of heavy rainfall in the watershed, the water level rises again along with the ammonium and nitrite concentrations. The next period of stability is accompanied by a new decline in the amounts of free nitrogenous compounds available. Only after the *cheia* has begun and the water level has ceased to rise do the concentrations of nitrogen compounds decrease below the detectable limits for the analyses used (Section 6). The *cheia* is characterized by stable and uniform conditions, both in the concentrations of these nutrients and related phenomena that manifest themselves in the oxygen concentration, pH, and humic matter content of the water, as determined by color on the Hazen scale.

The next period of rapid change occurs during the *vazante*, when low but detectable concentrations of ammonium are again encountered in the water. These continue to rise gradually until the *seca*, when radical increases to high concentrations follow the seasonal mortality of the aquatic biota and precede the complete disappearance of the standing water bodies.

Phosphorus cycle

The phosphorus cycle is much less complex than the nitrogen cycle because of the limitation on the supply to compounds available from mineral sources (Fig. 11.2). It must be assumed that nearly all natural phosphate-bearing minerals in the uppermost rock layers of the watershed were long ago dissolved and carried downstream in the streams and rivers toward the Pantanal (Heckman, 1995). These phosphates and any that have arrived through the atmosphere with dust are now being utilized by the living organisms, and the availability of phosphorus-bearing compounds is one of the most important factors regulating the biotic processes in the wetland.

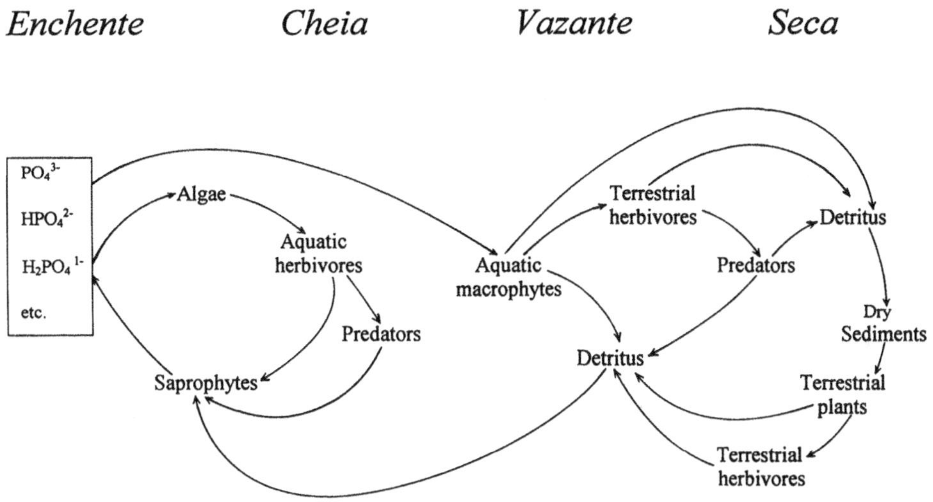

Figure 11.2: A simplified diagram of the phosphorus cycle in the Pantanal.

Due to the shallowness of the water in the Pantanal and the great activity of the microbiota, the existence of a considerable dormant phosphate reserve in the sediments of the rivers and deeper standing water bodies of the Pantanal is unlikely. The only phosphates withdrawn from the system for a relatively long time are those incorporated in the trees and smaller woody plants in the terrestrial forests, which were limited in their extent in the Pantanal, generally being located only on hummocks and along the high river banks of the large and medium-sized rivers.

Many of the seasonal characteristics of the phosphorus cycle are similar to those of the nitrogen cycle. This includes the rapid cycling of inorganic compounds leached from the newly inundated areas of the floodplain through the food chain from algae to microscopic herbivores and small predators. As in the case of the nitrogen compounds, this cycling ends after the available phosphates have been almost entirely assimilated by the aquatic macrophytes.

Carbon cycle

As in the case of the nitrogen cycle, the carbon supply in the Pantanal can be supplemented from atmospheric sources (Fig. 11.3). In contrast, minerals containing carbon in the form of carbonates and bicarbonates in the substrate are almost non-existent in the northern part of the Pantanal watershed, as demonstrated by the carbonate hardness values determined for the water (Section 6).

Simple carbon sources are absorbed by the plants directly from the atmosphere or through the ambient water. The great predominance of aquatic plants with emergent leaves suggests that direct absorption of carbon dioxide from the air is a more efficient way for the plants to meet their carbon needs for photosynthesis than the absorption of carbonate or bicarbonate molecules from the water during most of the rainy season. In contrast, during the dry season, the breakdown of large amounts of organic material by microorganisms releases large quantities of carbon dioxide and simple organic compounds that are available as carbon sources for the many developing algae. Euglenophytes and diatoms are known to consume simple organic substances to live heterotrophically or to use the carbon for photosynthesis. Nevertheless, *Euglena sanguinea* and *E. gracilis* move onto the surface tension layer of shallow standing water bodies, where they establish direct contact with the atmosphere for gas exchange. While this also may facilitate the release of oxygen at times that the underlying water is supersaturated with the gas, it also permits the rapid absorption of CO_2, which may have been depleted in the water during the daytime, judging by the steep rises in the pH (Section 6).

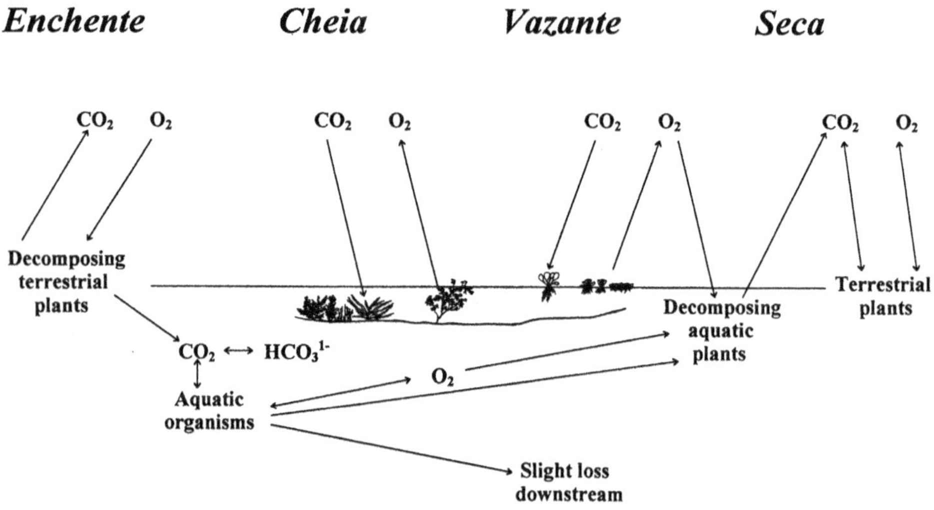

Figure 11.3: A simplified diagram of the carbon cycle in the Pantanal.

An indication of possible scarcities of carbon sources for photosynthesis in the water on the floodplain during the *cheia* are the remarkably low electrical conductivity and obviously poor buffer capacity of the water. With a conductivity approximating that of rainwater and a rather low carbonate hardness value, it is not a rich source of carbon dioxide or carbonates. The rather rapid but small changes in pH during the *cheia* accompanied by little increase in the oxygen concentration also suggest that the buffer capacity of the water in the Pantanal is low. This is not surprising in view of the extremely low buffer capacity of the water in the upland streams flowing toward the wetland (Heckman, 1995). Since the presence of dissolved CO_2, carbonates, and bicarbonates in the water would considerably increase the buffer capacity of the water beyond that observed levels, it must be assumed that these substances are largely depleted.

One result of the depletion of carbon sources during the *cheia* is the great dominance of emergent plants, which can exchange gases directly with the atmosphere. The reasons for this were just discussed. While the submerged species are often predominant during the *enchente*, the emergent species come to dominate the Pantanal later in the rainy season. Of course, some of this dominance can be explained by the direct exposure to sunlight enjoyed by the emergent species and the increased shading of the submerged flora. However, enough light is available to permit much more photosynthesis than actually takes place, as evidenced by the small changes in the diurnal curves for the oxygen concentration and pH recorded during the *cheia* (Section 6). Plants with deep root systems in the sediment and emergent leaves are thus at an advantage over the submerged species that can take their nutrients only from the ambient water. While the deep roots can tap new sources of phosphorus and nitrogen compounds, the emergent leaves ensure that enough carbon will be available, as well. This completely liberates the emergent plants from dependence upon the ambient water for life support.

Sulfur cycle

The sulfur cycle in the Pantanal is considerably simplified because the anaerobic breakdown of organic substances with the release of hydrogen sulfide does not seem to occur. Hence, sulfur is present in small amounts as sulfates or sulfites or as a component of organic substances. Hydrogen sulfide was almost never detected, even in detritus-rich sediments beneath the permanent lakes in the Pantanal, where methane but not H_2S is apparently produced under the prevailing anaerobic conditions. As reported in Section 6, hydrogen sulfide at small but detectable concentrations was encountered only after a massive mortality of fishes in the Rio Bento Gomes during the dry season.

Silicon cycle

The silicon cycle in the Pantanal is simple. The sediments are extremely rich in silicates, and there are generally large amounts of these compounds in the water, apparently mainly in colloidal form, judging from the electrical conductivity of the water (Section 6). The organisms that require this element for the formation of the frustules or spicules, namely the diatoms and the sponges, are limited in abundance by the concentrations of other inorganic nutrients and by the length of the high water period, respectively. Hence, silica is always present in superabundance, and there is an enormous supply in the sediment that can replenish the silicates carried out of the region by the water currents.

The empty frustules and the spicules of the desiccated sponges join the sediment and can gradually be broken down into colloidal particles. This breakdown is apparently rapid, judging from the scarcity of these structures in the sediment samples examined. Although such a rapid dissolution of silicates aroused the suspicion that fluorides might be abundant in the ground water, analyses of well and spring water upstream from the wetland as well as of the surface waters in the Pantanal itself seldom revealed considerable concentrations of fluoride, and these were almost always detected during the dry season together with high concentrations of chloride.

11.1.4. Successions of primary producers

There are marked seasonal changes in the floral composition on the floodplain, not only those involved in the general seasonal succession from wet to dry and dry to wet but also within both the aquatic and the terrestrial community. These are shown in diagram form in Figure 11.4. Again it must be stressed that, except during the *cheia*, the exact times of year during which the events in the succession occur vary according to the elevation of the individual site on the floodplain and to the differences in the water level changes from year to year.

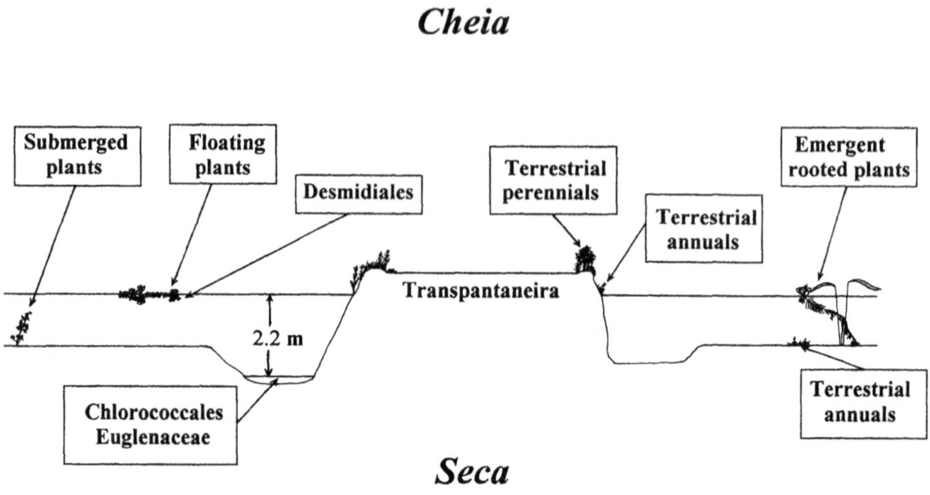

Figure 11.4: A diagrammatic depiction of the differences in vegetation during the high water period, the *cheia*, and the dry season, or *seca*.

Apparently, the main factor that determines which organisms will be dominant during a given season is the capacity to assimilate and store the available nutrients (Section 11.1.2). In general, the algae can absorb free nutrients more rapidly than the macrophytes and are therefore able to develop large populations more rapidly after the dry areas of the floodplain become inundated again during the early rainy season. These algae, however, are less capable of storing the nutrients in well protected structures and gradually disappear as the macrophytes develop. However, among the algae, there are also species with quite different seasonal patterns of development related to their abilities to absorb and store nutrients (De-Lamonica-Freire and Heckman, 1996).

In Figure 11.4, the communities shown to be dominant at the time of the freshet consist mainly of fast growing species of algae, mainly members of the Chlorococcales and Euglenophyta. These develop during the late dry season in the pools of water left on the lowest parts of the floodplain in which the temperature sometimes exceeds 40°C (Heckman *et al.*, 1993). The appearance of massive populations follows the death and decomposition of the aquatic macrophytes and sometimes fishes, as well. Some of these algae persist into the early rainy season and are joined by other species, as the dry parts of the floodplain gradually begin to be covered by water again, and the detritus from the drowned plants decomposes. Shortly thereafter, aquatic macrophytes begin to grow extremely rapidly in the continually deepening, nutrient-rich water. This fast growth continues until the supply of inorganic nutrients becomes exhausted, and then it tapers off. At the same time, the algae belonging to the Chlorococcales and Euglenophyta become continually scarcer (De-Lamonica-Freire and Heckman, 1996). By the beginning of the *cheia*, the dry season species have become rare, and the predominant algal species are desmids, which appear in sparse numbers but with a great diversity of species.

After the beginning of the *vazante*, a gradually increasing concentration of inorganic nutrients accompanies the death of many aquatic macrophytes in the shallow water on the floodplain. At this time, various dry season algal species begin to appear again, particularly euglenophytes and some diatoms. In addition, wetland plants begin to grow on the exposed ground as the water recedes, and these are then replaced by terrestrial species as the floodplain continues to dry out. An ecotone forms between the terrestrial and aquatic community, and this moves with the receding water (Heckman, 1997b). In this ecotone, as already mentioned, are the terrestrial forms of some wetland plant species which have both aquatic and terrestrial forms (Prado *et al.*, 1994). These include *Ludwigia inclinata, Hydrolea spinosa, Eleocharis minima*, and *Nymphoides grayana*. Together with these are some typical wetland species that develop rapidly along the ecotone and die off as the transition zone desiccates. These species include *Diodia kuntzei, Polygonum punctatum*, and *Echinodorus tenellus*. This information is repeated here as the basis for discussing the fundamental concept of "production" as presently calculated by ecologists. In the past, productivity studies were often undertaken as "applied science" research to estimate what was thought to be the absolute maximum amount of "biomass" that could potentially be produced each year. At one time, this was considered important to determine how much food could be produced in a particular kind of habitat. Some of the results found their way into the popular press and led to the formulation of such myths as the great productivity of irrigated deserts and the future use of marine plankton to feed the world. Before these myths were recognized for what they were, enormous areas of the world were ruined for agriculture and some very costly projects brought nothing but disappointment. Nevertheless, some of the methods developed for these studies are still occasionally used, and misleading results are still obtained.

Giving the events on the floodplain careful scrutiny, it is possible to see how some of the misconceptions about production arise. As the water recedes from the floodplain, the first moisture-loving plants appear, grow quickly, produce seeds, and die off. The productivity, if calculated by harvesting the plants at the time the seeds are being produced, would be extremely great because the plants grow in such a short time. However, the total amount of organic material produced is small. The next plants that appear as the ecotone moves over the sampling site also have a high rate of production but produce little material. Finally, the fully terrestrial plants grow slower but produce considerably more material. After the flooding, there is a "production" of algae. Then, the aquatic macrophytes display a high productivity at first, but almost cease to grow as the mineral nutrients are exhausted.

A "quantitative" study of the primary production during this succession can yield high values when the material produced during each of the individual steps is totaled. However, a grand total of nothing is "produced," except perhaps for a few more seeds than had existed the year before. Each species is simply securing its own survival by sharing nutrients that had already been used by other species in the succession. In the past, when inflated estimates of the potential productivity were used as the bases of calculations, ambitious agricultural projects were undertaken, which demonstrated again and again that most tropical and subtropical ecosystems cease to be productive as soon as the nutrient pool is reduced by removal of a crop. It is therefore vital to understand how the nutrient cycles during the seasonal succession actually function.

The succession not only involves an alternation between terrestrial and aquatic, it also encompasses short-term successional events within each of the main phases. Terrestrial species soon displace the wetland biota at the upper edge of the ecotone as the water continues to recede, while the aquatic plants at the lower end die off at an increasing rate as the water level on the floodplain continues to fall. As the aquatic plants and animals die off, decompose, and release their incorporated nutrients, the algal blooms of the dry season begin to develop. These include many species of Chlorococcales and Euglenophyta, as shown in Sections 7 & 10.

The drying pools of water filled with decomposing detritus and masses of microorganisms are familiar features of the dry season. Recapitulating the information provided above, the succession of primary producers is found not only to be a result of the physical and chemical changes in the water bodies but also a major factor determining to what extent these changes occur (Heckman, 1994a). In some of the drying pools of water, many fishes survive almost until the flood plain has completely dried up, while in others, mass mortalities begin early, greatly increasing the amount of decomposing organic matter and mineral nutrients for algae. The great variety of algae in the Pantanal is apparent from the distinctiveness of each pool of water during this season. For example, along the Transpantaneira Highway during the dry season of 1994, one pool supported a flora dominated by *Microcystis aeruginosa* and covered by a thin layer of *Euglena sanguinea* on the surface tension layer. The next pool along the road was dominated by various flagellates, including common chrysophyte species and euglenophytes, and the one beyond that, near the Rio Bento Gomes, was rich in chlorococcal green algae. Not only are there great spatial differences in the algal communities that develop, there are also considerable differences from year to year, brought about mainly by differences in the minimum water level reached. Thus, some locations remained flooded throughout one year and dried up completely during the next. In places that maintained enough water to permit the survival of the fishes until the next *enchente*, fewer inorganic nutrients were released into the water, and intensive algal blooms did not occur. Elsewhere in Brazil, blooms of these species are not infrequent, particularly in man-made habitats. *Microcystis aeruginosa* frequently blooms in the impoundments of eastern Brazil, such as the Barra Bonita Reservoir along the Tietê River in São Paulo (Calijuri and Dos Santos, 1996), and *Euglena sanguinea* has been found to bloom in fish breeding tanks in the same Brazilian state (Xavier *et al.*, 1991).

11.1.5. The conservative action of the aquatic vegetation

From studies of the upland streams and rivers feeding the Pantanal floodplain (Heckman, 1995), it became obvious that the entire region of southern Mato Grosso is subject to leaching and erosion processes that have been continuing for hundreds of millions of years (Section 3). The ancient kraton on which the Pantanal is located has been little influenced by the mountain-building events that resulted in the development of the Andes, and the mineral content of the streams has therefore not been enhanced by uplifted marine sediments during their known geological history. In short, all substances that are washed out of the system are forever lost, and the only inputs of minerals, brought by winds and rainfall, could easily be lost again with the water that rapidly flows through the system of the Rio Paraguay toward the Atlantic Ocean.

Under these circumstances, the ecosystem, to ensure its own continued existence, must act as a conservative agent to resist the natural leaching and erosion processes that threaten to remove the limited supplies of vital nutrients from the geographical region. This is effected through the seasonal cycle of plant growth, whereby appreciable amounts of mineral nutrients are present in soluble form only during the dry season, when there is no flow whatsoever in the tributaries, and the current in the Rios Cuiabá and Paraguay themselves is also at a minimum. During the *enchente*, the runoff from the rainwater moves from the rivers onto the floodplain, carrying with it the few nutrients from the rain as well as large amounts of inorganic substances leached from the soil over which it passes. As mentioned above, this material is rapidly incorporated into the tissues of the developing aquatic macrophytes, which eliminates the nutrient supply also utilized by the chlorococcal green algae and diatoms. These algae rapidly disappear from the water during the season of maximum macrophyte growth. By the beginning of the *cheia*, during which there is a large volume of water passing through the river, including much runoff from the floodplain, the mineral nutrients have been almost entirely absorbed from the water by anchored plants and associated algal species, mainly desmids, which are well secured by various spiny structures to the macrophytes. The submerged roots of these plants efficiently remove the last traces of nutrients carried by the water and greatly minimize the loss from the wetland. Characteristically, the species present are those noted for being able to live in nutrient-poor waters and grow with minimum amounts of nitrogen and phosphorus compounds in their tissues, such as the species of the family Pontederiaceae.

Some primary producers are capable of growing while the concentrations of nutrients in the water are extremely low. This indicates that they have the capability to efficiently absorb these substances and retain them while the supply remains low in the ambient water. Nevertheless, growth of the plants stagnates during the *cheia* because there is great competition among the individuals to absorb the last traces of inorganic substances. It is evident when observing the Pantanal that the maximum amount of living aquatic plant material is present during the *cheia*. This has been confirmed by monthly determinations of the dry weight of *Eichhornia azurea* and *Ludwigia helminthorrhiza* in lakes located near Barão de Melgaço, even though these lakes contain considerable amounts of water throughout the year (da Silva & Esteves, 1993). Similar studies along the floodplain of the Rio Paraná also revealed that the greatest amount of living plant material is present during the *cheia* (Sabatini, 1985). It is safe to assume that the maximum amounts of most nutrient elements are also present in the plant tissues during that season.

The rivers contain water with conductivity values hardly greater than the rainwater that feeds them during the *cheia*, permitting a loss of nutrients from the ecosystem approximating zero. Thus, under normal conditions, only the substances carried into the wetlands by the rain or their ionic equivalents can be carried out of the system by the rivers when the volume of water flowing through them is maximum. The plants first begin to die off and decompose again as the water level is falling rapidly and runoff into the rivers has practically ceased.

Obviously, this system is threatened by the activities of man, which includes uprooting the aquatic plants during the high water period and burning the vegetation on the floodplain during the dry season. Burning results in a greatly increased leaching by the first rains after the dry season has ended and the loss of the mineral nutrients into the main rivers before the growth of the aquatic macrophytes has begun. In some parts of the world, burning the vegetation may function to eliminate undesired underbrush, to favor certain plant species, and to produce a desired structure within the plant community. While fires may occur regularly due to natural causes in the Pantanal during the dry season, the great number of these fires that occur at the present time in due to deliberate arson in southern Mato Grosso has produced a real danger of extensive mineral nutrient loss during the first few weeks of the rainy season. Although these fires are usually reported to be accidental, on many occasions while returning from the Pantanal to Cuiabá in the evenings, persons were seen setting the fires near the roads using bottles of flammable fluids. Thus, deliberate arson rather than discarded cigarettes seems to be the cause of most fires.

There is a fundamental difference between the effects of fires on the upland regions of the watershed and in the wetland itself. In both cases, the conservative action of the plants in preventing nutrient loss is eliminated by their destruction. However, in the upland regions, the result will become apparent on the areas that are burned, which will be exposed to excessive erosion and leaching and begin to lose their fertility after several fires. In the wetland, however, the effect of leaching is collective in that the nutrients lost to a burned area will be replaced by the inflow of water and sediments from more elevated adjacent regions of the floodplain. Hence, the mineral loss will not become apparent where the fires have occurred. Rather, the loss must be subtracted from the entire nutrient pool in the northern Pantanal and will therefore be nearly impossible to detect.

11.1.6. Primary production by terrestrial plants

As mentioned in Section 11.1.3 above, the nutrient budget of the terrestrial vegetation is more difficult to determine with precision than that of the aquatic plants. Furthermore, it is more varied due to the diversity of the terrestrial communities. The total amount of organic material produced by the terrestrial herbs on the floodplain can obviously not exceed that which is present in the aquatic community at the same site during the *cheia*. Actually, due to the scarcity of water and characteristics of the soil, the amount of material produced is probably much less than that present during the flood period. However, local differences can be assumed to be considerable making any estimates for large areas of the Pantanal premature.

A different situation is encountered in the hummocks, that is, the permanent stands of woodlands on the areas of the Pantanal located above the high water level of the *cheia*. While a study of the regional terrestrial communities are beyond the scope of this book, several points should be mentioned with regard to the interrelationships between the wetland and adjacent terrestrial ecosystems.

The most obvious relationship is the contribution of allochthonous material from the terrestrial plants and animals to the aquatic community. This occurs through the decomposition of fallen leaves and other vegetable detritus in the water as well as through the consumption of the fruits and seeds of the terrestrial plants by aquatic animals. In addition, large herbivores, including the domestic cattle as well as the native capybaras, deer, and peccaries, act as transporters of nutrient materials from one community to another. Nutrient elements may move in the other direction, as well, being carried into the wooded areas during periods of particularly heavy rainfall and incorporated into the woody vegetation.

Another important role of the hummock and gallery forest vegetation is the shelter they provide for many species that interact primarily with the aquatic community, such as snail and fish-eating birds, giant otters, capybaras, aquatic reptiles, and the adults of many insects with aquatic larvae. Excluding the insects, nearly all such species reproduce in terrestrial habitats rather than water, and most prefer rather sheltered nesting sites.

11.2. Secondary production

Secondary production is here defined as the production of organisms at trophic levels higher than that of the primary producers without regard to whether they consume plants or other animals. A more differentiated discussion will be presented below in sections on primary and secondary consumers.

Much interest in secondary production has been generated by its relationship with the growth of fishes and other commercially valuable animals. In the Pantanal, some problems with the conventional conceptualization of this relationship arise. In this respect, water bodies in the tropical wet-and-dry climatic zone show considerable differences from those in more stable climatic regions, particularly the equatorial zones. A characteristic of the Pantanal is an extremely great seasonal mortality among the aquatic animals that must be compensated for by high rates of reproduction during the early rainy season. This mortality is sometimes extreme, as during the mass death of fishes in the Rio Bento Gomes during 1992 (Section 10.4.2), but it is not high each year at all locations. Hence, the spatial and temporal variation in the dynamics of the fish stocks on the floodplain and in the small rivers must be statistically considered in determining fish survival (Burnham *et al.*, 1987), and under the prevailing circumstances, an intensive monitoring program over a period of many years will be necessary to obtain accurate estimations of the fish production.

A simplistic view that is often tacitly accepted by economic planners in the absence of more detailed information is that more rapid secondary production leads to the presence of a greater stock. Since it is the standing stock that is harvested, fast-growing fishes with high rates of reproduction have frequently been propagated as ideal candidates for aquaculture, even when such fishes are not particularly prized for their flesh and often bring a poor price on the local markets. Actually, in equatorial regions where relatively stable conditions exist, as well as in the temperate zones where inorganic and organic nutrients are often present in excess in the water bodies, this simplistic concept equating fast growth and great productivity may hold true. In the Pantanal, however, the standing stock at any particular time of year is more or less the same as it had been one year earlier, in spite of an extremely high rate of secondary production during certain times of the year. Therefore, the rate of production and the amount of harvestable fish flesh produced are not correlated with one another. It is the annual mortality of the fishes through heavy predation, desiccation of the standing water bodies, and probably microbial toxins in the water that make the high rates of production possible by promoting a recycling of the nutrient elements. This is most certainly why the fish yield decreased considerably when the piscivorous caimans were slaughtered in the Amazon region (Fittkau, 1970).

The concept of a high productivity through a rapid recycling of the nutrient materials provides a plausible explanation for the rapid loss of productivity in the Pantanal and the upland parts of its watershed when anything is harvested and removed from the system. In this respect, the production of fishes and other aquatic animals parallels closely the production of macrophytes, discussed above. Nutrients are in short supply, and when they are completely removed from the region, the production nearly ceases. There is simply nothing left to recycle. Hence, at the present time, anthropogenic eutrophication is not a hazard to the Pantanal. The ecosystem can assimilate considerably more nutrients than are now present without changing the structure of its biotic communities. Unfortunately, experience shows that organic pollutants alone are not discharged from modern cities and towns. The eutrophication is almost always accompanied by toxic industrial wastes and agrotoxins, which have a considerable effect on susceptible elements of the flora and fauna and thereby negatively influence the structure of the biotic communities.

Returning to the concept of recycling nutrients, it is also obvious that eliminating the present environmental instability produced by the climate would greatly reduce secondary production. It is also certain that any change to make the hydrological conditions more stable would not increase the total standing stock but rather cause fundamental changes in the structure of the biotic communities. Old, slow-growing fishes would replace the young, fast growing ones. In contrast, aquaculture would enable the fast growth potential of the species in the Pantanal to be utilized because the time of harvest can be timed to the completing of the fastest growth stages of the fishes (Section 13).

The exploitation of the fishes by traditional fishing methods in impoundments formed after dams or dikes have regulated rivers to prevent the normally large changes in the water level from taking place does not always increase fish production over the long term, but rather it eliminates the highly productive floodplain species in favor of a few slower-growing fishes normally encountered in large rivers and lakes. Furthermore, the rates of production and growth of these fishes may become even slower because many of the small, fast growing fishes on which these larger species often feed are largely eliminated through the destruction of their shallow water breeding grounds. The reasons for these conclusions are discussed in detail below. Hence, if a market were found for the fishes of the Pantanal, which is presently lacking due to the low human population density in Mato Grosso, large-scale production would be recommended by intensive aquaculture in the upland parts of the watershed rather than by attempting to use inefficient traditional fishing methods in impounded water bodies constructed in the wetland itself. Any attempt to intensify the effort to catch members of the wild stocks and remove them from the system would be disastrous to the entire ecosystem. It would not only eliminate many of the abundant fish species, it would also spell disaster for the large numbers of caimans, water birds, and piscivorous mammals presently inhabiting the wetland.

11.2.1. Primary consumers

The term primary consumer is construed to refer to those species that feed on living green plants, including algae. Actually, the living macrophyte species are of considerable importance in the terrestrial and pseudoterrestrial habitats of the Pantanal but only of minor significance beneath the surface in the water bodies. Water plants, however, are often consumed by the terrestrial grazers, thereby contributing nutrients to the terrestrial food web. Domestic cattle are among the most prominent grazers on the dry floodplain, while a great variety of birds, monkeys, coatis, and even the iguana feed on fruit, leaves, and flowers of living plants. Among the species that feed on water plants, the most prominent are mammals, such as the Pantanal deer and the capybara, as well as the insects of the pseudoterrestrium, including grasshoppers, aphids, and beetles. The most prominent aquatic invertebrates that serve as herbivores in the food web are species of large pilid snails.

The positions of the largest and the most abundant primary consumers in the food webs are shown in Figure 11.5. It must be stressed that the consumers of small algae are generally restricted to a short season of activity at the end of the dry and beginning of the rainy season, when the nutrient supply in the water is sufficient to support large numbers of them. During the high water period, the large, spiny desmids that predominate among the algae are unsuitable as food for most of the small herbivores (Brook, 1981).

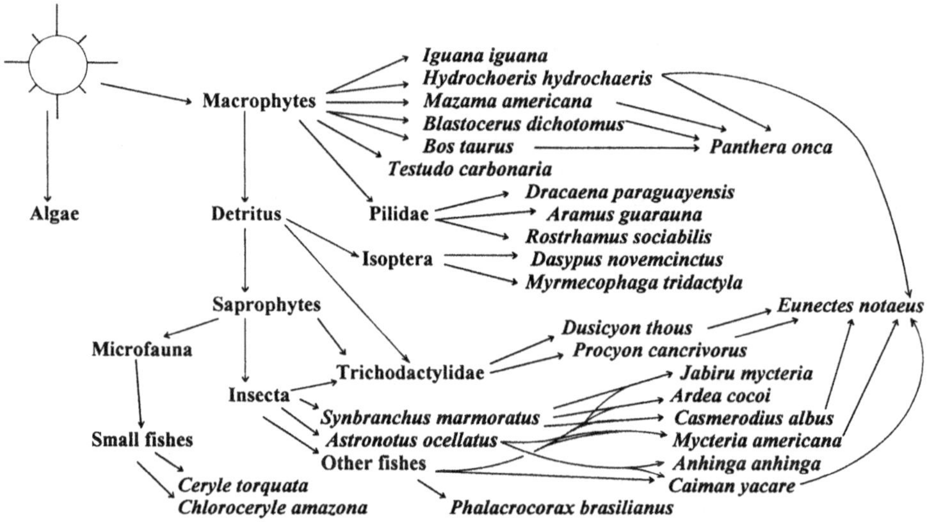

Figure 11.5: The positions of some of the largest and most abundant primary consumers in the food web.

A fundamental difficulty encountered when making quantitative estimates of secondary production results from the gaps in our knowledge of diet specialization. Some insects apparently feed on only one or a few plant species. Others graze on whatever is available. However, there are probably no animals that feed on all plants present in the Pantanal. The best known dietary habits are those of the cattle because these are of great economic importance to the ranchers. The plants that the cattle to not like are already well known, and much interest has been expressed in eliminating these species from the pastures. For example, species of *Andropogon* are not as palatable to the cattle as other grasses and therefore gradually displace the more valuable grasses on many areas of the floodplain. Toxic plants are not uncommon in Brazil (Schvartsman, 1979), and these may have an economic impact on animal husbandry. They may cause losses among the domestic animals and, at the same time, become more abundant as a result of selective grazing.

The marsh or Pantanal deer, *Blastocerus dichotomus*, feeds on many aquatic plants, including *Eichhornia azurea*, as personally observed on many occasions. As mentioned in Section 8, the diet of the capybara, *Hydrochoerus hydrochaeris*, includes many plant species, and littoral macrophytes, especially various grasses, seem to be preferred (Gonzáles-Jiménes, 1978).

There is a considerable difference between the spectrum of plants some animals will eat when they are available and those that they actually eat under natural circumstances. The pseudoterrestrial grasshopper, *Cornops aquaticum*, for example, fed on almost every plant species it was offered in the laboratory (Ferreira, pers. comm.), but in the Pantanal, it remains on the beds of emergent plants dominated by *Pontederia lanceolata* or *Eichhornia azurea*. Only when pursued and allowed no avenue of retreat will these grasshoppers leave the beds of plants and take refuge temporarily on shore. However, it will generally dive beneath the surface of the water rather than leave these beds of emergent aquatic plants. Thus, because of its close natural association with the water plant species, it feeds almost exclusively upon them under natural conditions, in a sense, not through preference but for convenience.

While there are relatively many terrestrial and pseudoterrestrial herbivores that subsist on the aquatic macrophytes, few aquatic species seem to do so. Exceptions, however, are found among the pilid snails, which are extremely abundant on the floodplain and serve as important links between the productive macrophytes and predatory vertebrates, especially lizards and birds. *Pomacea linearis* and *P. scalaris* are both general herbivores, which were observed to feed on the widest variety of water plants. They consume the emergent parts as well as the underwater roots and leaves. *Marisa planogyra* has similar feeding habits but apparently confines its consumption more to the underwater parts of the plants. Unlike the relatively small planorbids and ancylids, which browse chiefly on aufwuchs organisms and detritus, the pilids are mainly macrophyte consumers, at least after they have attained the size of adults.

Other animals that generally remain underwater and consume macrophytes, such as herbivorous insect larvae, are seldom observed and apparently not abundant. Among these, the burrowing larvae that live completely concealed in the stems of the host plants are apparently the most abundant. Such species include the curculionid larvae of *Neochetina eichhorniae*, which inhabit the stems of *Eichhornia* spp., and *Bagoidellus vernicatus* Hustache, 1929.

The greatest difficulties with the classification of the trophic position of species is encountered among the microorganisms, especially those that are non-selective filter feeders. Because many of these species consume all kinds of particles but can digest only a few of them, observations of feeding and examination of the stomach contents help little in determining the actual nutrition of the animals. Because the organisms consumed include saprophytes, algae, and small animals in proportions that change from day to day, the same filter feeders may be classified either as members of the saprobic food web or as primary or secondary consumers. In the past, species found in zooplankton have simply been classified as primary consumers, or secondary producers, although their position in the food web may be totally or in part quite different.

Although phytoplankton has often been studied in many parts of the world to provide a quantitative measure of productivity, there is still relatively little known about how much the common species can really contribute to the nutrition of animals at higher trophic levels. Even methodological difficulties, such as the extraction of the proteins from algae (Rausch, 1981) have led to greatly disparate estimates. It is therefore strongly recommended that estimates of potential secondary production by primary consumers made in the field not be used to calculate the fishery potential of the Pantanal. The long-term productivity depends upon the total supply of nutrients available to the system, which is low, rather than the instantaneous production rates of primary or secondary producers, which are extremely high, seasonally.

Studies downstream from the Pantanal have led researchers to conclude that the middle and upper sections of the Paraná and Paraguay River Systems are poor producers of photosynthetic microorganisms. The findings made during the studies of the biotic communities near Poconé confirm this but also suggest that part of the reason for the low algal productivity downstream in the system is the ability of the biotic community in the upstream wetlands to conserve the available nutrients, as already discussed. An example illustrating the effects of the scarcity of nutrients released downstream was provided by Quirós and Baigún (1985), who concluded that most of the fishes in the Rio da la Plata System of Argentina are illiophagous or detritivorous, reflecting, in their opinion, the low productivity by phytoplankton. They found a correlation between fish growth and the amount of organic matter in the water in the middle stretches of the Rio Paraná, as well. It might thus be concluded that the macrophyte vegetation of the

Pantanal and apparently that along the upper stretches of the Rio Paraná, as well, produces an overwhelmingly large proportion of the nutrients for species at higher trophic levels and that the organic material they produce reaches the secondary producers to a large part through the action of the saprobic community after the plants have died due to conditions provoked by the annual climatic cycle (Section 11.2.3).

11.2.2. Secondary consumers

The term "secondary consumer" refers to all predators, whether they feed on herbivores, making them secondary consumers *sensu stricto*, or whether they feed on other predators, placing them at tertiary trophic levels or higher. Among the secondary consumers are the many predators in both the aquatic and the terrestrial community. In addition, there are many mainly terrestrial species that live by hunting aquatic animals, mainly fishes, snails, and crabs. Although the food webs in most parts of the Pantanal are complex, the fauna includes several species with remarkably selective feeding habits. These have been shown in Table 8.2 for the birds, the behavior of which is best known among all of the animal groups in the Pantanal. Obviously, most such species are specialized for feeding on fishes, generally in one size class for each species. As mentioned above, the caiman lizard, limpkin, and snail kite feed mainly on the large pilid snails. The caimans seem to subsist almost entirely on fishes, and they were never observed to attempt to capture living birds or capybaras, although there are reports that they may do so. Only the diets of the "top predators" include groups of vertebrates with relatively few natural enemies. The anaconda feeds mainly on the moderately large mammals and birds, as well as on caimans. Adult anacondas at a length of three to four meters are capable of swallowing great white herons, small to medium sized peccaries, and other mammals similar in size. However, this snake has often been reported to swallow caimans, and these reptiles may well be the mainstay of the snake's diet.

The jaguar is the predator capable of consuming the largest prey, although it apparently feeds most frequently on mammals of moderate size, such as capybaras and coatis (Schaller & Vasconcelos, 1978a). Its habits of feeding occasionally on cattle has made the jaguar a target of poachers, who are hired by some ranchers to illegally kill this protected species (Section 13).

The role of the secondary consumers in the ecosystem dynamics is important. Without their control of the large herbivore and medium-sized predator populations, both the vegetation and the fishes might be locally decimated to the extent that structural changes would occur in the aquatic communities similar to those that occur on the pastures maintained for the grazing of cattle. As discussed above, such changes in the vegetation have already created problems for the ranchers on parts of the floodplain that are heavily grazed. It is the presence of the top predators, even when not as abundant as they would be under completely natural conditions, that induces the consumers of the fishes, invertebrates, and plants to assume defensive behavior patterns and not move with complete freedom through the wetland. The movements of cattle, capybaras, peccaries, and coatis in herds or large family groups as well as the selection of common trees for mass overnight roosting by various water bird species are forms of defensive behavior. Encountering many coatis and capybaras alone and the wandering of cattle far from the herds are signs that the predator pressure is presently light due to the decimation of the top predators.

Capybaras would seem to have a potential to produce populations dangerous to the well-being of the plant communities in a relatively short period of time. Adult capybaras are apparently relatively safe from attacks by smaller predators. However, Lord (1994) reported that young individuals in Venezuela were eaten by the northern spectacled caiman species, *Caiman crocodylus*, and even by caracaras and black vultures.

Most of the fishes seem to live mainly at the secondary consumer level, since even the omnivorous species subsist mainly on animals, at least during part of their life cycles. The seed-eating species frequently encountered in the Amazon region are not predominant among the ichthyofauna in the Pantanal, although some, such as *Colossoma brachypomum*, come to the Pantanal as migratory fishes in the large rivers. Unfortunately, almost nothing could be found in the literature about the specific feeding of most of the Pantanal fishes at the beginning of the study. Only the piranhas seemed to have been observed enough for the specific diets and feeding habits to have been included in ichthyological literature (Section 8.13.1). The fact that these fishes feed overwhelmingly on large morsels of food, mainly those bitten off other animals, has greatly aided the investigations, which have been conducted mainly on fishes in aquaria. The natural diets of fishes that consume microscopic food items were poorly known. Interestingly, the studies of the stomach contents of small fishes in the Pantanal by Silva Neto *et al.* (pers. comm.) revealed that two species, *Moenkhausia sanctaefilomenae* and *Hyphessobrycon callistus*, subsist to a large part on testaceans. Examinations of the fishes in September, 1995, revealed the presence of five *Difflugia* species, *Netzelia tuberculata* Ogden and Meisterfeld, 1989, and *Suiadifflugia* cf. *multipora*, a rare species described from Mato Grosso and found only in the stomachs of the fishes, never in plant or detritus samples.

In calculating the total productivity of the Pantanal, a large but undetermined proportion of the fish caught is not "biomass" that was produced in the Pantanal. About 80% of the fishery production in southern Mato Grosso comes from the Rio Cuiabá (Lima, 1987). These fishes belong overwhelmingly to the migratory species which apparently travel far downstream from the Pantanal to take advantage of the greater food supply while they are growing and maturing and return to the rivers in and upstream from the Pantanal to spawn (Lowe-McConnell, 1987). The travels and life cycles of these river fishes is an important field for future research because of their importance as a protein-rich food for the local human population.

The Pantanal, while not particularly rich in fish species, is extremely productive of the fishes that have successfully adapted to the wetland. These form enormous food reserves for secondary consumers, which is evident from the discussions of the caiman and the many piscivorous bird species in Section 8.13. It is well worth noting that the Brazilians in southern Mato Grosso have long observed fundamental difference between the "black" and the "white" water birds, which subsist mainly on fishes. Willis (1995) provided a summary of the distinctions between these two general groups. The "black" water birds are so-called because their feathers are generally dark in color. They are solitary, usually remaining concealed in the vegetation and waiting on or near the shore for fishes to swim within range. Many of the species are crepuscular. The "black" water birds nest in secluded spots, often in the gallery forests, during the *vazante*. These species belong to a variety of families, and a number of them belong to the Ardeidae.

The "white" water birds are white or mainly so, as the name implies. They flock together, often forming extremely dense feeding congregations, which move from one feeding ground to another during the dry season. Their habits are diurnal, and in the evening, they return to certain trees in which they roost together at night. They nest later than the "black" water birds, and the nests of many species are not at all concealed. Among the "white" water birds are the species of the family Ciconiidae. some of the ardeids, and the rose-colored roseate spoonbill in the family Threskiornithidae. However, because of its habits, the all-black Neotropic cormorant should also be classified as a "white" water bird.

11.2.3. The saprobic community

The liberation of nutrients and energy into the aquatic food web from the macrophytes and filamentous algae occurs mainly after these organisms or parts of them have died and been decomposed through the action of fungi or bacteria. This situation differs considerably from that in the terrestrial community, where many plants are subject to heavy grazing. The terrestrial saprobic species are of less importance than the aquatic ones because their action is often retarded greatly by the dryness of the soil. Nevertheless, various species of the Basidiomycetes as well as many kinds of bacteria are observed on the floodplain, especially after periods of rainfall. They appear to be most active in breaking down wood and the faecal matter of the large grazers. In contrast, the herbaceous plants generally dry up and show little sign of decomposition until after the flood waters have again inundated the seasonal wetlands.

As already mentioned, there are two periods during which the activity of the saprobic community is greatest. The first is during the *enchente*, when the submerged terrestrial vegetation decomposes rapidly beneath the flood waters. This decomposition is manifested in the physical and chemical data, including low dissolved oxygen concentrations, increasing concentrations of ammonium and phosphate, and the extremely dark color of the water, which at times resembles weak black coffee, its color being far greater than 150 Hazen, the maximum value readable on the test equipment used.

The second period begins in the late *vazante* after the aquatic plants have begun dying off in the receding water. It continues through the *seca* until dryness causes the process to nearly cease. While the water volume on the floodplain is continuously increasing during the *enchente*, moderating the increases in the concentrations of inorganic nutrients, it is rapidly decreasing due to evaporation during the *seca*, resulting in steady increases in the concentrations of inorganic substances.

In the aquatic milieu, the rapid rates of decomposition mediated by many microorganisms with peak rates of activity at the high prevailing water temperatures result in high rates of oxygen consumption. Thus, much of the decomposition of organic materials obviously occurs through anaerobic metabolism during the *enchente* and *vazante*. During the *enchente*, evidence for this is found in the color of the water, which far exceeded 150 Hazen, revealing the presence of large amounts of humic substances. These are carried slowly southwestward toward the Rio Paraguay with the flowing water, breaking down slowly as the dissolved oxygen concentrations increase due to the activity of the fast-growing submerged macrophytes. During the *cheia*, the oxygen concentration remains rather high and stable.

During the *vazante*, the oxygen concentration begins to decrease as the aquatic macrophytes progressively die off in the decreasing volumes of water on the floodplain. During this period, anaerobic decomposition process again begin to assume importance. Generally, such processes result in the release of hydrogen sulfide or methane, but rarely, if ever, both in the same milieu. In some of the baias, some gases evolved from the sediments, mainly along the wooded shores and beneath the sudd, or floating islands of vegetation. From its smell, this gas was assumed to be methane. In the Pantanal of Mato Grosso do Sul, Hamilton *et al.* (1995) demonstrated that methane is an important product of anaerobic metabolism and was present as a dissolved gas in the water. In contrast, the presence of hydrogen sulfide was detected only in exceptional situations and at low concentrations after mass mortalities of fishes.

During the *seca*, anaerobic metabolism again loses its importance for the breakdown of organic detritus. First, the decreasing depth of the water increases the surface area to volume ratio, permitting greater amounts of atmospheric oxygen per unit volume to be dissolved. Then, the development of large numbers of algae in response to increasing mineral nutrient concentrations promotes the release of large amounts of oxygen through photosynthesis. This resulted in the oxygen saturation reaching several hundred percent during daylight hours (Section 6). Finally, as the water completely evaporates, atmospheric oxygen comes directly in contact with the sediment, and anaerobic conditions are no longer encountered. Only in the permanent water bodies with little seasonal water flow, particularly the baias, are patches anaerobic sediments encountered throughout the year.

11.2.4. Recycling of nutrients

The nutrient cycles, discussed above (Section 11.1.3), result from the actions of various species which break down organic substances and release simple organic compounds and inorganic salts. This process is accomplished overwhelmingly by microorganisms, the most prominent of which are bacteria and fungi. These consume the detritus in the water, on moist soil, and even within the digestive tracts of metazoans. Protozoans also take part in the decomposition processes by breaking down simple organic materials released by bacteria and digesting the bacteria themselves. The roles of some of the organisms encountered in the Pantanal most responsible for the release of mineral salts from organic substances are depicted in Figure 11.6.

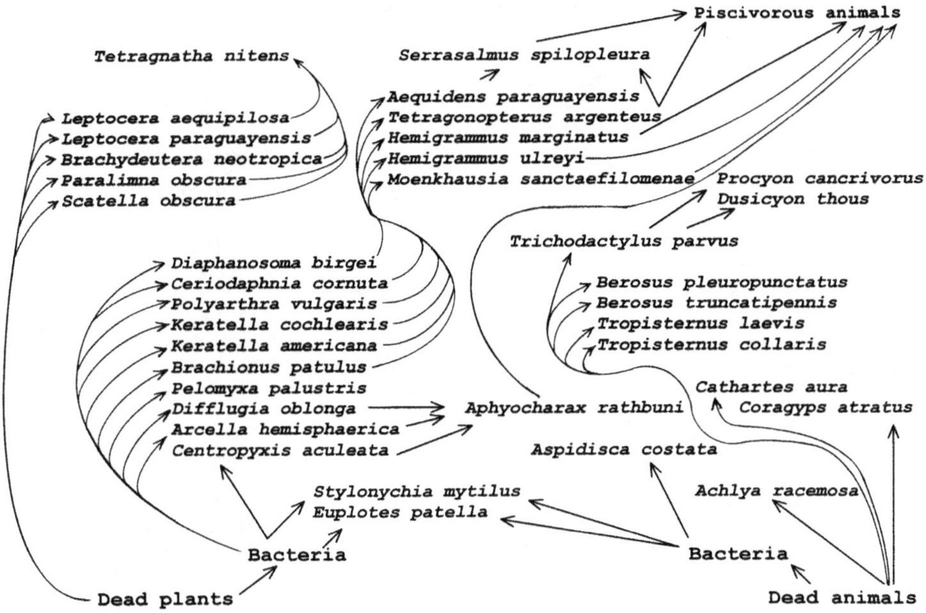

Figure 11.6: The roles in the food web of some saprobic organisms most responsible for breaking down detritus in the Pantanal. The species shown are representative for many others that perform more or less the same services.

A second source of inorganic nitrogen salts in the water is the action of nitrogen-fixing organisms, particularly the nodule bacteria in the roots of legumes and the cyanobacteria, including the symbiont, *Anabaena azollae* in the *Azolla* species. The great variety of plants in the family Leguminosae (Section 7) provide a good indication of the importance of these processes in the Pantanal.

Rain is another important source of inorganic nutrients which were carried into the atmosphere at other locations or fixed by the electrical discharge of lightning (Jones and Bromfield, 1970). At the time of the freshet, the air is usually extremely hazy from smoke and dust carried high into the air by the many dry season brush and forest fires. The rain is a much richer source of inorganic ions than the ground water at the headwaters of the rivers flowing into the Pantanal (Section 5.3.1.). Hence, the electrical conductivity in the upstream sections of the rivers generally rises rapidly after heavy rainfall (Heckman, 1995), and the water in the Pantanal during the *cheia* generally has an electrical conductivity similar to rainwater.

The next step in the recycling is the reincorporation of the inorganic nutrients into living organisms. This occurs mainly through the action of photosynthetic species, as previously discussed.

11.3. Food webs, dynamics of "skipping steps"

The interactions of the species within communities and between the communities themselves are naturally complex. However, the effort to learn some of the trophic relationships is rewarding because through a detailed analyses of the food web, it is possible to learn why the species occur where and when they do and why some become abundant while others nearly disappear. Figures 11.7 through 11.10 show typical food webs during the different seasons and in different kinds of habitat.

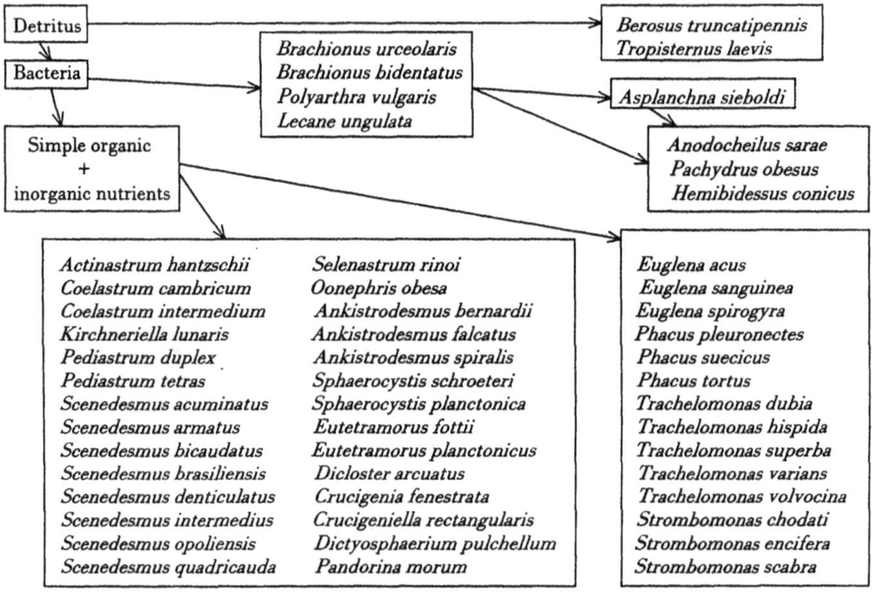

Figure 11.7: A simplified food web involving the microorganisms and small invertebrates in the shallow pools of water on the floodplain near the end of the dry season. There are no consumers of most algal species, which reproduce at a rate controlled only by space in the water body. This microcosm disappears when the water dries up or when the areas are again flooded after the freshet.

Basically, the aquatic species in the Pantanal belong to three distinct groups on the basis of habitat preference. The main group investigated and most thoroughly discussed in this book includes the species of the small streams, ponds, and flooded wetlands that undergo fundamental changes during the course of the annual cycle. The second group encompasses those species inhabiting the two large rivers draining the Pantanal of Poconé, the Rios Paraguay and Cuiabá. These species are potamal elements which may move long distances in the rivers, entering or leaving the region of the Pantanal without being required to change any of their typical behavior patterns. This group consists predominantly of fishes that can swim strongly generally are able to reach considerable size (Section 8). The members of the third group inhabit lentic waters intermediate between the two extremes. They prefer the large permanent lakes along the main rivers at the edge of the Pantanal, many of which are large backwaters or oxbows of the rivers. These species are adapted to standing or slowly moving waters, and some of them may leave the lakes and venture onto the flooded wetlands or into the main rivers at times, thus creating a link between the two otherwise separate systems.

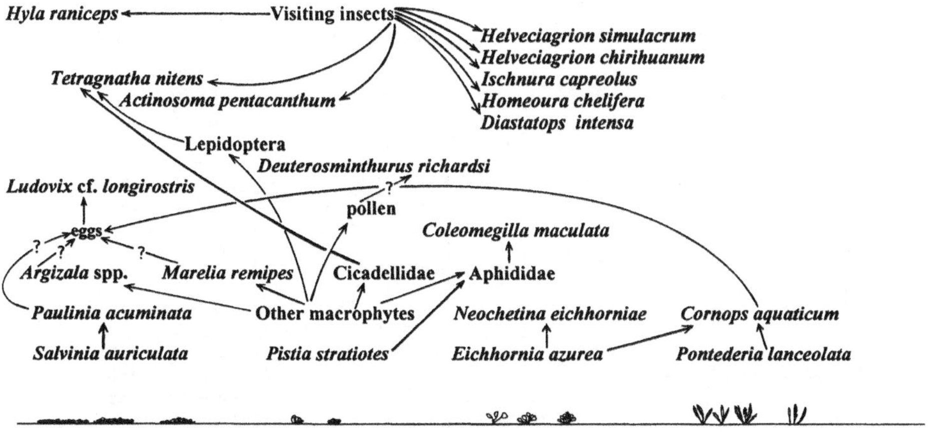

Figure 11.8: The food web of the species inhabiting the pseudoterrestrium. Some of the links are questionable, as indicated, because the species are still unidentified and their ecological roles in the community have not yet been elucidated.

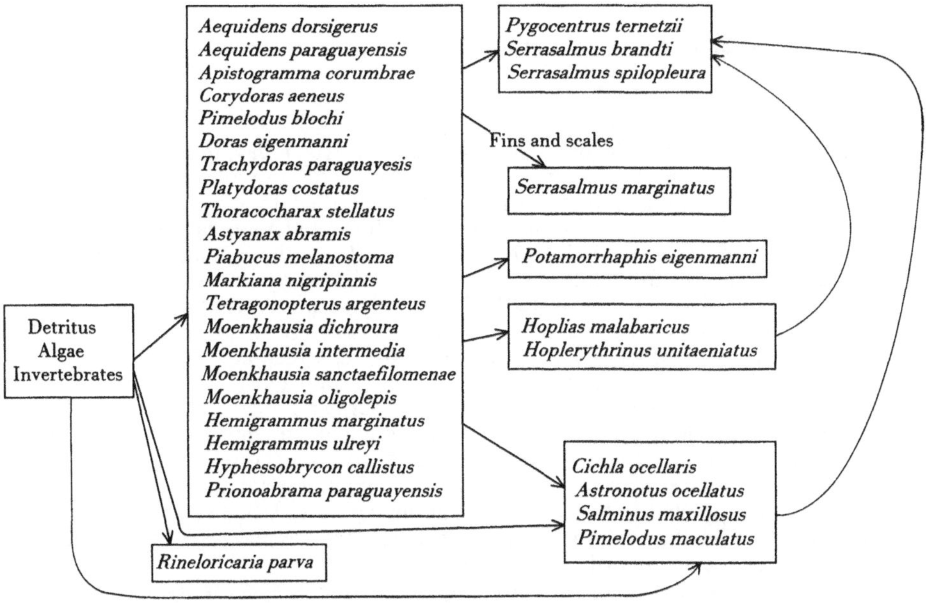

Figure 11.9: Partial food webs involving some of the fishes in the Pantanal. This diagram is highly simplified because many of the fishes are omnivores that have been observed to feed on almost anything thrown into the water.

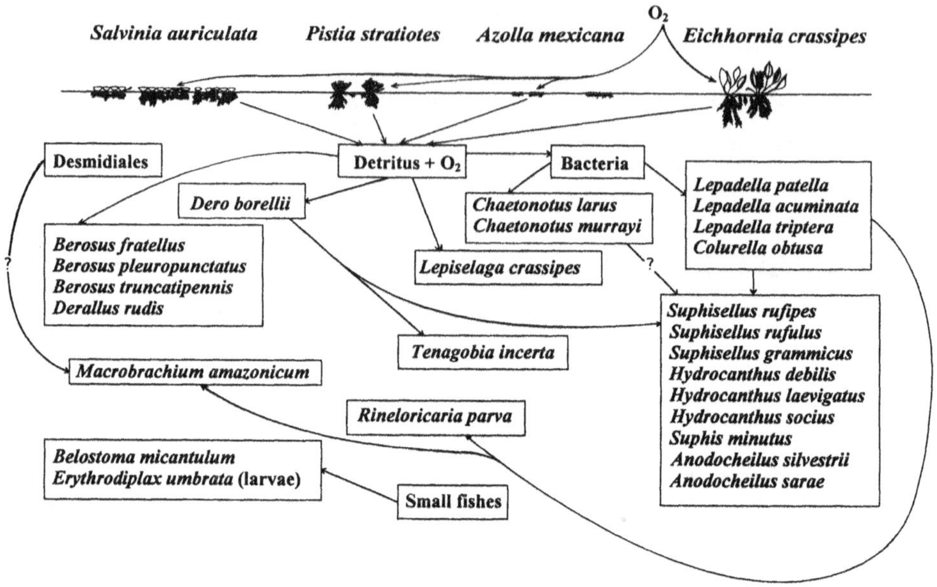

Figure 11.10: A simplified food web of the interrhizon community and some of the organisms that visit it to obtain food.

In comparison to most wetlands of the world as they exist today, the Pantanal supports many relatively large species. This is due to the inaccessibility of much of the region, which has greatly limited hunting and fishing and thus far discouraged grandiose construction projects that have eliminated the large birds and mammals in so many other wetlands throughout the world (Heckman, 1997a). Thus, the Pantanal has food webs that are quite different from those encountered by limnologists in other parts of the world. For example, in shallow water bodies, the large piscivorous birds and caimans eliminate a great many of the larger fishes, many of which prey on smaller fishes. This results in the presence of large schools of small fishes, which consume virtually all of the planktonic and most of the benthic organisms during most of the year. In most water bodies, this would result in the production of large numbers of bacteria and algae, which are free from grazing pressure from the zooplankters and benthic filter-feeders eliminated by the small fishes. However, there are not enough nutrients available during the rainy season to support more than a minimal algal growth because these nutrients are incorporated in the aquatic macrophytes. The lack of "top down" control can only manifest itself during the rainless seasons, when the plants die and release their nutrients through the activity of saprophytes.

This pattern of abundance and scarcity may be referred to as a "skipping a step" form of dynamics. Every other size group in the food web is abundant because the next larger size group of predatory organism is eliminated or nearly so. Perhaps the birds and caimans are overly dominant because the natural top predators, the jaguars and anacondas, have been reduced in numbers. If construction teams or hunters were to eliminate the birds and caimans, too, then large fishes would be expected to become overly abundant, eliminating the small fishes and making the waters suitable for the development of zooplankton and zoobenthos. These, in turn, would reduce the numbers of algae and bacteria present during the dry season, and it would seem likely that the conditions would come to resemble those which confront limnologists throughout most of the rest of the world, where large birds, piscivorous crocodilians, and the largest fish species have been eliminated for decades or centuries.

The food web pattern encountered in the Pantanal also provides insight into the need for top predators. These keep the predators at the upper intermediate levels from excessively reducing the numbers of their food species, including some of the fishes with relatively low reproduction rates, such as *Synbranchus marmoratus*, from the waters of the Pantanal. As a general rule, some check on the populations of species at all trophic levels is required. The toll the anacondas take on the caimans and large water birds hardly affects the size of the populations of these animals at the present time, but it provides a modest check on the activity of these piscivorous species, preventing them from reducing their own food supply to the extent that starvation occurs. The jaguars provide similar checks on various small mammals and birds, which could cause disturbances to the homeostasis of the ecosystem if they became too numerous. In most parts of the world, such top predators have already been eliminated, and many species that they formerly controlled have often become pests because of their overabundance. Classical ecological texts provide ample examples (Odum, 1971).

The step below the caimans and large piscivorous water birds are moderate-sized fishes, including the piranhas. Such predatory fishes have been largely eliminated in the shallow waters of the floodplains and small rivers, but they are rather numerous in the lentic oxbows and moderate-sized rivers in the Pantanal. Sazima and Machado (1990) observed the behavior of the piranhas and the fishes on which they prey and noted that the presence of these predators produced a considerable alteration in the behavior of smaller fishes, which are forced to devote more time to defensive behavior and often seek places of concealment. It can be surmised that many fishes too small to be eaten by the caimans seek out shallow waters in which they are present and thereby obtain some relief from the attacks of predatory fishes that are large enough to be on the caimans′ menu. It is noteworthy that waters in which large numbers of caimans congregate, such as the section of the Rio Bento Gomes near the Transpantaneira Highway, are characteristically inhabited by large schools of small fishes.

This provides some explanation for the observation that there is virtually never any zooplankton present in the shallow waters on the floodplain or in the small streams and rivers. The large schools of hungry fishes are obviously able to eliminate all edible invertebrates in the zooplankton size range that are not able to effectively conceal themselves. In the deeper sections of the medium-sized rivers and the *baias*, however, a moderately rich zooplankton is able to develop during the dry season, after the current in these water courses has ceased. The piranhas and other predatory species effectively eliminate the small fishes, which would otherwise have decimated the zooplankton. Only after the freshet, when tiny offspring of the piscivorous fishes appear in great numbers, are the zooplankters effectively eliminated from these water bodies, as well. Before they disappear, they provide the new generation of medium-sized piscivorous fishes with a large supply of suitable "first food."

Hence, it can be expected that where caimans and large piscivorous birds are present, large fishes will be scarce, small fishes abundant, zooplankton and other tiny invertebrates, scarce, and bacteria and algae, abundant. Where caimans and large piscicorous birds are absent, large fishes will be abundant, small fishes, rare, zooplankton, abundant, and bacteria and single-celled algae, reduced in numbers by grazing. This has a significant impact on fish production. Since some of the small fishes grow into large fishes, an abundance of small fishes means a maximal fish productivity. Juvenile fishes grow considerably faster than large ones of the same species, especially in warm habitats like the Pantanal. Where many large, senescent fishes decimate the juveniles of their own species, fish production is considerably slower than where large numbers of juveniles are permitted to grow to maturity with little mortality. This provides additional insight into the catastrophic decrease in fish productivity in the Amazon system after the caimans were eliminated (Fittkau, 1973).

12. Human impact

There has still been relatively little human activity in the Pantanal, but recently a good deal of interest has been expressed in its "economic development". This bodes ill for the region, if the same sort of economic development is practiced as in Amazonia and many other parts of the tropics. Until recently, however, the legal exploitation of the Pantanal has generally been limited to minor local enterprises and relatively harmless activities, consisting mainly of cattle ranching on lands at higher elevations and around the periphery of the wetlands (Fig. 12.1). Some other activities that have been limited mainly to the periphery of the Pantanal are more problematic. These include gold mining employing mercury (Lacerda *et al.* 1990, 1991; Tümpling *et al.*, 1995), the release of water contaminated by organic pollutants, and the promotion of erosion by various semi-industrial agricultural activities. Although the degradation of the quality of the Pantanal ecosystem due to industry and agriculture is still in its incipient phases, it has become evident enough to warrant concern (Sucksdorff, 1985; Alho *et al.*, 1988; Fereira *et al.*, 1994).

Figure 12.1: During the dry season, cattle are frequently driven along the roads in the Pantanal.

Fishing is another activity that influences the biotic communities in the Pantanal. Although it has become a popular activity there, the still rather small number of fishermen, the vast area of the region, and the great reproductive ability of the fish species minimizes the human impact due to this activity. However, a word of caution is necessary because attention has recently been drawn to the "economic development" of the Pantanal by expanding the fishery. Unfortunately, such programs in the past, especially those involving mass resettlement, have often included provisions to exploit local fish populations to provide almost all of the necessary protein to large numbers of human beings. Generally, there had been few assessments of the maximum sustainable yield, and any assessments that had been made were disregarded, resulting in many more fishermen being put to work than the water body was capable of supporting. One of the measures taken in the Amazon region to increase the amount of fish available for the fishermen was the elimination of the local caimans. As already mentioned, a study of the consequences showed that the disappearance of these reptiles resulted in a further decrease in the fish populations due to the role the caimans play in recycling the available nutrients (Fittkau, 1973). From the experience in the Amazon region and many other parts of the tropics, it would be well to plan now for the development of aquaculture for production of native fish species rather than trying to exploit any of the fish stocks inhabiting natural water bodies for a commercial fishery. Modern methods are available that permit fishermen to catch virtually all of the fishes of value. Thereafter, there is nothing left to catch, and a valuable renewable resource has been destroyed, perhaps permanently.

Still of minor importance, the tourist industry may well become the mainstay of the local economy in the Pantanal. The great attractiveness of the region, its flora, and its fauna has been documented in recent publications (Bertelli, 1984; Ravazanni *et al.*, 1991). These, together with various television programs, have served to awaken the interest of the Brazilians in their own natural wonders. A number of tourists from other continents also visit the Pantanal, and multilingual guides are in demand. If properly managed, tourism can be developed without causing any notable disturbance to the wetland, and the presence of tourists can help to protect the natural populations of wildlife species from decimation due to poaching. The money brought in by the visitors can also help pay for guarding the region against predatory economic activities. The only precautions that must be taken are the protection of the water bodies near the quarters of the tourists from organic pollution and the prevention of large-scale contruction projects within the wetland itself.

There have also been a considerable number of illegal activities in the Pantanal which represent a much more acute threat to the wildlife. Poaching is one of the most serious of these. The caimans are presently the chief victims of this activity. Although the skin of the caiman is reputed to be much less valuable than that of the alligator because of the bits of bone contained in its scales, the energetic protection of the North American alligator has resulted in the market demand shifting to the less well protected caiman. The skin can be used for shoes and watchbands after various treatments, so poaching has become more and more popular in recent years.

As a species of relatively large size that often leaves the water, the caiman is particularly vulnerable to large-scale decimation. Killing caimans acts directly against the tourist industry because being able to see 50 or more of these large reptiles sunning themselves along one short stretch of river bank during the dry season is an attraction that brings paying guests to the region. When these animals become so rare that the tourists can drive for the whole day without seeing one, a prime reason for visiting the wetland will have been eliminated.

In addition to the caimans, many of the relatively rare vertebrate species are exceedingly vulnerable to increased human pressure. Some of the birds, such as the hyacinthine macaw, have been eliminated throughout much of their former ranges due to intensive capture efforts for the pet trade (Plate 16a). Some of the mammals, such as the jaguar, are hunted because they occasionally kill cattle. The giant otter is rare enough to be eliminated rapidly. Moreover, because these animals move around together in large family groups, they are vulnerable to rapid slaughter by poachers using automatic weapons or explosives. Finally, slow-moving terrestrial animals, such as the giant anteater and the large tortoises, are endangered by the frequent fires that are set during the dry season. Although destructive human activities such as these are presently rather limited and generally illegal, there is a danger that they will increase rapidly in intensity if pressure to economically develop the Pantanal continues (Section 13).

To provide a detailed description of the human impact on the Pantanal in historical times, we must go back to the period prior to the colonization of Mato Grosso by Portuguese *bandeirantes* from São Paulo.

12.1. Pre-Columbian civilizations

Mato Grosso was and still is the home of many indigenous tribes, which inhabited the region since long before the arrival of Columbus. These tribes formerly lived in the region of the Pantanal, had large villages near Cuiabá, and hunted and fished throughout the present states of Mato Grosso and Mato Grosso do Sul, as well as in the neighboring parts of Bolivia and Paraguay. There were as many as 70 nations of native people in the Pantanal by the time the Europeans arrived (Bertelli, 1988).

The basic needs of these people were traditionally provided by the natural environment, which they modified no more than necessary to grow food, fish, and hunt. In prehistoric times, it may well be that the ancestors of these people wiped out some of the large mammal and bird species that are known today only from fossil remains. However, by the time the Portuguese arrived, they seem to have been better adapted to meet their needs without disturbing the natural equilibrium of the region than the newcomers from Europe.

There is little knowledge available today about the kind of civilization the indigenous people of southern Mato Grosso had developed by the time the first Europeans arrived because many of the tribes have disappeared with scarcely a trace. There is much historical information about the activities of the indigenous tribes during the colonial period, but under the stress of fighting for survival against technologically superior invaders, the cultures of these tribes must have changed considerably.

It was the presence of these native people that brought the first Europeans to the region of Cuiabá in 1713. They were known as the *bandeirantes* and traveled in groups from São Paulo looking for slaves. This brought them into conflict with the tribes, whose villages they plundered and burned. Over the next few decades, there were several fierce wars between the indigenous people and the European colonists. The colonists were forced to establish a series of strongholds along the rivers to protect their routes of travel (Por, 1995). These wars ended in the gradual elimination of the tribes from the area, although they were still strong enough as late as the period from 1864 to 1870 to be recruited to fight in the Chaco War (Valverde, 1972; Gomes de Souza, 1986; Por, 1995). Although some of the native population was assimilated into the newly developing civilization, later to become Brazilian, the languages and cultures of many of them failed to survive. Whether they were similar culturally to the natives in the Amazon watershed to the north is therefore difficult to say. The most dominant of the nations apparently spoke a language belonging to the Tupi-Guarani group. The dominant tribes were referred to collectively as the Mbayá and reportedly had a form of caste system (Por, 1995). The people they subjugated included those in the Arawakian farming groups, collectively called the Guaná. In addition, tribes from other ethnic and language groups inhabited the region of the Pantanal, including the Bororós to the east of the Pantanal and the primitive Nambikwaras to the north (Por, 1995).

According to Bertelli (1988), some of the native inhabitants of the Pantanal region used fire during deer hunts to drive the animals together. There is, however, some controversy about how much of the woodland was burned intentionally by man during the Pre-Columbian period.

Some native tribes cultivated *Oryza perennis*, a wild rice species (Zerries, 1968), and some of the artificial islands they built (Hoehne, 1936) can still be identified today (Por, 1995). For want of a written history, more detailed information about the native populations of southern Mato Grosso is available only for the periods after the arrival of the Europeans, during which the way of life of these people was altered fundamentally.

However, the native inhabitants quickly passed on the information about the use of the native plants, which had a great impact not only on the modern Brazilian diet but also on the development of the Brazilian economy through agricultural exports, which brought the plantation owners and commercial shippers great wealth. Much information about some of the edible plants in South America was provided by Zurlo and Brandão (1989). Some edible and medicinal species that are still not widely known have potential for becoming major export crops in the future (van den Berg, 1982, 1986; Almeida, 1993).

12.2. Early colonization and the development of ranching

The effects of the first arrival of European settlers in the Pantanal must be confined to those that directly affected the Pantanal ecosystem. The greatest initial impact was the introduction of cattle, horses, and other domestic animals to a region in which they had never before been found. These animals had apparently been brought to the Spanish colony of Paraguay from the La Plata region during the 16th century, and the first settlers in the Pantanal captured some from the wild herds and introduced them to ranches in the Pantanal, apparently in the middle of the 18th century. This race of cattle is called *tucuru* and is well adapted to the local conditions, being resistant to pests (Dourojeanni, 1980). The arrival of large ungulate herbivores naturally had an impact on the vegetation, the extent of which is difficult to say because the *status quo ante* is almost unknown. However, some of the concomitant activities of man have certainly had a much greater effect on the wetland. These include the introduction of grass species from other continents, especially Africa, and the arrival of many "weeds" that were apparently introduced with the grass seeds. These introduced plants have come to overgrow large areas of the pasture land on the floodplain, and some of them are causing the ranchers great problems. A number of the grasses are not liked by the cattle, and they displace the preferred species due to selective grazing (Section 7). The second activity related to ranching is the hunting of the large predators and various other mammals in the region to near extinction. This was practiced for many years to protect the cows, and although it is now strictly illegal, it is still practiced by some ranchers today.

By the 19th century, cebu cattle had reached the Pantanal. Two main races of cebu, the nelore and the gir, are presently abundant in the Pantanal, but the nelore seems to be the principal variety of Pantanal cattle at the present time.

During the 18th century, the settlers also brought firearms, which improved greatly in range and accuracy with the years. In the absence of legal protection of the wildlife, the colonists were able to kill many animals to meet their needs for food or clothing or for entertainment, and because the human population was sparse and the animals abundant, there was no thought of conservation. This produced a sort of hunting-cult which still exists today.

The settlement of Mato Grosso by the Europeans involved not only the introduction of a fully alien culture, it also brought about profound changes in the way of life of the native populations. During the early colonial period, the two dominant tribes in the region of the Pantanal were the Payaguas, who were expert boatmen, and the Guaicurús, who became skilled horsemen after horses were introduced (Por, 1995). Obviously, at least the Guaicurús must have had a different life style before the coming of the Europeans. Today, some of the Guaicurús still survive on reservations outside of the Pantanal. Por (1995) reported that the largest such group is the Cadiveus tribe along the Serra da Bodoquena. The Guaicurús apparently have been accorded a preferred status by the Brazilian government in recognition of their service against the Paraguayans in the Chaco War.

During the colonial period, deforestation and drastic modifications of the environment together with unlimited hunting caused the populations of many animals to drastically decline throughout the populated regions of Brazil, but the vast areas of Amazonia and the Pantanal were spared this decimation because of their inaccessibility. The only notable changes in the Pantanal and adjacent parts of southern Mato Grosso may well have been the reduction of the jaguar populations due to hunting, the decline of the tortoises and anteaters due to the frequent burning of the vegetation, and the displacement of the deer due to competition with the cattle and horses. However, definite proof that these changes have actually occurred cannot be provided because there is no information whatsoever available about the *status quo ante* of these populations. What is clear, however, is that the large-scale destruction of the ecosystems in the remote parts of Brazil have only begun very recently due to the input of large amounts of funds for "economic development" by international agencies.

12.3. Settlements along the rivers

During the early history of southern Mato Grosso, which was the central part of the state until the political independence of Mato Grosso do Sul was declared in 1979, overland travel to and from the cities in the East was difficult and almost never attempted. The cities of Cuiabá and Cáceres were reached almost exclusively by river boats, which made their way up the Rio Paraguay through the southern part of the Pantanal. The boats to Cáceres would continue up the Rio Paraguay, while those to Cuiabá entered the river of the same name at its confluence with the Rio Paraguay in the northern half of the wetland. Thus, both Cáceres and Cuiabá have extensive port facilities, most of which are no longer in use. After a railroad was constructed to Corumbá and on into Bolivia, the trip was somewhat shortened. Travelers were able to disembark in Corumbá and travel northward through the Pantanal in river boats. After the highway system was extended to Cuiabá and Cáceres, the river boat traffic all but ceased.

Old human settlements still exist along the rivers. They are difficult to reach by road. Many have become nearly isolated from the outside world since the demise of the river boat traffic, while other have flourished as vacation and tourist centers. The major town closest to Cuiabá, Santo Antonio de Leverger, can be reached by a paved road and has thus become a popular place for the people of Cuiabá to visit on weekends and holidays. The town of Barão de Melgaço is somewhat farther downstream and can only be reached over unpaved roads, which are sometimes impassable during the rainy season. It is nevertheless a popular place for residents of Cuiabá and tourists to visit. The towns farther downstream along the Rio Cuiabá have remained small and are almost completely dependent upon sport fishermen, who purchase their fuel and sometimes stay overnight in these settlements. Porto Cercado can be reached over an unpaved road from Poconé, while Porto Jofre is located at the southern terminus of the Transpantaneira Highway, which is also unpaved.

From these locations and from many even smaller settlements along the rivers, local fishermen set out to catch fish to meet their own needs and to sell on local markets. From the larger settlements, sport fishermen from all over Brazil set out to catch the large fishes that inhabit the rivers passing through the Pantanal. As mentioned in Section 8.13.1., a great many of these fishes are not Pantanal residents but rather species that undertake long migrations throughout the entire Rio Paraná System. Hence, the present fish production cannot be attributed to the productivity of the Pantanal alone, but the presence of this great wetland in its present condition seems to be a vital factor in this production. The upstream migration of many fishes seems to be mainly for spawning, and the presence of an enormous shallow water region along the downstream migration route of the hatchlings permits them maximum survival rates by offering concealment from predators and a rich supply of edible microorganisms at the end of the dry and beginning of the rainy season, just when the hatchlings require them.

In Brazil, Mato Grosso has a reputation as a rich source of freshwater fish. Lima (1987) reported that 80% of the fish caught there comes from the Cuiabá River. Hence, there is a concentration of the fishery along that section of the river settled during the period that the main route to Cuiabá was the river system. This explains several potential problems which may become acute in the near future. The few studies conducted to the south, mainly in Argentina (Lowe-McConnell, 1991), have provided only sketchy information about the fish migrations, and virtually nothing is known about the routes taken by the various species in the upper sections of the Rio Paraguay System. Some of the fishes are known to enter tributaries draining the Andes foothills in Bolivia to spawn (Bayley, 1973). Whether the fish species distribute themselves in the various tributaries in some recognizable pattern or whether some or all of them swim upstream and randomly enter any tributary is unknown. Without knowing anything about the sizes of the migrating populations, the main spawning sites, or the seasons the runs begin and end, it is not possible at this time to predict the impacts of the fishery on the most important commercial species.

The size of the towns and villages along the Cuiabá River and the local fishing methods used suggest that the local fishery has had little impact on the fish populations. The numbers caught from the dugout canoes and small rowboats have apparently had no negative impact on recruitment. The minor fishery effort and the nearly complete ignorance of the spawning habits of the fishes have led the local fisherman to regard the fishes as an inexhaustible resource, and there is a general disregard for the government regulations prohibiting fishing during the months of the early rainy season, when the fishes in the rivers have been observed to spawn intensively. The sporadic enforcement efforts generally lead to large numbers of arrests and seizures of equipment, indicating the magnitude of the violations. In many cases, the fisherman deny even knowing that fishing is seasonally prohibited.

The problem caused by the local fishermen is minor compared to that of the sport fishermen, many of whom have new and extremely powerful motorboats. It is often not clear how much of their fishery efforts can be classified as "sport" and how much of it is commercial. Furthermore, with such equipment, it is easy to escape the local forestry police, whose boats are smaller and slower. Thus, this group of fishermen have a potential for massive overfishing, as well as for poaching, if they so choose. In the Amazon region, an increased fishery effort has been shown to have caused substantial reductions in the populations of popular food fishes (Goulding, 1981). Acting in ignorance of the population dynamics of the species in question, it is difficult to judge the value of the various measures employed to protect the fishery resources.

12.4. Recent changes

As mentioned above, the Pantanal itself is influenced by relatively few human activities. It is fortunate that the great construction craze of the 1970s that ruined the economies and despoiled the resources of so many countries and so many other regions of Brazil somehow spared the Pantanal. Except for the abortive attempt to build the highway across it, no major projects to modify the environment were attempted in the wetland during that period.

Small scale changes that have influenced the conditions in the Pantanal include a resurgence of gold mining to take advantage of the high prices for the metal on the world market that prevailed early in the 1980s and introductions of new species, mainly herbaceous plants, in the hopes of improving the pastures for the cattle. The problem of weeds introduced with the plants has already been discussed.

Other introductions include the water buffalo, which has only recently arrived in the Pantanal (See below). It has been raised for a much longer period of time in the state of Amazonas, as described by Ohly (1985).

12.4.1. Agriculture

In the past, the small to medium-sized farm was important for the production of food in Brazil. In the State of Mato Grosso, small farms were rather widely dispersed because of the sparseness of the population. Nevertheless, many such farms were present near the cities and towns, for which they produced much of the food. In the rest of the state, extremely large areas of land, including much of the Pantanal, was occupied by vast ranches, which produced beef by methods that did not gravely affect the natural biota.

Recently, as in much of the rest of the world, there has been a mass migration from the rural areas to the cities, which has been promoted by a transition to mechanized monoculture. The small farmers have been displaced, and their work made unprofitable by the resulting economic changes. This has most gravely affected the *cerrado* region on the plateau to the east of the Pantanal. The most intensive agricultural activity in southern Mato Grosso is taking place to the southeast of Cuiabá, and even larger areas have been occupied by monocultures in the State of Mato Grosso do Sul, to the south. Unfortunately, the soil loses its fertility only a few years after it has been cleared for agriculture, and the insatiable appetite for new *cerrado* areas to slash and burn has resulted in a steady northward movement of the monoculture. This can have an impact of the Pantanal indirectly due to the increased erosion and leaching of nutrients, which is increasing the amounts of materials entering the wetland through the streams and rivers. The Pantanal itself is not an attractive region for cultivation due to the periodic flooding. This situation could change if an interest in rice farming develops, but an oversupply of this crop already exists in Brazil.

The leaching and erosion due to the stripping of the native *cerrado* vegetation from vast areas of the plateau has become severe, as evident throughout the region subject to monoculture. The formation of *voça rochas*, or erosion canyons cut through the soft sandstone layers, is evident along most of the water courses, and the results of the leaching invariably becomes evident when the productivity of the land decreases greatly after only a few years of cultivation. Under the methods presently used, the mechanized farmers simply abandon the stripped fields and cut more of the *cerrado* vegetation to develop new sites for cultivation. The native vegetation, which is adapted to surviving under the prevailing wet-and-dry climatic conditions by anchoring the soil and thereby interfering with the processes of leaching and erosion, also included species which the former rural population knew how to utilize for food (Siqueira, 1981). Many of these species and the knowledge of their usefulness may well disappear.

An even more insidious development associated with the monoculture is the application of enormous amounts of pesticides. This problem is still more acute in Mato Grosso do Sul, where the monocultures have been established longer. Yamaciro (1989) estimated that during the years from 1985 to 1987, 765,520 l of liquid and 115,427 kg of solid biocides were sold in Mato Grosso do Sul for application in the drainage area of the Rio Paraguay. Most of these are herbicides. Ferreira *et al.* (1994) reported that the most common substances sold include trifluralin, arachlor, monocrotophos, methyl parathion, and chlorates, mainly Mirex. However, reviewing the results of other authors, they concluded that the official consumption rates may amount to only 10% to 30% of the actual amounts used. As in many parts of the world, there is little awareness of the dangers to human health and the environment from these toxic substances. In 1988, one of my early impressions of Cuiabá was the sight of an agricultural supply store in a main shopping section of the city with the sidewalks in front of it white with pesticide powders of all kinds that had leaked out of the large sacks being offered for sale. Nobody, including the proprietor of the shop, seemed the least bit concerned with the danger of contact with these substances, and the general public throughout the world is largely ignorant of the toxic, carcinogenic, and teratogenic effects of such substances. The problematic methods of marketing these substances in the tropics were reviewed in a separate work (Heckman, 1993).

In contrast to the monoculture, the present methods of ranching in the Pantanal are extensive and have scarcely changed since the time of the early colonists. The only negative impacts of lasting consequence have been introductions of species, both grasses and new domestic animals. However, bad experiences from the introduction of noxious weeds that have spread rapidly may have encouraged the ranchers to desist from seeking new grass varieties abroad.

One recent development that may have long-range consequences for the Pantanal is the introduction of the water buffalo, *Bubalus bubalus*, which is far better adapted to the wetland environment than are cattle (Fig. 12.2). Although this animal is much better suited to life in the wetland and would require little modification of the natural system to survive, it is a cause of concern for many environmentalists. If its culture were to partially displace cattle ranching, however, less modification of the wetland would be carried out. For example, the harmful practice of burning large areas during the dry season would probably be discontinued because the water buffaloes forage mainly in and beside shallow water bodies rather than on dry pasture land. However, it is these foraging habits that have caused some of researchers who study the caiman to register some concern that buffaloes may interfere with the nesting of the caimans and cause some mortality among the young (Campos, 1993).

Figure 12.2: The most newly introduced large mammal to the Pantanal, the water buffalo, *Bubalus bubalus*. It is a domestic animal ideally suited for life in the wetland, but some zoologists are concerned about the impact it may have on native species, particularly the caiman, the nests of which may be threatened by the buffaloes.

The greatest impediments to buffalo ranching are the lack of acceptance of its flesh and milk on the local markets and the lack of experience of the cowboys in dealing with these animals, which react quite differently than cattle to the long-practiced methods of moving the herds.

The success of this venture, as in the case of previous introductions in Amazonia, is dependent upon the acceptance of buffalo meat by the consumer and the adoption of effective methods for handling these animals, which do not respond at all to the usual methods employed for handling cattle. It was learned in the Amazon region that although the buffalo is much more suitable to the habitat than cattle, which suffer from the heat and do not adapt well to flooding, the cowboys cannot handle them well, and there is a general rejection of buffalo meat by the consumers. Further objections come from naturalists who oppose any new introductions as a falsification of the fauna. It must, however, be admitted that the introduction of cattle was also such a falsification, and if the terrestrial ungulates could be reduced in abundance to make room for the buffalo, it would simply be a case of trading off one introduced species for another. In the tropical wet-and-dry climatic zone of Southeast Asia, both buffaloes and cattle are raised, but buffaloes are generally dominant in the temporarily flooded lowlands. If approached by a horseman trying to drive them like cattle, they do not flee to the herd but rather take a defensive stance and may become aggressive. However, buffaloes can be led or ridden, even by small children. Their milk is richer in fats, and their meat is somewhat more tender than beef. Because they are so easily handled by people familiar with them, they are unexcelled as work animals in muddy places.

12.4.2. Increasing organic pollution and disease

In recent years, the city of Cuiabá has grown faster than any other major city in Brazil. Like many other large cities in South America, it has no sewage treatment plant. The main method of collective sewage disposal is the channeling of raw sewage into the rivers, which has made urban pollution a problem of immense proportions in Brazil (Homewood, 1991). The entire bulk of the waste material, industrial and domestic, finds its way into the Rio Cuiabá upstream from the Pantanal.

Not only has there been a steady quantitative increase in sewage input, there have also been several important qualitative changes. One particularly serious one for the human population is the arrival of *Vibrio cholerae* in South America from Asia. The first outbreaks of cholera on the continent occurred in Peru at the beginning of the 1990s. From there the disease spread rapidly to the neighboring countries and arrived in Mato Grosso, probably in 1992, when a few isolated cases were reported. After a period of more than a year, the disease reached Cuiabá, and several minor epidemics occurred, causing a number of deaths. This disease is a particularly serious threat to the people living in the towns along the river downstream from Cuiabá because the people have traditionally obtained their water directly from the river, which is now contaminated with the vibrio. Not surprisingly, one of the most publicized outbreaks of cholera occurred among guests at a festival at Santo Antonio de Leverger, the first town downstream from Cuiabá.

A second problem has occurred due to the arrival of yellow and dengue fever from the East. These diseases had been rare or absent from southern Mato Grosso until about 1991, when a few cases of yellow fever appeared. About a year later, dengue fever arrived, and many people were infected. While yellow fever can be prevented by vaccination, there is no effective vaccine against dengue fever. To combat this disease, spraying campaigns were begun in the cities and towns, including Cuiabá, Varzea Grande, and Poconé. Both of these diseases are transmitted mainly by *Aedes (Stegomyia) aegypti* (Linnaeus, 1762), a cosmopolitan species that lives in close association with man. This urban mosquito has been subject to pesticide control programs worldwide for decades, and it is apparent that it has become resistant to almost every pesticide known to man. Predictably, the first results of the spraying campaign in Cuiabá was an extraordinary increase in the number of mosquitos, their natural enemies having been killed off by the sprays. Although the small epidemic soon ran its course as a result of seasonal factors, the disease returned with even greater severity the following year.

This disease and the ineffective methods used to control it will add another qualitative factor to the river pollution problem. From past experience worldwide, it is safe to assume that as the mosquitos increase in numbers, so will the amounts of pesticides used to control them. As the spraying increases, so will the populations of those urban mosquito species that are resistant to the chemicals. The result will be an ever increasing amount of toxic chemicals entering the Rio Cuiabá and finding their way into the Pantanal. One of the substances used in Mato Grosso against *Aedes aegypti* is DDT, to which these mosquitos have long ago become resistant. Unfortunately, DDT is one of the slowest of all pesticides to decompose (Kuhr *et al.*, 1974; Heckman, 1981), and before it does, some of it is transformed to DDE, which is harmful to bird populations by weakening the egg shells and thereby reducing the breeding success (Ames, 1966).

12.4.3. Gold mining

The bulk of the discussion about the impact of gold mining must be reserved for the next section because this is a major concern in planning conservation measures to protect the Pantanal. The reawakening of the interest in gold mining, which occurred during the 1980s, has resulted in the construction of large numbers of open pits, deep and shallow, in and near Poconé and Cangas. The gold is present as dust mixed with sand and silt. To separate the precious metal, mercury is employed. The amalgam is placed in cloth and squeezed to eliminate the excess mercury. The remainder is heated to drive off the remaining mercury as vapor. The final purification with a blow torch is generally completed in the shops of the gold buyers, which are common in Poconé and Cuiabá. The methods used usually fail to include an efficient recycling of the mercury, mainly due to disinterest on the part of the miners. The obvious results of the mining include large abandoned pits filled with rainwater, great masses of soft mud disposed of as tailings and contaminated to various degrees with mercury (Tümpling *et al.*, 1995), and miners and gold buyers displaying the symptoms of chronic mercury poisoning of the central nervous system, which eventually makes them into cripples. The amount of gold smuggling to evade paying taxes to the Brazilian government is reflected by the considerable amount of the metal exported from Paraguay, which does not produce any itself but is one of the main receivers of the smuggled gold. The mercury is not only be transported as vapor through the air, it also contaminates streams feeding the Pantanal when the contaminated tailings are washed into them during heavy rainstorms. Lacerda *et al.* (1991) reported as much as 0.61 mg $Hg \cdot kg^{-1}$ of dry weight in the fine material carried by stream water after a storm. They found that molluscs accumulated this metal, while macrophytes and fishes did not. The ambient water after removal of the suspended material was never found to carry as much as the minimum detectable amount of 0.04 $\mu g \cdot l^{-1}$, while the stream sediments contained from <0.02 to 0.18 $mg \cdot kg^{-1}$ dry weight, with the greatest amounts being found near the sources of the mine tailings.

For the long term economic development of Mato Grosso, the gold mining plays no role at all. When the gold is gone, the miners will depart without having founded any permanent industries. They will leave a scarred and eroded landscape and deposits of mercury behind, and they themselves will face crippling disabilities at a rather young age. During 1993 and 1994, the police began closing some of the mines, most of which were illegal. This incited protests by the miners. The lack of sufficient financial support from the government for enforcement of bans on mining permits the miners to continue their activities at night or at other locations not yet discovered by the police.

13. Conservation

In this section, the control of human activities to prevent the loss of natural resources will be discussed. In a sense, it parallels the material in the last section, but it is more theoretical in nature. Section 12 treated the anthropogenic changes that have actually occurred, while this section concentrates on potential activities that could cause damage in the future and the countermeasures necessary to prevent them. Generally speaking, the conservation laws in Brazil are good. Unfortunately, they are seldom effective because practically no funds are available for their enforcement. Thus, almost all of the destructive activities about to be discussed are already illegal, so no action from the lawmakers is required. Sometimes, perhaps, the politicians and enforcement officials look the other way when they should step in and take action, but most of the time, there are simply no enforcement officials present where the illegal acts are committed. In addition, there have been strong pressures brought on Brazil and other tropical nations from the mighty international development lobby to undertake mammoth construction projects, ostensibly to industrialize these countries and improve the economy. None of these projects have brought anything but debt to the recipient nations, but still new ones are being thought up. Many examples can be found of the severe problems created by poorly planned alterations of Brazilian river systems (Fearnside, 1990; Noda and Noda, 1990; Tundisi *et al.*, 1991; Tundisi and Straskraba, 1994; Petrere and Ribeiro, 1994). That the Pantanal or its river systems become one of the targets of "development" is probably the main danger to this unique wetland and its rich wildlife as well as to the human population in rural Mato Grosso, who stand to lose their livelihood and homes to make way for whatever mammoth project is envisioned.

Because of an extreme dearth of substantiated information in a form available to planners (Coutinho *et al.*, 1994), sensible developmental planning for the region of the Pantanal is still impossible, as it is in other parts of the tropics. Typically, follow-up studies of projects demonstrate that one expected negative effect has not been as bad as predicted, while another that was not predicted at all before the project, produces enormous problems. In Brazil, as well as elsewhere, the unexpected problems often have to do with public health (Petrere and Ribeiro, 1994). Bonetto and Wais (1992) noted that the system of the Rios Paraguay and Paraná drain two considerably large wetland areas. In addition to the vast Pantanal, a second system of floodplain wetlands borders the lower sections of the system. Obviously, major alterations of the Pantanal will significantly affect the extensive wetland system to the south in Argentina, as well. They stress the great need for research on this system before any modifications are undertaken.

13.1. Physical and ecological change

13.1.1. Ranching

As discussed in the last section, the most extensive anthropogenic change observed in the Pantanal has resulted from the conversion of relatively large areas at higher elevations to grasslands for raising cattle. This change involved considerable changes in the flora, and naturally, with the associated fauna, as well. There is no practical way to undo this falsification of the biota, and the introduced species are mostly there to stay. Major losses from the native fauna are not yet detectable, however. The great compulsion of many international development schemers, starting at about the time of Columbus's first trip to America, to introduce species to new continents has done much more damage with species exported from South America than with those imported. It is only necessary to consider the disastrous economic impact *Eichhornia crassipes* is having on Asia and Africa to realize that South America has still been spared the worst.

In fact, Columbus himself may have been one of the only intercontinental transporters of species who achieved mainly positive results, having imported the potato and other useful crops to Europe. Most people trying to imitate his success have accomplished little or nothing or have caused major economic catastrophes.

Aside from the introduction of new species in the past, cattle and horses included, the ranchers have done little to fundamentally change the Pantanal. For the purposes of conservation, restrictions need be placed only on several activities which are showing a tendency to become more widespread and could have severe negative effects on the flora and fauna if practiced too often.

The first of these is the uncontrolled burning of large areas to eliminate shrub growth and promote the development of grassland. Ironically, one of the factors responsible for the spread of large numbers of unwanted plants that the cattle do not like to eat was the introduction of these plants from other continents as or together which species that were thought to be preferred by cattle. Some of the large grasses, such as *Andropogon bicornis*, are not liked by the cattle, and as the more palatable plants species are grazed away, the "weeds" come to form nearly monospecific stands on the pasture land (Section 7). These and tough, woody shrubs soon change the more elevated parts of the wetland to shrubland into which the cattle no longer venture. When fires break out in such areas of undergrowth during the dry season, either in the Pantanal or in the upland regions surrounding it, it is officially claimed that they were started accidently. However, it is general knowledge that such fires are started deliberately, and in the evening, laborers can often be seen starting them with the bottles of flammable liquid at regular distances along the highway, as personally witnessed during the study on many occasions.

Burning is somewhat more widespread in the dry upland regions near the borders of the Pantanal. Much of the natural dry deciduous forest and *cerrado* on the plateau east of the Pantanal has already been destroyed to gain land for soybean and sugar cane plantations, and many remaining areas are burned each year to produce pasture. During the dry season in 1991, for example, over 40 major fires were counted in the region near Cuiabá in a single day, the airport had to be temporarily closed due to reduced visibility from the smoke. In 1993, particularly large fires destroyed rather large tracts of forest near the Rio Cuiabá at Porto Jofre. As a result of many decades of annual burning of the vegetation, the *cerrado* in the regions elevated somewhat above the level of the wetlands is sometimes called a "fire climax" (Section 2). There have been some studies of the effects of such fires on the vegetation of other continents, such as that of Menges and Kohlfeldt (1995) in Florida, but the real consequences of this phenomenon in South America are still poorly known. Many animals that cannot move rapidly and lack underground retreats are selectively eliminated by fire. Tortoises and large anteaters, as well as arboreal species, are decimated. Studies of mammals on other continents have shown that most escape the fire but later lack food and places of concealment, which negatively affects their rates of survival (Crowner and Barret, 1979).

A general survey of the effects fire has on the *cerrado* was provided by Coutinho (1990), while the relative resistance of various plant species in Mato Grosso and Mato Grosso do Sul to fire has been elucidated to a large degree by Lorenzi (1992) and Pott and Pott (1994).

After the unwanted vegetation is burned, grasses grow rapidly, and attractive pasture land is obtained for a short period of time, but the introduced plants that are unloved by the cattle develop more quickly than before, and while they are developing, much of the sparse nutrient supply in the soils is lost through leaching and erosion. Furthermore, in the Pantanal National Park, this activity is even more out of place because of the serious losses of rare species and nesting birds, which cannot escape the flames. To protect the Pantanal, this practice has to be eliminated, but it has become so common that this might require rather drastic enforcement methods.

The second constraint on ranching in the Pantanal is the requirement to protect the surviving large predators, mainly the jaguar. Brazilian laws strictly protect this and other endangered species, and the economic interests of the ranchers provide no grounds for a legal dispensation from the requirements of conservation. The problem lies in enforcement in the vast area of the Pantanal guarded by a ridiculously small number of forestry police. Many ranchers take their responsibility to guard the natural resources of the Pantanal seriously. Others do not. Generally, those profiting from tourism by renting quarters to visitors to the Pantanal and organizing guided tours realize that the jaguar is a great attraction, even if it seldom seen. This kind of awareness, that endangered natural resources also have economic value, is increasing among the ranchers and helping them understand that poaching is a counterproductive activity.

A third constraint to ranching that may become a problem in the future is the use of pesticides, particularly insecticides and herbicides (Heckman, 1983b). At the present time, ranchers in the Pantanal of Poconé generally claim that insecticides are used mainly in baths or as sprays applied directly to cattle in corrals. This type of use does minimum environmental damage, although it does not seem to accomplish much in combating the pests. Furthermore, the beef prices are not high, and profits can be made only when the production costs are kept low. In the Pantanal, the cattle roam free, and the cost of producing beef in minimal. They are not well protected, and losses are great, but there is presently an enormous surplus of beef, and production costs using the present methods are nearly non-existent.

A danger of considerable magnitude is the active advertising of these biocides by the chemical manufacturers, who have shown talent in the past for peddling many products that do not work but cause grave harm to the natural biotic communities as well as to the agricultural workers who are forced to work with them (Heckman, 1993). As these substances are increasingly forbidden or rejected by consumers in the industrial countries, increasing efforts are made to dump them on new customers in lands where they were never used before. In Cuiabá, there is a considerable amount of advertising on television and in the newspapers for agrotoxins, and much effort is put into winning over the government agronomists who are responsible for advising the farmers and ranchers on new and better methods. In fact, Brazilian scientists have complained that entomology symposia are often paid for entirely by multinational corporations that manufacture and distribute insecticides, and that their salesmen blanket these meetings trying to win entomologists over to help them peddle their products. It is also noteworthy that some of the pesticides offered for sale are not only prohibited in the countries where they are manufactured, their use is also forbidden by law in Brazil. However, no provisions have ever been made for law enforcement in these matters.

A final problem that may appear in the future involves the falsification of the native biota. An increasing number of ranchers are expressing interest in aquaculture. Indeed, Brazil has a vast potential to develop this industry, and at the present time, South America has the poorest aquaculture production of all continents except for Africa. In terms of quantity, only 1.8% of the world's yield comes from South America. However, this corresponds to 4% of the total worldwide value of the products (FAO, 1994). While vast areas of Africa suffer from a lack of water, South America has some of the greatest unpolluted freshwater resources in the world and an amazingly varied fish fauna. Nevertheless, due to a lack of expert help in domesticating local fish species, many persons are turning to the cultivation of introduced species, such as common carp and tilapias. Carp is especially unsuited for cultivation due to the high local water temperatures, and in the warmer regions of Brazil, the results are invariably worse than disappointing. Apparently, only in the southernmost state of Rio Grande do Sul can the common carp develop well. Near the Pantanal of Mato Grosso, the results would be disastrous due to the high water temperatures and low quantities of food for the carp during the high water period. The kind of eutrophic waters rich in benthic organisms that the carp prefers are simply not available.

Tilapia culture would have better chances for success, but why should the culture of African cichlids be encouraged when there are a variety of native fishes in this same family that grow faster and larger and are tastier to eat? Because of the similarities of their general habits to those of the native cichlids of the Pantanal, tilapias could prove to be serious competitors if they succeeded in escaping the culture facilities and colonizing the Pantanal. The second great danger of tilapia introductions is the possible transport of a serious cichlid pathogen or parasite that could decimate the native species that have not yet become resistant to them. It is not unusual in such cases for a virus, bacteria, or invertebrate parasite that lives on the introduced species without causing much harm to suddenly invade a native species previously unexposed to it with fatal results for a large percentage of the population.

Studies of a lake in Brasilia used for the culture of the African *Tilapia rendalli* (Boulenger, 1896) and a Chinese cyprinid, *Hypophthalmichthys molitrix* (Cuvier & Valenciennes, 1844), demonstrated that the presence of these fishes causes changes in the functioning of the ecosystem. The presence of tilapia resulted in a great increase in the populations of algae and rotifers. Starling (1993), who conducted the study, concluded that much more needs to be known about the ecological changes caused by the introduced fishes to preclude negative impacts on the ecosystem. Past experience has demonstrated, however, that aquaculture is often promoted in a reckless manner, particularly due to the inability of "experts" in economic development to learn that species other than Asian cyprinids and African cichlids can be successfully cultivated.

13.1.2. Mining

Mining, mainly for gold, has had a much more serious impact on the Pantanal. Not only has much pollution resulted from these activities (Section 13.2), a great deal of erosion has also been initiated, which has resulted in the deposition of large amounts of soft, muddy sediment downstream from the sites of the erosion. Fortunately, these activities have not yet become widespread, but some of the erosional sites at the edge of the Pantanal are causes for alarm.

Prospecting for gold dust is a particularly poor industry for the economic development the Pantanal region. The miners are able to earn a living only when the price of gold on the world market remains relatively high. The steady decline of the gold prices in recent years has therefore made this industry less lucrative. When prices become too low or the gold supply diminishes, the miners move to other regions, leaving nothing behind but large holes in the ground and large fields covered with mining sludge. The economic infrastructure in the towns of the gold mining region disintegrates, and the amassed wealth is transferred to other parts of the country or, more frequently, to foreign banks. The miners become unemployed, often with their health permanently impaired by the long-term contact with mercury.

The main environmental problem created by gold mining is the release of mercury, a heavy metal that is extremely toxic to mammals and birds. This will be discussed in detail below.

13.1.3. Public works

Another activity that has changed the Pantanal landscape is the construction of roads and highways. The Transpantaneira Highway itself is one of the best examples. After a road has been built, the continued disadvantages for the aquatic and terrestrial communities include losses of rare animals, which are run over on the road, sometimes deliberately, as well as an increased accessibility of the region to man, which often reduces the number of hiding places available to the wildlife species for rest and rearing the young. It also makes the area accessible to poachers, as discussed below.

The positive side of highway construction is the opening of the region to tourists, which often calls attention to the rich biota of the region and the need for conservation. In the case of the Transpantaneira Highway, the scooping of fill from the sides of the road during construction formed depressions on either side of the embankment. These hold water long after the floodplain has dried out during the *seca*. The flooded ditches provide additional refuges for the aquatic species during the driest period of the year, facilitating an increased survival and bringing many interesting species near enough to the road to serve as tourist attractions for the visitors.

Fortunately, practically no public works other than the limited amount of road and bridge construction have been undertaken in the Pantanal, and other than the dredging of the rivers, discussed below, none are presently planned. Sometimes, a shortage of public funds can be beneficial.

13.2. Pollution

Because the human population in the Pantanal is sparse, the amount of anthropogenic pollution of the surface waters is presently of minor importance. A local exception is the contamination of the sediments with mercury, which is used in the gold mines, as discussed below. The various kinds of pollution must be discussed separately. However, as long as pressure from international development organizations for the intensification of agriculture continues, a great potential for pollution exists, particularly toxic pollution from pesticides, which are presently being promoted by intensive advertising campaigns sponsored by the international chemical concerns in Mato Grosso. As discussed in Section 12, this problem is becoming acute in Mato Grosso do Sul, but it is still in its incipient phases in Mato Grosso.

In the temperate zones, pollution of the surface waters is usually associated with the eutrophication of the water bodies. Planning for applied environmental research in the Pantanal by European and North American institutes has therefore often included provisions for studying the effects of eutrophication. In fact, the Pantanal is a relatively nutrient-poor region that is not threatened at all by eutrophication, except locally where small water bodies receive massive inputs of organic substances from domestic sewage or the wastes of cattle ranches. Contagion is apparently the greatest danger associated with this pollution, as recent cholera epidemics in Cuiabá and neighboring towns has recently demonstrated. This problem will be discussed in detail in Section 13.2.5. below.

13.2.1. Heavy metals

Although heavy metal pollution is still only a local problem in the region of Poconé, where there are many gold mines, it is a form of pollution that is serious because of the extreme toxicity of the mercury released from the mines to birds and mammals as well as the fact that this and other heavy metals persist in the sediment for extremely long periods of time and may become concentrated in the food web whenever they are liberated through chemical action, sometimes mediated by microorganisms (Skerfving, 1972; Heckman, 1993). Mercury can also be transported as vapor from one place to another through the atmosphere and may distribute itself in sediments in an irregular and patchy pattern. This problem can become serious because of the tendency of heavy metals to become "biomagnified", whereby certain organisms accumulate mercury or other metals at concentrations one or more orders of magnitude greater than that in the ambient water. Such organisms can then pass the metals along to species at the next higher trophic level in the food web, where a further biomagnification can occur. Piscivorous raptors in North America have often been found to suffer population declines from a combination of mercury and DDE in their food, the mercury crippling or killing the birds of the present generation and the DDE weakening the egg shells so that the survival rate of the following generation remains poor (Section 12.4.2).

Analyses of the water and sediment samples revealed that the natural background levels of most heavy metals are low (Heckman, 1995). The concentrations of selected toxic metals in the sediments of the Pantanal and headwater streams feeding the Pantanal system are no more than slightly above the detectable limits and usually well below. Among the less dangerous metals, only iron, an element not physiologically toxic to most forms of life, is present in considerable amounts in the sediments of some water bodies. Manganese often accompanies the iron, but considerably smaller amounts are present (Section 4). Thus, elevated levels of toxic metals found in sediments of the northern Pantanal can with great probability be attributed to human activities.

The unusually low water hardness of the water feeding the Pantanal is a factor that must be considered in evaluating the potential impact of a heavy metal contamination on the aquatic organisms. Toxic effects of metals seem to be inhibited by calcium (Jeffree *et al.*, 1995), and such metals as zinc (Everall *et al.*, 1989), divalent trace metals (Markich and Jeffree, 1994), and alum sludge (George *et al.*, 1995) were all found to be more toxic in soft water. Furthermore, synergistic effects of more than one trace metal on daphnia were enhanced at low water hardness (Tomasik *et al.*, 1995). Under these circumstances, it would be wise to take precautions to prevent heavy metals from ever being disposed of in the streams and rivers of the Pantanal catchment area.

The primary industry responsible for heavy metal pollution in the Pantanal of Poconé at the present time is assumed to be gold mining, which is an activity dating back to the founding of Poconé in 1781. Most recent mining activity, however, dates from the early 1980s, when sharp rises in the gold price on the world market prompted the revival of this industry on a large scale. This has resulted in perhaps the most negative anthropogenic influence on the Pantanal: the contamination of the water bodies and sediment with mercury.

There are considerable deposits of gold dust in the drainage area of the Rio Cuiabá and its tributary, the Rio Bento Gomes. Recapitulating the information from previous sections, the technically primitive form of purifying the gold dust is briefly described. To extract the precious metal from the remaining sediment, mercury is presently used. This permits profitable extraction from rather poor quality gold-bearing sand. The mercury used is gradually lost into the river sediments, where it remains for long periods of time, or it is vaporized to purify the gold and transported through the air to various other locations in the Pantanal watershed. At some time in the future, it will almost certainly become incorporated into living organisms. It can then be passed along through the food web to vertebrates, including man. Much of the mercury could be recovered and reused, but few miners are interested in taking the trouble to do so.

The pollution of the sediments with mercury is the most alarming negative aspect of gold mining, as described below. This can have long-lasting consequences if the Pantanal becomes a site for the future development for aquaculture. The potential for a dangerous contamination of the fish produced will exist for a long time to come. It is a particularly difficult danger to assess because under favorable conditions, no mercury will find its way into the food produced, and analysis of random samples often fails to detect any problems. However, subtle changes in the chemistry of the upper sediment layer can lead to a mobilization of the mercury and the sudden appearance of large quantities in the food web.

In the Amazon Basin to the north, the problems arising from contamination of many rivers with mercury have already become so serious that fish taken from some of the rivers must be considered dangerous for human consumption. Nriagu *et al.* (1992) reported that detritivorous, omnivorous, and carnivorous fishes in the Rio Madeira and its tributaries in Rondônia contained 0.12 ± 0.06, 0.13 ± 0.08, and 1.01 ± 0.64 $\mu g \cdot g^{-1}$ of mercury, respectively, quantities that could easily result in a person consuming far more mercury per day than the 0.43 μg per kg of body weight recommended as the maximum intake by the World Health Organization. This is particularly serious in view of the great importance that fish has in the diets of the local human populations in the mining areas.

Studies on the present whereabouts of the mercury used for the extraction of gold in Poconé were undertaken by Tümpling *et al.* (1995). They assumed that from 10 to 15 metric tons of mercury had been used there during the decade prior to their study. Of this amount, they were able to account for an estimated 1.6 tons in the tailings from the mines. They discussed the difficulties in decontaminating the tailings because the mercury is unevenly distributed, with much being encountered in "hot spots" while the rest occurs at minimal concentrations. These tailings are stored in mounds around the edges of the mining areas, and from there, they can be washed into nearby streams by the heavy rainstorms.

Much of the remaining mercury finds its way into the atmosphere, as detected by analyses of the air in and near Poconé (Tümpling *et al.*, 1995). The mercury vapor, which is present in amounts significantly higher than the normal background values in Mato Grosso, reaches the atmosphere through evaporation from the tailings as well as from the process of heating the amalgam to drive off the mercury and thereby purify the gold. The mercury is then distributed over large areas through the air and will accumulate in the water bodies over a long period of time to come.

Because most of the mines have been illegally opened, they are frequently closed down by the police. Each time the police move in, there are massive protests from the miners, who can earn more money mining gold than they can by working in any other kind of industry. After the mines are closed, the miners continue their activities illegally elsewhere, often at night until the police units are withdrawn.

Actually, the danger from mercury poisoning is much more acute for the miners, gold buyers, and other persons who make their living in this industry. Many of these persons can be seen in the streets, apparently intoxicated, but actually suffering from advanced, permanent damage to the central nervous system. Buyers often show the symptoms most rapidly because they heat the amalgam with blow torches to drive off the mercury vapor, which they inhale. These people often become invalids within about ten years. They are often ignorant of the danger of their work, or they repress the thought that their own health could be impaired.

The long-term economic advantages of gold mining are minimal, as ghost towns all over the world attest to. When the gold reserves are depleted or the market price falls, the gold miners move away, leaving nothing behind. The miners have earned little more than a subsistence income for a few years and pay for their contact with mercury with ruined health, as discussed in Section 12.4.3. A few dealers become rich, often through smuggling the gold out of the country to avoid paying taxes. The tax paid on the gold produced and reported to the government is not enough to pay the interest on the debts accrued to pay for the colossal developmental projects of the 1970s, which have already proven to be economic catastrophes (Section 13.4). Thus, the gold mining with all of its environmental and public health side effects is perceived as necessary to pay for the mistakes of past developmental programs that went wrong.

It would be best for all concerned to give up this activity and instead develop industries with more promise of providing a long-term livelihood for the people without destroying their health.

13.2.2. Pesticides

The main form of agriculture in the region is cattle ranching. Thus, insecticides may be applied on the ranches to kill external parasites, but there has still been no effort to introduce large-scale spraying programs of any kind into the wetland itself. These would, in any case, be disastrous, mainly for reasons of public health. At the present time, although seasonally numerous, mosquitos present no risk whatever as disease vectors. In the northern Pantanal, only three species occur: *Psorophora varipes*, *Anopheles albitarsis*, and *Culex maxi*. None of these act as vectors of any known disease, although *A. albitarsis* is often reported to be one of the malaria vectors (Pessôa and Martins, 1988). However, as reported in Section 8, Lane (1953) did not think that the form present in the Pantanal is capable of spreading the disease. The survival of the mosquitos is possible only because their larvae are adapted to escaping the numerous predators in the water bodies. Nevertheless, their numbers are controlled by the many fish and invertebrate predators in the Pantanal. It was observed that the numbers of *Psorophora varipes*, for example, a species that attacks in swarms during the daytime, were quickly decimated when large numbers of dragonflies arrived. The dragonflies follow warm blooded animals and apparently are accustomed to using them for bait to lure the mosquitos out of their hiding places. The natural biological controls evidently still function in the Pantanal, and no vector of human diseases is able to establish itself there at the present time.

In Cuiabá, however, the mosquito species associated with human habitations are numerous, as evidenced by periodic epidemics of dengue and yellow fever. *Aedes aegypta* or congeneric disease vectors are required to spread these two diseases, and these species have long ago become resistant to all commonly used pesticides. Fogging programs carried out on a large scale in Cuiabá and other cities in Mato Grosso have visibly increased the numbers of mosquitos present within a short period of time (Section 12), and the diseases have continued to spread. Evidently, the removal of biological controls contributes greatly to future outbreaks of the diseases in the region. Because the spread of diseases is dependent on weather conditions conducive to the activity of adult mosquitos, contagion is sporadic, with large numbers of cases appearing periodically during warm, windless periods and with little spread of the diseases when cold, windy, or stormy conditions prevail. When new cases of the disease cease to appear, the authorities are quick to attribute this to pesticide fogging rather than the weather changes.

If insecticides were introduced into the Pantanal on a large scale, for any reason, the natural controls provided by the many invertebrate predators would be removed or greatly weakened, allowing the region to be colonized by the disease vectors, which are now confined to urban regions. It might also allow species of *Anopheles* capable of transmitting malaria to colonize the Pantanal. Such species are abundant in northern Mato Grosso, which is in the Amazon drainage area, and a great many persons who have travelled in this part of the state have contracted malaria. Increases in the natural mosquito populations could be expected, as well. The rapid increase of mosquitos in response to insecticide applications in the tropics has been amply demonstrated quantitatively by studies in other parts of the world (Service, 1977a, b; Heckman, 1993). It is therefore imperative that pesticides be kept out of the Pantanal, and the discontinuation of the spraying in the cities is also recommended.

13.2.3. Eutrophication

There are no sources of eutrophication within the central region of the Pantanal that are a cause for concern. The situation is quite different around the periphery of the wetlands, however. The city of Cuiabá, the capital of Mato Grosso, has been the fastest growing city in Brazil for the past few years. It has no sewage treatment whatsoever, and most of the wastewater from the city winds up in the Cuiabá River. Evidence of the great impact of this sewage on public health was amply provided by the rapid spread of cholera since its introduction to Mato Grosso in 1991 (Section 13.2.5.).

Throughout the world, the practice of dumping sewage into rivers has been and still is the cheapest way of removing the wastes and their effects from the urban environment. It has the advantage of keeping substances that cause eutrophication and sources of contagion from the doorstep because the river rapidly carries them away. It is particularly advantageous when the river carries the problems across a political border and into the jurisdiction of another government. Invariably, the problems that develop do not affect those causing the pollution but rather only those who live downstream along the water course or on the coast near the river mouth.

Because the Rio Cuiabá flows into the Pantanal, and during the rainy season, is the source of the water that enters the many bays, blind arms, and oxbows and floods vast expanses of the wetland, the nutrient input must influence the Pantanal ecosystem. Because of the present shortage of nutrients in the water during the rainy season, the question arises whether the effects of this nutrient input are to be considered harmful. The organic components of the pollution originating from man and other living organisms are quickly broken down at the relatively high water temperatures prevailing in the region, and the inorganic nutrients might well be eliminated largely by incorporation into the tissues of living organisms, such as the trees in the gallery forests and aquatic macrophytes and algae that supply nutrients to the food webs supporting the growth of the local fishes. Despite the fact that a large-scale pollution of the Rio Cuiabá from the cities in southern Mato Grosso is not presently a possibility, collections of pollutants in shallow arms of the river during the dry season might well cause local mortality among fishes and other aquatic organisms. For example, the toxicity of ammonia to amphipods was found to be enhanced appreciably by low water hardness (Ankley *et al.*, 1995), indicating that such a contamination might be particularly harmful in the very soft water of the Pantanal.

An cheap alternative to disposal of treated or untreated sewage through the river system is its burial in cesspools. This practice can have long-term consequences for public health. In the region of the Pantanal, it was noted that some wells contain excessive concentrations of nitrite. Subsequent testing revealed that the nitrate concentrations of this water were also extremely high, while the pH was often quite low, with one value as low as 4.2 being recorded. Many wells in which the pH was less than 6 were encountered (Heckman *et al.*, 1997). As reported in Section 6.2, a much higher chlorosity characterized many of the wells and, in combination with the other analytical values, suggested that the well water from the uppermost aquifer had been subject to contamination from domestic sewage, probably long buried beneath the town. Since Poconé was settled in 1781, it can be assumed that large amounts of organic wastes have been gradually broken down beneath the surface of the ground, releasing large amounts of simple nitrogenous compounds that can be oxidized to nitrate by microorganisms in the ground water. A scarcity of oxygen, however, retards the final oxidation process, causing much nitrite to accumulate. The absence of more than a trace of nitrite from both the deep wells used by the waterworks in Poconé and from the surface waters of the rivers reinforces the assumption that domestic wastes are the source of the nitrogenous compounds.

High concentrations of nitrate or nitrite in drinking water can have consequences for human health and render the well water unsuitable to use for raising animals. Galvanized pails were formerly used in the wells of Poconé, facilitating a reduction of the nitrate to nitrite in the slightly acid water. While the interference with oxygen transport caused by high concentrations of nitrite in drinking water may seldom produce serious illness in adults, it can be fatal to children (Johnson and Kross, 1990). Furthermore, it may react with proteins to produce nitrosamines, particularly in an acid medium. These substances have long been thought to be carcinogenic, and recently, they have actually been shown to be mutagenic and to cause neoplasms in man (Balaehandran *et al.*, 1991; Saravanan and Nagarajan, 1991; Zou *et al.*, 1994). Similar respiratory problems interfere with fish culture when the nitrite concentration in aquaculture facilities becomes excessive (Almendras, 1987; Ansari, 1987), and neoplasms may be induced in the fishes, as well (Bunton, 1989).

Therefore, while there are scarcely any problems to be anticipated from eutrophication of the surface waters, some protection of the ground water from an accumulation of organic pollutants is called for. An alternative to the contamination of vital drinking water resources is sewage treatment, although the contamination from past years will remain a problem for a long time to come.

While sewage treatment is urgently needed in Cuiabá for the sake of public health (Part 13.2.5), the avoidance of eutrophication is not a problem that warrants special measures at this time. A possible recycling of the wastewater nutrients for the benefit of agriculture would also be desirable, as it would be in most other parts of the world. A real danger related to the output of sewage from cities, however, is the almost universal mixture of domestic with industrial wastewater. While the eutrophication may not be sufficient to cause problems for man or wildlife, toxic chemicals certainly are. Presently, the two gravest problems are created by the two classes of pollutants discussed above: the heavy metals and the pesticides. The relatively few studies of this problem in the tropics have invariably shown that the toxic chemicals accompanying the domestic wastes present a serious problem, as in the case of cadmium found in municipal wastewater in Indonesia (Saeni *et al.*, 1980). and outbreaks of the Minamata (Skervfing, 1972) and itai-itai diseases, as in Japan (Friberg *et al.*, 1974), can be expected, although possibilities to diagnose them correctly in most tropical countries are lacking.

An additional problem is the fixation on eutrophication by engineers and scientists from the countries in the temperate zones. Methods employing various organisms as "indicators" of water quality (Gannon and Stemberger, 1978) are invariably oriented toward detecting increases in the trophic levels of lakes and rivers because this is the factor most salient for monitoring water quality in the countries where they usually work. Application of these methods in Mato Grosso would yield little information of value. What presently is a more urgent problem than eutrophication is the loss of nutrients with the vast volumes of water that pass through the Pantanal annually. This problem warrants special discussion.

13.2.4. Oligotrophication and erosion

The headwaters of the streams feeding the Rios Cuiabá and Paraguay fall into the "clear water" category of Sioli (1984). Chemical analyses of the waters in the springs feeding these streams and in the rivers into which they flow have demonstrated the great poverty of organic and inorganic nutrients they contain (Section 6.2; Heckman, 1995). An electrical conductivity even less than that found in the waters of the Rio Negro system, which have long been considered as an extreme case worldwide, is recorded in the headwaters of most of the streams.

The surface waters of the Pantanal itself are generally similar to rainwater in their mineral nutrient content during much of the year. They increase radically during the dry season, however, as discussed in Section 6. The aquatic macrophytes have a strong conservative effect on the nutrient balance in the Pantanal, quickly incorporating the dissolved substances immediately after the first flooding. This holds them in place and prevents their loss downstream during the *cheia*, when the greatest volumes of water are passing through the river systems. A lack of free mineral nutrients limits primary production by algae, thereby reducing the total material in the food webs that supply many fishes with nutrients. Shuter *et al.* (1989) discovered that starvation of some organisms occurs more rapidly in soft water, suggesting that the low water hardness in the Pantanal may stress the already hungry aquatic animals. A progressive loss of the mineral nutrients due to careless promotion of the leaching and erosion processes can therefore be expected to have negative consequences for the local fishery as well as for the wildlife species that consume aquatic prey.

Among the poor agricultural practices that have become increasingly popular in recent years is the burning of large areas of shrub growth and dry forest to gain more grassland for the cattle. Naturally, the ashes left by the fires provide the nutrients necessary to promote a rapid growth of grasses and herbs, which enhance the value of the land for grazing over the short term. However, large amounts of those ashes are carried away by the wind or washed to the rivers with the runoff from the heavy local thundershowers, robbing the land forever of its limited nutrient supply. The long-term productivity of the land is therefore greatly reduced, and after an area is burned a few times, it becomes too poor to be of much value for anything.

During the periods that large areas of the Pantanal are flooded, a corresponding reduction in the production of water plants, invertebrates, and fishes can also be expected due to nutrient shortages. This situation is still not fully comprehended by advisory personnel sent from industrialized countries to improve agriculture or protect water quality, because it runs contrary to all of the textbook knowledge they have accumulated. Agricultural experts in Europe, for example, must often exhort farmers to use less fertilizer in order to protect the ground water, while the main problem faced by hydrobiologists in those lands is the control of eutrophication.

In the catchment area of the Pantanal, and especially in the most elevated parts of its watershed, eutrophication is no problem at all, and it is rather a major problem of the farmers to apply fertilizer to the land faster than the nutrient elements are washed away by the heavy rainfall. Furthermore, high production costs cannot be compensated for by raising the prices of the crops because domestic consumers are generally not prosperous enough to pay more than minimum prices, and the international market can usually obtain its goods elsewhere when the price in any given country becomes too high. Thus, the agriculture tends to become a form of "slash and burn" farming, where crops are produced at a low cost as long as the soil holds out, and when it becomes too leached of nutrients to be productive anymore, the farmers move to another tract of virgin land to begin the process of burning and planting on the ashes anew. This completely robs the land of its productivity since there is no known way to make such land productive again after the thin topsoil and subsoil layers have been eroded away, leaving nothing behind except exposed layers of sandstone with isolated formations porous metamorphic rock. The productivity of the streams and rivers draining the area is also permanently destroyed because the natural sediments become covered by thick coats of fine silica transported downstream through the river systems from the exposed terrestrial subsoil layers. In effect, vast areas of *cerrado* are converted to desert to give speculators three or four years of large profits.

13.2.5. Pathogens and parasites

A major pollution problem from the human aspect is the transfer of diseases through contaminated water. With a rapid increase in the population of the state capital, Cuiabá, the raw sewage entering the Cuiabá River has tremendously increased over the past few years. The water in this river flows rather rapidly into the Pantanal, reaching it first in the Municipal District of Barão de Melgaço and proceeding from there through the central part of the wetland.

The problems arising from this pollution are not just of academic interest as a future danger, they have already become acute. Cholera first reached the state of Mato Grosso in 1991, and it began to assume epidemic proportions during the dry season of 1994, when a particularly large number of victims were infected during a festival in Santo Antônio Leverger, the first large town downstream from Cuiabá. Of course, many persons were infected along the river in Cuiabá itself, as well. This disease will tend to break out again and again during the dry seasons, because at this time, the water flow in the river is slowest, and the wastewater from the city tends to collect in shallow areas along the shores of the river, from which many of the poorer members of the population obtain their water for drinking and cooking. The spread of diseases through the water is a sure sign of a wastewater impact upon the ecosystem. The sporadic cases of cholera reported from the drainage area of the Rio Cuiabá during the other seasons are also showing a tendency to increase in frequency.

Of course, the obvious solution to the problem of contagion through the river is the construction of sewage treatment plants. This kind of facility, however, has traditionally received a low priority in the economic development programs that are generally supported by international, intergovernmental organizations. The lack of these facilities will probably have no significant effect on the wildlife of the Pantanal, but it certainly will on public health of the human population along the river.

The potential for the outbreak of a good many other diseases will also be encountered in the Pantanal if they are transported there by man (Pessôa and Martins, 1988). This potential is created when new species are introduced and their spread promoted by radical treatment of the environment, such as through chemical applications to the water bodies. For example, pesticide applications will certainly promote the colonization of the area by mosquito species, such as *Aedes aegypti*, which is not in the wetland presently but will be able to establish itself after the predatory arthropods are eliminated by insecticide applications. *Aedes aegypti* is an urban mosquito, already present in Cuiabá and responsible for the spread of yellow and dengue fever, which first became problems in Mato Grosso at the beginning of the 1990s. These mosquitos are resistant to all common insecticides and increase in abundance when insecticides are applied, due to the elimination of its natural enemies. This population increase was dramatic after a fogging program was begun in the cities of southern Mato Grosso in response to the appearance of yellow and dengue fever for the first time.

The potential for other diseases can also be created through the introduction of trematode parasites with snails as intermediate hosts. Schistosomiasis, which is common elsewhere in Brazil, is presently unknown in the Pantanal. This disease can be introduced with little difficulty through contamination of the water by infected persons and the establishment of conditions conducive to the spread of suitable planorbid snails, which serve as the intermediate hosts of the parasites. At least one such snail species, *Biomphalaria straminea*, already occurs in the Pantanal, making the appearance of the disease dependent only on the introduction of schistosome eggs from an infected person.

The spread of many digestive tract parasites is also facilitated through exposure to sewage in the river water. *Entamoeba histolytica* Schaudinn, 1903, and many nematode parasites can be transmitted directly through contaminated water as well as by flies that develop in the sewage.

The constant introduction of raw sewage into the Cuiabá River and other waterways also facilitates the spread of less dangerous pathogens, such as the infectious species of bacteria belonging to the genera *Salmonella*, *Shigella*, and *Vibrio*; parasitic flagellates including *Pentatrichomonas hominis* (Davaine, 1860) and *Giardia lamblia* Stiles, 1915; the ciliate, *Balantidium coli* (Malmstem, 1857); and various species of parasitic worm, which have not been given sufficient attention by the public health authorities. Furthermore, some sites along the small tributaries of the Cuiabá River in which excessive amounts of sewage collect also serve as breeding places for mosquitos and other insects capable of spreading diseases, such as the yellow and dengue fever which recently appeared for the first time. Being typically associated with polluted urban environments throughout the tropics, *Aedes aegypti* benefits from the heavy organic pollution which eliminates the fishes and predatory insects that would quickly eliminate its larvae in natural water bodies.

13.3. Poaching

One of the most serious problems for maintaining the larger wildlife species is poaching. The caiman is presently the chief victim of this form of illegal activity, but all other species are endangered, as well. The chief problem is the lack of funds and manpower for the forestry police who are in charge of protecting the Pantanal and the other natural resources of Mato Grosso from wanton destruction.

Poaching takes on two extreme forms with a variety of intergrades between them. The first is the organized international form that seeks to export large amounts of an animal product to the richer countries of the world, where people are willing to pay a great deal of money for them. This is the form of poaching that most endangers the caiman, the hide of which is used to meet the demand for "alligator leather". Mourão *et al.* (1996) reported that the poachers preferred to kill large individuals, based on a total of 2978 hides and 1358 skulls confiscated from poachers or found in their camps in the State of Mato Grosso do Sul. Apparently the number taken into custody by the authorities is only a small fraction of the total taken by the poachers, making the problem a serious one. Another major supporter of poaching is the pet industry, which peddles illegally captured rare species to unwary animal lovers in rich lands. The chief victim of this practice elsewhere has been the amethystine macaw, which has already been wiped out throughout much of its former range. It is a potential victim in the Pantanal, where at any time, the international poaching Mafia could appear.

The second form of poaching can be more or less described as killing animals for fun. This is practiced by many fishermen, tourists, and even by some of the people hired to cut the vegetation along the highways. This kind of poaching does not produce the caches of thousands of dead caimans, which are sometimes encountered by the police, but it does leave huge numbers of dead animals in those places most easily visited by the tourists and therefore, for economic reasons, most in the need of a rich fauna of exotic animals where they are easy to observe. While the populations of some species are less endangered by wanton destruction, the habits of the large wading birds to roost in vast numbers together in certain trees during the night (Yamashita and Vale, 1990) makes them particularly vulnerable to large-scale slaughter by vandals. The anaconda and other snakes, while harder to find, have a special attraction for thrill killers and are therefore particularly endangered. Two anacondas killed by the knives of fishermen were found in the Rio Bento Gomes near the Transpantaneira Highway on a single day, and smaller snakes and caiman lizards displaying scars from various injuries obviously caused my persons with knives were observed at various times.

Not only are the hunting regulations ignored, sound fishing practices are seldom abided by. The forestry police constantly encounter fishermen working hard on the rivers during the season the fishes are breeding, even though the law prohibits this. Many of the persons apprehended express surprise and deny knowing that they were doing anything illegal. Limitations on the amount of fuel allotted to the police preclude any kind of systematic law enforcement and even campaigns to educate the public about the hunting and fishing laws.

A root cause of the organized international poaching is the existence of an industry that processes and sells the products in countries where a high price is obtained. If the processing and sale were completely forbidden, there would be no more profit in the poaching. Nevertheless, even ecologists who should know better have frequently proposed a "limited exploitation" of the caiman populations. One way this is done is the collection and raising of the young caimans for later sale of their skins with a certificate of origin. Not only does this rob the Pantanal of some of the recruitment to the caiman population, it also supports an industry that, as experience teaches, develops a growing appetite for caiman hides. In any country of the world, not just developing countries, it is possible to buy a certificate for almost anything from a corrupt government official. How can a customs inspector determine which certificate for caiman hides was properly issued and which was obtained in return for a bribe? Both of them are exactly the same and perhaps even signed by the same person. If there is a quota of 2000 per year, how can an ordinary inspector determine whether 2,000, 20,000, or 200,000 are actually being exported with official documents from the country of origin.

Even if the Brazilian officials display absolute integrity in this matter, it is no problem for the poachers to smuggle their illegal wares over the long, unguarded borders into Paraguay or Bolivia and export them from there. Thus, the proposals for a "limited exploitation" of the caiman have to be rejected as naive and dangerous. The entire caiman skin market would bring great prosperity for perhaps five or six people in the world and a subsistence income for 20 or 30 others, as well as small earnings for transporter and the retail market. In such a case, the caiman would become extinct or so rare that the industry would disappear by itself. Thus, proposals for any kind of industrial "use" of the caimans in the Pantanal must be rejected as dilettantic. The best "use" the caimans serve is simply by their presence in large numbers, which makes the Pantanal an attractive goal for tourists (Plate 16b).

As a postscript, it is necessary to add that there is only one way in which the caimans might be utilized without endangering the species. This will depend upon whether a way is found to raise them in an aquaculture system permitting them to breed in captivity at a rate that precludes young animals from being taken from their natural habitat. A growing industry could be supplied by a increasing the number of caiman farms. The market value of the animals would be so reduced in this way that the profit from poaching would disappear, and a captive population would at least assure the survival of the species, no matter what fate befell the Pantanal. Of course the question arises whether so many consumers would like to have caiman-skin shoes and handbags. Furthermore, the whole industry would certainly be plagued by complaints from a growing "animals' rights" movement, which seeks to protect individual animals rather than the species.

The same kind of breeding industry might protect the hyacinthine macaw. If enough of these large parrots were bred in captivity, the pet industry could become saturated with the birds, causing their price to fall drastically and taking the profit out of catching wild ones. However, it is doubtful whether the resources will be available for macaw breeding farms anytime in the near future, so the protection of the wild species lies in eliminating the trade with the birds completely.

The key to eliminating the other kind of poaching, which is widespread but small in scale, is public education and on the spot enforcement. The organized poachers are able to operate without interference from the police because they know they are acting illegally and carefully prepare for their crimes. The "Sunday poacher" is generally just ignorant and usually rather stupid, so he often falls into the enforcement net. More checks by police, signs explaining why the animals should not be killed, and information in the media on how people should behave in a national park help to reduce this sort of vandalism.

At the present time, there is an opposite trend propagating bad behavior. From one bridge along the Transpantaneira, I witnessed a family of tourists from São Paulo having a good time stoning the caimans. The grandparents were smiling and looked particularly proud as their five or six-year-old grandson rushed over with a large handful of gravel to throw at the nearest head emerging from the water. At another bridge, some fishermen were killing the time throwing the largest rocks they could find at another caiman.

Another day, two medium-sized dead anacondas were pulled from the Rio Bento Gomes. Both had been hacked by large knives. Some boys fishing from the bridge said that there was also a third cadaver nearby, killed by the same fishermen.

On still another occasion, a caiman was found at the edge of the road without a tail. It was still alive, but it was badly hacked up. The tail is considered a gourmet item by some fishermen. Discussing this matter with various people revealed that even a tour guide with much experience in the Pantanal was convinced that the tail could be eaten in this way without killing the caiman. He was convinced that, like the tail of a gecko, the caiman's tail would simply regenerate after a short time.

This and other forms of vandalism are commonplace in the Pantanal because it is a vast area in which anyone can find places to do anything unobserved that he wants to do. Thus, fishermen from all over Brazil can find places in the Pantanal to chop off a caiman's tail and cook it on a fire made using wood chopped off one of the highway bridges, kill the anacondas and stone the caimans for amusement, and then dump all of their garbage into the water. In fact, some do this. This has given fishermen a bad reputation among tour guides, ranchers, and other persons who spend much time in the Pantanal. Tourists generally behave better, but this is again a matter of education. Those who come from far away to see the animals have generally learned something about them and don't do their best to kill them or drive them away.

On one bridge not far from a tourist hotel, four feet that had been chopped off a capybara were drying in the sun. Whoever had left them there obviously hoped to sell them as souvenirs. Tourists who buy this kind of kitsch help to finance the destruction of the wildlife in the park.

Road workers who are regularly sent to clear the vegetation along the Pantanal are often no better than the worst of the fishermen. After they had cleared the plants on one occasion, half of a small iguana was left on the road, chopped through by one of the machetes used to cut the plants. Because these animals are not eaten in southern Mato Grosso as they are in Central America, this act was simply one of vandalism. Snakes and small mammals meet similar fates whenever encountered by the workers, and the dried, cut vegetation is frequently burned in a way that the flames spread to neighboring dry grassland or forest. The large fires that result even burn the signs along the highway, as well.

Many of the ranchers have become conscious of their environment and the need to preserve it. On the Fazenda Ypiranga, where several experimental stations for the ecology project were established, the visit by jaguars meant the loss of one or two of the cattle, but it was tolerated as an event that brings added income through oral advertising, which attracts more tourists. Elsewhere, ranchers are not so progressive. While a German doctoral candidate was buying fuel for the project bus in Cáceres, a gas station attendant noticed his T-shirt with a picture of a jaguar on it and remarked that he had shot 17 of them. Some ranchers had paid him to kill each jaguar that appeared on the ranch. Of course, the killing of jaguars is absolutely forbidden in Mato Grosso, regardless of the circumstances.

The only practical way to prevent damage to the Pantanal and its wildlife from poaching is to greatly increase the number of persons assigned to guard and protect the wetland. Some of the money paid for this increased protection could be raised by heavy fines on those who kill or stone animals, spread garbage, destroy the wooden bridges, set fires, and fish out of season. An increased presence of responsible tourists and tour guides would also increase the chances of persons committing illegal acts being discovered.

The private property in the Pantanal typically consists of large ranches that are still minimally modified for cattle ranching. The cattle are more or less on their own, and mortality seems to be high. It is presently important to gain the cooperation of the ranchers to protect the wetland. Most of them seem knowledgeable about the problems, but there is considerable disagreement about how much tourism should be encouraged. Because the ranchers have considerable political power in the District of Poconé, it would be difficult to accomplish any program of protection without their cooperation. With their help, however, many of the harmful activities could be kept under control through increased law enforcement.

13.4. Mammoth projects

The most prominent economic factors in the general economy of Brazil today resulted from the massive developmental aid programs promoted by international developmental agencies during the years of the military dictatorship. These projects were all conceived on a scale far grander and anything previously attempted in human history. They include the construction of gigantic dams, some of the longest highways in the world, and sprawling industrial complexes. The Transpantaneira Highway was a product of these times, although it was never completed. The schemes also involved the resettlement of hundreds of thousands of people to remote regions of the country. At the time they were planned, these projects were touted to be absolutely indispensable for the economic development of Brazil, and achieving them was worth the sacrifice of large areas of natural environment together with their renewable resources. Not only Brazil, but a number of other countries around the world were heavily afflicted by these developmental programs.

Nearly every one of these projects has had similar economic effects, which range in degree from bankruptcy to catastrophe. Brazil now has such an enormous debt that it has great difficulties in raising the annual interest payments. The payments are presently met through a program of "readjustment" planned by the creditor agencies and involving saving money at the cost of the poorest members of the population. Although at the present time, there is no hope of ever being able to begin paying off the principle of the debts, the "readjustment" continues to be propagated as a solution to the dilemma. The regular diversion of the interest payments from the government's working capital results in the need to cut costs, generally at the expense of education, public health, and public services, that is, those outputs which are absolutely necessary to develop the human resources of the country.

What the country has gotten through the mammoth projects of the 1970s includes the Transamazonica Highway, an incredibly long unpaved road through the Amazonian rainforest, which remained passable for about two years after it was completed. For the past two decades, it has been a ruin of colossal proportions, with the approaches to the bridges washed out and deep erosional gullies cut completely across it. Repairing it would certainly cost more than the price of the original construction.

Elsewhere, some of the dams are more expensive to maintain than the electricity they produce is worth, and they are so far from the nearest cities that most of the power is lost in transmission through hundreds of kilometers of cables. Many settlers from the cities to the Amazonian region are poorer now than they had been before, and they are infected with malaria, as well. There have been at least 19 malaria epidemics in parts of Brazil where the disease had not previously been known after settlers in Amazonian regions came home to visit their relatives and friends.

Rather than learning from these experiences, the international organizations are planning new mammoth projects to "develop" Brazil. These are touted as having been planned more carefully than the ones in the past to avoid the mistakes. However, sober evaluation of their probable effects show them to be no better and perhaps even worse than what was done before. However, all of them have one thing in common: they promise to bring personal fortunes to the entrepreneurs who build them and power and higher salaries to the bureaucrats who plan, finance, and manage them (Hancock, 1989).

A major problem in discussing these projects is that their planning is now done in secret to prevent any protest against them from being organized before the decision has been made to implement them. It is therefore difficult to know exactly what the plans entail. The feasibility studies generally concentrate on the construction rather than economic need and environmental impact. There have been proposals to construct dams to produce hydroelectric power in the vicinity of Cuiabá (Alvarenga *et al.*, 1984), which would have an impact on the Pantanal, although Ferreira *et al.* (1994) pointed out that no consideration at all was given to this aspect.

One project that could have an enormous effect on the Pantanal is the "Hidrovia". At a time that the world, for better or for worse, is looking to highways or modern rail systems for the transportation of persons and goods, some planners have developed a concept of modifying the entire Paraná, Paraguay, and Cuiabá River System into a long waterway. It is claimed that this will permit large quantities of goods to be transported cheaply to and from Mato Grosso. At the present time there is only one main highway leading directly to São Paulo, over which enormous numbers of trucks pass each day.

Three objections to this plan come to mind immediately. The first two are purely economic, and the third, ecological with strong economic aspects. The grounds for the first economic objection can be found in simple hydrodynamics. Because the level in the river system is too low for some of the ships during the dry season, it is proposed to dredge the river channel to make the water deeper. The water flow from the Pantanal is considerably slowed down by the shallow water zones in the Rio Paraguay at the southern borders of the Pantanal. If this "sill" is removed, the water will flow much faster out of the wetland, and during the dry season, the water level in the rivers will become even shallower in spite of their deeper beds. Similar modification of various German rivers have resulted in increased flooding of cities and towns during high water periods and more frequent suspension of the river traffic during dry periods of the year. The length of the low water periods is increased rather than decreased, as desired (Nachtnebel, 1996). Unfortunately, conservationists who object to the project because of the damage it will do to the wetland biota often tacitly accept the argument that by deepening the channel, more shipping traffic will be able to use the river system (Bucher & Huszar, 1995), although a look at the hydrodynamics of the seasonal river flow indicates that the opposite will be true.

The reasons for the second economic objection are obvious from a glance at the map. Any passengers or goods to be transported overland to Cuiabá from the eastern part of Brazil must presently traverse 1600 to 2000 km of paved highway, which requires from one to two days. Goods brought in by ship from São Paulo or Rio de Janeiro would have to sail southward to the mouth of the Rio de la Plata between Argentina and Uruguay, traverse the lower Rios Paraná and Paraguay through Argentina and Paraguay, and pass through the Pantanal to reach the planned river ports of Cuiabá and Cáceres. Even if canals were built to connect the river systems, the tortuous trips would still take extremely long periods of time. For most consumer goods on the domestic market, this route is simply too long, even when the possible customs formalities are disregarded. The construction of this waterway would therefore bring no returns comparable to the enormous costs of its construction. Furthermore, because of the rapid siltation that could be expected during the rainy season, the construction costs would be dwarfed by the funds needed to keep the waterway usable.

The only good to be expected from such a waterway would be cheap transportation of sugar cane, soybeans, and other agricultural products presently being raised on the plateau to the east and southeast of Cuiabá. However, a decrease in the number of automobiles being manufactured to run on alcohol and a recent collapse in the world market price of soybeans has made this agriculture uneconomical. Therefore, even this possible advantage of the waterway has been eliminated.

The ecological grounds for not recommending the construction of this waterway are many (Bucher & Huszar, 1995). The most important at the present time is that the attractiveness of the Pantanal for tourists will be greatly reduced by any form of water level stabilization as well as by the draining of the wetland. If international funds are made available for this crackpot scheme, the results are certain to be disastrous, as they have been for the similar schemes carried out during the 1970s.

It is difficult to predict what other programs to "develop" the Pantanal will be thought up in the next few years. There seems to be no limit to the inventiveness of the international developers in finding reasons to build or modify the environment on a colossal scale of broadcast vast amounts of toxic substances indiscriminately over large areas. Although none of the proposals would stand up to a careful economic evaluation, they can be agreed upon in secret and bring enormous profits to the builder from the tax money collected in the industrial nations, leaving the debts to future generations of taxpayers in the developing countries. For each of these projects, enormous amounts of money are also made available for advertising, leading gullible members of the public to believe that enormous benefits will be derived from them. This has happened many times in the past and will continue to happen until more public scrutiny is given to this problem.

A major reason for the "failure" of the developmental programs is the lack of a technological infrastructure in Brazil capable of taking advantage of the many opportunities for development, and this infrastructure is lacking because the education system is not given proper support. Brazilian estimates on productivity have equated 60 man-hours in Brazil with only 15 in Germany and 7 in Japan. When 70% of the labor force is illiterate and most factory workers cannot read simple instructions written on the machines, it is no wonder that they take almost nine hours to produce what a Japanese worker can produce in one. Since this situation is well-known to the international development organizations, it seems that it would be hard for them to justify the deliberate exclusion of help for education from their aid programs.

The only feasible way of protecting the fabulous resources of the Pantanal from the grasp of the international aid planners would be to make a large area of the wetland a national park and international wildlife reserve with a large staff of guards to oversee a limited tourist traffic and protect the animals from poachers. The entire watershed should be placed under protection, as well, and the form of extensive ranching now practiced should be encouraged in the marginal areas of the wetland. The use of fertilizers should be monitored, and the introduction of pesticides, forbidden. Priority should be given to the construction of sewage treatment plants to minimize contagion.

The resources that should be protected in the watershed include the springs at the headwaters of the streams and rivers, which are presently among the purest water sources in the world, and the rich fish fauna. It would be wise to domesticate some of the fast-growing native fish species and raise them in aquaculture facilities near the cities in order to provide a cheap and reliable source of protein for the local population and eliminate the temptation to employ destructive fishing practices in the rivers. A modern railroad connecting Cuiabá with São Paulo to supplement the highway would be worth considering as an alternative to the foolish plan of building a waterway through Paraguay and Argentina. This waterway would effectively drain the Pantanal, not only depriving the wildlife of their habitat but also making the present form of cattle ranching impractical.

13.5. General concept for the protection of the Pantanal

13.5.1. Protection of the habitats

Before any practical concept for protecting the rich flora and fauna of southern Mato Grosso can be developed, it is necessary to find a way of protecting the natural habitats. Figure 13.1 shows a simple example of how some of the less abundant plant species in the Pantanal are able to survive in a limited number of special habitats along the Rio Bento Gomes System. Elimination of any of these distinct kinds of environment would doom certain plant specialized to inhabiting it to local extinction. Similarly, some animal species are generalists capable of living almost anywhere, while others are highly specialized in the habitat they require, making their local elimination certain if their particular kind of habitat, no matter now small, is destroyed.

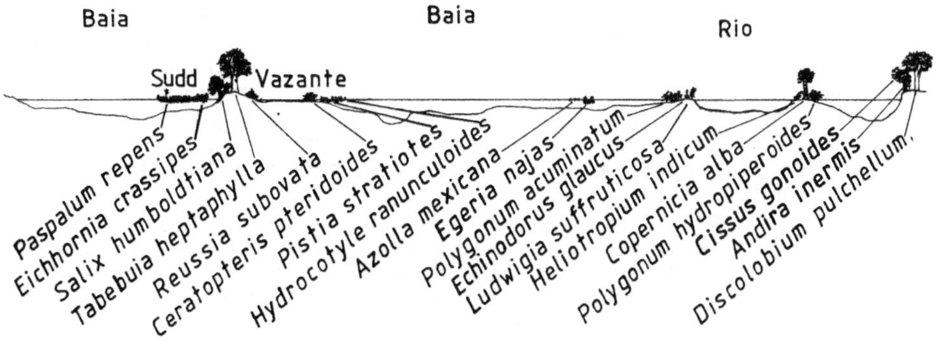

Figure 13.1: A cross section of various kinds of habitat found in or along the Rio Bento Gomes. Characteristic plant species of each individual kind of habitat are shown. Some of these are locally rare. The ubiquitous species found all over the Pantanal are omitted because they would be little affected by the elimination of any particular kind of habitat.

Without a suitable habitat, it is not possible for the species to survive, except in captivity or in small wildlife preserves where colonies of individual species are maintained, partly by artificial feeding systems. For terrestrial species, maintenance in captivity is much more costly than setting aside land for national parks in which the flora and fauna are protected within intact natural habitats. However, several well-known species have been saved from extinction by maintenance in captivity until the people living in or near their natural ranges became civilized enough to permit the recolonization of their former habitat by the progeny of captive stock. Small herds of the American bison returned to the great plains after captive individuals from zoos were brought to some of the national parks through programs promoted by President Theodore Roosevelt. Recently, oryx antelopes from herds maintained in the United States were returned to their home range on the Arabian Peninsula, where they had been fully eliminated.

For the aquatic and wetland species of the Pantanal, maintenance in captivity or in small preserves is much more difficult. Suitable preserves for the species require wetlands not subject to winter frosts, which, in contrast to terrestrial grasslands and forests, are not at all abundant in the world. Large aquatic animals are also more expensive to maintain in zoos than are terrestrial ones. Therefore, it is less practical to try to protect many or the Pantanal species in captivity than in their natural habitat.

We are confronted with the problem of how to protect a gigantic wetland system from encroachment by people looking for short-term economic gains at the expense of the long-term value of the system and, worse still, from the grasp of enormously powerful and influential international development organizations with a long record of habitat destruction accompanied by massive economic loss both for the donor and for the recipient countries.

To maintain a large natural system, it is necessary to integrate it into the economic structure of the human society assigned to protect it. The concept of a region maintained only for the animals is both impractical and counterproductive. Not only are people prevented from enjoying the aesthetic qualities of a wetland rich in natural wonders, the plants and animals lack protection from clandestine destruction by environmental criminals, which only an active rather than a passive form of protection can prevent. Mittermeier *et al.* (1990) summarized some of the reasons for protecting the Pantanal and stressed the fact that there are still many unspoiled areas of this enormous wetland. To keep them from becoming secret bases for poachers, an effective infrastructure must be set up, preferably based on economic interests of local residents, which they are willing to actively defend. Tourism is one obvious industry that is compatible with a protected environment, if it is properly managed. Other industries that might be promoted in the catchment area range from the export of its ground water in moderate amounts for washing and industrial purposes in other parts of Brazil to the cultivation of small aquarium fishes in special aquaculture facilities upstream from the wetland. The strong insolation of the entire region might also make it an outstanding site for producing solar energy.

Furthermore, it has often been observed by naturalists that various human activities, especially those that increase food supplies, produce an increase in the number of animals of all kinds near sites of human activity. Even a slight eutrophication of the water brings about an increase in the numbers of animals and plants without markedly changing the structure of the biotic community. Problems arise because anthropogenic eutrophication almost never remains slight. In the Pantanal, an example of the great value human activities can have in promoting wildlife survival is the excavation of the ditches along the Transpantaneira Highway. The extension of the time these artificial depressions remain under water during the dry season provides fishes and other aquatic animals with an opportunity of surviving longer and thereby prolonging the time they remain available for caimans, birds, and carnivorous mammals as food. When the drought is not too severe, the ditches also increase the number of aquatic species that survive to replenish the stock after the rainy season begins. These artificial water bodies not only benefit the animals, they also assure that a large number of attractive mammal, bird, and reptile species remain near the Transpantaneira Highway for the passing tourists to admire and photograph.

The question that must most urgently be addressed is how to protect the Pantanal from the slash and burn economic interests which have destroyed so much of the world's most attractive natural environments, especially wetlands worldwide and tropical regions, in general. Obviously, this protection is only possible if all forms of "development" that change the *status quo* are forbidden, not only in the wetland but also in the watershed. Experience elsewhere has shown that a wetland habitat can be grievously damaged even if fully set aside as a protected area if the watershed is not also protected. For example, the Florida Everglades is a wetland that suffers from severe drought because so much ground water is used to supply Miami and is subject to severe eutrophication and because pollution from agricultural areas upstream in the watershed, which are the sources of manure and pesticides. Obviously, protection of the region does not preclude the development of agriculture, industry, and commerce in southern Mato Grosso. It does, however, require that no internationally funded mammoth projects be permitted in the region.

The Pantanal should be declared a national park and become an internationally protected environment, in order to partially relieve the Brazilian government of the large costs of protecting it. The park should include the wetlands and the upland stretches of the streams and rivers feeding it as far as their sources on the plateau to the east and north of Cuiabá. International funds, if available for developmental projects, might best be used to build a sewage treatment plant for the cities of Cuiabá and Varzea Grande, which could also serve to provide some of the electric power for the cities.

The park should include a central part from which no animals or plants can be introduced or removed and in which tourists are permitted only under supervision. Hunting and fishing should be strictly prohibited, and the entry of potential poachers must be carefully prevented by continuous monitoring of the waterways. Animals and plants should not be removed from the park, and species from other parts of Brazil should not be introduced into the catchment area. Refuse should not be left by the tourists, and raw sewage released upstream from this protected reserve should be treated in suitable plants to preclude the spread of diseases.

The parts of the Pantanal presently used for cattle ranching can continue to be utilized as they have been for over a century. However, the introduction of pesticides should be prohibited. Since they are presently not used, this would require no change in the ranching methods. The construction of small tourist hostels on the ranches also presents little danger to the wetland.

13.5.2. Protection of the species

The Pantanal supports considerable populations of species that are generally considered rare or even endangered of extinction. These species are presently not subject to serious local reductions in numbers because the habitat is still intact, and the forms of agriculture and animal husbandry have still not become intensive.

The large and medium-sized fishes in the main rivers flowing through the Pantanal are of great importance, not only from the viewpoint of species protection, *per se*, put also for their value to the local fishermen. It was noted that 80% of the catch in the region comes from the Rio Cuiabá, where such migratory species are predominant (Lima, 1987). It was mentioned in the previous sections that a great many of these fishes travel long distances in the river system, and the greatest periods of growth seem to occur far downstream, where the availability of nutrients is considerably greater than in Mato Grosso. It is therefore safe to say that a large part of the "fish production" so important for maintaining the fishery in southern Mato Grosso occurs far to the south, and the fishes bring their "biomass" with them when they come upstream to spawn. As a corollary to this, it can be stated with reasonably certainty that an interruption in the free movement of the fishes due to construction projects along the rivers would have disastrous effects on the fishery production and would eliminate or greatly reduce the populations of many migratory species, even as far downstream as Argentina. Such consequences of modifying the water courses have already been noted in many parts of the world (Lelek, 1987; Holcík *et al.*, 1989).

Another grave danger to the native fish species is the introduction of fishes form other continents for aquaculture. This practice has been much promoted by international development organizations in the past because the culture methods for a few species are well known, and the aquaculture experts are too inflexible to find suitable methods for raising native fishes. Although excellent candidates for aquaculture can be found among the native species (Saint-Paul, 1986), common and tropical carp and tilapias have been introduced to Brazil and are accepted by a small number of gourmets as substitutes for the native species. The practice of introducing such fishes should be strongly condemned, not because purists might object to "falsifying" the South American fish fauna but because the practice has been shown numerous times in the past to be the cause of grave economic damage. This has come about occasionally because an introduced species displaced more valuable native fishes. Villwock (1972) described the disaster in Lake Titicaca caused by the introduction of fishes that the local people found unpalatable. These fishes effectively eliminated the highly prized endemic species and ended the formerly flourishing fishery on the lake. The introduced species could be described as "trash fishes" of the kind that have been deliberately poisoned by fishery organizations on occasion in North America, and the consumers in Peru and Bolivia rejected them completely. The international planners had failed to understand that a rapid growth rate does not make a fish more palatable on the table.

An even more devastating result of transporting fishes from other geographical areas is the unintentional introduction of diseases, which decimate less resistant native stocks. Because tilapias are African cichlids and much of the South American ichthyofauna belongs to the same family, the chance of introducing bacteria and viruses with a great pathogenicity to the local species is considerable.

Sometimes, herbivorous invertebrates have been introduced to South America for various reasons, such as the control of undesired plants (Maddox, 1970). These are even more dangerous than fishes because of their ability to fly or be transported by animals overland from one water body to another, and because of their high rates of reproduction. On other continents, such introductions have already had catastrophic effects, including great losses of rice caused by the South American snail, *Pomacea canaliculata* (Lamarck, 1822), which was recently introduced from South America to Southeast Asia (Halwart, 1994). It is hoped that South American scientists will learn from these examples not to import invertebrates from other regions of the world.

In addition to the general protection of plant and animal stocks from the consequences of foolish manipulations of the ecosystem, certain species require special measures to ensure their survival. Among the mammals, attention should be given to the giant otter, *Pteronura brasiliensis*, which inhabits the Rio Bento Gomes and apparently other small rivers in the Pantanal. This species is considered endangered of extinction (Vidal, 1993) as it survives in only a few locations in the Amazon region, as well as in the Pantanal. Apparently, the intensification of hunting and fishing by the impoverished settlers who were moved to the Amazon Valley as part of internationally funded resettlement programs has strongly reduced the giant otter populations, although these animals have little commercial value. As top predators in the aquatic system, they are sometimes regarded as unwanted competitors for the fishes or as "dangerous animals", and are therefore killed off.

For the protection of the giant otters, which live in small family groups, it is necessary to maintain isolated stretches of rivers free of human activity. This is difficult as long as fishing is tolerated, or even encouraged, in the Pantanal.

Among the most endangered terrestrial mammals in the Pantanal is the jaguar, *Panthera onca*, which is presently suffering from poaching at the behest of some of the ranchers, who wish to kill these large cats to protect their cattle. Remedies for this include not only better enforcement but also the augmentation of tourism. Like the lions and leopards in the national parks of Africa, the jaguar could become a great attraction if there were more of them to see. It must be remembered, however, that considerable resourcefulness by the tour guides will be required for the visitors to get a glimpse of this secretive cat.

The Pantanal deer, *Blastocerus dichotomus*, was formerly found throughout much of Brazil. Because it can be approached within only about 20 m before it flees, it was an easy species for hunters to kill, even before firearms were as well developed as they are today. As a result, within its entire former range, it is presently abundant only in the Bananal wetland along the Rio Araguaia and in the Pantanal. Although it is quite common in the Pantanal, its range has continued to decrease, and it is vulnerable to further reductions in abundance if poaching is not controlled. For its continued survival, both habitat protection and defenses against poaching will be necessary.

Among the birds, one of the most spectacular species demanding protection is the largest species in the parrot family, the hyacinthine macaw, *Anodorhynchus hyacinthinus*, which is threatened chiefly by the international pet industry (Plate 16a). Unlike most of the macaws and parrots, this species is not a resident of the rainforests but rather of the dry woodlands, such as those that predominate in the region of the Pantanal. Their great popularity as pets, especially in North America, has eliminated the species from much of its former range. The fact that one of these macaws can fetch a price of several thousand dollars makes the species an especially attractive victim for the poachers.

Other birds of the Pantanal that require protection include the stork jabiru, which is presently plentiful in the wetland but, unlike the herons, egrets, and wood storks, is not as widely distributed in other parts of South America. For many of the most abundant birds, the protection of the special fish fauna in the wetland is vital for survival. This is another reason to prohibit schemes tampering with the water level in the Pantanal. The avifauna is responsible for much of the value of the wetland as a future attraction for tourists, so it demands protection (Fig. 13.2).

Figure 13.2: Neotropic cormorants taking their places on a dead tree to dry off after gorging themselves on fishes. There is presently no other wetland in the world inhabited by such large numbers of so many water bird species.

One important step that could reduce the losses among the large water birds is the replacement of the power lines running parallel to the Transpantaneira Highway with a modern system of fiberglass conductors within a pipe running slightly above ground level. This would eliminate the great mortality among large birds, which frequently fly into these power lines and break their necks at night (Section 8).

The common reptiles of the Pantanal are not presently endangered, but their chances for survival would become precarious if the habitat were modified, and the caiman stands to become endangered if poaching is not controlled. The loss of caimans due to frequent poaching reduces the attractiveness of the wetland for tourists, as previously discussed (Section 8). The Pantanal anaconda, *Eunectes notaeus*, is confined mainly to the region of the wetland. This snake is smaller than the congeneric species in the Amazon region, but it still reaches an impressive size of over 4 m. Snakes are often killed wantonly by visitors to the Pantanal, and anacondas that had been hacked to death with knives by fishermen were found from time to time in the wetland, as mentioned in Section 8 (Fig. 13.3). Public education and enforcement measures are urgently needed to protect these reptiles. Among the many reasons their protection is particularly important for the regional economy is their attractiveness for tourists, who will generally name these giant snakes among the wildlife species that they would especially like to see and photograph. At present, they are still common enough for visitors to observe rather frequently.

Figure 13.3: One of two anacondas that was killed in a single day by a fisherman along the Rio Bento Gomes in the Pantanal. The knife wound is evident in the body.

13.5.3. Protection of the human population

Unfortunately, the relationship between environmental destruction and the impoverishment and sometimes elimination of the local human populations is seldom discussed. In the Amazon region, for example, the migration policy of the 1970s had increased the number of settlers of European origin but simultaneously decreased the numbers of Indios, who once inhabited these regions. Reports of massacres are frequent enough to have alarmed international organizations. In addition, many of the Indios die of "natural" causes as a result of exposure to diseases to which they had never before been exposed. The settlers are not at all prosperous, and it is doubtful if they will remain in most of the areas they have settled, considering how rapidly the soil loses its richness and the fish populations decline. Several million of the settlers engaged in gold mining, the only industry that offers them a moderate income, will most certainly leave the region after the precious metal supplies have been exhausted or if the gold price on the international market falls.

The Pantanal has been thinly settled since pre-Columbian times, and its area is presently divided up among large ranches, most of which are partially exploited for raising cattle. Not only the Pantanal but also the rest of Mato Grosso is thinly settled, with most of the permanent residents being engaged in ranching. A few small farmers and fishermen also inhabit the region, mainly along the rivers, but the great bulk of the population is urban. The gold miners constitute a considerable proportion of the population in rural regions, but they move from place to place and cannot be counted among the permanent residents. In addition, the life span of many of them is greatly limited by their constant exposure to mercury.

The native population of southern Mato Grosso is relatively prosperous, and the severe poverty observed in some of the other regions of Brazil is not encountered. In spite of the uncomfortably hot and humid climate, the city of Cuiabá has recently seen a rapid growth, probably faster than that of any other city in Brazil.

The basis of the long-term prosperity in the Pantanal region is agriculture. Gold mining brings only a brief boom because little of the wealth produced remains in Mato Grosso. The greatest problem presently faced by the ranchers is the worldwide overproduction of beef and the nationwide shortage of money. The many hungry people in the eastern part of the country lack the finds to buy much meat, and the export market has been largely eliminated by the oversupply of high quality beef in Europe and North America. Nevertheless, the present oversupply of food ensures the people of Mato Grosso an economical nutrition, and many people find a steady income through employment on the ranches and through transportation of the agricultural products.

The protection of the Pantanal also protects the way of life and prosperity of the inhabitants of the region. Except for the gold, the resources of the Pantanal are renewable, and the agricultural production is sustainable, as shown by the history of the region. Furthermore, there are various ways of improving the agriculture, avoiding the mistakes repeatedly made in other parts of the world. Among many suggestions to improve the lot of the local population economically are a development of the aquaculture of native fish species using methods developed in Asia, moderate increases in tourism, polyculture of fruit and vegetables, and an intensification of scientific research at permanent facilities in the wetland to catalog its living resources and find better ways to utilize them without causing environmental damage.

14. Summary and Conclusions

The Pantanal is one of the world's major wetlands, and it is apparently the largest such environment in the tropical wet-and-dry climatic zone that has been only marginally modified by man and therefore remains largely in its natural state. The fauna is moderately rich, but much less so than the neighboring equatorial region. Its species diversity must be considered relatively low for the tropics since vast numbers of individuals belonging to relatively few species dominate most of the wetland. The net annual production of living organisms approximates zero because nearly all of the organic material produced is consumed again by the living organisms. The gross production rate, however, is extremely high, and large amounts of plant material can be produced during short periods of time at the beginning of the flood period. This situation exists because the extreme annual variations in rainfall result in alternating periods of flood and desiccation, thereby causing seasonal successions of fully terrestrial and fully aquatic biotic communities at some locations and successions of lentic and lotic aquatic communities at others. Where terrestrial and aquatic species successively occupy areas of the floodplain during each annual cycle, the available mineral nutrients required by plants and animals, such as compounds of nitrogen, phosphorus, and calcium, are fully recycled at least twice each year. After the supply has been exhausted by the primary producers, growth stagnates.

The catchment area of the Pantanal is located on the Central Brazilian Shield, one of the oldest continental land masses on earth. Due to millions of years of leaching, few mineral nutrients are left in the upper rock layers. The soil consists mainly of silicates and is poor in most other minerals. The streams feeding it are in the "clear water" class and are extremely poor in ions, accounting for the extremely low electrical conductivity near the headwaters. The mineral content of the water in the northern Pantanal during the rainy season approximates that of rainwater, but during the dry season, high concentrations of substances such as ammonium and phosphate can be encountered due to the mass death of the aquatic biota and a rapid decomposition of the detritus be the saprobic community at water temperatures sometimes exceeding 40°C. Because the Pantanal seems to have been a desert during the last ice age, it has been colonized by its present aquatic biotic relatively recently, and hardly any species are endemic to the region.

The importance of the individual biotic communities to the ecosystem as a whole is quite distinctive. The seasonal development of these communities follows nearly the same pattern each year, although they may appear at different times due to variations in the dates of the first heavy rains, the overflow of the rivers, and the drying up of specific areas of the floodplain. Furthermore, during a series of "wet years", certain low-lying areas may remain flooded throughout the year, but during consecutive dry years, they may become completely desiccated, and local changes in the structure of the vegetation may occur.

For the protection of the wetland, it is important to note the enormous significance of the water level changes for the maintaining the integrity of the ecosystem. These changes promote the development of a flora and fauna with characteristically high rates of recruitment and growth. The plant and invertebrate species on the floodplains typically grow and reach maturity quickly, generally within one year, and suffer high rates of mortality during at least one of the seasons. Among the vertebrates, the r-strategy is similarly important to assure survival of the species. Exceptions to this are the birds and some of the mammals, which are seasonally present in such large numbers because the fast-growing plants, invertebrates, and fishes offer them enormous food supplies. During the seasons that food is scarce, these animals can fly or travel overland to remote feeding grounds far outside of the wetland. The fishes, amphibians, and reptiles must contend with a scarcity of water during one season of the year and a lack of food during another. It is obvious that an artificial stabilization of the water level would result in a displacement of the r-strategists by a richer biota dominated by k-strategists, but the concomitant reduction in food production would rob the large water birds and some of the mammals and reptiles of their superabundant food supply, and the wetland would lose its exceptional character as a haven for its rich and attractive avifauna, for the caimans, and for a great many aquatic animals with the capability of aestivating in the dried sediment.

As a means for judging anthropogenic impacts, much interest has been expressed in searching for indicator species to provide an early warning system to prevent environmental degradation. However, the population dynamics of most animal species precludes their selection as indicators. Due to the harshness of the physical changes during the course of the year, many species are regularly decimated and are even subject to frequent local extinction. Their resilience permits quick recoveries when conditions again become favorable. Among the only stable populations that might be selected for a monitoring program to detect negative environmental impacts are those of the large water birds. It is these species will be the first to disappear if the ecological conditions begin to change. The birds, however, are the animals generally regarded, rightly of wrongly, as the most desirable to protect. Hence, it would be counterproductive to wait for their disappearance to warn of environmental deterioration.

At the present time and in the foreseeable future, the Pantanal offers its greatest potential as an attraction for national and international tourism. To preserve the Pantanal for this purpose, however, the tourism must be controlled to prevent deleterious effects from construction of too many hotels, roads, and artificial waterways and to minimize the introduction of pathogens, pollutants, and garbage, particularly plastics that are not degraded by the natural action of microorganisms.

The greatest danger to the Pantanal, both as a natural wonder and as a source of livelihood for the human population of southern Mato Grosso, is international development. The leaders of industrial countries have long displayed a compulsion to use their superior wealth and technology to interfere with the natural environment of and in the lives of people belonging to other cultures around the world. Judging from past performance, any of a large number of proposed "developmental projects" for the Pantanal would lead to windfall profits to a few construction companies and local politicians and the impoverishment of a large segment of Mato Grosso's population. Brazil cannot afford to finance any form of massive destruction of the wetland at the present time, but it is almost certain that megalomanic bureaucrats in one of the international development agencies will think up some new construction project, prepare plans covertly for its execution, declare it absolutely necessary for the economic development of Brazil, and try to gain the permission of the Brazilian government officials by public relations methods involving, among other things, personal attacks on the reputations of anyone opposing the project. By the time it becomes clear that the project has been an economic failure, its proponents will be gone from public view, and the Brazilian government will have to come up with the money to pay the debts. The successors of the proponents in the international agencies will claim to have learned from the mistakes and propose new developmental project equally unlikely to produce any economic benefits and with a great probability of producing another economic and ecological disaster.

The priorities of the funding agencies are reflected in the projects receiving funding for scientific research. During the research project in the Pantanal, it was repeatedly stressed that only "applied scientific research" was desired. How can this possibly yield any positive results in a wetland as large as major European country about which even the most basic information is lacking? Many bureaucrats in the international development agencies do not see it as their task to give the people in the recipient countries what they need or want but rather to tell them what they are supposed to need and want. Simplistic thinking and clever propaganda to make the latest developmental dogma palatable to the taxpayers who have to finance its propagation mark the route of the international aid organizations. For example, scenes of hungry "street children" have often been used to promote projects to increase food production. In fact, it is an overproduction rather than a shortage of food that forces the children to go hungry in the city streets. Enormous food surpluses severely depress food prices and make it impossible for small farmers to earn enough from their harvests to support their families. The impoverishment of the rural population leads to crowding of economic refugees in the cities and consequent hunger. The children are hungry not because the food they need has not been produced but rather because they lack money to buy it, and it is exactly those funds that they need which must be siphoned off to pay foreign banks the interest on the debts incurred through the developmental follies of past decades.

The dredging of the Rio Paraguay to make it into an international waterway is an example of a proposal that could produce both economic and environmental damage on a major scale. It would be an economic disaster because after it is completed, there will still be little or nothing to ship. The highway is presently sufficient to meet the needs of southern Mato Grosso, and a supplemental railroad, if it were built, would carry the freight presently sent by highway at a fraction of the cost.

The waterway will certainly bring short-term employment for a small group of workers and enormous profits to the construction firms that carry it out. Local politicians can also expect to benefit considerably. Over the long term, the construction firms will continue to profit from the continual dredging necessary to remove new deposits of silt from the channel. These profits will come from international development funds since revenues from shipping will be minimal, at best. As learned from similar projects in the developed countries, dredging the channel of the upstream section of rivers permits the water to drain faster, decreasing rather than increasing the depth of the water during long dry periods.

Even if the work could be completed in the way its proponents wish, the valuable freshwater that the Pantanal presently holds for more than half of the year and releases slowly to the downstream sections of the river system will be drained much more quickly, thereby eliminating much of the aquatic vegetation, reducing the local evaporation and rainfall, and eventually turning the entire region into a hot, semi-arid or arid zone, as it was during its recent geological history. The duration of dry season conditions will increase, and the rainy season will produce unprecedented erosion, destroying the productivity of the wetland. It may well turn southern Mato Grosso into an uninhabitable desert, as the great developmental projects around the Aral Sea have already done in that part of the world.

Actually, such drastic changes cannot be expected because it is doubtful whether the river channels can be kept deep enough to permit passage of the ships as planned. In this case, the work will simply be a means of placing public funds in private pockets without producing any kind of recognizable change at all. In short, the money will simply be thrown away leaving the participating nations with even greater debts than they presently have.

For any real development of the economies in tropical countries, a prerequisite is an educated population. This is particularly true as the technological level of the world civilization increases. Unfortunately, international assistance to the development of education by many of the international aid organizations is specifically forbidden. The assistance that is given for training is so limited that it has to be classified as a token aid, and basic education is scarcely supported at all.

For the Pantanal to be conserved for future generations, it is vital that a permanent organization be established for that purpose. In addition to the protection of the wetland and watershed and regulation of the tourist industry by a group with broad police powers similar to those of the forest rangers in North American national parks and the game wardens in the African parks and game preserves, basic research to identify the resources and elucidate the ecological processes by scientists with facilities for conducting research on a permanent basis should be conducted. The local agriculture and industry should be developed only in ways that do not eliminate any of the renewable resources.

To guarantee the preservation of the Pantanal, it should be declared, together with its watershed, a national park. The present forms of agriculture should be permitted where it has long been practices by methods that do not cause leaching and erosion or require the use of pesticides. The main water courses and central part of the wetland should remain unchanged, with a general prohibition against removing or introducing plants and animals. Aquaculture in upland areas of the watershed should be encouraged to permit fish production without any fishery on the rivers. Tourism should be encouraged to supply funds for the protection of the park and to make it difficult for poachers to go about their work unobserved. Lodgings for the tourists should be maintained at a limited number of sites at which sewage treatment facilities are available and from which refuse is periodically removed. With careful management of the tourist traffic, a negative impact of the biota can be almost fully prevented.

The Pantanal is one of the last intact wetlands in the world supporting one of the largest unmodified ecosystems of any kind. In view of the many past mistakes, which have eliminated large tracts of diverse natural ecosystems together with their flora and fauna in order to make short-term economic gains for the few and cause long-term losses for the many, it is hoped that the Pantanal will not be changed in any way but rather conserved as it is to preserve the vast water supplies is contains, protect the valuable fishery resources of the entire river system, permit a continued production of beef at a sustainable level, and allow future generations to enjoy seeing its rich natural wildlife.

15. Acknowledgements

This book was produced with the cooperation of the limnologists, botanists, zoologists, and ecologists who worked in the Projeto Ecologia do Pantanal during the years 1991 through 1994 at the Federal University of Mato Grosso, through members of the staff of that university who were not included in the project, and numerous other experts, who provided much valuable information and advice. In particular, help was given by the scientists who collaborated in the field work, assisted with the analyses, and jointly wrote the scientific papers from which some of the information in this book was taken. Assistance was also provided by experts in other parts of the world who identified specimens of the flora and fauna. First, I thank those persons working within the project: Profa. Dr. E. L. Hardoim for collaboration in the work on the bacteria and protozoans, Mr. Benedito Rogeirio Santana Trindade for assistance with the study on the Euglenida as well as for help in collecting the field data. Profa. Anajde Lomes do Prado for collaboration in the study of the herbaceous macrophytes, Dipl. Biol. Axel U. Kretzschmar for providing information on the rotifers and molluscs, and Profa. Soraya Ferreira for help with the pseudoterrestrial grasshoppers.

In addition to the participants in the Project, other members of the faculty at the Universidade Federal de Mato Grosso contributed information and provided critical comments on the manuscript. Profa. Dr. Ermalinda De-Lamonica-Freire collaborated in the work with the algae, Dr. William Severi of the Universidade Federal de Mato Grosso critically read and provided valuable information on the section of the ichthyofauna. Prof. M. Carvalho read and corrected the section of the manuscript on the amphibians and reptiles of the Pantanal. Prof. Dalci Mauricio Oliveira was kind enough to read and suggest improvements in the section on the birds.

For considerable assistance with the chemical procedures for analyzing the water and for the confirmation of the accuracy of the results obtained in the field, I owe a debt of gratitude to Dipl. Biol. Joachim Kaiser of the Institut für Umweltschutz KLS GmbH in Pinneberg, Schleswig-Holstein. In addition, he generously provided the analytical results of analyses by atom absorption spectrometry for the metals in the sediments and overlying water of streams and lakes in and near the Pantanal.

Special thanks are also due to Dr. Arnildo Pott of the Empresa Brasileira de Pesquisa Agropecuária do Pantanal (EMBRAPA) in Corumbá, Mato Grosso do Sul, a foremost authority on the flora of the Pantanal, for initially identifying many of the specimens and photographs of the local flora. Later, he was kind enough to critically read the section on the macrophytes, correct the orthographic errors, and bring the taxonomic information up to the latest standards.

Additional information on macrophytes in the Pantanal was provided by Prof. Dr. F. Martins of the University of Campinas, SP. Specimens of *Hyptis* spp. were identified by Dr. Raymond Harley, while a positive identification of *Setaria parviflora* was provided by Mr. Steve Renvoise, both of the Royal Botanic Gardens, Kew. Dr. D. C. Zappi of the same institute provided additional information on the taxonomic status of *Diodia kuntzei*. I am also grateful to Dr. Henri Dumont of the State University of Ghent and editor of the series for providing valuable information on the cladocerans and for his critical reading of the manuscript.

Special thanks are due to Prof. Dr. H. Kausch of the Institut für Hydrobiologie und Fischereiwissenschaft for providing material support for the preparation of the manuscript. I am also grateful to Dr. A. Faubel of the same institute for providing information on the flatworms.

Valuable help in the search for literature, often in old or obscure journals, was provided by Frau Gerhild Lechner of the library of the Max-Planck-Institut für Limnology in Plön.

The staff of the Fundação Centro Tecnológico de Minas Gerais (CETEC) in Belo Horizonte also provided support by permitting me to produce some of the graphs using their computers and software and providing me with a spacious working place during my subsequent visits to Brazil.

Thanks are also due to the German Ministry of Research and Technology (BMBT), the Conselho Nacional de Pesquisa e Tecnologia (CNPq), and Instituto Brasileiro de Meio Ambiente e Recursos Naturais Renováveis (IBAMA) for financing the project through their SHIFT program. In addition, I express my special gratitude to the Johanna and Fritz Buch Gedächtnisstiftung in Hamburg for financing the printing of the color plates.

16. Literature

Ab'Saber, A. N. (1988) O Pantanal Mato-Grossense e a teoria dos refúgios. - Rev. Bras. Geogr. 50:9-57.

Adámoli, J. (1981) O Pantanal e suas relações fitogeográficas com os cerrados. Discussão sobre o conceito "Complexo do Pantanal". - Soc. Bras. Bot. 32:109-119.

Adámoli, J. (1986a) Fitogeografia do Pantanal. In: A. Boock (Ed.), Anais 1° Simpósio Recursos Naturais e Sócio-Econômico do Pantanal. EMBRAPA-CPAP, Corumbá. pp. 105-106.

Adámoli, J. (1986b) A dinâmica das inundações no Pantanal. In: A. Boock (Ed.), Anais 1° Simpósio Recursos Naturais e Sócio-Econômico do Pantanal. EMBRAPA-CPAP, Corumbá. pp. 51-61.

Adámoli, J. (1992) Research and management in the Pantanal of Brazil. In: E. Maltby, P. J. Dugan, & J. C. Lefeuvre (Eds.), Conservation and Development: The Sustainable Use of Wetland Resources. - IUCN, Gland, Switzerland. pp. 61-67.

Adis, J. (1984) Seasonal *igapó*-forests of Central Amazonian blackwater rivers and their terrestrial arthropod fauna. In: H. Sioli (Ed.), The Amazon. - W. Junk, Dordrecht. pp. 245-268.

Adis, J., & G. Righi (1990) Mass migration and life cycle adaptation - a survival strategy of terrestrial earthworms in Central Amazonian inundation forests. - Amazoniana 11:23-30.

Agostinho, A. A., & M. Zalewski (1995) The dependence of fish community structure and dynamics on floodplain and riparian ecotone zone in Parana River, Brazil. - Hydrobiologia 303:141-148.

Albuquerque, B. W. P. de (1981) Plantas forrageiras da Amazônia. I - Aquáticas flutuantes livres. - Acta Amazonica 11(3):457-471.

Alexander, C. P., & M. M. Alexander (1970) Family Tipulidae. In: N. Papavero (Ed.), Catalogue of the Diptera of the Americas south of the United States. Vol. 4. Mus. Zool., Univ. São Paulo. pp. 1-259.

Alfonsi, R. R., & M. B. Paes de Camargo (1986) Condições climáticas para a região do Pantanal Mato-Grossense. In: A. Boock (Ed.), Anais 1° Simpósio Recursos Naturais e Sócio-Econômico do Pantanal. EMBRAPA-CPAP, Corumbá. pp. 29-42.

Alho, C. J. R.; T. E. Lacher Jr. & H. C. Gonçalves (1988) Environmental degradation in the Pantanal ecosystem. - Bioscience 38(3):164-171.

Allanson, B. R. (1979) The physico-chemical limnology of Lake Sibaya. In: B. R. Allanson (Ed.), Lake Sibaya. - W. Junk, The Hague. pp. 42-74.

Allem, A. C., & J. F. M. Valls (1987) Recursos forrageiros nativos do Pantanal Mato-Grossense. - Dept. Dif. Tecnol., Brasilia. 339 pp.

Almeida, E. R. (1993) Plantas medicinais brasileiras: conhecimentos populares e científicos. - Hemos, São Paulo. 341 pp.

Almeida, V. L. L., E. K. Resende, M. S. Lima, & C. J. A. Ferreira (1993) Dieta e atividade alimentar de *Prochilodus lineatus* (Characiformes, Prochilodontidae) no Pantanal do Miranda-Aquidauana, Mato Grosso do Sul, Brasil. - Rev. Unimar 15(Suppl.):125-141.

Almendras, J. M. E. (1987) Acute nitrite toxicity and methemoglobinemia in juvenile milkfish (*Chanos chanos* Forsskal). - Aquaculture 61:33-40.

Alonso-Mejia, A., & R. A. Medellín (1991) *Micronycteris megalotis.* Mammalian Species 376:1-6.

Altafin, I. G., S. P. Mattos, C. G. B. Cavalcanti, & V. R. Estuqui (1995) Paranoá Lake - Limnology and recovery program. In: J. G. Tundisi, C. E. M. Bicudo, and T. M. Tundisi (Eds.) Limnology in Brazil. - Braz. Acad. Sci., Braz. Limnol. Soc., Rio de Janeiro. pp. 325-349.

Alvarenga, S. M., A. E. Brasil, R. Pinheira & H. J. H. Kux (1984) Estudo geomorfológico aplicado à bacia do Alto Rio Paraguai e Pantaneis Matogrossenses. - Bol. Tec. Proj. RADAM/BRASIL, Ser. Geomorphologia, Salvador 187:89-183.

Amaral, A. do (1977) Serpentes do Brasil. - Univ. São Paulo. - 246 pp.

Amaral Filho, Z. P. (1986) Solos do Pantanal Matogrossense. In: Anais 1° Simpósio Recursos Naturais e Sócio-Econômico do Pantanal. EMBRAPA-CPAP, Corumbá. pp. 29-42.

Ames, P. L. (1966) DDT residues in the eggs of the osprey in the northeastern United States and their relation to nesting success. - J. Appl. Ecol. 3 (Suppl.):87-97.

Anderson, D. J. (1986) Ecological succession. In: J. Kikkawa & D. J. Anderson (Eds.), Community Ecology: Pattern and Process. - Blackwell, Melbourne. pp. 269-285.

Andrade, M. A. de (1992) Aves Silvestres: Minas Gerais. - CIPA, Belo Horizonte. 176 pp.

Ankley, G. T., M. K. Schubauer-Berigan, & P. D. Monson (1995) Influence of pH and hardness on toxicity of ammonia to the amphipod *Hyalella azteca.* - Can. J. Fish. Aquat. Sci. 52:2078-2083.

Ansari, I. A. (1987) Acute toxicity of sodium nitrite on *Channa punctatus* (Bloch) and *Mystus (Mystus) vittatus* (Bloch). - Acta Hydrochim. Hydrobiol. 15:175-178.

Aranha, C, H. F. Leitão Filho & C. A. Yahn (1988) Sistemática de plantas invasoras. - Inst. Campineiro de Ensino Agrícola, Campinas. 179 pp.

Araujo, A. F. B. de (1994) Comunidade de lagartos Brasileiros. In: L. B. Nascimento, A. T. Bernades & G. A. Cotta (Eds.), Herpetologia no Brasil. - Fund. Biodiversitas, Fund. Ezequil Dias, Belo Horizonte, MG.

Araujo, R. L. (1958a) Contribuição à biogeografia dos térmitas de São Paulo, Brasil (Insecta Isoptera). - Arch. Inst. Biol., São Paulo 25:187-217.

Araujo, R. L. (1958b) Contribuição à biogeografia dos térmitas de Minas Gerais, Brasil (Insecta Isoptera). - Arch. Inst. Biol., São Paulo 25:219-236.

Arcifa, M. S., O. Froehlich & T. G. Northcote (1988) Distribution and feeding ecology of fishes in a tropical Brazilian reservoir. - Mem. Soc. Cien. Nat. La Salle 48(Suppl.):301-326.

Ashton, P. J., & D. S. Mitchell (1989) Aquatic plants: patterns and modes of invasion, attributes of invading species and assessment of control programmes. In: J. A. Drake, H. A. Mooney, F. di Castri, R. H. Groves, F. J. Kruger, M. Rejmanek & M. Williamson (Eds.), Biological Invasions: a Global Perspective. John Wiley, Chichester. pp. 111-154.

Askew, G. P., D. J. Moffatt, R. F. Montgomery & P. L. Searl (1970), Interrelationships of soils and vegetation in the savanna-forest boundary zone of northeastern Mato Grosso. - Geogr. J. 136:370-376.

Auvinen, H. (1988) Distribution and food of vendace [*Coregonus albula* (L.)] larvae in Lake Pyhaejaervi (Karelia, SE Finland). - Finn. J. Fish. Res. 1988:107-115.

Axelrod, H. R., W. E. Burgess, C. W. Emmens, N. Pronjek, J. G. Walls & R. Hunziker (1987) Dr. Axelrod's Mini-Atlas of Freshwater Aquarium Fishes. Mini-ed. - T. F. H., Neptune, N.J. 992 pp.

Azcárate, T. (1980) Sociobiología y manejo des capibara (*Hydrochoerus hydrochaeris*). - Doñana Acta Vert. 7:1-228.

Bachmann, A. O. (1969) Notas sobre Hydrophilini de Venezuela (Coleoptera, Hydrophilidae). - Mem. Soc. Cien. Nat. "La Salle" 28(81):281-300.

Bachmann, A. O. (1995) Insecta Lepidoptera. In. E. C. Lopretto & G. Tell (Eds.), Ecosistemas de aguas continentales. Vol. 3. - Ed. Sur, La Plata. pp. 1239-1242.

Balaehandran, B., S. N. Sivaswami, & V. M. Sivaramakrishnam (1991) Mutagenicity of nitrosated food items. - Indian J. Exp. Biol. 29:676-678.

Balousova, V. V. (1968) (On the taxonomy of the genus *Gallionela*.). - Microbiologiya 37:715-722. (in Russian).

Banarescu, M. (1990) Zoogeography of Fresh Waters. Vol. 1. - AULA, Wiesbaden. pp. 1-511

Banarescu, M. (1995) Zoogeography of Fresh Waters. Vol. 3. - AULA, Wiesbaden. pp. 1099-1617.

Barros, M. (1938) Ciperáceas Argentinas. III. Generos: *Androtrichum, Lipocarpha, Ascolepis, Cyperus*. - An. Mus. Argentina Cienc. Nat. "Bernardino Rivadavia" 85:253-381.

Barros Neto, J. (1979) A criação empirica de bovinos no Pantanal da Nhecolândia. - Recenha Tributária, São Paulo. 158 pp.

Barroso, G. M. (1952) Scrophulariaceae indígenas e exóticas no Brasil. - Rodriguésia 27:9.108.

Barthlott, W., K. Riede & M. Wolter (1994) Mimicry and ultrastructural analogy between the semi-aquatic grasshopper *Paulinia acuminata* (Orthoptera, Pauliniidae) and its foodplant, the water-fern *Salvinia auriculata* (Filicatae, Salviniaceae). - Amazoniana 13:47-58.

Bayley, P. B. (1973) Studies on the migratory characin *Prochilodus platensis* Holmberg, 1889. - J. Fish Biol. 5:25-40.

Bayo, V., & E. Cordiviola de Yuan (1992) Acidos grasos de juveniles de *Prochilodus lineatus* (Valenciennes) (Pisces, Curimatidae). - Rev. Hydrobiol. Trop. 25:153-157.

Beek, K. J., & D. L. Bramao (1968) Nature and geography of South American soils. In: E. J. Fittkau, J. Illies, H. Klinge, G. H. Schwabe & H. Sioli (Eds.), Biogeography and Ecology in South America. Vol. 1. - W. Junk, Dordrecht. pp. 82-112.

Beltzer, A. H. (1983a) Nota sobre fidelidad y participacion trofica del "bigua comun" (*Phalacrocorax olivaceus*) en ambientes del rio Parana medio (Pelicaniformes: Phalacrocoracidae). - Rev. Asoc. Cienc. Nat. Litoral 14:111-114.

Beltzer, A. H. (1983b) Alimentación de la <<garcita azulada>> (*Butorides striatus*) en el valle aluvial del rio Paraná medio (Ciconiiformes: Ardeidae). - Rev. Hydrobiol. trop. 16(2):203-206.

Beltzer, A. H. (1989) Fluctuaciones anuales en las poblaciones de garzas (Aves: Ardeidae) en la llanura aluvial del rio Parana medio, Argentina. - Rev. Asoc. Cienc. Nat. Litoral 20:111-114.

Beltzer, A. H. (1990a) Notes on the food of the rufescent tiger heron, *Tigrisoma lineatum* (Aves: Ardeidae) in the middle Paraná River floodplain, Argentina. - Stud. Neotrop. Fauna Environ. 25:93-96.

Beltzer, A. H. (1990b) Biologia alimentaria del halconcito comun *Falco sparverius* en el valle aluvial del rio Parana medio, Argentina. - Hornero 13:133-136.

Beltzer, A. H. (1991) Aspects of the breeding biology and the death rate of the *Butorides striatus* (Aves: Ardeidae). - Rev. Asoc. Cienc. Nat. Litoral 22:35-40.

Beltzer, A. H., E. B. de C. de Ara, E. D. Mosso, & A. T. R. de Montaner (1987) Alimentacion otoñal, anatomia macroscopica e histologica del tubo digestivo de la garcita bueyera *Bubulcus ibis ibis* (Aves: Ardeidae). - Rev. Asoc. Cienc. Nat. Litoral 18:155-165.

Beltzer, A. H., & E. D. Mosso (1992) Alimentacion de algunos patos (Aves: Anatidae) en el valle aluvial del rio Parana, Argentina. - Rev. Asoc. Cienc. Nat. Litoral 23:37-40.

Beltzer, A. H., & G. Paporello (1983) Alimentación de aves en el valle aluvial del rio Paraná. IV. *Agelaius cyanopus cyanopus* Vieillot, 1819 (Passeriformes, Icteridae). - Iheringia Ser. Zool. 62:55-60.

Belwood, J. J., & G. K. Morris (1987) Bat predation and its influence on calling behavior in Neotropical katydids. - Science 238:64-67.

Benson, W. W. (1985) Amazon ant-plants. In: G. T. Prance and T. E. Lovejoy (Eds.), Amazonia. - Pergamon, Oxford. pp. 239-266.

Berry, A. J. (1974) The anatomy of West Malaysian snails of parasitological significance. Malay Nature J. 27:131-165.

Berta, A. (1982) *Cerdacyon thous*. Mammalian Species 186:1-4.

Berta, A. (1986) *Atelocynus microtis*. Mammalian Species 256:1-3.

Bertelli, A. P. (1984) O paraíso das espécies vivas: Pantanal de Mato Grosso. - Cerifa, São Paulo. 337 pp.

Bertelli, A. P. (1988) O Pantanal: Mar dos Xaráes. Ed. Siciliano, São Paulo. 102 pp.

Berthold, T. (1991) Vergleichende Anatomie, Phylogenie und historische Biogeographie der Ampullariidae (Mollusca, Gastropoda). - Abh. nat.-wiss. Verein Hamburg (NF) 29:1-256.

Bican, J., K. Drbal, D. Dykyjová, J. Kvet, P. Marvan, J. P. Ondok, P. Pokorny, K. Pribán, & P. Smid (1986) Ecology of fishpond vegetation. In: S. Hejny, I. M. Raspopov, & J. Kvet (Eds.), Studies on Shallow Lakes and Ponds. Czech. Acad. Sci., Prague. pp. 171-229.

Bick, H., & S. Kunze (1971) Eine Zusammenstellung von autökologischen und saprobiologischen Befunden an Süßwasserciliaten. - Int. Revue ges. Hydrobiologie 56:337-384.

Bicudo, C. E. de M. (1969) Contribution to the knowledge of the desmids of the state of São Paulo, Brazil (including a few from the state of Minas Gerais). - Nova Hedwigia 17:433-549.

Bicudo, C. E. de M., & M. T. P. Azevedo (1977) Desmidioflórula paulista, 1: gênero *Arthrodesmus* Ehr. ex Ralfs emend Arch. - Bibl. Phycol., Vaduz 36:iii + 1-105.

Bicudo, C. E. de M., & A. A. J. de Castro (1994) Desmidioflórula paulista 4. Gêneros *Closterium, Spinoclosterium*. - Biblio. Phycol. 95. J. Cramer, Berlin. 191 pp.

Bicudo, C. E. de M., & E. M. De-Lamonica-Freire (1993) *Trachelomonas armata* (Euglenophyceae): an evaluation of the diagnostic features in the species. - Arch. Hydrobiol., Alg. Stud. 69:57-66.

Bicudo, C. E. de M., & I. M. Samanez (1984) Desmidioflórula paulista 3. Gêneros *Bambusina, Desmidium, Groenbladia, Hyalotheca, Onychonema, Phymatodocis, Spondylosium* & *Teilingia*. J. Cramer, Vaduz. pp. 1-138 + pl. 448 - 470.

Bicudo, C. E. de M., & L. Sormus (1982) Desmidioflórula paulista 2. Gênero *Micrasterias* C. Agardh ex Ralfs. - Biblio. Phycol., Vaduz. 57:v + 1-230.

Bicudo, C. E. de M., & R. M. Ventrice (1968) Algas do Brejo da Lapa, Parque Nacional do Itatiaia, Brasil. - Anais XIX Congr. nac. Bot., Fortaleza. pp. 1-30. 128 Figs.

Bicudo, R. M. T. (1968) An annotated list of Charophyceae already cited for Brazil. - Rickia 3:221-238.

Bisbal, F. J. (1993) Impacto humano sobre los carnívoros de Venezuela. - Stud. Neotrop. Fauna Environ. 28(3):145-156.

Bishop, J. E. (1973) Limnology of a small Malayan River Sungai Gombak. - W. Junk, The Hague. 485 pp.

Bistoni, M. de los A., J. G. Haro & M. Gutiérrez (1992) Ictiofauna del Rio Dulce en la Provincia de Cordoba (Argentina) (Pisces, Osteichthys). - Iheringia Sér. Zool. 72:105-111.

Blackwelder, R. E. (1944) Checklist of the Coleopterous Insects of Mexico, Central America, the West Indies, and South America. Parts 1 and 2. - Bull. U. S. Natl. Mus. 185:1-341.

Bohlin, K. (1897) Die Algen der ersten regnell'schen Expedition I. Protococcoideen. - Bih. K. Svenska Vetensk. Handl. 23(7):1-47.

Bok, A. H. (1989) Rearing of artificially spawned southern mullet. - S. Afr. J. Wildl. Cons. 19:31-34.

Bokermann, W. C. A. (1968) Uma nova espécie de '*Leptodactylus*' de Mato Grosso (Anura, Leptodactylidae). - Rev. bras. Biol. 29:13-16.

Bonaccorso, F. J. (1978) Foraging and reproductive ecology in a Panamanian bat community. - Bull. Fla. State Mus., Biol. Sci. 24:359-408.

Bonetto, A. A. (1975) Hydrologic regime of the Paraná River and its influence on ecosystems. In: D: Hassler (Ed.), Coupling of Land and Water Systems. - Springer, New York. pp. 175-197.

Bonetto, A. A., C. Bonetto & Y. Zalocar (1981) Contribución al conocímiento limnológico del río Paraguay en su tramo inferior. - Ecosur 8(16):55-88.

Bonetto, A. A., & A. M. de Ferrato, 1969a: Limnological investigations on biotic communities in the middle Paraná River Valley. - Verh. Int. Verein Theor. Angew. Limnol. 17:1035-1050.

Bonetto, A. A., & A. M. de Ferrato, 1969b: Ciclos hidrologicos del Rio Paraná y las poblaciones de peces contenidas en las cuencas temporarias de un valle de inundacion. Physis, Buenos Aires 29:213-223.

Bonetto, A., & C. Pignalberi (1964) Nuevos aportes al conocimiento de las migraciones de los peces en los rios mesopotamicos de la Republica Argentina. - Comm. Inst. Nac. Limnol., Santo Tome 1:1-14.

Bonetto, A., C. Pignalberi, E. Cordiviola, & O. Oliveros (1971) Informaciones complementarias sobre migraciones de peces en la cuenca del Plata. - Physis 30(81):505-520.

Bonetto, A. A., & I. R. Wais (1990) The Paraná River in the framework of modern paradigms of fluvial systems. - Limnol. Bras. 3:139-172.

Bonetto, A. A., & I. R. Wais (1992) Las grandes planicies aluviales del sistema potamico Parana-Paraguay. - Rev. Mus. Argent. Cienc. Natl. Bernardino Rivadavia Inst. Nac. Invest. Cienc. Nat. Argent. Hidrobiol. 6(4):1-29.

Bonetto, C. A. (1983) Fitoplancton y producción primaria del Paraná medio. - Ecosur 10(19-20):79-102.

Bonetto, C. A., Y. Zalocar, P. Caro & E. Vallejos (1979) Producción primaria des fitoplancton del río Paraná en el área de su confluencia con el río Paraguay. - Ecosur 6(12):207-227.

Borge, O. (1903a) Die Algen der ersten regnellschen Expedition, 2: Desmidiaceen. - Ark. Bot. 1:71-138.

Borge, O. (1903b) Die Algen der ersten regnellschen Expedition, 3: Zygnemaceen und Mesocarpaceen. - Ark. Bot. 1:277-285.

Borge, O. (1925) Die von Dr. F. C. Hoehne während der Expedition Roosevelt-Rondon gesammelten Süßwasseralgen. Ark. Bot. 19(17):1-56.

Bourrelly, P., & A. Couté (1982) Quelques algues d'eau douce de la Guyane Française. - Amazoniana 7:221-292.

Bovee, E. C. (1957) Protozoa of Amazonian and Andean waters of Colombia, South America. - J. Protozool. 4:63-66.

Bozelli, R. L. (1992) Composition of the zooplankton community of Batata and Massurá Lakes and of the Trombetas River, State of Pará, Brazil. - Amazoniana 12:239-261.

Brady, C. (1979) Observations on the behavior of the crab-eating fox (*Cerdocyon thous*). In: J. Eisenberg (Ed.), Vertebrate Ecology in the northern Neotropics. Smithson. Inst. Press, Washington. pp. 161-171.

Braga, R. A. (1952) Ovo, larva e alevino de tucunaré pinima *Cichla temensis* Humboldt (Actinopterygii, Cichlidae). Rev. Ind. Anim. 13:141-144.

Brako, L., & L. Zarucchi (1993) Catalogue of the flowering plants and gymnosperms of Peru. - Monogr. Syst. Bot. Missouri Bot. Gard. 45:i-xl, 1-1286.

Branco, C. C. Z., & O. Necchi Jr. (1996) Distribution of stream macrophytes in the eastern Atlantic rainforest of São Paulo State, Southeastern Brazil. - Hydrobiologia 333:139-150.

Brandorff, G. -O. (1976) The geographic distribution of the Diaptomidae in South America (Crustacea, Copepoda). Rev. Brasil. Biol. 36:613-627.

Brandt, R. A. M. (1974) The non-marine Mollusca of Thailand. Arch. Molluskenkunde. 105:1-423.

Braum, E. (1983) Beobachtungen über eine reversible Lippenextension und ihre Rolle bei der Notatmung von *Brycon* sp. (Pisces, Characidae) und *Colossoma macropomum* (Pisces, Serrasalmidae). - Amazoniana 7:355-374.

Braum, E., & R. Bock (1985) Funktionsmorphologische Untersuchungen über die Bärteln von *Osteoglossum bicirrhosum* (Pisces, Osteoglossidae) während der Notatmung. - Amazoniana 9:353-370.

Braun, A. (1882) Fragment einer Monographie der Characeen. Nach den hinterlassenen Manuskripten A. Braun's herausgegeben von Dr. Otto Nordstedt. - Abh. Akad. Wiss. Berlin 1882(1):1-211.

Britski, H. A. (1969) Lista dos tipos de peixes dos coleções do Departemento de Zoologia da Secretaria da Agricultura da São Paulo. - Pap. Avul. Zool., São Paulo 22:197-215.

Britski, H. A. (1972) Peixes de água doce do Estado de São Paulo: Systemática. In: Poluição e Piscicultura. - Commissão Interestadual da Bacia Paraná-Uruguay, São Paulo. pp. 79-108.

Brock, T. D. (1978) Thermophilic Microorganisms and Life at High Temperatures. - Springer Verlag, New York. 465 pp.

Brook, A. J. (1965) Planktonic algae as indicators of lake types with special reference to the Desmidiaceae. - Limnol. Oceanogr. 10:403-411.

Brook, A. J. (1981) The Biology of Desmids. - Blackwell, Oxford. 276 pp.

Brown, E. H., G. P. Askew, J. B. Thornes, A. Young, J. R. G. Townshend & S. G. Daultrey (1970) Geographical research on the Royal Society/Royal Geographical Society's expedition to North-Eastern Mato Grosso, Brazil: a symposium. - Geogr. J. 136:365-409.

Brown, K. S., Jr. (1986) Zoogeografia da região do Pantanal Mato-Grossense. In: A. Boock (Ed.), Anais 1° Simpósio Recursos Naturais e Sócio-Econômico do Pantanal. EMBRAPA-CPAP, Corumbá. pp. 137-178.

Brown, L., & D. Amadon (1968) Eagles, Hawks and Falcons of the World. 2 Vols. - Country Life, Feltham, Middlesex. 945 pp. + 165 pl.

Brücher, H. (1977) Tropischer Nutzpflanzen. - Springer, Berlin. 529 pp.

Brum, P. A. R. de & J. C. de Sousa (1985) Níveis de nutrientes minerais para gado em lagoas ("baías" e "salinas") no Pantanal Sul Matogrossense. - Pesq. Agropec. Bras. 20(12):1451-1454.

Brundin, L. (1966) On the real nature of Transantarctic relationships. - Evolution 19:496-505.

Brundin, L. (1967) Transantarctic relationships and their significance, as evidenced by chironomid midges. - Kungl. Svenske Vetenskap. Handl. 4:Ser. 11/1:1-472.

Bucher, E. H., & P. C. Huszar (1995) Critical environmental costs of the Paraguay-Paraná waterway project in South America. - Pestic. Sci. 45(3):3-9.

Bunton, T. E. (1989) Diethylnitrosamine (DEN)-induced degenerative, proliferative and neoplastic lesions in the liver of the medaka (*Oryzias latipes*) following short-term exposure. In: M. N. Moore & J. J. Stegeman (Eds.), Responses of Marine Organisms to Pollutants. Vol. 28:369-374.

Burkart, A. (1952) Las leguminosas Argentinas silvestres y cultivadas. - ACME, Buenos Aires. 476 pp.

Burkart, A., J. A. Caro, K. A. Okada, R. A. Palacios, Z. E. R. de Agrasar, E. S. de García, M. Toursarkissian & N. S. T. de Burkart (1969) Gramineas. In: A. Burkart (Ed.), Flora Ilustrada de Entre Rios (Argentina). - I.N.T.A., Buenos Aires. 551 pp.

Burmeister, E. -G., K. Göttlich, T. Grospietsch & G. Kaule (1980) Begriffsbestimmungen anhand der Moortypen Mitteleuropas. In: K. Göttlich (Ed.), Moor- und Torfkunde. - Schweizerbart, Stuttgart. pp. 1-46.

Burnham, K. P., D. R. Anderson, G. C. White, C. Brownie & K. H. Pollock (1987) Design and analysis methods for fish survival experiments based on release-recapture. - Amer. Fish. Soc. Monogr. 5. - Bethesda, Md. 437 pp.

Cabrera, A., & J. Yepes (1960) Mamiferos Sud Americanos. Vol. 1. - Librart, Buenos Aires. 187 pp.

Cadima, M. M. (1990) Fitoplancton de una laguna de varzea en el Valle del Sajta (Cochabamba - Bolivia). - Acta Limnol. Brasil. 3:419-437.

Calijuri, M. C., & A. C. A. Dos Santos (1996) Short-term changes in the Barra Bonita reservoir (São Paulo, Brazil): emphasis on the phytoplankton communities. Hydrobiologia 330:163-175.

Camargo, A. F. M., & F. A. Esteves (1995a) Biomass and productivity of aquatic macrophytes in Brazilian lacuatrine ecosystems. In: J. G. Tundisi, C. E. M. Bicudo, & T. Matsumura Tundisi (Eds.) Limnology in Brazil. ABC/SBL, Rio de Janeiro. pp. 137-149.

Camargo, A. F. M., & F. A. Esteves (1995b) Influence of water level variation on biomass and chemical composition of the aquatic macrophyte *Eichhornia azurea* (Kunth) in an oxbow lake of the Rio Mogi-Guaçu (São Paulo, Brazil). - Arch. Hydrobiol. 135:423-432.

Campos, Z. (1993) Effect of habitat on survival of eggs and sex ratio of hatchlings of *Caiman crocodilus yacare* in the Pantanal, Brazil. - J. Herpetol. 27:127-132.

Carbonell, C. S. (1957) Observaciones bio-ecológicas sobre *Marellia remipes* Uvarov (Orthoptera, Acridoidea) en el Uruguay. - Invest. Estud., Fac. Humanid. Cienc. (Montevideo). 21 pp. + 6 pl.

Carboneras, C. (1992a) Family Anhimidae (screamers). In: J. del Hoyo, A. Elliott, and J. Sargatal (Eds.), Handbook of the Birds of the World. Vol. 1. - Lynx, Barcelona. pp. 528-535.

Carboneras, C. (1992b) Family Anatidae (ducks, geese and swans). In: J. del Hoyo, A. Elliott, and J. Sargatal (Eds.), Handbook of the Birds of the World. Vol. 1. - Lynx, Barcelona. pp. 536-568.

Carozza, C., & E. Cordiviola de Yuan (1991) Estudios ictiológicos en la laguna La Cuarentena (Isla Carabajal), rio Paraná, Argentina: Edad y crecimiento del «sábalo» *Prochilodus lineatus* (Val.), período 1984-1985 (Pisces, Curimatidae). - Rev. Hydrobiol. Trop. 24:119-129.

Carvalho, J. C. (1986) Fauna terrestre e aquática. In: A. Boock (Ed.), Anais 1° Simpósio Recursos Naturais e Sócio-Econômico do Pantanal. EMBRAPA-CPAP, Corumbá. pp. 179-182.

Caspers, H., & C. W. Heckman (1981) Ecology of orchard drainage ditches along the freshwater section of the Elbe Estuary. - Arch. Hydrobiol./Suppl. 43 (Elbe-Aestuar 4):347-486.

Castellanos, A. (1959) Las Pontederiaceae de Brasil. - Arq. Jard. Bot. Rio de Janeiro 16:147-236.

Castro, A. L. de, 1986: Branchiura. - Manual Ident. Invert. Limn. Bras., No. 30. MCT and CNPq, Brasilia. 23 pp.

Catto, J. B., & J. F. R. Amato (1993a) Digenetic trematodes (Cryptogonimidae, Acanthostominae) parasites of the caiman, *Caiman crocodilus yacare* (Reptilia, Crocodylia) from the Pantanal Matogrossense, Brazil, with the description of a new species. - Mem. Inst. Oswaldo Cruz 88:435-440.

Catto, J. B., & J. F. R. Amato (1993b) Two new species of *Pseudotelorchis* (Digenea, Telorchidae), parasites of the caiman, *Caiman crocodilus yacare,* (Reptilia, Crocodylia) from the Pananal Mato-Grossense, Brazil. - Mem. Inst. Oswaldo Cruz 88:561-566.

Cei, J. M. (1980) Amphibians of Argentina. - Monit. zool. Ital., N.S., Monogr. 2:1-609.

Cherrie, G. K. (1916) A contribution to the ornithology of the Orinoco Region. - Brooklyn N.Y. Mus. Inst. Arts Sci. Bull. 2:133-374.

Cintra, R. (1985)Nascimento de filhotes de *Caiman yacare* (Daudin, 1802) (Crocodylia: Alligatoridae) em condições semi naturais no Pantanal Matogrossense. - Pap. Avul. Zool. 36(10):91-101.

Cintra, R. (1988) Nesting ecology of the Paraguayan caiman *(Caiman yacare)* in the Brazilian Pantanal. J. Herpetol. 22:219-222.

Cintra, R. (1989) Maternal care and daily pattern of behavior in a family of caimans, *Caiman yacare* in the Brazilian Pantanal. J. Herpetol. 23:320-322.

Cintra, R., & C. Yamashita (1990): Habitats, abundância e ocorrência das espécies de aves do Pantanal de Poconé, Mato Grosso, Brasil. - Pap. Avul. Zool. 37(1):1-21.

Cole, J. A., & S. C. Rittenberg (1971) A comparison of respiratory processes in *Spirillum volutans*, *Spirillum intersonii* and *Spirillum serpens*. - J. Gen. Microbiol. 69:375-383.

Collart, O. O., & L. C. Moreira (1993) Potencial pequeiro de *Macrobrachium amazonicum* na Amazônia Central (Ilha do Careiro: variação da abundância e do comprimento. - Amazoniana 12:399-413.

Colli, G. R. (1991) Reproductive ecology of *Ameiva ameiva* (Sauria, Teiidae) in the *cerrado* of Central Brazil. - Copeia 1991:1002-1012.

Conceição, C. de A., & J. E. de Paula (1986) Contribuição para o conhecimento da flora do Pantanal Mato-Grossense e sua relação com a fauna e o homen. In: Anais 1° Simpósio Recursos Naturais e Sócio-Econômico do Pantanal. EMBRAPA-CPAP, Corumbá. pp. 107-130.

Consoli, R. A. G. B., M. H. Pereira, A. L. Melo, & L. H. Pereira (1989) *Belostoma micantulum* Stål, 1858 (Hemiptera: Belostomatidae) as a predator of larvae and pupae of *Aedes fluviatilis* (Diptera: Culicidae) in laboratory conditions. - Mem. Inst. Oswaldo Cruz 84:577-578.

Cook, C. D. K. (1974) Water Plants of the World. - Dr. W. Junk, The Hague. 561 pp.

Cordiviola, E. (1966) Neuvos aportes al conocimiento de la biologia pesquera del "surubi" (*Pseudoplatystoma coruscans*) en el Parana medio (Pisces, Siluriformes). - Physis 26(71):237-244.

Cordiviola de Yuan, E., & C. Pignalberi de Hassan (1989) Icthyological studies in La Cuarentena Lagoon (Carabajal Island), Paraná River: density of *Prochilodus platensis* Holmberg (Curimatidae). - Hydrobiologia 183:43-46.

Cordiviola de Yuan, E., & O. B. Oliveros (1979) Campaña "Keratella I" a lo largo del rio Parana medio. I. Peces de ambientes leniticos. Acta Zool. Lilloana 35:629-642.

Costa, C. A. F., & J. B. Catto (1994) Helminth parasites of capibaras (*Hydrochoerus hydrochaeris*) in the sub-region of Nhecolândia, Pantanal, State of Mato Grosso do Sul. - Rev. Bras. Biol. 54(1):39-48.

Costa e Silva, P. P., & J. C. V. Ferreira (1994) Breve Historia de Mato Grosso e de Seus Municipios. Editora, Cuiabá. 263 pp.

Coutinho, L. M. (1990) Fire in the ecology of the Brazilian cerrado. In: Goldammer, J. G. (Ed.), Fire in the Tropical Biota. - Springer, Berlin. pp. 82-105.

Coutinho, M. E., G. M. Mourão, M. P. Silva & Z. Campos (1994) The sustainable use of natural resources and the conservation of the Pantanal wetland, Brazil. Acta Limnol. Bras. 5:165-176.

Cranston, P. S., & U. Nolte (1996) *Fissimentum*, a new genus of drought-tolerant Chironomini (Diptera: Chironomidae) from the Americas and Australia. - Ent. News 107(1):1-15.

Crawshaw, P. J., Jr. (1991) Effects of hunting on the reproduction of the Paraguayan Caiman (*Caiman yacare*) in the Pantanal of Mato Grosso, Brazil. In: J. G. Robinson & K. H. Redford (Eds.), Tropical Wildlife, Use and Conservation. - Univ. Chicago, Chicago. pp. 145-154.

Crespo, F. A. (1984) Nuevas citas de *Limnogonus (Limnogonus) aduncus* y *L. (Neogerris) lubricus* (Heteroptera, Gerridae). - Physis B 42(102):76.

Cresson, E. T., Jr. (1917) Studies in the American Ephydridae (Diptera) II. A revision of the species of the genera *Notiphila* and *Dichaeta*. - Trans. Amer. Ent. Soc. 43:27-66.

Crowner, A., & G. Barret (1979) Effects of fire on a small mammal component of an experimental grassland community. - J. Mammal. 60:803-812.

Cruz, J. A., J. R. Verani, L. Girardi & C. E. C. Torloni (1990) Levantamento da ictiofauna e aspetos da dinamica de população de algumas espécies do reservatorio de Promissão, SP (1a etapa). - Comp. Ener. São Paulo/Univ. Fed. São Carlos. Pesquisa e Desenvolvimento 52:1-78.

Cunha, A. M. da (1916) Contribuição para o conhecimento da fauna de protozoarios do Brasil. - Mem. Inst. Oswaldo Cruz 8:66-73.

Cunha, J. (1943) Análise química das águas. Cobre do Jauru, lagoas alcalinas do Pantanal. - Boll. Lab. Prod. Mineira 6:18-19.

Cunha Vieira, C. D. da (1951) Notas sobre os mamíferos obtidos pela expedição do Instituto Butantã ao Rio das Mortes e Serra do Roncador. - Pap. Avul. Dept. Zool. 10:105-125.

Currier, M. J. P. (1983) *Felis concolor*. - Mammalian Species 200:1-7.

Dabés, M. B. G. S. (1995) Composição e descrição do zooplâncton de 5 (cinco) lagoas marginais do rio São Francisco, Pirapora/Três Marias/ Minas Gerais/ Brasil. - Rev. Brasil. Biol. 55:831-845.

Daday, E. (1904) Süßwasser-Mikrofauna Paraguays. Protozoa. - Bibl. Zool. 44:4-46.

Da Silva, C. J., & F. de A. Esteves (1993) Biomass of three macrophytes in the Pantanal of Mato Grosso, Brazil. - Int. J. Ecol. Environ. Sci. 19:11-23.

Da Silva, C. J., & V. Pinto-Silva (1989) Macrófitas aquáticas e as condições físicas e químicas dos "alagados", "corixos" e rios, ao longo da rodovia Transpantaneira - Pantanal Matogrossense (Poconé - MT). - Rev. Brasil. Biol. 49:691-697.

Davenport, L. J. (1988) A monograph of *Hydrolea* (Hydrophyllaceae). - Rhodora 90:169-208.

Davis, W. B. (1976) Geographic variation in the lesser noctilio, *Noctilio albiventris* (Chiroptera). - J. Mamm. 57:687-707.

Davis, W. B., D. C. Carter & R. H. Pine (1964) Noteworthy records of Mexican and Central American bats. - J. Mamm. 45:375-387.

Dawson, G. (1973) Flora Argentina. Vol. 13. Lentibulariaceae. - Univ. Nac. La Plata, Bot. No. 70. 59 pp.

De Carlo, J. A. (1938) Los belostómidos americanos. An. Mus. Arg. Cienc. Nat. Bernardino Rivadavia 39:189-252.

De Carlo, J. A. (1964) Los Ranatridae de America. - Rev. Mus. Arg. Cienc. Nat. (Entomología) 1:133-215.

Decloitre, L. (1981) Le genre *Trinema* Dujardin, 1841. Arch. Protistenkunde 124:193-218.

Deflandre, G. (1929a) Le genre *Arcella* Ehrenberg. - Arch. Protistenkunde 64:152-287.

Deflandre, G. (1929b) Le genre *Centropyxis* Stein. - Arch. Protistenkunde 67:322-375.

Deflandre, G. (1937) Etude monographique sur le genre *Nebela* Leidy (Rhizopoda, Testacea). - Ann. Protistol. 5:201-286.

Deflandre, G. (1953) Thecamoebiens. In: P. -P. Grassé (Ed.), Traité de Zoologie 1(2):97-148.

De-Lamonica-Freire, E. M. (1989) Catalogo das algas referidas para o Estado de Mato Grosso, Brasil, 2. Rev. Bras. Biol. 49(3):679-689.

De-Lamonica-Freire, E. M. (1992a) Desmídias filamentosas (Zygnemaphyceae, Desmidiales) da estação ecológica da Ilha de Taiamã, Mato Grosso, Brasil. - Acta Limnol. Bras. 4:315-325.

De-Lamonica-Freire, E. M. (1992b) O gênero *Xanthidium* (Zygnemaphyceae, Desmidiales) na estação ecológica da Ilha de Taiamã, Estado de Mato Grosso, Brasil. - Bol. Inst. Biocienc. UFMT 1:1-9.

De-Lamonica-Freire, E. M., & C. L. Sant'Anna (1993) Chlorococcales (Chlorophyceae) da estação ecológica da Ilha de Taiama, Estado de Mato Grosso, Brasil. Hoehnea 20 (1/2):107-118.

De-Lamonica-Freire, E. M., C. E. de M. Bicudo & A. A. J. de Castro (1992) Ficoflórula do Pantanal de Poconé, Estado de Mato Grosso, Brasil, 1:Euglenaceae. - Rev. Brasil. Biol. 52:141-149.

De-Lamonica-Freire, E., & C. W. Heckman (1996) The seasonal succession of biotic communities in wetlands of the tropical wet-and-dry climatic zone: III. The algal communities in the Pantanal of Mato Grosso, Brazil. - Int. Revue ges. Hydrobiol. 81:255-282.

DeLoach, C. J., & H. A. Cordo (1978) Life history of the moth, *Sameodes albiguttalis*, as a candidate for biological control of waterhyacinth. - Environ. Ent. 7(2):309-321.

DeLoach, C. J., D. J. DeLoach & H. A. Cordo (1978) Observations on the biology of the moth, *Samea multiplicalis*, on waterlettuce in Argentina. - J. Aquatic Plant Management 17:42-44.

Dias, I. C. A. (1986) Zygnemaceae (Zygnemaphyceae) da Chapada dos Guimarães e Arredores, Mato Grosso, Brasil: uma contribuição ao seu conhecimento. - Rickia 13:69-75.

Dias, I. C. A. (1989) Chlorophyta filamentosas do municipio de Cáceres e arredores, Mato Grosso, Brasil: uma contribuição ao seu conhecimento. - Acta Bot. Bras. 3(2):3-12.

Dietz, J. M. (1985) *Chrysocyon brachyurus*. - Mammalian Species 234:1-4.

Dioni, W. (1967) Investigacion preliminar de la estructura basica de las asociaciones de la micro y meso fauna de las raizes de las plantas flotantes. - Acta Zool. Lilloana 23:111-132.

Dixon, J. R. (1980) The neotropical colubrid snake genus *Liophis*. The generic concept. - Milw. Public Mus. Contrib. Biol. Geol. 31:1-40.

Dixon, J. R. (1989) A key and checklist to the Neotropical snake genus *Liophis*. - Smithson. Herpetol. Info. Service, Publ. 79. 28 pp. + maps.

Donner, J. (1965) Ordnung Bdelloidea (Rotatoria, Rädertiere). - Akademie-Verlag, Berlin. 297 pp.

Dourojeanni, M. J. (1980) Desarrollo y conservacion en el Pantanal Matogrossense (Brasil) con especial referencia al manejo de la fauna silvestre. - Report Instituto Brasileiro de Desinvolvimento Florestal (IBDF), Brasilia.

Dourado, O. F. (1981) Principais peixes e crustáceos dos açudes controlados pelo DNOCS. - Dept. Nac. de Obras contra as Secas, Fortaleza. 40 pp.

Dragesco, J. (1980) Les Protozoaires. In: J. R. Durand & C. Lévêque (Eds.), Flore et Faune aquatique de l'Afrique Sahelo-Soudanienne. - T. I. Orstom, Paris. pp. 160-192.

Drake, C. J., & H. M. Harris (1934) The Gerrinae of the western hemisphere (Hemiptera). - Ann. Carnegie Mus. 23:179-241.

Drouet, F. (1938) The Brazilian Myxophyceae, 2. - Amer. J. Bot. 25:657-666.

Dubs, B. (1983) Vögel des südlichen Mato Grosso. - Verbandsdruckerei-Betadruck, Bern. 144 pp.

Dubs, B. (1992) Catalogue and Guide to the Birds of the Pantanal of Mato Grosso and its Border Areas. - Betrona, Küsnacht. 143 pp. + 58 plates.

Dubs, B. (1993) Catalogue of flowering plants from Mato Grosso and Mato Grosso do Sul, Brazil, represented in the herbarium collection of the Institute of Systematic Botany, University of Zürich (Z) and Regnellian Herbarium at Stockholm (S). - Betrone, Küsnacht.

Dubs, B. (1994) Differentiation of woodland and wet savanna habitats in the Pantanal of Mato Grosso, Brazil. - Betrone, Küsnacht. 103 pp.

Duda, O., 1925: Die außereuropäischen Arten der Gattung *Leptocera* Olivier = Limnsina Macquart (Dipteren). - Arch. Naturges., Abt. A, 11:1-215 + 4 pl.

Duke, A. (1960) Flora of Panama (Polygonaceae). - Ann. Misouri Bot. Gard. 47:330-331.

Dumont, H. J. (1969) A quantitative method for the study of periphyton. - Limnol. Oceanogr. 14:303-307.

Dumont, H. J., & J. Pensaert (1983) A revision of the Scapholeberinae (Crustacea: Cladocera). - Hydrobiologia 100:3-45.

Dunn, L. H. (1934) Notes on the water lettuce as a nursery of insect life. - Ecology 15:329-331.

Easterla, D. A. (1973) Ecology of 18 species of Chiroptera at Big Bend National Park, Texas. Northwest Missouri State Univ. Stud. 34:1-165.

Easterla, D. A., & J. O. Whitaker (1973) Food habits of some bats from Big Bend National Park, Texas. J. Mamm. 53:887-890.

Eidt, R. C. (1968) The climatology of South America. In: E. J. Fittkau, J. Illies, H. Klinge, G. H. Schwabe & H. Sioli (Eds.), Biogeography and Ecology in South America. Vol. 1. - W. Junk, Dordrecht. pp. 54-81.

Eigenmann, C. H. (1918) The American Characidae. - Mem. Mus. Comp. Zool. 43(2):103-208, pl. 9-93.

Eigenmann, C. H., & G. S. Myers (1929) The American Characidae. Part 5. - Mem. Mus. Comp. Zool. 43:428-558.

Eiten, G. (1972) The cerrado vegetation of Brazil. - Bot. Rev. 38:201-341.

Elliott, J. M., & K. H. Mann (1979) A Key to the British Freshwater Leeches. Freshw. Biol. Assoc. Sci. Publ. No. 40. Titus Wilson, Kendal. 72 pp.

Emmons, L. H. (1990) Neotropical Rainforest Mammals. - Univ. Chicago Press, Chicago. 281 pp.

Ernest, K. A. (1986) *Necromys squamipes*. - Mammalian Species 265:1-5.

Erwin, T. L., & J. Adis (1982) Amazonian inundation forests. Their role as short-term refuges and generators of species richness and taxon pulses. In: G. T. Prance (Ed.), Biological Diversification in the Tropics. - Columbia Univ. Press, New York. pp. 358-371.

Esteves, A. L., & J. A. Schnack (1980) Subfamilia Gelastocorinae. Revision de las especies de America del Sur (Hemiptera, Gelastocoridae). - Obra Centen. Mus. La Plata 6:75-89.

Esteves, F. (1988) Fundamentos da Limnologia. - Interciências, Rio de Janeiro. 585 pp.

Everall, N. C., N. A. A. Macfarlane, & R. W. Sedgwick (1989) The effects of water hardness upon the uptake, accumulation and excretion of zinc in the brown trout, *Salmo trutta* L. - J. Fish Biol. 35:881-892.

Ewel, K. C. (1991) Diversity in wetlands. - Evol. Trends in Plants 5(2):90-92.

Ewer, R. F. (1973) The Carnivores. - Weidenfeld & Nicolson, London. 494 pp.

Ezcurra de Drago, I., 1993: Distribucion geografica de las esponjas Argentinas (Porifera: Spongillidae, Potamolepidae y Metaniidae). Relaciones zoogeograficas, vias de poblamiento. In: A. Boltovskoy and H. L. López (Eds.), Conferencias de Limnologia. - Inst. Limnol. "Dr. R. A. Ringuelet", La Plata. pp. 115-125.

Fairchild, G. B. (1940) A note on the early stages of *Lepiselaga crassipes* Fabr. - Psyche 47:8-13.

Fairchild, T. R. (1978) Evidências paleontológicas de uma possível idade "Ediacarana" ou Cambriana inferior, para parte do grupo Corumbá (Mato Grosso do Sul). - Bol. Soc. Bras. Geol., XXX Congresso de Geologia. p. 181.

FAO (1994) Aquaculture Production 1986-1992. FAO Fisheries Circular 815:1-216.

Faria, S. M., G. P. Lewis, J. I. Sprent & J. M. Sutherland (1989) Occurrence of nodulation in the Leguminosae. - New Phytol. 111(607):607-619.

Fassett, N. C. (1953) A monograph of *Cabomba*. - Castanea 18:116-128.

Favre, T. C., T. H. P. Bogéa, L. Rotenberg, H. S. da Silva & O. S. Pieri (1995) Cercarial emergence of *Schistosoma mansoni* from *Biomphalaria glabrata* and *Biomphalaria straminea*. - Mem. Inst. Oswaldo Cruz 90(5):565-567.

Fearnside, P. M. (1990) Balbina: lições trágicas na Amazônia. - Ciência Hoje 11:34-40.

Fereira, C. J. A., B. M. A. Soriano, S. Galdino & S. K. Hamilton (1994) Anthropogenic factors affecting waters of the Pantanal wetland and associated rivers in the Upper Paraguay River Basin of Brazil. - Acta Limnol. Bras. 5:135-148.

Fernando, C. H. (1977) The ecology of the aquatic fauna of rice fields with special reference to South East Asia. - Geo-Eco-Trop. 1:169-188.

Fernando, C. H. (1980) The species and size composition of tropical freshwater zooplankton with special reference to the Oriental Region (South East Asia). - Int. Revue ges. Hydrobiol. 65:411-426.

Fernando, C. H. (1984) Freshwater invertebrates: some comments. In: C. H. Fernando (Ed.), Ecology and Biogeography in Sri Lanka. - Dr. W. Junk, The Hague. pp. 145-148.

Ferraz-de-Lima, J. A., G. Barbieri, & J. R. Verani (1984a) Periódo de reprodução, tamanho e idide de primeira maturação gonadal do pacú, *Colossoma mitrei,* em ambiente natural (Rio Cuiabá - Pantanal de Mato Grosso). - Ann. 3° Simp. Bras. Aquicult., pp. 477-497.

Ferraz-de-Lima, J. A., C. L. B. Ferraz-de-Lima, & G. Barbieri (1984b) Crescimento do pacú, *Colossoma mitrei* , em ambiente natural (Rio Cuiabá - Pantanal de Mato Grosso). - Ann. 3° Simp. Bras. Aquicult., pp. 499-521.

Ferreira, C. J. A., B. M. A. Soriano, S. Galdino, & S. K. Hamilton (1994) Anthropogenic factors affecting waters of the Pantanal wetland and associated rivers in the Upper Paraguay River Basin of Brazil. - Acta Limnol. Bras. 5:135-148.

Ferrell, C. S., & D. E. Wilson (1991) *Platyrrhinus helleri*. Mammalian Species 373:1-5.

Filgueiras, T. S. (1995) Flora dos Estados de Goiás e Tocantins Coleção Rizzo. Vol. 17. Gramineae (Poaceae). - Univ. Fed. Goiás, Goiânia. 143 pp.

Finlayson, C. M. (1993) Vegetation changes and biomass on an Australian monsoonal floodplain. In: B. Gopal, A. Hillbricht-Ilkowska & R. G. Wetzel (Ed.), Wetlands and Ecotones: Studies on Land-Water Interactions. Int. Sci. Publ., New Delhi. pp. 157-171.

Fitch, H., P. Goodrum & C. Newman (1952) The armadillo in the southeastern United States. J. Mamm. 33:21-37.

Fittkau, E. J. (1969) The Fauna of South America. - In: E. J. Fittkau, J. Illies, H. Klinge, G. H. Schwabe & H. Sioli (Eds.), Biogeography and Ecology in South America. Vol. 2. Dr. W. Junk, The Hague. pp. 624-658.

Fittkau, E. J. (1970) Role of caimans in the nutrient regime of mouth-lakes of Amazon affluents. Biotropica 2:138-142.

Fittkau, E. J. (1971) Ökologische Gliederung des Amazonasgebietes auf geochemischer Grundlage. - Münster. Forsch. Geol. Paläontol. 20/21:35-50.

Fittkau, E. J. (1973) Crocodiles and nutrient metabolism of Amazonian waters. Amazoniana 4:103-133.

Fittkau, E. J. (1974) Zur ökologischen Gliederung Amazoniens. I. Die erdgeschichtliche Entwicklung Amazoniens. - Amazoniana 5:77-134.

Fleming, T. H., E. T. Hooper & D. C. Wilson (1972) Three Central American bat communities: structure, reproductive cycles, and movement patterns. - Ecology 53: 555-569.

Foissner, W., & I. Foissner (1988) Stamm: Ciliophora. Catalogus Faunae Austriae, Part 1c. - Österr. Acad. Wiss., Vienna. 150 pp.

Fontenele, O. (1955) Contribuição para o conhecimento da biologia dos tucunarés (Actinopterygii, Cichlidae), em cativeiro. Aparelho de reprodução, hábitos de desouva e incubação. - Rev. Bras. Biol. 10:503-519.

Ford, L. S., & R. S. Hoffmann (1988) *Potos flavus*. - Mammalian Species 321:1-9.

Förster, K. (1969) Amazonische Desmidieen. 1 Teil: Areal Santarém. - Amazoniana 2:5-116 + 56 pl.

Förster, K. (1974) Amazonische Desmidieen. 2 Teil: Areal Maués - Abacaxis. - Amazoniana 5:135-242.

Fowler, H. W. (1948) Os Peixes de Agua Doce do Brasil. Vol. 6. Part 1. - Arq. Zool. 1-329.

Fowler, H. W. (1950) Os Peixes de Agua Doce do Brasil. Vol. 6. Part 2. - Arq. Zool. 330-625.

Franke, H. J. (1980) *Aphyocharax rathbuni*. Aquar. Terr. 27:22-23, 62-64.

Freude, H., K. W. Harde, & G. A. Lohse (1971) Die Käfer Mitteleuropas, 3. Adephaga 2: Palpicornia, Histeroidea. - Goecke & Evers, Krefeld. 365 pp.

Frey, R. (1995) Flora and vegetation of "Las Piedritas" and the margin of Laguna Caceres, Puerto Suarez, Bolivian Pantanal. - Bull. Torrey Bot. Club 122:314-319.

Friberg, L., & M. Piscator, G. F. Nordberg, & T. Kjellström (1974) Cadmium in the Environment. 2nd ed. CRC Press, Cleveland. 248 pp.

Frost, D. R. (1985) Amphibian Species of the World. - Assoc. Syst. Coll., Lawrence, Kans. 732 pp.

FUEM (Fundação Universidade Estadual de Maringá) (1987) Relatório anual do projeto "Ictiofauna e Biologia Pesqueira". Núcleo de Pesquisas em Limnologia, Ictiologia e Aquicultura. 2 Vols. 683 pp.

Fukuhara, H., & R. Henry (1987) Standing crop of zoobenthos in a small lake in northern Pantanal. In: Y. Saijo & J. G. Tundisi (Eds.), Limnological Studies in Rio Doce Valley Lakes and Pantanal Wetland, Brazil. - Water Res. Inst., Nagoya Univ., Nagoya. pp. 183-185.

Fukuhara, H., & O. Mitamura (1985) Standing crop of zoobenthos in the Pantanal. In: Y. Saijo & J. G. Tundisi (Eds.), Limnological Studies in Central Brazil. - Water Res. Inst., Nagoya Univ., Nagoya. pp. 197-201.

Furch, B., A. F. F. Corrêa, J. A. S. N. de Mello & K. -R. Otto (1985) Lichtklimadaten in drei aquatischen Ökosystemen verschiedener physikalisch-chemischer Beschaffenheit. 1. Abschwächung, Rückstreuung und vergleich zwischen Einstrahlung, Rückstrahlung und sphärisch gemessener Quantenstromdichte (PAR). - Amazoniana 9:411-430.

Furch, K. (1984a) Hydrogeochemie von Fließgewässern im Bereich der Transamazônica (Nordbrasilien). - Amazoniana 9:371-409.

Furch, K. (1984b) Water chemistry of the Amazon Basin: the distribution of chemical elements among freshwaters. In: H. Sioli (Ed.), The Amazon. - W. Junk, Dordrecht. pp. 167-199.

Galetti, P. M., Jr., K. E. Esteves, N. R. W. Lima, C. A. Mestriner, M. M. Cavallini, A. C. G. Cesar & C. S. Miyazawa (1990) Aspectos comparativos da ictiofauna de duas lagoas marginais do Rio Mogi-Guaçu (Alto Paraná - estação ecológica de Jitaí, SP). - Acta Limnol. Bras. 3:865-885.

Gallardo, J. M. (1962) A propósito de *Bufo variegtatus* (Günther), sapo del bosque húmedo Antartándico y otras especies de *Bufo* neotropicales. - Physis 23(64):93-102, 1 pl.

Gallardo, J. M. (1979) Composición, distribución y origen de la herpetofauna Chaqueña. - Monogr. Mus. Nat. Hist. Univ. Kans. 7:299-307.

Gambell, A. W., & D. W. Fisher (1964) Occurrence of sulfate and nitrate in rainfall. - J. Geophys. Res. 69:4203-4210.

Gannon, J. W., & R. S. Stemberger (1978) Zooplankton (esp. crustaceans and rotifers) as indicators of water quality. - Trans. Amer. Microsc. Soc. 97(1):16-35.

Gannon, M. R., M. R. Willig & J. K: Jones Jr. (1989) *Sturnira lilium*. - Mammalian Species 333:1-5.

Gans, C. (1967) A check list of recent Amphisbaenians (Amphisbaenia, Reptilia). - Bull. Amer. Mus. Nat. Hist. 135(2):61-105.

Gans, C. (1969) Amphisbaenians - reptiles specialized for a burrowing existence. - Endeavour 28:146-151.

Gans, C., & M. A. Amdur (1966) Rediscription of *Amphisbaena vermicularis* Wagler, with comments on its range and synonymy (Amphisbaenia: Reptilia). - Proc. Calif. Acad. Sci., 4th Ser. 33:69-90.

Gardner, A. L. (1977) Feeding habits. In: R. J. Baker, J. K. Jones Jr. & D. C. Carter (Eds.), Biology of Bats of the New World Family Phyllostomatidae. Part II. - Spec. Publ. Mus. Texas Tech. Univ. 13:293-350.

Gauthier-Lièvre, L., & R. Thomas (1958) Le genre *Difflugia*, *Pentagonia*, *Marghrebia*, et *Hoogenraadia* (Rhizopodes testacés) en Afrique. - Arch. Protistenkd., 103: 241-370.

George, D. B., S. G. Berk, V. D. Adams, R. S. Ting, R. O. Roberts, L. H. Park, & R. G. Lott (1995) Toxicity of alum sludge extracts to freshwater alga, protozoan, fish, and marine bacterium. - Arch. Environ. Contam. Toxicol. 29:149-158.

Géry, J. (1969) The fresh-water fishes of South America. In: E. J. Fittkau, J. Illies, H. Klinge, G. H. Schwabe & H. Sioli (Eds.), Biogeography and Ecology in South America. Vol. 2. - W. Junk, Dordrecht. pp. 828-848.

Géry, J. (1977) Characoids of the World. - TFH Publ., Neptune, N.J. 672 pp.

Géry, J., V. Mahnert & C. Dlouhy (1987) Poissons characoides non Characidae du Paraguay (Pesces, Ostariophysi). - Rev. Suisse Zool. 94(2):357-464.

Giani, A., & I. M. Leonardo (1988) Distribuição vertical de algas fitoplanctônicas no reservatório da Pampulha (Belo Horizonte, MG). - Acta Limnol. Brasil. 2:387-404.

Gibbs, R. J. (1967) The geochemistry of the Amazon River system. I. The factors that control the salinity and the composition and concentration of suspended solids. - Geol. Soc. Amer. Bull. 78:1203-1232.

Gils, J. van, & P. Wiersma (1996) Family Scolopacidae (sandpipers, snipes and phalaropes). In: J. del Hoyo, A. Elliott, and J. Sargatal (Eds.), Handbook of the Birds of the World. Vol. 3. - Lynx, Barcelona. pp. 488-499.

Ginn, H. (1973) Little, pygmy and elf owls. *Athene, Speotyto, Micrathene, Glaucidium, Aegolius, Pseudoscops*. In: J. A. Burton (Ed.), Owls of the World. - E. P. Dutton, New York. pp. 164-185.

Godoi Filho, J. D. (1986) Aspetos geológicos do Pantanal Matogrossense e de sua área de influèndia. In: Anais 1° Simpósio Recursos Naturais e Sócio-Econômico do Pantanal. EMBRAPA-CPAP, Corumbá. pp. 63-76.

Godoy, M. P. de (1965) Criação de peixes. - Est. Exp. Biol. Piscicult. Pirassununga, São Paulo. 26 pp.

Goldstein, R. J. (1973) Cichlids of the World. - TFH Publ., Hong Kong. 382 pp.

Gomes de Souza, L. (1986) Retrospectiva histórica do Pantanal. Anais 1° Simpósio Recursos Naturais e Sócio-Econômico do Pantanal. EMBRAPA-CPAP, Corumbá. pp. 199-203.

Gonzáles-Jiménes, E. (1978) Digestive physiology and feeding of capybara (*Hydrochoerus hydrochaerus*). In: M. Rechcigl (Ed.), Handbook Series in Nutrition and Food. - CRC Press, Cleveland. Sect. 9, 1:163-177.

Goodland, R., & M. G. Ferri, 1979: Ecologia de Cerrado. - Liv. Itatiaia Ed., Belo Horizonte. 193 pp.

Goodwin, G. G., & A. M. Greenhall (1961) A review of the bats of Trinidad and Tobago. - Bull. Amer. Mus. Nat. Hist. 122:191-301.

Gopal, B. (1986) Vegetation dynamics in temporary and shallow freshwater habitats. - Aquatic Botany 23:391-396.

Gopal, B. (1987) Water hyacinth. - Elsevier, Amsterdam. 471 pp.

Goulding, M. (1981) Man and Fisheries on an Amazon Frontier. - W. Junk, The Hague. 137 pp.

Goulding, M., M. L. Carvalho & E. G. Fereira (1988) Rio Negro, Rich Life in Poor Waters. SPB Acad. Publ., The Hague. 200 pp.

Grabert, H. (1964) Klimazeugen im Paläozoicum Brasiliens. - Geol. Rundschau 54:165-192.

Graça Salomão, M. da, S. M. A. Santos & G. Puorto (1995) Activity pattern of *Crotalus durissus* (Viperidae, Crotalinae): feeding, reproduction and snakebite. Stud. Neotrop. Fauna Environ. 30(2):101-106.

Grantsau, R. (1989) Os beija-flores do Brasil. 2nd ed. - Expressão e Cultura, Rio de Janeiro. 233 pp.

Green, J. (1970) Freshwater ecology in the Mato Grosso, Central Brazil. I. The conductivity of some natural waters. - J. Nat. Hist. 4:289-299.

Green, J. (1975) Freshwater ecology in the Mato Grosso, Central Brazil, IV: Associations of testate Rhizopoda. - J. Nat. Hist. 9:545-560.

Greene, H. W. (1988) Antipredator mechanisms in reptiles. - Biol. Reptilia 16:1-152.

Greenhall, A. M., G. Joermann, U. Schmidt & M. R. Seidel (1983) *Desmodus rotundus*. - Mammalian Species 202:1-6.

Greenhall, A. M., U. Schmidt & G. Joermann (1984) *Dyphylla ecaudata*. - Mammalian Species 227:1-3.

Grospietsch, T. (1965) Wechseltierchen (Rhizopoden). 2 ed. - Kosmos, Stuttgart. 87 pp.

Grospietsch, T. (1982) Der Thermalsee von Heviz (Westungarn) und seine Testaceen-Fauna. - Arch. Hydrobiol. 95:93-105.

Grzenda, A. R., & M. L. Brehmer (1960) A quantitative method for the collection and measurement of stream periphyton. Limnol. Oceanogr. 5:190-194.

Guarim Neto, G. (1984) Plantas do Brasil - Angiospermas do Estado de Mato Grosso. - Rodriguesia 36(59):105-121.

Guarim Neto, G. (1991) Plantas do Brasil - Angiospermas do Estado de Mato Grosso - Pantanal. Acta Bot. Bras. 5:25-47.

Guarim Neto, G. (1992) Biodiversidade do ecossistema pantaneiro: a vegetação do Pantanal. - An. Cong. Nac. Essências Nativas. pp. 106-110.

Guignot, F. (1961) Contribution à la connaissance des dytiscides et gyriniens sud-américaines (3e série). - Bull. Mus. Natl. Hist.
Nat. 33:311-317.

Haas, F. (1969) Superfamilie Unionacea. In: Das Tierreich 88: 1-663. - Walter de Gruyter, Berlin.

Halwart, M. (1994) The golden apple snail *Pomacea canaliculata* in Asian rice farming systems: present impact and future threat. - Int. J. Pest Managem. 40:199-206.

Hamilton, S. K., S. J. Sippel & J. M. Melack (1995) Oxygen depletion and carbon dioxide and methane production in waters of the Pantanal wetland of Brazil. - Biogeochemistry 30:115-141.

Hamilton, S. K., S. J. Sippel & J. M. Melack (1996) Inundation patterns in the Pantanal wetland of South America determined from passive microwave remote sensing. - Arch. Hydrobiol. 137:1-23.

Hancock, G. (1989) Lords of Poverty. - Macmillan, London. 234 pp.

Hancock J., & J. Kushlan, 1984: The Herons Handbook. - Harper & Row, New York. 288 pp.

Handley, C. O. (1976) Mammals of Smithsonian Venezuelan Project. - Brigham Young Univ. Sci. Bull., Biol. Ser., 10:1-91.

Hardoim, E. L., & C. W. Heckman (1996) The seasonal succession of biotic communities in wetlands of the tropical wet-and-dry climatic zone: IV. The free-living sarcodines and ciliates of the Pantanal of Mato Grosso, Brazil. - Int. Revue ges. Hydrobiol. 81:367-384.

Hardy, E. R., B. Robertson & W. Koste (1984) About the relationship between the zooplankton and fluctuating water level of Lago Camaleão, a Central Amazonian varzea lake. - Amazoniana 9:43-52.

Häusler, J. (1982) Schizomycetes. In: H. Ettl, J. Gerloff & H. Heynig (Eds.), Süßwasserflora von Mitteleuropa. Vol. 20. - Gustav Fischer, Stuttgart. 588 pp.

Haynes, R. R., & L. B. Holm-Nielsen (1994) The Alismataceae. - Flora Neotropica 64:1-112.

Heckman, C. W. (1974) Seasonal succession of species in a rice paddy in Vientiane, Laos. - Int. Rev. ges. Hydrobiol. 59:489-507.

Heckman, C. W. (1979) Rice Field Ecology in Northeastern Thailand. - W. Junk, The Hague. 228 pp.

Heckman, C. W. (1981) Long-term effects of intensive pesticide applications on the aquatic community in orchard drainage ditches near Hamburg, Germany. - Arch. Environ. Contam. Toxicol. 10:393-426.

Heckman, C. W. (1982a) Ecophysiological and phylogenetic characterization of a wintertime biotic community in shallow freshwater habitats near Hamburg. - Int. Rev. ges. Hydrobiol. 67:361-386.

Heckman, C. W. (1982b) Ecological functions of aquatic tracheophytes in lentic ecosystems. A comparative study of artificial urban water bodies in Hamburg. - Int. Revue ges. Hydrobiol. 67:187-207.

Heckman, C. W. (1983a) Comparative morphology of arthropod exterior surfaces with the capability of binding a film of air underwater. - Int. Revue ges. Hydrobiol. 68:715-736.

Heckman, C. W. (1983b) Reactions of aquatic systems to pesticides. In: J. O. Nriagu (Ed.): *Aquatic Toxicology*. Vol. 13 in the series Advances in Environmental Science and Technology. - John Wiley & Sons, New York. pp. 355-400.

Heckman, C. W. (1984) The ecological importance of wetlands along streams and rivers and the consequences of their elimination. - Int. J. Ecol. Environ. Sci. 10:11-29.

Heckman, C. W. (1985) The development of vertical migrations patterns in the sediments of estuaries as a strategy for the algae to resist drift with tidal currents. - Int. Revue ges. Hydrobiol. 70:151-164.

Heckman, C. W. (1986) Tidal influence on the wetland community structure behind the dike along the Elbe Estuary. - Arch. Hydrobiol./Suppl. 75 (Untersuch. Elbe Aestuar 6):1-117.

Heckman, C. W. (1989) The use of electronics in mapping the distribution and abundance of flora and fauna. In: Proceedings of the Regional Seminar on Methods of Biological Inventory and Cartography for Ecosystem Management (BICEM), Tokyo, Oct. 9-14 (1989. UNESCO/MAB, Tokyo. pp. 140-161.

Heckman, C. W. (1990) The fate of aquatic and wetland habitats in an industrially contaminated section of the Elbe Floodplain in Hamburg. - Arch. Hydrobiol./Suppl. 75 (Untersuch. Elbe-Aestuar 6):133-245.

Heckman, C. W. (1991) Necesidad de una caracterización limnologica nueva para el "Pantanal", Mato Grosso, Brasil. - Biologia Acuatica 15(1):58-59.

Heckman, C. W. (1993) The fate of pesticides in heavily polluted aquatic systems. In: M. M. Tilzer & M. Khondker (Eds.) Hypertrophic and Polluted Freshwater Ecosystems: Ecological Bases for Water Resource Management. - Dept. Botany, Univ. of Dhaka, Dhaka, Bengladesh. pp. 39-90.

Heckman, C. W. (1994a) The seasonal succession of biotic communities in wetlands of the tropical wet-and-dry climatic zone: I. Physical and chemical causes and biological effects in the Pantanal of Mato Grosso, Brazil. - Int. Revue ges. Hydrobiol. 79:397-421.

Heckman, C. W. (1994b) New limnological nomenclature to describe ecosystem structure in the tropical wet-and-dry climatic zone. - Arch. Hydrobiol. 130:385-407.

Heckman, C. W. (1995) The chemistry of headwater streams in the Rio das Mortes System and its effect on the structure of the biotic community. In: P. R. Seidl, O. R. Gottlieb & M. A. C. Kaplan (Eds.), Chemistry of the Amazon. - Amer. Chem. Soc., Washington. pp. 248-264.

Heckman, C. W. (1997a) Ecoclimatological survey of the wetland biota in the tropical wet-and-dry climatic zone. Global Ecol. Biogeogr. Letters 6::97-114.

Heckman, C. W. (1997b) Description of a dynamic ecotone in the Pantanal of Mato Grosso, Brazil. - Ecotropica 3:1-7.

Heckman, C. W. (1998) The Seasonal Succession of Biotic Communities in Wetlands of the Tropical Wet-and-Dry Climatic Zone: V. Climatological influences on the life cycles of the invertebrates in the Pantanal of Mato Grosso, Brazil. - Int. Rev. ges. Hydrobiol. 83:31-63.

Heckman, C. W., J. L. E. dos Campos & E. L. Hardoim (1997) Nitrite concentration in well water from Poconé, Mato Grosso, and its relationship to public health in rural Brazil. - Bull. Environ. Contam. Toxicol. 58:8-15.

Heckman, C.W., E. L. Hardoim, S. A. Ferreira & A. U. Kretzschmar (1993) Preliminary observations on some cosmpolitan algae in ephemeral water bodies of the Pantanal, Mato Grosso, Brazil. In: B. Gopal, A. Hillbricht-Ilkowska & R. G. Wetzel (Eds.), Wetlands and Ecotones. - Internat. Sci. Publ., New Delhi. pp. 279-292.

Heckman, C. W., & R. Schade (1989) Biotische Verhältnisse im Harburger Binnenhafen. In: Umweltbehörde Hamburg (Ed.), Der Hafen. Eine Ökologische Herausfforderung. pp. 207-212.

Heckman, C.W., B. R. S. Trindade & E. L. Hardoim (1996) Environmental conditions in the Pantanal of Mato Grosso conducive to natural euglenophyte populations. - Japan. J. Limnol. 58:21-34.

Hegewald, E. (1984) Hochtemperaturstämme in der Gattung *Scenedesmus* Sektion *Desmodesmus*. - Arch. Hydrobiol./Suppl. 67 (Algological Studies 37):441-446.

Hekstra, G. P. (1973) Scops and screech owls. *Otus, Lophostrix*. In: J. A. Burton (Ed.), Owls of the World. - E. P. Dutton, New York. pp. 94-115.

Herd, R. M. (1983) *Pteronotus parnellii*. - Mammalian Species 209:1-5.

Heyer, W. R., A. S. Rand, C. A. G. da Cruz, O. L. Peixoto & C. E. Nelson (1990) Frogs of Boracéia. - Arq. Zool., São Paulo 31(4):231-410.

Hilty, S. L., & W. L. Brown (1986) Birds of Colombia. Princeton Univ. Press, Princeton, N.J. 836 pp.

Hirn, K. (1900) Monographie und Iconographie der Oedogoniaceen. - Acta Soc. Sci. Fenn. 27(1):1-394.

Hoehne, F. C. (1923) Phytophysionomia do Estado do Mato Grosso. - Melhoramentos, São Paulo 1923:53-58.

Hoehne, F. C. (1936) O grande Pantanal de Mato Grosso. - Bol. Agric., São Paulo:443-447.

Hoge, A. R., & S. A. R. W. L. Romano-Hoge (1981) Sinopse das serpentes poçonhentas do Brasil. 2nd ed. Mem. Inst. Butantan 42/43:373-496.

Holčík, J., R. Kinzelbach, L. I. Sokolov & V. P. Vasil'ev (1989) *Acipenser sturio* Linnaeus, 1758. In: J. Holcík (Ed.), The Freshwater Fishes of Europe, Vol. 1, Part II. - Aula, Wiesbaden. pp. 367-394.

Höll, K. (1968) Wasser. 4th ed. - Walter de Gruyter, Berlin. 393 pp.

Holst, R. W., & J. H. Yopp (1979) Studies of the *Azolla-Anabaena* symbiosis using *Azolla mexicana*. I. Growth in nature and laboratory. - Amer. Fern J. 69:17-25.

Holthuis, L. B. (1952) A general revision of the Palaemonidae (Crustacea, Decapoda, Natania) of the Americas. II. The subfamily Palaemoninae. Occ. Pap. Allan Hancock Foundation, Publ. 12:1-396.

Holthuis, L. B. (1980) Shrimps and prawns of the world. - FAO Species Catalogue. FAO Fisheries Synopsis No. 125, Vol. 1. 271 pp.

Homewood, B. (1991) Dirty cities head Brazil's environmental ills. - New Scientist 7(1785):21.

Hood, C. S., & J. K. Jones Jr. (1984) *Noctilio leporinus*. - Mammalian Species 216:1-7.

Hood, C. S., & J. Pitocchelli (1983) *Noctilio albiventris*. - Mammalian Species 197:1-5.

Hubendick, B. (1964) Studies on Ancylidae. The subgroups. - Medd. Goteb. Mus. Zool. Avd. 137:3-72.

Huber-Pestalozzi, G. (1955) Das Phytoplankton des Süßwassers. 4. Euglenophyceen. - Schweizerbart, Stuttgart. 606 pp. + 114 pl.

Huber-Pestalozzi, G. (1961) Das Phytoplankton des Süßwassers. 5. Ordnung Volvocales. - Schweizerbart, Stuttgart. 744 pp. + 158 pl.

Huber-Pestalozzi, G. (1968) Das Phytoplankton des Süßwassers. 3. Cryptophyceae, Chloromonadophyceae, Dinophyceae. 2nd ed. - Schweizerbart, Stuttgart. 322 pp.

Hueck, K. (1966) Die Wälder Südamerikas. - Gustav Fischer, Stuttgart. 422 pp.

Hungerford, H. B. (1948) The Corixidae of the western hemisphere. - Univ. Kans. Sci. Bull. 32:1-827 + 11 pl.

Hynes, H. B. N. (1970) The Ecology of Running Water. - Liverpool Univ. Press. 555 pp.

Illies, J. (1965) Die wegenersche Kontinentalverschiebundstheorie im Lichte der modernen Biogeographie. - Naturwissenschaften 52:505-511.

Illies, J. (1966) Neue Theorien der Kontinentalverschiebung. - Zool. Anz. 176:46-50.

Illies, J. (1969) Biogeography and ecology of Neotropical freshwater insects, especially those from running waters. In: E. J. Fittkau, J. Illies, H. Klinge, G. H. Schwabe & H. Sioli (Eds.), Biogeography and Ecology in South America. Vol. 2. - W. Junk, Dordrecht. pp. 685-708.

Irwin, H. S., & R. C. Barneby (1981) Tribe 2. Cassieae Bronn (1822). In: R. M. Polhill & P. H. Raven (Eds.): Advances in Legume Systematics. Part 1. - Roy. Bot. Gardens, Kew. pp. 97-106.

Irwin, H. S., & R. C. Barneby (1982) The American Cassiinae. Parts 1 and 2. - Mem. N. Y. Bot. Soc. 35:1-918.

Jaeckel, S. G. A., Jr. (1969) Die Mollusken Südamerikas. In: E. J. Fittkau, J. Illies, H. Klinge, G. H. Schwabe & H. Sioli (Eds.), Biogeography and Ecology in South America. Vol. 2. - W. Junk, Dordrecht. pp. 794-827.

Jedicke, A., B. Furch, U. Saint-Paul & U. -B. Schlüter (1989) Increase in the oxygen concentration in Amazon waters resulting from the root exudation of two notorious water plants, *Eichhornia crassipes* (Pontederiaceae) and *Pistia stratiotes* (Araceae). - Amazoniana 11(1):53-69.

Jeffree, R. A., S. J. Markich, & P. L. Brown (1995) Australian freshwater bivalves: their applications in metal pollution studies. - Australas. J. Ecotoxicol. 1:33-41.

Jenni, D. A. (1996) Family Jacanidae (jacanas). In: J. del Hoyo, A. Elliott, and J. Sargatal (Eds.), Handbook of the Birds of the World. Vol. 3. - Lynx, Barcelona. pp. 276-291.

Johnson, C. J. (1990) Continuing importance of nitrate contamination of groundwater and wells in rural areas. - Agricultural Occupational and Environmental Health Policy Strategies for the Future Scientific Basis. Part 3, Vol. 18(4):449-456.

Johnson, D. S. (1967) On the chemistry of freshwaters in southern Malaya and Singapore. - Arch. Hydrobiol. 63:477-496.

Joly, A. B. (1980) Conheça a vegetação brasiliera. - USP, São Paulo. 165 pp.

Jones, C., & S. Anderson (1978) *Callicebus moloch*. - Mammalian Species 112:1-5.

Jones, J. K., Jr., & J. Arroyo-Cabrales (1990) *Nyctinomops aurispinosus*. - Mammalian Species 350:1-3.

Jones, M. J., & A. R. Bromfield (1970) Nitrogen in the rainfall at Samaru, Nigeria. - Nature (London) 227:86.

Judziewicz, E. J. (1990) Poaceae (Gramineae). In: A. R. A. Görts-van Rijn (Ed.), Flora of the Guianas. Ser. A, Fasc. 8. 727 pp.

Julien, M. H. (1987) Biological control of weeds: a world catalogue of agents and their target weeds. 2nd ed. - CAB International, Wallingford. 144 pp.

Kadlubowska, J. Z. (1984) Conjugatophyceae I. Chlorophyta VIII. Zygnemales. In: H. Ettl, J. Gerloff & H. Heynig (Eds.), Süßwasserflora von Mitteleuropa. Vol. 16. Gustav Fischer, Stuttgart. 532 pp.

Kahl, A. (1935) Urtiere oder Protozoa I. Wimpertiere oder Ciliata (Infusoria). In: M. Dahl & H. Bischoff (Eds.), Die Tierwelt Deutschlands 18, 21, 25, and 30. Gustav Fischer, Jena. 886pp.

Kato, T., M. F. Watanabe & M. Watanabe (1991) Allozyme divergence in Microcystis (Cyanophyceae) and its taxonomic interference. - Arch. Hydrobiol./Suppl. 92 (Algological Studies 64):129-140.

Kaufmann, J., & A. Kaufmann (1965) Observations on the behavior of tayras and grisons. - Z. Säugertierk. 30:146-155.

Kaul, V., & U. Bakaya (1976) The noxious, floating, lemnid-*Salvinia* weed complex in Kashmir. - In: C. K. Varshney and J. Rzóska (Eds.), Aquatic Weeds in S. E. Asia. - Dr. W. Junk, The Hague. pp. 183-192.

Kehr, A. I., & N. Basso (1992) Consideraciones sobre la estructura de edades, supervivencia y tiempo de metamorfosis en los estados larvales de *Hyla pulchella pulchella* (Anura: Hylidae). - Acta Zool. Lilloana 41:125-133.

Kensley, B., & Walker, I. (1982) Palaemonid shrimps from the Amazon Basin, Brazil (Crustacea: Decapoda: Natantia). - Smithsonian Cont. Zool. 362. 28 pp.

Kevern, N. R., J. L. Wilhm & G. M. van Dyne (1966) Use of artificial substrates to estimate the productivity of periphyton. - Limnol. Oceanogr. 11:499-502.

Killeen, T. J., & P. N. Hinz (1992) Grasses of the Precambrian Shield region of eastern lowland Bolivia: I. Habitat preferences. - J. Trop. Ecol. 8(4):389-407.

Klammer, G. (1982) Die Paläowüste des Pantanal von Mato Grosso und die pleistozäne Klimageschichte der brasilianischen Randtropen. - Z. Geomorph. N.F. 26:393-416.

Klammer, G. (1984) The relief of the extra-Andean Amazon basin. In: H. Sioli (Ed.), The Amazon. - W. Junk, Dordrecht. pp. 47-83.

Knöppel, H. -A. (1972) Zur Nahrung tropischer Süßwasserfische aus Südamerika. - Amazoniana 3(2):231-246.

Komárek, J., & A. Comas González (1982) Taxonomical definition of the genera and several species of *Ankistrodesmus* and *Selenastrum* (Chlorococcales). - Arch. Hydrobiol./Suppl. 63 (Algological Studies 32):259-277.

Komárek, J., & B. Fott (1983) Das Phytoplankton des Süßwassers. 7. First Half. Chlorophyceae (Grünalgen), Ordnung Chlorococcales. - Schweizerbart, Stuttgart. 1044 pp.

Komárková-Legnerová, J., & G. Cronberg (1994) Planktic blue-green algae from lakes in South Scania, Sweden. Part 1. Croococcales. - Arch. Hydrobiol./Suppl. 102 (Algological Studies 72):13-51.

Koste, W. (1972) Rotatorien aus Gewässern Amazoniens. - Amazoniana 3:258-505.

Koste, W. (1974) Zur Kenntnis der Rotatorienfauna der "schwimmenden Wiese" einer Uferlagune in der Várzea Amazoniens, Brasilien. - Amazoniana 5:25-59.

Koste, W. (1975) Uber den Rotatorienbestand einer Mikrobiozonose in einem tropischen aquatischen Saumbiotop, der *Eichhornia-crassipes*-Zone im Litoral des Bung Borapet, einem Stausee im Zentralthailand. - Gewässer und Abwässer 57/58:43-58.

Koste, W. (1978) Rotatoria. 2 Vols. - Borntraeger, Berlin and Stuttgart. 673 pp. + 234 pl.

Koste, W., & K. Böttger (1992)Rotatorien aus Gewässern Ecuadors II. - Amazoniana 12:263-303.

Koste, W., & B. Robertson (1983) Taxonomic studies on the Rotifera (Phylum Aschelminthes) from a Central Amazonian varzea lake, Lago Camaleão (Ilha de Marchanteria, Rio Solimões, Amazonas, Brazil). - Amazoniana 8:225-254.

Koste, W., B. Robertson & E. Hardy (1984) Further taxonomical studies on the Rotifera from Lago Camaleão, a Central Amazonian varzea lake (Ilha de Marchanteria, Rio Solimões, Amazonas, Brazil). - Amazoniana 8:555-576.

Kramer, K. U., & P. S. Green, Eds. (1990) Pteridophytes and Gymnosperms. Vol. 1. The Families and Genera of Vascular Plants. - Springer, Berlin. 404 pp.

Krammer, K., & H. Lange-Bertalot (1986) Bacillariophyceae. 1. Naviculaceae. In: H. Ettl, J. Gerloff & H. Heynig (Eds.), Süßwasserflora von Mitteleuropa. Vol. 2, Part 1. Gustav Fischer, Stuttgart. 876 pp.

Krammer, K., & H. Lange-Bertalot (1988) Bacillariophyceae. 2. Bacillariaceae, Epithemiaceae, Surirellaceae. In: H. Ettl, J. Gerloff & H. Heynig (Eds.), Süßwasserflora von Mitteleuropa. Vol. 2, Part 2. Gustav Fischer, Stuttgart. 596 pp.

Krammer, K., & H. Lange-Bertalot (1991a) Bacillariophyceae. 3. Centrales, Fragilariaceae, Eunotiaceae. In: H. Ettl, J. Gerloff, H. Heynig & D. Mollenhauer (Eds.), Süßwasserflora von Mitteleuropa. Vol. 2, Part 3. Gustav Fischer, Stuttgart. 576 pp.

Krammer, K., & H. Lange-Bertalot (1991b) Bacillariophyceae. 4. Achnanthaceae. Kritische Ergänzungen zu *Navicula* (Lanceolatae) and *Gomphonema*. In: H. Ettl, G. Gärtner & J. Gerloff (Eds.), Süßwasserflora von Mitteleuropa. Vol. 2, Part 4. Gustav Fischer, Stuttgart. 437 pp.

Kretzschmar, A. U., S. A. Ferreira, E. L. Hardoim & C. W. Heckman (1993) Peak growth of the *Asplanchna sieboldi* (Leydig, 1854) rotifer aggregation in relation to the seasonal wet and dry cycle in the Pantanal, Mato Grosso, Brazil. In: B. Gopal, A. Hillbricht-Ilkowska & R. G. Wetzel (Eds.), Wetlands and Ecotones. - Internat. Sci. Publ., New Delhi. pp. 293-301.

Kretzschmar, A. U., & C. W. Heckman (1995) Estratégias de sobrevivência das espécies de Ampullariidae durante mudanças das condições ambientais extremas do ciclo sazonal sob o clima tropical úmido-e-seco. - Acta Limnol. Brasil. 7:60-66.

Krieg, H. (1936) Tiergeographische Wirkungen der jährlichen Überschwemmungen im Stromgebiete des Paraguay-Paraná. - Isis 1934-1936:5-7.

Krieger, W. (1937) Die Desmidiaceen. Rabenhorst's Kryptogamenflora von Deutschland, Österreich und der Schweiz. Vol. 13. Part 1. - Akad. Verlagsges., Leipzig. 712 pp.

Krieger, W. (1939) Die Desmidiaceen. Rabenhorst's Kryptogamenflora von Deutschland, Österreich und der Schweiz. Vol. 13. Part 2. - Akad. Verlagsges., Leipzig. 117 pp.

Kudo, R. R. (1966) Protozoology. 5th ed. Charles C. Thomas, Springfield, Ill. 1174 pp.

Kuhr, R. J., A. C. Davis, & J. B. Bourke (1974) DDT residues in soil, water, and fauna from New York apple orchards. - Pesticide Monit. J. 7:200-203.

Lacerda, L. D., W. C. Pfeifer, O. Malm, C. M. M. Souza, E. G. Silveira, & W. R. Bastos (1990) Mercúrio no meio ambiente: risco potential das áreas garimpeiras no Brasil. - Acta Limnol. Bras. 3:969-977.

Lacerda, L. D., W. C. Pfeifer, R. V. Marins, S. Rodrigues, C. M. M. Souza, & W. R. Bastos (1990) Mercury dispersal in water, sediments and aquatic biota of a gold mining tailing deposit drainage in Poconé, Brazil. - Water Air Soil Pollut. 55:283-294.

Lacerda, L. D., W. Salomons, W. C. Pfeiffer, & W. R. Bastos (1991) Mercury distribution in sediment profiles from lakes in the high Pantanal, Mato Grosso State, Brazil. Biogeochemistry 14(2):91-98.

Lajmanovich, R. C. (1994) Contribucíon al conocimiento de la alamentacíon de larvas de la rana criolla *Leptodactylus ocellatus* (Amphibia, Leptodactylidae) en el Paraná medio. - Stud. Neotrop. Fauna Environ. 29:55-61.

Lallana, V. H. (1980) Productividad de *Eichhornia crassipes* (Mart.) Solms in una laguna isleña de la cuenca del río Paraná medio. II: Biomasa y dimánica de población. - Ecología 5:1-16.

Landolt, E. (1986) Biosystematic investigations in the family of duckweeds (Lemnaceae). Vol. 2. - Veröff. Geobot. Inst. Eidg. Techn. Hochschule, Stift. Rübel No. 71:1.566.

Landsberg, H. E.; H. Lippmann; K. H. Paffen & C. Troll (Eds.) (1965) Weltkarten zur Klimakunde. - Springer, Berlin. 28 pp. + Maps.

Lane, J. (1953) Neotropical Culicidae. 2 Vols. Univ. São Paulo, São Paulo. 1112 pp.

Lang, J. W. (1989) Sex determination. In: C. A. Ross (Ed.), Crocodiles and Alligators. - Golden Press, Silverwater, NSW. pp. 120-121.

Lansac Tôha, F. A., A. F. Lima, S. M. Thomaz, & M. C. Roberto (1993) Zooplâncton de uma planície de inundação do Rio Paraná. II. Variação sazonal e influência dos níveis fluviométricos sobre a comunidade. - Acta Limnol. Bras. 6:42-55.

Lanzer-de Souza, M. E. (1991) Nova espécie do gênero *Lethocerus* Mayr, 1853 (Heteroptera, Belostomatidae, Lethocerinae). - Iheringia 71:139-143.

Lassier, S., & D. E. Wilson (1989) *Lonchorhina aurita*. - Mammalian Species 347:1-4.

Laubmann, A. (1939) Die Vögel von Paraguay. Vol. 1 - Strecker und Schröder, Stuttgart. 246 pp.

Laubmann, A. (1940) Die Vögel von Paraguay. Vol. 2 - Strecker und Schröder, Stuttgart. 229 pp.

LaVal, R. K., & M. L. LaVal (1980) Prey selection by a Neotropical foliage-gleaning bat, *Micronycteris megalotis*. - J. Mamm. 61:327-330.

Lelek, A. (1987) The Freshwater Fishes of Europe. Vol. 9, Threatened Fishes of Europe. - Aula, Wiesbaden. 343 pp.

Lellinger, D. B. (1989) The ferns and fern-allies of Costa Rica, Panama, and the Chaco (Part 1: Psilotaceae through Dicksoniaceae). - Amer. Fern Soc., Washington, D.C. 364 pp.

Levi, H. W. (1981) The American orb-weaver genera *Dolichognatha* and *Tetragnatha* north of Mexico (Araneae: Araneidae, Tetragnathinae). - Bull. Mus. Comp. Zool. 149:271-318.

Levi, H. W. (1995) Orb-weaving spiders *Actinosoma, Spilasma, Micrepteira, Pronous*, and four new genera. - Bull. Mus. Comp. Zool. 154:153-213.

Lewis, G. P. (1987) Legumes of Bahia. - Royal Bot. Gard., Kew. 369 pp.

Lichtenstein, J. S. de (1939) Las especies argentinas del género *Salvinia*. - Darwiniana 3(2):93-110.

Lima, A. F., F. A. Lansac-Tôha, & C. C. Bonecker (1996) Zooplankton in the floodplains of a tributary to the Paraná River in Mato Grosso do Sul, Brazil. Stud. Neotrop. Fauna & Environm. 31:112-116.

Lima, J. A. F. de (1987) A pesca no Pantanal de Mato Grosso (Rio Cuiabá: importância dos peixes migradores). - Acta Amazonica 16/17:87-94.

Link, D., & M. E. L. de Souza (1983) Captura de Belostomatidae (Hemiptera, Heteroptera) com armadilha luminosa. 1. *Lethocerus annulipes* (Herrich-Schaffer, 1846). - Iheringia Ser. Zool. 62:121-131.

Loczy, L. de (1969) Tectônismo transversal na constituição de America do Sul e a importância econômica das falhas transcorrentes. - Mineração Metalurgia 50:273-280.

Loczy, L. de (1973) Some problems of the tectonic framework of the Guiana Shield with special regard for the Roraima Formation. - Geol. Rdsch. 62:318-342.

Löffler, H. (1973) Tropical high mountain lakes of New Guinea and their zoogeographical relationship compared with other tropical high mountain lakes. - Arctic and Alpine Res. 5, Part 2:A193-A198.

Longfield, C. (1929) A list of Odonata of the State of Matto Grosso, Brazil. - Trans. Ent. Soc. Lond. 77:125-139 + pl. 12.

Lopes, R. M. (1993) Zooplankton spatial and seasonal distribution in the Tibagi River (Paraná State, Brazil). - Semina 14:95-102.

Lopez-Ruf, M. L. (1991) The genus *Pelocoris* in Argentina (Heteroptera-Limnocoridae). 1. Introduction, diagnosis, key to species and redescription of *P. (P.) binotulatus nigriculus* Berg, *P. (P.) binotulatus binotulatus* and *P. (P.) impicticollis*. - Physis B:49(115/117): 13-22.

Lopretto, E. C. (1995) Crustaceos. In: E. C. Lopretto & G. Tell (Eds.), Ecosistemas de Aguas continentales. Vol. 3.

Lord, R. D. (1994) A descriptive account of capybara behaviour. - Stud. Neotrop. Fauna Environ. 29:11-22.

Lorenzi, H. (1991) Plantas daninhas do Brasil. 2nd ed. - Ed. Fransciscana, Piracicaba, São Paulo. 425 pp.

Lorenzi, H. (1992) Arvores Brazileiras. - Ed. Plantanum, Nova Odessa, SP. 368 pp.

Loureiro, M. F., E. K. James, J. I. Sprent, & A. A. Franco (1995) Stem and root nodules on the tropical wetland legume *Aeschynomene fluminensis*. - New Phytol. 130:531-544.

Lowe-McConnell, R. H. (1987) Ecological Studies in Tropical Fish Communities. - Cambridge Univ. Press, Cambridge. 382 pp.

Lowe-McConnell, R. H. (1991) Natural history of fishes in Araguaia and Xingu Amazonian tributaries, Serra do Roncador, Mato Grosso, Brazil. - Ichthyol. Explor. Freshwaters 2:63-82.

Ludwig, G. M. (1994) Tank culture of sunshine bass *Morone chrysops* x *M. saxatilis* fry with freshwater rotifers *Brachyonus calyciflorus* and salmon starter meal as first food sources. - J. World Aquacult. Soc. 25:337-347.

Lutz, B. (1946) A notable frog chorus in Brazil. - Copeia 1946(3):153-155.

Lutz, B. (1972) Geographical and ecological notes on Cisandine to Platine frogs. - J. Herpetol. 6(2):83-100.

Lutz, B., & G. A. Lutz (1973) Brazilian species of *Hyla*. - Univ. Texas, Austin. 260 pp + Appendix.

Lynch, J. D. (1970) Systematic status of the American Leptodactylid frog genera *Engystomops*, *Eupemphix*, and *Physalaemus*. - Copeia 1970:488-496.

Lynch, J. D. (1971) Evolutionary Relationships, Osteology, and Zoogeography of Leptodactyloid Frogs. - Univ. Kans. Mus. Nat. Hist., Misc. Publ. 53. 238 pp.

MacDonald, D. W. (1981) Dwindling recources and the social behaviour of capybaras (*Hydrochoreus hydrochaeris*) (Mammalia). - J. Zool. London 194:371-391.

Machado, A. (1980) *Helveciagrion* N. G., com descrição de uma nova espécie do Parque Estadual do Rio Doce, Minas Gerais (Odonata Coenagrionidae). - Lundiana 1:59-87.

Maddox, D. M. (1970) The bionomics of a stem borer, *Vogtia malloi* (Lepidoptera: Phycitidae) on alligatorweed in Argentina. - Ann. Ent. Soc. Amer. 63(5):1267-1273.

Magalhães, A. C. de (1931) Monographia Brazileira de Peixes Fluviaes. - Graphicars, São. 260 pp.

Magalhães, N. W. de (1992) Conheça o Pantanal. - Terragraph, São Paulo. 390 pp.

Magnusson, W. E., K. A. Vliet, A. C. Pooley, & R. Whitaker (1989) Reproduction. In: C. A. Ross (Ed.), Crocodiles and Alligators. - Golden Press, Silverwater, NSW. pp. 118-119, 121, 123-135.

Mago, F. M. L. (1970) Estudios preliminares sobre la ecologia de los peces de los llanos de Venezuela. - Acta Biol. Venez. 7:71-102.

Maldonado, M., & E. Goitia (1993) Primer aporte al conocimiento de las poblaciones de bivalvos del arroyo aguas negras en la Estación Biológica Beni (Beni-Bolivia). - Acta Limnol. Bras. 6:106-113.

Mansur, M. C. D., & R. M. Valer (1992) Bivalve molluscs from Rio Uraricoera and Rio Branco, Roraima, Brasil. - Amazoniana 12(1):85-100.

Mara, D. (1976) Sewage Treatment in Hot Climates. - John Wiley, London. 168 pp.

Marcus, E. (1945) Sobre microturbelários do Brasil. - Com. Zool. Mus. Hist. Nat. Montevideo 25:1-74 + 11 pl.

Margalhães, C. (1985) Desenvolvimento larval obtido en laboratório de palaemonídeos da Região Amazônica. I. *Macrobrachium amazonicum* (Heller, 1862) (Crustacea, Decapoda). - Amazoniana 9:247-274.

Margalhães, C., & M. Türkay (1996) Taxonomy of the neotropical freshwater crab family Trichodactylidae, I. The generic system with description of some new genera (Crustacea: Decapoda: Brachyura). - Senckenberg biol. 75:63-95.

Marglianesi, R. (1973) Principales caracteristicas químicas y físicas de las águas del Alto Paraná y Paraguay inferior. Phisis 32(85):185-197.

Marins, R. V., P. N. Conceição & J. A. F. Lima (1981) Estudos ecológicos das principais espécies de peixes de interesse comercial, esportivo e ornamental da Bacia do Alto Paraguai. - SEMA, Brasilia.

Markich, S. J., & R. A. Jeffree (1994) Absorption of divalent trace metals as analogues of calcium by Australian freshwater bivalves: an explanation of how water hardness reduces metal toxicity. - Aquat. Toxicol. 29:257-290.

Marshall, L. G. (1978) *Chironectes minimus*. Mammalian Species 109:1-6.

Mathews, A. G. A. (1977) Studies on termites from the Mato Grosso State, Brazil. - Acad. Bras. Ciênc., Rio de Janeiro. 267 pp.

Matsumura-Tundisi, T. (1986) Latitudinal distribution of Calanoida copepods in freshwater aquatic systems of Brazil. - Rev. Bras. Biol. 46(3):527-553.

Mayer, J. J., & R. M. Wetzel (1987) *Tayassu pecari*. - Mammalian Species 293:1-7.

Mazzeo, N., D. Crosa, & R. Sommaruga (1993) Productividad y variación estacional de la biomasa de *Pistia stratiotes* L. en le reservorio del cisne, Uruguay. - Acto Limnol. Bras. 6:186-195.

McBee, K., & R. J. Baker (1982) *Dasypus novemcinctus*. - Mammalian Species 162:1-9.

McLachlan, A. J. (1979) The aquatic environment: I. Chemical and physical characteristics of Lake Chilwa. - In: M. Kalk, A. J. McLachlan & C. Howard-Williams (Eds.), Lake Chilwa. - W. Junk, The Hague. pp. 59-78.

Medellín, R. A. (1989) *Chrotopterus auritus*. - Mammalian Species 343:1-5.

Medellín, R. A., & H. T. Avrita (1989) *Tonatia evotis* and *Tonatia sylvicola*. - Mammalian Species 334:1-5.

Menezes, M. (1986) Ficoflórula da Chapada dos Guimarães e arredores, Mato Grosso, Brasil: Euglenaceae pigmentadas (Euglenophyceae). - Rickia 13:87-95.

Menezes, M., & V. O. Fernandes (1987) Euglenaceae pigmentadas do municipio de Cáceres e arredores, Mato Grosso, Brasil: uma contribuição ao seu conhecimento. - Rickia 14:53-71.

Menezes, M., & V. O. Fernandes (1989) Euglenaceae (Euglenophyceae) pigmentadas do noroeste do Estado de Mato Grosso, Brasil: municípios de Barra do Bugres, Cáceres, Juina e Porto Esperid_o. - Hoehnea 16:35-55.

Menges, E. S., & N. Kohfeldt (1995) Life history strategies of Florida scrub plants in relation to fire. - Bull. Torrey Bot. Club 122(4):282-297.

Meschiatti, A. J. (1995) Alimentação da comunidade de peixes de uma lagoa marginal do rio Mogi-Guaçu, SP. - Acta Limnol. Bras. 7:115-137.

Mickel, J. T., & J. M. Beitel (1988) Pteridophyte flora of Oaxaca, Mexico. - Mem. N. Y. Bot. Gard. 46:1-568.

Miller, F. W. (1930) Notes on some mammals of southern Mato Grosso, Brazil. - J. Mamm. 11:10-22.

Milner, J., C. Jones & J. K. Jones Jr. (1990) *Nyctinomops macrotis*. - Mammalian Species 351:1-4.

Milstead, W. W. (1963) Notes on Brazilian frogs of the genera *Physalaemus* and *Pseudopaludicola*. - Copeia 1963:565-566.

Mitamura, O., K. Hino, Y. Saijo, J. G. Tundisi, T. Matsumura-Tundisi, I. Ikusima, T. Sunaga, N. Nakamoto & H. Fukuhara, 1985: Physico-chemical feature of the Pantanal water system. In: Y. Saijo & J. G. Tundisi (Eds.), Limnological Studies in Central Brazil. - Water Res. Inst., Nagoya Univ., Nagoya. pp. 189-196.

Mitchell, D. S. (1985) African aquatic weeds and their management. In: P. Denny (Ed.), The Ecology and Management of African Wetland Vegetation. - W. Junk, Dordrecht. pp. 177-202.

Mittermeier, R. A., I. de Gusmão-Camara, M. T. Jorge-Padua, & J. Blanck (1990) Conservation in the Pantanal of Brazil. - Oryx 24:103-112.

Moreira, L. C., & O. O. Collart (1993) Migração vertical nictemeral das larvas de *Macrobrachium amazonicum* num lago de *várzea* na Amazônia Central, Ilha do Careiro, Brasil. - Amazoniana 12:385-398.

Mores, A., & J. Ojasti (1986) *Hydrochoerus hydrochaerus*. - Mammalian Species 264:1-7.

Moschini-Carlos, V., L. M. L. Pompêo & J. J. Soares (1995) Fenologia de *Scirpus cubensis* Poepp. & Kunth na Lagoa do Infernão (Estação Ecológia do Jataí, Luiz Antônio - SP. - Acto Limnol. Bras. 7:1-9.

Mosso, E. D., A. T. R. de Montaner, A. H. Beltzer, & E. B. de Carlo (1994) Aspectos anatómicos e histológicos del aparato digestivo de la garza bruja *Nycticorax nycticorax hoactli* Gmelin, 1789 (Aves: Ardeidae). Rev. Ceres 41(234):113-125.

Mothes de Moraes, B. (1983) Revisão do gênero *Drulia* Gray, 1867 (Porifera, Spongillidae). - Iheringia, Ser. Zool. 62:13-36.

Mouchamps, R. (1957) Contribution à la connaissance des Cybisterini (Col. Dytiscidae) du Musée d'Histoire Naturelle de Vienne. - Ann. Nat. Hist. Mus. Wien 61:278-287.

Mourão, G., Z. Campos, M. Coutinho, & C. Abercrombie (1996) Size structure of illegally harvested and surviving caiman *Caiman crocodilus yacare* in Pantanal, Brazil. - Biol. Conserv. 75:261-265.

Mourão, G. M. de, T. H. Ishii & Z. M. S. Campos (1988) Alguns fatores limnológicos relacionados com a ictiofauna de baías e salinas do Pantanal da Nhecolândia, Mato Grosso do Sul, Brasil. - Acta Limnol. Bras. 2:181-198.

Müller, L., & W. Hellmich (1936) Amphibian und Reptilien. I. Teil: Amphibia, Chelonia, Loricata. In: H. Krieg (Ed.), Wissenschaftliche Ergebnisse der Deutschen Gran Chaco-Expedition. - Strecker und Schröder, Stuttgart. pp. 1-120.

Müller, P., & G. Weimer (1976) Bemerkungen zu den Verbreitungszentren der südamerikanischen Callichthyiden und Cichliden. - Amazoniana 6:105-121.

Munz, P. A. (1947) Onagraceas. Flora Brasílica, Vol. 41(1). Sec. Agric., São Paulo. 62 pp + 51 pl.

Murata, N., & I. Nishida (1987) Lipids of blue-green algae (Cyanobacteria). In: P. K. Stumpf (Ed.), The Biochemistry of Plants. Vol. 9. Lipids: Structure and Function. - Academic Press, Orlando. pp. 315-347.

Nachtnebel, H. P. (1996) Begradigung, Uferverbau und Stauhaltungen. In: J. L. Lozan & H. Kausch (Eds.), Warnsignale aus Flüssen und Ästuaren. - Parey, Berlin. pp. 152-161.

Napier, J., & P. Napier (1967) A Handbook of Living Primates. - Academic Press, New York. 456 pp.

Nascimento, M. T., & C. N. Cunha (1989) Estrutura e composição florística de um cambarazal no Pantanal de Poconé, MT. - Acta Bot. Bras. 3(1):3-23.

Naumburg, E. M. B. (1930) The birds of Matto Grosso, Brazil. - Bull. Amer. Mus. Nat. Hist. 60:1-432.

Neal, E. C., B. C. Patten & C. E. Depoe (1967) Periphyton growth on artificial substrate in a radioactively contaminated lake. - Ecology 48:918-924.

Neiff, J. J. (1990) Aspects of primary productivity in the Lower Paraná and Paraguay Rivers. - Acta Limnol. Bras. 3:77-113.

Neto, M. S. S., R. Alves, A. C. Figueiredo & K. Hino (1993) Caracterização hydrogeoquímica da Bacia do Rio Manso - Cuiabá, Mato Grosso. - Acta Limnol. Bras. 6:230-244.

Neves, L. D., R. M. D. S. Isaias & L. E. D. Mello Filho (1993) Anatomic study of the leaf of *Ficus elliotiana* Moore. Bradea 6(22):196-205.

Newcomer, R. J., & J. W. Crenshaw (1967) Electrophoretic comparison of blood proteins of two closely related species of South American tortoises. - Copeia 1967:481-483.

Nieschulz, O. (1928) Zoologische Beiträge zum Surraproblem. XVII. Weitere Surraübertragungsversuche mit *Tabanus rubidus* Wied. - Arch. Protistenk. 61:92-118.

Nieser, N. (1975) The water bugs (Heteroptera: Nepomorpha) of the Guyana Region. - Studies on the fauna of Suriname and other Guyanas 59:1-310 + 24 plates.

Nieser, N. (1977) A revision of the genus *Tenagobia* Bergroth (Heteroptera: Corixidae). - Stud. Neotrop. Fauna Environ. 12:1-56.

Nijssen, H. (1970) Revision of the Surinam catfishes of the genus *Corydoras* Lacépède, 1803 (Pisces, Siluriformes, Callichthyidae). - Beaufortia 18:1-75.

Noda, S. N., & H. A. Noda (1990) A água envenenada. - Ciência Hoje 11:40-42.

Nolte, U. (1995) From egg to imago in less than seven days: *Apedilum elachistus* (Chironomidae). In: P. S. Cranston (Ed.), Chironomids. From Genus to Ecosystems. - CSIRO Publications, East Melbourne. pp. 177-184.

Nolte, U., R. S. Tietboehl, & W. P. McCafferty (1996) A mayfly from tropical Brazil capable of tolerating short-term dehydration. - J. N. Amer. Benthol. Soc. 15:87-94.

Noreña-Janssen, C, 1995a: Studies on the taxonomy and ecology of the Turbellaria (Plathelminthes) in the floodplain of the Paraná River (Argentina) I. biotope and research area. Arch. Hydrobiol./Suppl. 107:211-262.

Noreña-Janssen, C, 1995b: Studies on the taxonomy and ecology of the Turbellaria (Plathelminthes) in the floodplain of the Paraná River (Argentina) II. taxonomy and ecology of the Turbellaria. Arch. Hydrobiol./Suppl. 107:211-262.

Nriagu, J. O., W. G. Pfeiffer, O. Malm, C. M. M. de Souza & G. Mierle (1992) Mercury pollution in Brazil. - Nature (London) 356:389.

Nusch, E. A., & G. Palme (1975) Biologische Methoden für die Praxis der Gewässeruntersuchung. I. Bestimmung des Chlorophyll-a- und Phaepigmentgehaltes in Oberflächenwasser. - Gas- und Wasserfach 116(2):562-565.

Nyonje, B., & J. Radull (1991) The effects of feeding freshwater chlorella, bakers yeast and culture selco on the culture of rotifers (*Brachionus* sp.). - Spec. Publ. Eur. Aquacult. Soc. No. 15:106-109.

Ochoa, J. G., H. Castellanous & C. Ibañez (1988) Records of bats and rodents from Venezuela. - Mammalia 52:175-180.

Ochs, G. (1964) Zur Kenntnis der Gyriniden (Col.) von Suriname und vom Rio Parú in benachbarten Brasilien. - Studies on the fauna of Suriname and other Guyanas 7(27):82-110.

Ochs, G. (1965) Die südliche Artengruppe um *Gyretes pygmaeus* Rég. (Col., Gyrinidae). - Mitt. Münch. Entomol. Ges. 55:278-312.

Ochs, G. (1966) Über neue und interessante Gyriniden aus den brasilianischen Staaten Parana und São Paulo (Col.). - Stud. Entom. 9(1-4):441-464.

Odum, E. P. (1971) Fundamentals of Ecology. 3rd ed. - Saunders, Philadelphia. 574 pp.

Ohly, J. J. (1985) Die Wasserbüffelwirtschaft am mittleren Amazonas aus der Sicht der jüngsten Entwicklungen. - Amazoniana 9:441-457.

Ojasti, J. (1973) Estudio biológico del chigüire o capibara. - Fondo Nac. Invest. Agropec., Caracas. 275 pp.

Ojasti, J. (1983) Ungulates and large rodents of South America. In: F. Burliere (Ed.), Ecosystems of the World. 13. Tropical Savannas. Elsevier, Amsterdam. pp. 427-439.

Oliva, A. (1989) El genero <*Berosus*> (Coleoptera: Hydrophilidae) en America des Sur. - Rev. Mus. Arg. Cienc. Nat. <Bern. Rivadavia> 6(4):57-236.

Oliveira-Filho, A. T., & F. R. Martins (1991) A comparative study of five cerrado areas in southern Mato Grosso, Brazil. - Edinb. J. Bot. 48(3):307-332.

Olrog, C. C. (1963) List y distribucion de las aves Argentinas. - Opera Lilloana 9:1-377.

Oniki, Y., & E. O. Willis (1983) Breeding records of birds from Manaus, Brazil: 5. Icteridae to Fringillidae. - Rev. bras. Biol. 43(1):55-64.

Opuszynski, K. K., & J. V. Shireman (1993) Strategies and tactics for larval culture of commercially important carp. - J. Appl. Aquacult. 2:189-219.

Orchard, A. E., & C. Kasselmann (1992) Notes on *Myriophyllum mattogrossense* (Haloragaceae). - Nord. J. Bot. 12(1):81-84.

Orchymont, A. d' (1939) Notes sur quelques *Palpicornia* de la Republique Argentine. - Rev. Soc. Entomol. Arg. 10:253-264.

Orioli, A. L., Z. P. Amaral & A. B. Oliveira (1982) Pedologia. In: Projeto Radambrasil. Vol. 27. Rio de Janeiro.

Page, F. C. (1976) An Illustrated Key to Freshwater and Soil Amoebae. Freshwater Biol. Assoc. Publ. No. 34. - Titus Wilson, Kendal. 155 pp.

Paggi, J. C. (1989) A new species of *Ilyocryptus* (Crustacea: Anomopoda) from the River Paraná Basin, Argentina. - Hydrobiologia 182:239-248.

Paggi, J. C., & S. Jose de Paggi (1990) Zooplàncton de ambientes lóticos e lênticos do Rio Paraná médio. - Acta Limnol. Bras. 3:685-719.

Pain, T. (1972) The Ampullariidae, an historical survey. - J. Conch. 27:453-462.

Paiva, M. P., & F. H. Nepomuceno (1989) On the reproduction in captivity of the oscar, *Astronotus ocellatus* (Cuvier), according to the mating methods (Pisces - Cichlidae). - Amazoniana 10:361-377.

Parma de Croux, M. J. (1983) Nivel de onígeno y mínimo de supervivencia de *Hoplias malabaricus malabaricus* (Bloch, 1794) (Pisces, Erythrinidae). - Iheringia Ser. Zool. 63:91-101.

Partzsch, M. (1987) Wachstum and Entwicklung von *Scenedesmus acutus* Chodat, *Kirchneriella lunaris* (Kirchn.) Maob. und *Monoraphidium contortum* (Thur.) Kom.-Legn. in Abhängigkeit von Licht und Temperatur. - Arch. Hydrobiol./Suppl. 78 (Algological Studies 46):89-98.

Patrick, R., & C. W. Reimer (1966) The Diatoms of the United States, Vol. 1. Monogr. Acad. Nat. Sci. Phila. Sutter House, Lititz, Pa. 688 pp.

Patrick, R., & C. W. Reimer (1975) The Diatoms of the United States, Vol. 2. Monogr. Acad. Nat. Sci. Phila. Sutter House, Lititz, Pa. 213 pp.

Pauly, D. (1984) Fish population dynamics in tropical waters: a manual for use with programmable calculators. - ICLARM, Manila. 325 pp.

Payer, H. D., Y. Cheimvichak, K. Hosakul, C. Kongpanichkul, L. Kraidej, M. Nguitragul, S. Reungmaniphytoon & P. Buri (1980) Temperature as an important climatic factor during mass production of microscopic algae. In: G. Shelef & C. J. Soeder (Eds.), Algal Biomass: Production and Use. - Elsevier Biomedical, Amsterdam. pp. 389-399.

Pearson, N. E. (1937) The fishes of the Beni-Mamoré and Paraguay Basins, and a discussion of the origin of the Paraguayan fauna. - Proc. Calif. Acad. Sci. (4)23:99-114.

Penney, J. T., & A. A. Racek (1968) Comprehensive revision of a worldwide collection of freshwater sponges (Porifera: Spongillidae). - Bull. U. S. Natl. Mus. 272:1-184.

Perfil'ev, B. V., & D. R. Gabe (1961) (The capillary methods for inoculating microorganisms.). - Izdat. Ak. Nauk SSSR. Moscow. (in Russian).

Perotti de Jorda, N. M. P. de (1984) Estudios limnologicos en una seccion transversal del tramo medio del rio Parana. - Rev. Asoc. Cinec. Nat. Lit. 15:117-133.

Perraca, M. G. de (1904) Viagio del Dr. Borelli nel Matto Grosso Brasiliano e nel Paraguay, 1899. - Rettili ed Abfibi. - Bol. Mus. Anat. Comp. Univ. Torino 19(460):1-15.

Persia, D. H. di (1980) The aquatic Oligochaeta of Argentina: current status of knowledge. In: R. O. Brinkhurst & D. G. Cook (Eds.), Aquatic Oligochaete Biology. - Plenum, New York. pp. 79-113.

Pessôa, S. B., & A. V. Martins (1988) Parasitologia Médica. 11th ed. Guanabara, Rio de Janeiro. 872 pp.

Peters, J. A., & R. Donoso-Barros (1970) Catalogue of the Neotropical Squamata. Part II. Lizards and Amphisbaenians. - U. S. Natl. Mus. Bull. 297. 293 pp.

Peters, J. A., & A. Orejas-Miranda (1970) Catalogue of the Neotropical Squamata. Part I. Snakes. - U. S. Natl. Mus. Bull. 297. 347 pp.

Petr, T. (1983) Limnology of the Purari basin. Part 1. The catchment above the delta. In. T. Petr (Ed.), The Purari - tropical environment of a high rainfall river basin. - W. Junk, The Hague. pp. 141-177.

Petr, T., & G. Irion (1983) Geochemistry of soil and sediments of the Purari River basin. In. T. Petr (Ed.), The Purari - tropical environment of a high rainfall river basin. - W. Junk, The Hague. pp. 109-122.

Petrere, M., Jr., & M. C. L. de Brito Ribeiro (1994) The impact of a large tropical hydroelectric dam: the case of the Tucuruí in the middle River Tocantins. - Acta Limnol. Bras. 5:123-133.

Petri, S., & V. J. Fulfaro (1988) Geologia do Brasil. In: T. A. Queiroz Ed. USP. São Paulo.

Piedade, M. T. F. (1993) Biologia e ecologia de *Echinochloa polystachya* (H.B.K.) Hitchcock (Gramineae = Poaceae), capim semi-aquático de várzea Amazônica. - Acta Limnol. Bras. 6:173-185.

Pignalberi, C., & E. Cordiviola de Yuan (1973) Biometria del "sabalo" (*Prochilodus platensis* Holmberg) de diversos ambientes del Parana medio (Pisces, Tetragonopteridae). - Physis, Section B 32(84):29-40.

Pilsbry, H. A. (1933) Zoological results of the Mato Grosso Expedition to Brazil in 1931. II. Mollusca. Proc. Acad. Nat. Sci. Phila. 85:67-76.

Pinder, L., & A. P. Grosse (1991) *Blastocerus dichotomus.* - Mammalian Species 380:1-4.

Pine, R. H., I. R. Bishop & R. L. Jackson (1970) Preliminary list of mammals of the Xavantina/Cachimbo Expedition (central Brazil). - Trans. Roy. Soc. Trop. Med. Hyg. 64:668-670.

Pinto, C. (1925) Protozoários observados no Brasil. Memórias do Instituto Oswaldo Cruz, Tomo XVIII: 211-327.

Pinto, O. M. de O. (1938) Catálogo das Aves do Brasil (Non-Passeriformes). Rev. Mus. Paulista 12:1-566.

Pinto, O. M. de O. (1939) Duas formas novas da avifauna de Pernambuco. - Bol. Biol. São Paulo 4:189-195 + 1 pl.

Pinto, O. M. de O. (1940) Aves do Pernambuco. - Arq. Zool. Estado São Paulo 1:219-282 + 6 pl.

Pinto, O. M. de O. (1944) Catálogo das Aves do Brasil (Passeriformes). Dept. Zool. Sec. Agric., São Paulo. 700 pp.

Pinto, S. Y. (1974) Peixes de água doce do Estado da Guanabara. - Arq. Mus. Hist. Nat., Belo Horizonte 1:49-125.

Poglayen-Neuwall, I. (1978) Breeding, rearing and notes on the behavior of tayras, *Eiras barbata.* - Int. Zoo Yearbook, London 18:134-140.

Poi-de-Neiff, A. C. (1983) Observaciones comparativas de la mesofauna asociada a *Pistia stratiotes* L. (Araceae) en algunos ambientes acuaticos permanentes y temporarios (Chaco, Argentina). - Physis B:41(101):95-102.

Pompêo, M. L. M., & V. Moschini-Carlos (1995) Zonação e biomassa das macrófitas aquáticas na Lagoa Dourada (Brotas, SP), com ênfase na *Utricularia gibba* L. - Acta Limnol. Bras. 7:78-86.

Ponce, V. M., & C. N. da Cunha (1993) Vegetated earthmounds in tropical savannas of Central Brazil: a synthesis with special reference to the Pantanal of Mato Grosso. - J. Biogeogr. 20:219-225.

Pope, B. (1966) Population characteristics of howler monkeys (*Alouatta caraya*) in northern Argentina. - Am. J. Phys. Anth. 24:361-370.

Popovsky, J., & L. A. Pfiester (1990) Dinophyceae (Dinoflagellida). In: H. Ettl, J. Gerloff, H. Heynig & D. Mollenhauer (Eds.), Süßwasserflora von Mitteleuropa. Vol. 6. Gustav Fischer, Stuttgart. 272 pp.

Por, F. D. (1995) The Pantanal of Mato Grosso (Brazil). - Kluwer, Dordrecht. 122 pp.

Posadas V. C. I. G. (1966) Rb-Sr whole rock age in the Imatacá Complex, Venezuela. - Mass. Inst. Technol. Ann. Prog. Rpt. 14:17-24.

Pott, A., & V. J. Pott (1994) Plantas do Pantanal. - EMBRAPA, Brasilia. 320 pp.

Pott, E., & A. Pott (1989) Níveis de nutrientes em plantas não-gramíneas pastejadas por bovinos na subregião dos Paiaguás, do Pantanal Mato-Grossense. - Pesq. Agropec. Bras., Brasilia 22(11-12):1293-1299.

Pott, E., & A. Pott (1990) Flórula ruderal da cidade de Corumbá, MS. In: Anais do Congresso Nacional de Botânica 36 (1985), Curitiba. - IBAMA, Brasilia. pp. 519-535.

Pott, V. J., N. C. Bueno, R. A. C. Pereira, S. M. de Salis & N. L. Vieira (1989) Distribuição de macrófitas aquáticas numa lagoa na Fazenda Nhumirim, Nhecolândia, Pantanal, MS. - Acta bot. bras. 3(2):153-168.

Pott, V. J., & A. C. Cervi, in press: A familia Lemnaceae no Pantanal (Mato Grosso e Mato Grosso do Sul), Brasil. - Rev. Bras. Bot. (in press).

Prado, A. L. do, C. W. Heckman & F. R. Martins (1994) The seasonal succession of biotic communities in wetlands of the tropical wet-and-dry climatic zone: II. The aquatic macrophyte vegetation in the Pantanal of Mato Grosso, Brazil. - Int. Revue ges. Hydrobiol. 79:569-589.

Prado, C. H. B. A., & J. A. P. V. Moraes (1990) Estudo do pH da rizosfera de *Pistia stratiotes* L. condicionada a duas fontes de nitrogênio: nitrato e amônio. Acta Limnol. Bras. 3:675-683.

Prado, D. E., P. E. Gibbs, A. Pott & V. J. Pott (1992) The Chaco-Pantanal transition in southern Mato Grosso, Brazil. In: P. A. Furley, J. Proctor & J. A. Ratter (Eds.), Nature and Dynamics of Forest-Savanna Boundaries. - Chapman & Hall, London. pp. 451-470.

Prance, G. T. (1980) A note on the pollination of *Nymphaea amazonum* Mart. & Zucc. (Nymphaeaceae). - Brittonia 32(4):505-507.

Prance, G. T., & G. B. Schaller (1982) Preliminary study of some vegetation types of the Pantanal, Mato Grosso, Brazil. - Brittonia 34(2):228-251.

Prescott, G. W., C. E. de M. Bicudo & W. C. Vinyard (1982) A. Synopsis of North American Desmids, 2: Desmidiaceae: Placodermae, Section 4. - Univ. Nebr. Press, Lincoln. 700 pp.

Prescott, G. W., H. T. Croasdale & W. C. Vinyard (1972) Desmidiales, 1: Secodermae, Mesotaeniaceae. North American Flora, Ser. 2, Part 6. N. Y. Bot. Garden, New York. 84 pp.

Prescott, G. W., H. T. Croasdale & W. C. Vinyard (1975) A. Synopsis of North American Desmids, 2: Desmidiaceae: Placodermae, Section 1. - Univ. Nebr. Press, Lincoln. 275 pp.

Prescott, G. W., H. T. Croasdale & W. C. Vinyard (1977) A. Synopsis of North American Desmids, 2: Desmidiaceae: Placodermae, Section 2. - Univ. Nebr. Press, Lincoln. 413 pp.

Prescott, G. W., H. T. Croasdale, W. C. Vinyard & C. E. de M. Bicudo (1981) A. Synopsis of North American Desmids, 2: Desmidiaceae: Placodermae, Section 3. - Univ. Nebr. Press, Lincoln. 720 pp.

Prohaska, F. (1976) The climate of Argentina, Paraguay & Uruguay. In: W. Schwerdtfeger (Ed.), Climates of Central and South America. - Elsevier, Amsterdam. pp. 13-112..

Putzer, H. (1962) Die Geologie von Paraguay. - Borntraeger, Berlin. 183 pp.

Putzer, H. (1968) Überblick über die geologische Entwicklung Südamerikas. In: E. J. Fittkau, J. Illies, H. Klinge, G. H. Schwabe & H. Sioli (Eds.), Biogeography and Ecology in South America. Vol. 1. - W. Junk, Dordrecht. pp. 1-24.

Putzer, H. (1984) The geological evolution of the Amazon Basin and its mineral resources. In: H. Sioli (Ed.), The Amazon. - W. Junk, Dordrecht. pp. 15-46.

Quirós, R., & C. Baigún (1985) Fish abundance related to organic matter in the Plata River Basin, South America. - Trans. am. Fish. Soc. 114:377-387.

Rácenis, J. (1958) Los odonatos neotropicales en la coleccion de la faculdad de agronomia de la Universidad Central de Venezuela. - Acta Biol. Venez. 2(19):179-226.

Rai, H., & G. Hill (1984) Primary production in the Amazonian aquatic ecosystem. In: H. Sioli (Ed.), The Amazon. - W. Junk, Dordrecht. pp. 311-335.

Ramamoorthy, T. P., & E. M. Zardini (1987) The systematics and evolution of *Ludwigia* sect. *Myrtocarpus* sensu lato (Onagraceae). - Missouri Bot. Gard. Monogr. 19:1-120.

Rand, A. S. (1968) A nesting aggregation of iguanas. - Copeia 1968:552-561.

Rataj, K. (1972) Revision of the genus *Sagittaria*. Part II. (The species of West Indies, Central and South America). - Ann. Zool. Bot. No. 78. 61 pp.

Rataj, K. (1975) Revision of the genus *Echinodorus* Rich. - Czech. Akad. VED, Prague. 156 pp.

Ratisbona, L. R. (1976) The Climate of Brazil. In: W. Schwerdtfeger (Ed.), Climates of Central and South America. - Elsevier, Amsterdam. pp. 219-293.

Ratter, J. A., G. P. Askew, R. F. Montgomery & D. R. Gifford (1978) Observations on the vegetation of northeastern Mato Grosso II. Forests and soils of the Rio Suiá - Missu area. - Proc. R. Soc. Lond. B 203:191-208.

Ratter, J. A., A. Pott, V. J. Pott, C. N. da Cunha & M. Haridasan (1988) Observations on woody vegetation types in the Pantanal and around Corumbá. - Notes Roy. Bot. Gardin Edinburgh 45(3):503-525.

Rausch, T. (1981) The estimation of micro-algal protein content and its meaning to the evaluation of algal biomass. I. Comparison of methods for extracting protein. - Hydrobiologica 78:237-251.

Ravazzani, C., H. Wiederkehr Filho, J. P. Fagnani, S. da Costa (1991) Pantanal. - Edibran, Curitiba. 180 pp.

Redford, K. H., & R. M. Wetzel (1985) *Euphractus sexcinctus*. - Mammalian Species 252:1-4.

Regen, C. T. (1912) A revision of the South American siluroid fishes of the genus *Corydoras*, with a list of the specimens in the British Museum (Natural History). - Ann. Mag. Nat. Hist. London, ser 8, 10:209-220.

Reid, J. W. (1985) Chave de identificação e lista de referências bibliográficas para as espécies continentais Sulamericanas de vide livre da ordem Cyclopoida (Crustacea, Copepoda). - Bolm. Zool., Univ. São Paulo 9:17-143.

Reid, J. W. (1993) The harpacticoid and cyclopoid copepod fauna in the cerrado region of Central Brazil. I. Species composition, habitats, and zoogeography. - Acta Limnol. Bras. 6:56-68.

Reid, J. W., & I. H. Moreno (1990) The Copepoda (Crustacea) of the southern Pantanal, Brazil. - Acta. Limnol. Brasil. 3:721-739.

Resende, E. K. (1992) Bioecologia do curimbatá, *Prochilodus lineatus* no Pantanal do Miranda-Aquidauana, Mato Grosso do Sul, Brasil. - Acta Limnol. Bras. 4:261-276.

Richmond, E. A. (1920) Studies on the biology of the aquatic Hydrophilidae. - Bull. Amer. Mus. Nat. Hist. 42:1-93 + 16 pl.

Riedel-Lorjé, J. C. (1981) Untersuchungen über den Indikationswert von Aufwuchs in Süß- und Brackwasserzonen des Elbe-Aestuars unter Berücksichtigung industrieller Einleitungen. - Arch. Hydrobiol./Suppl. 61 (Untersuch. Elbe-Aestuar 5):153-226.

Riehl, R., & H. A. Baensch (1991) Aquarien Atlas. 8th ed. - Mergus, Melle. 992 pp.

Ringuelet, R. A. (1981) Hirudinea. In: S. H. Hurlbert, G. Rodríguez & N. D. dos Santos (Eds.) Aquatic Biota of Tropical South America, Volume 2, Part 1. Anarthropoda. - San Diego State Univ., San Diego. pp. 191-196.

Ringuelet, R. A., R. de Aramburu & A. A. de Aramburu (1967) Los peces argentinos de aqua dulce. - Com. Inv. Cienc., Prov. Buenos Aires, La Plata. 602 pp.

Ris, F. (1913) Neuer Beitrag zur Kenntnis der Odonatenfauna von Argentina. - Mem. Soc. Ent. Belg. 22:55-102.

Rizzini, C. T., & E. P. Heringer (1962) Preliminares acêrca das formações vegetais e do reflorestamento no Brasil Central. - Serviço de Informações Agrícola, Min. Agric., Rio de Janeiro. 79 pp.

Robertson, B. A., & E. R. Hardy (1984) Zooplankton of Amazonian lakes and rivers. In: H. Sioli (Ed.), The Amazon. Limnology and landscape ecology of a mighty tropical river and its basin. - W. Junk, The Hague. pp. 337-352.

Rodina, A. G. (1953) (The development of iron bacteria in an iron-enriched fish pond.). - Mikrobiologiya 12:171-178. (in Russian).

Rodriguez, G. (1992) The freshwater crabs of America. Family Trichodactylidae and supplement to the family Pseudothelphusidae. - Faune trop., ORSTROM, Paris 31:1-109.

Roessler, E. W. (1995) Review of Colombian Conchostraca (Crustacea) - ecological aspects and life cycle - family Cyclestheriidae. - Hydrobiologia 298:113-124.

Rolla, M. E., M. B. G. S. Dabés, R. C.França & E. M. V. M. Ferreira (1990) Aspectos limnológicos do Reservatório de Volta Grande, Minas Gerais/ São Paulo. Acta Limnol. Bras. 3:219-244.

Rose, F. (1953) A survey of the ecology of the British lowland bogs. - Proc. Linn. Soc. Lond. 164:186-211.

Rossi, L. M. (1989) Alimentación de larvas de *Salminus maxillosus* Val. 1840 (Pisces Characidae). - Iheringia Sér. Zool. 69:49-59.

Roth, P. (1989): Der Hyazinthara, *Anodorhynchus hyacinthinus*. - Papageien 1/89:20-24.

Rott, E. (1981) Some results from phytoplankton counting intercalibrations. - Schweiz. Z. Hydrologie 43:34-62.

Roze, J. (1983) New World coral snakes (Elapidae): a taxonomic and biological summary. Mem. Inst. Butantan 46:305-338.

Rudescu, L. (1967) Fauna Republicii Socialiste România, 2, Trochelminthes. Part 3, Gastrotricha. - Ed. Acad. Rep. Soc. Romênia, Bucharest. 289 pp.

Ruschi, A. (1982) Aves do Brasil. Vols. 4 and 5. Villa Rica Ed., Rio de Janeiro. 206 pp and 452 pp.

Růžička, J. (1977) Die Desmidiaceen Mitteleuropas. Vol. 1. - Schweizerbart, Stuttgart. pp. 1-291.

Růžička, J. (1981) Die Desmidiaceen Mitteleuropas. Vol. 2. -Schweizerbart, Stuttgart. pp. 293-736.

Sabatini, R. A. (1985) Dinamica y productividade de *Paspalum repens* Bergies (canutillo) en um ambiente lenitico del valle aluvial del rio Parana. - Rev. Hydrobiol. Trop. 18:3-11.

Saeni, M. S.; R. T. M. Sutamihardja & J. Sukra (1980) Water quality of the Musi River in the city area of Palembang. In: J. I. Furtado (Ed.), Tropical Ecology and Development. - Int. Soc. Trop. Ecol., Kuala Lumpur. pp. 717-724.

Sahai, R., & A. B. Sinha (1970) Contribution to the ecology of Indian aquatics. I. Seasonal changes in biomass of water hyacinth (*Eichhornia crassipes*). - Idrobiologia 35(3-4):376-382.

Saijo, Y., O. Mitamura, I. Ijusima, J. G. Tundisi, T. Sunaga, N. Nakamoto, H. Fukuhara, F. A. R. Barbosa, R. Henry, C. T. Seixas Filho & V. P. Silva (1987) Physico-Chemical features of small lakes near Porto Jofre in northern Pantanal. In: Y. Saijo & J. G. Tundisi (Ed.), Studies in Rio Doce Lakes and Pantanal Wetland, Brazil. Vol. 2. - Water. Res. Inst., Nagoya Univ., Nagoya. pp. 177.182.

Saint-Paul, U. (1984) Physiological adaptation to hypoxia of a Neotropical characoid fish, *Collosoma macropomum*, Serrasalmidae. - Environ. Biol. Fishes 11(1):53-62.

Saint-Paul, U. (1986) Potential for aquaculture of South American freshwater fishes: a review. - Aquaculture 54:205-240.

Sanchez, P. A. (1989) Soils. In: H. Lieth & M. J. A. Werger (Eds.), Ecosystems of the World 14B. Tropical Rainforest Ecosystems. - Elsevier, Amsterdam. pp. 73-88.

Sant'Anna, C. L. (1984) Chlorococcales (Chlorophyceae) do Estado do São Paulo, Brasil. - Biblio. Phycol. 67. J. Cramer, Vaduz. 348 pp.

Saravanan, K., & B. Nagarajan (1991) Carcinogenic potential of catechol and nitrite-treated fish. - Med. Sci. Res. 19:781-783.

Sarmiento, G. (1983) The savannas of tropical America. In: F. Bourlière (Ed.), Ecosystems of the World 13. Tropical Savannas. - Elsevier, Amsterdam. pp. 245-288.

Sazama, I. (1986) Similarities in feeding behaviour between some marine and freshwater fishes in two tropical communities. - J. Fish Biol. 29:53-65.

Sazama, I. (1988) Territorial behaviour in a scale-eating and a herbivorous Neotropical characiform fish. - Rev. Bras. Biol. 48(2):189-194.

Sazama, I., & E. Pelligrini Caramaschi (1989) Comportamento alimentar de duas espécies de *Curimata*, sintópicas no Pantanal de Mato Grosso (Osteichthyes, Characiformes). - Rev. Bras. Biol. 49(2):325-333.

Sazima, I., & F. A. Machado (1990) Underwater observations of paranhas in western Brazil. - Environ. Biol. Fish.28:17-31.

Sazima, I., & C. Zamprogno (1984) Use of water hyacinths as shelter, foraging place, and transport by young piranhas *Serrasalmus spilopleura*. - Environ. Biol. Fishes 12(3):237-240.

Schaden, R. (1978) Zur Diversität und Identität amazonischer Rotatorienzoome. - Amazoniana 6:347-371.

Schaller, G. B. (1983) Mammals and their biomass on a Brazilian ranch. - Arq. Zool., São Paulo 31:1-36.

Schaller, G. B., & P. G. Crawshaw (1980) Movement patterns of jaguar. Biotropica 12:161-168.

Schaller, G. B., & P. G. Crawshaw (1981) Social organization of a capypara population. - Säugetierkund. Mitt. 29:3-16.

Schaller, G. B., & P. G. Crawshaw (1982) Fishing behavior of the Paraguayan caiman (*Caiman crocodilus*). - Copeia 1982:66-72.

Schaller, G. B., & J. M. C. Vasconcelos (1978a) Jaguar predation on capybara. Z. Säugetierkunde 43:296-301.

Schaller, G. B., & J. M. Vasconcelos (1978b) A marsh deer census in Brazil. - Oryx 14:345-354.

Schauensee, R. M. de & W. H. Phelps (1978) Birds of Venezuela. - Princeton Univ. Press. 424 pp.

Scheßl, M. (1997) Flora und Vegetation des nördlichen Pantanal von Mato Grosso, Brasilien. - Arch. naturwiss. Dissertationen 4:1-276 + 2 pl.

Schmidle, W. (1901) Algen aus Brasilien. Hedwigia 15:45-54.

Schmidt, G. W. (1973) Primary production of phytoplankton in the three types of Amazonian waters. - Amazoniana 4:379-404.

Schnack, J. A., G. R. Spinelli, A. L. Esteves, & E. A. Domizi (1986) Aspectos de la ecologia reproductiva de poblaciones Bonaerensis de *Belostoma elegans* (Mayr) y *B. micantulum* (Stål) (Hemiptera, Belostomatidae). - Rev. Asoc. Cienc. Nat. Litor. St. Tome 17:31-37.

Schnell, R. (1987) La Flore et la végétation de L'Amérique Tropicale. Vol. 2. - Masson, Paris. 448 pp.

Schneller, S. J. (1990) Salviniaceae. In: K. U. Kramer & P. S. Green (Eds.), Pteridophytes and Gymnosperms. Vol. 1. The Families and Genera of Vascular Plants. - Springer, Berlin. pp. 256-258.

Schvartsman, S. (1979) Plantas venenosas. - Sarvier, São Paulo. 176 pp.

Schwank, P., & I. Bartsch (1990) Gastrotricha and Nemertini. Süßwasserfauna von Mitteleuropa 3/1+2. - Gustav Fischer, Stuttgart. 262 pp.

Schweizer, J. (1992) Ariranhas no Pantanal. - Edibran, Curitiba. 202 pp.

Schwerdtfeger, W. (1976) Introduction. In: W. Schwerdtfeger (Ed.), Climates of Central and South America. - Elsevier, Amsterdam. pp. 1-12.

Scott, H. (1912) A contribution to the knowledge of the fauna of Bromeliaceae. - Ann. Mag. Nat. Hist. (Ser. 8) 10:424-438.

Sendacz, S. (1993) Distribuiç_o geográfica de alguns organismos zooplanctônicos na America do Sul. - Acta Limnol. Bras. 6:31-41.

Serruya, C., & U. Pollingher (1983) Lakes of the Warm Belt. - Cambridge Univ. Press, Cambridge. 569 pp.

Service, M. W. (1977a) Mortalities of the immature stages of species B of the *Anopheles gambiae* complex in Kenya: comparison between rice fields and temporary pools, identification of predators, and effects of insecticidal spraying. - J. Med. Entomol. 13:535-545.

Service, M. W. (1977b) Some problems in the control of malaria. In: F. H. Perring & K. Mellanby (Eds.), Ecological Effects of Pesticides. - Academic Press, London. pp. 151-164.

Seymour, K. L. (1989) *Panthera onca*. - Mammalian Species 340:1-9.

Seymour, R. L. (1970) The genus *Saprolegnia*. - Nova Hedwigia 19:1-124.

Sharp, D. (1882) On aquatic carnivorous Coleoptera or Dytiscidae. - Alex. Thom, Dublin. 1004 pp. + 18 pl.

Shiobara, Y., S. Inoue, K. Kato, Y. Oischi, N. Nishimoto, F. Oliveira, G. Akisue, M. K. Akisue & G. Hashimoto (1993) A nortriterpenoid, triterpenoids and ecdysteroids from *Pfaffia glomerata*. - Phytochemistry 32(6):1527-1530.

Short, L. L. (1975) A zoogeographic analysis of the South American chaco avifauna. - Bull. Am. Mus. Nat. Hist. 154:165-352.

Shump, K. A. Jr. (1982) *Lasiurus cinereus*. - Mammalian Species 185:1-5.

Shump, K. A., Jr. & A. U. Shump (1982) *Lasiurus borealis*. - Mammalian Species 183:1-6.

Shuter, B. J., P. E. Ihssen, D. L. Wales, & E. J. Snucins (1989) The effects of temperature, pH and water hardness on winter starvation of young-of-the-year smallmouth bass, *Micropterus dolomieni* Lacepede. - J. Fish. Biol. 35:765-780.

Sick, H. (1988) Ornitologia Brasileira. 2 Vols. 3rd ed. - Ed. Univ. Bras. & Linha Gráfica Ed., Brasilia. 827 pp.

Silva, J. M. C., & Y. Oniki (1988) Lista preliminar da avifauna da estação ecológica Serra das Araras, Mato Grosso, Brazil. - Bol. Mus. Goeldi, Zool. 4:123-143.

Silva, N. M. F. de & M. da C. Valente (1996) Flora dos Estados de Goiás e Tocantins Coleção Rizzo. Vol. 19. Combretaceae. - Univ. Fed. Goiás, Goiânia. 59 pp.

Silva, T. R. dos S. (1996) Flora dos Estados de Goiás e Tocantins Coleç_o Rizzo. Vol. 18. Droseraceae. - Univ. Fed. Goiás, Goiânia. 16 pp.

Silva Neto, E. V. da, E. L. Hardoim, & L. M. K. Borges (1997) Diversidade de Testacea (Protozoa: Rhizopoda) em conteudo estomical de *Hyphessobrycon callistus* (Boulenger, 1900) (Pisces: Characidae) e *Aphyocharax rathbuni* (Eigenmann, 1907) (Pisces: Characidae) da vazante Birici, Pantanal de Poconé - MT. In: II Simpósio sobre Recursos Naturais e Socio-economicos do Pantanal - Manejo e Conservação, Corumbá, in press.

Simone, L. R. L. de (1994) Anatomical characters and systematics of *Anodontites trapesialis* (Lamarck, 1819) from South America (Mollusca, Bivalvia, Unionioda, Muteloidea). - Stud. Neotrop. Fauna Environ. 29:169-185.

Sioli, H. (1957) Tropical rivers as expressions of their terrestrial environments. In: F. B. Golley & E. Medina (Eds.), Tropical Ecosystems, Trends in Terrestrial and Aquatic Research. - Springer, New York. pp. 275-287.

Sioli, H. (1984) The Amazon and its main affluents: hydrography, morphology of the river courses, and river types. In: H. Sioli (Ed.), The Amazon. - W. Junk, Dordrecht. pp. 127-165.

Siqueira, J. (1981) Utilização popular das plantas do cerrado. - Loyola, São Paulo. 60 pp.

Siqueira, J. (1989) Flora do Estado de Goiás Coleção Rizzo. Vol. 12. Amaranthaceae. - Univ. Fed. Goiás, Goiânia. 44 pp.

Skerfving, S. (1972) Organic mercury compounds - relation between exposure and effects. In: L. Friberg & J. Vostal (Eds.), Mercury in the Environment. - CRC Press, Cleveland. pp. 141-168.

Sládecek, V., & A. Sládecková (1964) Determination of the periphyton production my means of the glass slide method. Hydrobiologia 23:125-158.

Smith, C. (1977) Feeding behavior and social organization in howler monkeys. - In: T. Clutton-Brock (Ed.), Primate Ecology. - Academic Press, London. pp. 97-126.

Smith, H. M. (1963) Revival of Bonaparte's subfamily Heterodontinae of colubrid snakes. Herpetologica 19:288-291.

Smith, L. B., D. C. Wasshausen & R. M. Klein (1982) Gramíneas. 1 Parte. In: R. Reitz (Ed.), Flora Ilustrada Catarinense. 1407 pp.

Soderstrom, T. S. (1981) Observations on a fire-adapted bamboo of the Brazilian cerrado, *Actinocladium verticillatum* (Poaceae: Bambusoideae). - Amer. J. Bot. 68:1200-1211.

Solheim, W. G. II (1971) New light on a forgotten past. - Natl. Geogr. 139:330-339.

Sophia, M. G., & L. H. S. da Silva (1989) Considerações sobre a flora de desmidias filamentosas (Zygnemaphyceae) do noroeste de Mato Grosso e sudeste de Rondônia, Brasil. - Rev. Bras. Biol. 49 (4):943-956.

Sorokin, C. (1967) New high temperature *Chlorella*. - Science 158:1204-1205.

Sota, E. R. de la (1962) Contribución al conocimiento de las Salvinaceae neotropicales, II *Salvinia auriculata* Aublet. - Darwiniana 12(3):499-513.

Spangler, P. J. (1981) Coleoptera. In: S. H. Hurlbert, G. Rodríguez & N. D. dos Santos (Eds.) Aquatic Biota of Tropical South America, Volume 2, Part 1. Arthropoda. - San Diego State Univ., San Diego. pp. 129-220.

Starling, F. L. R. M. (1993) Análise experimental dos efeitos da tilápia do Congo (*Tilapia rendalli*) e carpa prateada (Hypophthalmichthys molitrix) sobre a estrutura de comunidade planctônica do lago Paranoá, Brasilia (DF). Acta Limnol. Bras. 6:144-156.

Starmach, K. (1985) Chrysophyceae und Haptophyceae. In: H. Ettl, J. Gerloff & H. Heynig (Eds.), Süßwasserflora von Mitteleuropa. Vol. 1. Gustav Fischer, Stuttgart. 515 pp.

Steinbacher, J. (1962) Beiträge zur Kenntnis der Vögel von Paraguay. - Abh. senckenb. naturforsch. Ges. 502:1-106.

Sterba, G. (1977) Süßwasserfische aus aller Welt. - Melsungen, Berlin. 558 pp.

Sterba, G. (1987) Süßwasserfische der Welt. - Eugen Ulmer, Stuttgart. 914 pp.

Strickman, D. (1981) Number of eggs per male giant water bug *Belostoma mecantulum* in Paraguay. - Mosq. News 41:375.

Strussmann, C., & I. Sazima (1990) Esquadrinhar com a cauda: uma tática de caça da serpente *Hydrodynastes gigas* no Pantanal, Mato Grosso. - Mem. Inst. Butantan 52:57-61.

Strussmann, C., & I. Sazima (1993) The snake assemblage of the Pantanal at Poconé, western Brazil: faunal composition and ecological summary. - Stud. Neotrop. Fauna Environ. 28(3):157-168.

Stymne, S., & A. K. Stobart (1987) Triacylglycerol biosynthesis. In: P. K. Stumpf (Ed.), The Biochemistry of Plants. Vol. 9. Lipids: Structure and Function. - Academic Press, Orlando. pp. 175-214.

Sucksdorff, A. (1985) Pantanal, um paraíso perdido? - Siciliano, Rio de Janeiro. 159 pp.

Sucksdorff, A. (1989) Desenvolvimento e preservação. II. Vis_o nacional. - An. I. Cong. Int. Cons. Pantanal. - Campo Grande. pp. 113-115.

Svenson, H. K. (1944) The New World species of *Azolla*. - Am. Fern J. 34:69-84.

Tarifa, J. R. (1986) O sistema climático do Pantanal. Da compreensão do sistema à definição de prioridade de pisquisa climatológica. In: Anais 1° Simpósio Recursos Naturais e Sócio-Econômico do Pantanal. EMBRAPA-CPAP, Corumbá. pp. 9-27.

Taylor, E. H. (1968) The Caecilians of the World. - Univ. Kans., Lawrence. 848 pp.

Taylor, J. A., & D. Tullock (1985) Rainfall in the wet-dry tropics: extreme events at Darwin and simularities between years during the period 1870-1983. - Australian J. Ecol. 10:281-295.

Taylor, P. (1989) The Genus *Utricularia* - a Taxonomic Monograph. Kew Bull. Additional Ser. 14. 724 pp.

Teiling, E. (1967) The desmid genus *Staurodesmus*: a taxonomic study. - Ark. Bot., Ser. II, 6(11):467-629.

Tell, G., & V. Conforti (1986) Euglenophyta Pigmentadas de la Argentina. - Biblio. Phycol. 75. J. Cramer, Berlin. 301 pp.

Tell, G., I. Izaguirre & I. O'Farrell (1994) Ecological and taxonomical remarks on the desmid flora of the Lower Uruguay River Basin (Argentina). - Bibliotheca Phycologica 96. J. Cramer, Berlin. 141 pp.

Thatcher, V. E. (1991) Amazonian fish parasites. - Amazoniana 11:263-572.

Thérézien, Y. (1989a) Algues d'eau douce de la partie Amazonienne de la Bolivia. 1. Cyanophycees, Euglenophyycees, Chrysophycees, Xanthophycees, Dinophycees. - Bibliotheca Phycologica 82. J. Cramer, Berlin. 52 pp. + 5 pl.

Thérézien, Y. (1989b) Troisieme contribution a l'etude des Chlorophytes des eaux douces de la partie Amazonienne de la Bolivie. - Bibliotheca Phycologica 82. J. Cramer, Berlin. 62 pp. + 6 pl.

Thomasson, K. (1963) Araucarian lakes. - Acta phytogeogr. suec. 47:1-139.

Thompson, F. C. (1972) A contribution to a generic revision of the Neotropical Milesinae (Diptera: Syrphidae). - Arq. Zool. 23:73-215.

Thornback, J., & M. Jenkins (1982) The IUCN mammal red data book. Part 1. - Gresham Press, Old Woking. 516 pp.

Todd, E. L. (1955) A taxonomic revision of the family Gelastocoridae (Hemiptera). - Univ. Kans. Sci. Bull. 37:277-475.

Tokarnia, C. H., & J. Döbereiner (1982) Intoxicação de bovinos por *Vernonia rubricaulis* (Compositae) em Mato Grosso. - Pesq. Vet. Bras. 2(4):143-147.

Tomás, W. M., 1995: Seasonality of the antler cycle of pampas deer (*Ozotoceros bezoarticus leucogaster*) from the Pantanal wetland, Brazil. - Stud. Neotrop. Fauna Environ. 30(4):221-227.

Tomasik, P., C. H. D. Magadza, S. Mhizha, & A. Chirume (1995) The metal-metal interactions in biological systems. Part 3. *Daphnia magna*. - Water, Air, Soil Pollut. 82:695-711.

Torgon, L. C., M. Garcia-Baptista, C. Odebrecht & O. O. Möller Jr. (1995) Distribuição vertical do fitoplâncton na Laguna dos Patos, Rio Grande do Sul, Brasil (Verão, 1986). - Acta Limnol. Brasil. 7:67-77.

Treasurer, J. W. (1990) The food and daily food consumption of lacustrine 0+ perch, *Perca fluviatilis* L. - Freshwater Biol. 24:361-374.

Trémouilles, E. R., A. Oliva & A. O. Bachmann (1995) Insecta Coleoptera. In: E. C. Lopretto & G. Tell (Eds.), Ecosistemas de Aguas continentales. Vol. 3. pp. 1133-1197.

Tricart, J. (1982) El Pantanal, un ejemplo del impacto de la geomorphologia sobre el medio ambiente. - Geografia 7(13-14):37-50.

Troll, C. (1965) Jahreszeitenklimate der Erde. In: H. E. Landsberg, H. Lippmann, K. H. Paffen & C. Troll (Eds.). World Maps of Climatology. - Springer, Berlin. pp. 7-18.

Tryon, R. M., A. F. Tryon & K. U. Kramer (1990) Pteridaceae. In: K. U. Kramer & P. S. Green (Eds.), Pteridophytes and Gymnosperms. Vol. 1. The Families and Genera of Vascular Plants. - Springer, Berlin. pp. 230-256.

Tümpling, W. von, Jr., P. Zeilhofer, U. Ammer, J. Einax & R.-D. Wilken (1995) Estimation of mercury content in tailings of the gold mine area of Poconé, Mato Grosso, Brazil. - Environ. Sci. Pollut. Res. 2(4):225-228.

Tundisi, J. G., & T. Matsumura-Tundisi (1985) The "Pantanal" wetlands of Western Brazil. In: Y. Saijo & J. G. Tundisi (Eds.), Limnological Studies in Central Brazil. - Water Res. Inst., Nagoya Univ., Nagoya. pp. 177-188.

Tundisi, J. G., T. Matsumura-Tundisi, M. C. Calijuri & E. M. L. Novo (1991) Limnology of six reservoirs in the Middle Tietê River, São Paulo State, Brazil. - Verh. Verein. Internat. Limnol. 24:1489-1496.

Tundisi, J. G., & M. Straskraba (1994) Ecological basis for the application of ecotechnologies to watershed/reservoir recovery and management. - Acta Limnol. Bras. 5:49-72.

Turner, P. N., & C. da Silva (1992) Littoral rotifers from the State of Mato Grosso, Brazil. - Stud. Neotrop. Fauna Environ. 27(4):227-241.

Uherkovich, G. (1966) Die *Scenedesmus*-Arten Ungarns. - Akademia Kiado, Budapest. 173 pp.

Uherkovich, G. (1976) Algen aus den Flüssen Rio Negro und Rio Tapajós. - Amazoniana 5:465-515.

Uherkovich, G. (1981) Algen aus einigen Gewässern Amazoniens. - Amazoniana 7:191-219.

Uherkovich, G., & M Franken (1980) Aufwuchsalgen aus zentralamazonischen Regenwaldbächen. Amazoniana 7:49-79.

Uherkovich, G., & H. Rai (1979) Algen aus dem Rio Negro und seinen Nebenflüssen. - Amazoniana 6:611-638.

Uherkovich, G., & G. W. Schmidt (1974) Phyhtoplanktontaxa in dem zentralamazonischen Schwemmlandsee Lago do Castanho. - Amazoniana 5:243-283.

Valverde, D. (1972) Fundamentos geográficos de planejamento rural do Município de Corumbá. - Rev. Bras. Geogr. 34(1):49-144.

van den Berg, M. E. (1982) Contribuição ao conhecimento da flora medicinal do Mato Grosso. - Ciên. Cult. 34:163-170.

van sen Berg, M. E. (1986) Formas atuais e potenciais de aproveitamento das espécies nativas e exóticas do Pantanal Mato-Grossense. In: Anais 1° Simpósio Recursos Naturais e Sócio-Econômico do Pantanal. EMBRAPA-CPAP, Corumbá. pp. 131-136.

Vanzolini, P. E. (1968a) Environmental temperature and number of body annuli in *Amphisbaena alba*: notes of a cline (Sauria: Amphisbaenidae). - Pap. Dept. Zool. São Paulo 21(23):231-241.

Vanzolini, P. E. (1968b) Lagartos Brasileiros da familia Gekkonidae (Sauria). - Arq. Zool. 17:1-84.

Vanzolini, P. E. (1968c) Geography of the South American Cekkonidae (Sauria). - Arq. Zool. 17:85-112.

Vanzolini, P. E., & J. Valença (1966) The genus *Dracaena*, with a brief consideration of microteiid relationships (Sauria: Teiidae). - Arq. Zool. 13:7-35.

Varela, M. E. (1990) Notas taxonómicas y ecológicas sobre algunos oligoquetos delceacuicolas des nordeste artentino. I. Naididae. - Stud. Neotrop. Fauna Environ. 25(4):223-233.

Vari, R. P. (1989a) A phylogenetic study of the Neotropical characiform family Curimatidae (Pisces: Ostariophysi). - Smithsonian Cont. Zool. 471. 71 pp.

Vari, R. P. (1989b) A phylogenetic study of the Neotropical characiform genus *Psectrogaster* Eigenmann and Eigenmann (Pisces: Characiformes). - Smithsonian Cont. Zool. 481. 43 pp.

Vari, R. P. (1992a) Systematics of the Neotropical characiform genus *Curimatella* Eigenmann and Eigenmann (Pisces: Ostariophysi), with summary comments on the Curimatidae. - Smithsonian Cont. Zool. 533. 48 pp.

Vari, R. P. (1992b) Systematics of the Neotropical characiform genus *Cyphocharax* Fowler (Pisces: Ostariophysi). - Smithsonian Cont. Zool. 529. 137 pp.

Vassalo, M. C. M., & L. A. Kieffer (1984) Esdudios limnologicos en una seccion transversal del tramo medio del rio Parana. VII. Química del agua. - Rev. Asoc. Cien. Nat. Litoral 15:97-108.

Velho, L. F. M., F. A. Lansac-Tôha & M. Serafim-Junior (1996) Testate amoebae (Rhizopodea-Sarcodina) from zooplankton of the high Paraná River floodplain, State of Mato Grosso do Sul, Brazil: I. Families Arcellidae and Centropyxidae. - Stud. Neotrop. Fauna & Environm. 31:35-50.

Veloso, H. P. (1947) Considerações gerais sobre a vegetação do Estado do Mato Grosso. II. Notas preliminares sobre o Pantanal e e zonas de transição. - Mem. Inst. Oswaldo Cruz 45:253-272.

Veloso, H. P. (1972) Aspetos fitoecológicos da Bacia do Alto Rio Paraguai. - Biogeografia 7:1-31.

Vidal, O. (1993) Aquatic mammal conservation in Latin America: problems and perspectives. - Conservation Biol. 7:788-795.

Villwock, W. (1972) Gefahren für die endemische Fischfauna durch Einführungsversuche und Akklimatisation von Fremdfischen am Beispiel des Titicaca-Sees (Peru-Bolivien) und des Lanao-Sees (Mindanao/Philippinen). - Verh. Internat. Ver. Limnol. 18:1227-1234.

Vitt, L. J., & G. R. Colli (1994) Geographical ecology of a Neotropical lizard: *Ameiva ameiva* (Teiidae) in Brazil. - Can. J. Zool. 72:1986-2008.

Volkmer-Ribeiro, C, R. de Rosa-Barbosa & H. Fonseca Mello (1983) The unexpected occurrence of *Drulia browni* (Bowerbank, 1863) (Porifera, Spongillidae) in an oxbow lake at the extreme South of Brazil. - Iheringia, Ser. Zool. 63:3-10.

Vucetich, M. C. (1970) Algunos Tecamebianos de la Provincia de Formosa. - Neotropica 16:42-48.

Vucetich, M. C. (1972) Tecamebianos del eupleuston de cuorpos de agua de la Provincia de Buenos Aires. - Acta Zool. Lilloana 29:271-287.

Vucetich, M. C. (1976) Tecamebianos del Lago San Roque y de un ambiente lentico artificial vinculado as mesmo (Cordoba, Argentina). - Limnobios 1:29-34.

Vucetich, M. C. (1980) Tecamebianos de las cumbres Calchaquies, Tucuman, Argentina (Rhizopods Testacea). - Limnobios 1(2):397-402.

Vucetich, M. C. (1989) Descripcion y ultraestructura tecal de dos especies nuevas des genero *Difflugia* Leclerc. - Limnobios 2:740-743.

Walker, I. (1982) The thecamoebae (Protozoa, Rhizopoda) of small Amazonian forest streams and their possible use as indicator organisms for waterquality. - Acta Amazonica, Suppl. 12(3):79-105.

Walter, H., & S. W. Breckle (1984) Spezielle Ökologie der tropischen und subtropischen Zonen. - Ökologie der Erde. Vol. 2. Gustav Fischer, Stuttgart. 461 pp.

Wanderley, M. das G. L. (1989) Xiridaceae. Flora do Estado de Goiás. Coleç_o Rizzo. Vol. 11. - Univ. Fed. Goiás, Goiânia. 81 pp.

Webster, W. D., & R. D. Owen (1984) *Pygoderma bilabiatum*. - Mammalian Species 220:1-3.

Welcomme, R. L. (1985) River Fisheries. FAO Fish. Tech. Paper 262:1-330.

Wermuth, H., & R. Mertens (1961) Schildkröten. Krokodile. Brückenechsen. - VEB Gustav Fischer, Jena. 424 pp.

Werner, F. (1933) Loricata (Reptilia). Das Tierreich, Leipzig. 62:1-40.

Whitaker, J. O., Jr. & J. S. Findley (1980) Foods eaten by some bats from Costa Rica and Panama. - J. Mamm. 61:540-544.

Wiebach, F. (1970) Amazonische Moostiere (Bryozoa) II. - Amazoniana 2(3):353-362.

Wiebach, F. (1974) Amazonische Moostiere (Bryozoa) III. - Amazoniana 5(2):293-303.

Wilhelmy, H. (1958) Das Grosse Pantanal. Umshau 18:555-559.

Williams, E. E. (1960) Two species of tortoises in northern South America. - Breviora Mus. Comp. Zool. 120:1-13.

Willig, M. R., & R. R. Hollander (1987) *Vampyrops lineatus*. - Mammalian Species 275:1-4.

Willig, M. R., & J. K. Jones Jr. (1985) *Neoplatymops matogrossensis*. - Mammalian Species 204:1-3.

Willis, E. O. (1995) Black versus white waterbird colonies (Aves) in the Bolivian-Brazilian Pantanal. - Iheringia, Ser. Zool. 78:95-97.

Willis, E. O., & Y. Oniki (1990a) Levantamento preliminar das aves de inverno em dez areas do sudoeste de Mato Grosso, Brasil. - Ararajuba 1:19-38.

Willis, E. O., & Y. Oniki (1990b) Nomes gerais para as aves brasileiras. "Grafica da Região" - Américo Brasiliense, São Paulo. 246 pp.

Willis, N. E. (1976) Effect of a cold wave on the Amazonian avifauna in the Upper Paraguay drainage, western Mato Grosso, and suggestions on oscine-suboscine relationships. - Acta Amazonica 6:279-294.

Wilson, D. E., & R. K. LaVal (1974) *Myotis migricans*. - Mammalian Species 39:1-3.

Wimsatt, W. A. (1962) The responses of captive vampires to cold and warm environments. - J. Mamm. 43:185-191.

Wirth, W. W., & F. Cavalieri (1977) Ceratopogonidae. In: S. H. Hurlbert (Ed.), Aquatic Biota of Southern South America. - San Diego State Univ., San Diego. pp. 280-283.

Wolf, A. (1990) Observations on the vegetation of some shallow lakes near the mouth of the Rio Ypané, Paraguay. - Amazoniana 11:167-184.

Wolters, H. E. (1975-1982) Die Vogelarten der Erde. - Paul Parey, Hamburg. 745 pp.

Wooldridge, D. P. (1973) New *Paracymus* from South America. - J. Ent. Soc. Kans. 46(1):116-123.

Worbes, M. (1985) Structural and other adaptations to long-term flooding by trees in Central Amazonia. - Amazoniana 9:459-484.

Wu, I. C. H., & C. R. Curds (1979) A guide to the species of the genus *Aspidisca*. - Bull. Brit. Mus. Nat. Hist. (Zool.) 36:1-34.

Xavier, M. B., C. S. R. Mainardes-Pinto & M. Takino (1991) *Euglena sanguinea* Ehrenberg bloom in a fish-breeding tank (Pindamenhangaba, São Paulo, Brazil). - Arch. Hydrobiol./Alg. Stud. 62:133-142.

Yamaciro, R. G. (1989) Macrozonamento do Mato Grosso do Sul quanto ao uso de agrotóxicos. - SEMA/MS, Campo Grande. 37 pp.

Yamashita, C., & M. de P. Vale (1990) Sobre ninhais de aves do Pantanal do Município de Poconé, Mato Grosso, Brasil. Vida Silv. Neotrop. 2(2):49-63.

Young, F. A. (1994) Food preferences in tropical marine fish larvae. - Aquacult. Mag. 20:57-59.

Young, J. O., & R. M. Procter (1986) Are the lake dwelling leeches, *Glossiphonia complanata* (L.) and *Helobdella stagnalis* (L.) opportunistic predators on molluscs and do they partition this food resource? - Freshwater Biol. 16:561-566.

Zaccagnini, M. E., & A. H. Beltzer (1982) Alimentacion de *Bubulcus ibis ibis* L. 1758 y su relacion trofica con *Egretta thula thula* (Molina 1782) en Leales, Tucuman (Ciconiiformes: Ardeidae). - Rev. Asoc. Cienc. Nat. Litoral 13:73-80.

Zafar, A. R. (1959) The taxonomy of lakes. Hydrobiologia 13:287-299.

Zakrys, B., & P. L. Walne (1994) Floristic, taxonomic and phytogeographic studies of green Euglenophyta from the Southeastern United States, with emphasis on new and rare species. - Arch. Hydrobiol./Alg. Stud. 72:71-114.

Zayas, J. C. (1968) La guayaba y outras frutas myrtaceas. - Inst. Libro, Havana. 87 pp.

Zerries, O. (1968) The South American Indians and their culture. In: E. J. Fittkau, J. Illies, H. Klinge, G. H. Schwabe & H. Sioli (Eds.), Biogeography and Ecology in South America. Vol. 1. - W. Junk, Dordrecht. pp. 329-388.

Ziliukiene, V. (1994) (Peculiarities of feeding of pike and bream juveniles reared in illuminated cages.). - Ekologiya 1994(3):23-28. (in Russian).

Zimmermann, A. (1921) Beiträge zur Kenntnis der südamerikanischen Schwimmkäferfauna nebst 41 Neubeschreibungen. - Arch. Naturgeschichte 87(3):181-206.

Zou, X. N., S. H. Lu, & B. Liu (1994) Volatile N-nitrosamines and their precursors in Chinese salted fish - a possible etological factor for NPC in China. - Int. J. Cancer 59:155-158.

Zurlo, C., & M. Brandão (1989) Ervas comestívais: descrição, ilustrações e receitas. - Globo, Rio de Janeiro.

Taxonomic index

—A—

—B—

—C—

—D—

—F—

—H—

—M—

—N—

—O—

—P—

—Q—

—S—

—T—

Subject index

MONOGRAPHIAE BIOLOGICAE

31. M. J. A. Werger (ed.): *Biogeography and Ecology of Southern Africa*, 2 vols. 1978 ISBN 90-6193-083-9
32. C. Serruya (ed.): *Lake Kinneret* [Lake of Tiberias / Sea of Galilee]. 1978 ISBN 90-6193-085-5
33. Ph. D. Mordukhaí-Boltovskoi (ed.): *The River Volga and Its Life*. 1979 ISBN 90-6193-084-7
34. Ch. W. Heckman: *Rice Field Ecology in Northeastern Thailand*. The Effect of Wet and Dry Seasons on a Cultivated Aquatic Ecosystem. 1979 ISBN 90-6193-086-3
35. M. Kalk, A. J. McLachlan and C. Howard-Williams (eds.): *Lake Chilwa*. Studies of Change in a Tropical Ecosystem. 1979 ISBN 90-6193-087-1
36. B. R. Allanson (ed.): *Lake Sibaya* [South Africa]. 1979 ISBN 90-6193-088-X
37. H. Löffler (ed.): *Neusiedlersee*. The Limnology of a Shallow Lake in Central Europe. 1979 ISBN 90-6193-089-8
38. J. Rzóska: *Euphrates and Tigris*. Mesopotamian Ecology and Destiny. With Contributions by J. F. Talling and K. E. Banister. 1980 ISBN 90-6193-090-1
39. K. A. Brodsky: *Mountain Torrent of the Tien Shan*. A Faunistic-Ecology Essay. Translated from Russian by V. V. Golosov. 1980 ISBN 90-6193-091-X
40. M. S. Mani and L. E. Giddings (eds.): *Ecology of Highlands*. 1980 ISBN 90-6193-093-6
41. A. Keast (ed.): *Ecological Biogeography of Australia*, 3 vols. 1981 ISBN 90-6193-092-8
42. J. L. Gressit (ed.): *Biogeography and Ecology of New Guinea*, 2 vols. 1982 ISBN 90-6193-094-4
43. H. Heathwole, T. Done and E. Cameron: *Community Ecology of a Coral Cay*. A Study of One-Tree Island, Great Barrier Reef, Australia. 1981 ISBN 90-6193-096-0
44. P. S. Maitland (ed.): *The Ecology of Scotland's Largest Lochs*. Lomond, Awe, Ness, Morar and Shiel. 1981 ISBN 90-6193-097-9
45. K. Müller (ed.): *Coastal Research in the Gulf of Bothnia*. 1982 ISBN 90-6193-098-7
46. G. K. Rutherford (ed.): *The Physical Environment of the Faeroe Islands*. 1982 ISBN 90-6193-099-5
47. J. I. Furtado and S. Mori (eds.): *Tasek Bera* [Malaysia]. The Ecology of a Freshwater Swamp. 1982 ISBN 90-6193-100-2
48. G. R. South (ed.): *Biogeography and Ecology of the Island of Newfoundland*. 1983 ISBN 90-6193-101-0
49. J. A. Thornton (ed.): *Lake McIlwaine*. The Eutrophication and Recovery of a Tropical African Man-made Lake. 1982 ISBN 90-6193-102-9
50. R. W. Edwards and M. P. Brooker: *The Ecology of the Wye* [U.K.]. 1982 ISBN 90-6193-103-7
51. T. Petr (ed.): *The Purari* [Papua New Guinea]. Tropical Environment of a High Rainfall River Basin. 1983 ISBN 90-6193-104-5
52. H. Kuhbier, J. A. Alcover and C. Guerau d'Arellano Tur (eds.): *Biogeography and Ecology of the Pityusic Islands*. 1984 ISBN 90-6193-105-3
53. J.-P. Carmouze, J.-R. Durand and C. Lévêque (eds.): *Lake Chad*. Ecology and Productivity of a Shallow Tropical Ecosystem. 1983 ISBN 90-6193-106-1
54. S. Horie (ed.): *Lake Biwa* [Japan]. 1984 ISBN 90-6193-095-2
55. D. R. Stoddart (ed.): *Biogeography and Ecology of the Seychelles Islands*. 1984 ISBN 90-6193-107-X
56. H. Sioli (ed.): *The Amazon*. Limnology and Landscape Ecology of a Mighty Tropical River and Its Basin. 1984 ISBN 90-6193-108-8

MONOGRAPHIAE BIOLOGICAE

57. C. H. Fernando (ed.): *Ecology and Biogeography in Sri Lanka.* 1984 ISBN 90-6193-109-6
58. S. J. Casper (ed.): *Lake Stechlin* [Germany]. A Temperate Oligotrophic Lake. 1985 ISBN 90-6193-512-1
59. U. Th. Hammer: *Saline Lake Ecosystems of the World.* 1985 ISBN 90-6193-535-0
60. B. R. Davies and K. F. Walker (eds.): *The Ecology of River Systems.* 1986 out of print
61. P. de Deckker and W. D. Williams (eds.): *Limnology in Australia.* 1986 ISBN 90-6193-578-4
62. Y. Yom-Tov and E. Tchernov (eds.): *The Zoogeography of Israel.* The Distribution and Abundance at a Zoogeographical Crossroad. 1988 ISBN 90-6193-650-0
63. F. D. Por: *The Legacy of Tethys.* An Aquatic Biogeography of the Levant. 1989 ISBN 0-7923-0189-7
64. B. R. Allanson, R. C. Hart, J. H. O'Keeffe and R. D. Robarts: *Inland Waters of Southern Africa.* An Ecological Perspective. 1990 ISBN 0-7923-0266-4
65. F. di Castri, A. J. Hansen and M. Debussche (eds.): *Biological Invasions in Europe and the Mediterranean Basin.* 1990 ISBN 0-7923-0411-X
66. R. W. Edwards, A. S. Gee and J. H. Stoner (eds.): *Acid Waters in Wales.* 1990 ISBN 0-7923-0493-4
67. S. K. Jain and L. W. Botsford (eds.): *Applied Populations Biology.* 1992 ISBN 0-7923-1425-5
68. C. Dejoux and A. Iltis (eds.): *Lake Titicaca* [Peru/Bolivia]. A Synthesis of Limnological Knowledge. 1992 ISBN 0-7923-1663-0
69. R.B. Wood and R.V. Smith (eds.): *Lough Neagh.* The Ecology of a Multipurpose Water Resource. 1993 ISBN 0-7923-2112-X
70. P.E. Ouboter (ed.): *The Freshwater Ecosystems of Suriname.* 1993 ISBN 0-7923-2408-0
71. M.A. Brunt and J.E. Davies (eds.): *The Cayman Islands.* Natural History and Biogeography. 1994 ISBN 0-7923-2462-5
72. V. Fet and K. Atamuradov (eds.): *Biogeography and Ecology of Turkmenistan.* 1994 ISBN 0-7923-2738-1
73. F.D. Por: *The Pantanal of Mato Gross (Brazil).* World's Largest Wetlands. 1995 ISBN 0-7923-3481-7
74. D.S. Edwards, W.E. Booth and S.C. Choy (eds.): *Tropical Rainforest Research – Current Issues.* Proceedings of the Conference held in Bandar Seri Begawan. 1996 ISBN 0-7923-4038-8
75. R.W. Schmieder (ed.): *Rocas Alijos.* 1996 ISBN 0-7923-4056-6
76. Z. Mianping: *An Introduction to Saline Lakes on the Qinghai-Tibet Plateau.* 1996 ISBN 0-7923-4098-1
77. C.W. Heckman: *The Pantanal of Poconé.* Biota and Ecology in the Northern Section of the World's Largest Pristine Wetland. 1998 ISBN 0-7923-4863-X

KLUWER ACADEMIC PUBLISHERS – DORDRECHT / BOSTON / LONDON

GPSR Compliance

The European Union's (EU) General Product Safety Regulation (GPSR) is a set of rules that requires consumer products to be safe and our obligations to ensure this.

If you have any concerns about our products, you can contact us on ProductSafety@springernature.com

In case Publisher is established outside the EU, the EU authorized representative is:

Springer Nature Customer Service Center GmbH
Europaplatz 3
69115 Heidelberg, Germany

Batch number: 10809436

Printed by Printforce, the Netherlands